Craig O'Donnell cro3
610-868-7785

Table A.2: Properties of Areas

A = Area
I = Area moment of inertia
J = Polar area moment of inertia
y = Centroidal distance

Rectangle

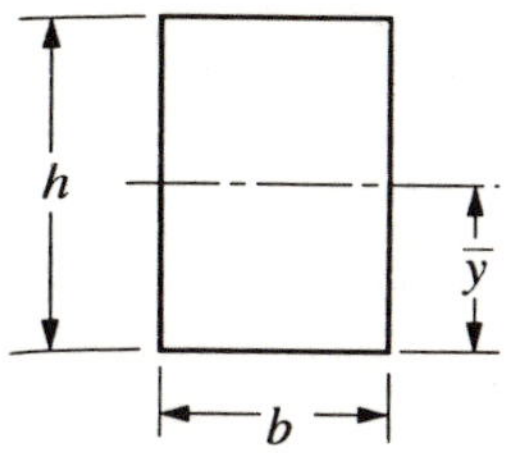

$$A = bh \qquad y = \frac{h}{2}$$
$$I = \frac{bh^3}{12}$$

Triangle

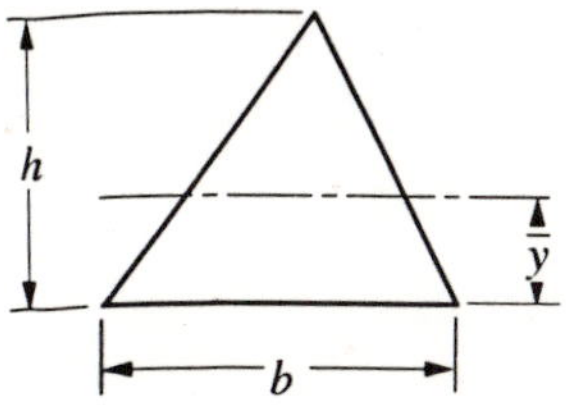

$$A = \frac{bh}{2} \qquad y = \frac{h}{3}$$
$$I = \frac{bh^3}{36}$$

Circle

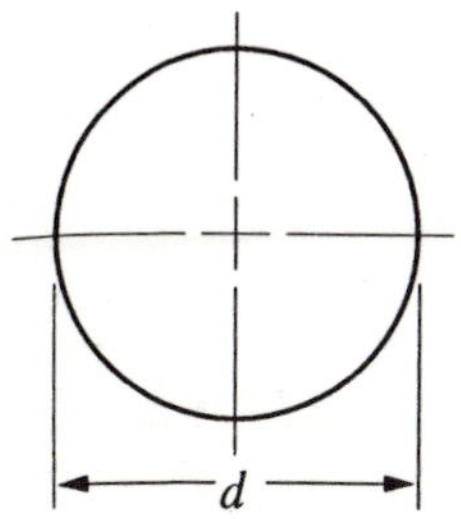

$$A = \frac{\pi d^2}{4} \qquad J = \frac{\pi d^4}{32}$$
$$I = \frac{\pi d^4}{64} \qquad y = \frac{d}{2}$$

Hollow Cylinder

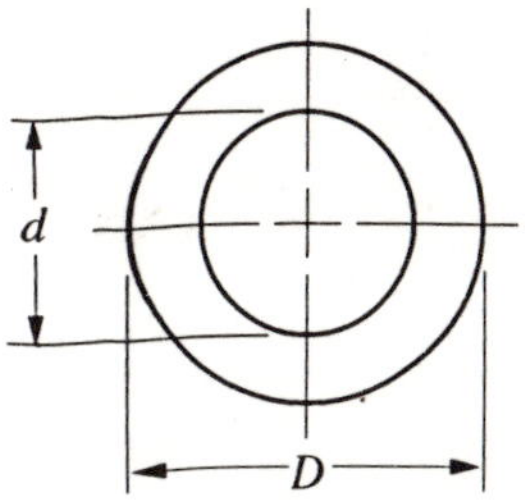

$$A = \frac{\pi(D^2 - d^2)}{4} \qquad J = \frac{\pi(D^4 - d^4)}{32}$$
$$I = \frac{\pi(D^4 - d^4)}{64} \qquad y = \frac{D}{2}$$

KINEMATICS, DYNAMICS, AND Design of Machinery

KINEMATICS, DYNAMICS, AND *Design of Machinery*

KENNETH J. WALDRON
GARY L. KINZEL

The Ohio State University

John Wiley & Sons, Inc.

New York • Chichester • Brisbane • Toronto • Singapore

ACQUISITIONS EDITOR: Regina Brooks

MARKETING MANAGER: Katherine Hepburn

PRODUCTION EDITOR: Patricia McFadden

OUTSIDE PRODUCTION MANAGEMENT: Ingrao Associates

COVER DESIGN: John McHale

COVER PHOTOGRAPH: © Gary L. Kinzel

PHOTO EDITOR: Kim Khatchatourian

ILLUSTRATION EDITOR: Sigmund Malinowski

ELECTRONIC ILLUSTRATIONS: ClarisCad files edited by Radiant

This book was set in 10/12 Times Roman by BI-COMP, Inc., and printed and bound by Hamilton Printing
The cover was printed by Phoenix Color Corp.

This book is printed on acid-free paper. ♾

The paper in this book was manufactured by a mill whose forest management programs include sustained yield harvesting of its timberlands. Sustained yield harvesting principles ensure that the numbers of trees cut each year does not exceed the amount of new growth.

Library of Congress Cataloging-in-Publication Data

Waldron, Kenneth J.
Kinematics, dynamics, and design of machinery / Kenneth J. Waldron, Gary L. Kinzel.
p. cm.
Includes index.
ISBN 0-471-58399-5 (cloth : alk. paper)
1. Machinery, Kinematics of. 2. Machinery, Dynamics of. 3. Machine design. I. Kinzel, G. (Gary) II. Title.
TJ175.W35 1999
621.8′1—dc21 98-25607
CIP

Printed in the United States of America

10 9 8 7 6 5 4 3 2

Preface

The design of mechanisms is a traditional core topic in mechanical engineering. In our university, the first curriculum in mechanical engineering, 125 years ago, included a project in which the students had to perform the kinematic and mechanical design and manufacture by casting and machining the metal parts of a model illustrating a principal of mechanism design. We still have some of those models.

The subject of mechanism design is also very unique to mechanical engineering. Most topics in the mechanical engineering are shared, at least in part, with other engineering disciplines, but mechanism design is one topic that mechanical engineers can claim as uniquely theirs.

Despite the long history of the subject, going back, in fact, at least to Roman times, mechanism design remains a vital component of modern machine design practice. As with all engineering subjects, the practice in this field is continually changing with the development of new technologies. New technologies that have dramatically changed the way we do mechanism design, also have made fundamental changes to the nature of the machines we design, and further have made substantial changes in the way students learn the subject. That is why it is appropriate to introduce a new textbook in this old, but vital and constantly developing subject.

Until relatively recently, mechanisms were designed using manual, graphical techniques. The advent of digital computers has revolutionized the methodology. Computer-aided-design software tools lend themselves to direct automation of some traditional graphical methods. However, in most cases it is more productive to go back to the fundamental kinematic geometry to develop analytical formulations on which solution algorithms can be based. This is the basis of special purpose software packages that have been developed to carry out the most used mechanism design operations.

More recently, there have been advances in the way engineers use computers to do routine calculations. Traditional, procedural programming languages have been largely superseded by specialized numerical computation packages, and symbolic mathematics packages. It is appropriate that the methods presented in this book reflect this change.

The influence of computers has also been influential in two other respects. First, it has become much more practicable to design three-dimensional mechanisms since solid modeling packages and three-dimensional simulators overcome the difficulty of visualizing a three-dimensional system from traditional machine drawings. Second, advances in actuator technology and in digital control and communication techniques have freed the designer from the traditional machine in which everything is mechanically coordinated off a single prime mover. Thus, whereas an old fashioned printing press, or packaging machine, might have dozens of functions mechanically coordinated off a single, large motor, it may now be possible to use a mechanically simpler machine with multiple actuators that are electronically coordinated. This strategy offers great functional flexibility, but is not always appropriate. Obviously, it is important that the theory and methods used in designing such systems be introduced in this text.

In some situations, the integration of computers into manufacturing machinery has produced fundamental changes in the way we design and manufacture machines. The advent of computers for design and numerically controlled machine tools for manufacture has led to the use of polynomial cam profiles, rather than the traditional parabolic, harmonic, and cycloidal types. It is important that this book include a strong treatment of numerically based cam design.

As with everything else, there have been advances in computer programming tools. The first software for mechanism synthesis and analysis was written in assembly language. Shortly thereafter, traditional, procedural languages like FORTRAN were used in a batch processing, mainframe environment. In today's distributed computing environments, with powerful workstations and personal computers that are much more powerful than those early mainframes, it is possible for the user to interact intensively with the computer. This has led to the use of different computing strategies for both specialized packages, and general use programming languages. In particular, we now have software systems that combine the universal flexibility of a procedural language with powerful, inbuilt numerical functions. These software systems have become the common means of solving mechanical engineering design problems. Accordingly, we have chosen to arrange the material presented in this text to take full advantage of the capabilities of such integrated problem solving software systems. The one that we use is MATLAB, but there are several competing packages that could equally well be used.

We have chosen not to include computer code in this book, since that would limit its use to those programs that use the same software. However, the CD-ROM disk included with the book does include a library of functions written in MATLAB to perform many of the important computations in the text. The CD-ROM disk also includes downloadable copies of all line figures in the text, as well as supplementary problem sets for most chapters.

The presentation used here will not suit all instructors in the subject, since there are several different representational philosophies in use in the professional community, and because different instructors have different pedagogical approaches. Nevertheless, the material is based in long experience in teaching the subject and has been thoroughly tested in the classroom. Between us we have over fifty years of classroom experience in teaching this subject. We look forward to learning how the text fares as an instructional tool in the hands of other teachers, and to improving the product in the light of that feedback.

We express our sincere thanks to the colleagues and students who have contributed to this book. We would especially like to acknowledge Dr. George Sutherland whose teaching approach and lectures inspired several sections of the book. Dr. Joe Davidson and Dr. Michael Stanisic made many suggestions on an early draft of the book and suggested some exercise problems. Dr. Necip Berme also suggested corrections for some examples in an early draft. Two graduate students, Sven Esche and Jim Schmiedeler, did outstanding jobs of proofreading manuscript and page proofs. Other graduate students making significant contributions were Vasfi Omurlu and Wei Kang who developed solutions to some exercise problems.

Kenneth J. Waldron and Gary L. Kinzel
June 15, 1998

Contents

Chapter 4: Planar Linkage Design . . . 236

KINEMATICS, DYNAMICS, AND *Design of Machinery*

Chapter 1

Introduction

1.1 HISTORIC PERSPECTIVE

A mechanism is a machine composed of rigid members that are jointed together. The members interact with one another by virtue of the joints. The joints are formed by portions of the surfaces of the members joined that contact one another. The geometries of the contacting surface segments determine the properties of each joint.

The design of mechanisms is a technical area that is unique to mechanical engineering. Its history stretches back to prehistoric times. Artisans such as blacksmiths and carpenters also functioned as the designers of mechanisms. One of the original functions of engineers was the design of mechanisms both for warfare and for peaceful uses. In renaissance times, we find Leonardo da Vinci depicting a sophisticated variety of mechanisms, mostly for military purposes. Sometime thereafter the distinction between civil engineering and military engineering appeared. The modern era in mechanism design, along with the history of mechanical engineering as a distinct discipline, can be viewed as starting with James Watt.

Figure 1.1 The Adaptive Suspension Vehicle. Each leg is a planar pantograph mechanism hinged to the body about an axis parallel to the longitudinal axis of the vehicle.

That is not to say that the subject has remained static. In fact, there have been dramatic changes in the practice of mechanism design in recent years. Traditionally, machines have been designed to be powered by a single "prime mover," with all functions mechanically coordinated. That tradition certainly predates Watt. Recent developments in computer technology, coupled with improvements in electric motors and other actuators, have made it possible to use a different approach. This is an approach in which machines are powered by multiple actuators coordinated electronically. The resulting machines are simpler, less expensive, more easily maintained, and more reliable. Another major change is in the techniques used in mechanism design. The use of interactive computer graphics has had a dramatic impact on design practice. One of our motivations in producing this book, even when there are a number of excellent texts available in mechanism kinematics, is to provide a treatment that reflects these changes in practice.

1.2 KINEMATICS

Kinematics is the study of position and its time derivatives. Specifically, we are concerned with the positions, velocities, and accelerations of points and with the angular positions, angular velocities, and angular accelerations of solid bodies. Together these entities are sufficient to describe the positions, velocities, and accelerations of solid bodies. The position of a body can be defined by the position of a nominated point of that body combined with the angular position of the body. In some circumstances we are also interested in the higher time derivatives of position and angular position.

The subject of kinematics is a study of the geometry of motion. This is an accurate title because kinematics is geometry with the element of time added. The bulk of the subject matter of this book is often referred to as the kinematics of mechanisms. Our objective is to present techniques that can be used to design mechanisms to meet specific motion requirements. That is why the subject matter is approached from a mechanical designer's perspective.

1.3 DESIGN: ANALYSIS AND SYNTHESIS

The material in this book falls into two classifications. The first consists of techniques to determine the positions, velocities, and accelerations of points in the members of mechanisms and the angular positions, velocities, and accelerations of those members. These are kinematic analysis techniques. The second type of material comprises methods for mathematically determining the geometry of a mechanism to produce a desired set of positions and/or velocities or accelerations. These are rational synthesis techniques.

The activity that distinguishes engineering from science is design. Science is the study of what is; engineering is the creation of what is to be. This creative activity is design or, more formally, synthesis. The rational synthesis techniques developed by kinematicians offer a rather direct route to mechanism design that lends itself well to automation using computer graphics workstations. However, these techniques do not represent the only way to design mechanisms, and they are relatively restrictive: rational synthesis techniques exist only for specific types of mechanism design problems, and many practical mechanism design problems do not fit within the available class of solutions. An alternative is to use informal synthesis. This is a methodology used by engineers to solve design problems in many technical areas, not just in mechanism design. The basic procedure is to "guess" a set of dimensions and then use analysis to check the resulting performance. The dimensions are then adjusted to attempt to more closely match the performance specifications and the mechanism is analyzed again. The process is repeated until an acceptably close match to the specifications is achieved. Thus, a primary use of the analysis material is also in mechanism design.

From an engineering point of view, it is not possible to treat mechanism design solely in terms of kinematics. The motivation for performing an acceleration analysis is often to enable inertia forces on the links to be calculated, allowing, in turn, computation of the forces transferred between links and the internal forces, or stresses, within the links. Mechanisms must usually drive loads, as well as generate motions. Of course, as soon as we introduce the concept of force, we leave the domain of pure kinematics and enter that of kinetics. Insofar as the largest forces in many mechanisms are inertia forces created by motion, it is convenient to study them within the general framework of kinematic techniques. There is also an important symmetry between the geometry of the force distribution and that of the velocity distribution which is particularly useful when working with spatial mechanisms. Thus, it is entirely appropriate to treat mechanism statics or kinetics within the general geometry of motion framework constructed to study mechanism kinematics. Such a treatment is presented in the later chapters of this book.

1.4 MECHANISMS

Mechanisms are assemblages of solid members connected together by joints. Mechanisms transfer motion and mechanical work from one or more actuators to one or more "output" members. For the purposes of kinematic design, we idealize a mechanism to a kinematic linkage in which all the members are assumed to be perfectly rigid and are connected by kinematic joints. A kinematic joint is formed by direct contact between the surfaces of two members. One of the earliest codifications of mechanism kinematics was that of Reuleaux (1876),[1] and some of the basic terminology we use originated with him. He called a kinematic joint a "pair." He further divided joints into "lower pairs" and "higher pairs." A lower pair joint is one in which contact between two rigid members occurs at every point of one or more surface segments. A higher pair is one in which contact occurs only at isolated points or along line segments.

Joints are the most important aspect of a mechanism to examine during an analysis. They permit relative motion in some directions while constraining motion in others. The types of motion permitted are related to the number of degrees of freedom (dof) of the joint. The number of degrees of freedom of the joint is equal to the number of independent coordinates needed to specify uniquely the position of one link relative to the other constrained by the joint.

Lower pair joints are necessarily restricted to a relatively small number of geometric types, since the requirement that surface contact be maintained constrains the geometry of the contacting surfaces. It can be shown that there are only six fundamentally different types of lower pair joints, classified by the types of relative motion that they permit. There is, in contast, an infinite number of possible higher pair geometries. The lower pair joint types are shown in Table 1.1. Some important examples of higher pair joints are shown in Table 1.2.

Lower pair joints are frequently used in mechanism design practice. They give good service because wear is spread out over the contact surface and because the narrow clearance between the surfaces provides good conditions for lubrication and a tight constraint on the motion. The change in the geometric properties of the joint with wear is slow for a lower pair. At least as important are the simple geometries of the relative motions that these joints permit.

Higher pair joints that involve pure rolling contact, or that approximate that condition, are also used frequently. In pure rolling contact, the points in one of the two joint surfaces

[1] Reuleaux, Franz, *The Kinematics of Machinery* (Translated and edited by Alexander B. W. Kennedy), Dover Publications, New York, 1963.

Table 1.1 Lower Pair Joints

Connectivity (No. of Degrees of Freedom)	Names	Letter Symbol	Typical Form	Sketch Symbol
1	Revolute Hinge Turning pair	R	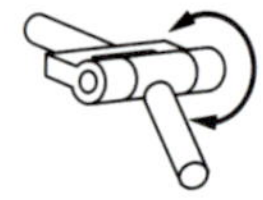	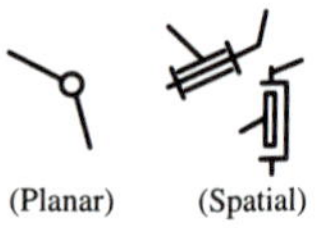
1	Prismatic joint Slider Sliding pair	P	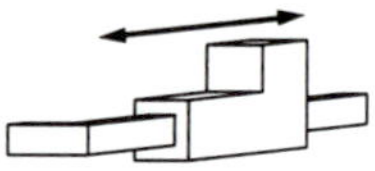	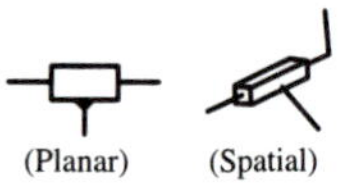
1	Screw joint Helical joint Helical pair	H	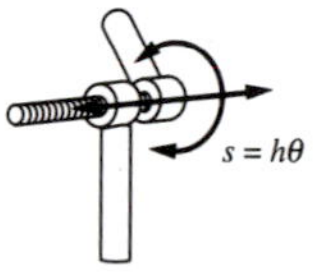	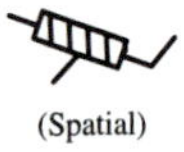
2	Cylindric joint Cylindric pair	C	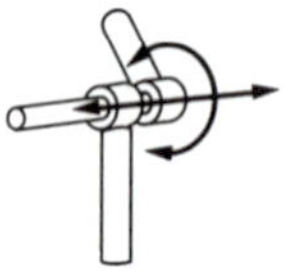	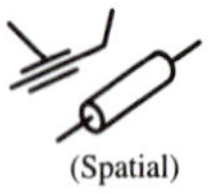
3	Spherical joint Ball joint Spherical pair	S	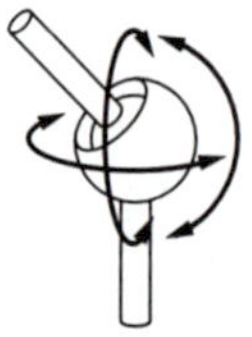	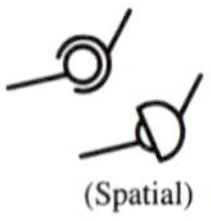
3	Planar joint Planar pair	P_L	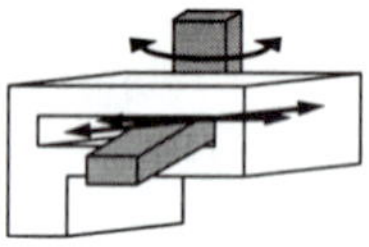	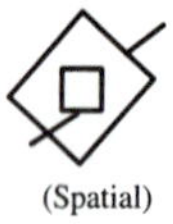

that are actually in contact with the other surface at any instant are at rest relative to that surface. Hence there is no relative sliding of the surfaces and joint friction and wear are minimized. Physically, the limitation of this kind of joint is the stress intensity that the material of the contacting bodies can support. Stresses are necessarily high because of the very small contact areas. If the bodies were perfectly rigid, contact would occur only at discrete points or along a line, the contact area would be zero, and the stresses would be locally infinite!

Lower pair joints such as revolute joints and cylindrical joints are also often simulated by mechanisms such as ball or roller bearings in which there are actually many elements acting in parallel. The actual contact joints in a ball bearing are rolling contacts, which are higher pairs. In this way, the low-friction properties of rolling contacts are exploited to obtain a joint with lower friction and higher load and relative speed capabilities than would be possible with a plain revolute joint. At the same time, the simple overall relative motion geometry of the revolute joint is retained. This is one example of a compound joint in which the joint is actually a complex mechanism but is regarded as kinematically equivalent to a simple revolute. Several examples of compound joints are shown in Table 1.3.

Table 1.2 Some Higher Pair Joints

Connectivity (No. of Degrees of Freedom)	Names	Typical Form	Comments
1	Cylindrical roller	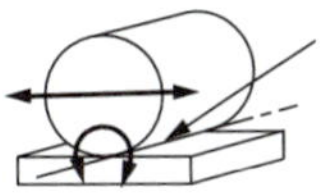	Roller rotates about this line at this instant in its motion. Roller does not slip on the surface on which it rolls.
2	Cam pair	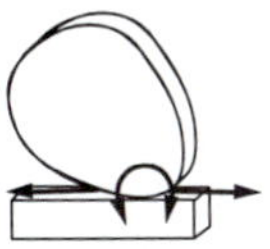	Cam rolls and slides on follower.
3	Rolling ball	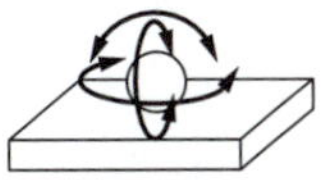	Ball rolls without slipping.
4	Ball in cylinder		Ball can rotate about any axis through its center and slide along cylinder axis.
5	Spatial point contact	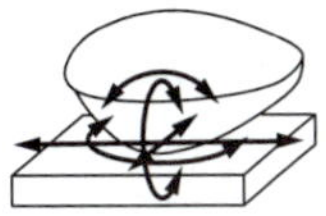	Body can rotate about any axis through the contact point and slide in any direction in the tangent plane.

Conversely, higher pairs are sometimes replaced by equivalent lower pair joints (Fig. 1.2). For example, a pin-in-a-slot joint becomes a combination of a revolute joint and a prismatic joint. Note that this involves adding extra members to the mechanism. In both the case in which a lower pair is replaced by a rolling contact bearing, or compound joint, and this case, the two mechanisms are said to be *kinematically equivalent.* This means that the relative motions that are permitted between the bodies in the two cases are the same, even though the joint is physically quite different.

The number of degrees of freedom of a joint is the minimum number of independent parameters required to define the positions of all points in one of the bodies it connects relative to a reference frame fixed to the other. The term *connectivity* is used to denote this freedom of the body, even though the "joint" may be something very elaborate such as the antifriction bearing shown in Table 1.3 on p. 6 and Fig. 1.3 on p. 7. If motion is

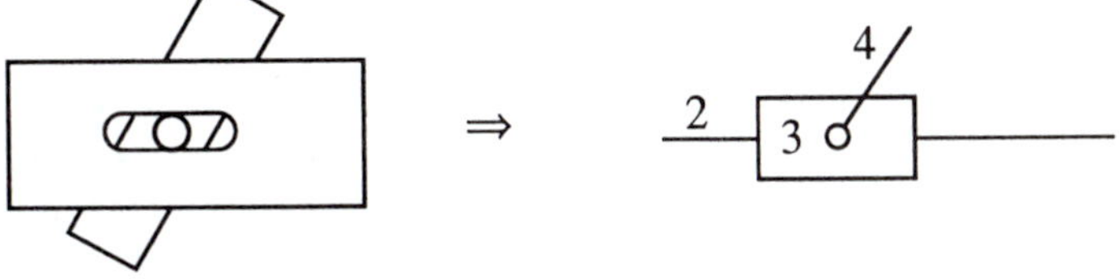

Figure 1.2 Replacement of a higher-pair joint by a kinematically equivalent combination of lower-pair joints.

Table 1.3 Some Examples of Compound Joints

Connectivity	Names	Form
1	Ball bearing Antifriction bearing Rolling contact bearing	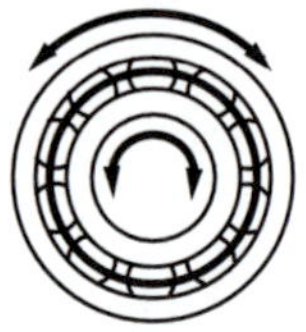
2	Universal joint Hooke joint Cardan joint	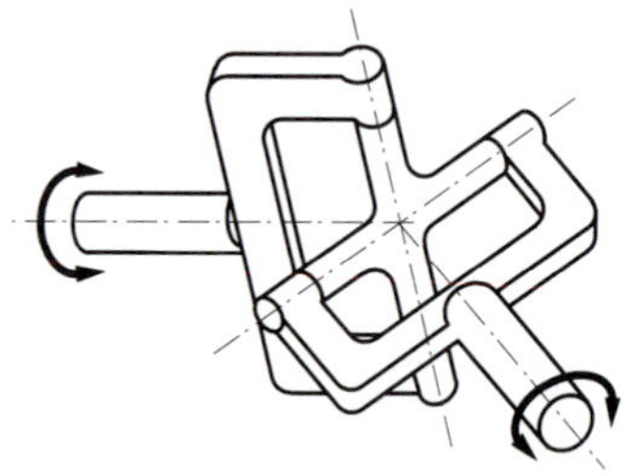
1	Roller slide Roller glide	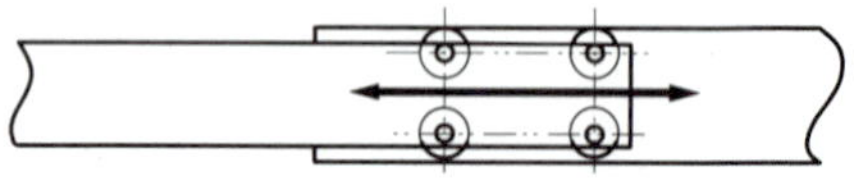

restricted to a plane, the maximum number of degrees of freedom is three. In general spatial motion, the maximum number is six. The number of degrees of freedom for each joint is listed in Tables 1.1, 1.2, and 1.3 in the first column.

1.5 PLANAR LINKAGES

A planar linkage is one in which the velocities of all points in all members are directed parallel to a plane, called the plane of motion. The only lower pair joints that are properly compatible with planar motion are revolute and prismatic joints. The axes of rotation of all revolute joints must be normal to the plane of motion because points would not move in parallel planes otherwise. The directions of sliding of all prismatic joints must be parallel to the plane of motion, since all points in a member connected to another by a prismatic joint move on lines parallel to the sliding direction relative to the second member. Occasionally other lower pair joints will appear in what is otherwise a planar linkage. However, they then function only as revolute or prismatic joints. For example, a spherical joint may be substituted for a revolute joint, but if the linkage is functionally planar, that spherical joint will operate as a revolute with rotation occurring about only the axis normal to the plane of motion. This type of situation will be discussed in more detail in the context of degrees of freedom and mobility.

A common schematic method of representing planar linkages is to represent revolute joints by small circles as shown in Table 1.1. Binary links, those that have two joints mounted on them, are represented as lines joining those joints. Ternary links, those that have three joints mounted on them, are represented as triangles with the joints at the vertices, and so on. Examples of the resulting representation are shown in Figs. 1.4–1.6. The link geometries may then be easily reproduced, giving an accurate view of the linkage in a specified position. Alternatively, the schematic may be used conceptually without accurate geometric data to indicate the topology of the linkage. Topology is the branch of geometry that deals with issues of connectedness without regard to shape. Links with

Figure 1.3 Various rolling-element and plain bearings.

three or more joints should be shaded or crosshatched. Otherwise, the schematic for a quaternary link, one with four joints, cannot be distinguished from the schematic for a four-bar linkage loop.

A kinematic chain is any assemblage of rigid links connected by kinematic joints. A closed chain is one in which the links and joints form one or more closed circuits. Each closed circuit is a loop in which each link is attached to at least two other links.

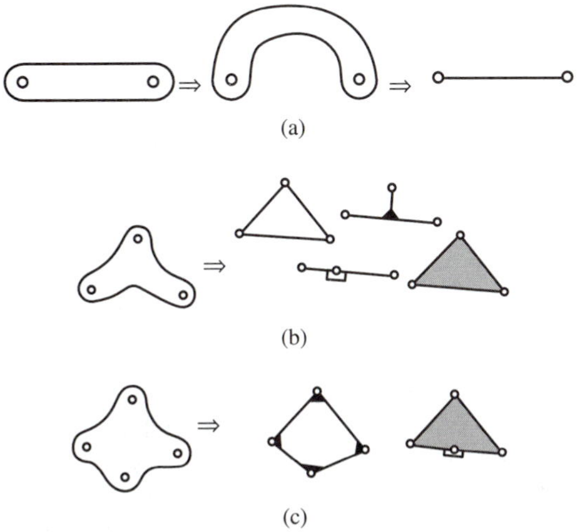

Figure 1.4 Representations of links. (a) Binary links: those to which two joints are mounted. (b) Ternary links and (c) quaternary links; these have three and four joints, respectively.

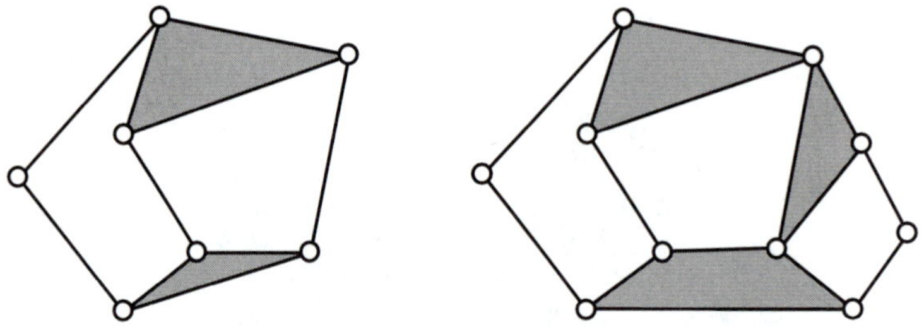

Figure 1.5 Conventional representations of planar linkages. Revolute joints are indicated by circles. Binary links, those with two joints mounted on them, are represented by line segments. Ternary links, with three joints, are represented by triangles, and so on.

Prismatic joints are represented by means of a line in the direction of sliding, representing a slide, with a rectangular block placed on it. This produces linkage representations such as those shown in Fig. 1.6.

A ***frame*** or base member is a link that is fixed. That is, it has zero degrees of freedom relative to the fixed coordinate system. A ***linkage*** is a closed kinematic chain with one link selected as the frame.

In cases in which it is necessary to distinguish the base member of a linkage, it is customary not to show the base as a link in the normal manner but to indicate joints to base by "mounts," as shown in Figs. 1.7 and 1.8.

The term ***mechanism*** is somewhat interchangeable with ***linkage.*** In normal usage, it is a somewhat more generic term encompassing systems with higher pairs, or combinations of lower and higher pair joints, whereas the term linkage tends to be restricted to systems that have only lower pair joints. Mechanisms or linkages are generally represented by their links and joints. The links are numbered with the frame link usually taken as link 1.

Simple, single-loop linkages are given a symbolic designation by a sequence of letters denoting joint types written in clockwise order beginning and ending with the joints mounted to the frame link as shown in Fig. 1.9. The letter designations for the different joints are given in Table 1.1.

The profiles of the contacting surfaces of higher pairs, such as cams and followers, are

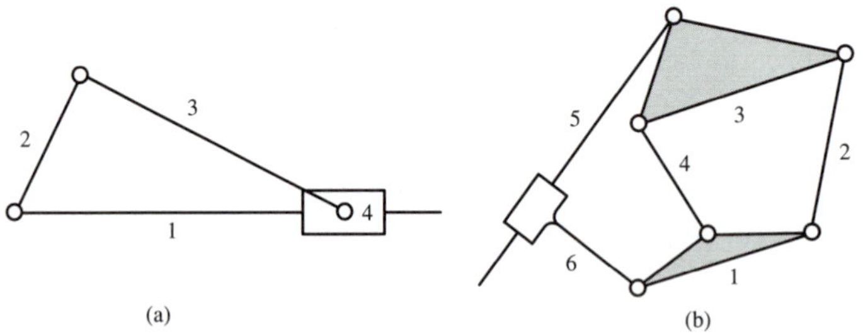

Figure 1.6 Representations of planar linkages with prismatic joints. (a) A four-bar slider-crank mechanism. Note that the sliding "block" is a binary member of the mechanism with a revolute joint and a prismatic joint providing the connections to adjacent members in the loop. The fillets connecting the block to a binary member represented by a line in (b) represent a rigid connection. Thus, the combination is, in this case, a binary member of the linkage.

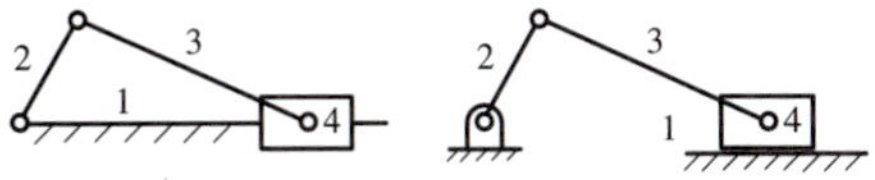

Figure 1.7 Selection of a frame member converts the chain of Fig. 1.5 into a linkage. This linkage is known as a slider-crank linkage.

drawn in planar linkages producing representations such as that shown in Fig. 1.10. Those surfaces must be general (not necessarily circular) cylinders whose straight-line generators are normal to the plane of motion. The profile drawn is, therefore, the generating curve of the cylinder shown in Fig. 1.11. The cylinder is generated by translating that curve along a straight line in the direction normal to its plane. The familiar cylinder with a circular generating curve is called a right circular cylinder.

1.6 VISUALIZATION

Because linkage motion is inextricably intertwined with geometry, it is always important to the designer to visualize the motion. In this respect, planar linkages are relatively easy to work with because their geometry, and loci representing their motion, can be drawn on a two-dimensional surface. Nevertheless, it can be very difficult to visualize successive positions of the links of a planar linkage from only a drawing of that linkage in a representative position. Yet this succession of positions and the relative locations of all the links in each of the positions are very important when trying to predict effects such as interference with each other and with other machine parts. Mechanism designers have traditionally solved this problem by constructing simple physical models with the links cut from cardboard and revolute joints formed by pins or grommets. Card cut from a manila folder with thumbtacks for revolute joints provides an acceptable material for quick visualization models. Prototyping kits (Fig. 1.12) or even children's construction toys (Fig. 1.13) provide an alternative that requires more construction time but gives a more functional model.

When mechanisms are designed using computer graphics systems, animation on a computer is often used to visualize the motion of the mechanism, rather than construction of a physical model. Animation should be used with caution, however. As will be seen in a later section of this chapter, there are important interference effects that do not lend themselves to planar representation but which, if present, are immediately apparent in a physical model.

Furthermore, adding realistic boundary profiles to the representations of links on computer graphic systems is often time consuming and simply not worth the effort when trying

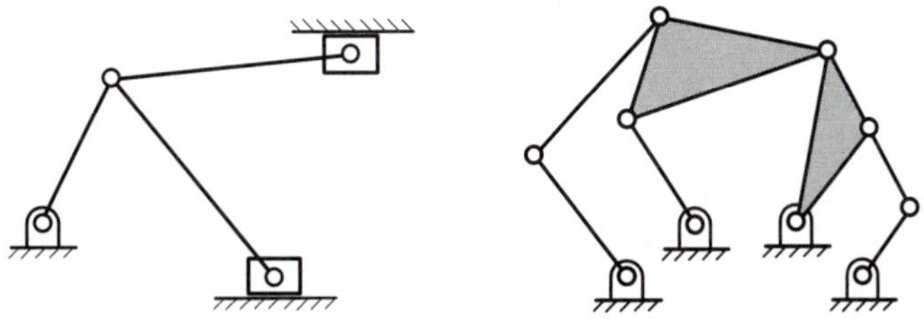

Figure 1.8 Representations of planar linkages with the base link not shown in the same form as the other links. The page can be thought of as representing the base link. The joints to the base link are indicated by hatched "mounts."

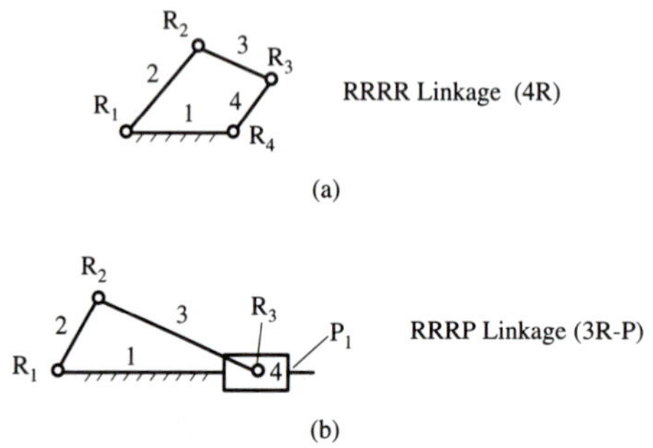

Figure 1.9 Designation of single-loop linkages by means of their joints. The joints are taken in clockwise order around the loop, starting and finishing with a joint to frame.

a variety of different alternative linkage configurations. Instead, quick physical visualization models may be a more efficient alternative. The reader is urged to get into the habit of constructing simple models to visualize the motions of linkages that are being designed, or analyzed, and to make use of computer animation when it is available.

Three-dimensional systems are much more difficult than planar systems to visualize because the depths of the positions for points on the links are no longer constant. Construction of an adequate physical model is often a major effort requiring machining to shape three-dimensional parts. In this case the most efficient solution is to use one of the solid modeling software packages that support linkage joint representations and animation of the linkage. Construction of the model involves a considerable effort since each link must be described as a three-dimensional solid. Nevertheless, the effort is usually much less than would be required for the construction of a physical model. Usually it is possible to change the viewpoint from which the representation is projected. This allows the motion to be viewed from several different directions. It also allows areas of interference to be identified and corrected.

1.7 CONSTRAINT ANALYSIS

The number of degrees of freedom of a body is the number of independent coordinates needed to specify uniquely the position of that body relative to a given reference frame. Similarly, we call the minimum number of coordinates needed to specify uniquely the positions of all of the members of a system of rigid bodies the number of degrees of freedom of that system. In fact, we will use the concept of the number of degrees of freedom in three distinct but closely related ways. The first is the number of degrees of freedom of a body relative to a specified reference frame, which is defined as above. The

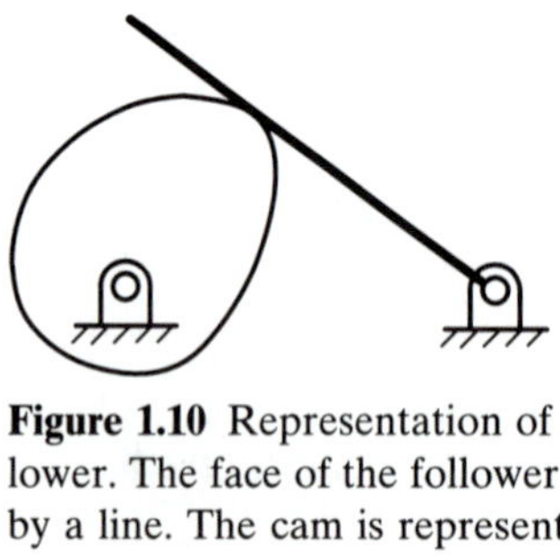

Figure 1.10 Representation of a plate cam with a rocker follower. The face of the follower is a plane, so it is represented by a line. The cam is represented by its profile curve.

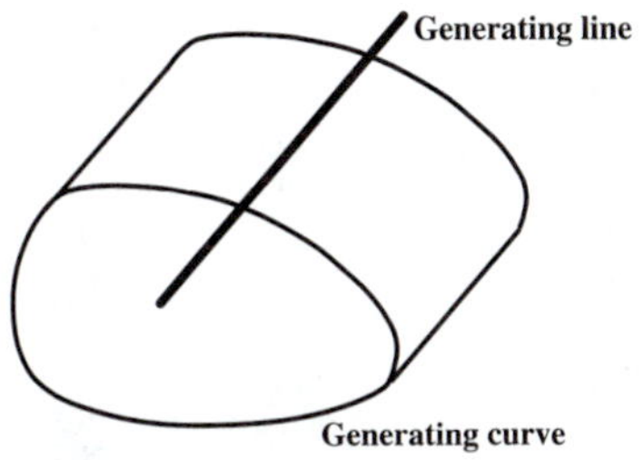

Figure 1.11 General cylinder. The generating curve is a plane curve. Its plane is normal to the generating line. The surface may be considered to be generated by moving the generating curve so that a point on it moves along the generating line. Alternatively, it may be generated by moving the generating line so that a point on it traverses the generating curve.

second is the number of degrees of freedom of a kinematic joint. The third is the number of degrees of freedom of a linkage or mechanism.

Both because "number of degrees of freedom" is such a mouthful and because we are using a distinct concept, we will refer to the number of degrees of freedom of a joint as its ***connectivity.*** In addition, the term will apply to the number of relative freedoms between two bodies. Likewise, we will refer to the number of degrees of freedom of a linkage as the ***mobility*** of that linkage. These terms may be formally defined as follows:

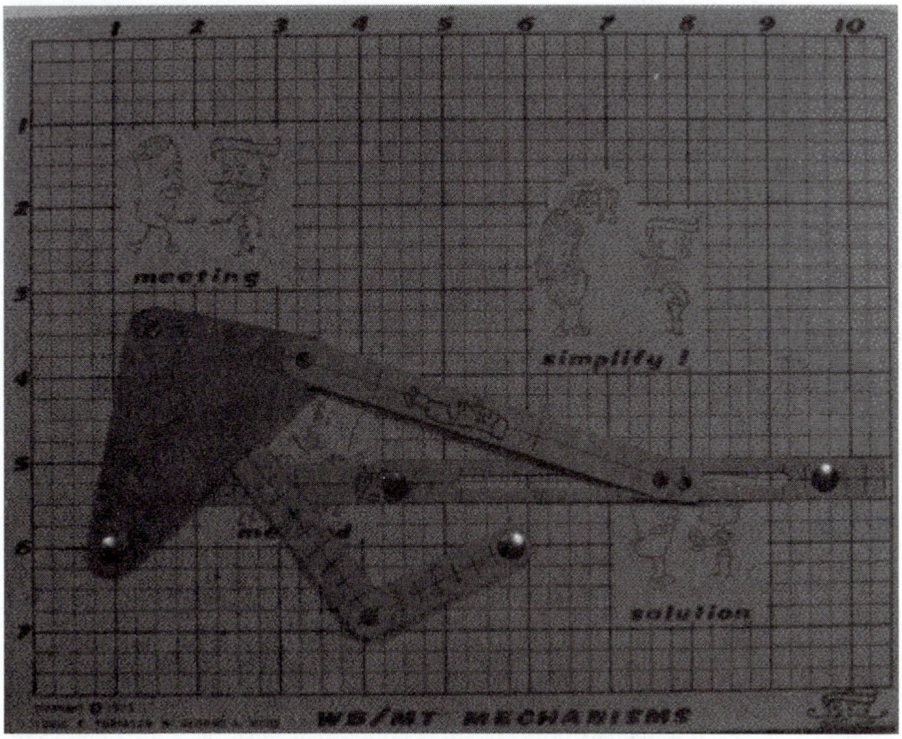

Figure 1.12 J. Woody Blockhead model.[2]

[2] G. .A. Wood and L. E. Torfason, *Mechanism Modeling,* Wood & Torfason, Lincoln, Massachusetts.

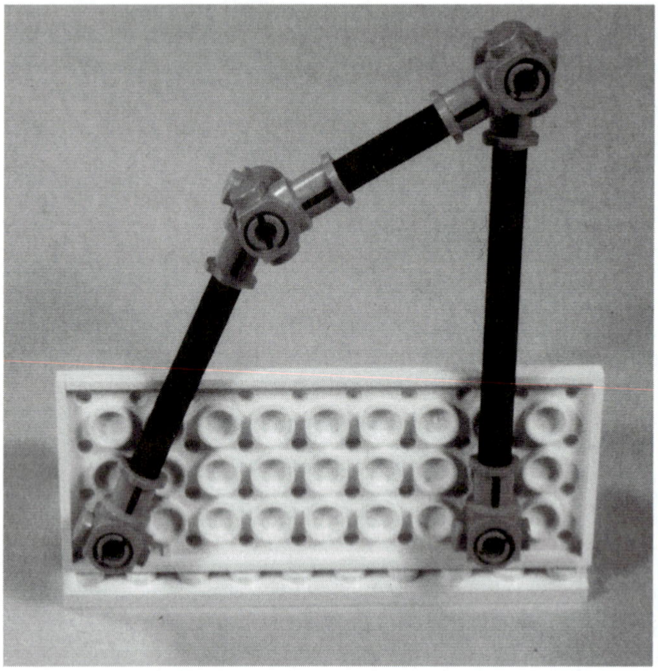

Figure 1.13 Model made with LEGOS Technics.[3]

If a kinematic joint is formed between two rigid bodies that are not otherwise connected, the ***connectivity*** of that joint is the number of degrees of freedom of motion of (either) one of the two bodies joined relative to the other.

The ***mobility*** of a mechanism is the minimum number of coordinates needed to specify the positions of all members of the mechanism relative to a particular member chosen as the base or frame.

The mobility, or number of degrees of freedom of a linkage, is used to determine how many pair variables must be specified before the positions of all of the points on all of the members of the linkage can be located as a function of time. A linkage has a mobility of one or more. Traditionally, almost all linkages had one degree of freedom. However, in modern design practice, linkages with two or more degrees of freedom are becoming more common. If the mobility is zero, or is negative, as determined by the constraint equations developed below, the assemblage is a structure. If the mobility is zero, the structure is statically determinate. If the mobility is negative, the structure is statically indeterminate.

To compute the mobility, let us consider the planar case first and then extend the results to the spatial case. As indicated in Fig. 1.14, in the plane, a body moving freely has three degrees of freedom. Suppose that in a given linkage, there are n links. If they are all free to move independently, the system has mobility $3n$. If one link is chosen as the frame link, it is fixed to the base reference frame and loses all of its degrees of freedom. Therefore the total mobility of the system is $3(n - 1)$ with no joints formed between the members.

If a joint with connectivity f_i (f_i degrees of freedom) is formed between two bodies, the mobility of the system is diminished since those two bodies originally had three degrees of freedom of motion relative to one another. After formation of the joint, they have only f_i degrees of freedom of relative motion. Hence the reduction in the system mobility is

[3] LEGO Systems, Inc., 555 Taylor Road, Enfield, Connecticut.

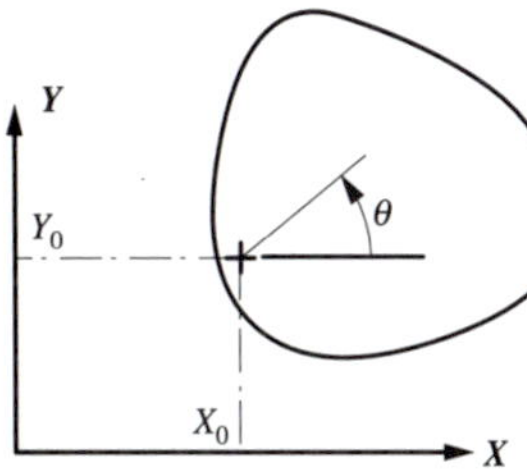

Figure 1.14 One set of three coordinates that can be used to describe planar motion. The number of degrees of freedom of a body is the number of independent coordinates needed to specify its position. Therefore, a body moving freely in a plane has three degrees of freedom.

$3 - f_i$. If joints continue to be formed until there are j joints, the loss of system mobility is

$$(3 - f_1) + (3 - f_2) + \cdots + (3 - f_j) = \sum_{i=1}^{j} (3 - f_i) = 3j - \sum_{i=1}^{j} f_i$$

Then the total mobility of the linkage will be

$$M = 3(n - 1) - \left(3j - \sum_{i=1}^{j} f_i \right) = 3(n - j - 1) + \sum_{i=1}^{j} f_i \qquad \textbf{(1.1)}$$

Equation (1.1) is called a constraint criterion. There are many different-appearing versions of this relationship to be found in the literature. They all, in fact, are equivalent to one another, except that some are restricted to a subset of the cases covered by Eq. (1.1).

A problem arises in some cases in which more than two members are apparently connected by the same joint. Typically, three or more members are pinned together by the same shaft and are free to rotate relative to one another about the same revolute axis. This difficulty is readily resolved if we recall that a kinematic joint is formed by contact between the surfaces of *two* rigid bodies. This is the reason for Reuleaux's name "pair" for what we here call a "joint." Considering the present case, we see that there is not one joint but several between the bodies. In fact, if p members are connected by a "common" joint, the connection is equivalent to $p - 1$ joints all of the same type. Inclusion of this number in j, and $(p - 1)f$ in the connectivity sum of Eq. (1.1) will ensure correct results (see Example 1.3).

EXAMPLE 1.1 **(Degrees of Freedom in a Simple Four-Bar Linkage)**

PROBLEM   Determine the mobility of the planar four-bar linkage shown in Fig. 1.15.

SOLUTION

$n = j = 4$

$\sum_{i=1}^{j} f_i = j \times 1 = 4$

$M = 3(4 - 4 - 1) + 4 = 1$

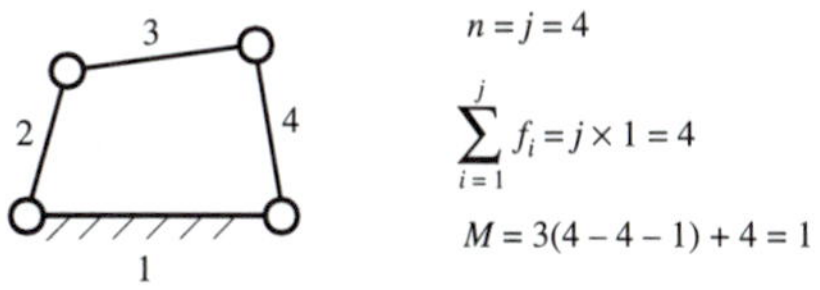

Figure 1.15 Mobility analysis of a planar four-bar linkage.

EXAMPLE 1.2 *(Degrees of Freedom in a Complex Mechanism)*

PROBLEM

Determine the mobility of the linkage shown in Fig. 1.16. The linkage is planar and all joints have connectivity one.

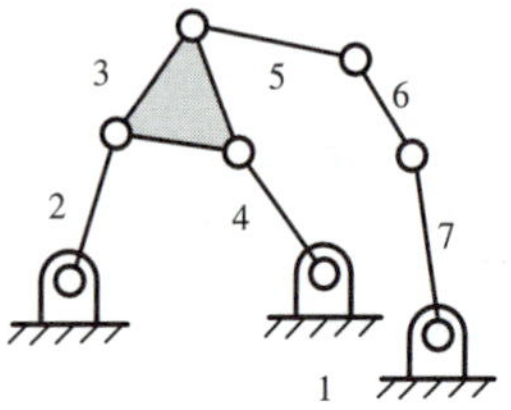

$$n = 7,\ j = 8$$

$$\sum_{i=1}^{j} f_i = j \times 1 = 8$$

$$M = 3(7 - 8 - 1) + 8 = 2$$

Figure 1.16 Mobility analysis of a two-loop planar linkage.

SOLUTION

Notice that the base member must always be counted even when it is not shown in the same way as the other members but just by a set of "bearing mounts." ■

EXAMPLE 1.3 *(Degrees of Freedom When Joints Are Coincident)*

PROBLEM

Determine the mobility of the linkage shown in Fig. 1.17. The linkage is planar and all joints have connectivity one. Links 3, 4, and 5 are connected at the same revolute joint axis.

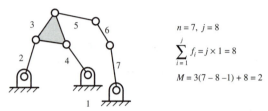

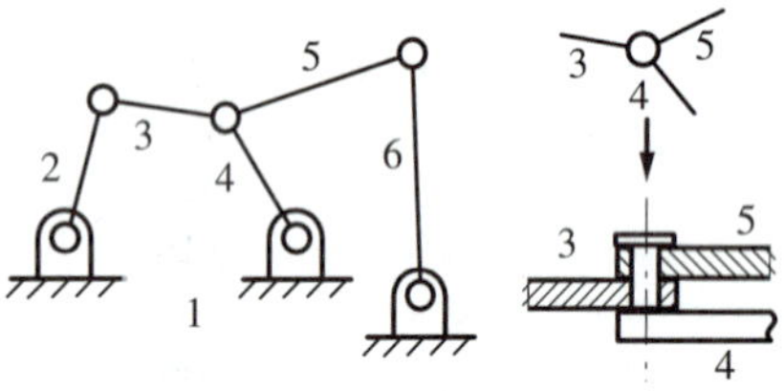

Figure 1.17 Mobility analysis of a linkage when more than two members come together at a single point location.

SOLUTION

$$n = 6, j = 7$$

$$\sum_{i=1}^{j} f_i = j \times 1 = 7$$

$$M = 3(6 - 7 - 1) + 7 = 1$$

As indicated above, when p members are connected at the same joint axis, then $p - 1$ joints are associated with the same axis. Hence the location where links 3, 4, and 5 come together counts as two revolute joints. As indicated in the figure, members 3 and 5 can be thought of as being connected to link 4 by two separate revolute joints that have the same axis of rotation. ■

EXAMPLE 1.4 *(Degrees of Freedom for a Mechanism Containing a Higher Pair)*

PROBLEM

Determine the mobility of the linkage shown in Fig. 1.18. The linkage is planar and not all of the joints have connectivity one.

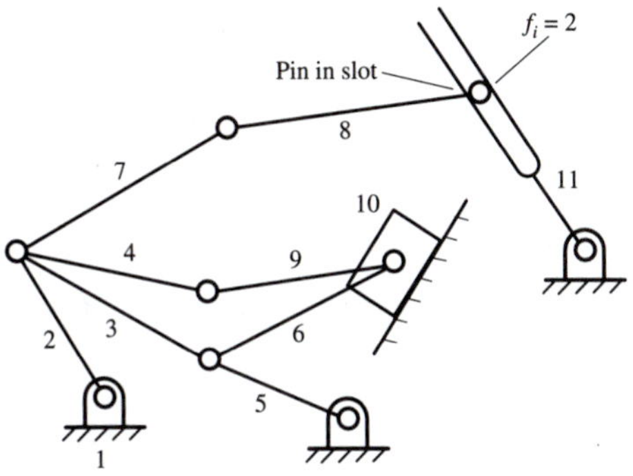

Figure 1.18 Mobility analysis of a linkage with various types of joints.

SOLUTION

In this mechanism, there are three places where more than two links come together at the same revolute joint location. In addition, there is a pin-in-a-slot joint that permits two degrees of freedom (connectivity equals 2). Therefore, the joints must be counted carefully. When this is done, we find n and j to be

$$n = 11, j = 14$$

and

$$\sum_{i=1}^{j} f_i = 13 \times 1 + 1 \times 2 = 15$$

Then,

$$M = 3(11 - 14 - 1) + 15 = 3$$ ■

A special case that deserves attention occurs when the mobility in Eq. (1.1) is set to one and all joints have connectivity one ($f_i = 1$). Then,

$$1 = 3(n - j - 1) + j$$

or

$$4 = 3n - 2j \quad \textbf{(1.2)}$$

Because n and j are integers, n must be even because 4 and $2j$ are both even numbers. This is an example of a Diophantine equation. That is one which admits only integral solutions. Written as an expression for j in terms of n, the equation becomes

$$j = 3n/2 - 2$$

Some of the possible solutions are listed in Table 1.4. In each case, the joints may be either revolute or prismatic joints, since they are the only lower pair joints that can properly be included in planar linkages.

Table 1.4 Different Integer Solutions to Eq. (1.2) for Mobility of One

Solution No.	n	j	Number of Configurations
1	2	1	1
2	4	4	1
3	6	7	2
4	8	10	16
5	10	13	230
6	12	16	6856

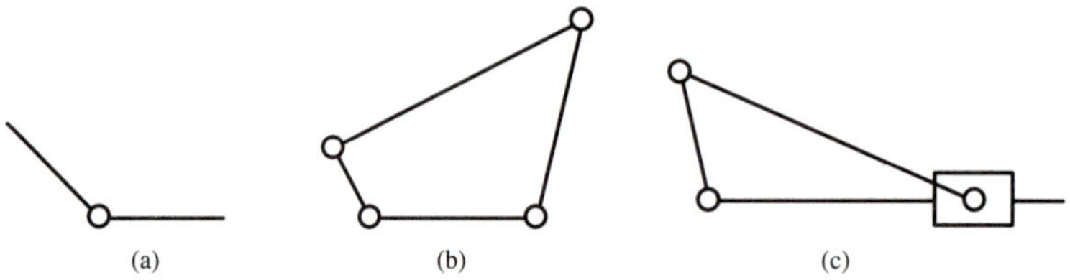

Figure 1.19 Solutions of the planar mobility equation for $M = 1$ when $n = 2$ and $n = 4$.

Solution 1 gives the rather trivial case of two bodies connected by a single revolute or slider joint. This is shown in Fig. 1.19a. Actually, this mechanism is very common. For example, a door, its hinges, and the door frame form an open kinematic chain and a mechanism of this type.

Solution 2 gives a single, closed loop of four members with four joints. Two forms are shown in Fig. 1.19b and c. The one in Fig. 1.19b is the planar four-bar linkage which forms a major element in planar linkage theory. The one in Fig. 1.19c is the slider-crank linkage, which is also extensively studied.

Solution 3 presents two new features. First, members with more than two joints mounted on them appear. Second, even when only revolute joints are included, there are two possible, topologically distinct, configurations of six members with seven joints. These are respectively named the Watt and Stephenson six-bar linkages and are shown in Fig. 1.20.

Solution 4 gives 16 possible different topological configurations, shown in Fig. 1.21, and solution 5 gives 230. The number increases very rapidly with larger numbers of members. For example, solution 6 gives 6856 configurations (Hunt, 1978).[4]

From the above, it should be apparent why we spend so much effort on the design of four-link mechanisms. The four-link arrangement is the simplest possible nontrivial linkage. It turns out that most design requirements can be met by four- or six-link mechanisms.

Note that, in the above, the type of the joints was not specified. All that was specified was that the joints have connectivity one and that the linkage is planar and has mobility one. Although the joints pictured in Figs. 1.19–1.22 are all revolute, rolling contact joints could be substituted for any of the joints, and prismatic joints could be substituted for some of them. Thus, even if the joints are confined to lower pairs, the four-link, four-joint solution represents the four different chains shown in Fig. 1.22. The Scotch Yoke, based on the 2R-2P chain, is shown in Fig. 1.23.

Furthermore, as discussed later in this chapter, the important concept of inversion generates several different linkages from any mechanism based on the 3R-P and 2R-2P

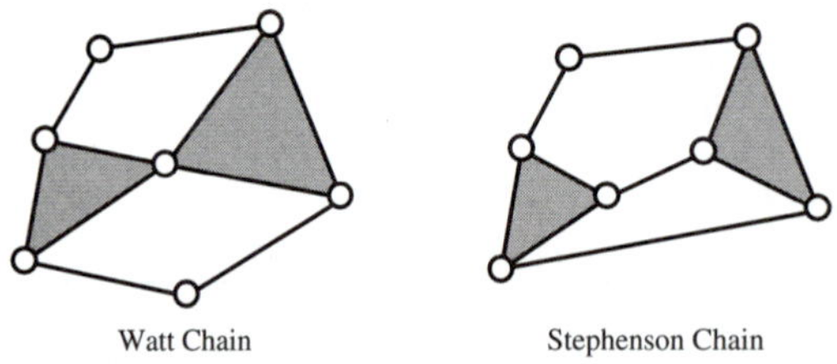

Figure 1.20 The two solutions of the planar mobility equation for seven revolute joints. $M = 1$ and each kinematic chain has six members.

[4] Hunt, K. H., *Kinematic Geometry of Mechanisms,* Oxford University Press, Oxford, p. 40, 1978.

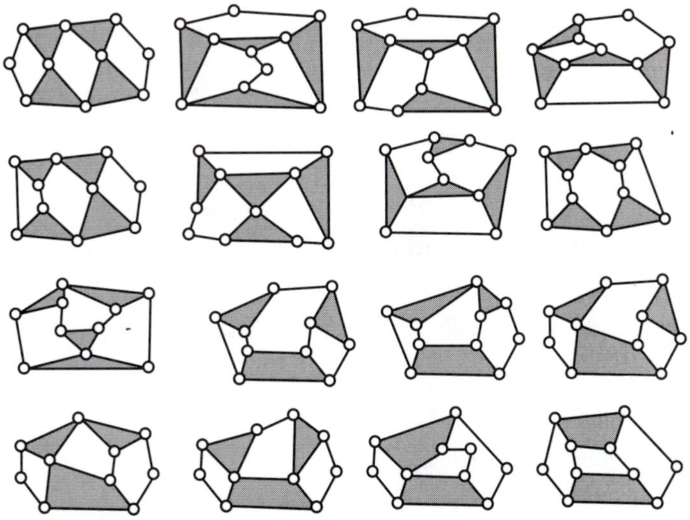

Figure 1.21 The 16 solutions of the planar mobility equation for ten revolute joints. $M = 1$ and each kinematic chain has eight members.

chains. An inversion is a different mechanism derived from a given mechanism or linkage. "Different" means that the motion relative to the frame that can be produced by the inversion is different from that provided in the original mechanism; that is, the inversion produces a different general form for the paths of points on the different links or a different input-output function.

A different inversion is produced for each choice of frame link. As a result, the 3R-P chain produces four different mechanisms. In the basic slider-crank mechanism, the frame member has one revolute and one prismatic joint mounted on it. We can also make the slider the frame. The other two inversions are turning block linkages in which the base has two revolutes mounted on it. The 2R-2P chain can produce three different mechanisms: the Scotch yoke has one revolute and one prismatic joint mounted on the frame; the double slider has two prismatic joints on the frame; and the third mechanism, the Oldham coupling, has both revolutes mounted on the frame.

1.8 CONSTRAINT ANALYSIS OF SPATIAL LINKAGES

In spatial motion, each body that moves freely has six degrees of freedom rather than three. Using exactly the same reasoning as was used in the planar case, the constraint criterion equation becomes

$$M = 6(n - j - 1) + \sum_{i=1}^{j} f_i \tag{1.3}$$

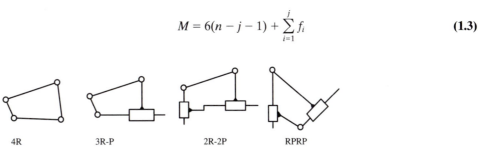

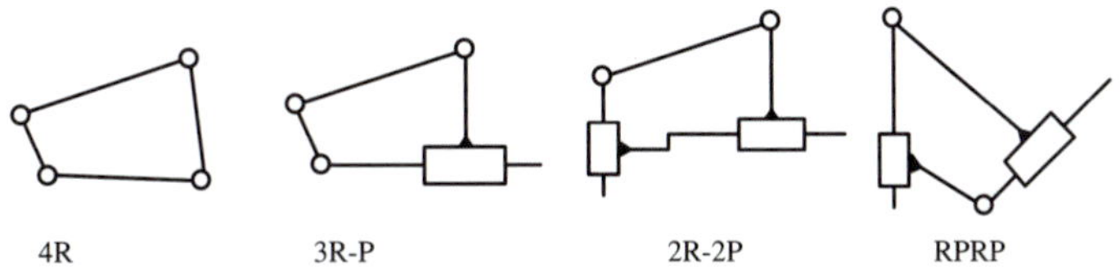

Figure 1.22 Four different forms of four-bar chains with combinations of revolute and prismatic joints.

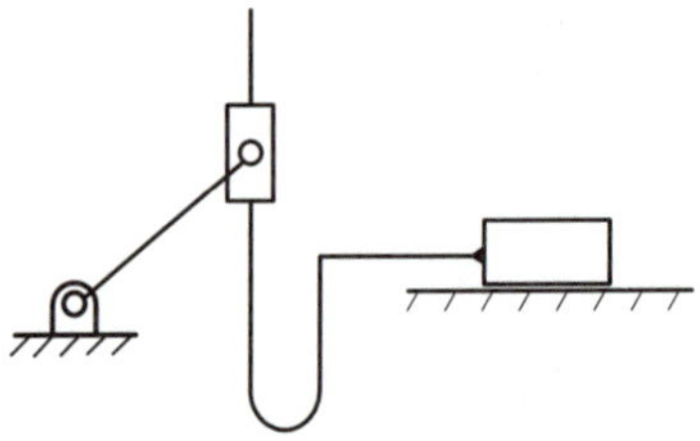

Figure 1.23 The 2R-2P chain as a Scotch yoke mechanism.

This is called the Kutzbach criterion. If only lower pair joints are involved, each with connectivity one, the equations become

$$M = 6(n - j - 1) + j = 6n - 5j - 6$$

If the linkage is required to have mobility one, this gives

$$6n = 7 + 5j \quad \textbf{(1.4)}$$

Equation (1.4) corresponds to Eq. (1.2) derived in the case of planar motion. Like that equation, it is a Diophantine equation that admits only integral values of the variables. Evidently, j must be odd because $5j$ must be odd to combine with the odd number 7 to produce the even number $6n$. The sum $7 + 5j$ must also be divisible by three. Solutions to Eq. (1.4) are a little harder to generate than those of Eq. (1.2). The simplest solution is given by $j = 1$ and $n = 2$. This is exactly the same as the simplest solution in the planar case depicted in Fig. 1.19a. The next allowable solution is $j = 7$ and $n = 7$. This is a single, closed loop with seven members and seven joints. It bears the same relationship to general spatial linkage topologies that the planar four-bar linkage does to planar ones. The next order solution is $j = 13$, $n = 12$. There are three distinct topological forms in this case. For spatial mechanisms, the complexity increases with the number of members and joints even more rapidly than it does for planar joints.

EXAMPLE 1.5 *(Degrees of Freedom in a Spatial Mechanism)*

PROBLEM

Determine the mobility of the linkage shown in Fig. 1.24. The linkage is spatial. The joints are lower pairs of the types labeled.

Note how the three-dimensional joints are drawn. There is no formalism that is more or less universally recognized for representing spatial mechanisms as there is for planar linkages; however, we will follow the symbols shown in Table 1.1.

SOLUTION

$$n = j = 4$$

$$\sum_{i=1}^{j} f_i = 2 \times 3 + 1 \times 1 + 1 \times 2 = 9$$

$$M = 6(4 - 4 - 1) + 9 = 3$$

■

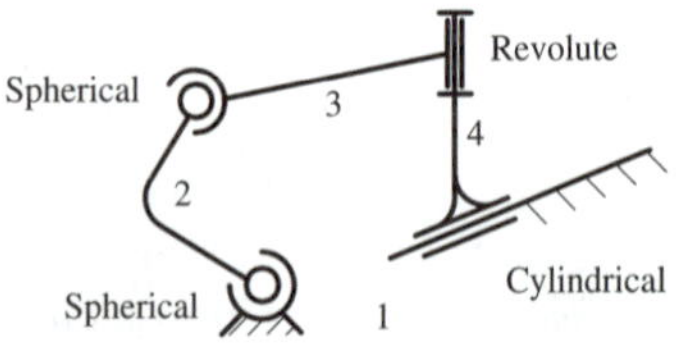

Figure 1.24 A four-member, single-loop, spatial linkage.

Another way of looking at the constraint criterion is in terms of closures. Imagine building up the linkage by starting with the base link and successively adding members and joints. If a joint connects an additional member to the system, the number of degrees of freedom is increased by f_i, if f_i is the connectivity of that joint, and the numbers of members and joints are both increased by one. If a joint is made between two members that are already part of the linkage, the total number of degrees of freedom is decreased by the number of constraints imposed by that joint. The number of constraints imposed by a joint is the number of degrees of freedom lost by the system when that joint is formed. For a spatial mechanism, it is $6 - f_i$ since two bodies have six degrees of freedom of motion relative to one another when they are free of each other and only f_i degrees of freedom of relative motion after the joint is formed. Also, in this case, the formation of the joint results in the formation of a closed loop of members and joints within the linkage. This is called a closure. Proceeding in this manner, the mobility of the linkage can be expressed as

$$M = \sum_{i=1}^{j} f_i - 6c$$

where c is the number of closures. Now, when a closure is formed, the number of members does not increase, whereas the number of joints increases by one. If there are no closures (open kinematic chain), the number of link members is given by

$$n = j + 1$$

the additional member being the base member. Therefore, if there are c closures in the linkage

$$c = j + 1 - n$$

Thus, substitution for c in the expression for the mobility leads to Eqs. (1.4). The relationship among c, j, and n is illustrated in Fig. 1.25.

The reason for looking at the constraint criterion from this viewpoint is that it relates to the position analysis of a spatial linkage. When a closure is formed, a set of six algebraic equations called closure equations can be written. The formulation of these equations will be briefly treated in Chapter 7, although their study lies largely beyond the scope of this book. The quantity $6c = 6(j + 1 - n)$ is therefore the number of equations available for position analysis of the mechanism. The variables in those equations are the joint parameters, the variables needed to fix the relative positions of the bodies connected by each

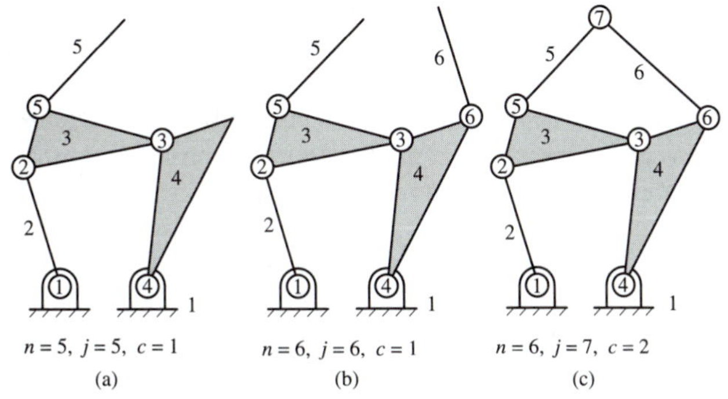

Figure 1.25 The effect of adding a member to a linkage together with a joint (b) and of adding a joint without an additional member (c). Adding a joint without a member always closes a loop within the linkage.

joint. There are f_i of these joint parameters for joint i. Therefore the total number of variables in the linkage is

$$\sum_{i=1}^{j} f_i$$

In this way, it may be seen that Eq. (1.4) expresses the mobility of the linkage as the number of variables less the number of equations for the system.

Yet another viewpoint on the constraint criterion that it is productive to pursue is that of static force analysis. Free-body diagrams can be drawn for all members except the base. Six static equilibrium equations can be written for each free body. Hence there are $6(n - 1)$ equations describing the system. At each joint there is a number of reaction force and torque components which is equal to the number of constraints of that joint. These force components are the variables in a static force analysis. Since the number of constraints at joint i is $6 - f_i$, the number of variables is

$$\sum_{i=1}^{j} (6 - f_i) = 6j - \sum_{i=1}^{j} f_i$$

Therefore, the difference between the number of variables and the number of equations is

$$6j - \sum_{i=1}^{j} f_i - 6(n - 1) = -M$$

Thus, the mobility is meaningful from the point of view of static force analysis also. If $M = 0$, the linkage is not movable and is a structure. The position problem can be solved to obtain the joint positions that cannot vary. The static equilibrium problem can be solved for all of the reaction force and torque components. The structure is statically determinate since there is a unique solution to the static equilibrium problem.

If the mobility is -1, the number of equations for the position problem exceeds the number of variables. Therefore, in general there is no solution to the position problem. For a solution to exist it is necessary for the equations to be dependent. This means that the geometry of the mechanism must satisfy the conditions needed for the equations to be dependent. Physically, this means that, in general, it is not possible to assemble the linkage. One or more of the closures cannot be made. However, if the link geometry is changed to bring the surfaces for the closing joint into alignment, the linkage may be assembled.

From the viewpoint of force analysis, the mobility is the number of static equilibrium equations less the number of force variables: the converse of the situation for position analysis. Thus, if $M = -1$ there is one more force variable than the number of force equations. Therefore, in this case solutions of the system exist, but there is no unique solution. The force problem cannot be solved without additional information relating the forces in the system. The linkage is a statically indeterminate structure. If the links are modeled as elastic, rather than rigid solids, compatibility of their deflections under load provides the necessary additional relationship.

Conversely, if the mobility is one or more, the number of position variables is greater than the number of position equations. Solutions to the system exist, but there is no unique solution. The number of force equations is greater than the number of force variables, so, in general, no solution to the static force problem exists. In practice, application of an arbitrary set of loads to the linkage would lead to rapid, uncontrolled acceleration, and the system behavior could not be described without writing dynamic equations. However, this invalidates the assumption of a static model.

Specification of the value of a joint parameter is equivalent to fixing that joint. Physically, it might be done by putting an actuator on that joint which would hold it in position. The joint can now support a force, or torque. The effect is to increase the number of unknown force variables by one. If a linkage has mobility one, fixing the position of a joint with connectivity one converts it into a structure. It also converts the static force problem from one in which there is one more equation than there are variables to one in which the number of variables is the same as the number of equations. That is, it is statically determinate.

Fixing the torque applied about a revolute joint, or the force applied by an actuator at a prismatic joint, has a quite different effect. It does not change the number of variables or the number of equations in either the position or the force problem. This is because having a passive joint is already equivalent to fixing the force or torque variable about that joint. The torque applied at a passive revolute joint is fixed to zero. Changing it to any other value does not affect the number of unknown variables. Of course, it does affect the values of the unknown force variables.

This is quite important in practical applications of multiply actuated mechanisms. Consider the manipulator arm shown in Fig. 1.26. It has seven members (italic numbers) and six joints. The heavy dashed lines with bold numbers indicate the joint axes. Joints 1, 2, 4, 5 and 6 are revolute joints. Joint 3 is a prismatic joint. The axes of joints 3 and 4 are the same. Member one is the base member.

Applying the constraint criterion to this mechanism, we have $n = 7, j = 6, \Sigma f_i = 6$, so

$$M = 6(7 - 6 - 1) + 6 = 6$$

If we actuate all of the joints so that we can specify their positions, the position of the mechanism is uniquely specified.

Consider now what happens if the manipulator grips an object that is fixed relative to the base member, as shown in Fig. 1.27. It is assumed that the gripper grasps the object tightly so that no relative motion is possible. The effect is to make link 7 a part of link 1. Therefore, application of the constraint equation gives $n = 6, j = 6, \Sigma f_i = 6$, so

$$M = 6(6 - 6 - 1) + 6 = 0$$

The mechanism is now a structure and we do not have the liberty of setting the joint variables to any value we wish. Attempting to control the mechanism by commanding joint positions, as done when the manipulator is moving freely, is not effective in this case.

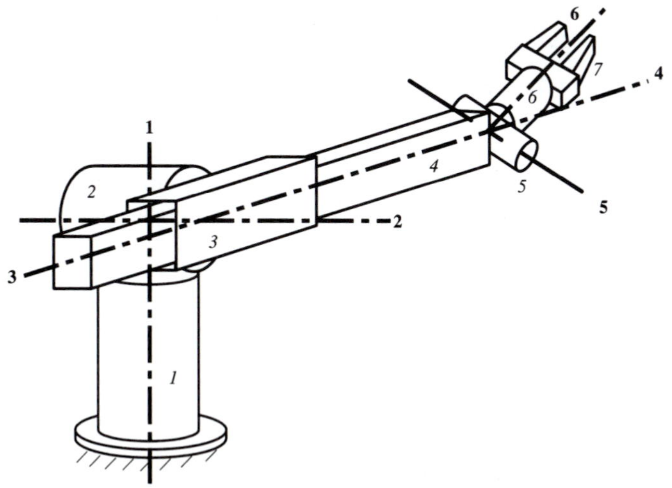

Figure 1.26 A robotic manipulator that is used to produce general spatial motions of its gripper. The mechanism has seven members, indicated by the italic numbers, and six joints. Joints 1, 2, 4, 5, and 6 are revolutes. Joint 3 is a prismatic joint. The heavy dashed lines indicate the joint axes. The axes of joints 3 and 4 are coincident.

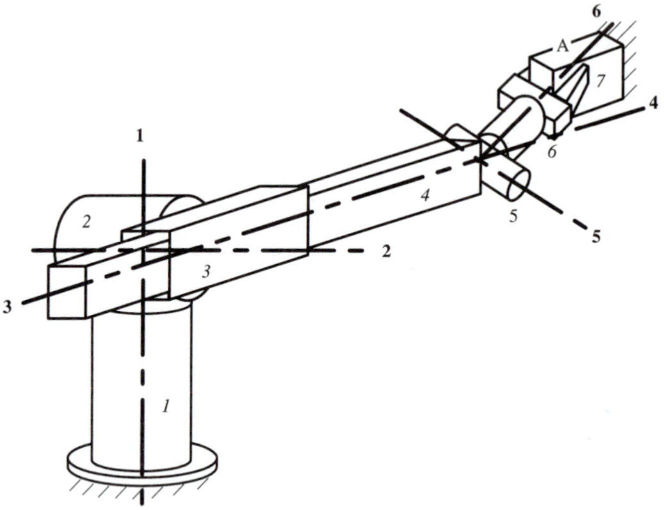

Figure 1.27 The robotic manipulator of Fig. 1.26 gripping a fixed object. If the gripper grasps the object so that no relative motion is possible, the gripper becomes fixed to member one. This reduces the number of members in the system to six and closes a loop.

Since most manipulator structures are very stiff, a small position error results in very large forces on the actuators. The usual result is that the actuator controllers become unstable, producing violent vibratory behavior. However, if the actuators are commanded to produce specified forces or torques, there is no problem. The actuator torques and forces can be set to any desired set of values. In this way it is possible to apply a specified force system to the fixed object A by means of the manipulator. Notice that commanding forces and torques all the time is not a solution. If actuator forces are commanded when the manipulator is moving freely, the number of static equilibrium equations exceeds the number of variables by six and the manipulator will perform rapid uncontrolled movements, violating the assumption of static stability.

1.9 IDLE DEGREES OF FREEDOM

Equation (1.4) sometimes gives misleading results. There are several reasons for this. One is the phenomenon of idle degrees of freedom. Consider the linkage shown in Fig. 1.28. This linkage has four members and four joints. Two of the joints are revolutes. The other two are spherical joints. This mechanism is quite often used in situations such as the steering mechanisms of automobiles. Applying the constraint criterion, we have $n = 4$, $j = 4$, $\Sigma f_i = 2 \times 1 + 2 \times 3 = 8$.

Therefore

$$M = 6(4 - 4 - 1) + 8 = 2$$

Nevertheless, practical experience with this mechanism shows that there is a unique value of the output joint angle, ϕ, for any given value of the input angle, θ. How can this be explained?

Examination of the mechanism reveals that the coupler member is free to spin about the line through the centers of the two spherical joints. This motion can take place in any position of the linkage without affecting the values of the input and output joint angles. It is what is termed an idle degree of freedom. That is, it is a degree of freedom which does not affect the input-output relationship of the linkage.

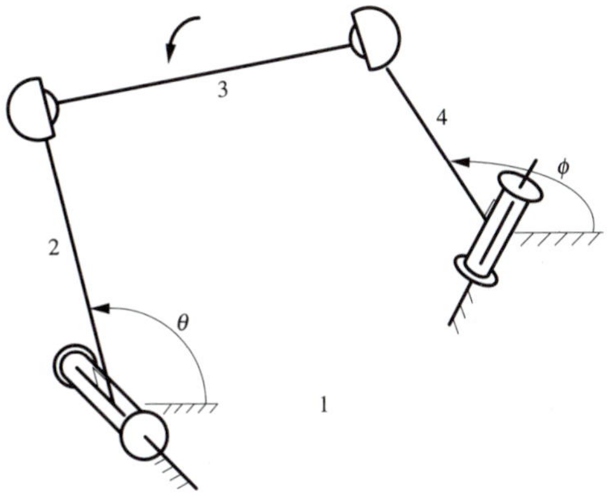

Figure 1.28 A spatial four-member, four-joint linkage. Two of the joints are revolutes. The other two are spherical joints. θ is the input joint angle and ϕ is the output joint angle. The linkage has an idle degree of freedom since member 3 can spin about the line joining the centers of the spherical joints without affecting the relationship between θ and ϕ.

The real problem here is that usually we are not really interested in the mobility of the entire linkage, that is, of all of its links. Rather, we are interested in the connectivity that the linkage provides as a joint between two of its members. This is a new use of the term connectivity. Previously we applied it only to simple joints at which the members contact each other directly. However, a mechanism constrains the number of degrees of freedom of relative motion of any two of its members. Therefore it can be regarded as forming a kinematic joint between any two of its members. We can define its connectivity as a joint between those members and as the number of degrees of freedom of relative motion which it permits between the members.

In the example of Fig. 1.28, the connectivity of the linkage as a joint between the input and output members is one, even though the mobility of the linkage is two, and the connectivity between links 3 and 1 is two. The mobility places an upper bound on the connectivity of the mechanism as a joint between any two of its members. There is no simple method of directly determining connectivity, so the mobility equation is used. If the mobility is one and the linkage is not overconstrained in some local region, there is no problem. The connectivity of the linkage as a joint between any two of its members is also one. If the mobility is greater than one, strictly speaking, all that can be said is that the connectivity between any given pair of members may be equal to the mobility or may be less than that number. Fortunately, idle degrees of freedom usually can be identified by inspection.

Another example is shown in Fig. 1.29. This is one form of the so-called Stewart platform mechanism. This mechanism is commonly used to produce general spatial motions in aircraft simulators for training pilots. The output member is connected to the base by six "limbs," each of which has an actuated prismatic joint in the middle and two spherical joints at either end. There are 14 members: 2 in each of the limbs plus the base and output members. There are 18 joints: 6 prismatic joints and 12 spherical joints. Hence $\Sigma f_i = 6 \times 1 + 12 \times 3 = 42$.

Therefore

$$M = 6(14 - 18 - 1) + 42 = 12$$

However, it is easily seen that each limb is free to spin about the line joining the centers of its spherical joints without affecting the position of the output member relative to the

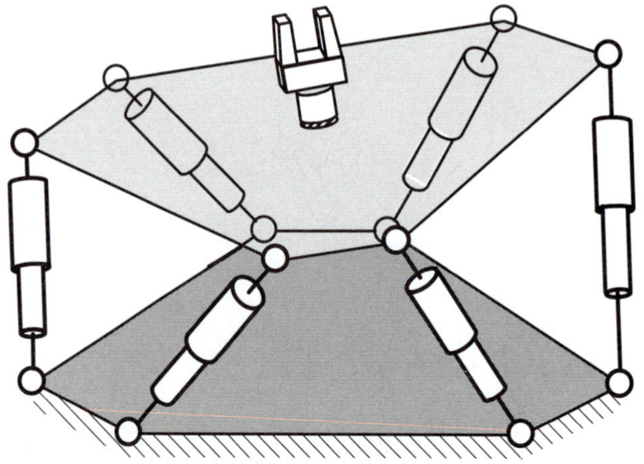

Figure 1.29 Stewart platform.

base. Therefore, the mechanism has six idle degrees of freedom, and its connectivity as a joint between base and output member is

$$C = M - 6 = 6$$

Therefore, by appropriately positioning the actuated prismatic joints, the output member can be placed in any position within its working volume.

While idle degrees of freedom are most common in spatial linkages, they can also occur in planar linkages. Typically, this occurs when cam roller followers are involved. For example, if the mobility of the linkage in Fig. 1.30 is computed, it will be found to be 1 if there is rolling contact between the roller (link 5) and the cam (link 6) at point C. However, if there is cam contact at C, the mobility will be 2. The extra degree of freedom is associated with the free rotation of link 5 relative to the frame. Usually, this rotation will be of no interest because the motion of all of the other links in the mechanism will be unaffected by this rotation.

1.10 OVERCONSTRAINED LINKAGES

A second reason why the constraint criteria (Eqs. 1.1 and 1.4) sometimes give misleading results is the phenomenon of overconstraint. A mechanism can be overconstrained either

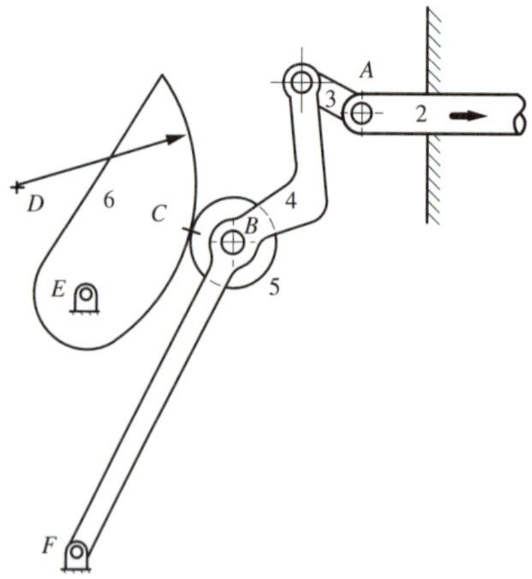

Figure 1.30 Planar mechanism with an idle degree of freedom.

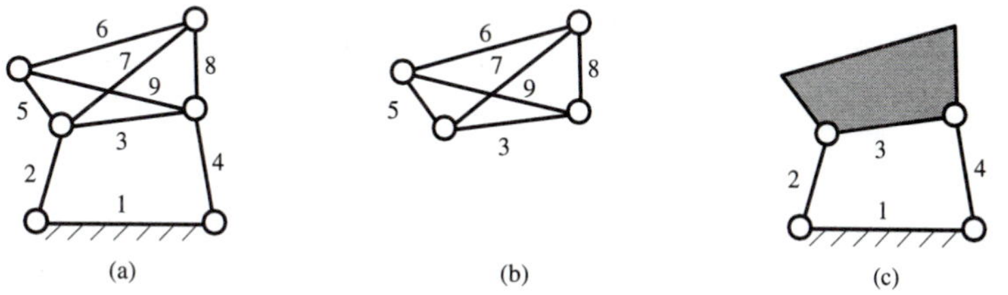

Figure 1.31 (a) A planar mechanism in which part of the mechanism is a structure, leading to a misleading value of mobility. All joints are revolutes. (b) The part of the mechanism which is a statically indeterminate structure. (c) A modified model of the linkage which gives the correct mobility value.

locally or generally. If the mechanism is overconstrained locally, a portion of the system may be a structure, but the entire mechanism can move. When this happens, we must replace that portion of the linkage with a single rigid body and recompute the mobility of the mechanism. An example is shown in the planar system of Fig. 1.31a.

Here $n = 9$, $j = 2 \times 1 + 2 \times 2 + 2 \times 3 = 12$. Note that there are two joints at which three members are connected and two at which four members are connected. $\Sigma f_i = j = 12$. Hence

$$M = 3(9 - 12 - 1) + 12 = 0$$

However, it can be observed that the portion of the linkage consisting of members 3, 5, 6, 7, 8, and 9 is a statically indeterminate structure. This portion is shown in Fig. 1.31b. Here $n = 6$, and because three members are connected at each joint location, $j = 4 \times 2 = 8$. Also, $\Sigma f_i = j = 8$. Therefore,

$$M = 3(6 - 8 - 1) + 8 = -1$$

revealing the statically indeterminate nature of the structure and the source of the error in the mobility value. A portion of the linkage that is a statically determinate structure does not cause an error in calculating mobility.

To get a correct value of mobility, the linkage is remodeled as shown in Fig. 1.31c with the portion that is a structure replaced by a single, rigid member. The linkage is now revealed to be a planar four-bar linkage for which the mobility is one.

Mechanisms and especially spatial mechanisms can also be generally overconstrained. Figure 1.32 shows a spatial linkage with four members and four revolute joints. It has a

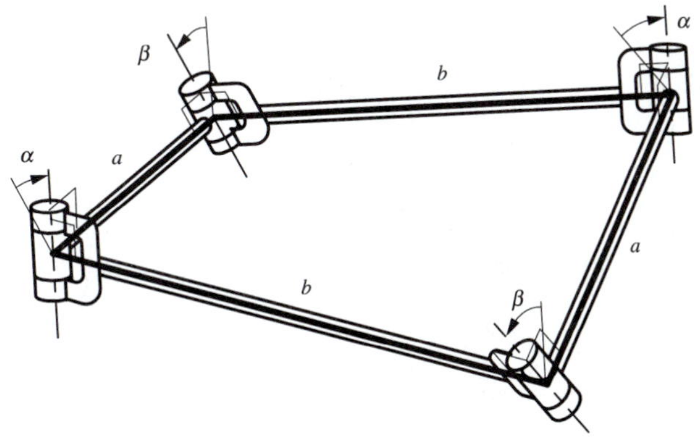

Figure 1.32 The Bennett mechanism. The side lengths and twist angles obey the relationship $a \sin \beta = b \sin \alpha$.

special geometry. The opposite members are identical, and the normals to the pairs of axes in the links intersect at the joint axes. The lengths of those normals (a and b) are related to the angles between successive axes (α and β) by the relationship

$$a \sin \beta = b \sin \alpha$$

As was demonstrated nearly a hundred years ago by Bennett, this linkage has mobility one. However, if we apply the constraint criterion with $n = j = 4$ and $\Sigma f_i = 4$, the result is

$$M = 6(4 - 4 - 1) + 4 = -2$$

In this case, because of the special geometry, the position equations of the linkage turn out to be dependent in all positions. For this reason, the effective number of equations is only three, rather than the six that would be expected for a single closed loop. Because the constraint criterion calculates the difference between the number of position variables and the number of available equations, it miscounts the mobility by three degrees of freedom. It turns out that rather a large number of linkages have anomalous mobility like the Bennett mechanism. They are called overconstrained linkages. Many of these are largely curiosities. However, there are several very important families of overconstrained linkages that are exceedingly common in engineering practice.

The most common example of overconstraint is the family of planar linkages. There is no *a priori* reason why planar linkages should not obey the general spatial mobility criterion. Nevertheless, they do not. Equation (1.4) gives a value for M that is always $3c$ less than the correct value, where c is the number of independent closure equations for the linkage. The fact that planar linkages obey Eq. (1.2), which has the same form as Eq. (1.4) but with the coefficient 6 replaced by 3, indicates that only three of the six equations produced by any closure are independent for a planar linkage.

Another common family of overconstrained linkages is the family of spherical linkages. These are linkages whose joints are all revolutes. The axes of those joints all pass through a single point. Figure 1.33 shows a spherical four-bar linkage.

Spherical linkages obey the same form of constraint criterion as planar linkages and the Bennett linkage. Thus, three of the equations resulting from each closure in a spherical linkage are always dependent.

Compared with properly constrained linkages (those that obey Eq. 1.4), overconstrained linkages have properties that are different in important and practical ways. They tend to be much stiffer and stronger in supporting loads, particularly those orthogonal to the direction of motion at the point of application. However, they are sensitive to dimensional

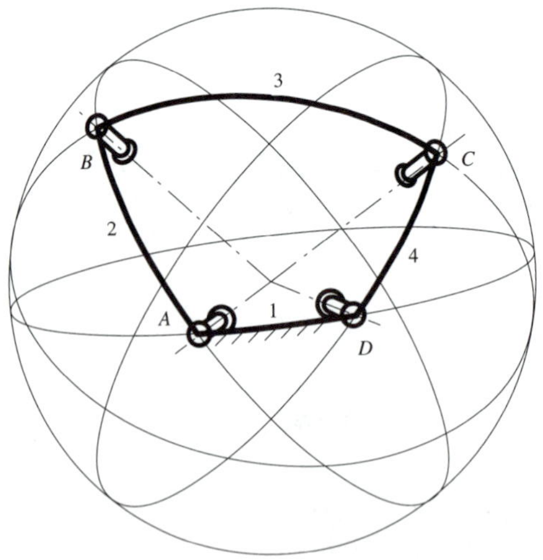

Figure 1.33 Spherical four-bar linkage.

Figure 1.34 Front-end loader. If analyzed using the planar mobility equations, the mechanism will be found to have fewer than one degree of freedom. Parallel actuators are used on both sides of the machine to balance the load and increase stiffness. The loader part of the machine has two degrees of freedom. (Courtesy of Deere & Company, Moline, Illinois.)

accuracy in their members. This requires manufacture to relatively tight tolerances, which can increase cost. Conversely, properly constrained linkages are completely insensitive to link geometry, as far as mobility is concerned. This means that, in lightly loaded situations, they can absorb abuse that deforms links and still function, at least after a fashion. This is an important property in situations such as the control linkages of agricultural machinery. In heavily loaded situations, the design engineer will often deliberately increase the degree of overconstraint to improve stiffness and strength. An example is the bucket support linkage of a front-end loader. A photograph of the loader is shown in Fig. 1.34, and one of the bucket support linkages is identified in Fig. 1.35.

In principle, only one of the two planar inverted, slider-crank linkages is needed to lift/support the bucket. In this case, we would have $n = j = \Sigma f_i = 4$ and

$$M = 6(4 - 4 - 1) + 4 = -2$$

Since the true mobility is 1, the degree of overconstraint is $1 - (-2) = 3$. However, the linkage is doubled up with identical linkages supporting each end of the bucket. This gives

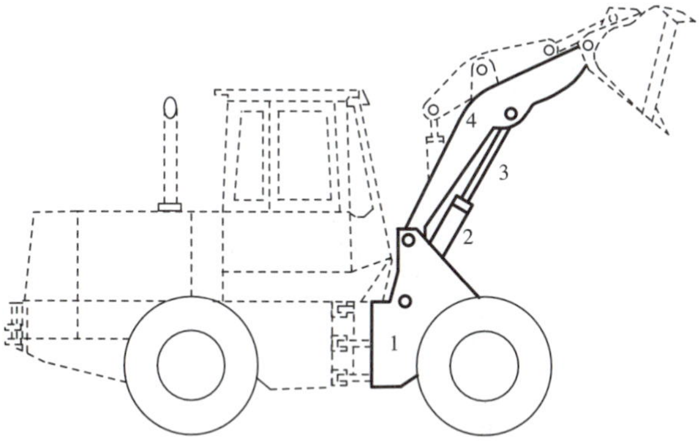

Figure 1.35 Schematic of the right-side bucket support linkage for the front-end loader in Fig. 1.34.

$n = 6$ and $j = \Sigma f_i = 8$. Thus

$$M = 6(6 - 8 - 1) + 8 = -10$$

Therefore, for the doubled linkage the degree of overconstraint is $1 - (-10) = 11$. The result is a much stronger mechanism since the individual planar loops do not have to support the large out-of-plane moments that a single linkage would have to support. The cost is that the axes of the corresponding joints on either side must be collinear to a high degree of accuracy, requiring careful manufacturing.

1.11 USES OF THE MOBILITY CRITERION

The mobility criterion is most useful to the engineer when examining an unfamiliar mechanical system. It allows a quick check to determine whether the links, joints, and actuators identified are consistent with system function. Inconsistency may indicate that some elements have been misidentified or that passive degrees of freedom are present. Of course, as already discussed, overconstraint may also need to be considered. In particular, if the linkage is planar or spherical, the appropriate form of the constraint equation should be used in place of the general form.

It is possible to formulate expressions for the mobility that accommodate overconstrained closures of arbitrary type. These expressions are equivalent to the form

$$M = \sum_{k=1}^{c} b_k + \sum_{i=1}^{j} f_i \tag{1.5}$$

where $c = n - j - 1$ is the number of closures of the linkage.

Unfortunately, unless the values of b_k associated with the different closures can be identified by inspection, such expressions have no value. The reason is that the mobility equation gives a quick check of the number of position variables and independent equations without the need to develop those equations. However, the only way to verify an overconstrained closure of a type not identifiable by inspection is to develop the closure equations and analyze them for dependency. Therefore the quick-check advantage of the mobility equation disappears, and there is no way to derive information about the linkage without performing a complete position analysis.

1.12 INVERSION

A commonly used tactic in studying mechanism kinematics is inversion. This is a change of the fixed reference frame from one link to another that causes different characteristics of the motion relative to the frame. For example, Fig. 1.36 shows the different inversions

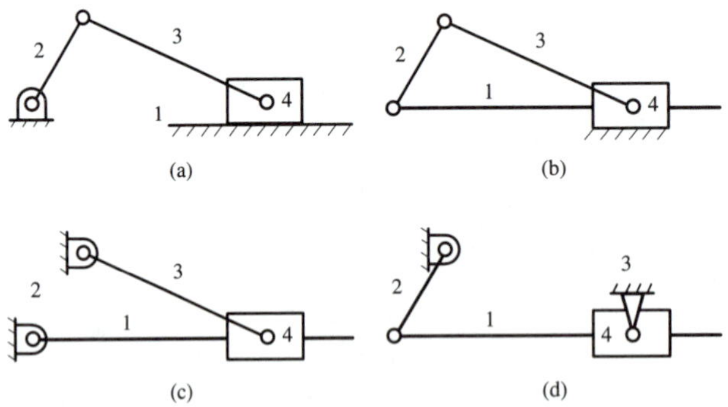

Figure 1.36 Inversions of the slider-crank linkage. The linkage in (a) is the original linkage and those in (b), (c), and (d) are the inversions.

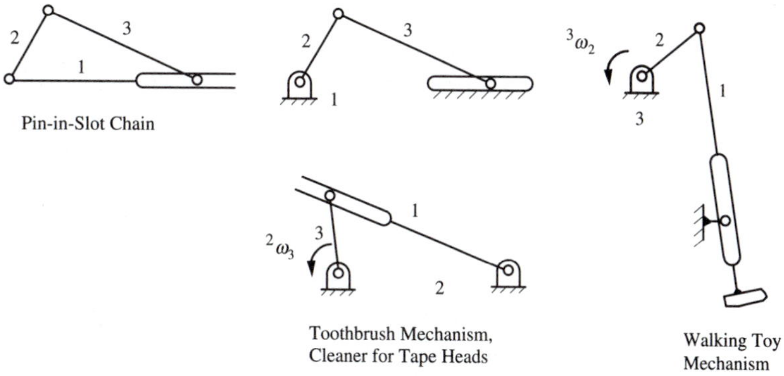

Figure 1.37 Uses of inversions of a pin-in-a-slot linkage.

of a slider-crank linkage, and Fig. 1.37 shows the inversions of a pin-in-a-slot mechanism. These inversions are often used as inexpensive substitutes for the slider-crank inversions. The motion characteristics of the coupler links for each of the mechanisms are all very different. Nevertheless, the linkage topology and the relative angular relationships among the links are the same in all cases. Therefore, useful information obtained from the study of the linkage in one inversion can be transferred to the study of other inversions. Note that in Fig. 1.36, the relative positions of all of the joints are the same for the position chosen. It is only when the mechanisms move that the different motion characteristics are revealed.

To determine the inversions of a mechanism, it is convenient to start with the chain from which the mechanism is formed. A different linkage results whenever a different link is selected as the frame.

1.13 REFERENCE FRAMES

It is necessary to be careful about reference frames when working with systems of many bodies. A reference frame can be attached to each body, and we can express positions, velocities, and accelerations relative to any or all of them.

As far as kinematics is concerned, there is no restriction on the use of reference frames. All frames are equally viable. We can invert from one frame to another without restriction.

It is only when we introduce forces and enter the realm of kinetics that a restriction appears. It is then that Newton's first and second laws, which relate motion properties to force, are true only if all motion properties are referred to a common reference frame. This common reference frame must be of a special type, called an inertial reference frame. For the purposes of mechanism design the inertial reference frame is almost always fixed to the earth. There are engineering problems, such as the design of mechanisms to be carried on spacecraft, for which the primary inertial reference frame must be used. The primary inertial reference frame is fixed relative to the fixed stars. A more complete discussion of inertial reference frames can be found in most texts on rigid-body dynamics. Einstein showed that in a space-time framework all reference frames are equally valid, thereby removing the Newtonian distinction between inertial reference frames and others. However, in the domain in which mechanical engineers usually operate, Newtonian mechanics provides a very accurate simplification of relativistic mechanics which is of great practical utility.

It is important to remember that position and motion properties can be expressed only relative to a reference frame. A habit has grown up in this subject of referring to a velocity or acceleration of a point relative to another point. This will be found to be convenient in some types of problems, particularly in graphical analysis, and there is no harm in using it provided that it is clearly understood that it is a shorthand expression for the velocity or acceleration of the first point relative to a reference frame in which the second point is fixed. The identity of that reference frame should always be kept in mind.

In many discussions in the following it will be convenient to have a notation that explicitly states the reference frame in which a particular vector is expressed. A method that is often used is to indicate the reference frame by means of a superscript placed in front of the symbol for the vector. Thus ${}^1\mathbf{v}_A$ indicates the velocity of point A relative to reference frame 1; ${}^2\omega_3$ indicates the angular velocity of member 3 relative to reference frame 2.

Usually, we will associate one reference frame with each member of a linkage and will number it to agree with the number of the linkage. Reference frame 1 will usually refer to the fixed link. Unfortunately, the use of superscripts to indicate reference frames complicates expressions and makes them more difficult to read. For this reason, the superscripts will usually be dropped when all vectors are referred to reference frame 1.

1.14 MOTION LIMITS

A member of a linkage which is connected to the base by a revolute joint and which rotates completely as the linkage moves through its motion cycle is called a crank. Usually, there will also be members in the linkage which look exactly like cranks because they are connected to the base by a revolute, but which cannot rotate completely.

Consider the four-bar linkage shown in Fig. 1.38a in which the link *AB* is a crank rotating fully about the revolute joint *A*. It will be assumed to rotate continuously in the counterclockwise direction. Complete revolution of this link requires that it pass through the positions shown in Fig. 1.38b and c. Now consider the motion of the revolute joint *D*. Prior to reaching the position of Fig. 1.38b link, *CD* was rotating counterclockwise about joint *D*. In the position of Fig. 1.38b further rotation of *CD* about *D* in the counterclockwise direction is not possible. *CD* comes to rest and reverses its direction of motion. Similarly, before entering the position of Fig. 1.38c, the link *CD* is rotating clockwise about the joint *D*. In this position, further rotation in this direction is not possible, and the link comes to rest and then reverses direction. The positions shown in Fig. 1.38b and c are called motion limit positions for the joint *D*. The link *CD* does not perform full rotation but simply oscillates between these positions. That is, it is not a crank; it is a rocker.

1.15 ACTUATION

At this point it is necessary to introduce some terminology to describe the different members of a four-bar linkage. The fixed link, that is, the member to which the frame of

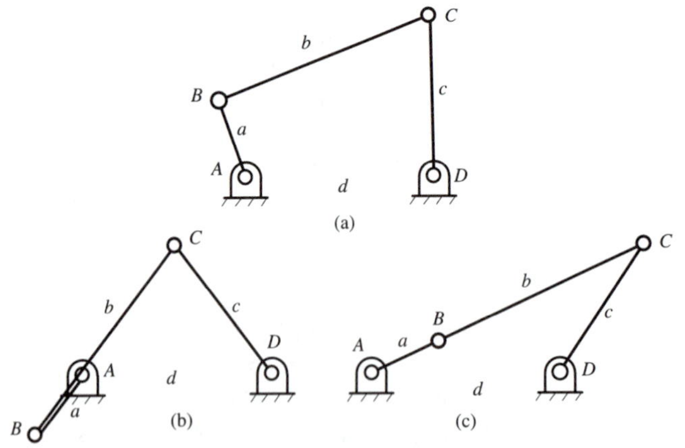

Figure 1.38 Limiting positions of joint *D* of a four-bar linkage.

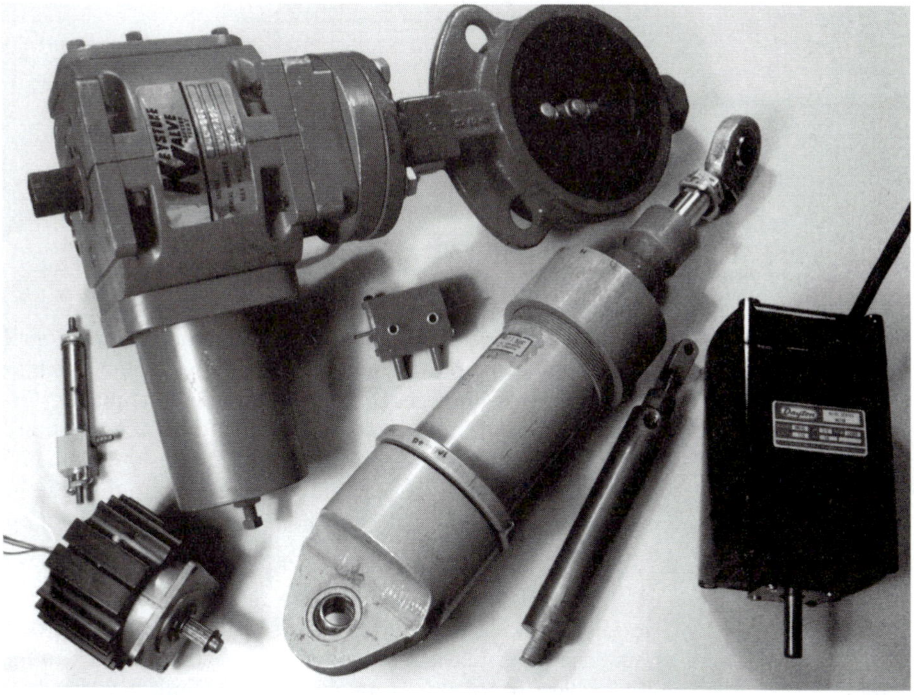

Figure 1.39 Photographs of a variety of actuators.

reference is attached, is called the base. The two members that are connected to the base by revolute joints are called turning links. The link that is jointed to both turning links and has no direct connection to base is called the coupler. The turning links may be further distinguished by the terms crank, for a link capable of complete revolution relative to the base, and rocker, for a link that is only capable of oscillating between motion limits.

A linkage is actuated, or driven, by applying force to one of its moving links. This may be done in a variety of ways, as is evident from the number of different types of commercial actuators (Fig. 1.39). It is frequently convenient for that powered link to be connected to the base by a revolute joint. The linkage may then be actuated by applying a torque to that link. In this case it is usually also preferable that the link be continuously rotatable since it may then be actuated by means of a continuously rotating motor. For this reason it is important to be able to identify four-bar linkages that have continuously rotatable joints and to locate those joints. This may be done by means of a simple set of rules called Grashof's rules.

Grashof distinguished two fundamentally different types of four-bar linkage by means of the inequality

$$s + l < p + q \tag{1.6}$$

where, as shown in Fig. 1.40, s is the length of the shortest side, l is the length of the longest side, and p and q are the lengths of the other two sides. Linkages that obey this

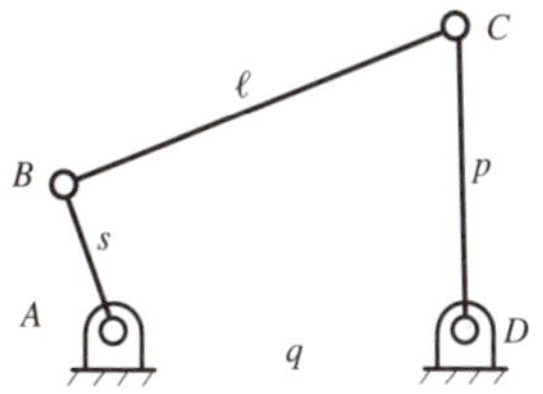

Figure 1.40 Nomenclature for Grashof's inequality.

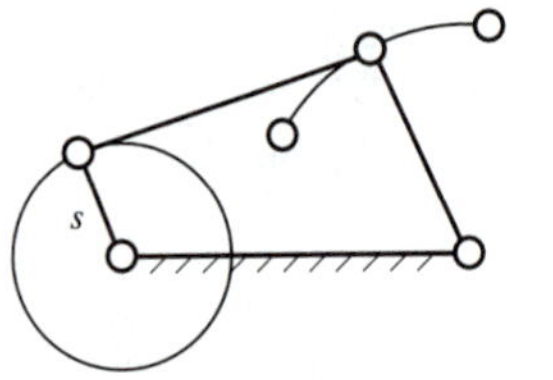

Figure 1.41 Crank-rocker subtype of Grashof type 1 linkage. This linkage type occurs when the shortest link is jointed to the base of the linkage.

inequality (Grashof type 1) have two joints that perform complete rotations and two that oscillate between motion limits. The two fully rotatable joints are those on either end of the shortest link. Linkages that do not obey the inequality (Grashof type 2) have no fully rotatable joints. All four joints then oscillate between motion limits.

The behavior of a linkage that obeys the Grashof inequality is strongly dependent on the locations of the fully rotatable joints relative to the base link. That is, it is dependent on the inversion of the linkage. The following additional rules distinguish three subtypes that have different behavior:

1. If the shortest link is jointed to the base, the linkage is a crank-rocker (Fig. 1.41). The joint between the shortest link and the base is fully rotatable. Hence that link is a crank. The other fully rotatable joint connects that crank to the coupler. Hence the other joint to base is not fully rotatable, and the link it connects to base oscillates. It is the rocker. A crank-rocker can be conveniently driven across the joint connecting the crank to base.

2. If the shortest link is the base, both joints at the base are fully rotatable, and so both links connected to the base are cranks (Fig. 1.42). The linkage is a double-crank, also known as a drag-link. It may be conveniently actuated at either of the base joints.

3. If the shortest link is the coupler, neither base joint is fully rotatable (Fig. 1.43). The linkage is a type 1 double-rocker. Its behavior is different from that of type 2 double-rockers, those that do not satisfy the inequality, since the two floating joints rotate completely. The result is that the coupler tumbles, performing a complete rotation relative to the base. The angular motion of the coupler of a type 2 double-rocker is an oscillation relative to the base.

The Grashof inequality may be proved as follows:

Consider the linkage shown in Fig. 1.44a. In order to perform a complete rotation it must pass through the positions shown in Fig. 1.44b and c. Let a be the length AB, b the

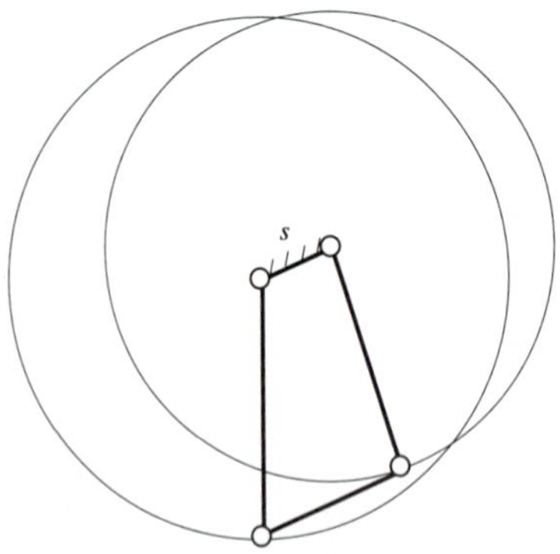

Figure 1.42 Double-crank subtype of Grashof type I linkage. This linkage type is also called a drag-link. It occurs when the shortest link is the base.

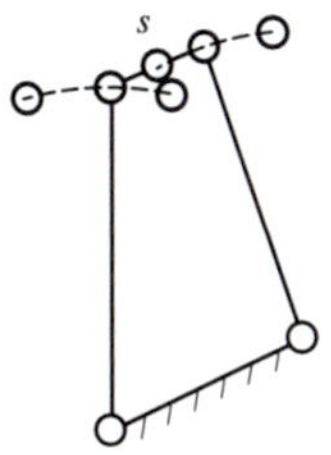

Figure 1.43 Type 1 double-rocker. This subtype occurs when the shortest link is the coupler.

length BC, c the length CD, and d the length DA. It is assumed that

$$a < d$$

The triangle inequality states that the sum of the lengths of any two sides of a triangle is greater than that of the third. It may be applied three times to Fig. 1.44b to give

$$a + d < b + c \quad \textbf{(a)}$$

$$b < c + a + d \quad \textbf{(b)}$$

$$c < b + a + d \quad \textbf{(c)}$$

It may also be applied three times to Fig. 1.44c to give

$$d - a < b + c \quad \textbf{(d)}$$

$$b < c + d - a \quad \textbf{(e)}$$

$$c < b + d - a \quad \textbf{(f)}$$

Examination of these inequalities reveals that if (e) is true then (b) is certainly true, because the right-hand side of (b) is that of (e) plus $2a$. We say that inequality (e) is stronger than inequality (b). Hence inequality (b) can be eliminated. Inequality (e) can be written in the form

$$a + b < c + d \quad \textbf{(e')}$$

by adding a to both sides of the inequality.

Similarly, inequality (c) is certainly true if inequality (f) is true. Once again, the right-hand side of inequality (c) is larger by $2a$. Inequality (f) assumes the form

$$a + c < b + d \quad \textbf{(f')}$$

if a is added to both sides.

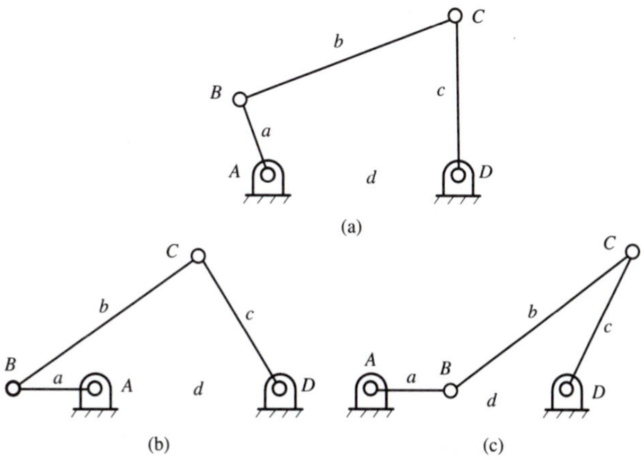

Figure 1.44 Extreme positions for a four-bar linkage.

Inequality (d) is certainly true if inequality (a) is true, since its left-hand side is less than that of inequality (a) by $2a$. Hence, the six inequalities are reduced to three: (a), (e′), and (f′). Addition of both sides of inequalities (a) and (e′) gives

$$2a + b + d < 2c + b + d$$

so that

$$a < c$$

Likewise, addition of both sides of inequalities (a) and (f′) gives

$$2a + c + d < 2b + c + d$$

so that

$$a < b$$

Since a has also been assumed to be less than d, it follows that a is the shortest link length. Now, whichever of the inequalities (a), (e′), and (f′) has the longest link length on the left added to a will be the strongest. That is, the left-hand side is largest and the right-hand side is smallest. Whichever one this is assumes the form

$$s + l < p + q$$

where $s = a$ is the shortest link length, l is the longest link length, and p and q are the two remaining link lengths.

It must be remembered that we assumed that a was less than d. It is also necessary to deal with the case in which a is larger than d. This can be handled by inverting the linkage so that AB becomes the base link and DB becomes the link jointed to it by the continuously rotatable joint. Pursuing the application of the triangle inequality then results in d being the shortest link length, and the Grashof inequality again results.

What we have shown so far is that the Grashof inequality is a necessary condition for the presence of a fully rotatable joint, and that joint is always at one end of the shortest link. Now, there can never be just one fully rotatable joint in a four-bar linkage. There must always be at least two. If there were just one fully rotatable joint, a topological contradiction would result when the rotation of AB relative to the other links after one cycle were to be considered. If that link were to perform a complete rotation about joint A, and joints B, C, and D were to oscillate back to their initial positions, AB would have performed a complete rotation relative to each of the other links. That is, it would have performed a complete revolution relative to BC. However, joint B has not performed a complete revolution but, rather, has performed zero net rotation. Hence there cannot be just one completely rotatable joint. Since we have shown that any completely rotatable joint must be at one end of the shortest link, it follows that there are two completely rotatable joints, and they are at either end of the shortest link. This completes the proof of Grashof's rules.

The shortest link of a type 1 linkage performs a complete revolution in each motion cycle relative to the other members. The net rotations of the fully rotatable joints on either end of that link cancel one another so that the net rotations of the remaining links relative to one another are zero for a complete motion cycle.

Of course, sometimes it is not necessary for the mechanism to perform a complete motion cycle. A restricted range of driving joint motion may be adequate. In that case linear actuators, such as hydraulic or pneumatic cylinders acting across the driving joint, may be used. However, it is still necessary that the driving joint not pass through a motion limit within the necessary range of motion. Grashof's rules are often useful in ensuring that this does not happen.

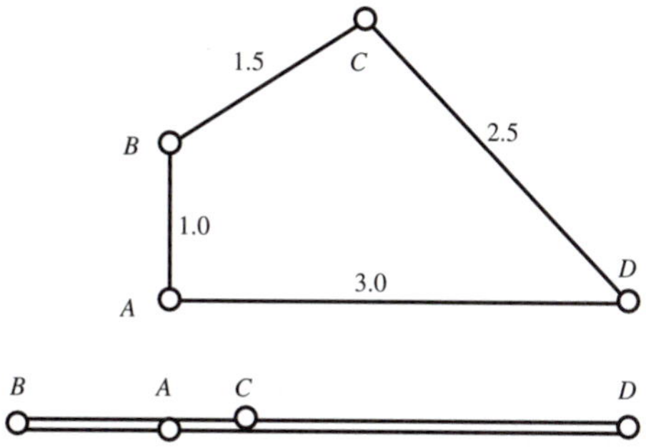

Figure 1.45 Grashof neutral linkage.

Occasionally it is necessary to drive a crank-rocker linkage by oscillating the rocker through a part of its motion range. In this case the linkage is usually referred to as a rocker-crank.

The reasons associated with the use of type 2 double-rocker linkages, or with the use of type 1 linkages driven by rockers rather than cranks, will be better understood after a discussion of linkage synthesis. Often, a linkage that is synthesized to produce a specific motion cannot be driven through the motion without the driving joint passing through a motion limit position. In that case, a solution might be to drive the other base joint.

A special case arises when

$$s + l = p + q$$

This is called a transition linkage or Grashof neutral linkage. In this case the linkage can assume a "flattened" configuration as shown in Fig. 1.45. When passing through this position, it can change from one to the other of the two configurations in which the linkage can be assembled with a given driving crank angle. In practice, this is usually undesirable, because it leads to unpredictable behavior and possibly large loads on the members and joints.

1.16 COUPLER-DRIVEN LINKAGES

In some applications linkages are actuated not by applying a force or torque to one of the links jointed to the base but rather by applying a force or torque to the coupler, the member that has no direct connection to the base. Everyday examples are not uncommon. Polycentric hinges for heavy doors or for automotive hood and trunk lids come to mind.

It is still important for a coupler-driven mechanism not to pass through a motion limit within the desired motion range. The motion limit positions for a coupler drive are quite different from those for a crank drive. They are the positions in which the two rotating links become parallel, as shown in Fig. 1.46. In these positions the angular motion of the coupler ceases and must reverse if motion is to continue. Elimination of these motion

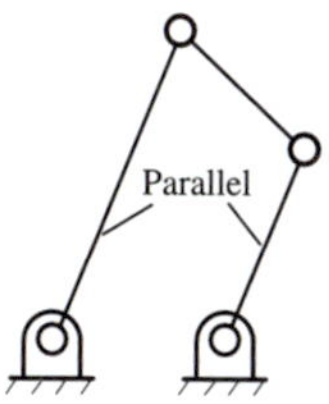

Figure 1.46 Motion limit for the coupler.

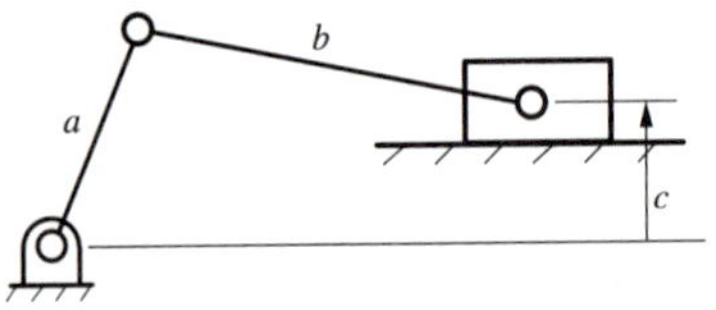

Figure 1.47 General slider-crank mechanism with offset dimension c.

limits produces a linkage whose coupler performs a complete revolution relative to the base link. Because in a type 1 linkage the shortest link rotates completely relative to the remaining links, that link must be either the coupler or the base. It follows that the Grashof subtypes for which complete rotation of the coupler relative to the base is possible are the type 1 double-rocker and drag-link subtypes.

1.17 MOTION LIMITS FOR SLIDER-CRANK MECHANISM

The limits for a slider-crank mechanism can be determined by considering the combinations of link lengths that will cause the linkage to lock up. A typical slider-crank is shown in Fig. 1.47.

The limit positions of the rotating link a (Fig. 1.47) are determined when the coupler link is perpendicular to the direction of slider travel. The limiting assembly position occurs for one of the four geometries shown in Fig. 1.48.

From the four limit positions shown in Fig. 1.48, it is apparent that the following relationships must be maintained if it is to be possible to drive the slider-crank for a full 360° rotation of the crank.

$$b > a$$

and

$$b - a > c$$

where

a = length of the crank
b = length of the coupler
c = offset distance from crank-ground pivot to slider pin (measured positive upward)

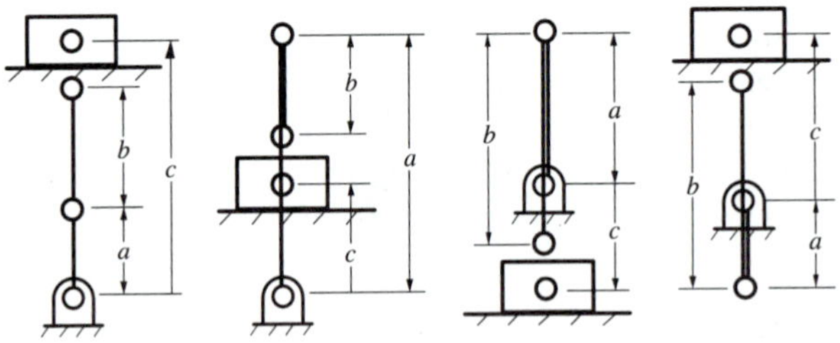

Figure 1.48 Positions for which the slider-crank mechanism cannot be assembled.

EXAMPLE 1.6 ***(Using Grashof's Equation)***

PROBLEM

In the Watt six-bar linkage shown in Fig. 1.49, the joint between links 5 and 6 must be placed on the arc indicated. Using Grashof's rule, determine the region for joint E that will allow full

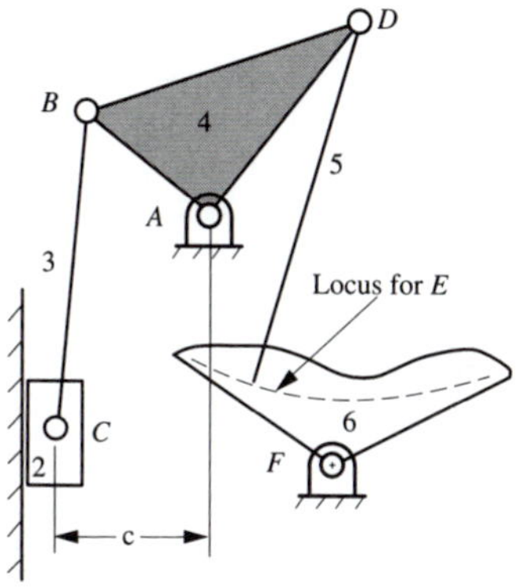

Figure 1.49 Mechanism for which point E is to be determined.

rotation of link 6. The critical dimensions are

$$AB = 1.14 \text{ in} \quad BC = 2.26 \text{ in} \quad AD = 1.74 \text{ in}$$
$$AF = 2.00 \text{ in} \quad DE = 2.68 \text{ in} \quad c = 1.09 \text{ in}$$

SOLUTION

Consider first the slider-crank mechanism (ABC) even though the crank AB does not rotate 360°. Clearly, if the crank AB can rotate for 360°, it will not lock up in any intermediate position. Based on the dimensions given,

$$BC > AB$$

and $BC - AB = 1.12$. Therefore,

$$BC - AB > c$$

and the crank of the slider-crank mechanism can rotate a full 360°.

Next consider the crank rocker mechanism ($ADEF$). For a crank rocker, link 6 must be the crank, which means that EF must be the shortest link. The longest link is DE. Therefore, based on Eq. (1.6), for $ADEF$ to be a crank-rocker mechanism,

$$EF + DE < AF + AD$$

or

$$EF + 2.68 < 2.00 + 1.74$$

or

$$EF < 1.06 \text{ in}$$

The allowable range for E is shown in the Fig. 1.50. ■

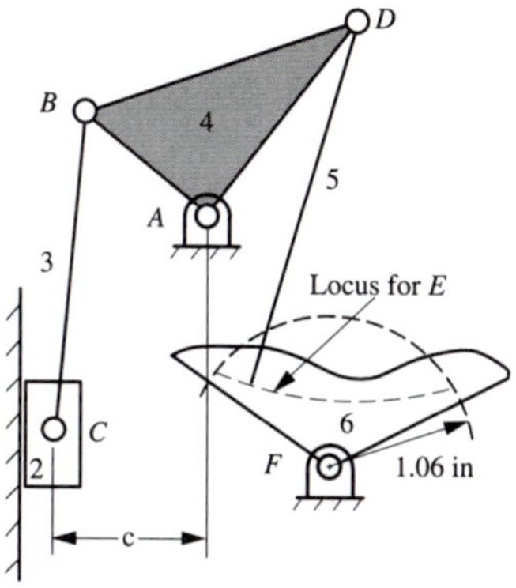

Figure 1.50 Allowable range for point E.

1.18 INTERFERENCE

This is a topic that is often ignored in courses and texts on mechanism design. That is unfortunate since a full cycle motion capability can be prevented by topological interference even when Grashof's rules indicate that it is possible. An understanding of topological interference is particularly important at the present time, when linkages are often designed using CAD systems and their functioning checked by animation rather than by construction of physical models. It is very difficult to represent topological interference adequately on a planar display. For this reason, the reader is urged to construct models using cardboard and thumbtacks, or whatever other appropriate materials are available, when reading this section. That is the best way to gain an understanding of the nature of topological interference. There is also a tendency to regard interference as a result of the physical shape of the links and as something that can be avoided if enough care is given to the design of the physical link geometry. That is not what we are talking about here. Topological interference is a fundamental property of a linkage configuration in the same way that Grashof type is. It cannot be avoided by simply reshaping the links.

Topological interference really affects only the capability of executing a complete motion cycle using a rotary input. If oscillatory motion over a partial cycle is all that is required, topological interference can usually be circumvented.

The topological and physical limitation that the links cannot pass through each other creates difficulties in arranging for input and output motion transfer to and from type 1 linkages. When a simple, type 1 four-bar linkage is viewed as a three-dimensional structure with revolute joint axes of finite length, there is only one way in which it can be assembled to avoid any of the links or joint axes having to pass through each other. This is shown in Fig. 1.51. The problem is the fully rotatable joints.

The fundamental nature of the problem can be thought of by considering a closed loop of string as shown in Fig. 1.52. A segment of the string may be rotated through an unlimited number of cycles relative to the rest of the loop as long as that segment does not cross the remainder of the loop. If it does cross, the remainder of the loop becomes twisted, and each cycle becomes increasingly different because it adds another twist to the string.

The oscillatory joints of a type 1 linkage never pass through positions in which their joint angles ϕ and ψ, shown in Fig. 1.51, become either zero or π. If either one did so, the joint diagonally opposite it would be at a motion limit, preventing it from a complete rotation. Consequently, the axes of these joints never cross the lines of the links BC and DA, so there is no interference. However, when joint A is fully rotated, the axis of joint

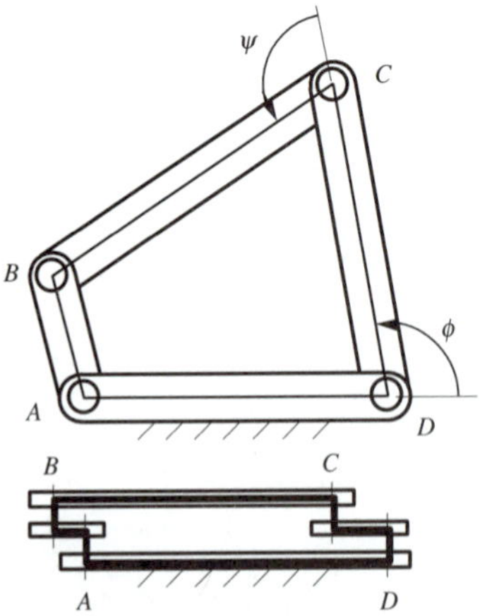

Figure 1.51 Assembly of type 1 linkage needed to avoid interference.

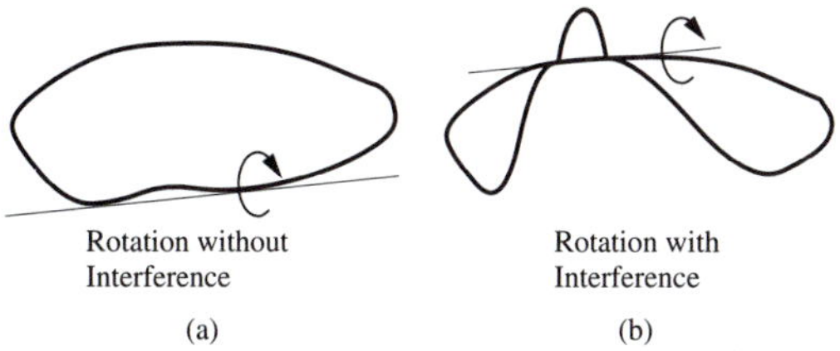

Figure 1.52 Interference illustrated with a string.

B must cross the line *DA* and, since *AB* is the shortest link, it must cross between *D* and *A*. Likewise, when joint *B* is fully rotated, the axis of joint *A* will cross the line *BC* between *B* and *C*. Viewing the linkage from a direction normal to the joint axes, it can be seen that, if joint *B* is on one side of the link *AB,* the link *DA* must be on the other side, otherwise the link will cut the joint axis. Similarly, joint *A* must be on the opposite side of *AB* to link *BC*. It follows that, in the direction along the link axes, *BC* and *DA* are on either side of *AB* and *CD*. This may seem to be dependent on the physical realization of the links, but it is, in fact, a fundamental topological property of the loop.

The simplest situation for motion transfer is when both input and output motions are rotary. Motion can then be transferred into and out of the linkage by means of shafts attached to the turning links. Interference constrains the arrangement of the input and output shafts, as shown in Fig. 1.53. If the linkage is a *crank-rocker* both shafts must enter from the same side in order to avoid interference between the shafts and the coupler link. Notice that the shafts must pass through the base link to get to the turning links, which are on the inside of the linkage in this inversion. Physically, the shafts are supported in bearings mounted in the base link.

If the linkage is a *drag-link,* the shafts may be attached directly to the turning links, one on either side, since those links are on the outside of the linkage in this inversion. However, if this is done, the fixed bearings may be moved to the outside, essentially turning the base link inside out. The base link becomes a pair of fixed bearing mounts on either side of the linkage, as is shown in Fig. 1.52b. A *drag-link* linkage must always be mounted

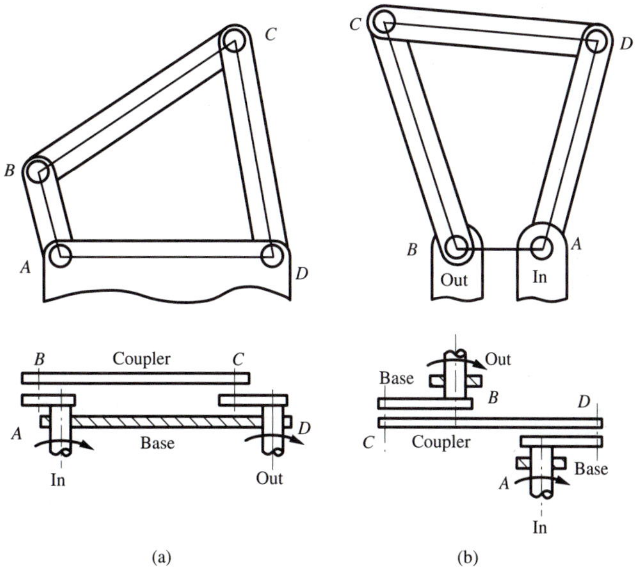

Figure 1.53 Shaft drive of crank-rocker (a) and drag-link (b) linkages.

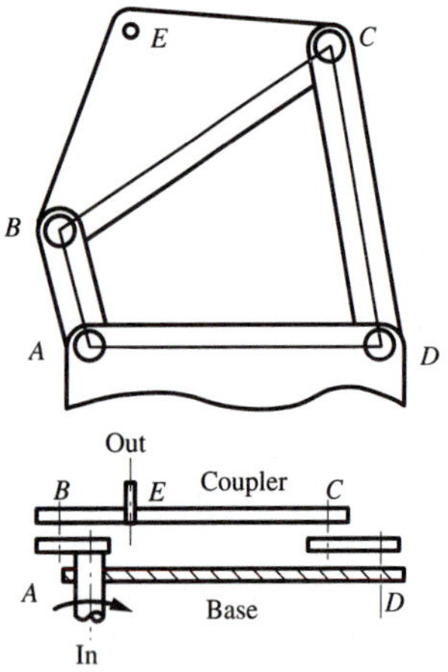

Figure 1.54 Transfer of motion from a point on the coupler of crank-rocker mechanism.

in this manner to achieve full cycle motion, regardless of the means of input or output, since otherwise the coupler must pass through the base.

The discussion of rotary input and output to a *type 1 double-rocker* linkage will be left until later since it is not possible to achieve full cycle motion with a crank drive in this type of linkage.

A more complex case is that in which the input is rotary and motion must be transferred from a point on the coupler link. This is easy enough to arrange in the *crank-rocker* case, as shown in Fig. 1.54, since the base and coupler are on the outside of the linkage.

Much more difficult is the case in which motion must be transferred from a point on a crank or on the coupler of a *drag-link.* Because the coupler moves between the cranks there is no way to avoid interference of a shaft coming off the coupler with those cranks in full cycle motion. Further, because the two parts of the base are outside the cranks, there is also a problem of interference with the base. This latter problem also affects the transfer of motion to or from points on the cranks.

It is possible to circumvent the interference problem for motion transfer from points on a crank by "doubling" the crank. This is shown in Fig. 1.55. The crank is essentially

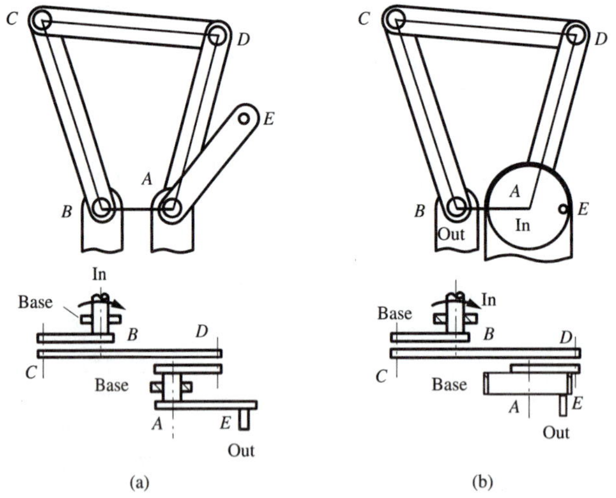

Figure 1.55 Motion transfer from a point on the crank of a drag-link.

duplicated on the outside of the base bearing. A shaft rigidly fixed to both the crank and the duplicate passes through the bearing forming the base joint and ensures that both move together. This effectively makes points on the crank available outside the base, where additional links can be attached at motion transfer joints. In this way, multiloop linkages such as that shown in Fig. 1.56 can be built up and driven by the driving crank of the master *drag-link* loop.

If the transfer point is reasonably close to the base joint of the crank, the result in Fig. 1.55a can be achieved by using a bearing of sufficiently large diameter to encompass the transfer point. This is shown in Fig. 1.55b.

There is no simple way to transfer motion from a point on the coupler of the *drag-link* without preventing full cycle mobility. It can be done by splitting one of the joints between crank and coupler, allowing the linkage loop to pass through itself. This requires the addition of at least one auxiliary link, so the mechanism, strictly speaking, is no longer a four-bar. The yoke shown in Fig. 1.56 carries the two bearings that replace the simple joint of the original linkage. It is undesirable to leave this link unconstrained, so it is usual to add a second link connecting it to base, as shown in the figure. This allows the coupler to be moved outside the crank. However, it is still not possible to transfer motion directly from a point on the coupler because of interference with the yoke. For this reason, the coupler is doubled in the same way that the crank was in Fig. 1.55, producing the six-bar arrangement shown. As can be seen, this is quite an extensive modification!

The situation for coupler drive, in which the driving torque is applied to the coupler link and the output motion is taken off that same link, is quite similar. The two linkage types that can, in principle, perform complete motion cycles in the coupler drive mode are the *type 1 double-rocker* and *drag-link* types. Both present a problem because the base and coupler are inside the cranks in the basic loop. The *type 1 double-rocker* can be made to allow full cycle motion, without interference, by doubling the coupler.

Once again, the *drag-link* presents additional problems because of the necessity of splitting the base, resulting in the base being outside the cranks. Coupler-driven full cycle motion of a simple *drag-link* is not possible because of interference. The six-bar arrangement of Fig. 1.56 can be used for full cycle motion which is identical to that of the *drag-link* with coupler drive.

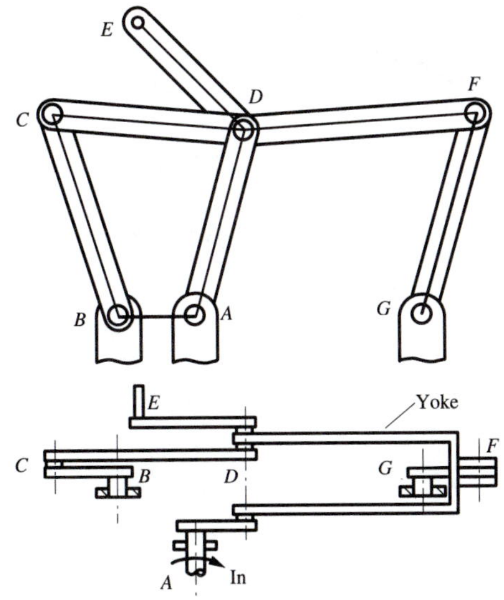

Figure 1.56 Six-bar modification to achieve motion transfer from the coupler of a drag-link linkage.

All of the foregoing discussion relates only to full-cycle motion, that is, to motion in which the driving link performs a complete rotation. If oscillation through a partial motion cycle is adequate for the application, interference can always be avoided by modifying the physical shapes of the links. This is true even for the *drag-link* type. Type 2 linkages can also be used in this mode. It is necessary only to ensure that it is not necessary for such linkages to pass through motion limit positions of the driving link when traversing the desired segment of the motion cycle.

1.19 CHAPTER 1 *Exercise Problems*

PROBLEM 1.1 Find a mechanism as an isolated device or in a machine and make a realistic sketch of the mechanism. Then make a freehand sketch of the kinematic schematics for the mechanism chosen.

PROBLEM 1.2 The drawings shown in the figure below are pictorial representations of real mechanisms that are commonly encountered. Make a freehand sketch of the kinematic schematic representation of each mechanism. Use this sketch to calculate the mobility, or number of degrees of freedom, of each of the three mechanisms.

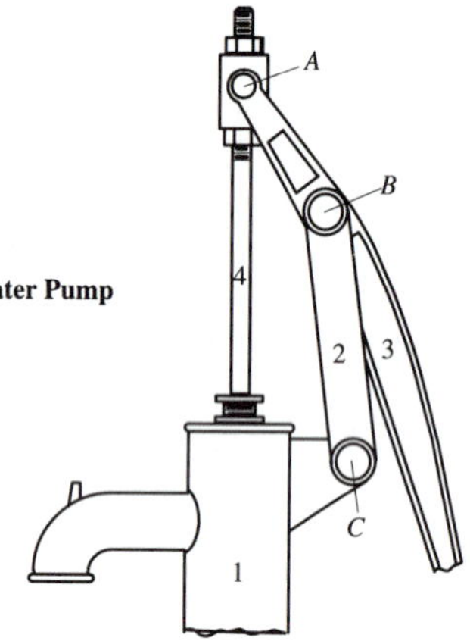

Water Pump

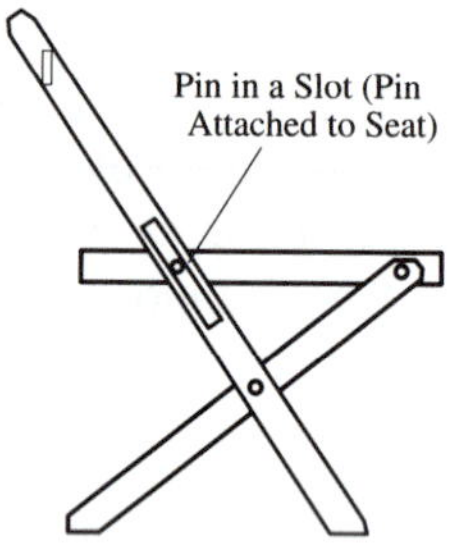

Folding Chair

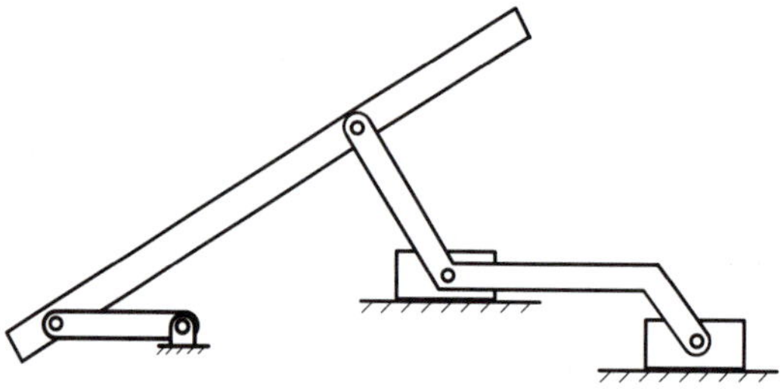
Casement Window Mechanism

PROBLEM 1.3 Find four different kinematic models for folding chairs. Make a freehand sketch of the kinematic schematics for the different chair mechanisms.

PROBLEM 1.4

What are the number of members, number of joints, and mobility of each of the planar linkages shown below?

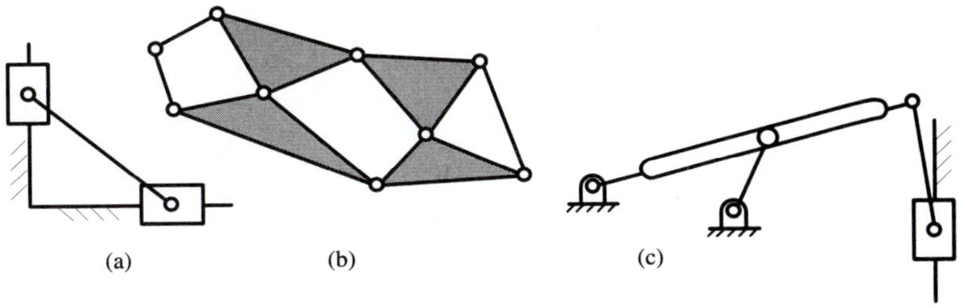

PROBLEM 1.5

Determine the mobility of each of the planar linkages shown below. Show the equations used to determine your answers.

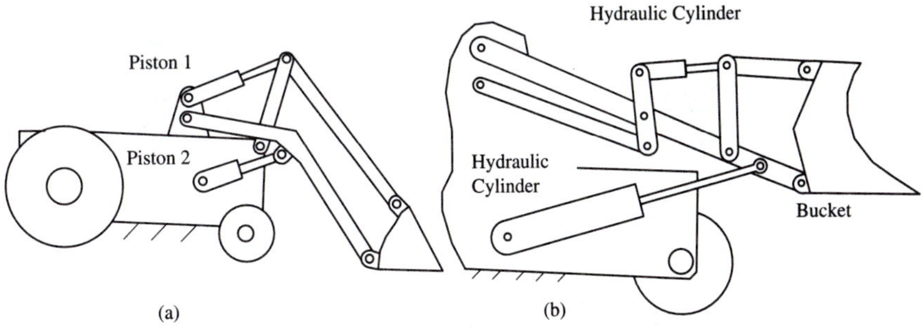

PROBLEM 1.6

Determine the mobility and the number of idle degrees of freedom of each of the planar linkages shown below. Show the equations used to determine your answers.

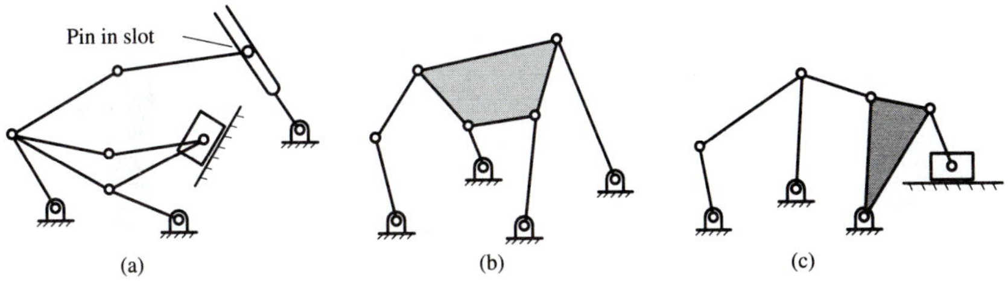

PROBLEM 1.7

Determine the mobility and the number of idle degrees of freedom of each of the planar linkages shown below. Show the equations used to determine your answers.

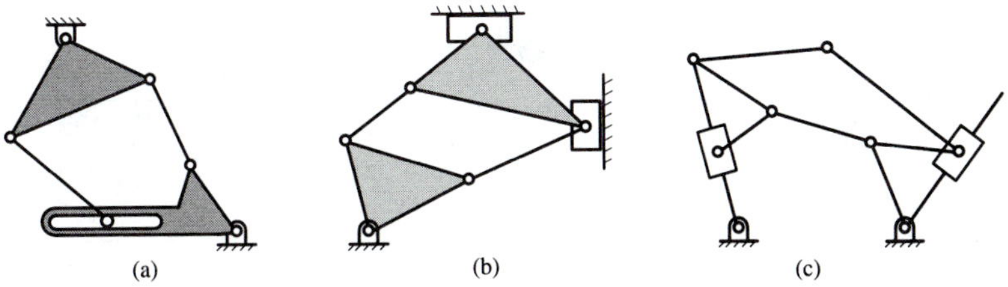

PROBLEM 1.8 Determine the mobility and the number of idle degrees of freedom of each of the planar linkages shown below. Show the equations used to determine your answers.

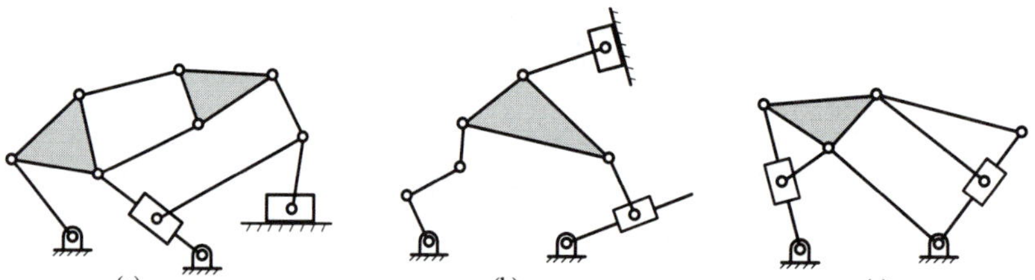

(a) (b) (c)

PROBLEM 1.9 What are the number of members, number of joints, mobility, and the number of idle degrees of freedom of each of the spatial linkages shown below?

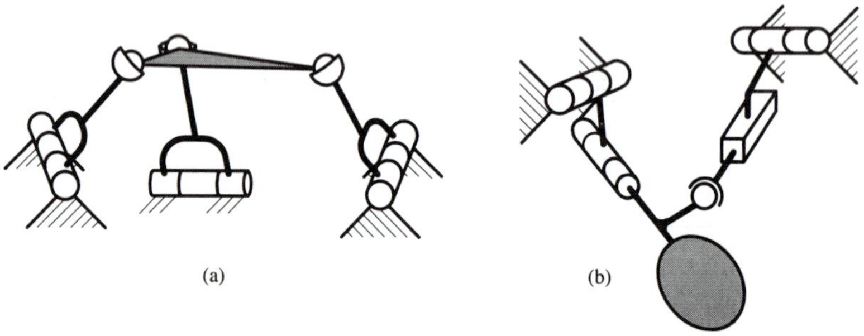

(a) (b)

PROBLEM 1.10 Determine the mobility and the number of idle degrees of freedom of the spatial linkages shown below. Show the equations used to determine your answers.

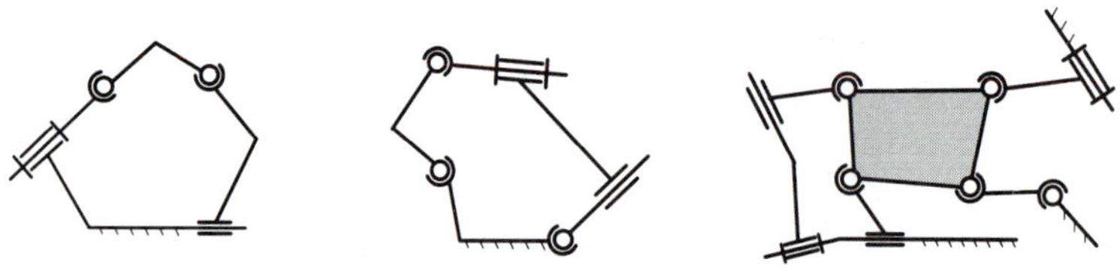

(a) (b) (c)

PROBLEM 1.11 Determine the mobility and the number of idle degrees of freedom of the spatial linkages shown below. Show the equations used to determine your answers.

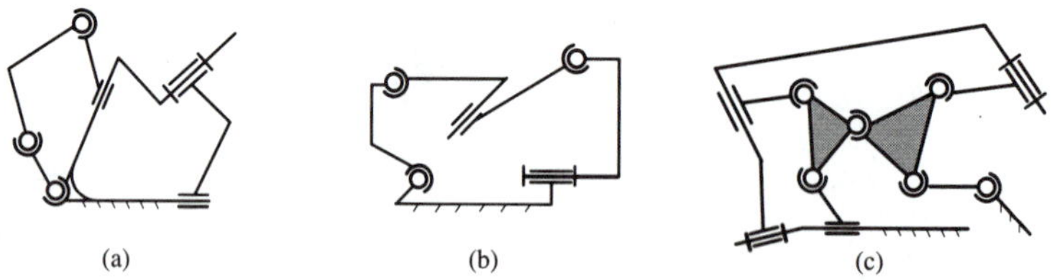

(a) (b) (c)

PROBLEM 1.12

If position information is available for all points in the planar linkage shown below, can all of the velocities be determined uniquely if the value of ω is given? Explain your answer.

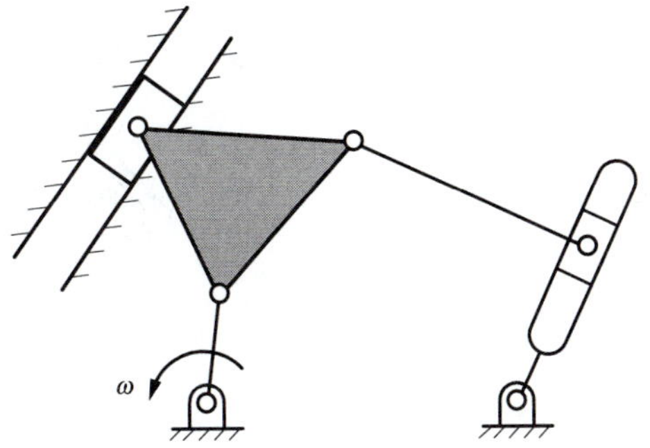

PROBLEM 1.13

Determine the mobility and the number of idle degrees of freedom of the linkages shown below. Show the equations used to determine your answers.

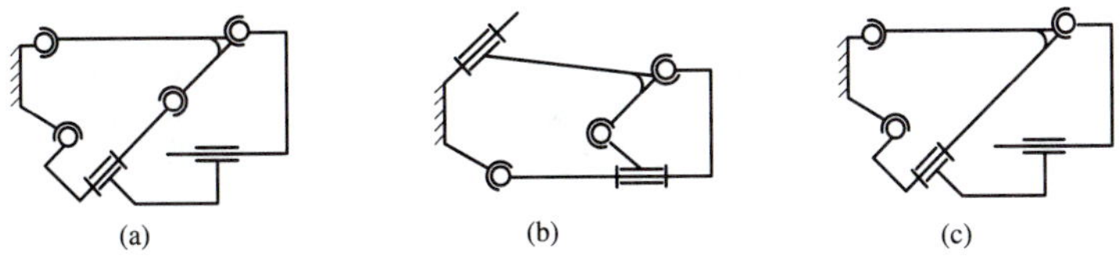

PROBLEM 1.14

Determine the mobility and the number of idle degrees of freedom for each of the mechanisms shown. Show the equations used to determine your answers.

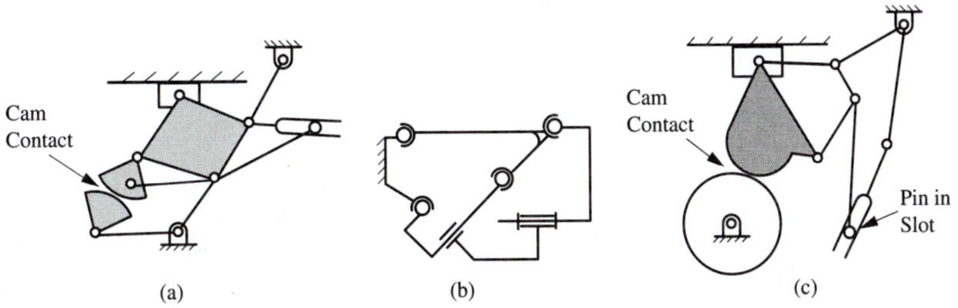

PROBLEM 1.15

Determine the mobility and the number of idle degrees of freedom for each of the mechanisms shown. Show the equations used to determine your answers.

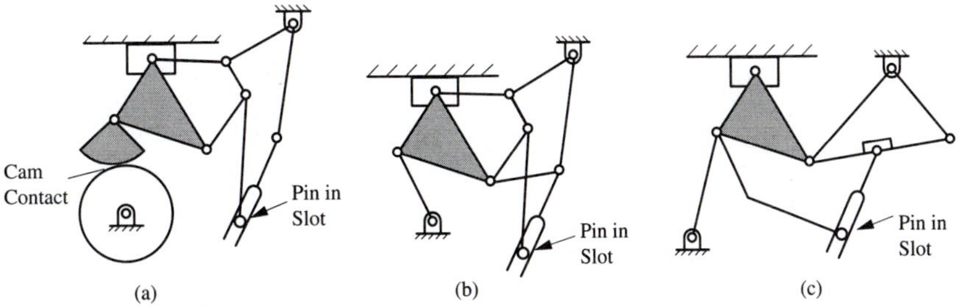

PROBLEM 1.16

Determine the mobility and the number of idle degrees of freedom for each of the mechanisms shown. Show the equations used to determine your answers.

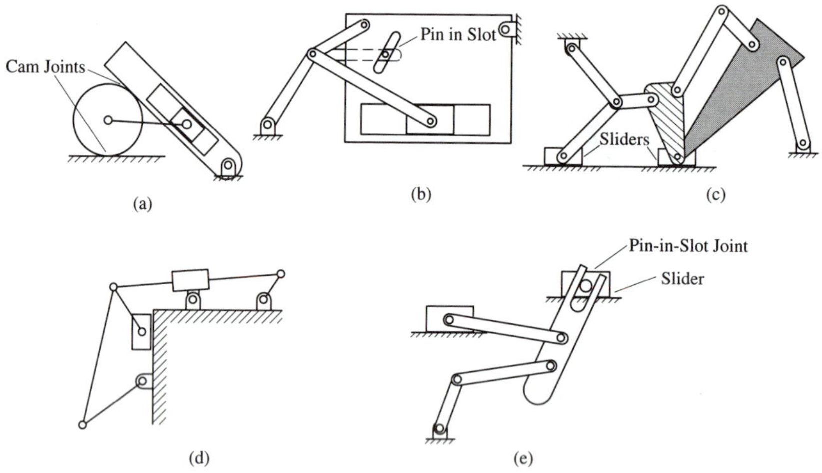

PROBLEM 1.17

Determine the mobility and the number of idle degrees of freedom associated with each mechanism.[5] Show the equations used to determine your answers.

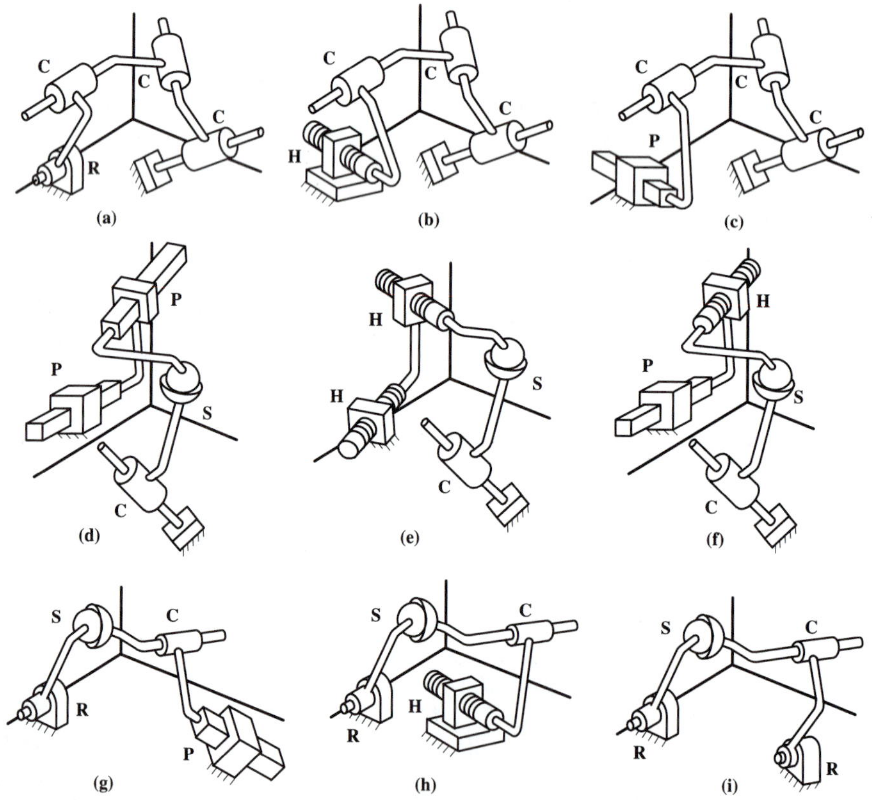

[5] Problem based on paper entitled "A Number Synthesis Survey of Three-Dimensional Mechanisms" by L. Harrisberger, *Trans. ASME, J. of Eng. for Ind.*, May, 1965, pp. 213–220.

PROBLEM 1.18[6]

Determine the mobility and the number of idle degrees of freedom associated with the mechanism shown below. The mechanism is a side-dumping car that consists of body 2 and truck 3 connected together by two six-bar linkages, *ABCDEF* and *AGHKLMN*. Link *NM* is designed as a latch on its free end (see left drawing). When jack 1 is operated, body 3 is lifted to the dumping position shown in the right-hand drawing. Simultaneously, the six-bar linkage *AGHKLMN* opens the latch on link *NM* and raises link *GH*. Linkage *ABCDEF* swings open side *BC* and the load can be dumped at some distance from the car (see right-hand drawing). Show the equations used to determine your answers.

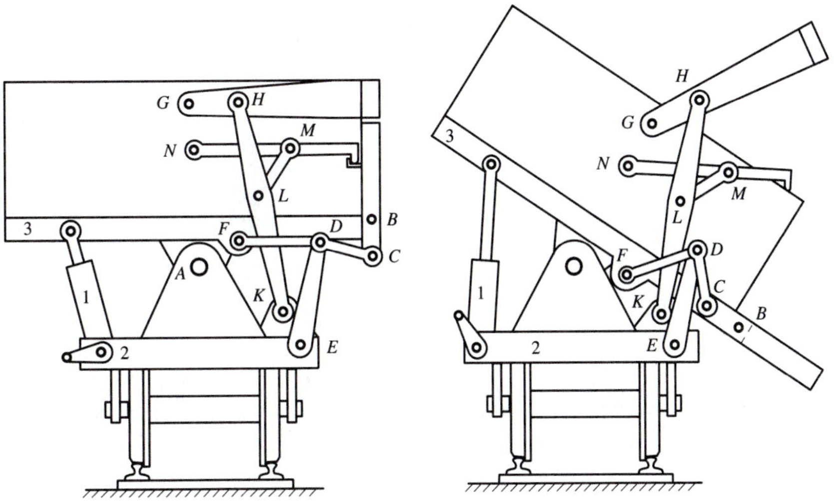

PROBLEM 1.19

Determine the mobility and the number of idle degrees of freedom associated with the mechanism below. The round part rolls without slipping on the pieces in contact with it.

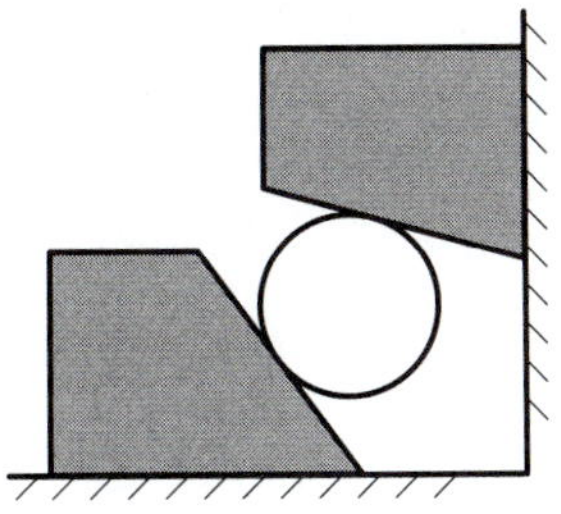

[6] Problem courtesy of Joseph Davidson, Arizona State University.

PROBLEM 1.20 [7]

Determine the mobility and the number of idle degrees of freedom associated with the mechanism below. The figure is a schematic of the entire linkage for a large power shovel used in strip mining. It can cut into a bank 20 m high and can dump to a height of 14.5 m. Link 7 is connected to link 8 with a revolute joint.

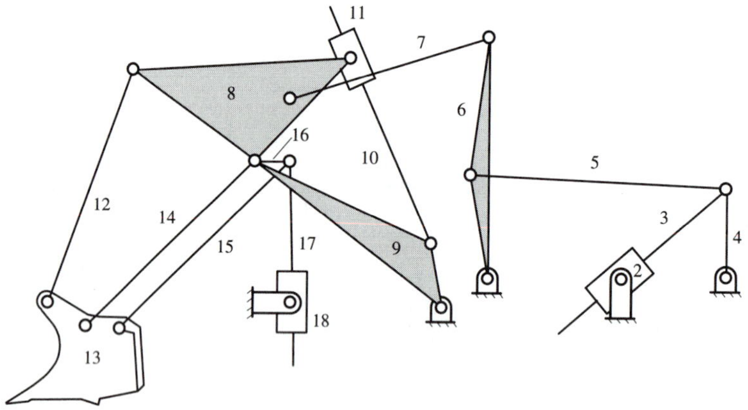

PROBLEM 1.21

In the figure is a portion of the support mechanism for the dipper on a large earth-moving machine used in removing overburden in strip mining operations. The fixed centers for the portion of the mechanism really move, but useful information can be obtained by observing the dipper motion relative to the "frame" as shown in the sketch. Both links 4 and 5 are mounted at O_4. Links 4 and 6 are parallel and of equal length. The dipper is moved by a hydraulic cylinder driving crank 5 about its fixed cylinder. Determine the number of degrees of freedom of the mechanism.

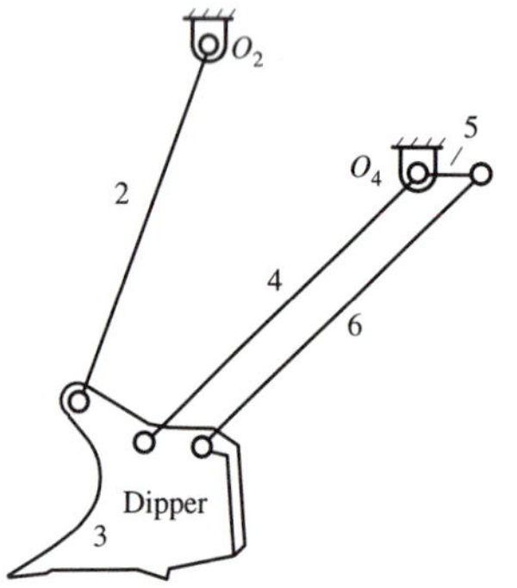

[7] Problem courtesy of Joseph Davidson, Arizona State University.

PROBLEM 1.22 Determine which (if either) of the following linkages can be driven by a constant-velocity motor. For the linkage(s) which can be driven by the motor, indicate the driver link.

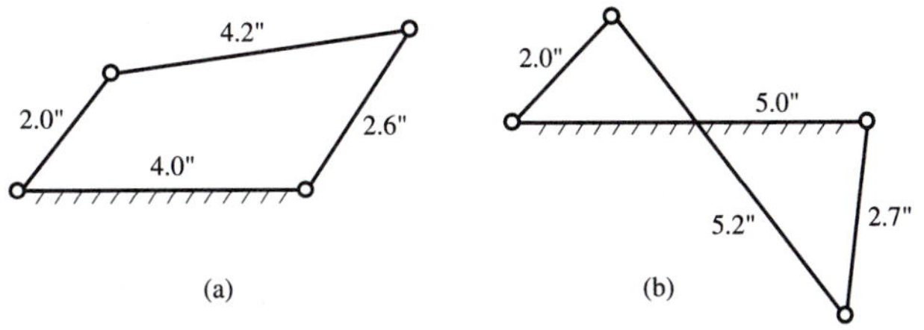

PROBLEM 1.23 Assume that you have a set of links of the following lengths: 2 in, 4 in, 5 in, 6 in, 9 in. Design a four-bar linkage that can be driven with a continuously rotating electric motor. Justify your answer with appropriate equations, and make a scaled drawing of the linkage. Label the crank, frame, coupler, and rocker (follower).

PROBLEM 1.24 Assume that you have a set of links of the following lengths: 20 mm, 30 mm, 45 mm, 56 mm, 73 mm. Design a four-bar linkage that can be driven with a continuous-rotation electric motor. Justify your answer with appropriate equations, and make a freehand sketch (labeled) of the resulting linkage. Label the crank, frame, coupler, and rocker (follower).

PROBLEM 1.25 For the four-bar linkages below, indicate whether they are Grashof type 1 or 2 and whether they are crank-rocker, double-crank, or double-rocker mechanisms.

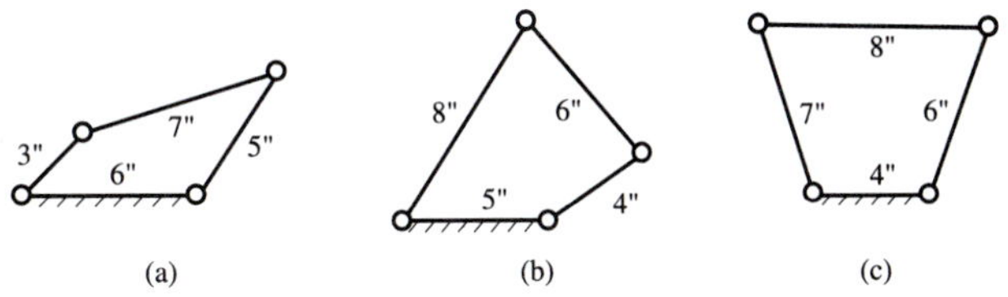

PROBLEM 1.26 You are given a set of three links with lengths 2.4 in, 7.2 in, and 3.4 in. Select the length of a fourth link and assemble a linkage that can be driven by a continuously rotating motor. Is your linkage a Grashof type 1 or Grashof type 2 linkage? (Show your work.) Is it a crank-rocker, double-rocker, or double-crank linkage? Why?

PROBLEM 1.27 You have available a set of eight links from which you are to design a four-bar linkage. Choose the links such that the linkage can be driven by a continuous-rotation motor. Sketch the linkage and identify the type of four-bar mechanism resulting.

$$L_1 = 2'', L_2 = 3'', L_3 = 4'', L_4 = 6'', L_5 = 7'', L_6 = 9.5'', L_7 = 13'', \text{ and } L_8 = 9''$$

PROBLEM 1.28

Determine the number of fully rotating cranks in the planar mechanisms shown below. Show your calculations.

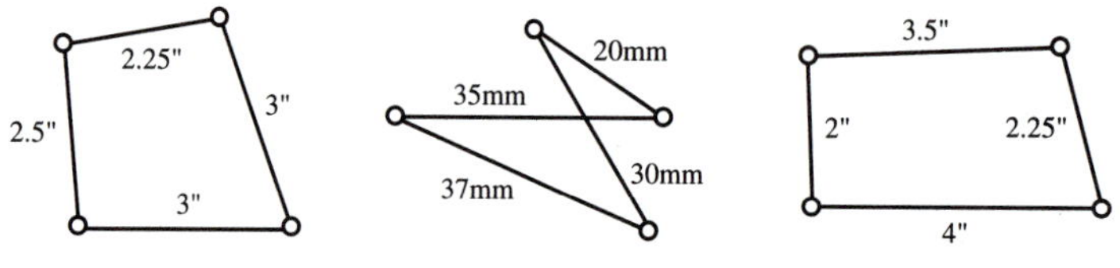

(a) (b) (c)

PROBLEM 1.29

The mechanisms shown below are drawn to scale.

(a) Sketch kinematic schematics showing the relationships between the members and joints.
(b) Determine the Grashof type of each four-bar linkage in each mechanism.

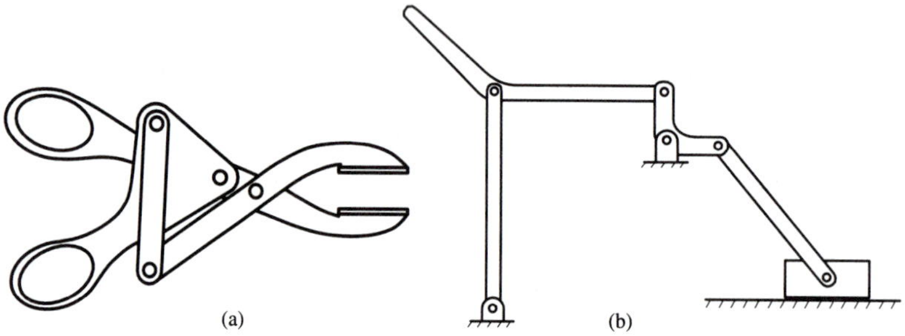

(a) (b)

Chapter 2

Graphical Position, Velocity, and Acceleration Analysis

2.1 INTRODUCTION

Historically, planar linkage analysis problems were solved graphically using drafting equipment. In recent years computer techniques have offered a viable and attractive alternative. Some teachers of the subject now prefer to concentrate their time on analytical approaches. Nevertheless, there are still many situations in which graphical techniques offer the most efficient solution, and the insight into the problem obtained by an understanding of the graphical approach is, we feel, essential. For this reason we have chosen to present both approaches. In this chapter, we present the graphical approach, and in Chapter 3, we present analytical approaches.

There is a tendency to discard traditional graphical techniques in favor of numerical solutions based on the analytical formulations presented in Chapter 3. However, there are many situations in which graphical techniques are useful. For example, it is necessary to check and debug computer programs. This is done most effectively by comparing the numerical solutions of sample problems with solutions to the same problems obtained using completely different techniques. Graphical techniques are ideal for providing these alternative solutions. At other times, a quick answer to a problem is needed, and no suitable program is available. Rather than writing and debugging a program specifically to solve the problem, it is often more efficient to use the graphical approach. Most important, insight into the kinematic geometry that governs all mechanism behavior is obtained by an understanding of the graphical approach.

In the present chapter, graphical techniques for the analysis of a variety of mechanisms will be developed. It is necessary to develop methods for the analysis of linkages with multiple closed loops, as well as for linkages with prismatic joints and higher pairs. It is also necessary to develop techniques that allow the analysis of linkages that are driven via members that are not cranks. An important class of simple mechanisms is driven via the coupler.

In this chapter, we will use two different approaches to a graphical kinematic analysis. The first approach is based on solving the position, velocity, and acceleration equations with vector polygons. For the case in which only velocities are of interest, an alternative method for a graphical velocity analysis based on instant centers of relative motion is introduced. When two laminae are moving relative to one another there exists, at every instant, a point in one lamina that is at rest relative to the other, and vice versa. This is the instant center of relative motion of those laminae. The technique of velocity analysis based on instant centers presents advantages when solving certain types of problems. Therefore, it is advantageous for the engineer to be familiar with this technique, as well as the vector polygon technique.

2.2 GRAPHICAL POSITION ANALYSIS

Regardless of what procedure is used for a linkage analysis, it is necessary to determine the angular positions of the links before it is possible to perform a velocity analysis. Likewise, it is necessary to know the link angular velocities before an acceleration analysis can be performed. That is, the kinematic analysis of a linkage must *always* proceed in this sequence: position analysis, then velocity analysis, then acceleration analysis. If the linkage has one degree of freedom and the driver is a crank, it is necessary that the angular position, angular velocity, and angular acceleration of a driving link be specified for a solution to be possible. If the driving member is connected to the base by a prismatic joint, the linear position, velocity, and acceleration of any point in that link must be specified.

When working graphically, the position analysis consists of simply drawing the linkage to scale. Usually this is so straightforward that it tends to be forgotten as an important step in the solution process. The representation used is a geometric skeleton of the linkage: links connected by revolute joints are represented by the line, or lines, joining the joint axes. Prismatic joints are represented by lines in the direction of sliding. Revolute joints are usually represented only by the points that are the intersections of their axes with the plane of motion. The way the method works in the analysis of a simple linkage is illustrated in the examples.

As will be shown in Chapter 3, the position equations for mechanisms are inherently nonlinear. In many cases, the mechanism can be assembled (or drawn) in two possible configurations. It is necessary to know before the analysis is conducted which solution is desired. This will be illustrated in the examples that will be discussed after the equations for velocity and acceleration are developed.

We will begin the analysis of velocities and accelerations with a relatively simple case involving two points fixed to the same rigid link. The equations for this case are commonly developed in courses in mechanics using the procedure we shall use here. The equations developed will be directly applicable to mechanisms with revolute joints and/or sliders on fixed lines. We will illustrate the use of the procedure with several examples.

For more complex joints, a more rigorous and general approach will be used to develop the velocity and acceleration equations. This will entail identifying the coordinate systems relative to which each of the vectors is described and relative to which the time derivatives are desired. It will be shown that the velocity and acceleration equations developed for the case of two points on a rigid link are special cases of the more general equations.

2.3 PLANAR VELOCITY POLYGONS

Velocity analysis is the determination of the angular velocities of different links in a mechanism and of the velocities of points on the links, given either the angular velocity of some member or the velocity of some point on the link designated as the input. The vector polygon technique will be used here to solve the velocity and acceleration equations. The method facilitates the solution of a large variety of velocity and acceleration problems and also has the advantage that the acceleration polygon solution has a strong similarity to that of the velocity polygon, which makes it relatively straightforward to learn and remember. Almost all practical problems can be solved by this approach.

In theory, however, the technique is not general. It is possible to formulate problems

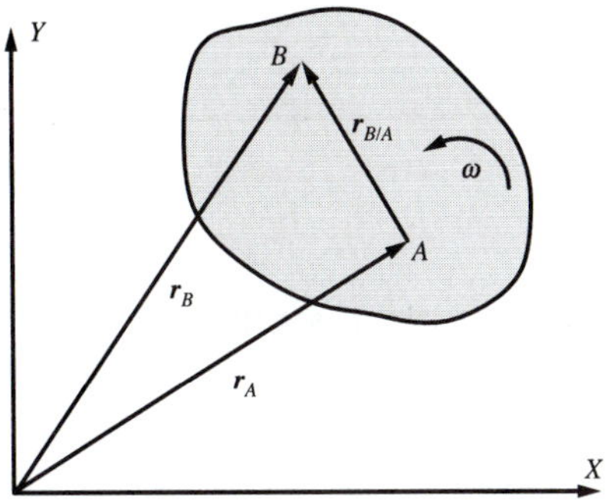

Figure 2.1 Position relationships of two points embedded in a moving lamina.

that cannot be solved by the methods presented here. Special techniques have been developed that allow treatment of some of the simpler cases that are not amenable to the vector polygon method; however, it is possible to formulate problems that cannot be solved by even these embellished techniques. The reader is referred to books by Hirschhorn,[1] Hall,[2] and Holowenko[3] for the auxiliary–point technique and other methods of handling more general mechanisms. It should be emphasized, however, that problems that cannot be solved by the methods presented in this chapter are rarely encountered in practice.

The key to the graphical velocity analysis of most linkages is the relationship between the velocities of any two points embedded in a rigid body. This relationship is

$$\boldsymbol{v}_B = \boldsymbol{v}_A + \boldsymbol{\omega} \times \boldsymbol{r}_{B/A} \tag{2.1}$$

where A and B are points fixed in a moving lamina (rigid body) as shown in Fig. 2.1, $\boldsymbol{v}_A$ and $\boldsymbol{v}_B$ are the respective velocities relative to the frame of those points, $\boldsymbol{r}_{B/A}$ is the vector $\overline{AB}$, and $\boldsymbol{\omega}$ is the angular velocity of the lamina relative to the frame.

To prove this relationship, consider the two points A and B fixed in the lamina shown in Fig. 2.1. The lamina is moving with general planar motion. Let the position of point A relative to a fixed reference frame be $\boldsymbol{r}_A$ and that of point B be $\boldsymbol{r}_B$. The vector $\overline{AB}$ is $\boldsymbol{r}_{B/A}$ and is pointed from A to B. Therefore

$$\boldsymbol{r}_B = \boldsymbol{r}_A + \boldsymbol{r}_{B/A} \tag{2.2}$$

Differentiating Eq. (2.2) with respect to time gives

$$\boldsymbol{v}_B = \boldsymbol{v}_A + d\boldsymbol{r}_{B/A}/dt$$

Now, since points A and B are fixed in the moving lamina, vector $\boldsymbol{r}_{B/A}$ is fixed in that lamina and moves with it. It has constant length, so only its direction changes. Let the change in direction in a small time interval δt be $\delta\boldsymbol{\theta}$ as shown in Fig. 2.2. The magnitude of the change in $\boldsymbol{r}_{B/A}$ is

$$\delta r = r_{B/A}\, \delta\theta$$

As δt and hence $\delta\theta$ approach zero, the angle between the vectors $\delta\boldsymbol{r}_{B/A}$ and $\boldsymbol{r}_{B/A}$ approaches 90°. If ω is the magnitude of the angular velocity of the lamina,

$$\delta\theta = \omega\, \delta t$$

[1] Hirschhorn, J., *Kinematics and Dynamics of Plane Mechanisms,* McGraw Hill Book Co., New York, 1962.
[2] Hall, A., *Kinematics and Linkage Design,* Balt Publishers, West Lafayette, IN, 1966.
[3] Holowenko, A. R., *Dynamics of Machinery,* John Wiley & Sons, Inc., New York, 1955.

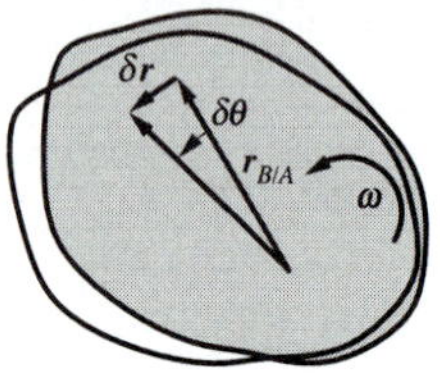

Figure 2.2 Successive positions of the lamina separated by a small time interval, δt.

Therefore

$$\delta r / \delta t = r\omega$$

so, in the limit as δt approaches zero,

$$|d\boldsymbol{r}_{B/A}/dt| = r_{B/A}\omega$$

If $\boldsymbol{\omega}$ is considered to be a vector normal to the plane of motion, clockwise (CW) if directed away from the observer and counterclockwise (CCW) if directed toward the observer, the direction of $d\boldsymbol{r}_{BA}/dt$ is normal to $\boldsymbol{\omega}$ and to $\boldsymbol{r}_{B/A}$ and obeys the right-hand screw rule with respect to those vectors. Therefore $d\boldsymbol{r}_{B/A}/dt$ can be represented by the expression

$$d\boldsymbol{r}_{B/A}/dt = \boldsymbol{\omega} \times \boldsymbol{r}_{B/A}$$

Thus

$$\boldsymbol{v}_B = \boldsymbol{v}_A + \boldsymbol{\omega} \times \boldsymbol{r}_{B/A} \tag{2.1}$$

As will be shown in Section 2.11, this expression is actually valid for general, spatial motion, although the derivation above applies only to planar motion.

It is convenient to write Eq. (2.1) in the form

$$\boldsymbol{v}_B = \boldsymbol{v}_A + \boldsymbol{v}_{B/A} \tag{2.3}$$

where

$$\boldsymbol{v}_{B/A} = \boldsymbol{\omega} \times \boldsymbol{r}_{B/A} \tag{2.4}$$

The vector $\boldsymbol{v}_{B/A}$ is usually called the velocity of B relative to A, although, strictly speaking, it is meaningless to talk of a velocity relative to a point. Velocities are vectors and are measured relative to reference frames. Therefore, $\boldsymbol{v}_{B/A}$ would be the velocity of point B relative to a reference frame which has its origin at point A. The reference frame moves so as to be always parallel to the fixed frame.

The basic technique used in a graphical linkage analysis is to work from one or more points with known velocity to one of unknown velocity using the relationship in Eq. (2.1)

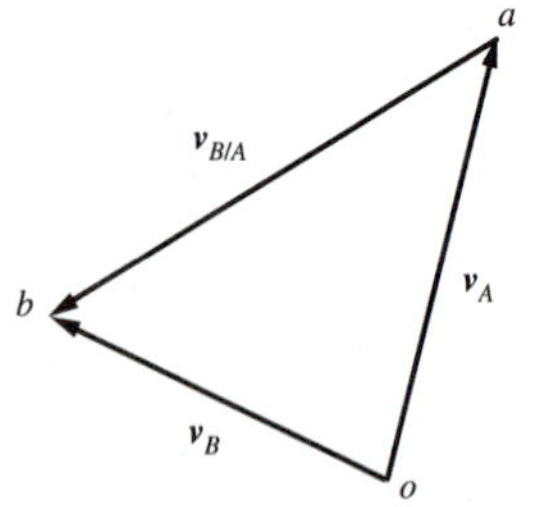

Figure 2.3 Velocities of two points embedded in a lamina.

Figure 2.4 The direction relationship among the vectors $\boldsymbol{v}_{B/A}$, $\boldsymbol{\omega}$, and $\boldsymbol{r}_{B/A}$ for planar motion.

between the velocities of two points fixed in the same lamina. The intersections of the axes of revolute joints with the plane of motion form transfer points because they are actually coincident points fixed in two different links. Thus, the velocity of a revolute point can be obtained by considering it to be a point in one of the links it connects. That information can then be used by considering it to be fixed in the other link.

Equation (2.3) can be represented graphically as the vector triangle shown in Fig. 2.3. This triangle can always be solved given the direction and magnitude of one of the three vectors and the directions of the remaining two. This is the normal situation in planar velocity analysis. Again, the way in which this works will be illustrated in several examples after all of the necessary equations have been developed.

Based on Eq. (2.3), to find the angular velocity, $\boldsymbol{\omega}$, for a given link, we must compute the relative velocity between two points on the link, and the velocity must be given relative to the desired reference frame. For example, the relative velocity relationship for points B and A can be written as

$$\boldsymbol{v}_{B/A} = \boldsymbol{\omega} \times \boldsymbol{r}_{B/A} \tag{2.5}$$

The vectors will be mutually orthogonal as indicated schematically in Fig. 2.4. Because we will know the lines along which each of the vectors must lie, the main problem is to determine the directions along the lines and the magnitudes of each of the vectors. Given any two of the vector directions, we can find the direction of the third by observing the directions given by the right-hand screw rule. Two examples are shown in Fig. 2.4.

Notice that $\boldsymbol{v}_{B/A}$ and $\boldsymbol{r}_{B/A}$ are always perpendicular to each other. Also, visually, we can determine the direction of $\boldsymbol{v}_{B/A}$ by rotating $\boldsymbol{r}_{B/A}$ 90° in the direction of $\boldsymbol{\omega}$. Similarly, if we know the directions of $\boldsymbol{r}_{B/A}$ and $\boldsymbol{v}_{B/A}$, we can determine the direction of $\boldsymbol{\omega}$ by visualizing the direction in which we must rotate $\boldsymbol{r}_{B/A}$ to obtain the direction of $\boldsymbol{v}_{B/A}$.

Because the three vectors in Eq. (2.4) are orthogonal, their magnitudes are related by

$$|\boldsymbol{v}_{B/A}| = |\boldsymbol{\omega}||\boldsymbol{r}_{B/A}| \tag{2.6}$$

Given any two of the three magnitudes in Eq. (2.6), we can easily solve for the third magnitude.

2.4 GRAPHICAL ACCELERATION ANALYSIS

Just as was the case for velocity analysis, the key to most graphical acceleration analyses is the relationship between the accelerations of two points fixed in the same rigid lamina or link. This relationship can be derived by differentiating the velocity relationship with respect to time. Rewriting Eq. (2.1)

$$\boldsymbol{v}_B = \boldsymbol{v}_A + \boldsymbol{\omega} \times \boldsymbol{r}_{B/A} \tag{2.1}$$

Differentiating

$$\boldsymbol{a}_B = \boldsymbol{a}_A + d\boldsymbol{\omega}/dt \times \boldsymbol{r}_{B/A} + \boldsymbol{\omega} \times d\boldsymbol{r}_{B/A}/dt$$

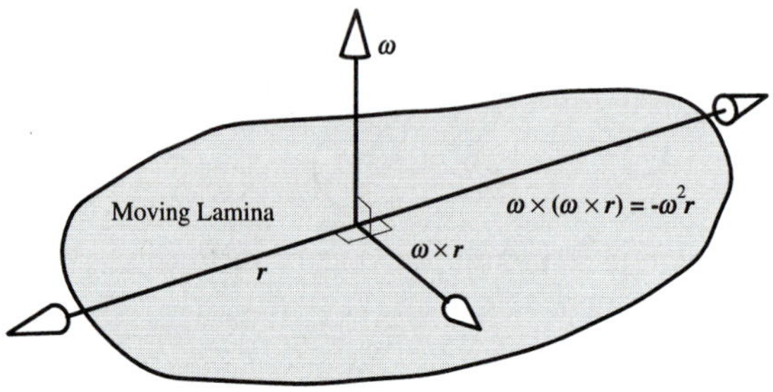

Figure 2.5 The derivation of the relationship $\boldsymbol{\omega} \times (\boldsymbol{\omega} \times \boldsymbol{r}) = -\omega^2 \boldsymbol{r}$ which is valid for planar motion.

As was shown in Section 2.3,

$$d\boldsymbol{r}_{B/A}/dt = \boldsymbol{\omega} \times \boldsymbol{r}_{B/A}$$

Also, $\boldsymbol{\alpha}$ is defined to be $d\boldsymbol{\omega}/dt$. Hence

$$\boldsymbol{a}_B = \boldsymbol{a}_A + \boldsymbol{\alpha} \times \boldsymbol{r}_{B/A} + \boldsymbol{\omega} \times (\boldsymbol{\omega} \times \boldsymbol{r}_{B/A}) \tag{2.7}$$

As will be demonstrated in Section 2.11, this expression is generally valid for three-dimensional motion, although it has been derived here only in the planar motion context. For planar motion, it is possible to simplify the expression by noting that, in this case, $\boldsymbol{\omega}$ and $\boldsymbol{r}_{B/A}$ are orthogonal, as shown in Fig. 2.5. Also, $\boldsymbol{\omega} \times \boldsymbol{r}_{B/A}$ has the magnitude $\omega r_{B/A}$ and is normal to both $\boldsymbol{\omega}$ and $\boldsymbol{r}_{B/A}$ in the sense given by the right-hand screw rule. Then, $\boldsymbol{\omega} \times (\boldsymbol{\omega} \times \boldsymbol{r}_{B/A})$ has the magnitude $\omega^2 r_{B/A}$ and is orthogonal to both $\boldsymbol{\omega}$ and $\boldsymbol{\omega} \times \boldsymbol{r}_{B/A}$. Applying the right-hand screw rule, it can be seen that this vector $\boldsymbol{\omega} \times (\boldsymbol{\omega} \times \boldsymbol{r}_{B/A})$ is always in the negative $\boldsymbol{r}_{B/A}$ direction. Therefore it can be written as $-\omega^2 \boldsymbol{r}_{B/A}$, and the relationship between the accelerations of points A and B is

$$\boldsymbol{a}_B = \boldsymbol{a}_A + \boldsymbol{\alpha} \times \boldsymbol{r}_{B/A} - \omega^2 \boldsymbol{r}_{B/A} \tag{2.8}$$

It is usual to write

$$\boldsymbol{a}^r_{B/A} = -\omega^2 \boldsymbol{r}_{B/A} \quad \text{and} \quad \boldsymbol{a}^t_{B/A} = \boldsymbol{\alpha} \times \boldsymbol{r}_{B/A} \tag{2.9}$$

with $\boldsymbol{a}^r_{B/A}$ called the radial component of the acceleration of B relative to A and $\boldsymbol{a}^t_{B/A}$ called the transverse component of the acceleration of B relative to A. As was noted in the case of velocities, it is not really proper to talk about the velocity or acceleration of one point relative to another point. The vector $\boldsymbol{a}_{B/A}$ is really the acceleration of point B relative to a reference frame with origin at A. The reference frame moves so that it is always parallel to the fixed frame.

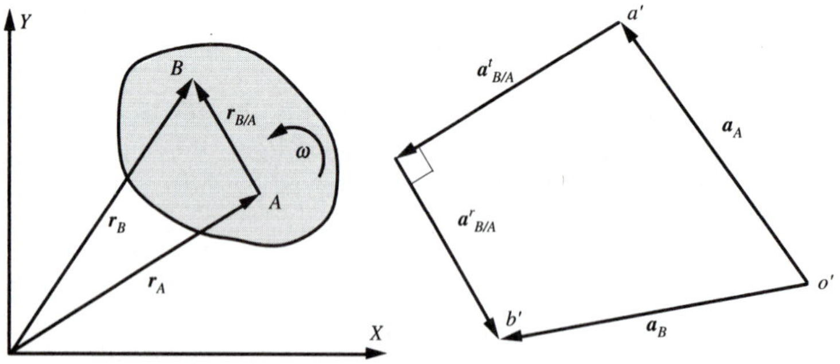

Figure 2.6 Accelerations of two points embedded in a moving lamina.

Figure 2.7 The direction relationship among the vectors $\boldsymbol{a}^t_{B/A}$, $\boldsymbol{\alpha}$, $\boldsymbol{r}_{B/A}$ for planar motion.

The vector polygon corresponding to Eq. (2.7) is shown in Fig. 2.6. If a velocity analysis of the linkage has been performed, the angular velocities of all the links are known, and so the radial component $\boldsymbol{a}^r_{B/A} = -\omega^2\boldsymbol{r}_{B/A}$ can always be calculated and plotted. Hence, if one of the other vectors is known, and the directions of the remaining two are also known, the polygon can be solved in very much the same way as the vector triangle was used in velocity analysis. This is the normal procedure for a graphical acceleration analysis.

The angular acceleration for a given link is obtained in the same manner as the angular velocity except that the *tangential* component of relative acceleration is used instead of the linear velocity. To find a value of $\boldsymbol{\alpha}$, we must know the tangential component of the relative acceleration between any two points on the link. That tangential component of acceleration must be given relative to the desired reference frame. For example, the relative tangential acceleration relationship for points B and A can be written as

$$\boldsymbol{a}^t_{B/A} = \boldsymbol{\alpha} \times \boldsymbol{r}_{B/A}$$

Because we will know the lines along which the vectors must lie, the main problem again is to determine the directions along the lines and the magnitude of each of the vectors. Given any two of the vector directions, we can find the direction of the third by observing the directions given by the right-hand screw rule. Two examples are shown schematically in Fig. 2.7.

Notice that these relationships are exactly the same as for the velocity expressions if $\boldsymbol{\omega}$ is replaced by $\boldsymbol{\alpha}$ and $\boldsymbol{v}_{B/A}$ is replaced by $\boldsymbol{a}^t_{B/A}$. In particular, notice that $\boldsymbol{a}^t_{B/A}$ and $\boldsymbol{r}_{B/A}$ are always perpendicular to each other. Also, we can determine the direction of $\boldsymbol{a}^t_{B/A}$ by visually rotating $\boldsymbol{r}_{B/A}$ 90° in the direction of $\boldsymbol{\alpha}$. Similarly, if we know the directions of $\boldsymbol{r}_{B/A}$ and $\boldsymbol{a}^t_{B/A}$, we can determine the direction of $\boldsymbol{\alpha}$ by visualizing the direction in which we must rotate $\boldsymbol{r}_{B/A}$ to obtain the direction of $\boldsymbol{a}^t_{B/A}$.

Because the three vectors in Eq. (2.9) are orthogonal, their magnitudes are related by

$$|\boldsymbol{a}^t_{B/A}| = |\boldsymbol{\alpha}||\boldsymbol{r}_{B/A}| \tag{2.10}$$

Given any two of the three magnitudes in Eq. (2.10), we can easily solve for the third magnitude.

2.5 GRAPHICAL ANALYSIS OF A FOUR-BAR MECHANISM

Having derived the basic equations for relative velocities and accelerations between two points on a rigid link, we will illustrate the use of the equations for the graphical analysis of several mechanisms. The first example involves the position, velocity, and acceleration analysis of the four-bar mechanism given in Fig. 2.8. The analysis for this example will be conducted in detail, but in subsequent examples, less detail will be given. In all of the examples, subscripts will be used to identify the links to which the points are attached. This is necessary because *the equations derived in Sections 2.3 and 2.4 apply only when the points (A and B) are fixed to the same link or lamina.*

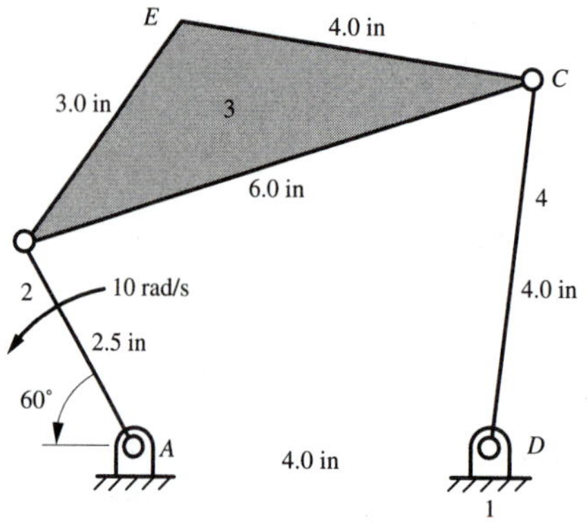

Figure 2.8 The four-bar linkage of Example 2.1. (Note that the figure has been reduced in size for printing.)

EXAMPLE 2.1 *(Graphical Analysis of a Four-Bar Mechanism)*

PROBLEM

Determine the angular positions, angular velocities, and angular accelerations of all members of the linkage shown when link AB is at 60° to the horizontal as shown in Fig. 2.8. Also find the position, velocity, and acceleration of point E in the coupler member of the linkage. Link AB is driven at a constant angular velocity of 10 rad/s counterclockwise (CCW).

SOLUTION

(a) Position Analysis

We will first address the graphical determination of the link positions. The first step is to choose a scale. The larger the scale, the more accurate the results. Therefore, it is best to use a drawing table with a drafting machine and B- or C-sized drawing paper if accurate results are desired. A CAD system that supports the construction of lines and arcs and locates intersections of lines and arcs may also be used. In the present case, we want to fit the figure onto a regular book page, so the construction will be described when it is drawn at half-scale (1 inch on the drawing corresponds to 2 inches on the actual mechanism). The reader is encouraged to draw the figures in this, and following examples, at full scale when working through them.

The construction is shown in the position diagram in Fig. 2.9. A horizontal line representing the base link is drawn first, and the two points bounding an interval of 2 inches (half-scale) are marked to represent A and D. Next locate point B, which is where the driver link (link 2) is joined to the coupler (link 3). A line through point A at an angle of 120° to AD is drawn, and a point on that line at a distance of 1.25 inches is marked to represent point B. Next locate point C, which is where the coupler is joined to the rocker (link 4).

To locate point C, a compass is set to a radius of 3.0 inches, and an arc is drawn with center point B. The compass is then reset to a 2.0 inch radius, and a second arc is drawn with center D. C is at the intersection of the two arcs. Actually, there are two possible intersection points corresponding to the two assembly modes of the mechanism. This is a common situation with many mechanisms, and it is necessary for the designer to know which assembly mode is desired. In the present case, the correct one is easily located by referring to Fig. 2.8.

Point E can now be located in a similar manner because we know the distance from point E to point B and to point C. The compass is set to a radius of 1.5 inches (the scaled distance from E to B), and an arc is drawn with center B. The compass is reset to a radius of 2.0 inches (the scaled distance from E to C), and an arc is drawn with center C. E is at an intersection of the two arcs. Once again two intersections are possible (one below BC and one above BC), and the correct intersection can be identified by referring to Fig. 2.8.

This completes the construction of the scale drawing of the linkage and hence completes the solution of the position analysis problem. The resulting construction is shown in Fig. 2.9. The angular positions of the links may be measured from the drawing. Likewise, the position of point E can be measured and the coordinates multiplied by the scale factor of 2. In practice,

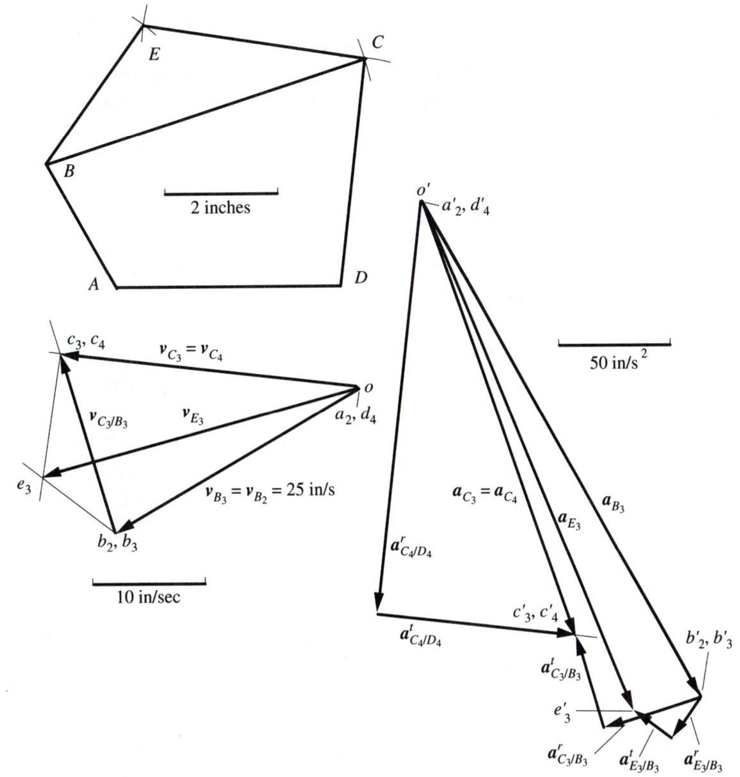

Figure 2.9 Position, velocity, and acceleration polygons for the four-bar linkage of Fig. 2.8. Note that the position solution is necessary to draw the velocity polygon, and the velocity polygon is needed to draw the acceleration polygon. (Note also that the size of the figure has been reduced in the printing process.)

if the position analysis is being performed solely as a preliminary step to a velocity analysis, the angular positions of the links and the position of point C would not need to be measured directly. Rather, that information would be directly transferred to the velocity diagram using a drawing machine or drafting triangles to construct normal or parallel lines. For the acceleration analysis, however, the linear distances would be required.

(b) Velocity Analysis

In the velocity analysis, we will typically use the same points in the same order that we used for the position analysis. We will first compute the velocity of point B, then the velocity of C, and finally the velocity of E. Location B is actually the location of two points, B_2 and B_3 in links 2 and 3, respectively, and to be rigorous, we need to identify which of the points we are considering. Start with B_2, which is the point on the driver link 2. We want to compute $\boldsymbol{v}_{B_2}$, which is the absolute velocity of point B_2.

This absolute velocity can be expressed as the relative velocity between B_2 and any point that has zero velocity. The point we shall use is A_2. It has zero velocity because it is always coincident with A_1, which is fixed to frame 1. All points in frame 1 have zero velocity. Then the velocity expression in Eq. (2.1) can be written as

$$\boldsymbol{v}_{B_2} = \boldsymbol{v}_{A_2} + \boldsymbol{v}_{B_2/A_2} = \boldsymbol{\omega}_2 \times \boldsymbol{r}_{B/A}$$

because points A_2 and B_2 are both on the same link or lamina (link 2). Note that we do not need to identify the link associated with A and B in $\boldsymbol{r}_{B/A}$ because all of the A's have the same coordinates, and all of the B's also have the same coordinates.

Note also that we know the directions and magnitudes for both $\boldsymbol{\omega}_2$ and $\boldsymbol{r}_{B/A}$, and we know that the two vectors are orthogonal to each other. Therefore, by the cross-product, the velocity $\boldsymbol{v}_{B_2/A_2}$ will be orthogonal to $\boldsymbol{\omega}_2$ and $\boldsymbol{r}_{B/A}$, and the direction will be given by the right-hand screw rule. The relationship among the three vectors is represented by a diagram similar to that shown in Fig. 2.5. We can compute the magnitude of the velocity of B_2 from an equation similar to Eq. (2.5). The magnitude is given by

$$|\boldsymbol{v}_{B_2/A_2}| = |\boldsymbol{\omega}_2||\boldsymbol{r}_{B/A}| = (10 \text{ rad/s})(2.5 \text{ in}) = 25 \text{ in/s}$$

The direction for $\boldsymbol{v}_{B_2/A_2}$ is given by using the right-hand screw rule or by rotating $\boldsymbol{r}_{B/A}$ 90° in the direction of $\boldsymbol{\omega}_2$.

It is now necessary to select a scale to plot $\boldsymbol{v}_{B_2/A_2}$. We used a scale of 10 in/s to 1 inch in the velocity diagram shown in Fig. 2.9. The direction of $\boldsymbol{v}_{B_2/A_2}$ may be obtained by placing one of the orthogonal edges of a triangle along $\overrightarrow{AB}$ and drawing a line along the other edge, since $\boldsymbol{v}_{B_2/A_2}$ is normal to $\boldsymbol{r}_{B/A}$ or $\overrightarrow{AB}$. Two points, o and b_2, separated by an interval of 2.5 in are marked as shown in Fig. 2.9. On the polygon, $\overrightarrow{ob_2}$ may be labeled as the vector $\boldsymbol{v}_{B_2/A_2}$. Here we are using the convention of labeling points on the velocity polygon with lowercase letters and the corresponding points on the position polygon with uppercase letters. Thus, the absolute velocity of point B_2 given above by $\boldsymbol{v}_{B_2/A_2}$ would be represented on the velocity polygon by $\overrightarrow{ob_2}$ or $\overrightarrow{a_2b_2}$.

Next we want to compute the velocity of point C. We know that B_2 and B_3 are permanently pinned together so that

$$\boldsymbol{v}_{B_3} = \boldsymbol{v}_{B_2} = 25 \text{ in/s}$$

in the direction shown in Fig. 2.9. Similarly, C_3 and C_4 are permanently pinned together. Therefore,

$$\boldsymbol{v}_{C_3} = \boldsymbol{v}_{C_4}$$

Because B_3 and C_3 are both fixed to link 3, we can write a relative velocity equation similar to Eq. (2.1). That is,

$$\boldsymbol{v}_{C_3} = \boldsymbol{v}_{B_3} + \boldsymbol{v}_{C_3/B_3} \qquad \textbf{(2.11)}$$

In Eq. (2.11), the vector $\boldsymbol{v}_{B_3}$ is entirely known. Also, because B_3 and C_3 are on the same rigid link, we know that $\boldsymbol{v}_{C_3/B_3}$ is given by

$$\boldsymbol{v}_{C_3/B_3} = \boldsymbol{\omega}_3 \times \boldsymbol{r}_{C/B}$$

Therefore, we know that the vector $\boldsymbol{v}_{C_3/B_3}$ is perpendicular to $\boldsymbol{r}_{C/B}$ or $\overrightarrow{BC}$. We can then construct a line through point b_3 on the velocity polygon in a direction perpendicular to $\overrightarrow{BC}$ on the position diagram. This is easily done with the help of drafting triangles. This defines one locus of c_3. To find another locus for c_3, we need to find the direction of the vector $\boldsymbol{v}_{C_3}$. We know that $\boldsymbol{v}_{C_3} = \boldsymbol{v}_{C_4}$, and we can identify the direction of the velocity of C_4 by inspection (C can only move on a circle about point D, and therefore $\boldsymbol{v}_{C_4}$ must be perpendicular to the line $\overrightarrow{DC}$) or we can write a relative velocity equation for $\boldsymbol{v}_{C_4}$. Again, the velocity $\boldsymbol{v}_{C_4}$ is an absolute velocity, and it can be expressed as the relative velocity between C_4 and any point that has zero velocity. If we choose D_4 as that point, the velocity equation becomes

$$\boldsymbol{v}_{C_4} = \boldsymbol{v}_{D_4} + \boldsymbol{v}_{C_4/D_4} = \boldsymbol{v}_{C_4/D_4} = \boldsymbol{\omega}_4 \times \boldsymbol{r}_{C/D} \qquad \textbf{(2.12)}$$

Because of the cross-product, it is clear that $\boldsymbol{v}_{C_4}$ must be perpendicular to $\boldsymbol{r}_{C/D}$ or the line $\overrightarrow{DC}$. To locate c_4 on the velocity polygon, draw a line through point o in a direction perpendicular to $\overrightarrow{DC}$ on the position diagram. Once again, this is most easily done with drafting triangles. Because c_3 and c_4 are located at the same point, this gives a second locus for c_3 that can be determined as shown in Fig. 2.9. The vectors $\boldsymbol{v}_{B_3}$, $\boldsymbol{v}_{C_4}$, and $\boldsymbol{v}_{C_3/B_3}$ are as shown. The magnitude of $\boldsymbol{\omega}_3$ may be found from the expression for the relative velocity $\boldsymbol{v}_{C_3/B_3}$. Then,

$$|\boldsymbol{v}_{C_3/B_3}| = |\boldsymbol{\omega}_3||\boldsymbol{r}_{C/B}|$$

or

$$|\omega_3| = |v_{C_3/B_3}|/|r_{C/B}|$$

To get v_{C_3/B_3}, measure the length of $\overrightarrow{b_3c_3}$ on the velocity polygon and multiply by the scale factor. In the present case,[4] $\overrightarrow{b_3c_3} = 1.65$ in, so $v_{C_3/B_3} = 1.65 \times 10 = 16.5$ in/s. Then

$$|\omega_3| = |v_{C_3/B_3}|/|r_{C/B}| = 16.5/6.0 = 2.75 \text{ rad/s}$$

To get the direction, visualize the direction in which we would have to rotate $r_{C/D}$ to obtain the direction of v_{C_3/B_3}. This is the counterclockwise (CCW) direction.

Next compute the angular velocity ω_4 from Eq. (2.12). The magnitude can be found from an expression for the relative velocity v_{C_4/D_4}. Then

$$|v_{C_4/D_4}| = |\omega_4||r_{C/D}|$$

or

$$|\omega_4| = |v_{C_4/D_4}|/|r_{C/D}|$$

To obtain the velocity v_{C_4/D_4}, measure the distance c_4d_4 on the velocity polygon and multiply by the scale factor. In the present case, $c_4d_4 = 2.69$ in, so $v_{C_4/D_4} = 2.69 \times 10 = 26.9$ in/s. Then

$$|\omega_4| = |v_{C_4/D_4}|/|r_{C/D}| = 26.9/4.0 = 6.73 \text{ rad/s}$$

To get the direction, visualize the direction in which we would have to rotate $r_{C/D}$ to obtain the direction of v_{C_4/D_4}. The direction is CCW.

The velocity of point E_3 may be obtained by considering first the point pair E_3 and B_3 and then the pair E_3 and C_3. Both pairs are fixed to member 3. The relative velocity expressions are

$$v_{E_3} = v_{B_3} + v_{E_3/B_3} = v_{B_3} + \omega_3 \times r_{E/B} \tag{2.13}$$

and

$$v_{E_3} = v_{C_3} + v_{E_3/C_3} = v_{C_3} + \omega_3 \times r_{E/C} \tag{2.14}$$

The velocities v_{C_3} and v_{B_3} are known and have been plotted as $\overrightarrow{ob_3}$ and $\overrightarrow{oc_3}$ on the velocity polygon. We can compute $|v_{E_3}|$ two different ways as implied by Eqs. (2.13) and (2.14). One way is to compute the cross-product in Eq. (2.13) and add the resulting vector to v_{E_3}. We could make similar calculations using Eq. (2.14). A second way is to solve both equations simultaneously. Using Eq. (2.13), we know that v_{E_3/B_3} lies on a line through b_3 on the velocity diagram and is perpendicular to $\overrightarrow{EB}$ on the position diagram. Similarly, v_{E_3/C_3} lies on a line through c_3 on the velocity diagram and is perpendicular to $\overrightarrow{EC}$ on the position diagram. The point e_3 lies on the intersection of the two lines, and v_{E_3} is the vector from o to the point e_3.

The magnitude of v_{E_3} can be obtained by measuring $\overrightarrow{oe_3}$ and multiplying by the scale factor. The distance $\overrightarrow{oe_3} = 2.93$ in, so $|v_{E_3}| = 29.3$ in/s. Its direction may be measured from the diagram with a protractor. The direction is $-164.9°$ with the zero angle reference being horizontal and positive to the right. This completes the velocity analysis of the linkage.

(c) *Acceleration Analysis*

The acceleration analysis can be conducted using the points that were used in the velocity analysis. We will first compute the acceleration of point B_2 (and B_3), then the acceleration of C_3 (and C_4), and finally the acceleration of E_3. The acceleration of B_2 can be expressed as the absolute acceleration between B_2 and A_2. Because two points on the same rigid link are involved,

[4] The distances identified refer to the original drawings developed for this book. Because the drawings were reduced when the book was printed, the distances reported here cannot be measured directly from the book pages. However, the results can be verified by making measurements from the drawings and using the small scales included with each of the drawings.

an acceleration expression similar to Eq. (2.7) can be written as

$$\boldsymbol{a}_{B_2} = \boldsymbol{a}_{A_2} + \boldsymbol{a}_{B_2/A_2} = \boldsymbol{\alpha}_2 \times \boldsymbol{r}_{B/A} + \boldsymbol{\omega}_2 \times (\boldsymbol{\omega}_2 \times \boldsymbol{r}_{B/A}) = \boldsymbol{a}^t_{B_2/A_2} + \boldsymbol{a}^r_{B_2/A_2}$$

Note that we know the directions and magnitudes for $\boldsymbol{\omega}_2$, $\boldsymbol{\alpha}_2$, and $\boldsymbol{r}_{B/A}$, and therefore we can compute each of the vectors in the equation. Because of the cross-product, the acceleration $\boldsymbol{a}^t_{B_2/A_2}$ will be orthogonal to $\boldsymbol{\alpha}_2$ and $\boldsymbol{r}_{B/A}$, and the direction will be given by the right-hand screw rule. The direction of $\boldsymbol{a}^r_{B_2/A_2}$ will be opposite to the direction of $\boldsymbol{r}_{B/A}$. We can compute the magnitude of $\boldsymbol{a}^t_{B_2/A_2}$ from an equation similar to Eq. (2.9). The magnitude is given by

$$|\boldsymbol{a}^t_{B_2/A_2}| = |\boldsymbol{\alpha}_2||\boldsymbol{r}_{B/A}| = (0)(2.5\text{ in}) = 0$$

The magnitude of the radial component can be computed by using Eq. (2.7). Then,

$$\boldsymbol{a}^r_{B_2/A_2} = \boldsymbol{\omega}_2 \times (\boldsymbol{\omega}_2 \times \boldsymbol{r}_{B/A}) = |\boldsymbol{\omega}_2|^2|\boldsymbol{r}_{B/A}| = 10^2(2.5) = 250\text{ in/s}^2$$

and the direction is opposite $\boldsymbol{r}_{B/A}$. The direction of $\boldsymbol{a}^r_{B_2/A_2}$ is therefore 60° below the horizontal (down and to the right). It is now necessary to choose a scale and plot the acceleration of point B_2. We will use a scale of 50 in/s^2 to 1 inch to ensure that the diagram will fit on a quarter-sized page. The acceleration $\boldsymbol{a}_{B_2/A_2} = \boldsymbol{a}^r_{B_2/A_2}$ is plotted in Fig. 2.9 as $\overrightarrow{o'b_2'}$. Here, we are using the convention that a lowercase letter with a prime (′) indicates the acceleration of the corresponding point on the position diagram. The most convenient way to plot a line parallel to $\overrightarrow{AB}$ is to place a drafting triangle with one of the two orthogonal sides along $\overrightarrow{ob_2}$ on the velocity diagram and draw a line through o' along the other side of the triangle. Since $\overrightarrow{ob_2}$ is normal to $\overrightarrow{AB}$, this results in a line parallel to $\overrightarrow{AB}$. Once again, $\overrightarrow{o'b_2'}$ is directed down and to the right because it is in the negative $\overrightarrow{AB}$ direction.

Next we want to compute the acceleration of point C. Recall that B_2 and B_3 are permanently pinned together. Therefore,

$$\boldsymbol{a}_{B_2} = \boldsymbol{a}_{B_3} = 250\text{ in/s}^2$$

in the direction shown in Fig. 2.9. Similarly, C_3 and C_4 are permanently pinned together. Therefore,

$$\boldsymbol{a}_{C_3} = \boldsymbol{a}_{C_4}$$

Because B_3 and C_3 are both fixed to link 3, we can write the following relative acceleration equation:

$$\boldsymbol{a}_{C_3} = \boldsymbol{a}_{B_3} + \boldsymbol{a}_{C_3/B_3} = \boldsymbol{a}_{B_3} + \boldsymbol{a}^r_{C_3/B_3} + \boldsymbol{a}^t_{C_3/B_3} \tag{2.15}$$

In Eq. (2.15), the vector $\boldsymbol{a}_{B_3}$ is entirely known. Also, the radial component of the acceleration, which is a function of position and velocity only, can be computed directly from the following:

$$\boldsymbol{a}^r_{C_3/B_3} = \boldsymbol{\omega}_3 \times (\boldsymbol{\omega}_3 \times \boldsymbol{r}_{C/B})$$

From the velocity analysis, we computed the magnitude of the angular velocity to be $|\boldsymbol{\omega}_3| =$ 2.75 rad/s. The radial acceleration is a vector from C to B on the position diagram (opposite $\boldsymbol{r}_{C/B}$), and the magnitude is given by

$$|\boldsymbol{a}^r_{C_3/B_3}| = |\boldsymbol{\omega}_3|^2|\boldsymbol{r}_{C/B}| = 2.75^2(6) = 45.4\text{ in/s}^2$$

A convenient way to draw a line parallel to $\overrightarrow{BC}$ is, again, to place a triangle with one of the orthogonal sides along $\overrightarrow{b_3c_3}$ on the velocity polygon and draw a line along the other orthogonal side through point b_3'. The direction is down and to the left because this component is in the negative $\boldsymbol{r}_{C/B}$ direction.

The tangential component $\boldsymbol{a}^t_{C_3/B_3}$ is given by

$$\boldsymbol{a}^t_{C_3/B_3} = \boldsymbol{\alpha}_3 \times \boldsymbol{r}_{C/B}$$

The magnitude of $\boldsymbol{a}^t_{C_3/B_3}$ is unknown because $\boldsymbol{\alpha}_3$ is unknown. However, this vector will be normal to $\overrightarrow{BC}$. Hence a line is drawn through the tip of the $\boldsymbol{a}^r_{C_3/B_3}$ vector to represent this direction.

This defines one locus of c_3'. To find another locus for c_3', we need to find another equation for the vector $\boldsymbol{a}_{C_3}$. We know that $\boldsymbol{a}_{C_3} = \boldsymbol{a}_{C_4}$, and we can write an equation for the acceleration of C_4. Again, the acceleration $\boldsymbol{a}_{C_4}$ is an absolute acceleration, and it can be expressed as the relative acceleration between C_4 and any point that has zero acceleration. If we choose D_4 as that point, the acceleration equation becomes

$$\boldsymbol{a}_{C_4} = \boldsymbol{a}_{C_4/D_4} = \boldsymbol{a}^r_{C_4/D_4} + \boldsymbol{a}^t_{C_4/D_4} \tag{2.16}$$

The radial component of the acceleration is a function of position and velocity only and can be computed directly from the following:

$$\boldsymbol{a}^r_{C_4/D_4} = \boldsymbol{\omega}_4 \times (\boldsymbol{\omega}_4 \times \boldsymbol{r}_{C/D})$$

From the velocity analysis, we computed the magnitude of the angular velocity to be $|\boldsymbol{\omega}_4| =$ 6.73 rad/s. The radial acceleration is a vector from C to D on the position diagram (opposite $\boldsymbol{r}_{C/D}$), and the magnitude is given by

$$|\boldsymbol{a}^r_{C_4/D_4}| = |\boldsymbol{\omega}_4|^2|\boldsymbol{r}_{C/D}| = 6.73^2(4) = 181.2 \text{ in/s}^2$$

This vector is plotted from o' in Fig. (2.9).

The tangential component $\boldsymbol{a}^t_{C_4/D_4}$ is given by

$$\boldsymbol{a}^t_{C_4/D_4} = \boldsymbol{\alpha}_4 \times \boldsymbol{r}_{C/D}$$

The magnitude of $\boldsymbol{a}^t_{C_4/D_4}$ is unknown because $\boldsymbol{\alpha}_4$ is unknown. However, this vector will be normal to $\overrightarrow{BD}$. Hence a line is drawn through the tip of the $\boldsymbol{a}^r_{C_4/D_4}$ vector to represent this direction. This defines a second locus for c_3' and c_4'. The points c_3' and c_4' are located where the two loci intersect as shown in Fig. 2.9. The vectors $\boldsymbol{a}^t_{C_3/B_3}$, $\boldsymbol{a}^t_{C_4/D_4}$, and $\boldsymbol{a}_{C_4/D_4}$ are as shown. The magnitude of $\boldsymbol{\alpha}_3$ may be found from the expression for the tangential component of the relative acceleration between C_3 and B_3. Then,

$$|\boldsymbol{a}^t_{C_3/B_3}| = |\boldsymbol{\alpha}_3||\boldsymbol{r}_{C/B}|$$

or

$$|\boldsymbol{\alpha}_3| = |\boldsymbol{a}^t_{C_3/B_3}|/|\boldsymbol{r}_{C/B}|$$

To get $|\boldsymbol{a}^t_{C_3/B_3}|$, measure the length of the vector on the acceleration polygon and multiply by the scale factor. On the acceleration polygon, the length of the line corresponding to $|\boldsymbol{a}^t_{C_3/B_3}|$ is 0.847 in. Therefore, $|\boldsymbol{a}^t_{C_3/B_3}| = 0.847 \times 50 = 42.2$ in/s^2. Then

$$|\boldsymbol{\alpha}_3| = |\boldsymbol{a}^t_{C_3/B_3}|/|\boldsymbol{r}_{C/B}| = 42.2/6.0 = 7.06 \text{ rad/s}^2$$

To get the direction, visualize the direction in which we would have to rotate $\boldsymbol{r}_{C/B}$ to obtain the direction of $\boldsymbol{a}^t_{C_3/B_3}$. The direction is CCW.

Next compute the angular velocity $\boldsymbol{\alpha}_4$. The magnitude can be found from an expression for the tangential component of the relative acceleration $\boldsymbol{a}^t_{C_4/D_4}$. Then,

$$|\boldsymbol{a}^t_{C_4/D_4}| = |\boldsymbol{\alpha}_4||\boldsymbol{r}_{C/D}|$$

or

$$|\boldsymbol{\alpha}_4| = |\boldsymbol{a}^t_{C_4/D_4}|/|\boldsymbol{r}_{C/D}|$$

To obtain the magnitude of the tangential component of acceleration, measure $|\boldsymbol{a}^t_{C_4/D_4}|$ on the acceleration diagram and multiply by the scale factor. From Fig. 2.9, $|\boldsymbol{a}^t_{C_4/D_4}| = 1.816 \times 50 =$ 90.8 in/s^2. Then

$$|\boldsymbol{\alpha}_4| = |\boldsymbol{a}^t_{C_4/D_4}|/|\boldsymbol{r}_{C/D}| = 90.8/4.0 = 22.7 \text{ rad/s}^2$$

To get the direction, visualize the direction in which we would have to rotate $\boldsymbol{r}_{C/D}$ to obtain the direction of $\boldsymbol{a}^t_{C_4/D_4}$. The direction is CW.

The acceleration of point E_3 may be obtained by considering first the point pair E_3 and B_3

and then the pair E_3 and C_3. Both pairs are fixed to member 3. The relative acceleration expressions are

$$\boldsymbol{a}_{E_3} = \boldsymbol{a}_{B_3} + \boldsymbol{a}_{E_3/B_3} = \boldsymbol{a}_{B_3} + \boldsymbol{a}^r_{E_3/B_3} + \boldsymbol{a}^t_{E_3/B_3} = \boldsymbol{a}_{B_3} + \boldsymbol{\omega}_3 \times (\boldsymbol{\omega}_3 \times \boldsymbol{r}_{E/B}) + \boldsymbol{\alpha}_3 \times \boldsymbol{r}_{E/B} \quad \textbf{(2.17)}$$

and

$$\boldsymbol{a}_{E_3} = \boldsymbol{a}_{C_3} + \boldsymbol{a}_{E_3/C_3} = \boldsymbol{a}_{C_3} + \boldsymbol{a}^r_{E_3/C_3} + \boldsymbol{a}^t_{E_3/C_3} = \boldsymbol{a}_{C_3} + \boldsymbol{\omega}_3 \times (\boldsymbol{\omega}_3 \times \boldsymbol{r}_{E/C}) + \boldsymbol{\alpha}_4 \times \boldsymbol{r}_{E/C} \quad \textbf{(2.18)}$$

The accelerations $\boldsymbol{a}_{B_3}$ and $\boldsymbol{a}_{C_3}$ are known and have been plotted as $\overrightarrow{o'b'_3}$ and $\overrightarrow{o'c'_3}$ on the acceleration polygon. As in the corresponding case of velocities, we can compute $\boldsymbol{a}_{E_3}$ two different ways as implied by Eqs. (2.17) and (2.18). One way is to compute the cross-products in Eq. (2.17) and add the resulting vectors to $\boldsymbol{a}_{B_3}$. We could also make similar calculations using Eq. (2.18). A second way is to solve both equations simultaneously as was done in the velocity analysis. We will use the first procedure here. To determine $\boldsymbol{a}_{E_3}$ using Eq. (2.17), we must compute $\boldsymbol{a}^r_{E_3/B_3}$ and $\boldsymbol{a}^t_{E_3/B_3}$. The direction of the radial component is opposite $\boldsymbol{r}_{E/B}$, and the magnitude is given by

$$|\boldsymbol{a}^r_{E_3/B_3}| = |\boldsymbol{\omega}_3|^2 |\boldsymbol{r}_{E/B}| = 2.75^2(3) = 22.7 \text{ in/s}^2$$

This vector is added to $\boldsymbol{a}_{B_3}$ in Fig. 2.9.

The direction of $\boldsymbol{a}^t_{E_3/B_3}$ is found using the right-hand screw rule or by turning $\boldsymbol{r}_{E/B}$ 90° in the direction of $\boldsymbol{\alpha}_3$. Recall that $\boldsymbol{\alpha}_3$ is CCW. The magnitude of $\boldsymbol{a}^t_{E_3/B_3}$ is given by

$$|\boldsymbol{a}^t_{E_3/B_3}| = |\boldsymbol{\alpha}_3||\boldsymbol{r}_{E/B}| = 7.06(3) = 21.2 \text{ in/s}^2$$

This vector is plotted in Fig. 2.9. The point e'_3 is located at the tip of $\boldsymbol{a}^t_{E_3/B_3}$. The acceleration of E_3 is located by the vector from o' to e'_3 on the acceleration diagram. To determine the magnitude, measure $\overrightarrow{o'e'_3}$ and multiply by the scale factor. The result is

$$|\boldsymbol{a}_{E_3}| = 4.85 \times 50 = 2421.5 \text{ in/s}^2$$

The vector is pointed in a direction that is 67° below the horizontal and to the right. A much more efficient way to locate point e'_3 will be presented later in this chapter. ■

2.6 GRAPHICAL ANALYSIS OF A SLIDER-CRANK MECHANISM

The analysis of a slider-crank mechanism depends on whether the crank or the slider is the driver. If the crank is the driver, we need to know the angular position, velocity, and acceleration of the crank. If the slider is the driver, we need to know the position, velocity, and acceleration of some point on the slider. Each point on the slider will have a unique position, but all points will have the same velocity and the same acceleration.

We will analyze the slider-crank mechanism shown in Fig. 2.10, where the crank is the

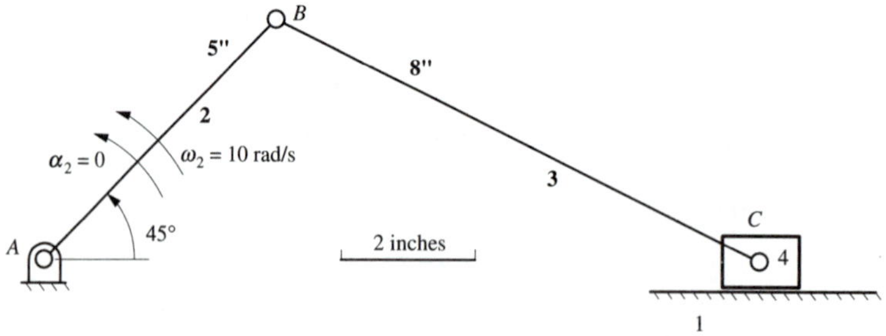

Figure 2.10 The slider-crank linkage to be analyzed in Example 2.2.

driver. As was the case for the four-bar linkage, the key to the acceleration analysis of this mechanism is the relationship between the velocities and accelerations of two points on the same rigid body.

EXAMPLE 2.2 (*Graphical Analysis of a Slider-Crank Mechanism*)

PROBLEM

Find $\boldsymbol{a}_C$ and $\boldsymbol{\omega}_3$ for the slider-crank linkage in the position shown in Fig. 2.10. The crank *AB* (link 2) is driven at a constant angular velocity of 10 rad/s CCW. *C* is the axis of the revolute joint connecting the connecting rod, link 3, to the slider, link 4. In the position shown, *AB* is at 45° to *AC*, and the link lengths are shown on the drawing.

SOLUTION

(a) Position Analysis

The linkage is first drawn to scale to establish the direction of member *BC*. To do this, first locate the horizontal line through *A* on which *C* lies. Next, draw member *AB* to scale. Then draw an arc scaled to represent 8 inches and centered at *B*. The arc intersects the horizontal line through *A* at two locations. The desired location is to the right of *A* as indicated in Fig. 2.10. The scaled drawing is shown in Fig. 2.11.

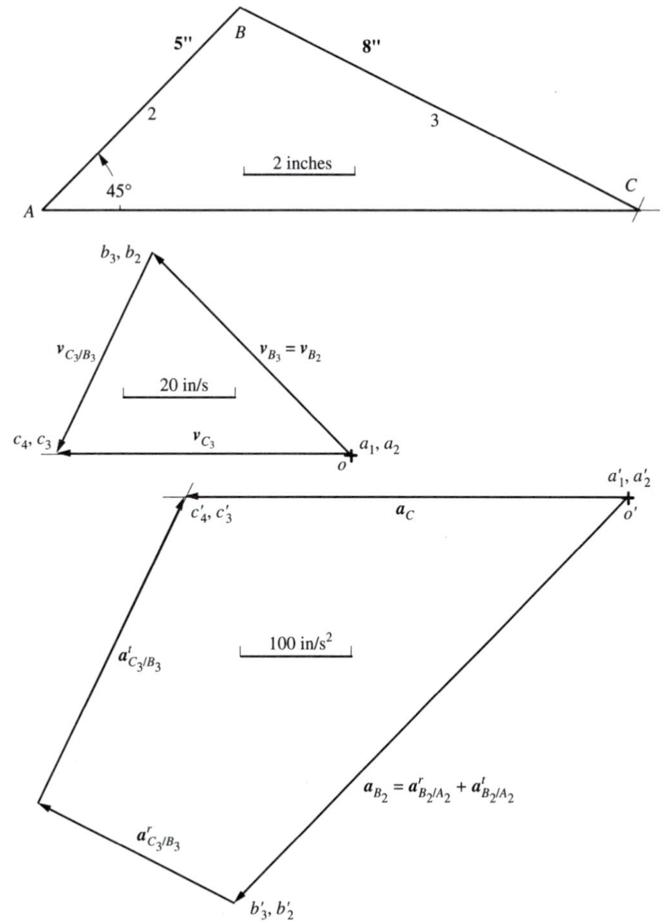

Figure 2.11 Velocity and acceleration polygons for Example 2.2.

(b) Velocity Polygon

The basic equation to be solved is

$$\boldsymbol{v}_{C_3} = \boldsymbol{v}_{B_3} + \boldsymbol{v}_{C_3/B_3} = \boldsymbol{v}_{B_2} + \boldsymbol{v}_{C_3/B_3} = \boldsymbol{v}_{B_2/A_2} + \boldsymbol{v}_{C_3/B_3}$$

From the given data:

$$\boldsymbol{v}_{B_2} = \boldsymbol{v}_{B_2/A_2} = \boldsymbol{\omega}_2 \times \boldsymbol{r}_{B/A} = 10 \times 5 = 50 \text{ in/s (normal to } \boldsymbol{r}_{B/A})$$

The direction for the velocity of C_3 must be horizontal. This lets us solve the basic velocity equation as shown in Fig. 2.11. By measurement in Fig. 2.11:

$$|\boldsymbol{v}_{C_3/B_3}| = 1.98 \times 20 = 39.6 \text{ in/s}$$

Then

$$|\boldsymbol{\omega}_3| = |\boldsymbol{v}_{C_3/B_3}|/|\boldsymbol{r}_{C/B}| = 39.6/8 = 4.95 \text{ rad/s CW}$$

(c) Acceleration Polygon

The basic acceleration equation to be solved is

$$\boldsymbol{a}_{C_3} = \boldsymbol{a}_{B_2/A_2} + \boldsymbol{a}_{C_3/B_3} = \boldsymbol{a}^r_{B_2/A_2} + \boldsymbol{a}^t_{B_2/A_2} + \boldsymbol{a}^r_{C_3/B_3} + \boldsymbol{a}^t_{C_3/B_3}$$

From the given data:

$$|\boldsymbol{a}^r_{B_2/A_2}| = |\boldsymbol{\omega}_2|^2|\boldsymbol{r}_{B/A}| = 10^2(5) = 500 \text{ in/s}^2$$
$$|\boldsymbol{a}^t_{B_2/A_2}| = |\boldsymbol{\alpha}_2||\boldsymbol{r}_{B/A}| = 0(5) = 0$$

Using information from the velocity analysis,

$$|\boldsymbol{a}^r_{C_3/B_3}| = |\boldsymbol{\omega}_3|^2|\boldsymbol{r}_{C/B}| = 4.95^2(8) = 196 \text{ in/s}^2$$

The direction for the acceleration of C_3 must be horizontal, and the basic acceleration equation can now be solved for the acceleration of C_3 as shown in Fig. 2.11. By measurement in Fig. 2.11

$$|\boldsymbol{a}_{C_3}| = 3.98(100) = 398 \text{ in/s}^2$$

and

$$|\boldsymbol{a}^t_{C_3/B_3}| = 2.98 \times 100 = 298 \text{ in/s}^2$$

Then

$$|\boldsymbol{\alpha}_3| = |\boldsymbol{a}^t_{C_3/B_3}|/|\boldsymbol{r}_{C/B}| = \frac{298}{8} = 37.3 \text{ rad/s}^2 \text{ CCW}$$

The steps for the total analysis are summarized in the following, and the results are shown in Fig. 2.11

1. Draw the linkage to scale.
2. Construct the velocity polygon and compute $\boldsymbol{\omega}_3$.
3. Compute the magnitudes of $\boldsymbol{a}^r_{B_2/A_2}$, $\boldsymbol{a}^t_{B_2/A_2}$, and $\boldsymbol{a}^r_{C_3/B_3}$ and identify their directions.
4. Choose a suitable scale and plot $\boldsymbol{a}^r_{B_2/A_2}$ opposite to $\boldsymbol{r}_{B/A}$. Put the tail of the vector at the acceleration pole, o'.
5. Plot $\boldsymbol{a}^t_{B_2/A_2}$ (zero in this case) normal to $\boldsymbol{r}_{B/A}$ and through the tip of $\boldsymbol{a}^r_{B_2/A_2}$. The tip of $\boldsymbol{a}^t_{B_2/A_2}$ gives the point b'_2. Here, the direction for $\boldsymbol{a}_{B_2/A_2}$ is in the direction of $-\boldsymbol{r}_{B/A}$.
6. Plot vector $\boldsymbol{a}^r_{C_3/B_3}$ opposite to $\boldsymbol{r}_{C/B}$ with its tail at point b'_2.
7. Draw a line through the tip of vector $\boldsymbol{a}^r_{C_3/B_3}$ normal to line $\overrightarrow{BC}$.
8. Draw a line through o' parallel to line $\overrightarrow{AC}$. The intersection of the lines drawn in steps 7 and 8 gives point c'_3.
9. Measure the magnitude $\boldsymbol{a}_C$ as $\overrightarrow{o'c'_3}$ and note its direction.
10. Measure $\boldsymbol{a}^t_{C_3/B_3}$ and compute $|\boldsymbol{\alpha}_3| = |\boldsymbol{a}^t_{C_3/B_3}|/|\boldsymbol{r}_{C/B}|$. Note that the sense of $\boldsymbol{\alpha}_3$ is found by visualizing C rotating about B so that it moves in the $\boldsymbol{a}^t_{C_3/B_3}$ direction. ■

2.7 THE VELOCITY IMAGE THEOREM

To conduct a graphical analysis of a linkage with more than one loop, it is necessary to obtain the velocities of additional points on a rigid link once the kinematic properties of the first two points are known. After the velocities and accelerations of two points are known, the angular velocity of the body can be determined. Knowing the velocity of a point and the angular velocity of the body, the velocity of any other point on the rigid body can be computed using Eq. (2.3). Similarly, if the velocity analysis has been conducted, and the acceleration of a point and the angular acceleration of the body are known, the acceleration of any other point on the body can be found using Eq. (2.8). An alternative method for determining the velocity and acceleration of a third point on a rigid body is to use the concept of velocity and acceleration image. The velocity image theorem will be discussed first.

Notation. As indicated previously, a convenient means of labeling the velocity polygon is to use a lowercase letter to identify the absolute velocity of each point on the position diagram. Then a vector from the velocity pole to the point will represent the absolute velocity of the point. A vector between any two points will correspond to the relative velocity between the points. For example, in Fig. 2.12, $\boldsymbol{v}_{C/B} = \overrightarrow{bc}$.

The velocity image theorem states that, if PQR is a triangle fixed in a rigid body, the triangle pqr in the velocity diagram is similar to triangle PQR, is rotated from PQR by 90° in the positive $\boldsymbol{\omega}$ direction, and is magnified by the factor ω. This theorem, stated here for triangles, can be extended to apply to any polygon, because any polygon can be broken down into triangles, or indeed to any shape, since any shape can be approximated by a polygon to any desired degree of accuracy. Thus, the velocity image of any member is similar to that member, is rotated relative to that member through 90° in the positive $\boldsymbol{\omega}$ direction, and is magnified by a factor ω.

The proof of the theorem can be developed using Fig. 2.13. In that figure, the position diagram for the rigid link is PQR, and the velocity diagram is pqr. Using $\boldsymbol{v}_{Q/P}$ as an example,

$$\boldsymbol{v}_{Q/P} = \boldsymbol{\omega} \times \boldsymbol{r}_{Q/P}$$

Therefore, $\boldsymbol{v}_{Q/P}$ is normal to PQ and has the magnitude ωPQ. Hence pq has magnitude ωPQ and is rotated 90° in the $\boldsymbol{\omega}$ direction. Similarly, $qr = \omega RP$ and is rotated 90° in the $\boldsymbol{\omega}$ direction. Hence, triangle pqr is similar to triangle PQR, is rotated from triangle PQR through 90° in the $\boldsymbol{\omega}$ direction, and is magnified over triangle PQR by a factor ω.

Note that the velocity image can be used to determine directly the velocity of any point in the rigid body given the position of the point and the velocity diagram. Conversely, the location of a point with a given velocity can be found by mapping points in the velocity diagram to points in the position diagram.

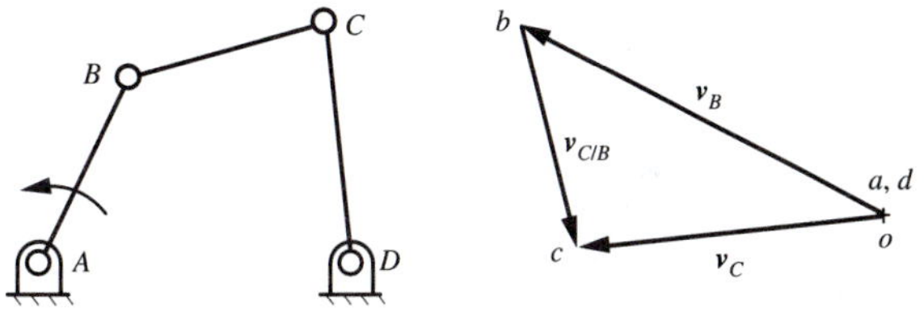

Figure 2.12 Notation used on the velocity polygon to facilitate drawing the velocity image. The lowercase letters on the velocity polygon correspond to the letters on the linkage. $\boldsymbol{v}_B = \overrightarrow{ob}$, $\boldsymbol{v}_C = \overrightarrow{oc}$, and $\boldsymbol{v}_{B/A} = \overrightarrow{ab}$. The point o corresponds to all fixed points. That is, A and D both map into point o.

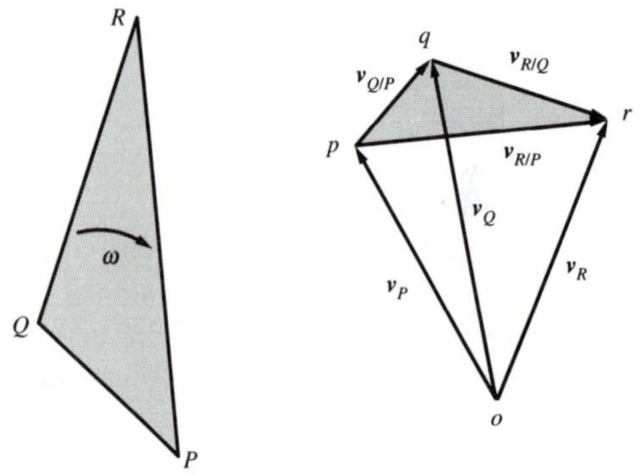

Figure 2.13 Link PQR and its velocity image in the velocity polygon. Triangle pqr is similar to triangle PQR and is rotated from it by 90° in the $\boldsymbol{\omega}$ direction.

The manner in which the velocity image is used to analyze multiloop linkages is illustrated in Example 2.3 involving a six-bar linkage.

EXAMPLE 2.3 ***(Graphical Velocity Analysis of Six-Bar Linkage)***

Develop a procedure for finding the angular velocities of links 3, 5, and 6 and the velocity of point B of the linkage shown in Fig. 2.14.

SOLUTION

The polygons for the analysis are shown in Fig. 2.14. The velocity analysis starts with the slider-crank part of the mechanism. The equations involved and the order in which they are solved

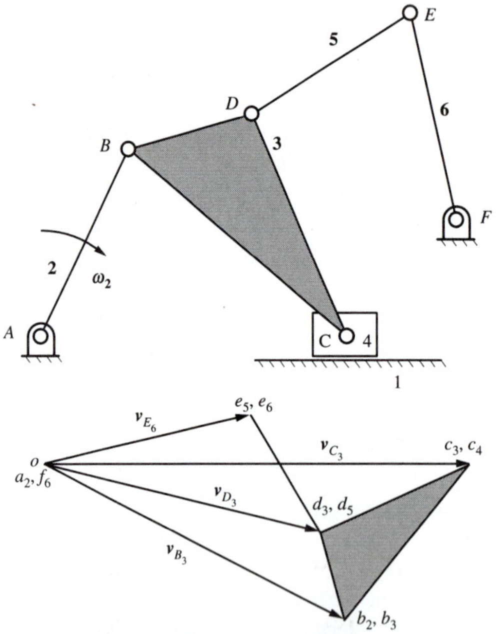

Figure 2.14 The linkage and velocity polygon of Example 2.3. The velocity image theorem is used to locate point d_3.

are given in the following:

$$\boldsymbol{v}_{B_3} = \boldsymbol{v}_{B_2} = \boldsymbol{v}_{B_2/A_2} = \boldsymbol{\omega}_2 \times \boldsymbol{r}_{A/B}$$

and

$$\boldsymbol{v}_{C_3} = \boldsymbol{v}_{B_3} + \boldsymbol{v}_{C_3/B_3}$$

Next we will find the velocity of point D_3 by image. Then the dyad (links 5 and 6) can be analyzed using

$$\boldsymbol{v}_{E_5} = \boldsymbol{v}_{E_6}$$

$$\boldsymbol{v}_{E_5} = \boldsymbol{v}_{D_5} + \boldsymbol{v}_{E_5/D_5} = \boldsymbol{v}_{D_5} + \boldsymbol{\omega}_5 \times \boldsymbol{r}_{E/D}$$

$$\boldsymbol{v}_{E_6} = \boldsymbol{v}_{F_6} + \boldsymbol{v}_{E_6/F_6} = \boldsymbol{\omega}_6 \times \boldsymbol{r}_{E/F}$$

STEPS

1. Draw the linkage to scale in the nominated position.
2. Select a suitable scale and plot $\boldsymbol{v}_{B_3} = \boldsymbol{v}_{B_2/A_2} = \overrightarrow{ob}$ normal to line AB. Point o represents the points on the fixed frame and all other points with zero velocity. That is, *all* points with zero velocity in the linkage map into point o, and all points at o map to the linkage as points with zero velocity.
3. Draw a line through point o parallel to line $\overrightarrow{AC}$. The velocity of C_3 must lie on this line.
4. Draw a line through point b_3 normal to line $\overrightarrow{BC}$. The intersection of the lines drawn in steps 3 and 4 gives point c_3.
5. Now find the velocity image of D_3. Start by drawing a line through point b_3 normal to line $\overrightarrow{BD}$.
6. Draw a line through point c_3 normal to line $\overrightarrow{CD}$. The intersection of the lines drawn in steps 5 and 6 is point d_3.
7. Next locate e_5 (and e_6). Start by drawing a line through point d_3 normal to line $\overrightarrow{DE}$.
8. Draw a line through point o normal to line $\overrightarrow{EF}$. The intersection of the lines drawn in steps 7 and 8 is point e_5.
9. Compute ω_3 from $|\boldsymbol{\omega}_3| = |\boldsymbol{v}_{C_3/B_3}|/|\boldsymbol{r}_{C/B}|$. Note that the sense is CCW. This is inferred by noting that C_3 must rotate CCW about B_3 in order to move in the direction of $\boldsymbol{v}_{C_3/B_3}$.
10. Compute ω_5 from $|\boldsymbol{\omega}_5| = \boldsymbol{v}_{D_5/E_5}|/|\boldsymbol{r}_{D/E}|$. The sense is CCW, since D_5 must rotate CCW about E_5 in order to move in the direction of $\boldsymbol{v}_{D_5/E_5}$.
11. Compute ω_6 from $|\boldsymbol{\omega}_6| = |\boldsymbol{v}_{E_6/F_6}|/|\boldsymbol{r}_{E/F}|$. The sense is CW, since E_6 must move CW about D_6 in order to move in the direction of $\boldsymbol{v}_{E_6/F_6}$. ■

The velocity image theorem is very useful for finding the velocity of a point on the coupler of a linkage at which an additional joint is placed. It is important to notice that the *shape* of any velocity polygon (i.e., all angles in it) is determined only by the dimensions of the linkage. See, for instance, Fig. 2.14. Further, the speed at which the linkage is operated can affect only the *size,* or scale, of the polygon and not the shape. This property will play a pivotal role in later sections (e.g., Section 2.10).

2.8 THE ACCELERATION IMAGE THEOREM

As was the case in the velocity analysis, an acceleration image theorem provides an easy way to obtain accelerations of additional points on a rigid body when the accelerations of two points are already known. This is useful when multiple loops are involved in the linkage. In the acceleration diagram we will use primed lowercase letters to indicate the absolute accelerations of various points. Thus $\boldsymbol{a}_{Q/P} = \overrightarrow{p'q'}$, $\boldsymbol{a}_{B/A} = \overrightarrow{a'b'}$, etc. Once again, o' on the acceleration diagram corresponds to the pole where all points with zero acceleration map.

The acceleration image theorem states that, if PQR is a triangle fixed in a rigid link in motion relative to the fixed frame, then triangle $p'q'r'$ is similar to triangle PQR.

Triangle $p'q'r'$ is magnified by a factor that is a function of $\boldsymbol{\alpha}$ and $\boldsymbol{\omega}$ and is rotated from triangle PQR by an angle that is also a function of $\boldsymbol{\alpha}$ and $\boldsymbol{\omega}$.

Proof. To prove the acceleration image theorem, we will use Fig. 2.15. Then

$$\boldsymbol{a}_{Q/P} = \overrightarrow{p'q'} = -\omega^2 \boldsymbol{r}_{Q/P} + \alpha|\boldsymbol{r}_{Q/P}|\boldsymbol{n}'$$

where $\boldsymbol{n}'$ is normal to $\boldsymbol{r}_{Q/P}$. Therefore the magnitude of the relative acceleration vector is given by

$$|\overrightarrow{p'q'}| = PQ\sqrt{\omega^4 + \alpha^2}$$

Similarly

$$|\overrightarrow{q'r'}| = QR\sqrt{\omega^4 + \alpha^2}$$

and

$$|\overrightarrow{r'p'}| = RP\sqrt{\omega^4 + \alpha^2}$$

Hence triangle $p'q'r'$ is similar to triangle PQR. The magnification factor is $|\overrightarrow{r'p'}|/RP = |\overrightarrow{q'r'}|/QR = |\overrightarrow{p'q'}|/PQ = \sqrt{\omega^4 + \alpha^2}$. Referring to Fig. 2.15, the angle of rotation is

$$\theta = \pi - \tan^{-1}(a^t_{R/P}/a^r_{R/P})$$

or

$$\theta = \pi - \tan^{-1}(\alpha/\omega^2)$$

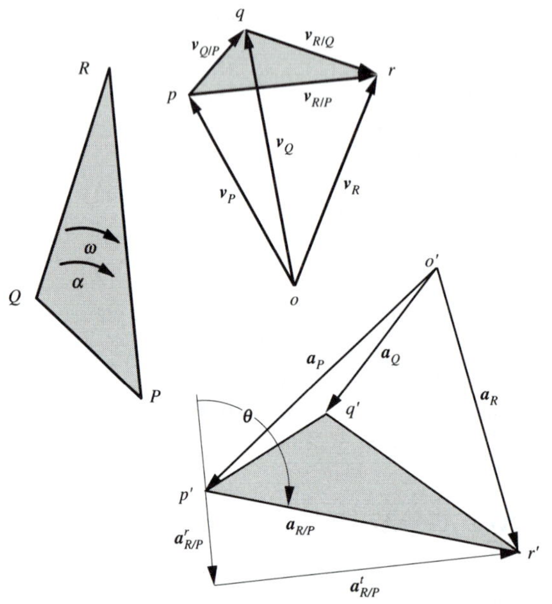

Figure 2.15 The acceleration image theorem. The example used is the same as for the velocity image in Fig. 2.15. Triangle $p'q'r'$ is similar to triangle PQR in the original lamina. Hence it is also similar to triangle pqr, which is the velocity image of PQR. If $\boldsymbol{a}_P$ is plotted, together with the radial and transverse components of the acceleration of R relative to $P(\boldsymbol{a}_{R/P})$ to locate points p' and r', q' can be located from the image to give $\boldsymbol{a}_Q$.

Once again, this result can be extended to cover members of any shape by noting that any polygon may be broken down into triangles, and any area bounded by a plane curve may be approximated by a polygon as closely as desired.

Because the angle of rotation in the acceleration image is not usually 90°, similar triangles must be constructed by making corresponding angles equal.

EXAMPLE 2.4 *(Graphical Acceleration Analysis of a Six-Bar Linkage)*

PROBLEM

Given the dimensions of the linkage shown in Fig. 2.16, find $\boldsymbol{a}_C$ and $\boldsymbol{\alpha}_6$ if $\boldsymbol{\omega}_2 = 60$ rpm CW and $\boldsymbol{\alpha}_2 = 0$.

SOLUTION

The steps of the analysis are shown in Fig. 2.17. The scales for position, velocity, and acceleration are shown with the polygons. The velocity analysis follows the procedure developed in Example 2.3. The initial equation to be solved is for the slider-crank mechanism. That is:

$$\boldsymbol{v}_{C_3} = \boldsymbol{v}_{B_3} + \boldsymbol{v}_{C_3/B_3}$$

where

$$\boldsymbol{v}_{B_3} = \boldsymbol{v}_{B_2} = \boldsymbol{v}_{B_2/A_2} = \boldsymbol{\omega}_2 \times \boldsymbol{r}_{B_2/A_2}$$

Next we will find the velocity of point D_3 by image. Then the dyad (links 5 and 6) can be analyzed using

$$\boldsymbol{v}_{E_5} = \boldsymbol{v}_{D_5} + \boldsymbol{v}_{E_5/D_5} = \boldsymbol{v}_{E_6} = \boldsymbol{v}_{F_6} + \boldsymbol{v}_{E_6/F_6}$$

STEPS

1. Draw the linkage to scale.
2. Compute the magnitude of $\boldsymbol{v}_B = \boldsymbol{v}_{B_2/A_2}$ and identify its direction. Plot it as the vector $\overrightarrow{ob}$.

$$\omega_2 = 60 \times 2\pi/60 = 6.283 \text{ rad/s}$$
$$\boldsymbol{v}_B = 6.283 \times 1.5 = 9.42 \text{ in/s normal to } \overrightarrow{AB}$$

3. Draw a line through point $\boldsymbol{b}$ normal to line $\overrightarrow{BC}$.
4. Draw a line through o parallel to $\overrightarrow{AC}$. The intersection of this line with that plotted in step 3 gives point c_3 (and c_4).
5. Construct triangle ***bcd*** similar to triangle *BCD*, thereby locating point d_3. This step is a use of the velocity image theorem.
6. Draw a line through point d normal to line $\overrightarrow{DE}$.
7. Draw a line through point o normal to $\overrightarrow{EF}$. The intersection of this line with that drawn in step 6 gives point e_5 (and e_6).

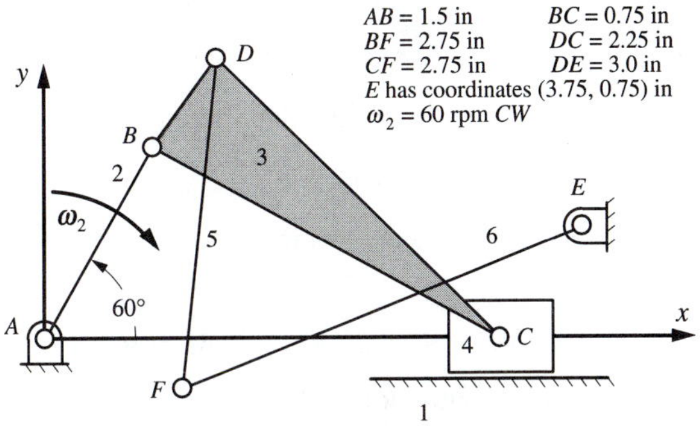

Figure 2.16 Problem statement for Example 2.4.

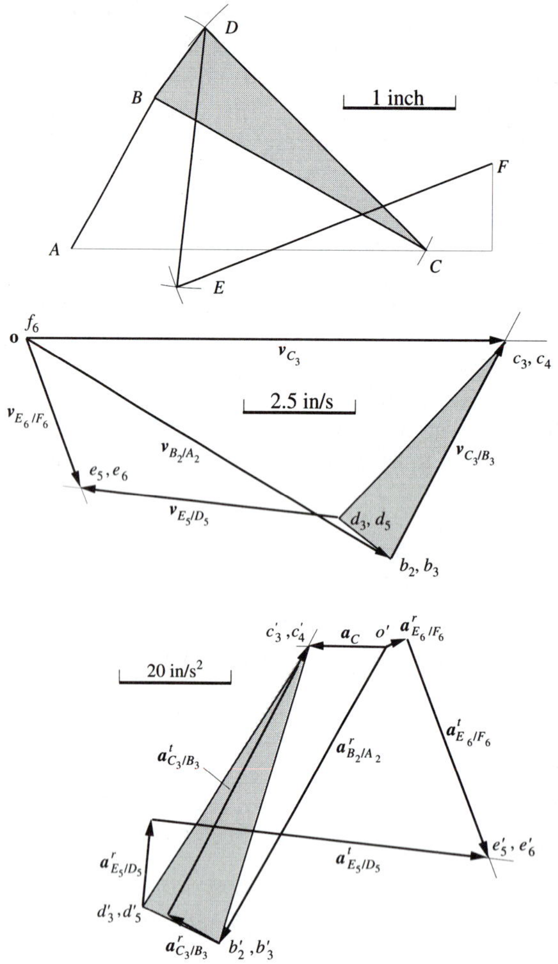

Figure 2.17 Position, velocity and acceleration polygons for Example 2.4.

8. Measure the magnitudes of $\boldsymbol{v}_{C_3/B_3} = \overrightarrow{c_3 b_3}$, $\boldsymbol{v}_{E_5/D_5} = \overrightarrow{e_5 d_5}$, and $\boldsymbol{v}_{E_6} = \overrightarrow{f_6 e_6}$.

$$v_{C_3/B_3} = 5.34 \text{ in/s}, \qquad v_{E_5/D_5} = 5.82 \text{ in/s}, \qquad v_{E_6} = 3.41 \text{ in/s}$$

9. Compute $|\boldsymbol{\omega}_3| = |\boldsymbol{v}_{C_3/B_3}|/|\boldsymbol{r}_{C/B}|$, $|\boldsymbol{\omega}_5| = |\boldsymbol{v}_{D_5/E_5}|/|\boldsymbol{r}_{D/E}|$, and $|\boldsymbol{\omega}_6| = |\boldsymbol{v}_{E_6/F_6}|/|\boldsymbol{r}_{E/F}|$. This completes the velocity analysis of the linkage.

$$\boldsymbol{\omega}_3 = 5.34/2.75 = 1.94 \text{ rad/s CCW}, \qquad \boldsymbol{\omega}_5 = 5.82/2.25 = 2.59 \text{ rad/s CW}$$

$$\boldsymbol{\omega}_6 = 3.41/3.0 = 1.137 \text{ rad/s CCW}$$

10. For the acceleration analysis, we must solve the equations

$$\boldsymbol{a}_{C_3} = \boldsymbol{a}^r_{B_2/A_2} + \boldsymbol{a}^t_{B_2/A_2} + \boldsymbol{a}^r_{C_3/B_3} + \boldsymbol{a}^t_{C_3/B_3}$$

Next we will find the acceleration of point D_3 (and D_5) by image. Then the dyad can be analyzed using

$$\boldsymbol{a}_{E_5} = \boldsymbol{a}_{D_5} + \boldsymbol{a}^r_{E_5/D_5} + \boldsymbol{a}^t_{E_5/D_5} = \boldsymbol{a}_{E_6} = \boldsymbol{a}_{F_6} + \boldsymbol{a}^r_{E_6/F_6} + \boldsymbol{a}^t_{E_6/F_6}$$

First compute $\boldsymbol{a}_{B_2} = \boldsymbol{a}^r_{B_2/A_2}$ and plot as the vector $\overrightarrow{o'b'_2}$.

$$\boldsymbol{a}_{B_2} = 1.5 \times 6.283^2 = 59.2 \text{ in/s}^2 \text{ in the } \overrightarrow{BA} \text{ direction}$$

11. Compute the magnitudes of vectors $\boldsymbol{a}^r_{C_3/B_3}$, $\boldsymbol{a}^r_{E_5/D_5}$, $\boldsymbol{a}^r_{E_6/F_6}$ and identify their directions.

$$\boldsymbol{a}^r_{C_3/B_3} = 2.75 \times 1.94^2 = 10.35 \text{ in/s}^2 \text{ in the } \overrightarrow{CB} \text{ direction}$$
$$\boldsymbol{a}^r_{E_5/D_5} = 2.25 \times 2.59^2 = 15.09 \text{ in/s}^2 \text{ in } \overrightarrow{ED} \text{ direction}$$
$$\boldsymbol{a}^r_{E_6/F_6} = 3.0 \times 1.137^2 = 3.87 \text{ in/s}^2 \text{ in } \overrightarrow{EF} \text{ direction}$$

12. Plot vector $\boldsymbol{a}^r_{C_3/B_3}$ in the $\overrightarrow{CB}$ direction with its tail at b'.
13. Draw a line normal to line $\overrightarrow{CB}$ through the tip of vector $\boldsymbol{a}^r_{C_3/B_3}$.
14. Draw a line through o' parallel to line $\overrightarrow{AC}$. The intersection of this line with that drawn in step 13 gives point c'_3.
15. Construct triangle $b'c'd'$ similar to triangle BCD to locate point d'_3. This step is a use of the acceleration image theorem.
16. Plot $\boldsymbol{a}^r_{E_5/D_5}$ in the $\overrightarrow{ED}$ direction with its tail at point d'.
17. Draw a line normal to $\overrightarrow{ED}$ through the tip of vector $\boldsymbol{a}^r_{E_5/D_5}$.
18. Plot $\boldsymbol{a}^r_{E_6/F_6}$ in the $\overrightarrow{EF}$ direction with its tail at o'.
19. Draw a line normal to $\overrightarrow{EF}$ through the tip of vector $\boldsymbol{a}^r_{E_6/F_6}$. The intersection of this line with that drawn in step 17 gives the point e'_5 (and e'_6).
20. Measure the magnitudes of $\boldsymbol{a}_{C_4}$ and $\boldsymbol{a}^t_{E_6/F_6}$.

$$|\boldsymbol{a}_{C_4}| = 13.9 \text{ in/s}^2, \qquad |\boldsymbol{a}^t_{E_6/F_6}| = 40 \text{ in/s}^2$$

21. Compute $\boldsymbol{\alpha}_6 = |\boldsymbol{a}^t_{E_6/F_6}|/|\boldsymbol{r}_{E/F}|$.

$$\boldsymbol{\alpha}_6 = 40/3.0 = 13.3 \text{ rad/s}^2 \text{ CCW}$$

22. The sense of $\boldsymbol{\alpha}_6$ is obtained by visualizing E rotating about F so as to move in the $\boldsymbol{a}^t_{E_6/F_6}$ direction. ■

EXAMPLE 2.5 (*Using Velocity and Acceleration Image*)

PROBLEM

The mechanism in Fig. 2.18 is drawn to scale. Also given is the velocity polygon for the slider-crank linkage, and the acceleration of point B on the round link is shown on the acceleration polygon. Use the image technique to determine the velocity and acceleration of point D_4. Then determine the velocity and acceleration images of link 4.

SOLUTION

To solve the problem, we need only find the image of point D_4 on both the velocity and acceleration diagrams. The images of link 4 will both be circles with centers at d and d', respectively, and with radii of bd and $b'd'$, respectively. We find the velocity image of D_4 by constructing triangle bdc similar to BDC to locate d and drawing the circle centered at d and

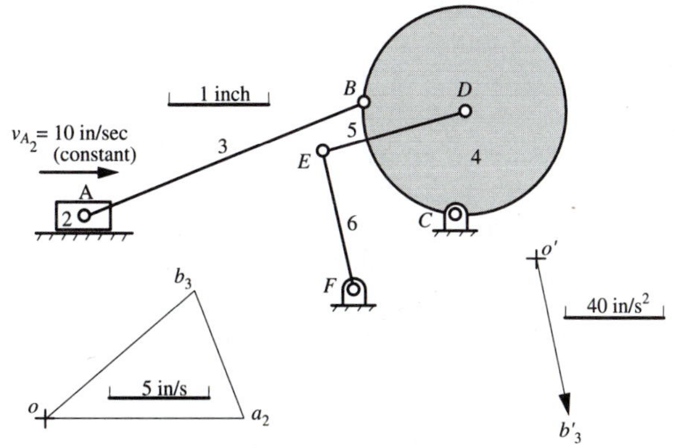

Figure 2.18 Figure for Example 2.5.

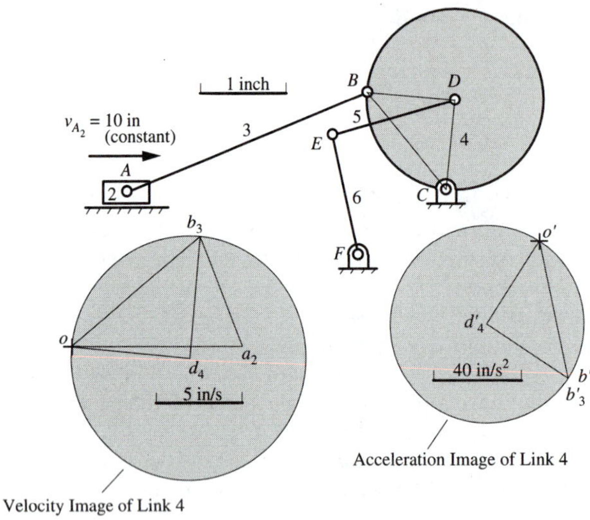

Figure 2.19 Solution to Example 2.5.

with radius *bd*. Similarly, the acceleration image is found by constructing the triangle $b'd'c'$ similar to *BDC* and drawing the circle centered at d' and with radius $b'd'$. The solution is shown in Fig. 2.19.

2.9 SOLUTION BY INVERSION

In general, if we have a linkage where the driver link is not part of a four-bar loop that contains the frame as one of the members, it is not possible to analyze the linkage directly using the vector polygon approach. The Stephenson six-bar linkage shown in Fig. 2.20 can be solved using the techniques in the previous sections *provided* the driving crank is O_AA or O_BB. However, if the linkage is driven by crank O_CC, the linkage cannot be analyzed using the techniques developed so far. This is because O_CC does not form a part of any four-bar loop in the linkage. Consequently, plotting the velocity, or acceleration, of point *C* does not provide enough information to close a velocity or acceleration polygon.

If the position of the linkage is known, however, a velocity solution can be achieved recognizing that all of the velocities in the linkage are linearly related to the velocity of the input member. Therefore, we can solve the velocity problem indirectly by first assuming the linkage to be driven by O_AA, rotating at 1 rad/s in a specified direction. The velocity polygon is completed and the angular velocity of O_CC is found. A scaling factor is then computed. It is the ratio of the actual angular velocity of O_CC to that calculated. It also carries a sign that is positive if both angular velocities are in the same direction and negative if they are opposed. All velocities and angular velocities are then multiplied by that scaling factor to complete the solution.

This solution technique is an example of *inversion.* The driving and driven cranks are interchanged in order to perform the solution. That is, the linkage is inverted by having the driver moved to a different location. This may seem different from inversion as described in Chapter 1, in which the base link is changed. However, it is closely related, as will be seen later when we deal with the case in which the mechanism is driven via a floating link. A detailed discussion of the issues involved in inversion is also given by Goodman.[5]

[5] Goodman, T. P. "An Indirect Method for Determining Accelerations in Complex Mechanisms," *Trans. ASME,* Nov., 1958, pp. 1676–1682.

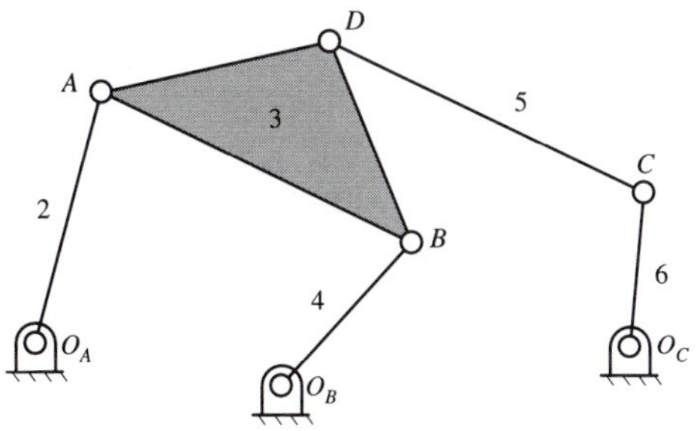

Figure 2.20 A simple linkage that can be analyzed using the techniques of the preceding sections if it is driven by crank O_AA, or by crank O_BB, but not by crank O_CC.

A serious situation arises in most problems requiring inversion. It was assumed above that the position of the linkage was known. Normally, that is not the case, and it is first necessary to determine the angular positions of all links by drawing the linkage to scale. Consideration of Fig. 2.20 reveals, however, that this is not straightforward when the position of crank O_CC is given. Again, the problem is that this crank does not form part of a four-bar loop but appears only in loops with five members. Therefore it is not possible to complete the loop when only the position of that crank is given.

One approach to the solution of this problem is to note that when the angular position of crank O_CC is specified, point D can lie anywhere on a circle with center point C and radius length CD. The position of point D is also constrained by the four-bar linkage O_AABO_B to lie on a unique curve, called a coupler curve. If the coupler curve is plotted, its intersection with the circle gives the location of point D. Unfortunately, the coupler curve is a complicated planar curve of degree six. The only reasonably efficient way to plot it is to construct successive positions of the linkage O_AABO_B as the angular position of the crank O_AA is incremented. Also, there may be as many as six intersections between the coupler curve and the circle, giving up to six different possible positions of the linkage with crank O_CC in the specified position. Each gives an acceptable assembly configuration for the linkage, so the designer must choose the proper one for a given application.

Another approach to the problem is to iterate for the location of the dyad made up of links D and C, and this technique works well when the linkage is drawn using a computer graphics package. For this approach, assume a position for link O_AA, draw the rest of the linkage, and note the position of link O_CC. If the position of link O_CC is not correct, select a different position for link O_AA and reconstruct the linkage again. Measure the position of link O_CC and continue changing O_AA and measuring the position of O_CC until O_CC is in the desired orientation. This may take a number of iterations; however, once the proper position for O_AA is bounded, the procedure will converge fairly rapidly.

If the entire range of motion for the linkage is of interest, then accurately locating the position of O_CC in specific positions is not necessarily an issue. Link O_AA can be located in representative positions in its range of motion and the analysis conducted for each position. Smooth curves can then be drawn through the results.

A procedure for the solution of problems that can be approached by inversion is detailed in Example 2.6.

EXAMPLE 2.6 *(Velocity Analysis by Inversion)*

The linkage shown in Fig. 2.21 is driven by crank O_CC. Find the angular velocities of all members of the linkage in the position in which θ_C is 135°. The angular velocity of O_CC is 10 rad/s CCW.

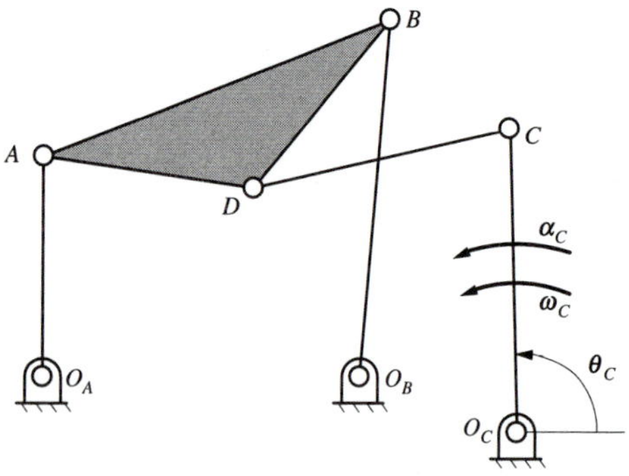

Figure 2.21 The linkage of Example 2.6. This is an example of a linkage that cannot be solved graphically without the use of inversion techniques.

$O_AA = 2$ in, $AB = 3.5$ in, $O_BB = 3.25$ in, $CD = 2.5$ in, $O_CC = 2.75$ in, and $AD = BD = 2.0$ in. With origin at O_A, O_B is the point (3.0, 0) and O_C is the point (4.5, −0.5).

SOLUTION

We must conduct the analysis by starting with the position analysis.

(*a*) *Position*

It is first necessary to construct the linkage in the specified position. The intersection of the coupler curve generated by point *D* with the circular locus of *D* centered on *C* is shown in Fig. 2.22. The coupler curve is plotted by constructing the four-bar O_AABO_B in successive positions with equal increments of the angle of the crank O_AA and plotting the corresponding positions of point *D*. This process is not shown on the figure, but the basic steps are as follows:

1. Plot O_A, O_B, and O_C.
2. Select the angle O_BO_AA and plot O_AA.
3. With center *A* and radius *AB*, draw an arc.
4. With center O_B and radius O_BB, draw an arc. Its intersection with the arc from step 3 is point *B*.
5. Construct the triangle *ABD* on line *AB* to locate point *D*.
6. Increment angle O_BO_AA and repeat steps 1 through 5.
7. Plot the coupler curve, that is, the locus of the successive positions of point *D*. The comma-shaped curve shown in Fig. 2.22 is the resulting coupler curve.

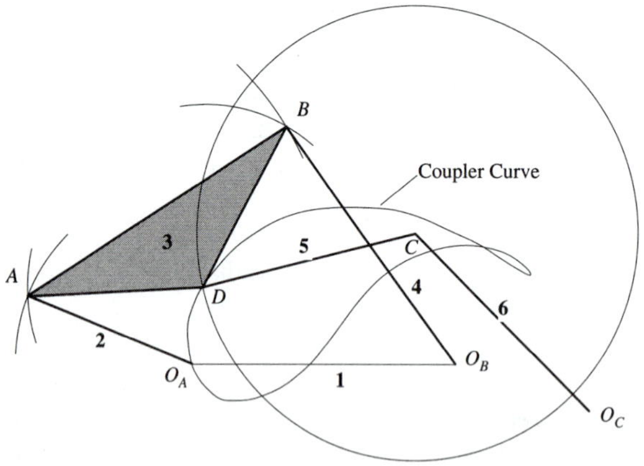

Figure 2.22 Position solution of the linkage of Example 2.6.

The configuration of the linkage can now be constructed as follows:

8. $O_C C$ is drawn at the specified angle, and a circle is drawn with center C and radius CD. Its intersections with the coupler curve give possible positions of point D for the specified value of θ_C. Notice that there are two possible positions for D in this case (there may be as many as six). We choose the position of D that gives the linkage configuration closest to that shown in Fig. 2.21.

Once D is located, point A is located as follows:

9. Set radius $O_A A$ and strike an arc centered on O_A.
10. Set radius DA and strike an arc centered on D. The intersection with the arc of step 9 is point A.
11. Construct triangle ABD on $\overrightarrow{AD}$ to locate point B. The linkage can now be drawn in the specified position.

(b) Velocities

The procedure for solving for the velocities is to draw the velocity polygon with the angular velocity of link 2 assumed to be $\Omega_2 = 1$ rad/s. The value of the angular velocity of link 6, Ω_6, is found for this assumption and a scaling factor is calculated to scale Ω_6 to the specified value of $\omega_6 = 10$ rad/s. The same scaling factor is then applied to all other velocities and angular velocities to give their correct values when $\omega_6 = 10$ rad/s. This is a solution *by inversion,* because it is necessary first to solve the problem with link 2 assumed to be the driving crank rather than working directly with the actual driving crank, which is link 6. The solution with the assumed value of ω_2 is inverted to that with the required value of ω_6 by scaling it.

For the velocity analysis, the basic equations that we will solve are:

$$\boldsymbol{v}_A = \boldsymbol{v}_{A/O_A} = \boldsymbol{\omega}_2 \times \boldsymbol{r}_{A/O_A}$$

$$\boldsymbol{v}_B = \boldsymbol{v}_A + \boldsymbol{v}_{B/A} = \boldsymbol{v}_{B/OB}$$

$$\boldsymbol{v}_D = \boldsymbol{v}_A + \boldsymbol{v}_{D/A} = \boldsymbol{v}_B + \boldsymbol{v}_{D/B}$$

$$\boldsymbol{v}_C = \boldsymbol{v}_D + \boldsymbol{v}_{C/D} = \boldsymbol{v}_{C/O_C}$$

STEPS

1. Compute the value of $\boldsymbol{v}_A$ with the assumption that $\Omega_2 = 1$ rad/s CCW, and plot $\boldsymbol{v}_A$ (as $\overrightarrow{od}$) normal to $\overrightarrow{O_A A}$ as shown in Fig. 2.23.

$$v_A = 2.0 \times 1 = 2 \text{ in/s}$$

2. Draw a line through a normal to $\overrightarrow{AB}$.
3. Draw a line through o normal to $\overrightarrow{O_B B}$. The intersection of this line with that of step 2 gives point b.
4. Draw a line through a normal to $\overrightarrow{AD}$.
5. Draw a line through b normal to $\overrightarrow{BD}$. The intersection of this line with that from step 4 gives point d (velocity image).
6. Draw a line through d normal to $\overrightarrow{CD}$.
7. Draw a line through o normal to $\overrightarrow{O_C C}$. The intersection of this line with that from step 6 gives point c.
8. Measure $\boldsymbol{v}_C = \overrightarrow{oc}$ and compute Ω_6.

$$v_C = 1.214 \text{ in/s}$$

$$\Omega_6 = v_C/O_C C = 1.214/2.75 = 0.441 \text{ rad/s CCW}$$

9. Compute scaling factor $\sigma = \omega_6/\Omega_6$ where ω_6 is the specified angular velocity of link 6.

$$\sigma = 10/0.441 = 22.7$$

Since both the calculated and specified values of $\boldsymbol{\omega}_6$ are CCW, σ is positive. If they had been in opposite directions, σ would be negative.

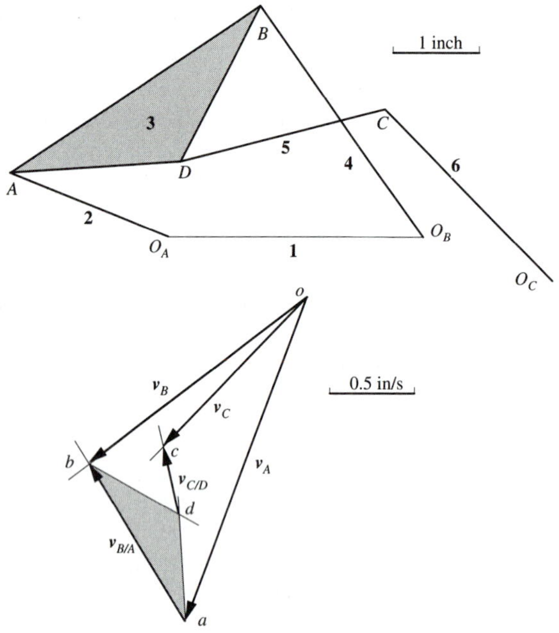

Figure 2.23 Velocity solution for Example 2.6.

10. Compute the angular velocities $\mathbf{\Omega}_2$, $\mathbf{\Omega}_3$, $\mathbf{\Omega}_4$, and $\mathbf{\Omega}_5$ and scale the results.

$$\boldsymbol{\omega}_2 = \sigma \times \mathbf{\Omega}_2 = 1 \times 22.7 = 22.7 \text{ rad/s CCW}$$
$$\boldsymbol{\omega}_3 = \sigma \times \mathbf{\Omega}_3 = \sigma \times v_{B/A}/AB = 22.7 \times 1.09/3.5 = 7.05 \text{ rad/s CCW}$$
$$\boldsymbol{\omega}_4 = \sigma \times \mathbf{\Omega}_4 = \sigma \times v_B/O_BB = 22.7 \times 1.62/3.25 = 11.32 \text{ rad/s CCW}$$
$$\boldsymbol{\omega}_5 = \sigma \times \mathbf{\Omega}_5 = \sigma \times v_{C/D}/CD = 22.7 \times 0.42/2.5 = 3.73 \text{ rad/s CCW}$$

This completes the velocity analysis. ■

2.10 REFERENCE FRAMES

If a linkage involves only revolute joints or sliders on fixed lines, the equations developed in Sections 2.3 and 2.4 are sufficient for conducting the kinematic analysis. However, for other types of joints, the equations become more complex, and it is necessary to use more than one reference frame for the velocities and accelerations. In general, each link may be assumed to have a reference frame attached to it. In fact, when each link is manufactured, the machine tool that is used to form the link geometry will be guided relative to the local coordinate system or reference frame fixed to the link.

When it is important to distinguish the reference frames to which positions, velocities, and accelerations are referred, we will use a superscript before the vector symbol to identify the relevant reference frame. Typically, we will use the link number or letter as the reference frame for that link. Thus if B is a general link that is moving relative to another link R, ${}^R\boldsymbol{\omega}_B$ is the angular velocity of the moving body, B, relative to frame R. ${}^R\mathbf{v}_Q$ is the absolute velocity of point Q relative to frame R, and ${}^B\mathbf{r}_Q$ is the absolute position of point Q relative to the reference frame fixed to body B. ${}^2\boldsymbol{\alpha}_3$ is the angular acceleration of member 3 relative to the reference frame fixed in member 2.

The vector ${}^R\mathbf{v}_{B/A}$ is usually called the velocity of B relative to A in reference frame R. However, as discussed earlier, this definition is technically incorrect. Vectors must be measured relative to reference frames. Therefore, ${}^R\mathbf{v}_{B/A}$ would be the velocity of point B relative to a reference frame R that has its origin at point A and moves so as always to be parallel to the fixed frame. Similarly, one would call ${}^R\mathbf{r}_{Q/P}$ the position of Q relative to a reference frame, with origin at P, which remains, at all times, parallel to the frame R. The complexity of this statement explains the widespread use of the term "position of Q relative to P" for ${}^R\mathbf{r}_{Q/P}$.

Often, when all vectors are referred to the same reference frame, R, we will drop the superscript R to simplify the notation. That is, $\boldsymbol{\omega}_B \equiv {}^R\boldsymbol{\omega}_B$. This was the case in Sections 2.3 and 2.4 when the fixed frame (link 1) was understood to be the reference frame for all vectors.

The basis of the velocity analysis of planar linkages is the relationship between the velocities of two different points when something about the motion of the two points is known relative to a moving coordinate system. To derive this relationship in a form suitable for the formulation of a velocity polygon, let us consider the points P and Q shown in Fig. 2.24. If ${}^R\mathbf{r}_P$ is the position of point P relative to reference frame R, ${}^R\mathbf{r}_Q$ is the position of Q relative to reference frame R, and ${}^B\mathbf{r}_{Q/P}$ is the vector from point P to point Q defined relative to the moving reference system B, then we can write

$$ {}^R\mathbf{r}_Q = {}^R\mathbf{r}_P + {}^B\mathbf{r}_{Q/P} \tag{2.2} $$

As indicated before, the vector ${}^B\mathbf{r}_{Q/P}$ is called the position of Q relative to P when the observer is fixed relative to reference system B.

Although P and Q may be fixed to body B, Eq. (2.2) is valid regardless of the link to which points P and Q are fixed (i.e., P and Q may be fixed to link B or some other link). To obtain the velocities, we must differentiate Eq. (2.2) when the observer is in reference frame R.

Note that in the position considered, the coordinate axes for systems B and R must be parallel. Otherwise we cannot add vector components as implied in Eq. (2.2). If the nominal coordinate systems attached to the two links are not parallel, we must use another set of coordinate systems that are momentarily parallel. The two coordinate systems fixed to a given link would be related by a simple coordinate transformation.

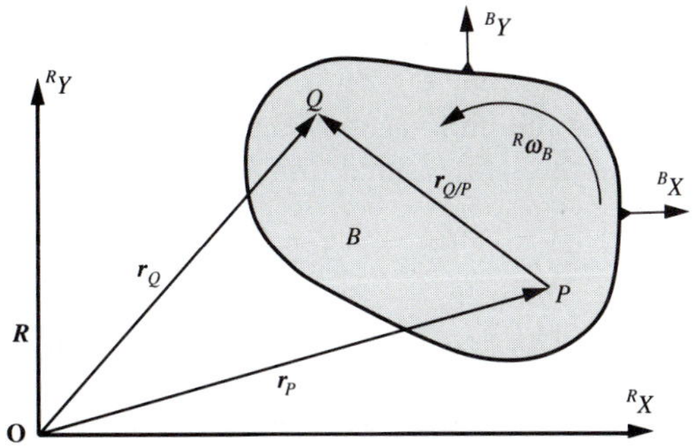

Figure 2.24 Positions of two points in the moving lamina, B. $\mathbf{r}_P$ and $\mathbf{r}_Q$ are the respective positions of points P and Q relative to the fixed reference frame, R. $\boldsymbol{\omega}_B$ is the angular velocity of B. Note that this figure is similar to Fig. 2.1.

2.11 GENERAL VELOCITY AND ACCELERATION EQUATIONS

2.11.1 Velocity Equations

When we differentiate Eq. (2.2) with the observer in the reference system R, we get

$$\frac{{}^{R}d}{dt}({}^{R}\mathbf{r}_{Q}) = \frac{{}^{R}d}{dt}({}^{R}\mathbf{r}_{P}) + \frac{{}^{R}d}{dt}({}^{B}\mathbf{r}_{Q/P}) \tag{2.19}$$

The derivatives of the position vectors defined relative to reference system R can be represented in a straightforward manner as velocities relative to reference system R because the reference axes relative to which the vectors are defined are fixed to R and do not move with time. Therefore, Eq. (2.2) becomes

$${}^{R}\mathbf{v}_{Q} = {}^{R}\mathbf{v}_{P} + \frac{{}^{R}d}{dt}({}^{B}\mathbf{r}_{Q/P}) \tag{2.20}$$

However, note that ${}^{B}\mathbf{r}_{Q/P}$ is a vector defined relative to the coordinate system fixed to body B and the reference axes of body B rotate relative to those of reference system R with an angular velocity ${}^{R}\boldsymbol{\omega}_{B}$. Therefore the derivative $({}^{R}d/dt)({}^{B}\mathbf{r}_{Q/P})$ must account for this rotation. In particular, the derivative involves two terms, one associated with the change in magnitude of the vector and one associated with the change in direction. This is apparent if we represent the vector ${}^{B}\mathbf{r}_{Q/P}$ as a general three-dimensional vector in terms of its components and unit vectors. Then,

$${}^{B}\mathbf{r}_{Q/P} = x^{B}\boldsymbol{i} + y^{B}\boldsymbol{j} + z^{B}\mathbf{k}$$

and

$$\begin{aligned}\frac{{}^{R}d}{dt}({}^{B}\mathbf{r}_{Q/P}) &= \frac{{}^{R}d}{dt}(x^{B}\boldsymbol{i} + y^{B}\boldsymbol{j} + z^{B}\mathbf{k}) \\ &= \left(\frac{{}^{R}dx}{dt}{}^{B}\boldsymbol{i} + \frac{{}^{R}dy}{dt}{}^{B}\boldsymbol{j} + \frac{{}^{R}dz}{dt}{}^{B}\mathbf{k}\right) + \left(x\frac{{}^{R}d^{B}\boldsymbol{i}}{dt} + y\frac{{}^{R}d^{B}\boldsymbol{j}}{dt} + z\frac{{}^{R}d^{B}\mathbf{k}}{dt}\right)\end{aligned} \tag{2.21}$$

The derivatives of the components correspond to the change in the length of the vector, and this is defined relative to the coordinate system fixed to body B. Therefore, this is just the velocity defined relative to body B. The second term accounts for the rotation of the coordinate axes of B relative to the reference frame R.

Because ${}^{B}\boldsymbol{i}$, ${}^{B}\boldsymbol{j}$, and ${}^{B}\mathbf{k}$ are unit vectors, only their directions can change with time. We can determine how to evaluate the derivatives if we look at an infinitesimal angular displacement $\delta\boldsymbol{\theta}$ of body B relative to R during an infinitesimal time increment, δt.

Because infinitesimal angular rotations are involved, we can treat $\delta\boldsymbol{\theta}$ as a vector with x, y, z components (i.e., $\delta\boldsymbol{\theta} = \delta\theta_{x}{}^{R}\boldsymbol{i} + \delta\theta_{y}{}^{R}\boldsymbol{j} + \delta\theta_{z}{}^{R}\mathbf{k}$) and determine how each component changes the directions of the unit vectors. The angular velocity will be the change in the angular position during the infinitesimal time increment, δt. That is,

$$\boldsymbol{\omega}_{x} = \frac{{}^{R}\delta\boldsymbol{\theta}_{x}}{\delta t}, \qquad \boldsymbol{\omega}_{y} = \frac{{}^{R}\delta\boldsymbol{\theta}_{y}}{\delta t}, \qquad \boldsymbol{\omega}_{z} = \frac{{}^{R}\delta\boldsymbol{\theta}_{z}}{\delta t}$$

and

$${}^{R}\boldsymbol{\omega}_{B} = \omega_{x}{}^{R}\boldsymbol{i} + \omega_{y}{}^{R}\boldsymbol{j} + \omega_{z}{}^{R}\mathbf{k}$$

To identify the trend, consider the effect of the angular components about the X axis. For the x direction (unit vector ${}^{B}\boldsymbol{i}$), the change in the unit vector is represented in Fig. 2.25. From the figure,

$${}^{R}\delta({}^{B}\boldsymbol{i}) = 1{}^{R}\boldsymbol{j}\,\delta\theta_{z} - 1{}^{R}\mathbf{k}\,\delta\theta_{y} \tag{2.22}$$

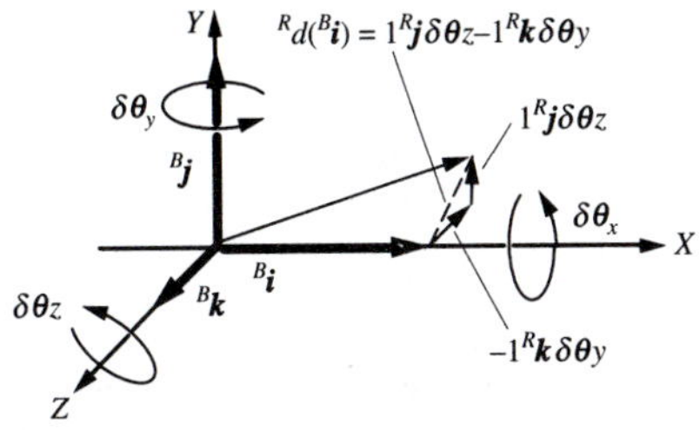

Figure 2.25 Change in $\boldsymbol{i}$ due to a rotation about the X, Y, and Z axes.

The change takes place during the time increment δt. Therefore, dividing Eq. (2.22) by δt, we get

$$\frac{^R d}{dt}(^B\boldsymbol{i}) = {}^R\boldsymbol{j}\omega_z - {}^R\boldsymbol{k}\omega_y = {}^R\boldsymbol{\omega}_B \times {}^B\boldsymbol{i} = \begin{vmatrix} ^R\boldsymbol{i} & ^R\boldsymbol{j} & ^R\boldsymbol{k} \\ \omega_x & \omega_y & \omega_z \\ 1 & 0 & 0 \end{vmatrix}$$

Similarly,

$$\frac{^R d}{dt}(^B\boldsymbol{j}) = {}^R\boldsymbol{\omega}_B \times {}^B\boldsymbol{j}$$

$$\frac{^R d}{dt}(^B\boldsymbol{k}) = {}^R\boldsymbol{\omega}_B \times {}^B\boldsymbol{k}$$

Therefore,

$$x\frac{^R d^B\boldsymbol{i}}{dt} + y\frac{^R d^B\boldsymbol{j}}{dt} + z\frac{^R d^B\boldsymbol{k}}{dt} = x^R\boldsymbol{\omega}_B \times {}^B\boldsymbol{i} + y^R\boldsymbol{\omega}_B \times {}^B\boldsymbol{j} + z^R\boldsymbol{\omega}_B \times {}^B\boldsymbol{k}$$
$$= {}^R\boldsymbol{\omega}_B \times (x^B\boldsymbol{i} + y^B\boldsymbol{j} + z^B\boldsymbol{k}) = {}^R\boldsymbol{\omega}_B \times {}^B\boldsymbol{r}_{Q/P}$$

Then,

$$\frac{^R d}{dt}(^B\boldsymbol{r}_{Q/P}) = \left(\frac{^R dx}{dt}{}^B\boldsymbol{i} + \frac{^R dy}{dt}{}^B\boldsymbol{j} + \frac{^R dz}{dt}{}^B\boldsymbol{k}\right) + \left(x\frac{^R d^B\boldsymbol{i}}{dt} + y\frac{^R d^B\boldsymbol{j}}{dt} + z\frac{^R d^B\boldsymbol{k}}{dt}\right) \tag{2.23}$$
$$= {}^B\boldsymbol{v}_{Q/P} + {}^R\boldsymbol{\omega}_B \times {}^B\boldsymbol{r}_{Q/P}$$

Now, Eq. (2.20) can be written as

$$^R\boldsymbol{v}_Q = {}^R\boldsymbol{v}_P + {}^B\boldsymbol{v}_{Q/P} + {}^R\boldsymbol{\omega}_B \times {}^B\boldsymbol{r}_{Q/P} \tag{2.24}$$

Before proceeding to the development of the acceleration equations, it is interesting to note that Eq. (2.23) is quite general. We could have derived a similar expression for the derivative of *any* vector that is defined relative to a moving coordinate system. For example, if $\boldsymbol{s}$ is any vector (e.g., position, velocity, acceleration) and if U and W are any two different coordinate systems,

$$\frac{^U d}{dt}(^W\boldsymbol{s}) = \frac{^W d}{dt}(^W\boldsymbol{s}) + {}^U\boldsymbol{\omega}_W \times {}^W\boldsymbol{s} \tag{2.25}$$

Note that angular velocity is a property of a body and linear velocity is a property of a point. In both cases, it is necessary to specify, or at least understand, which reference frame is used to define the quantity. Also note that if the vector is defined in the coordinate system in which the observer stands, the term involving the angular velocity will be zero. That is,

$$^{U}\boldsymbol{\omega}_U = 0$$

2.11.2 Acceleration Equations

The acceleration equations will involve the derivative of each angular velocity. In general, angular acceleration can be written as

$$^{R}\boldsymbol{\alpha}_B = \frac{^{R}d}{dt}(^{R}\boldsymbol{\omega}_B)$$

where again B is the moving body and R is the reference system. As in the case of the angular velocity, the angular acceleration is a property of the entire body. It is a vector and has a magnitude and direction.

If a velocity vector is defined in the coordinate system in which the observer is located, the corresponding acceleration can be expressed simply. For example, if the velocity vector is given by $^{R}\mathbf{v}_Q$, then the acceleration is given by

$$^{R}\boldsymbol{a}_Q = \frac{^{R}d}{dt}(^{R}\mathbf{v}_Q)$$

To obtain the linear acceleration relationship for the points P and Q in Fig. 2.25, we can differentiate Eq. (2.24). Differentiating term by term with the observer in reference system R gives

$$^{R}\boldsymbol{a}_Q = \frac{^{R}d}{dt}(^{R}\mathbf{v}_Q) = \frac{^{R}d}{dt}(^{R}\mathbf{v}_P) + \frac{^{R}d}{dt}(^{B}\mathbf{v}_{Q/P}) + \frac{^{R}d}{dt}(^{R}\boldsymbol{\omega}_B \times {}^{B}\mathbf{r}_{Q/P})$$

Considering each term and recognizing that vectors $^{B}\mathbf{v}_{Q/P}$ and $^{B}\mathbf{r}_{Q/P}$ are both defined relative to the moving coordinate system (B), we get after differentiation,

$$^{R}\boldsymbol{a}_Q = {}^{R}\boldsymbol{a}_P + \frac{^{B}d}{dt}(^{B}\mathbf{v}_{Q/P}) + {}^{R}\boldsymbol{\omega}_B \times {}^{B}\mathbf{v}_{Q/P} + {}^{R}\boldsymbol{\alpha}_B \times {}^{B}\mathbf{r}_{Q/P} + {}^{R}\boldsymbol{\omega}_B \times (^{B}\mathbf{v}_{Q/P} + {}^{R}\boldsymbol{\omega}_B \times {}^{B}\mathbf{r}_{Q/P})$$

and collecting terms gives

$$^{R}\boldsymbol{a}_Q = {}^{R}\boldsymbol{a}_P + {}^{B}\boldsymbol{a}_{Q/P} + 2{}^{R}\boldsymbol{\omega}_B \times {}^{B}\mathbf{v}_{Q/P} + {}^{R}\boldsymbol{\alpha}_B \times {}^{B}\mathbf{r}_{Q/P} + {}^{R}\boldsymbol{\omega}_B \times (^{R}\boldsymbol{\omega}_B \times {}^{B}\mathbf{r}_{Q/P})$$

Note that in the last term, the operation $(^{R}\boldsymbol{\omega}_B \times {}^{B}\mathbf{r}_{Q/P})$ must be carried out before the operation $^{R}\boldsymbol{\omega}_B \times (^{R}\boldsymbol{\omega}_B \times {}^{B}\mathbf{r}_{Q/P})$. Obviously, $(\boldsymbol{\omega} \times \boldsymbol{\omega}) \times \boldsymbol{r} \neq \boldsymbol{\omega} \times (\boldsymbol{\omega} \times \boldsymbol{r})$.

The term $2(^{R}\boldsymbol{\omega}_B \times {}^{B}\mathbf{v}_{Q/P})$ is called the Coriolis term and is a function of velocities only. The term $^{R}\boldsymbol{\alpha}_B \times {}^{B}\mathbf{r}_{Q/P}$ is the transverse or tangential component of acceleration identified before. This component of acceleration is perpendicular to the radius vector. The term $^{R}\boldsymbol{\omega}_B \times (^{R}\boldsymbol{\omega}_B \times {}^{B}\mathbf{r}_{Q/P})$ is the radial component of acceleration, and it points in the direction opposite to the radius vector. The term $^{B}\boldsymbol{a}_{Q/P}$ is the acceleration of Q relative to P when the observer is in the moving body B.

2.11.3 "Chain Rule" for Positions, Velocities, and Accelerations

When dealing with mechanisms with a relatively large number of members, it is helpful to have relationships between the relative velocities and accelerations of several points and between the relative angular velocities and angular accelerations of several members. These relationships are particularly relevant to the spatial chain mechanisms discussed in Chapter 7.

2.11.3.1 Positions, Velocities, and Accelerations of Points

Let A, B, C, D, and E be any arbitrary points moving with respect to the reference frame R as shown in Fig. 2.26. Then a position equation can be written as

$$^R\mathbf{r}_{E/D} + {^R\mathbf{r}_{D/C}} + {^R\mathbf{r}_{C/B}} + {^R\mathbf{r}_{B/A}} = {^R\mathbf{r}_{E/A}} \tag{2.27}$$

This type of equation is just a simple expression of vector addition, and it applies regardless of the number of points involved. For velocities, we can differentiate Eq. (2.27) with the observer in system R. Then,

$$^R\mathbf{v}_{E/D} + {^R\mathbf{v}_{D/C}} + {^R\mathbf{v}_{C/B}} + {^R\mathbf{v}_{B/A}} = {^R\mathbf{v}_{E/A}} \tag{2.28}$$

and the acceleration equation becomes

$$^R\mathbf{a}_{E/D} + {^R\mathbf{a}_{D/C}} + {^R\mathbf{a}_{C/B}} + {^R\mathbf{a}_{B/A}} = {^R\mathbf{a}_{E/A}} \tag{2.29}$$

Equations (2.27) through (2.29) are applicable to any set of points, and they are especially useful when determining the kinematic information for points on mechanisms after the basic kinematic information associated with each link is known. They are also useful when analyzing manipulators and robots.

The relationship among three *arbitrary* points (A, B, C) is

$$^R\mathbf{r}_{C/A} = {^R\mathbf{r}_{C/B}} + {^R\mathbf{r}_{B/A}}$$

Then,

$$^R\mathbf{r}_{C/B} = {^R\mathbf{r}_{C/A}} - {^R\mathbf{r}_{B/A}} \tag{2.30}$$

Because A is arbitrary, Eq. (2.30) indicates that we can find the relative position between two points by subtracting the relative position vectors between the two points and the same third point. Similarly, for velocities and accelerations,

$$^R\mathbf{v}_{C/B} = {^R\mathbf{v}_{C/A}} - {^R\mathbf{v}_{B/A}} \tag{2.31}$$

and

$$^R\mathbf{a}_{C/B} = {^R\mathbf{a}_{C/A}} - {^R\mathbf{a}_{B/A}} \tag{2.32}$$

Note the position of A, B, and C in each of the expressions.

2.11.3.2 Relative Angular Velocities

A chain rule for angular velocities works the same way as for linear velocities except that now reference systems are involved instead of points. Consider three coordinate systems (1, 2, and 3) that are *momentarily parallel.* Then,

$$^1\boldsymbol{\omega}_3 = {^1\boldsymbol{\omega}_2} + {^2\boldsymbol{\omega}_3} \tag{2.33}$$

and

$$^2\boldsymbol{\omega}_3 = {^1\boldsymbol{\omega}_3} - {^1\boldsymbol{\omega}_2} = {^1\boldsymbol{\omega}_3} + {^2\boldsymbol{\omega}_1} \tag{2.34}$$

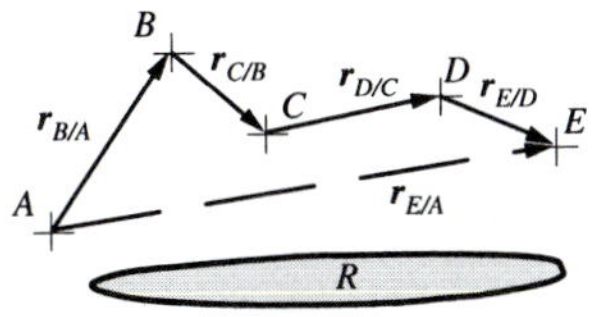

Figure 2.26 Relationship among the positions of a series of points.

This means we can find the relative angular velocity between any two bodies by computing the angular velocity difference between each of the bodies and the same third body (in this case body 2).

For n bodies, the relative angular velocities are related by

$$^{n}\boldsymbol{\omega}_1 = {}^{2}\boldsymbol{\omega}_1 + {}^{3}\boldsymbol{\omega}_2 + {}^{4}\boldsymbol{\omega}_3 + \cdots + {}^{n-1}\boldsymbol{\omega}_{n-2} + {}^{n}\boldsymbol{\omega}_{n-1}$$

2.11.3.3 *Relative Angular Accelerations*

For relative accelerations, we can differentiate the relative velocity equation, Eq. (2.33):

$$^{1}\boldsymbol{\omega}_3 = {}^{1}\boldsymbol{\omega}_2 + {}^{2}\boldsymbol{\omega}_3$$

or

$$\frac{^{1}d}{dt}(^{1}\boldsymbol{\omega}_3) = \frac{^{1}d}{dt}(^{1}\boldsymbol{\omega}_2) + \frac{^{1}d}{dt}(^{2}\boldsymbol{\omega}_3) \tag{2.35}$$

The first two terms are straightforward because the derivatives are both taken with respect to the reference system in which each vector is defined. That is,

$$\frac{^{1}d}{dt}(^{1}\boldsymbol{\omega}_3) = {}^{1}\boldsymbol{\alpha}_3, \qquad \frac{^{1}d}{dt}(^{1}\boldsymbol{\omega}_2) = {}^{1}\boldsymbol{\alpha}_2$$

The third term is a vector described in the second coordinate system (superscript 2). Therefore using Eq. (2.25), this term can be written as

$$\frac{^{1}d}{dt}(^{2}\boldsymbol{\omega}_3) = \frac{^{2}d}{dt}(^{2}\boldsymbol{\omega}_3) + {}^{1}\boldsymbol{\omega}_2 \times {}^{2}\boldsymbol{\omega}_3 = {}^{2}\boldsymbol{\alpha}_3 + {}^{1}\boldsymbol{\omega}_2 \times {}^{2}\boldsymbol{\omega}_3$$

The relative angular acceleration expression in Eq. (2.35) can then be written as

$$^{1}\boldsymbol{\alpha}_3 = {}^{1}\boldsymbol{\alpha}_2 + {}^{2}\boldsymbol{\alpha}_3 + {}^{1}\boldsymbol{\omega}_2 \times {}^{2}\boldsymbol{\omega}_3$$

This expression can be extended to n bodies using

$$^{1}\boldsymbol{\alpha}_n = {}^{1}\boldsymbol{\alpha}_{n-1} + {}^{n-1}\boldsymbol{\alpha}_n + {}^{1}\boldsymbol{\omega}_{n-1} \times {}^{n-1}\boldsymbol{\omega}_n$$

Then,

$$^{1}\boldsymbol{\alpha}_n = {}^{1}\boldsymbol{\alpha}_2 + {}^{2}\boldsymbol{\alpha}_3 + {}^{3}\boldsymbol{\alpha}_4 + \cdots + {}^{n-1}\boldsymbol{\alpha}_n + {}^{1}\boldsymbol{\omega}_2 \times {}^{2}\boldsymbol{\omega}_3 + {}^{1}\boldsymbol{\omega}_3 \times {}^{3}\boldsymbol{\omega}_4 + {}^{1}\boldsymbol{\omega}_4 \times {}^{4}\boldsymbol{\omega}_5 + \cdots + {}^{1}\boldsymbol{\omega}_{n-1} \times {}^{n-1}\boldsymbol{\omega}_n$$

Note that in the plane, all of the $\boldsymbol{\omega}$'s will be parallel, making the cross products all zero. Thus in *planar* problems, the chain rule for angular accelerations reduces to a scalar equation in the magnitudes of the angular accelerations with signs according to some rule (say + for CCW and − for CW).

$$^{1}\alpha_n = {}^{1}\alpha_2 + {}^{2}\alpha_3 + {}^{3}\alpha_4 + \cdots + {}^{n-1}\alpha_n \quad \text{(planar problems)}$$

2.12 SPECIAL CASES FOR THE VELOCITY AND ACCELERATION EQUATIONS

Equations (2.24) and (2.26) are the most general forms for the relative velocity and acceleration equations for points that we will encounter in the kinematic analysis of linkages. In most practical problems, some of the terms in the expressions are zero. Three special cases occur, and these will be discussed separately in the following.

2.12.1 Points *P* and *Q* Fixed to *B*

This is the most common situation that exists in the analysis of mechanisms. If P and Q are both fixed to B, we have

$$^{B}\boldsymbol{v}_{Q_B/P_B} = {}^{B}\boldsymbol{a}_{Q_B/P_B} = 0 \tag{2.36}$$

because P and Q do not have any motion relative to an observer in the moving body B. When Eq. (2.36) is used to simplify Eqs. (2.24) and (2.26), the results are

$$^{R}\boldsymbol{v}_{Q} = {}^{R}\boldsymbol{v}_{P} + {}^{R}\boldsymbol{\omega}_{B} \times \boldsymbol{r}_{Q/P} \tag{2.37}$$

which can be recognized as being the same as Eq. (2.1), and

$$^{R}\boldsymbol{a}_{Q} = {}^{R}\boldsymbol{a}_{P} + {}^{R}\boldsymbol{\alpha}_{B} \times \boldsymbol{r}_{Q/P} + {}^{R}\boldsymbol{\omega}_{B} \times ({}^{R}\boldsymbol{\omega}_{B} \times \boldsymbol{r}_{Q/P}) \tag{2.38}$$

which is the same as Eq. (2.7).

Here we have dropped the superscript on $\boldsymbol{r}_{Q/P}$ because all coordinate systems are assumed to be parallel, and $\boldsymbol{r}_{Q/P}$ will have the same coordinates in all coordinate systems. Note also that we could have rewritten Eqs. (2.24) and (2.26) relative to any other link; however, only the choice of the link (B) to which Q and P are attached simplifies the equation. Using the radial and tangential notation, we can also rewrite Eq. (2.38) as

$$^{R}\boldsymbol{a}_{Q} = {}^{R}\boldsymbol{a}_{P} + {}^{R}\boldsymbol{a}^{r}_{Q/P} + {}^{R}\boldsymbol{a}^{t}_{Q/P}$$

where

$$^{R}\boldsymbol{a}^{t}_{Q/P} = {}^{R}\boldsymbol{\alpha}_{B} \times \boldsymbol{r}_{Q/P} \tag{2.39}$$

and

$$^{R}\boldsymbol{a}^{r}_{Q/P} = {}^{R}\boldsymbol{\omega}_{B} \times ({}^{R}\boldsymbol{\omega}_{B} \times \boldsymbol{r}_{Q/P}) \tag{2.40}$$

We will use the radial and tangential notation extensively in mechanism analyses. For planar mechanisms, $^{R}\boldsymbol{a}^{t}_{Q/P}$ and $^{R}\boldsymbol{a}^{r}_{Q/P}$ will be orthogonal to each other because $^{R}\boldsymbol{\omega}_{B}$ and $^{R}\boldsymbol{\alpha}_{B}$ are both orthogonal to $\boldsymbol{r}_{Q/P}$. In spatial mechanisms, however, this will not always be the case.

2.12.2 *P* and *Q* Are Coincident

A second special case in kinematics is that in which P and Q belong to different bodies but are momentarily coincident. Then, $\boldsymbol{r}_{P/Q}$ is momentarily zero, and Eqs. (2.24) and (2.26) reduce to

$$^{R}\boldsymbol{v}_{Q} = {}^{R}\boldsymbol{v}_{P} + {}^{B}\boldsymbol{v}_{Q/P} \tag{2.41}$$

If Eq. (2.41) is considered carefully, it is apparent that the equation for the relative velocity remains the same regardless of the body chosen as the moving body. This means that the relative velocity term $^{B}\boldsymbol{v}_{Q/P}$ is independent of the coordinate system chosen for the "moving" body. Therefore,

$$^{i}\boldsymbol{v}_{Q/P} = {}^{B}\boldsymbol{v}_{Q/P} = {}^{R}\boldsymbol{v}_{Q/P}$$

where i and B are *any* systems.

The acceleration equation, Eq. (2.26), simplifies to

$$^{R}\boldsymbol{a}_{Q} = {}^{R}\boldsymbol{a}_{P} + {}^{B}\boldsymbol{a}_{Q/P} + 2{}^{R}\boldsymbol{\omega}_{B} \times {}^{B}\boldsymbol{v}_{Q/P} \tag{2.42}$$

Here, the Coriolis term is a function of velocities, so it can be computed as soon as the velocity analysis is completed. Only $^{B}\boldsymbol{a}_{Q/P}$ involves new information not available from the velocity analysis.

2.12.3 *P* and *Q* Are Coincident and in Rolling Contact

If points P and Q are not only momentarily coincident but also in rolling contact, Eqs. (2.41) and (2.42) can be simplified still further. If two points are in rolling contact, they have the same velocity, and their relative velocity must be zero. This means that

$$^{B}\boldsymbol{v}_{Q/P} = 0$$
$$^{R}\boldsymbol{v}_{Q} = {}^{R}\boldsymbol{v}_{P}$$

and

$$^{R}\boldsymbol{a}_{Q} = {}^{R}\boldsymbol{a}_{P} + {}^{B}\boldsymbol{a}_{Q/P}$$

Using logic similar to that used with Eq. (2.42), it is apparent that while the relative acceleration $^{B}\boldsymbol{a}_{Q/P}$ is not usually zero, it is independent of whatever coordinate system is used for reference. This means that the relative acceleration will be the same when observed from any of the links in the mechanism.

2.13 LINKAGES WITH ROTATING SLIDING JOINTS

Mechanisms in this class can have either a slider that slides on a line that is rotating or a pin-in-a-slot joint where the slot is straight and rotating. These cases are shown in Fig. 2.27.

Mechanisms with sliders that rotate are common in practice. Typical examples are door closers, the hydraulic cylinders on power shovels, and the power cylinders on some robots. The pin-in-slot joints, often with a free-spinning roller centered on the pin, are typically used as inexpensive substitutes for slider joints. They function where the transmitted loads are low. Examples are electric toothbrush mechanisms, audiotape cleaners, and walking-toy mechanisms.

The analysis of these mechanisms can be approached using the special case in Section 2.12.2 for relative velocities and accelerations of coincident points. The resulting velocity and acceleration equations for two coincident points P and Q are given by Eqs. (2.41) and (2.42) as

$$^{R}\boldsymbol{v}_{Q} = {}^{R}\boldsymbol{v}_{P} + {}^{B}\boldsymbol{v}_{Q/P}$$
$$^{R}\boldsymbol{a}_{Q} = {}^{R}\boldsymbol{a}_{P} + {}^{B}\boldsymbol{a}_{Q/P} + 2\,{}^{R}\boldsymbol{\omega}_{B} \times {}^{B}\boldsymbol{v}_{Q/P}$$

When the points are coincident, P and Q will share the same coordinates, and they will usually be designated by the same letter with a subscript identifying the link to which they are attached. For example, if 3 and 4 are the links to which the coincident points are attached, if body B corresponds to link 5, and if the frame is 1, the velocity and acceleration equations can be written as

$$^{1}\boldsymbol{v}_{P_3} = {}^{1}\boldsymbol{v}_{P_4} + {}^{5}\boldsymbol{v}_{P_3/P_4} \tag{2.43}$$

and

$$^{1}\boldsymbol{a}_{P_3} = {}^{1}\boldsymbol{a}_{P_4} + {}^{5}\boldsymbol{a}_{P_3/P_4} + 2\,{}^{1}\boldsymbol{\omega}_5 \times {}^{5}\boldsymbol{v}_{P_3/P_4} \tag{2.44}$$

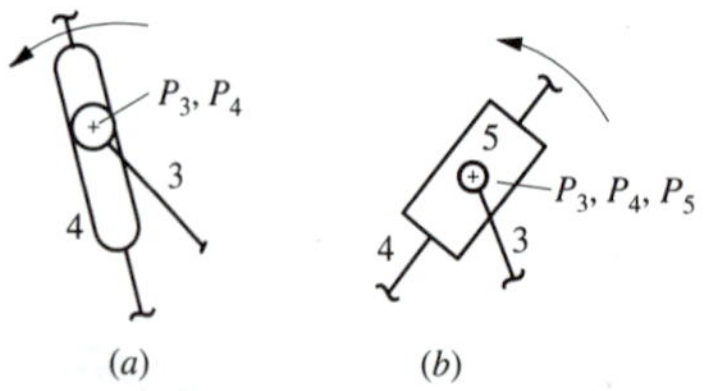

Figure 2.27 Joints that can be analyzed as rotating sliding joints. (a) Pin in straight slot; (b) rotating slider.

Once again, ${}^5\boldsymbol{a}_{P_3/P_4}$ is called the acceleration of P_3 relative to P_4 when the observer is in system 5. The term $2{}^1\boldsymbol{\omega}_5 \times {}^5\boldsymbol{v}_{P_3/P_4}$ is the Coriolis component of acceleration, and it can be written as $\boldsymbol{a}^C_{P_3/P_4}$. Equation (2.44) can then be written as

$${}^1\boldsymbol{a}_{P_3} = {}^1\boldsymbol{a}_{P_4} + {}^5\boldsymbol{a}_{P_3/P_4} + \boldsymbol{a}^C_{P_3/P_4}$$

for graphical analyses.

In planar motion, the Coriolis component is normal to ${}^5\boldsymbol{v}_{P_3/P_4}$ and has the magnitude $2({}^1\omega_5)({}^5v_{P_3/P_4})$. Its sense is obtained by imagining ${}^5\boldsymbol{v}_{P_3/P_4}$ to be rotating about its tail in the ${}^1\boldsymbol{\omega}_5$ direction. The direction of movement of the head of ${}^5\boldsymbol{v}_{P_3/P_4}$ gives the sense. To illustrate the manner in which Eqs. (2.43) and (2.44) are used in graphical linkage analysis, consider the following example.

EXAMPLE 2.7 ***(Velocity and Acceleration Analysis of a Quick-Return Mechanism)***

Find the sliding velocities of the two slides, the angular accelerations of links 3 and 4, and the acceleration of slide 5 for the quick-return mechanism of Fig. 2.28. The dimensions are as shown. Link 2 is driven with a constant angular velocity of 10 rpm CCW.

SOLUTION

Link 2 is the driver, so we will begin the analysis with point A_2. We will conduct the velocity analysis first. If that analysis is done carefully, we can proceed with the same points for the acceleration analysis. As in the previous examples, we will develop the basic equations first and then give the graphical procedure for solving them. The velocity of point A_2 is given by

$${}^1\boldsymbol{v}_{A_2} = {}^1\boldsymbol{v}_{A_2/O_A} = {}^1\boldsymbol{\omega}_2 \times \boldsymbol{r}_{A_2/O_A}$$

In the analysis of mechanisms of this type, it is important to identify the link in which the observer is located. Therefore, the left superscripts will be maintained. We must now use the coincident point A_3 in order to be able to develop an equation relating a point on link 2 to a point on link 3. We can write the relative velocity equation in one of two ways:

$${}^1\boldsymbol{v}_{A_2} = {}^1\boldsymbol{v}_{A_3} + {}^1\boldsymbol{v}_{A_2/A_3} = {}^1\boldsymbol{v}_{A_3} + {}^3\boldsymbol{v}_{A_2/A_3} \qquad \textbf{(2.45)}$$

or

$${}^1\boldsymbol{v}_{A_3} = {}^1\boldsymbol{v}_{A_2} + {}^1\boldsymbol{v}_{A_3/A_2} = {}^1\boldsymbol{v}_{A_2} + {}^2\boldsymbol{v}_{A_3/A_2} \qquad \textbf{(2.46)}$$

To solve the problem, we must be able to recognize the direction of the relative velocity defined in the moving coordinate system. Referring to the mechanism in Fig. 2.28, if the observer is fixed to link 2, it is not possible to identify directly the direction of the velocity ${}^2\boldsymbol{v}_{A_3/A_2}$; however, if the observer is in link 3, it is possible to identify the direction of the velocity ${}^3\boldsymbol{v}_{A_2/A_3}$ because the pin at A is constrained to move along the straight slot in link 3. Therefore, the direction of the velocity ${}^3\boldsymbol{v}_{A_2/A_3}$ must be along the slot. Because we can determine the direction of ${}^3\boldsymbol{v}_{A_2/A_3}$ by inspection, Eq. (2.45) is more useful than Eq. (2.46).

In problems such as this, it is important to identify clearly the links relative to which the velocity directions can be identified. The same links can be used for the subsequent acceleration analysis, and it is usually much easier to visualize velocities than it is to visualize accelerations.

After Eq. (2.45) is solved for the unknowns, ${}^1\boldsymbol{v}_{A_3}$ will be known. Then ${}^1\boldsymbol{v}_{B_3}$ can be found from the velocity image of link 3 using O_B, A_3, and B_3. Knowing ${}^1\boldsymbol{v}_{B_3}$, which is the same as ${}^1\boldsymbol{v}_{B_4}$, we can write the following equation for the velocity of C_4.

$${}^1\boldsymbol{v}_{C_4} = {}^1\boldsymbol{v}_{B_4} + {}^1\boldsymbol{v}_{C_4/B_4} \qquad \textbf{(2.47)}$$

Because the directions of ${}^1\boldsymbol{v}_{C_4}$ and ${}^1\boldsymbol{v}_{C_4/B_4}$ are known, we can solve Eq. (2.47) for the unknowns. After Eqs. (2.45) and (2.47) are solved, we can compute the angular velocities of links 3 and 4 from

$${}^1\boldsymbol{v}_{B_3/O_B} = {}^1\boldsymbol{\omega}_3 \times \boldsymbol{r}_{B_3/O_B}$$

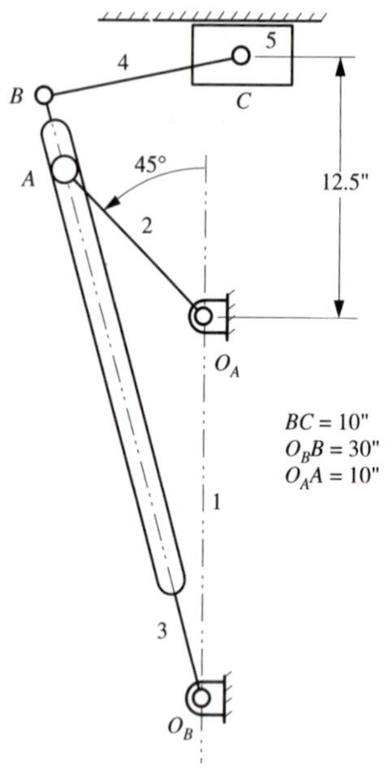

Figure 2.28 The quick-return linkage to be analyzed in Example 2.7.

and

$$^1\boldsymbol{v}_{C_4/B_4} = {}^1\boldsymbol{\omega}_4 \times \boldsymbol{r}_{C_4/B_4}$$

For the acceleration analysis, we need only differentiate Eqs. (2.45) and (2.47). The results are

$$^1\boldsymbol{a}_{A_2} = {}^1\boldsymbol{a}_{A_3} + {}^1\boldsymbol{a}_{A_2/A_3}$$

and

$$^1\boldsymbol{a}_{C_4} = {}^1\boldsymbol{a}_{B_4} + {}^1\boldsymbol{a}_{C_4/B_4}$$

Expanding the equations in terms of vectors relative to moving coordinate systems, we obtain

$$^1\boldsymbol{a}^r_{A_2/O_A} + {}^1\boldsymbol{a}^t_{A_2/O_A} = {}^1\boldsymbol{a}^r_{A_3/O_B} + {}^1\boldsymbol{a}^t_{A_3/O_B} + {}^3\boldsymbol{a}_{A_2/A_3} + 2({}^1\boldsymbol{\omega}_3 \times {}^3\boldsymbol{v}_{A_2/A_3}) \quad \textbf{(2.48)}$$

and

$$^1\boldsymbol{a}_{C_4} = {}^1\boldsymbol{a}_{B_3/O_B} + {}^1\boldsymbol{a}^r_{C_4/B_4} + {}^1\boldsymbol{a}^t_{C_4/B_4} \quad \textbf{(2.49)}$$

where

$^1\boldsymbol{a}^r_{A_2/O_A} = r_{A_2/O_A}\omega_2^2$ from A to O_A
$^1\boldsymbol{a}^t_{A_2/O_A} = {}^1\boldsymbol{\alpha}_2 \times \boldsymbol{r}_{A_2/O_A}$ perpendicular to $\overrightarrow{AO}_A$
$^1\boldsymbol{a}^r_{A_3/O_B} = r_{A_3/O_B}\omega_3^2$ from A to O_B
$^1\boldsymbol{a}^t_{A_3/O_B} = {}^1\boldsymbol{\alpha}_3 \times \boldsymbol{r}_{A_3/O_B}$. ${}^1\boldsymbol{\alpha}_3$ is unknown but the result is perpendicular to $\overrightarrow{AO}_B$
$^3\boldsymbol{a}_{A_2/A_3}$ magnitude is unknown but direction is along slot in link 3
$2({}^1\boldsymbol{\omega}_3 \times {}^3\boldsymbol{v}_{A_2/A_3})$ Coriolis acceleration perpendicular to ${}^3\boldsymbol{v}_{A_2/A_3}$
$^1\boldsymbol{a}_{C_4}$ along the slider path of link 5
$^1\boldsymbol{a}_{B_3/O_B}$ found by acceleration image of link 3
$^1\boldsymbol{a}^r_{C_4/B_4} = r_{C_4/B_4}\omega_4^2$ from C to B
$^1\boldsymbol{a}^t_{C_4/B_4} = {}^1\boldsymbol{\alpha}_4 \times \boldsymbol{r}_{C_4/B_4}$ magnitude is unknown but direction is perpendicular to $\overrightarrow{CB}$.

Based on the position and velocity analyses, there will be only two unknown magnitudes in Eqs. (2.48) and (2.49). All of the direction will be known. Therefore, the equations can be solved.

STEPS

1. Draw linkage to scale as shown in Fig. 2.29.
2. Compute $\boldsymbol{v}_{A_2}$, and plot $\boldsymbol{v}_{A_2}$ normal to $\overrightarrow{O_A A}$ as $\overrightarrow{oa_2}$.

$$^1\omega_2 = 10 \times 2\pi/60 = 1.0472 \text{ rad/s CCW}$$

$$v_{A_2} = 10 \times 1.0472 = 10.472 \text{ in/s}$$

3. Draw a line through a_2 parallel to $\overrightarrow{O_B B}$.
4. Draw a line through o normal to $\overrightarrow{O_B A_3}$. The intersection with the line from step 3 gives point a_3.
5. Locate b_3 using the velocity image $\dfrac{ob_3}{oa_3} = \dfrac{O_B B}{O_B A}$.

$$ob_3 = 7.26 \times (30/26.48) = 8.22 \text{ in/s}$$

6. Draw a line through b_3 normal to $\overrightarrow{BC}$.
7. Draw a line through o parallel to the slide. Its intersection with the line drawn in step 6 gives point c_4.
8. Measure $v_{C_4} = \overrightarrow{oc_4}$, $v_{A_2/A_3} = \overrightarrow{a_2a_3}$, and $v_{C_4/B_4} = \overrightarrow{c_4b_4}$.

$$v_{C_4} = 11.06 \text{ in/s}, \qquad v_{A_2/A_3} = 5.09 \text{ in/s}, \qquad v_{C_4/B_4} = 2.823 \text{ in/s}$$

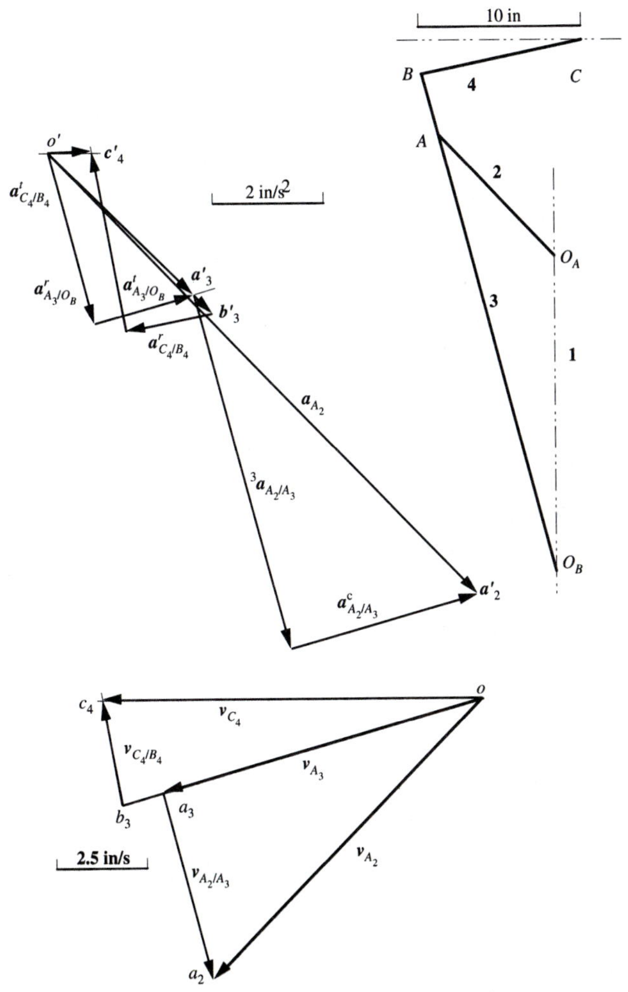

Figure 2.29 Solution of Example 2.7.

9. Compute ${}^1\boldsymbol{\omega}_3 = v_{A_3/O_B}/O_BA_3$ and ${}^1\omega_4 = v_{C_4/B_4}/BC$.

$${}^1\boldsymbol{\omega}_3 = (9.07)/(26.37) = 0.344 \text{ rad/s CCW.}$$

$${}^1\boldsymbol{\omega}_4 = 2.823/5 = 0.565 \text{ rad/s CCW.}$$

10. Get the senses of ${}^1\boldsymbol{\omega}_3$ and ${}^1\boldsymbol{\omega}_4$ by looking at the directions of rotation of $\boldsymbol{r}_{A_3/O_B}$ and $\boldsymbol{r}_{C_4/B_4}$ needed to get the respective relative velocity directions.

11. Compute $\boldsymbol{a}_{A_2}$ and plot it as $\overrightarrow{o'a_2'}$.

$$a_{A_2} = 10 \times 1.0472^2 = 10.97 \text{ in/s}^2$$

12. Compute $a^C_{A_2/A_3} = 2({}^1\omega_3)(v_{A_2/A_3})$ and get the sense of $\boldsymbol{a}^C_{A_2/A_3}$ by rotating $\boldsymbol{v}_{A_2/A_3}$ 90° in the ${}^1\boldsymbol{\omega}_3$ direction. Plot it with the tip at a_2'.

$$a^C_{A_2/A_3} = 2 \times 5.09 \times 0.344 = 3.50 \text{ in/s}^2$$

13. Draw a line normal to $\boldsymbol{a}^C_{A_2/A_3}$ and through the tail of $\boldsymbol{a}^C_{A_2/A_3}$. This line corresponds to ${}^3\boldsymbol{a}_{A_2/A_3}$, which is along the slot.

14. Compute ${}^1\boldsymbol{a}^r_{A_3/O_B}$ and plot it from o' in the $\overrightarrow{A_3O_B}$ direction.

$${}^1a^r_{A_3/O_B} = 26.37 \times 0.344^2 = 3.12 \text{ in/s}^2$$

15. Draw a line through the tip of ${}^1\boldsymbol{a}^r_{A_3/O_B}$ normal to $\overrightarrow{O_BA_3}$. This vector corresponds to ${}^1\boldsymbol{a}^t_{A_3/O_B}$. Its intersection with the line drawn in step 13 gives point a_3'.

16. Locate point b_3' using the acceleration image $\dfrac{o'b_3'}{o'a_3'} = \dfrac{O_BB}{O_BA}$.

$$o'b_3' = 3.62 \times (30/26.37) = 4.12 \text{ in/s}^2$$

17. Plot ${}^1\boldsymbol{a}^r_{C_4/B_4}$ from point b_3' parallel to $\overrightarrow{CB}$.

$${}^1a^r_{C_4/B_4} = 5 \times 0.608^2 = 1.85 \text{ in/s}^2$$

18. Draw a line through the tip of $\boldsymbol{a}^r_{C_4/B_4}$ normal to $\overrightarrow{CB}$. This vector corresponds to $\boldsymbol{a}^t_{C_4/B_4}$.

19. Draw a line through o' parallel to the slide. Its intersection with the line generated in step 18 gives point c_4'.

20. Measure ${}^1a^t_{A_3/O_B}$ and ${}^1a^t_{C_4/B_4}$.

$${}^1a^t_{A_3/O_B} = 1.870 \text{ in/s}^2 \quad \text{and} \quad {}^1a^t_{C_4/B_4} = 3.197 \text{ in/s}^2$$

21. Compute ${}^1\boldsymbol{\alpha}_3 = \dfrac{{}^1a^t_{A_3/O_B}}{r_{A_3/O_B}}$ and ${}^1\boldsymbol{\alpha}_4 = \dfrac{{}^1a^t_{C_4/B_4}}{r_{C_4/B_4}}$ and get the senses of these angular accelerations by considering the directions of rotation needed to rotate the position vectors in the directions of ${}^1\boldsymbol{a}^t_{A_3/O_B}$ and ${}^1\boldsymbol{a}^t_{C_4/B_4}$, respectively.

$${}^1\alpha_3 = 1.870/(26.37) = 0.071 \text{ rad/s}^2 \text{ CW}$$

$${}^1\alpha_4 = 3.197/5 = 0.640 \text{ rad/s}^2 \text{ CCW}$$

22. Measure a_{C_4}.

$$a_{C_4} = 0.7986 \text{ in/s}^2 \text{ directed to the right}$$

One of the useful features of the quick-return linkage is a long range of motion with relatively uniform velocity on the forward stroke. The small value of a_{C_4} is indicative of this property. ■

2.14 ROLLING CONTACT

Rolling contact is quite often used in practical linkages. In addition to the obvious case of a wheel rolling on a surface or a rail, rolling contact between a cam and a roller follower is a common example. Also, the pitch cylinders of spur and helical gear pairs and the pitch cones of bevel gear pairs can be considered to be in pure rolling contact. In that case, although the actual physical contact between the gear teeth is a general combination of rolling and sliding, the gear pair can be modeled as a pair of simple elements in pure rolling contact from the point of view of investigating gross kinematic properties.

Rolling contact can be approached in two different ways, depending on the level of detail desired. If the velocities and accelerations of the rolling elements themselves are immaterial, it is possible to solve for the velocities and accelerations of the other links in a rolling-contact problem by replacing the actual linkage with a virtual linkage in which the rolling elements are replaced by a single link with length equal to the sum of their radii of curvature. If the velocities and accelerations of all the links are important, then one or more additional relative velocity (or angular velocity) relations are necessary to obtain the angular velocities of one or more rolling links. Both approaches will be discussed.

2.14.1 Basic Kinematic Relationships for Rolling Contact

Figure 2.30 shows two rigid bodies in rolling contact. The bodies are arbitrarily taken as links 2 and 4. The contact location is B, and the centers of curvature of the two bodies corresponding to B_2 and B_4 are O_2 and O_4, respectively. At the point of contact for two bodies rolling on each other, there is no relative sliding between the two points (B_2 and B_4) at the location of contact. Because B_2 and B_4 are not only momentarily coincident but also in rolling contact, they have the same velocity, and their relative velocity must be zero. This means that

$$^1\boldsymbol{v}_{B_2} = {}^1\boldsymbol{v}_{B_4}$$

and

$$^1\boldsymbol{v}_{B_2/B_4} = {}^4\boldsymbol{v}_{B_2/B_4} = {}^4\boldsymbol{v}_{B_2} = 0$$

Note that this is exactly the same velocity condition as that for a revolute joint. Therefore, for velocities **ONLY,** the point of rolling contact can be treated as a revolute joint. However, this is not true for accelerations.

The relative acceleration $^1\boldsymbol{a}_{B_2/B_4}$ is not usually zero, but it is independent of coordinate system. Therefore,

$$^1\boldsymbol{a}_{B_2/B_4} = {}^4\boldsymbol{a}_{B_2/B_4} = {}^4\boldsymbol{a}_{B_2} \tag{2.50}$$

From Eq. (2.50), it is apparent that the direction of $^1\boldsymbol{a}_{B_2/B_4}$ is the same as the direction of $^4\boldsymbol{a}_{B_2}$, which is the absolute acceleration of point B_2 observed from link 4. Therefore, it is useful to determine the path that B_2 traces on 4 (or B_4 traces on 2) to determine the

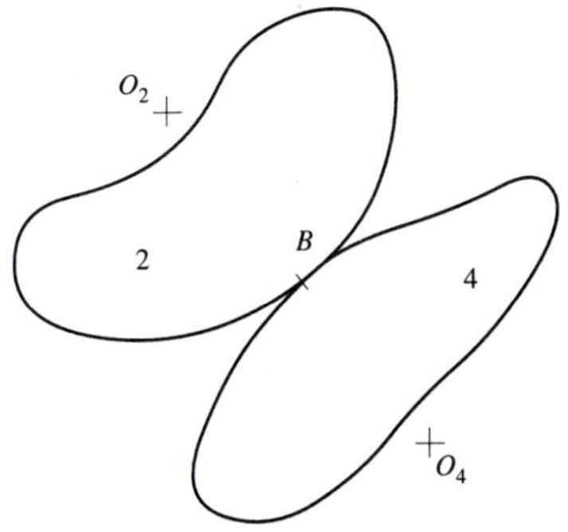

Figure 2.30 Two links in rolling contact.

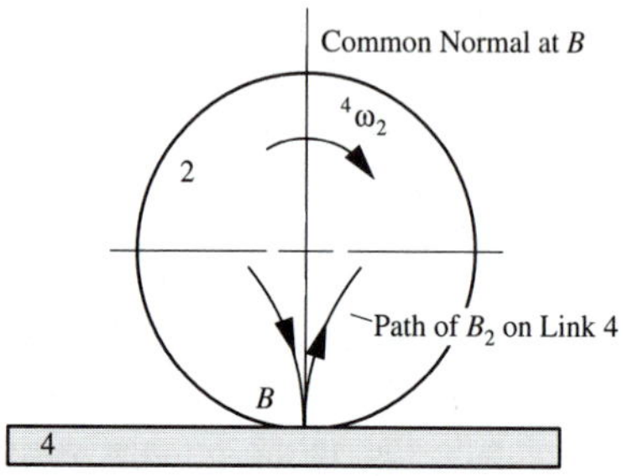

Figure 2.31 Path of motion of B_2 relative to link 4.

direction of the acceleration of ${}^4\boldsymbol{a}_{B_2}$. To do this, first imagine that link 2 is a circle and link 4 is a straight line. From experience (for example, from looking at a bicycle tire reflector at night) we know that the path of B_2 will look as shown in Fig. 2.31. That is, the path forms a cusp at the contact location. The cusp will approach the contact point in a direction which is tangent to the common normal at the contact point, and the cusp will also leave the contact point in a direction that is tangent to the common normal. This means that the acceleration must be along the common normal at the point of contact. The same kind of relationship also applies for general bodies.

To conduct an acceleration analysis of mechanisms involving rolling contact, it is necessary to determine both the magnitude and direction of the relative acceleration between the two contact points. Because we know that the direction of the relative accelerations will be along the common normal at the point of contact, we need only determine the magnitude.

To do this, first consider a general rigid body *R*. If the contour of the rigid body is known, which must be the case for a kinematic analysis, at any given point on the contour, the center of curvature, O_R, for that body can be found. If a circle of radius O_RB is drawn, that circle will be tangent to the contour at *B*, and it will share three points (separated by infinitesimal distances) with the curve *R*. This circle is called the osculating circle to the curve at point *B*, and the circle is a unique property of the curve for the point considered. An example is shown in Fig. 2.32.

If we consider two general links (2 and 4) in rolling contact, we can draw the osculating circle for each curve. As the two bodies roll together, the three points shared by the osculating circles will be in contact with each other. Therefore, for two differentially separated time periods, the curves could be replaced by their osculating circles. Because only two differentially separated time periods must be considered for accelerations, we can replace the original curves with their osculating circles, and the kinematic results for position, velocity, and acceleration will remain unchanged. If higher time derivatives than accelerations are desired, however, we cannot replace the original surfaces with their

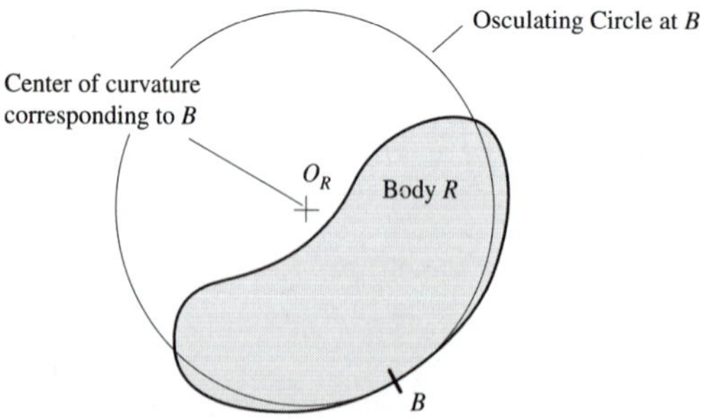

Figure 2.32 Osculating circle.

osculating circles. Obviously, a different osculating circle may be required for each contact position if the surface of body R is totally general. However, if a kinematic analysis is to be conducted, the geometry of the surface must be known, and therefore the osculating circle corresponding to each point on the periphery can be identified.

The replacement of the general surface with osculating circles is extremely useful in kinematics whenever higher pairs are involved. The special properties of circles make it relatively simple to analyze linkages with rolling and cam joints.

Because we can replace the two curves with their osculating circles, we can connect the two centers of curvature by a virtual link pinned to the two bodies at the centers of curvature, and the two bodies can still move relative to each other. This is precisely the condition existing when two gears in a standard transmission are meshed. For the sake of discussion, let the two bodies again be links 2 and 4 and the virtual link be designated as x as shown in Fig. 2.33. With this arrangement, we are now in a position to compute the relative acceleration ${}^4\boldsymbol{a}_{B_2/B_4}$. To do this, we will use Eq. (2.50) and compute ${}^1\boldsymbol{a}_{B_2/B_4}$, which is equal to ${}^4\boldsymbol{a}_{B_2/B_4}$.

As with any planar vector, the acceleration ${}^1\boldsymbol{a}_{B_2/B_4}$ can be resolved into two orthogonal components. It is convenient to resolve the vector into one component along the common normal to the two curves at B and another along the common tangent. That is,

$$ {}^1\boldsymbol{a}_{B_2/B_4} = {}^1\boldsymbol{a}^n_{B_2/B_4} + {}^1\boldsymbol{a}^t_{B_2/B_4} $$

However, we know from the discussion before that the relative acceleration must lie along the common normal. Therefore, the tangential component must be zero, and the *total* relative acceleration between B_2 and B_4 can be represented as

$$ {}^1\boldsymbol{a}_{B_2/B_4} = {}^1\boldsymbol{a}^n_{B_2/B_4} $$

We can compute the normal acceleration by writing the relative accelerations among the points B_2, B_4, O_2, and O_4, that is,

$$ {}^1\boldsymbol{a}_{B_2/B_4} = {}^1\boldsymbol{a}^n_{B_2/O_2} + {}^1\boldsymbol{a}^n_{(O_2)_x/(O_4)_x} + {}^1\boldsymbol{a}^n_{O_4/B_4} $$

Now consider individually each term on the right-hand side of the equation. Each term will be a function of velocities and can be computed in a variety of ways. For example,

$$ {}^1\boldsymbol{a}^n_{B_2/O_2} = -{}^1\omega_2^2(\boldsymbol{r}_{B/O_2}) = {}^1\boldsymbol{\omega}_2 \times {}^1\boldsymbol{v}_{B_2/O_2} = \frac{|{}^1\boldsymbol{v}_{B_2/O_2}|^2}{|\boldsymbol{r}_{B/O_2}|} \quad \text{(from } B \text{ to } O_2\text{)} $$

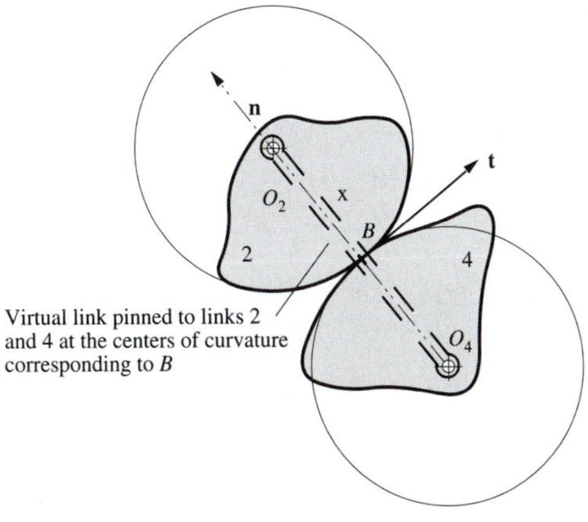

Figure 2.33 Virtual link pinned at the centers of curvature of the two bodies in rolling contact.

and

$$ {}^1\boldsymbol{a}^n_{O_4/B_4} = -{}^1\omega_4^2(\boldsymbol{r}_{O_4/B}) = {}^1\boldsymbol{\omega}_4 \times {}^1\boldsymbol{v}_{O_4/B_4} = \frac{|{}^1\boldsymbol{v}_{O_4/B_4}|^2}{|\boldsymbol{r}_{O_4/B}|} \quad \text{(from } O_4 \text{ to } B\text{)} $$

Similarly,

$$ {}^1\boldsymbol{a}^n_{(O_2)_x/(O_4)_x} = -{}^1\omega_x^2(\boldsymbol{r}_{O_4/O_2}) = {}^1\boldsymbol{\omega}_x \times {}^1\boldsymbol{v}_{O_4/O_2} = \frac{|{}^1\boldsymbol{v}_{O_4/O_2}|^2}{|\boldsymbol{r}_{O_4/O_2}|} \quad \text{(from } O_2 \text{ to } O_4\text{)} \tag{2.51} $$

To evaluate the first two expressions on the right-hand side of Eq. (2.51), we need to develop an expression for ${}^1\boldsymbol{\omega}_x$. To do this, we can derive relative velocity expressions among B_2, B_4, O_2, and O_4. Considering links 2, 4, and x,

$$ \boldsymbol{v}_{B_2} = \boldsymbol{v}_{O_2} + \boldsymbol{v}_{B_2/O_2} = \boldsymbol{v}_{O_2} + {}^1\boldsymbol{\omega}_2 \times \boldsymbol{r}_{B/O_2} $$

$$ \boldsymbol{v}_{B_4} = \boldsymbol{v}_{O_4} + \boldsymbol{v}_{B_4/O_4} = \boldsymbol{v}_{O_4} + {}^1\boldsymbol{\omega}_4 \times \boldsymbol{r}_{B/O_4} $$

$$ \boldsymbol{v}_{O_2} = \boldsymbol{v}_{O_4} + \boldsymbol{v}_{O_2/O_4} = \boldsymbol{v}_{O_4} + {}^1\boldsymbol{\omega}_x \times \boldsymbol{r}_{O_2/O_4} $$

Combining these equations and recognizing that $\boldsymbol{v}_{B_2} = \boldsymbol{v}_{B_4}$, we get

$$ {}^1\boldsymbol{\omega}_x \times \boldsymbol{r}_{O_2/O_4} = \boldsymbol{v}_{O_2} - \boldsymbol{v}_{O_4} = \boldsymbol{v}_{O_2/O_4} = {}^1\boldsymbol{\omega}_4 \times \boldsymbol{r}_{B/O_4} - {}^1\boldsymbol{\omega}_2 \times \boldsymbol{r}_{B/O_2} \tag{2.52} $$

Recognizing that $\boldsymbol{r}_{B/O_2} = -\boldsymbol{r}_{O_2/B}$, Eq. (2.52) can also be written as

$$ {}^1\boldsymbol{\omega}_x \times \boldsymbol{r}_{O_2/O_4} = {}^1\boldsymbol{\omega}_4 \times \boldsymbol{r}_{B/O_4} + {}^1\boldsymbol{\omega}_2 \times \boldsymbol{r}_{O_2/B} \tag{2.53} $$

The magnitude of ${}^1\boldsymbol{\omega}_x$ is given by

$$ {}^1\omega_x = \frac{|{}^1\boldsymbol{\omega}_4 \times \boldsymbol{r}_{B/O_4} + {}^1\boldsymbol{\omega}_2 \times \boldsymbol{r}_{O_2/B}|}{|\boldsymbol{r}_{O_2/O_4}|} = \frac{|{}^1\boldsymbol{\omega}_2 \times \boldsymbol{r}_{O_2/B} + {}^1\boldsymbol{\omega}_4 \times \boldsymbol{r}_{B/O_4}|}{|\boldsymbol{r}_{O_2/B} + \boldsymbol{r}_{B/O_4}|} $$

If the direction is of interest, it can be determined from the vectors in Eq. (2.53).

To summarize, in rolling-contact problems, we know that the two contact points (e.g., B_2 and B_4) have the same velocity. Also, given the acceleration of one of the points, say B_4, the acceleration of the other point can be computed from

$$ {}^1\boldsymbol{a}_{B_2} = {}^1\boldsymbol{a}_{B_4} + {}^1\boldsymbol{a}^n_{B_2/B_4} $$

where ${}^1\boldsymbol{a}^n_{B_2/B_4}$ can be computed using *any* of the following:

$$ {}^1\boldsymbol{a}^n_{B_2/B_4} = \underset{(B \text{ to } O_2)}{\frac{|{}^1\boldsymbol{v}_{B_2/O_2}|^2}{|\boldsymbol{r}_{B_2/O_2}|}} + \underset{(O_2 \text{ to } O_4)}{\frac{|{}^1\boldsymbol{v}_{O_2/O_4}|^2}{|\boldsymbol{r}_{O_2/O_4}|}} + \underset{(O_4 \text{ to } B)}{\frac{|{}^1\boldsymbol{v}_{O_4/B_4}|^2}{|\boldsymbol{r}_{O_4/B_4}|}} \tag{2.54} $$

or

$$ {}^1\boldsymbol{a}^n_{B_2/B_4} = \underset{(B \text{ to } O_2)}{-{}^1\omega_2^2(\boldsymbol{r}_{B_2/O_2})} - \underset{(O_2 \text{ to } O_4)}{{}^1\omega_x^2(\boldsymbol{r}_{O_2/O_4})} - \underset{(O_4 \text{ to } B)}{{}^1\omega_4^2(\boldsymbol{r}_{O_4/B_4})} \tag{2.55} $$

or

$$ {}^1\boldsymbol{a}^n_{B_2/B_4} = \underset{(B \text{ to } O_2)}{{}^1\boldsymbol{\omega}_2 \times {}^1\boldsymbol{v}_{B_2/O_2}} + \underset{(O_2 \text{ to } O_4)}{{}^1\boldsymbol{\omega}_x \times {}^1\boldsymbol{v}_{O_4/O_2}} + \underset{(O_4 \text{ to } B)}{{}^1\boldsymbol{\omega}_4 \times {}^1\boldsymbol{v}_{O_4/B_4}} \tag{2.56} $$

If one of the rolling surfaces is flat, the center of curvature will approach infinity, and the corresponding acceleration term will become zero. For example, if the rolling surface for link 2 is flat, then O_2 is at infinity, and the acceleration expressions reduces to

$$ {}^1\boldsymbol{a}^n_{B_2/B_4} = \frac{|{}^1\boldsymbol{v}_{O_4/B_4}|^2}{|\boldsymbol{r}_{O_4/B_4}|} = -{}^1\omega_4^2(\boldsymbol{r}_{O_4/B}) = {}^1\boldsymbol{\omega}_4 \times {}^1\boldsymbol{v}_{O_4/B_4}) \quad \text{from } (O_4 \text{ to } B) $$

EXAMPLE 2.8 (*Analysis of Linkage with a Rolling-Contact Joint*)

In the linkage shown in Fig. 2.34, link 4 is a gear, pivoted at O_B. Link 3 is a gear meshing with 4 and has a lever fixed to it that is hinged to link 2 at A. Link 2 is driven at constant angular velocity 10 rad/s CCW. Find the angular acceleration of gear 4.

In this instance we cannot ignore the acceleration of either of the two contacting bodies. The angular acceleration of gear 3 is the same as that of arm AQ to which it is rigidly affixed. The angular acceleration of gear 4 is the quantity to be found.

For the velocity analysis, the equations to be solved are

$$\boldsymbol{v}_{A_2/O_A} = {}^1\boldsymbol{\omega}_2 \times \boldsymbol{r}_{A_2/O_A}$$

$$\boldsymbol{v}_{P_4/O_B} = \boldsymbol{v}_{A_3} + \boldsymbol{v}_{P_3/A_3}$$

For the acceleration analysis, the corresponding equations are

$$\boldsymbol{a}_{A_2/O_A} = \boldsymbol{a}^r_{A_2/O_A} + \boldsymbol{a}^t_{A_2/O_A}$$

$$\boldsymbol{a}^r_{P_4/O_B} + \boldsymbol{a}^t_{P_4/O_B} = \boldsymbol{a}_{A_2/O_A} + \boldsymbol{a}^r_{P_3/A_3} + \boldsymbol{a}^t_{P_3/A_3} + \boldsymbol{a}^n_{P_4/P_3}$$

Here the unknowns are the magnitudes of the two transverse components, $\boldsymbol{a}^t_{P_3/A_3}$ and $\boldsymbol{a}^t_{P_4/O_B}$.

STEPS

1. Draw the linkage to scale as shown in Fig. 2.35. To do this, first draw link 2 and locate point A. Next find the center Q, knowing that it is on a circle of radius AQ centered at A and also on a circle of radius QO_B centered at O_B. After locating Q, draw the line AQ, and the circles corresponding to the pitch circles of the two gears.
2. Compute $\boldsymbol{v}_{A_2}$ and plot as $\overrightarrow{oa_2}$.

$$v_{A_2} = 1 \times 10 = 10 \text{ in/s}, \qquad \boldsymbol{v}_{A_2} \text{ is normal to } \overrightarrow{O_A A}$$

3. Draw a line through o normal to $\overrightarrow{O_B P}$.
4. Draw line $\overrightarrow{AP}$.
5. Draw a line through a_2, a_3 normal to $\overrightarrow{AP}$. The intersection of this line with the line generated in step 3 gives the point P. Notice that since the bar $\overrightarrow{AQ}$ is rigidly fixed to gear 3, line $\overrightarrow{AP}$ is fixed in member 3. Although at any instant, the point at P is fixed to member 3, a different P (and different point) is involved for each position of the linkage.
6. Draw a line through a_3 normal to $\overrightarrow{AQ}$. The intersection of this line with the line generated in step 3 gives the point q_3. Point q_3 could also have been located from points a_3 and p_3 by using the velocity image theorem.

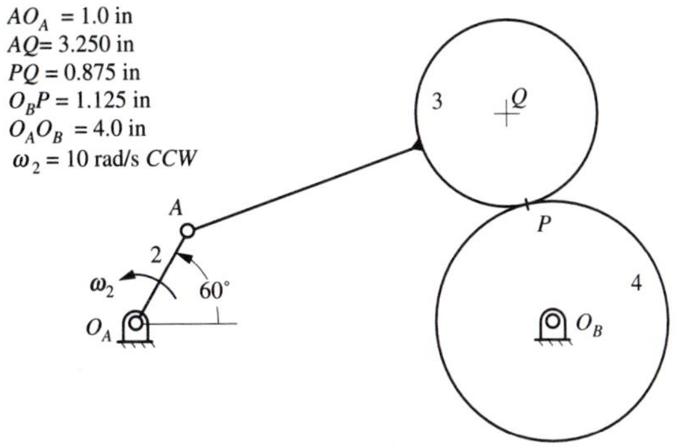

Figure 2.34 The linkage of Example 2.8. This is an example of a geared five-bar linkage. Geared five-bar and six-bar linkages are used quite frequently as alternatives to four-bar linkages. They allow more flexibility in synthesis than four-bar linkages because there are more dimensions that can be varied.

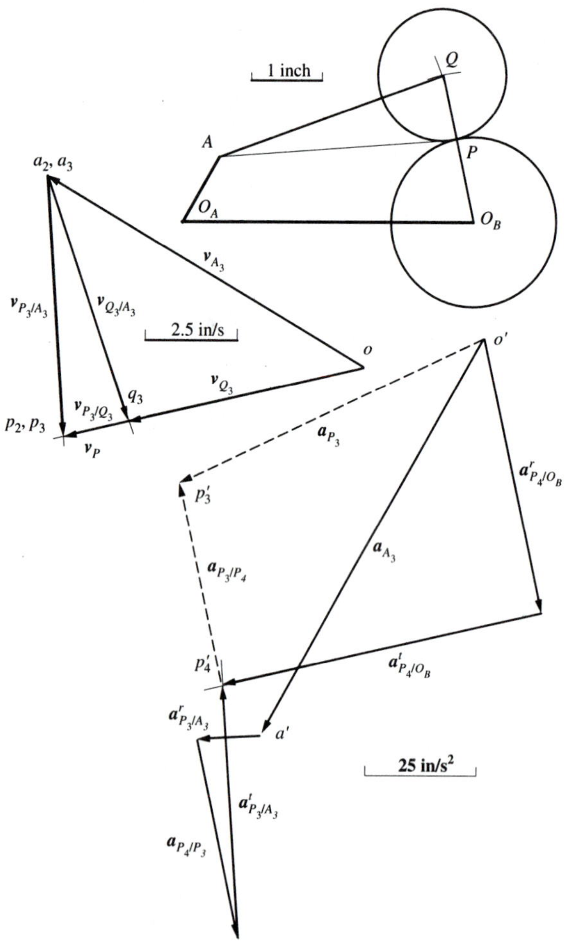

Figure 2.35 The graphical solution of Example 2.8.

7. Compute ${}^1\omega_3 = v_{P_3/A_3}/AP$ and ${}^1\omega_4 = v_{P_4/O_B}/PO_B$ and find the senses needed to give the directions of $\mathbf{v}_{P_3/A_3}$ and $\mathbf{v}_{P_4/O_B}$. Because of the pure rolling condition of the pitch circles of the gears, the velocity of point P is the same regardless of whether it is considered to be in member 3 or member 4.

$${}^1\boldsymbol{\omega}_3 = 6.85/3.258 = 2.10 \text{ rad/s CW}$$

$${}^1\boldsymbol{\omega}_4 = 8.32/1.125 = 7.39 \text{ rad/s CCW}$$

8. Compute $\boldsymbol{a}_{A_3} = \boldsymbol{a}^r_{A_3/O_A}$ and plot as $\overrightarrow{o'a'}$.

$$a_{A_3} = 1 \times 10^2 = 100 \text{ in/s}^2$$

9. Compute and plot $\boldsymbol{a}^r_{P_3/A_3}$ in the $\overrightarrow{PA}$ direction from point a'.

$$a^r_{P_3/A_3} = AP \times {}^1\omega_3^2 = 3.258 \times 2.10^2 = 14.39 \text{ in/s}^2$$

10. Compute $\boldsymbol{a}_{P_4/P_3}$ using the equation form given in Eq. (2.54). Then,

$$\boldsymbol{a}^n_{P_4/P_3} = \underset{(P \text{ to } O_B)}{\frac{|{}^1\mathbf{v}_{P_4/O_B}|^2}{|\mathbf{r}_{P_4/O_B}|}} + \underset{(O_B \text{ to } Q_3)}{\frac{|{}^1\mathbf{v}_{O_B/Q_3}|^2}{|\mathbf{r}_{O_B/Q_3}|}} + \underset{(Q_3 \text{ to } P_3)}{\frac{|{}^1\mathbf{v}_{Q_3/P_3}|^2}{|\mathbf{r}_{Q_3/P_3}|}} = \underset{(P \text{ to } O_B)}{\frac{|8.32|^2}{1.125}} + \underset{(O_B \text{ to } Q_3)}{\frac{|6.54|^2}{2}} + \underset{(Q_3 \text{ to } P_3)}{\frac{|1.92|^2}{0.875}}$$

Arbitrarily taking direction $\overrightarrow{PO_B}$ as positive, the signs of the individual terms can be identified. Then,

$$a_{P_4/P_3} = 61.53 - 21.38 + 4.21 = 44.36 \text{ in/s}^2$$

Note also that $\boldsymbol{a}_{P_4/P_3} = -\boldsymbol{a}_{P_3/P_4}$ and has direction from the center of wheel 3 toward the center of wheel 4. Plot $\boldsymbol{a}_{P_4/P_3}$ from the tip of $\boldsymbol{a}^r_{P_3/A_3}$.

11. Draw a line through the tip of vector $\boldsymbol{a}_{P_4/P_3}$ normal to $\overrightarrow{AP}$.

12. Compute $a^r_{P_4/O_B}$ and plot $\boldsymbol{a}^r_{P_4/O_B}$ from o' in the $\overrightarrow{PO_B}$ direction.

$$|\boldsymbol{a}^r_{P_4/O_B}| = |\boldsymbol{r}_{P_4/O_B} \times {}^1\boldsymbol{\omega}_4^2| = 1.125 \times 7.39^2 = 61.4 \text{ in/s}^2$$

13. Draw a line through the tip of vector $\boldsymbol{a}^r_{P_4/O_B}$ normal to $\overrightarrow{O_BP}$. The intersection of this line with that drawn in step 11 gives p_4'.

14. Compute ${}^1\alpha_4 = a^t_{P_4/O_B}/r_{P_4/O_B}$ and find the sense needed to give the direction of $\boldsymbol{a}^t_{P_4/O_B}$.

$${}^1\boldsymbol{\alpha}_4 = 74.25/1.125 = 66.0 \text{ rad/s}^2 \text{ CCW}$$

Note that this construction, with the vectors in the order shown, gives the correct position for p_4' but not for p_3'. This does not matter for the present purpose. However, if the correct position of p_3' were required, either to get the absolute acceleration or if the acceleration image was to be constructed, it would be obtained by plotting $\boldsymbol{a}_{P_3/P_4}$ from p_4' as shown in Fig. 2.35. ■

Although the acceleration of the contacting point in one body relative to that in the other has been worked out assuming circular contacting profiles, it can also be used if the profiles are not circular. The radius of the circular profile is simply replaced with the osculating circle of the profile at the point of contact.

2.14.2 Modeling Rolling Contact Using a Virtual Linkage

As a second example of rolling contact, we will consider the plate cam with roller follower shown in Fig. 2.36. In this mechanism, we are given the angular velocity and acceleration of link 2, and we wish only to know the angular velocity and acceleration of link 4. We are not interested in the velocity and acceleration of link 3. When this is the case, we can model the linkage with a virtual link between the centers of curvature of links 2 and 3

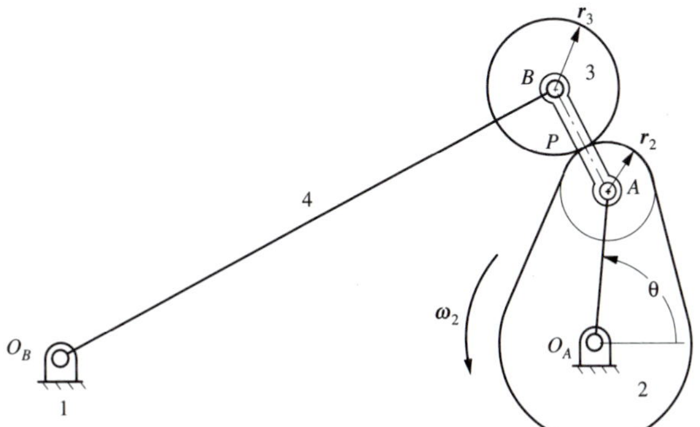

Figure 2.36 A plate cam with roller follower. For given angular velocity and acceleration of the cam, 2, the angular velocity and angular acceleration of the arm 4 can be found by replacing the linkage with the virtual four-bar O_AABO_B. Here point A is the center of curvature of the cam profile at the contact point P.

corresponding to the contact point P. The cam-follower mechanism can then be analyzed as the virtual four-bar linkage O_AABO_B. Line O_AA is fixed to link 2, so the angular velocity and acceleration of O_AA will be the angular velocity and acceleration for link 2.

EXAMPLE 2.9 (*Analysis of a Geared Linkage—Rolling Contact*)

Find the angular velocity and angular acceleration of the arm (link 4) of the linkage in Fig. 2.36 for $\boldsymbol{\theta} = 90°$ when the cam, 2, is rotating CCW with constant angular velocity 1000 rpm. The following basic dimensions are given for the mechanism:

$$O_AO_B = 4.0 \text{ in}; \quad O_BB = 4.25 \text{ in}; \quad r_2 = 0.5 \text{ in}; \quad r_3 = 2.5 \text{ in};$$
$$O_AA = 1.153 \text{ in}; \quad AB = 0.901 \text{ in}$$

The steps to analyzing this mechanism are the same as those required for the four-bar linkage in Example 2.1. The velocity equations that must be solved graphically are

$$\boldsymbol{v}_{A_2/O_A} = {}^1\boldsymbol{\omega}_2 \times \boldsymbol{r}_{A_2/O_A}$$

$$\boldsymbol{v}_{B_4/O_B} = \boldsymbol{v}_{A_2/O_A} + \boldsymbol{v}_{B_5/A_5}$$

where the virtual link is designated as link 5. The acceleration equations that must be solved are

$$\boldsymbol{a}_{A_2/O_A} = \boldsymbol{a}^r_{A_2/O_A} + \boldsymbol{a}^t_{A_2/O_A}$$

$$\boldsymbol{a}^r_{B_4/O_B} + \boldsymbol{a}^t_{B_4/O_B} = \boldsymbol{a}^r_{A_2/O_A} + \boldsymbol{a}^t_{A_2/O_A} + \boldsymbol{a}^r_{B_5/A_5} + \boldsymbol{a}^t_{B_5/A_5}$$

STEPS

1. Draw the mechanism to scale as shown in Fig. 2.37. Note that, for this analysis, we need draw only the virtual mechanism.
2. Compute ${}^1\omega_2$.

$${}^1\omega_2 = 1000 \times 2\pi/60 = 104.72 \text{ rad/s}$$

3. Compute and plot $\boldsymbol{v}_{A_2/O_A}$.

$$v_{A_2/O_A} = 1.153 \times 104.72 = 120.74 \text{ in/s}; \qquad \boldsymbol{v}_{A_2/O_A} \text{ is normal to } O_AA$$

4. Solve the velocity equation graphically and measure $\boldsymbol{v}_{B_4/O_B}$ and $\boldsymbol{v}_{B_5/A_5}$.

$$v_{B_5/O_B} = 38.95 \text{ in/s}; \qquad v_{B_5/A_5} = 108.2 \text{ in/s}$$

5. Compute ${}^1\boldsymbol{\omega}_5$ and ${}^1\boldsymbol{\omega}_4$ and determine their senses:

$${}^1\boldsymbol{\omega}_5 = \boldsymbol{v}_{B_5/A_5}/AB = 108.2/0.901 = 120.1 \text{ rad/s CW}$$

$${}^1\boldsymbol{\omega}_4 = \boldsymbol{v}_{B_4/O_B}/O_BB = 38.95/4.25 = 9.16 \text{ rad/s CCW}$$

Notice that ${}^1\boldsymbol{\omega}_5$ is the angular velocity of a *virtual* link containing the line AB.

6. Compute and plot $\boldsymbol{a}_{A_2/O_A} = \boldsymbol{a}^r_{A_2/O_A}$:

$$a^r_{A_2/O_A} = O_AA \times {}^1\omega_2^2 = 1.153 \times 104.7^2 = 12{,}640 \text{ in/s}^2; \boldsymbol{a}^r_{A_2/O_A} \text{ is in the } \overrightarrow{AO_A} \text{ direction}$$

7. Compute and plot $\boldsymbol{a}^r_{B_4/O_B}$ from point o' and $\boldsymbol{a}^r_{B_5/A_5}$ from point a':

$$a^r_{B_4/O_B} = O_BB \times {}^1\omega_4^2 = 4.25 \times 9.16^2 = 356.6 \text{ in/s}^2; \qquad \boldsymbol{a}^r_{B_4/O_B} \text{ is in the } \overrightarrow{BO_B} \textit{ direction}$$

$$a^r_{B_5/A_5} = AB \times {}^1\omega_5^2 = 0.901 \times 120.1^2 = 13{,}000 \text{ in/s}^2; \qquad \boldsymbol{a}^r_{B_5/A_5} \text{ is in the } \overrightarrow{BA} \text{ direction}$$

8. Measure $a^t_{B_4/O_B}$:

$$a^t_{B_4/O_B} = 25{,}290 \text{ in/s}^2$$

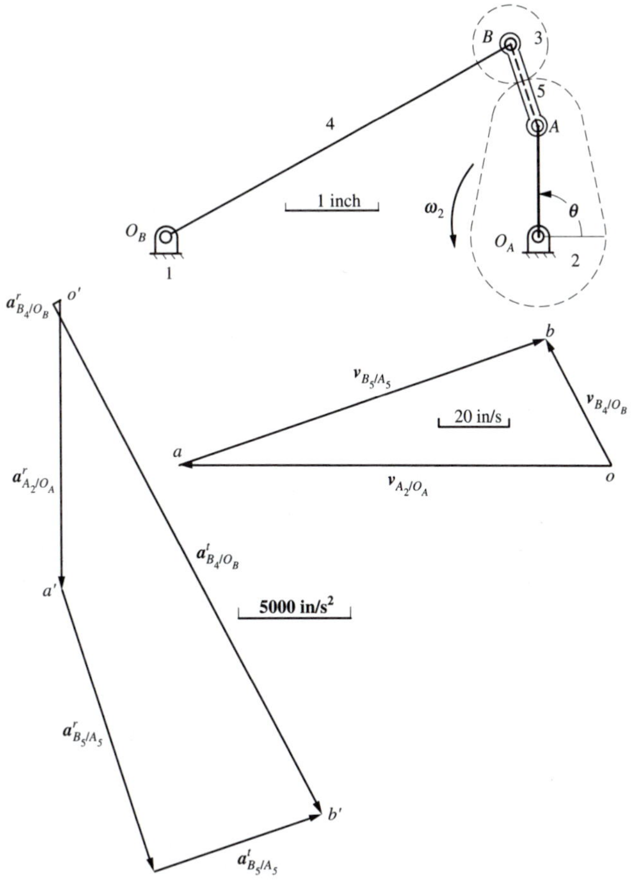

Figure 2.37 The velocity and acceleration polygons for Example 2.9.

9. Compute ${}^1\boldsymbol{\alpha}_4$ and determine its sense:

$$ {}^1\boldsymbol{\alpha}_4 = \boldsymbol{a}^t_{B_4/O_B}/O_B B = 25{,}290/4.25 = 4{,}440 \text{ rad/s}^2 \text{ CW} $$

Notice that the reason this relatively simple approach can be used is that we are not interested in the angular acceleration of the roller, link 3. This is definitely not equal to the angular acceleration of the line AB, which, for convenience, was called the virtual link 5. ■

2.15 CAM CONTACT

The analysis of mechanisms with cam joints can be conducted either directly or through the use of equivalent linkages. We will look at the direct approach first.

2.15.1 Direct Approach to the Analysis of Cam Contact

In the general case of cam contact, there will be both rolling and sliding at the contact point, and this is probably the most typical type of higher-pair contact between two bodies. If we look at two arbitrary bodies (e.g., 2 and 4) at the contact location B, we know B_2 and B_4 have the same coordinates.

$$ \boldsymbol{r}_{B_2} = \boldsymbol{r}_{B_4} $$

However,

$$^1\boldsymbol{v}_{B_2} \neq {}^1\boldsymbol{v}_{B_4}$$

or

$$^1\boldsymbol{v}_{B_2/B_4} \neq 0$$

We can obtain some information on $\boldsymbol{v}_{B_2/B_4}$ by recognizing that coincident points are involved and

$$^1\boldsymbol{v}_{B_2/B_4} = {}^4\boldsymbol{v}_{B_2B_4} = {}^4\boldsymbol{v}_{B_2}$$

Therefore, to analyze the velocity of B_2 relative to B_4 or link 4, it is convenient to represent the velocity in terms of components in the tangential ($\boldsymbol{t}$) and normal ($\boldsymbol{n}$) directions relative to the tangent at the contact point as shown in Fig. 2.38. Then,

$$^1\boldsymbol{v}_{B_2/B_4} = {}^4\boldsymbol{v}_{B_2/B_4} = {}^4\boldsymbol{v}_{B_2} = {}^4\boldsymbol{v}^n_{B_2} + {}^4\boldsymbol{v}^t_{B_2}$$

If the two bodies are rigid, there can be no component of velocity in the normal direction or the bodies would either penetrate each other or separate. Therefore, the normal component of the relative velocity must be zero, and the relative velocity direction must be along the common tangent to the two bodies at the point of contact. That is,

$$^4\boldsymbol{v}^n_{B_2} = 0$$

and

$$^1\boldsymbol{v}_{B_2/B_4} = {}^4\boldsymbol{v}^t_{B_2} = {}^1\boldsymbol{v}^t_{B_2/B_4}$$

We cannot determine anything more about $^1\boldsymbol{v}^t_{B_2/B_4}$; however, knowing the direction for the relative velocity usually provides sufficient information to conduct a velocity analysis.

We cannot determine anything about $^1\boldsymbol{a}_{B_2/B_4}$ directly; however, if we expand $^1\boldsymbol{a}_{B_2/B_4}$ into normal and tangential components, we can compute additional information about it. Then,

$$^1\boldsymbol{a}_{B_2/B_4} = {}^1\boldsymbol{a}^n_{B_2/B_4} + {}^1\boldsymbol{a}^t_{B_2/B_4}$$

Note that there is no Coriolis term because the acceleration is defined in link 1 and not link 4. This is directly analogous to the case of rolling contact except that now the tangential component is not zero. However, by definition, we know the direction of the tangential component.

Using the same nomenclature as in the case of rolling contact (see Fig. 2.33), the normal component of relative acceleration is given by Eqs. (2.54), (2.55), or (2.56). For example,

$$^1\boldsymbol{a}^n_{B_2/B_4} = \underset{(B \text{ to } O_2)}{\frac{|^1\boldsymbol{v}_{B_2/O_2}|^2}{|\boldsymbol{r}_{B_2/O_2}|}} + \underset{(O_2 \text{ to } O_4)}{\frac{|^1\boldsymbol{v}_{O_2/O_4}|^2}{|\boldsymbol{r}_{O_2/O_4}|}} + \underset{(O_4 \text{ to } B)}{\frac{|^1\boldsymbol{v}_{O_4/B_4}|^2}{|\boldsymbol{r}_{O_4/B_4}|}}$$

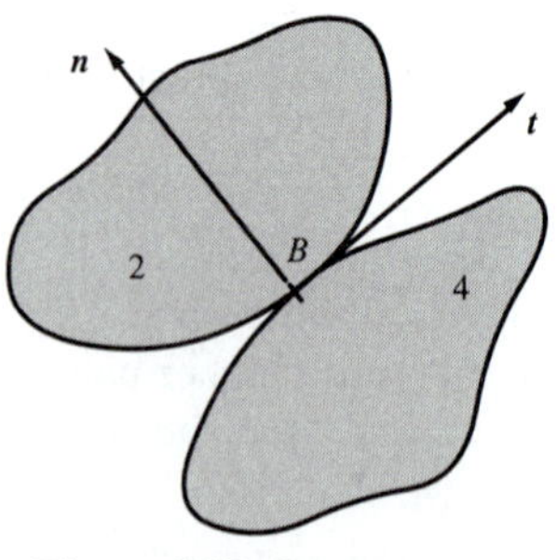

Figure 2.38 Cam contact.

If one of the rolling surfaces is flat, the relative position vector corresponding to the center of curvature will approach infinity, and the corresponding acceleration term will become zero. For example, if the surface for link 2 is flat, then O_2 is at infinity, and the normal component of acceleration expression reduces to

$$ {}^1\boldsymbol{a}^n_{B_2/B_4} = \frac{|{}^1\boldsymbol{v}_{O_4/B_4}|^2}{|\boldsymbol{r}_{O_4/B_4}|} \quad \text{from } O_4 \text{ to } B $$

EXAMPLE 2.10 (*Analysis of Mechanism with a Cam Joint*)

Find the velocity and acceleration of the cam follower (link 3) given in Fig. 2.39 if the cam is rotating at a constant angular velocity of 100 rad/s CCW.

SOLUTION

To analyze the problem, we can determine the velocity and acceleration of any point on link 3 because *all* points on link 3 have the same velocity and the same acceleration. The point to choose is the contact point C_3. To solve for the velocity and acceleration of C_3, first find the velocity of point C_2. Then write the relative velocity expression between points C_2 and C_3 and solve for the velocity of C_3. The relevant equations are

$$ {}^1\boldsymbol{v}_{C_2} = {}^1\boldsymbol{v}_{C_2/A_2} = {}^1\boldsymbol{\omega}_2 \times \boldsymbol{r}_{C/A} $$

and

$$ {}^1\boldsymbol{v}_{C_3} = {}^1\boldsymbol{v}_{C_2} + {}^1\boldsymbol{v}_{C_3/C_2} $$

Next solve for the velocity of B_2 either directly or by image. This will be needed for the acceleration analysis. The acceleration equations that must be solved are

$$ {}^1\boldsymbol{a}_{C_2} = {}^1\boldsymbol{a}_{C_2/A_2} = {}^1\boldsymbol{a}^n_{C_2/A_2} + {}^1\boldsymbol{a}^t_{C_2/A_2} $$

$$ {}^1\boldsymbol{a}_{C_3} = {}^1\boldsymbol{a}_{C_2} + {}^1\boldsymbol{a}_{C_3/C_2} = {}^1\boldsymbol{a}_{C_2} + {}^1\boldsymbol{a}^t_{C_3/C_2} + {}^1\boldsymbol{a}^n_{C_3/C_2} $$

and

$$ {}^1\boldsymbol{a}^n_{C_3/C_2} = {}^1\boldsymbol{a}^n_{C_3/D_3} + {}^1\boldsymbol{a}^n_{D_3/B_2} + {}^1\boldsymbol{a}^n_{B_2/C_2} = \frac{|{}^1\boldsymbol{v}_{C_3/D_3}|^2}{|\infty|} + \frac{|{}^1\boldsymbol{v}_{D_3/B_2}|^2}{|\infty|} + \frac{|{}^1\boldsymbol{v}_{B_2/C_2}|^2}{|\boldsymbol{r}_{B/C}|} = \frac{|{}^1\boldsymbol{v}_{B_2/C_2}|^2}{|\boldsymbol{r}_{B/C}|} $$

where D_3 is the center of curvature of the cam follower surface and is located at infinity. The steps in solving the equations are given in the following.

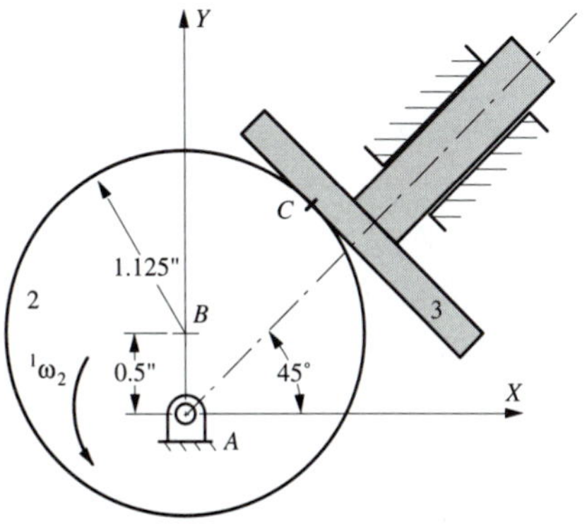

Figure 2.39 Cam and flat-faced follower.

STEPS

1. Draw the mechanism to scale as shown in Fig. 2.40. To do this, draw the cam circle centered at *B*. Next locate point *A* at 0.5 in below *B*. Then construct a line through *A* at an angle of 45°. This locates the direction of travel of the flat-faced follower. Finally, draw a line perpendicular to the 45° line and tangent to the cam. This locates point *C*.
2. Compute ${}^1\boldsymbol{v}_{C_2} = {}^1\boldsymbol{\omega}_2 \times \boldsymbol{r}_{C/A}$:

$$ {}^1\boldsymbol{v}_{C_2} = 100(1.52) = 152 \text{ in/s} $$

and it is perpendicular to AC and in the direction shown in Fig. 2.40. This will locate c_2.

3. Draw a line from o at an angle of 45° with the horizontal. Point c_3 will be on this line.
4. Draw a line through the tip of c_2 and perpendicular to the line at 45°. The intersection of this line with that drawn in step 3 locates c_3.

$$ \boldsymbol{v}_{C_3} = 35 \text{ in/s in the direction shown} $$

5. Locate b_2 by image.
6. Compute and plot $\boldsymbol{a}_{C_2/A_2} = \boldsymbol{a}^r_{C_2/A_2}$.

$$ \boldsymbol{a}^r_{C_2/A_2} = AC \times {}^1\omega_2^2 = 1.52 \times 100^2 = 15{,}200 \text{ in/s}^2; \qquad \boldsymbol{a}^r_{C_2/A_2} \text{ is in the } \overrightarrow{CA} \text{ direction} $$

7. Draw a line from o' at an angle of 45° with the horizontal. Point c_3' will be on this line.
8. Compute $\boldsymbol{a}^n_{C_3/C_2}$, determine its direction, and plot the resulting vector through the tip of $\boldsymbol{a}^r_{C_2/A_2}$.

$$ \boldsymbol{a}^n_{C_3/C_2} = \frac{|{}^1\boldsymbol{v}_{B_2/C_2}|^2}{|\boldsymbol{r}_{B/C}|} = \frac{|112.9|^2}{1.11} = 11{,}480 \text{ in/s}^2 \text{ from } B \text{ to } C $$

9. Draw a line through the tip of and perpendicular to $\boldsymbol{a}^n_{C_3/C_2}$. The intersection of this line with the line drawn in step 7 will be the point c_3'.
10. Measure a_{C_3}.

$$ \boldsymbol{a}_{C_3} = 3{,}250 \text{ in/sd}^2 \text{ in the direction shown} $$

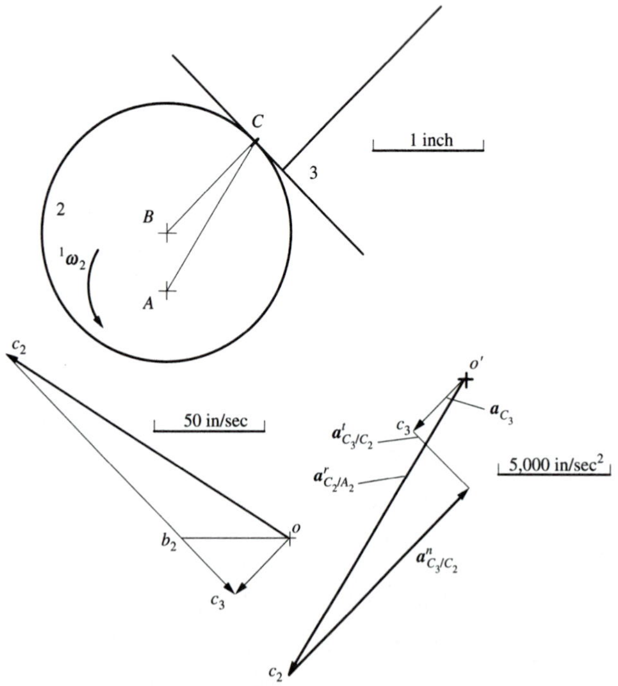

Figure 2.40 Position, velocity, and acceleration polygons for Example 2.10.

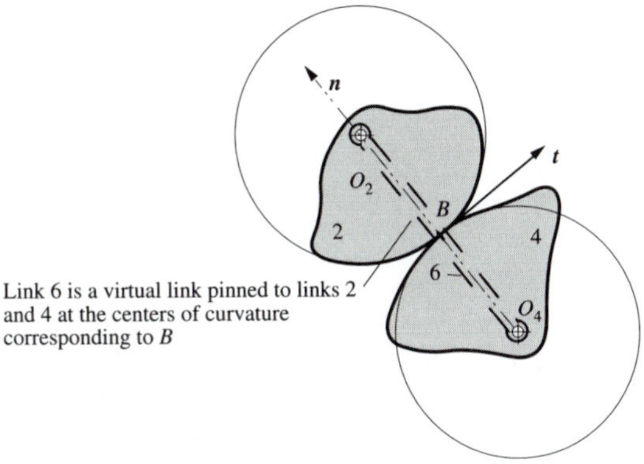

Figure 2.41 Virtual link pinned at the centers of curvature of the two bodies in cam contact.

2.15.2 Analysis of Cam Contact Using Equivalent Linkages

Another approach to determining the velocities and accelerations is to use the concept of equivalent linkages. For this we represent the two cam surfaces by their osculating circles and attach a binary link from one center of curvature to the other using revolute joints. As in the case of rolling contact, this technique can be used for velocities and acceleration, but it will not give correct results for higher derivatives. In Fig. 2.41, link 6 is a virtual link that usually changes length with each *finite* change in position. (It is constant for differential changes in position, however.)

The use of equivalent linkages usually simplifies the velocity and acceleration analyses because the equivalent linkages are usually standard four-bar linkages or one of the inversions of the common slider-crank mechanism. For example, a simple three-link cam mechanism becomes a four-bar linkage when replaced by its equivalent linkage as shown in Fig. 2.42. In the example in Fig. 2.42, the kinematic information for link 4 (virtual link) can be computed; however, this is usually of no interest. It is important to remember that the equivalent linkage is valid for one position only. The length of the virtual link usually changes with each position of interest.

If one of the surfaces is flat, the fictitious link becomes infinitely long, and the movement of the fictitious link can be represented by a slider. An example of this is shown in Fig. 2.43. The slider need not "slide" on the face of the flat cam surface through B. The only restriction is that it slide on a line that is parallel to the cam face.

The equivalent linkage is analyzed as any other linkage with pin and slider joints would be. The kinematic properties computed for links 2 and 3 will be the same for both the

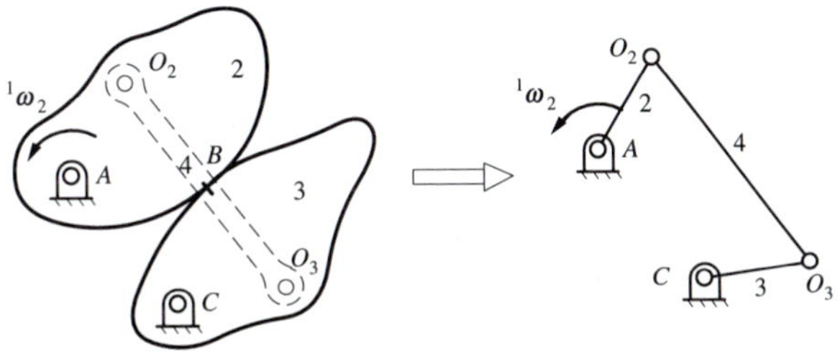

Figure 2.42 Equivalent linkage for cam mechanism with curved cam surfaces and revolute joints between the cams and the frame.

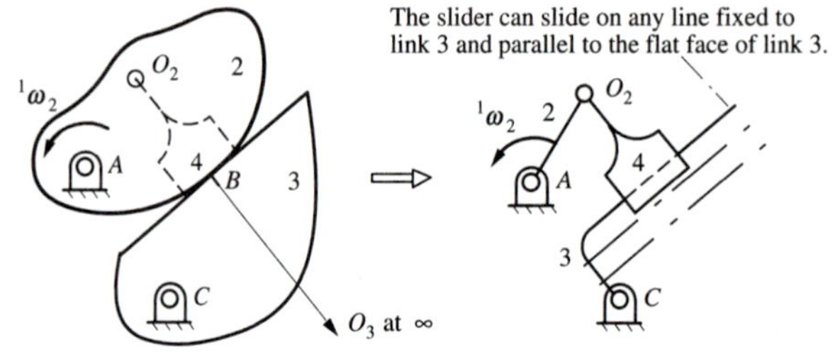

Figure 2.43 Equivalent linkage for cam mechanism with one flat-faced cam and revolute joints between the cams and the frame. The slider can slide on any line that is parallel to the cam face and fixed to link 3.

equivalent linkage and the actual linkage. The equivalent linkages for the other two types of three-bar cam linkages are given in Fig. 2.44.

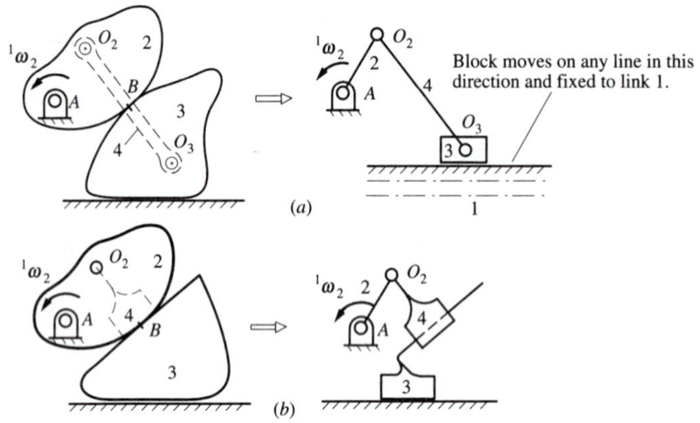

Figure 2.44 Equivalent linkage for cam mechanism. (a) Sliding joint between link 3 and frame. (b) Sliding between links 2 and 3 and between link 3 and frame.

EXAMPLE 2.11 ***(Mechanism Analysis Using an Equivalent Linkage)***

Use equivalent linkages to compute the velocity and acceleration of the cam follower (link 3) in Fig. 2.39 if the cam is rotating at a constant angular velocity of 100 rad/s CCW.

SOLUTION

The mechanism in Fig. 2.39 is of the type represented in Fig. 2.44b. Therefore, link 3, the follower, will have a sliding joint with the frame and with the virtual link (link 4). The resulting equivalent linkage is shown in Fig. 2.45. Notice that the location of the line on which link 4 must slide relative to link 3 is arbitrary as long as the line is fixed to link 3 and is parallel to the face of link 3. Therefore, the line that passes through B is chosen for simplicity. Similarly, the location of the line on which link 3 slides relative to the frame is arbitrary as long as the line is inclined at an angle of 45°.

For the equivalent linkage, we need only find the velocity and acceleration of point B_2. The velocity and acceleration of B_3 can then be found using the procedure given in Section 2.7. The velocity equations that must be solved are

$$^1\boldsymbol{v}_{B_2} = {}^1\boldsymbol{v}_{B_2/A_2} = {}^1\boldsymbol{\omega}_2 \times \boldsymbol{r}_{B/A}$$

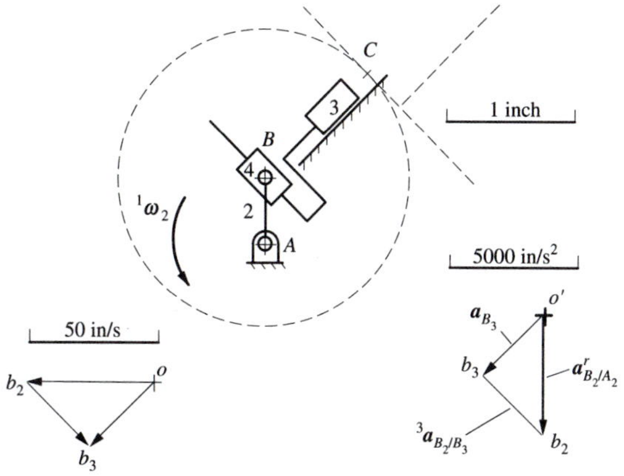

Figure 2.45 Position, velocity, and acceleration polygons for Example 2.11.

and

$$^1\boldsymbol{v}_{B_2} = {}^1\boldsymbol{v}_{B_3} + {}^1\boldsymbol{v}_{B_2/B_3} \tag{2.57}$$

Here we have written the velocity equation in terms of the velocity of B_2 relative to B_3 rather than vice versa because we can easily identify the direction of the velocity of B_2 relative to B_3. We also know the direction for the velocity and acceleration of B_3. The acceleration equations that must be solved are

$$^1\boldsymbol{a}_{B_2} = {}^1\boldsymbol{a}_{B_2/A_2} = {}^1\boldsymbol{a}^r_{B_2/A_2} + {}^1\boldsymbol{a}^t_{B_2/A_2}$$

$$^1\boldsymbol{a}_{B_2} = {}^1\boldsymbol{a}_{B_3} + {}^1\boldsymbol{a}_{B_2/B_3} = {}^1\boldsymbol{a}_{B_3} + \boldsymbol{a}^c_{B_2/B_3} + {}^3\boldsymbol{a}_{B_2/B_3}$$

and

$$\boldsymbol{a}^c_{B_2/B_3} = 2\,{}^1\boldsymbol{\omega}_3 \times {}^3\boldsymbol{v}_{B_2/B_3} = 0$$

The Coriolis term is a function of velocities only and can be computed; however, links 3 and 4 simply translate, making $^1\boldsymbol{\omega}_3 = 0$. Therefore, the acceleration expression becomes

$$^1\boldsymbol{a}_{B_2} = {}^1\boldsymbol{a}_{B_3} + {}^3\boldsymbol{a}_{B_2/B_3}$$

By geometry, $^3\boldsymbol{a}_{B_2/B_3}$ must move in the direction parallel to the face of the cam follower. Therefore, the equation has only two unknowns (once $^1\boldsymbol{a}_{B_2}$ is computed), and the equation can be solved for $^1\boldsymbol{a}_{B_3}$ and $^3\boldsymbol{a}_{B_2/B_3}$.

STEPS

1. Draw the mechanism to scale using the procedure given in Example 2.10. Then draw the equivalent mechanism.
2. Compute $^1\boldsymbol{v}_{B_2} = {}^1\boldsymbol{\omega}_2 \times \boldsymbol{r}_{B/A}$:

$$^1\boldsymbol{v}_{B_2} = 100(0.5) = 50\text{ in/s}$$

and it is perpendicular to AB and pointed in the direction shown in Fig. 2.45. This will locate b_2.

3. Draw a line from o at an angle of 45° with the horizontal. Point b_3 will be on this line.
4. Draw a line through the tip of b_2 and perpendicular to the line at 45°. The intersection of this line with that drawn in step 3 locates b_3.

$$\boldsymbol{v}_{B_3} = 35\text{ in/s in the direction shown}$$

5. Compute and plot $\boldsymbol{a}_{B_2/A_2} = \boldsymbol{a}^r_{B_2/A_2}$:

$$\boldsymbol{a}^r_{B_2/A_2} = AB \times {}^1\boldsymbol{\omega}_2^2 = 0.5 \times 100^2 = 5000\text{ in/s}^2; \qquad \boldsymbol{a}^r_{B_2/A_2}\text{ is in the }\overrightarrow{BA}\text{ direction}$$

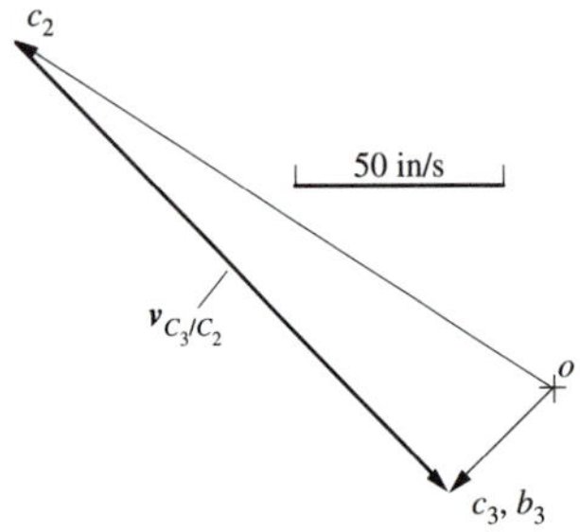

Figure 2.46 Calculation of the relative velocity v_{C_3/C_2}.

6. Draw a line from o' at an angle of 45° with the horizontal. Point b_3' will be on this line.
7. Draw a line through the tip of $a^r_{B_2/A_2}$ and parallel to the face of the cam follower (link 3). This is the direction of ${}^3a_{B_2/B_3}$. The intersection of this line with the line drawn in step 7 will be the point b_3'.
8. Measure a_{B_3}.

$$a_{B_3} = 3{,}250 \text{ in/s}^2 \text{ in the direction shown}$$

This is the acceleration of link 3. Note that this is the same solution as obtained in Example 2.10. All points in link 3 have the same velocity and the same acceleration. Therefore, points B_3 and C_3 have the same velocity and the same acceleration. Also note that considerably less work is required to obtain the final result when equivalent linkages are used. ■

When equivalent linkages are used, no information is used about the relative motion at the contact point. If the relative motions between the coincident points at contact are of interest, these can be computed directly after the basic analysis is completed. This velocity might be of interest for lubrication considerations.

EXAMPLE 2.12 ***(Analysis of Sliding Velocity in a Cam Mechanism)***

Find the sliding velocity at the point of contact for the mechanism in Example 2.11.

SOLUTION

The sliding velocity at the point of contact is the relative velocity between points C_2 and C_3. This velocity appears in the Eq. (2.57) and can be computed from

$$v_{C_3/C_2} = v_{C_3} - v_{C_2} \quad \textbf{(2.58)}$$

Because

$$v_{C_3} = v_{B_3}$$

we need only solve for v_{C_2} to determine v_{C_3/C_2} in Eq. (2.58). From Example 2.10, ${}^1v_{C_2} = {}^1\omega_2 \times r_{C/A}$, and it is perpendicular to $\overrightarrow{AC}$ as shown in Fig. 2.40. The vector v_{C_3/C_2} is shown in Fig. 2.46. Measurement of the vector gives $v_{C_3/C_2} = 148$ in/s and the direction is shown in the figure. ■

2.16 GENERAL COINCIDENT POINTS

In mechanisms, pin-in-slot joints are common, and occasionally the slots will be curved paths. Also, occasionally, sliders will be used on circular paths that rotate. Mechanisms employing these types of joints can be analyzed using general coincident points. In general, we can use any coincident points to help in a kinematic analysis if we can recognize the

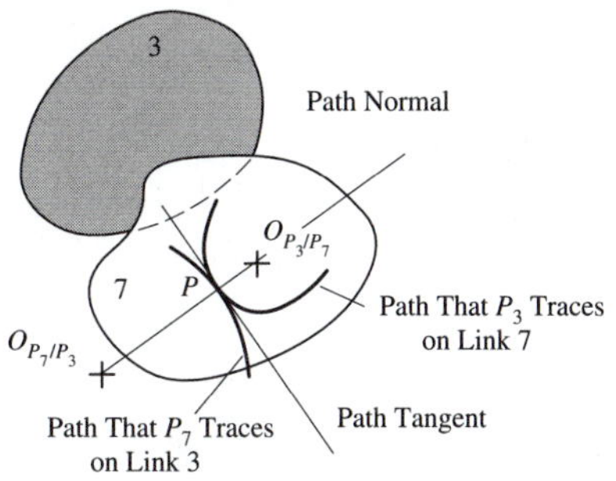

Figure 2.47 Geometric properties of relative paths traced by coincident points.

path that one of the points traces on the other link. For this, we must "stand" in one link and watch the coincident point on the other link move.

For the analysis, we need the center of curvature of the path and the corresponding tangent to the path. The tangent is normal to the line from the coincident points to the center of curvature of the path. For illustration, assume that the two bodies in question are links 3 and 7, and the coincident points are located at P as shown in Fig. 2.47. Then for any coincident points,

$$^1\boldsymbol{v}_{P_3/P_7} = {}^7\boldsymbol{v}_{P_3/P_7} = -{}^1\boldsymbol{v}_{P_7/P_3} = -{}^3\boldsymbol{v}_{P_7/P_3}$$

Two paths will be traced, and these can be designated as path P_3/P_7 and path P_7/P_3. The paths will share a common tangent vector, and the normal to the paths will contain the two coincident points and the two centers of curvature as shown in Fig. 2.47. The path that P_3 traces on link 7 will be fixed to link 7, and the path P_7/P_3 will be fixed to link 3.

To solve problems involving general coincident points, we must be able to recognize one of the relative paths by inspection. This means that we must be able to determine the center of curvature of the path. Sometimes, we can recognize one of the relative paths but not the other. This is still useful because of the relationships

$$^1\boldsymbol{v}_{P_7/P_3} = -{}^7\boldsymbol{v}_{P_3/P_7}$$

and

$$^1\boldsymbol{a}_{P_7/P_3} = -{}^7\boldsymbol{a}_{P_3/P_7}$$

This means that if we can recognize one of the paths, we can always rewrite the kinematic equations so that the information will appear in the correct form. Some examples of paths that are obvious are given in Fig. 2.48.

2.16.1 Velocity Analyses Involving General Coincident Points

The velocity analysis of mechanisms that involve general coincident points will generally require that the direction for the relative velocity vector ($\boldsymbol{v}_{P_3/P_7}$ or $\boldsymbol{v}_{P_7/P_3}$) be known. This direction can be determined by using the same technique as was used in the analyses using cam pairs. For this, we replace the path P_3/P_7 by its osculating circle at P. Recall that we can do this without compromising the accuracy of the solution as long as we are interested in only velocities and accelerations. Next connect P_3 to the center of curvature O of the path P_3/P_7 by a virtual link x. The geometry is represented schematically in Fig. 2.49.

The motion of P_3 relative to link 7 will be the same as for P_x relative to link 7 if the

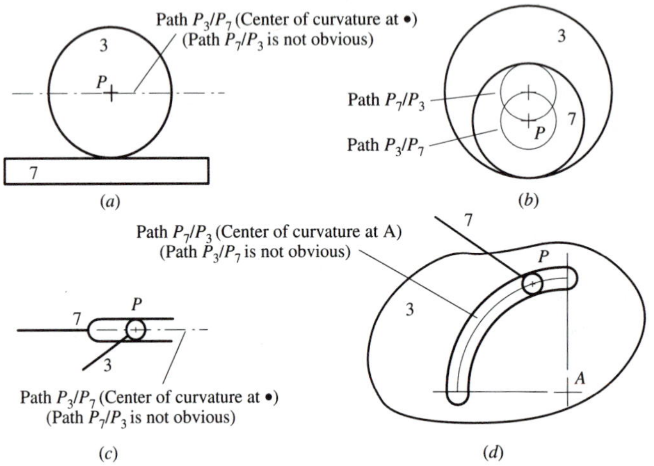

Figure 2.48 Obvious relative paths of general coincident points.

two points are pinned together. The relative velocity between P_3 and P_7 can then be written as

$$ {}^1\mathbf{v}_{P_3/P_7} = {}^7\mathbf{v}_{P_3/P_7} = {}^7\mathbf{v}_{P_3/P_x} + {}^7\mathbf{v}_{P_x/O_x} + {}^7\mathbf{v}_{O_x/O_7} + {}^7\mathbf{v}_{O_7/P_7} \tag{2.59} $$

or because points P_3 and P_x and O_x and O_7 are pinned together, and the last term involves the motion of two points in system 7 as observed from system 7,

$$ {}^1\mathbf{v}_{P_3/P_7} = {}^7\mathbf{v}_{P_3/P_7} = {}^7\mathbf{v}_{P_x/O_x} \tag{2.60} $$

Because two points on the same rigid link are involved, the term on the right-hand side of Eq. (2.60) can be written as

$$ {}^1\mathbf{v}_{P_3/P_7} = {}^7\mathbf{v}_{P_3/P_7} = {}^7\boldsymbol{\omega}_x \times {}^x\mathbf{r}_{P_x/O_x} $$

This vector is perpendicular to the line from the point P to the center of curvature of the path P_3/P_7, and it is therefore along the direction of the tangent to the path. Therefore, when the direction for the relative velocity is required, we need only determine the center of curvature of the path P_3/P_7 and draw a line perpendicular to it.

The magnitude of the angular velocity term will be required for the acceleration analysis, and it can be written as

$$ |{}^7\boldsymbol{\omega}_x| = \frac{|{}^1\mathbf{v}_{P_3/P_7}|}{|{}^x\mathbf{r}_{P_x/O_x}|} $$

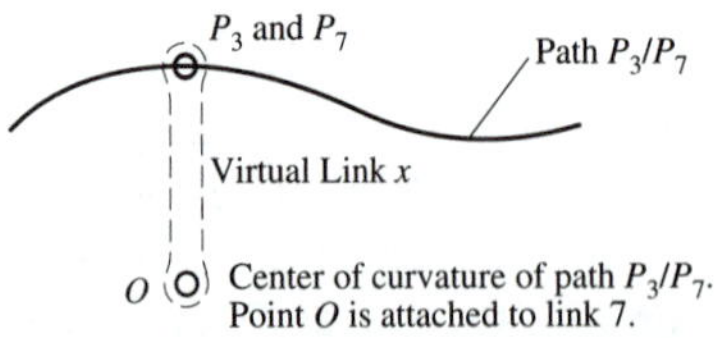

Figure 2.49 Connecting virtual link from point P_3 to center of curvature of path that P_3 traces on link 7.

2.16.2 Acceleration Analysis Involving General Coincident Points

The acceleration analysis is slightly more complex than the velocity analysis when general coincident points are involved. For the relative acceleration, again assume that the path that P_3 traces on link 7 is known. This means that the center of curvature of the path is also known. The development of the relative acceleration expression is similar to that used for the case of a rotating slider, and the relative acceleration expression can be written as

$$^1\boldsymbol{a}_{P_3/P_7} = {^7\boldsymbol{a}_{P_3/P_7}} + 2\,{^1\boldsymbol{\omega}_7} \times {^7\boldsymbol{v}_{P_3/P_7}} + {^1\boldsymbol{\omega}_7} \times ({^1\boldsymbol{\omega}_7} \times {^7\boldsymbol{r}_{P_3/P_7}}) + {^1\boldsymbol{\alpha}_7} \times {^7\boldsymbol{r}_{P_3/P_7}}$$

Because ${^7\boldsymbol{r}_{P_3/P_7}} = 0$ at the moment considered,

$$^1\boldsymbol{a}_{P_3/P_7} = {^7\boldsymbol{a}_{P_3/P_7}} + 2\,{^1\boldsymbol{\omega}_7} \times {^7\boldsymbol{v}_{P_3/P_7}} \tag{2.61}$$

The second term in the expression is the Coriolis term, which is a function of position and velocity only. Therefore, it can be computed as soon as the velocity analysis is completed. The direction of the Coriolis term is given by the cross-product. Graphically, we can get the direction by rotating ${^7\boldsymbol{v}_{P_3/P_7}}$ (which equals ${^1\boldsymbol{v}_{P_3/P_7}}$) 90° in the direction of ${^1\boldsymbol{\omega}_7}$.

The first term in Eq. (2.61) is simply the acceleration of P_3 as observed from system 7. This term can be written as

$$^7\boldsymbol{a}_{P_3/P_7} = {^7\boldsymbol{a}_{P_3/\text{any point in system 7}}} = {^7\boldsymbol{a}_{P_3}} \tag{2.62}$$

Unlike the case of the rotating slider, the direction for this acceleration component is not immediately obvious. However, by using the technique begun in the velocity analysis, we can determine a vector expression for this component that involves only one unknown.

To begin, replace the path P_3/P_7 by its osculating circle at P and rewrite the acceleration expression in Eq. (2.62) in terms of the virtual link x and the center of curvature of the path of P_3/P_7. This is similar to what was done with velocities in Eq. (2.59). The relative acceleration between P_3 and P_7 can the be written as

$$^7\boldsymbol{a}_{P_3} = {^7\boldsymbol{a}_{P_3/P_7}} = {^7\boldsymbol{a}_{P_3/P_x}} + {^7\boldsymbol{a}_{P_x/O_x}} + {^7\boldsymbol{a}_{O_x/O_7}} + {^7\boldsymbol{a}_{O_7/P_7}}$$

or because points P_3 and P_x and O_x and O_7 are pinned together, and the last term involves the motion of two points in system 7 as observed from system 7,

$$^7\boldsymbol{a}_{P_3} = {^7\boldsymbol{a}_{P_3/P_7}} = {^7\boldsymbol{a}_{P_x/O_x}} \tag{2.63}$$

Because two points on the same rigid link are involved, the term on the right-hand side of Eq. (2.63) can be written as

$$^7\boldsymbol{a}_{P_x/O_x} = {^7\boldsymbol{a}^r_{P_x/O_x}} + {^7\boldsymbol{a}^t_{P_x/O_x}}$$

The radial component is a function of velocities and position only and can be written as

$$^7\boldsymbol{a}^r_{P_x/O_x} = {^7\boldsymbol{\omega}_x} \times {^x\boldsymbol{v}_{P_x/O_x}} = \frac{|{^1\boldsymbol{v}_{P_x/O_x}}|^2}{|{^x\boldsymbol{r}_{P_x/O_x}}|} = \frac{|{^1\boldsymbol{v}_{P_3/P_7}}|^2}{|\boldsymbol{r}_{P/O}|} \quad \text{(from } P \text{ to } O\text{)}$$

The magnitude of the vector ${^7\boldsymbol{a}^t_{P_x/O_x}}$ cannot be computed directly; however, we know that the direction is perpendicular to the line from the point P to the center of curvature of the path P_3/P_7, and it is therefore along the direction of the tangent to the path. The total acceleration can now be represented as

$$^1\boldsymbol{a}_{P_3/P_7} = {^7\boldsymbol{a}^t_{P_x/O_x}} + {^7\boldsymbol{a}^r_{P_x/O_x}} + 2\,{^1\boldsymbol{\omega}_7} \times {^7\boldsymbol{v}_{P_3/P_7}}$$

or in terms of the original subscripts,

$$^1\boldsymbol{a}_{P_3/P_7} = {^7\boldsymbol{a}^t_{P_3/P_7}} + {^7\boldsymbol{a}^r_{P_3/P_7}} + 2\,{^1\boldsymbol{\omega}_7} \times {^7\boldsymbol{v}_{P_3/P_7}}$$

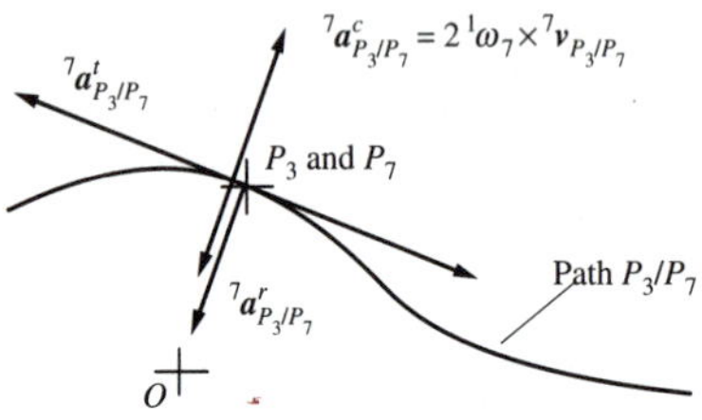

Figure 2.50 Acceleration components associated with the relative acceleration between P_3 and P_7.

Of the three vectors on the right-hand side of the equation, only the direction of ${}^7\boldsymbol{a}^t_{P_3/P_7}$ will be unknown after the velocity analysis. The directions for the individual terms are summarized in Fig. 2.50. Note that the radial and Coriolis terms are both normal to the tangent of the path of P_3/P_7; however, only the radial component always points from P to the center of curvature of the path of P_3/P_7. The direction of the Coriolis term will depend on the directions of both ${}^1\boldsymbol{\omega}_7$ and ${}^7\boldsymbol{v}_{P_3/P_7}$.

EXAMPLE 2.13 ***(Analysis of Mechanism with Pin-in-Slot Joint)***

In the mechanism shown in the Fig. 2.51, point B_2 moves on a curved slot in link 3. The radius of the slot is 3 m. Points C, B, and D are collinear, and the other distances between points are as given in Fig. 2.51. Link 2 rotates with an angular velocity of 2 rad/s CCW and an angular acceleration of 3 rad/s² CCW. For the position shown, find:

1. ${}^1\boldsymbol{\omega}_3$, ${}^1\boldsymbol{\alpha}_3$, ${}^1\boldsymbol{v}_{D_3}$, ${}^1\boldsymbol{a}_{D_3}$

2. The center of curvature of the path that B_3 traces on link 2

SOLUTION

Once the position of the linkage is drawn, the following vector quantities can be measured:

$$\boldsymbol{r}_{B/A} = 5\angle 60°$$
$$\boldsymbol{r}_{B/C} = 5\angle 120°$$
$$\boldsymbol{r}_{D/C} = 8\angle 120°$$

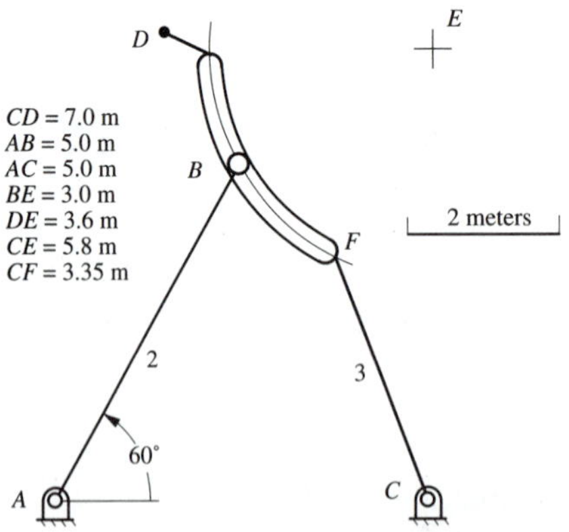

Figure 2.51 Mechanism for Example 2.13.

For the velocity analysis, we can first compute the velocity of point B_2. That is:

$$\boldsymbol{v}_{B_2} = \boldsymbol{v}_{B_2/A_2} = {}^1\boldsymbol{\omega}_2 \times \boldsymbol{r}_{B_2/A_2}$$

Next go to point B_3 on link 3.

$${}^1\boldsymbol{v}_{B_3} = {}^1\boldsymbol{v}_{B_3/C_3} = {}^1\boldsymbol{\omega}_3 \times {}^3\boldsymbol{r}_{B/C}$$

Because ${}^1\boldsymbol{\omega}_3$ is unknown, this term cannot be computed without another equation. Consider the two coincident points B_2 and B_3. Then,

$${}^1\boldsymbol{v}_{B_3} = {}^1\boldsymbol{v}_{B_2} + {}^1\boldsymbol{v}_{B_3/B_2}$$

This equation is technically correct; however, we cannot recognize the path that B_3 traces on link 2. Therefore, the equation cannot be differentiated to help us in the acceleration analysis. Therefore, write the equation of terms of ${}^1\boldsymbol{v}_{B_2}$. Then,

$${}^1\boldsymbol{v}_{B_2} = {}^1\boldsymbol{v}_{B_3} + {}^1\boldsymbol{v}_{B_2/B_3}$$

This equation is useful because we can pick out the path that B_2 traces on link 3 by inspection. This equation can be solved, although there are two unknown directions on the right-hand side of the equation. This is handled by beginning one vector at the velocity pole and ending the other vector at the end of ${}^1\boldsymbol{v}_{B_2}$.

After the velocity polygon is drawn, we can measure ${}^1\boldsymbol{v}_{B_3}$, and determine ${}^1\boldsymbol{v}_{D_3}$ by image. We can also measure ${}^1\boldsymbol{v}_{B_2/B_3}$, which will be required for the acceleration analysis.

The velocity analysis uses two basic equations:

$${}^1\boldsymbol{v}_{B_2} = {}^1\boldsymbol{v}_{B_2/A_2}$$

and

$${}^1\boldsymbol{v}_{B_2} = {}^1\boldsymbol{v}_{B_3/C_3} + {}^1\boldsymbol{v}_{B_2/B_3}$$

and these two equations show the solution path for the accelerations. Again start at B_2. Then

$${}^1\boldsymbol{a}_{B_2} = {}^1\boldsymbol{a}_{B_2/A_2} = {}^1\boldsymbol{a}^t_{B_2/A_2} + {}^1\boldsymbol{a}^r_{B_2/A_2} = {}^1\boldsymbol{\alpha}_2 \times {}^2\boldsymbol{r}_{B/A} + {}^1\boldsymbol{\omega}_2 \times {}^2\boldsymbol{v}_{B/A}$$

Now differentiate the velocity expression involving B_3.

$$\begin{aligned}{}^1\boldsymbol{a}_{B_2} &= {}^1\boldsymbol{a}_{B_3/C_3} + {}^1\boldsymbol{a}_{B_2/B_3} = {}^1\boldsymbol{a}^r_{B_3/C_3} + {}^1\boldsymbol{a}^t_{B_3/C_3} + {}^3\boldsymbol{a}^n_{B_2/B_3} + {}^3\boldsymbol{a}^t_{B_2/B_3} + 2\,{}^1\boldsymbol{\omega}_3 \times {}^1\boldsymbol{v}_{B_2/B_3} \\ &= {}^1\boldsymbol{\omega}_3 \times {}^1\boldsymbol{v}_{B_3/C_3} + {}^1\boldsymbol{\alpha}_3 \times {}^3\boldsymbol{r}_{B_3/C_3} + {}^3\boldsymbol{a}^n_{B_2/B_3} + {}^3\boldsymbol{a}^t_{B_2/B_3} + 2\,{}^1\boldsymbol{\omega}_3 \times {}^1\boldsymbol{v}_{B_2/B_3}\end{aligned} \quad \textbf{(2.64)}$$

This equation has only two unknowns and can be solved. We can compute the acceleration of D_3 using acceleration image.

To find the center of curvature of the path that B_3 traces on link 2, we must find an expression that involves the radius of curvature of the path. This term is ${}^2\boldsymbol{a}^n_{B_3/B_2}$, and it can be evaluated from the following.

$${}^1\boldsymbol{a}_{B_2/B_3} = -{}^1\boldsymbol{a}_{B_3/B_2}$$

Therefore,

$${}^1\boldsymbol{a}^t_{B_2/B_3} = -{}^1\boldsymbol{a}^t_{B_3/B_2}$$

and

$${}^1\boldsymbol{a}^n_{B_2/B_3} = -{}^1\boldsymbol{a}^n_{B_3/B_2}$$

Also,

$${}^3\boldsymbol{a}^n_{B_2/B_3} + 2\,{}^1\boldsymbol{\omega}_3 \times {}^1\boldsymbol{v}_{B_2/B_3} = -{}^2\boldsymbol{a}^n_{B_3/B_2} - 2\,{}^1\boldsymbol{\omega}_2 \times {}^1\boldsymbol{v}_{B_3/B_2}$$

and

$$^2a^n_{B_3/B_2} = \frac{|^1\boldsymbol{v}_{B_3/B_2}|^2}{|\boldsymbol{r}_{B/G}|}$$

or

$$r_{B/G} = \frac{|^1\boldsymbol{v}_{B_3/B_2}|^2}{^2a^n_{B_3/B_2}}$$

where G gives the location of the center of curvature of the path that B_3 traces on link 2. The location of G on the proper side of B is found by the direction of $^2\boldsymbol{a}^n_{B_3/B_2}$ because it points from B to the center of curvature of the path.

STEPS

1. Select a scale and draw link 2 at a scaled distance of 5 m at an angle of 60° to the horizontal. This will locate point B as shown in Fig. 2.52. Draw an arc of radius BE centered at B. Then draw a second arc of radius CE centered at C. The intersection of this arc with the first will locate point E. Next draw the arc of radius BE centered at E. Draw a line from point C through point B of length CD.
2. Compute $^1\boldsymbol{v}_{B_2} = {^1\boldsymbol{v}_{B_2/A_2}} = {^1\boldsymbol{\omega}_2} \times \boldsymbol{r}_{B_2/A_2}$ and draw the vector from o in the direction $\perp BA$. The sense of $^1\boldsymbol{v}_{B_2}$ is determined by rotation of $\boldsymbol{r}_{B_2/A_2}$ 90° in the direction of $^1\boldsymbol{\omega}_2$. This will locate b_2.

$$\boldsymbol{v}_{B_2/A_2} = {^1\boldsymbol{\omega}_2} \times \boldsymbol{r}_{B_2/A_2} = (2)(5) = 10\, m/s \perp BA$$

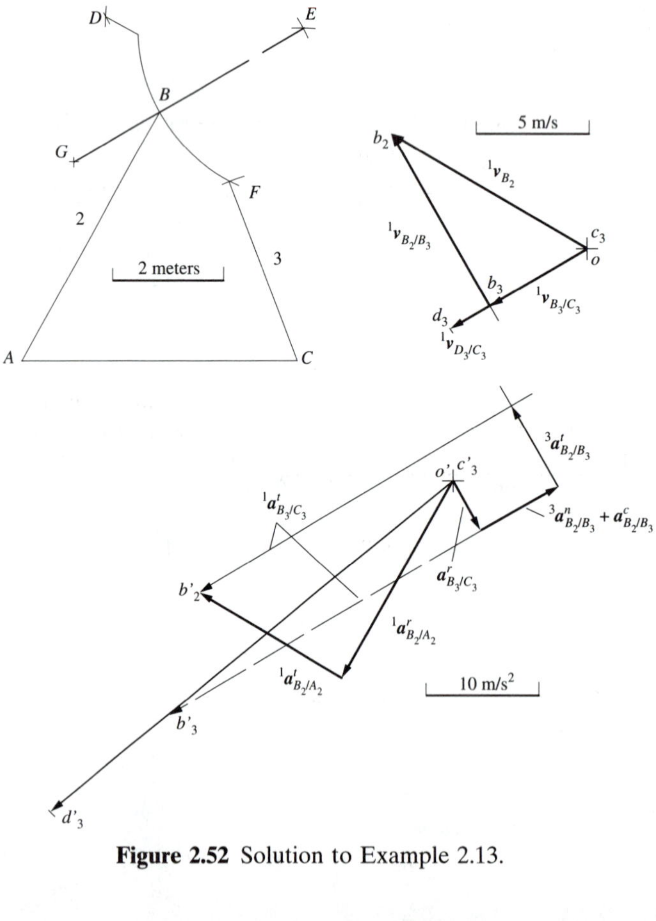

Figure 2.52 Solution to Example 2.13.

3. Draw a line through o in the direction perpendicular to CB.

4. Draw a line through b_2 in the direction tangent to the path that B_2 traces on link 3 (perpendicular to the radius BE). The intersection of this line with that drawn in step 3 will locate point b_3. Locate the arrowheads on the velocity polygon to conform with Eq. (2.48).

5. Locate point d_3 by image.

$$od_3 = \frac{CD}{CB} ob_3 = 7.0 \text{ m/s at an angle of } 210° \text{ to the horizontal}$$

6. Measure ${}^1\boldsymbol{v}_{B_2/B_3} = 8.7$ m/s at an angle of 120° to the horizontal.

7. Measure ${}^1\boldsymbol{v}_{B_3/C_3} = 5.0$ m/s at an angle of 210° to the horizontal and compute ${}^1\omega_3$. Determine the sense of ${}^1\boldsymbol{\omega}_3$ by rotating $\boldsymbol{r}_{B_3/C_3}$ 90° in the direction of ${}^1\boldsymbol{\omega}_3$ to get the direction of ${}^1\boldsymbol{v}_{B_3/C_3}$.

$${}^1\omega_3 = \frac{|{}^1\boldsymbol{v}_{B_3/C_3}|}{|\boldsymbol{r}_{B_3/C_3}|} = \frac{5}{5} = 1 \text{ rad/s CCW}$$

8. Compute ${}^1a^t_{B_2/A_2}$ and ${}^1a^r_{B_2/A_2}$ and, starting from o', draw the resulting vectors after scaling.

$${}^1\boldsymbol{a}^r_{B_2/A_2} = {}^1\boldsymbol{\omega}_2 \times \boldsymbol{v}_{B/A} = 2(10) = 20 \text{ m/s}^2 \text{ opposite } \boldsymbol{r}_{B/A}$$

$${}^1\boldsymbol{a}^t_{B_2/A_2} = {}^1\boldsymbol{\alpha}_2 \times \mathbf{r}_{B/A} = 3(5) = 15 \text{ m/s}^2$$

perpendicular to $\boldsymbol{r}_{B/A}$ in the direction given by rotating $\boldsymbol{r}_{B/A}$ 90° in the direction of ${}^1\boldsymbol{\alpha}_2$. This vector is added to ${}^1\boldsymbol{a}^r_{B_2/A_2}$ as shown in Fig. 2.52.

9. Compute ${}^1\boldsymbol{a}^r_{B_3/C_3}$, ${}^3\boldsymbol{a}^n_{B_2/B_3}$, and $\boldsymbol{a}^c_{B_2/B_3}$. All of these accelerations are functions of the velocity and position data.

$${}^1\boldsymbol{a}^r_{B_3/C_3} = {}^1\boldsymbol{\omega}_3 \times {}^1\boldsymbol{v}_{B_3/C_3} = 1(5) = 5 \text{ m/s}^2 \quad \text{from } B \text{ to } C$$

$${}^3\boldsymbol{a}^n_{B_2/B_3} = \frac{|{}^1\boldsymbol{v}_{B_2/B_3}|^2}{|\boldsymbol{r}_{B_3/E_3}|} = \frac{(8.7)^2}{3} = 25.23 \text{ m/s}^2 \quad \text{from } B \text{ to } E$$

$$\boldsymbol{a}^c_{B_2/B_3} = 2\,{}^1\boldsymbol{\omega}_3 \times {}^1\boldsymbol{v}_{B_2/B_3} = 2(1)(8.7) = 17.4 \text{ m/s}^2 \quad \text{from } E \text{ to } B$$

10. Note that ${}^3\boldsymbol{a}^n_{B_2/B_3}$ and $\boldsymbol{a}^c_{B_2/B_3}$ are in opposite directions. Therefore, determine the resultant before plotting.

$${}^3\boldsymbol{a}^n_{B_2/B_3} + \boldsymbol{a}^c_{B_2/B_3} = 25.23 - 17.4 = 7.83 \text{ m/s}^2 \quad \text{from } B \text{ to } E$$

11. Starting from o', add the vectors ${}^1\boldsymbol{a}^r_{B_3/C_3}$ and ${}^3\boldsymbol{a}^n_{B_2/B_3} + \boldsymbol{a}^c_{B_2/B_3}$ as shown in Fig. 2.52.

12. Draw a line through the tip of ${}^3\boldsymbol{a}^n_{B_2/B_3} + \boldsymbol{a}^c_{B_2/B_3}$ in the direction perpendicular to ${}^3\boldsymbol{a}^n_{B_2/B_3} + \boldsymbol{a}^c_{B_2/B_3}$ and to EB (i.e., tangent to the path that B_2 traces on link 3).

13. Draw a line through the tip of ${}^1\boldsymbol{a}^t_{B_2/A_2}$ in the direction perpendicular to $\overrightarrow{BC}$. The intersection of this line with that from step 12 will give ${}^3\boldsymbol{a}^t_{B_2/B_3}$ and ${}^1\boldsymbol{a}^t_{B_3/C_3}$. The locations of the arrowheads (directions) are given by Eq. (2.64).

14. Measure $\boldsymbol{a}^t_{B_3/C_3}$ and compute the magnitude of the angular acceleration, ${}^1\boldsymbol{\alpha}_3$.

$${}^1\alpha_3 = \frac{|{}^1\boldsymbol{a}^t_{B_3/C_3}|}{|\boldsymbol{r}_{B_3/C_3}|} = \frac{32.2}{5} = 11.2 \text{ rad/s}^2$$

The sense is given by ${}^1\boldsymbol{a}^t_{B_3/C_3}$ and $\boldsymbol{r}_{B_3/C_3}$. Namely, we rotate $\boldsymbol{r}_{B_3/C_3}$ 90° in the direction of ${}^1\boldsymbol{\alpha}_3$ to get the direction of ${}^1\boldsymbol{a}^t_{B_3/C_3}$. The direction is CCW.

15. Locate the acceleration of D_3 by acceleration image. To do this, determine the absolute acceleration of B_3. This is done by adding ${}^1\boldsymbol{a}^t_{B_3/C_3}$ to ${}^1\boldsymbol{a}^r_{B_3/C_3}$ to locate b'_3. Then find d'_3 using

$$o'd'_3 = \frac{CD}{CB} o'b'_3 = 45.9 \text{ m/s}^2 \quad \text{in the direction shown in Fig. 2.52}$$

16. Compute ${}^2\boldsymbol{a}^n_{B_3/B_2} = -({}^3\boldsymbol{a}^n_{B_2/B_3} + 2\,{}^1\boldsymbol{\omega}_3 \times {}^1\boldsymbol{v}_{B_2/B_3} + 2\,{}^1\boldsymbol{\omega}_2 \times {}^1\boldsymbol{v}_{B_3/B_2})$. Arbitrarily select the direction $\overrightarrow{BE}$ as positive. Then,

$$ {}^2a^n_{B_3/B_2} = -(25.23 - 17.4 + 34.8) = -42.63\ \text{m/s}^2 $$

The minus sign means that the direction of the center of curvature of the path is opposite $\overrightarrow{BE}$ or in the $\overrightarrow{EB}$ direction.

17. Compute the radius of curvature by locating G_2 using

$$ r_{B/G} = \frac{|{}^1\boldsymbol{v}_{B_3/B_2}|^2}{|{}^2\boldsymbol{a}^n_{B_3/B_2}|} = \frac{(8.7)^2}{42.63} = 1.78\ \text{m} $$

The center of curvature is shown in Fig. 2.52. ■

2.17 INSTANT CENTERS OF VELOCITY

2.17.1 Introduction

At every instant during the motion of a rigid body in a plane there exists a point that is instantaneously at rest. This point is called the instant center. The concept of an instant center of velocity for two bodies with planar motion was first discovered by Johann Bernoulli in 1742. This concept was later extended by Chasles (1830) to include general spatial motion using the instantaneous screw axis concept. However, although the instant center of velocity has proved to be very useful in general mechanism velocity analysis, the corresponding instant center of acceleration has found little use. This is because, in general, more calculations are required to find the acceleration center than would be required to find the accelerations of interest using methods previously outlined. Therefore, only instant centers of velocity will be considered here.

The instant center technique for velocity analysis is particularly useful when only two or three velocities, or angular velocities, are of interest. It can be a very efficient technique, for example, for finding input-output velocity relationships of very complex mechanisms. When combined with virtual work, or conservation of energy, it provides an efficient way to obtain input-output force or torque relationships. Instant centers are also very helpful when analyzing mechanisms with higher pairs, such as cam mechanisms, or gear trains. In principle, the instant center and velocity polygon techniques are *alternative* methods for solving the same set of problems. However, they are quite different techniques, and each is better suited to some situations than to others. Some experience is necessary to identify easily the most applicable technique for a particular problem.

It should be emphasized that instant centers of velocity are applicable to velocities only and are usually of little help if accelerations are ultimately of interest. If an acceleration analysis must be performed, then the velocity analysis should be conducted using one of the previously discussed traditional procedures based on vector methods.

2.17.2 Definition

Given two bodies B and C moving with planar motion relative to each other in a reference frame R, there is, in general, one and only one location P in the plane of motion where the coincident points at a given instant have the same velocity with respect to the reference frame R. One coincident point is in body B and the other is in body C. This location is called the instant center of velocity for bodies B and C and is represented by I_{BC} or I_{CB}. If P is the instant center, then

$$ {}^R\boldsymbol{v}_{P_B} = {}^R\boldsymbol{v}_{P_C} $$

or

$$^{R}\boldsymbol{v}_{P_B/P_C} = {}^{R}\boldsymbol{v}_{P_C/P_B} = 0$$

If the points are permanently attached to each other, they are called permanent instant centers. If the points are only momentarily coincident, the instant centers are called instantaneous instant centers.

2.17.3 Existence Proof

The existence of an instant center between arbitrary links B and R may be inferred, and its location found, by the use of the relationship between the velocities of two points in body B. In the following, all velocities are defined in system R, so the left superscript designating the coordinate system will be omitted for simplicity.

$$\boldsymbol{v}_{Q_B} = \boldsymbol{v}_{P_B} + \boldsymbol{\omega}_B \times \boldsymbol{r}_{Q/P}$$

Now, assume that Q is the instant center I_{RB}. Then, $\boldsymbol{v}_{Q_B} = 0 = \boldsymbol{v}_{I_{RB}}$ and

$$-\boldsymbol{v}_{P_B} = \boldsymbol{\omega}_B \times \boldsymbol{r}_{Q/P} \tag{2.65}$$

From Section 2.5, we know that the radial component of the relative acceleration between two points P and Q on the same rigid link B is

$$\boldsymbol{\omega}_B \times (\boldsymbol{\omega}_B \times \boldsymbol{r}_{Q/P}) = -\omega_B^2 \boldsymbol{r}_{Q/P}$$

Therefore, cross-multiplication of both sides of Eq. (2.65) by $\boldsymbol{\omega}_B$ gives

$$\boldsymbol{\omega}_B \times (\boldsymbol{\omega}_B \times \boldsymbol{r}_{Q/P}) = -\boldsymbol{\omega}_B \times \boldsymbol{v}_{P_B} = -\omega_B^2 \boldsymbol{r}_{Q/P}$$

or

$$\boldsymbol{r}_{Q/P} = \frac{\boldsymbol{\omega}_B \times \boldsymbol{v}_{P_B}}{\omega_B^2}$$

In planar motion $\boldsymbol{\omega}_B \times \boldsymbol{v}_{P_B}$ is normal to $\boldsymbol{v}_{P_B}$ and may be written $(\omega_B)(v_{P_B})\boldsymbol{n}$, where $\boldsymbol{n}$ is a unit vector normal to $\boldsymbol{v}_{P_B}$ with the sense given by visualizing $\boldsymbol{v}_{P_B}$ rotated about its tail in the $\boldsymbol{\omega}_B$ direction. Hence

$$\boldsymbol{r}_{Q/P} = \frac{v_{P_B}}{\omega_B}\boldsymbol{n}$$

Thus, the distance, $r_{P/Q}$, in Fig. 2.53 from the instant center, I_{RB}, to point P is v_{P_B}/ω_B and the line IP is normal to $\boldsymbol{v}_{P_B}$. Its sense is such that rotation of $\boldsymbol{r}_{P/Q}$ about I in the $\boldsymbol{\omega}_B$ direction produces $\boldsymbol{v}_{P_B}$.

If more than one location in the plane of motion is found to be an instant center for two bodies, then those two bodies, for velocity analysis purposes, can be considered to be instantaneously fixed to each other. That is, if more than one location is an instant center, then all locations are instant centers.

On the other hand, if no finite location can be found that qualifies as an instant center of relative motion of two bodies, then the two bodies are translating with planar motion with respect to each other. In this case, the instant center can be considered to be a location at infinity reached by a line drawn perpendicular to the relative velocity vector between two arbitrary coincident points in the two bodies considered.

Instant centers are useful because they permit velocities to be computed easily. For example, if we know the velocity of Point P_B by analysis, we know the velocity of P_R directly.

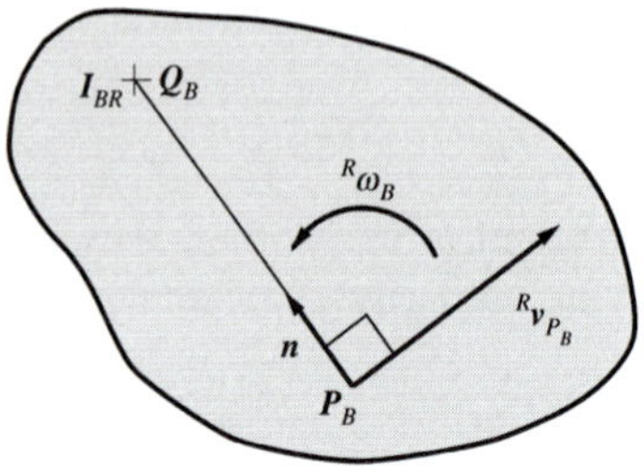

Figure 2.53 Proof of the existence of an instant center of velocity in planar motion.

2.17.4 Location of an Instant Center from the Directions of Two Velocities

Assume that we know the velocities of two points (P and Q) in body C where the velocities are defined relative to the coordinate system in a second body B. This condition is shown in Fig. 2.54. We can then search for some point in C that has zero velocity relative to body B. The location of this point is the instant center designated by I_{BC} or, in the development here, it can be represented simply as I. To find the instant center location, let I_B be the point in body B and I_C be the coincident point in body C. We can write relative velocity expressions for points P and Q as follows:

$$^B\mathbf{v}_{P_C} = {}^B\mathbf{v}_{P_C/I_B} = {}^B\mathbf{v}_{P_C/I_C} + {}^B\mathbf{v}_{I_C/I_B}$$

and

$$^B\mathbf{v}_{Q_C} = {}^B\mathbf{v}_{Q_C/I_B} = {}^B\mathbf{v}_{Q_C/I_C} + {}^B\mathbf{v}_{I_C/I_B}$$

However, by definition of the instant center,

$$^B\mathbf{v}_{I_C/I_B} = {}^B\mathbf{v}_{I_C} = 0$$

so that

$$^B\mathbf{v}_{P_C/I_B} = {}^B\mathbf{v}_{P_C/I_C} = {}^B\boldsymbol{\omega}_C \times \mathbf{r}_{P_C/I_C}$$

and

$$^B\mathbf{v}_{Q_C/I_B} = {}^B\mathbf{v}_{Q_C/I_C} = {}^B\boldsymbol{\omega}_C \times \mathbf{r}_{Q_C/I_C}$$

By definition of the cross product, $^B\mathbf{v}_{P_C/I_C}$ must be perpendicular to $\mathbf{r}_{P_C/I_C}$, and $^B\mathbf{v}_{Q_C/I_C}$ must be perpendicular to $\mathbf{r}_{Q_C/I_C}$. The location of the instant center (I_{BC}) is given by the intersection of the two perpendicular lines as shown in Fig. 2.54.

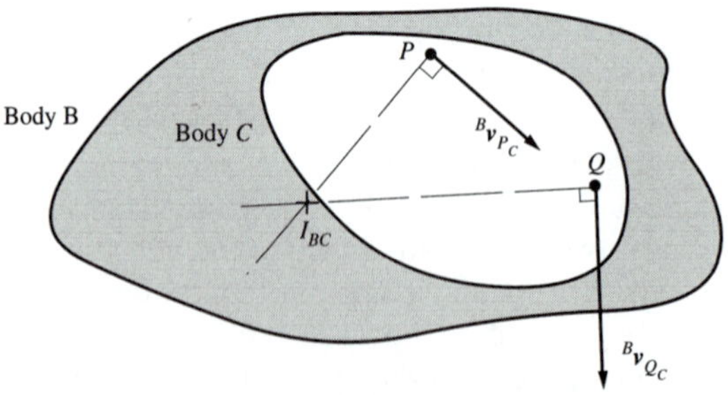

Figure 2.54 Location of the instant center given the directions of the velocities of two points.

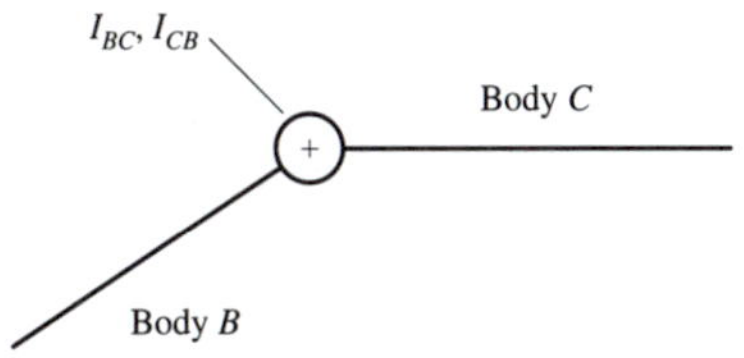

Figure 2.55 Permanent instant center.

2.17.5 Instant Center at a Revolute Joint

The center of rotation at a revolute joint, I_{BC}, has the same velocity whether it is considered to be part of link B or link C. Therefore, it qualifies as a permanent instant center. This is indicated in Fig. 2.55.

2.17.6 Instant Center of a Curved Slider

If body B is a block moving on a circular arc on body C as shown in Fig. 2.56, then the center of the arc is a stationary location common to both bodies. Therefore, this location qualifies as a permanent instant center. If the curve is not circular at the location of interest, the curve can be replaced by its osculating circle (for velocities and accelerations) and the center of the osculating circle or center of curvature of the path at the given point would be the instant center. Actually, a circular slider is kinematically equivalent to a revolute joint. The center of the equivalent revolute is the center of curvature. That is, it is the instant center. A noncircular slide is realizable only as a higher pair.

2.17.7 Instant Center of a Prismatic Joint

If the radius of curvature, ρ, in the case of the curved slider is allowed to become very large, the arc will approach a straight line. Also, the location of the instant center will tend toward infinity. However, the velocity of P relative to system B will still remain perpendicular to the line from P to the instant center. Therefore, if we know the direction of the velocity of *any* point P relative to system B, we can find one locus for the instant center; that is, it must lie on a line perpendicular to the velocity vector as shown in Fig. 2.57.

Note that the location of the line to infinity is unimportant; only the direction is defined by the velocity direction. This can be thought of as being the parallax phenomenon in which the direction to a distant object appears to remain the same, regardless of motion of the observer.

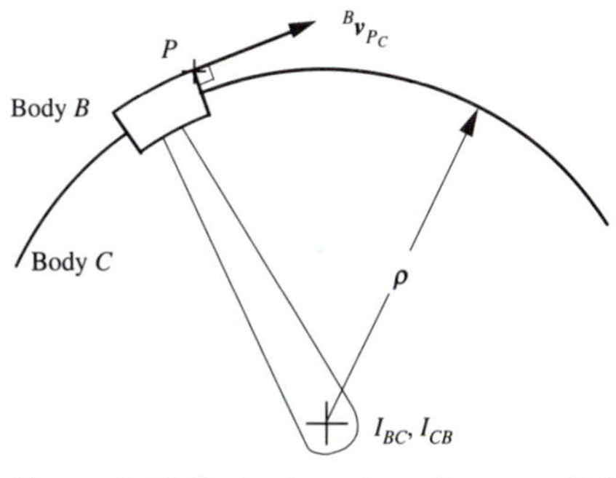

Figure 2.56 Instant center of a curved slide.

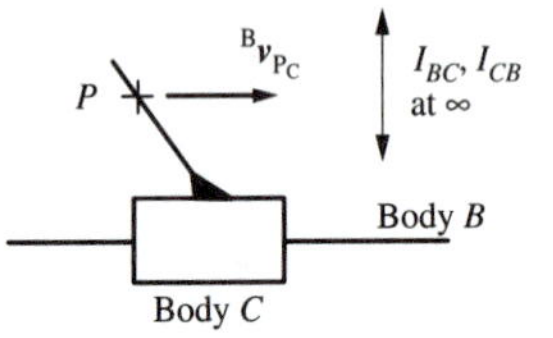

Figure 2.57 Instant center of a prismatic joint.

2.17.8 Instant Center of a Rolling Contact Pair

The instant center of pure rolling contact between two rigid bodies B and C is located at the point of contact of the two bodies as shown in Fig. 2.58. This is a direct consequence of the rolling condition that the two points in contact be at rest relative to one another. The instant center for the relative motion of involute spur gears is at the pitch point: the point of rolling contact between their pitch circles.

2.17.9 Instant Center of a General Cam-Pair Contact

When two planar bodies (B and C in Fig. 2.59) are held in general cam contact, it is assumed that the bodies will neither penetrate each other nor separate. In general, the bodies both roll on each other and slide over each other. If sliding is involved, the direction of relative sliding must be along the common tangent of the profiles of the two bodies as shown in Fig. 2.59. If P is the contact point location, then the velocity of point P_C, as well as the velocity of point P_B, will lie along the common tangent. Therefore, the instant center must be located on a line perpendicular to the common tangent at the contact point P. This means that the instant center must lie on a line through the centers of curvature (O_C and O_B) corresponding to P in each of the two bodies.

To locate precisely the position of I_{BC}, some further information about the relative motion of bodies B and C is required. For example assume that body B is link 2 and body C is link 3 and that links 2 and 3 are both connected to the frame by revolute joints as shown in Fig. 2.60.

If we arbitrarily pick the point A as a candidate for the instant center, we see that the velocities $\boldsymbol{v}_{A_2}$ and $\boldsymbol{v}_{A_3}$ cannot be equal because they are not in the same direction. The only location where they can be equal in direction is at the point Q on a line through the two pivots. Note that the two fixed pivots are the instant centers I_{12} and I_{13}.

2.17.10 Centrodes

As two bodies, B and C, move relative to each other, I_{BC} traces a path on each of the bodies (path BC_B on body B and BC_C on body C). These paths are the centrodes for the two bodies. At any instant, the two paths will be in contact with each other at the instant

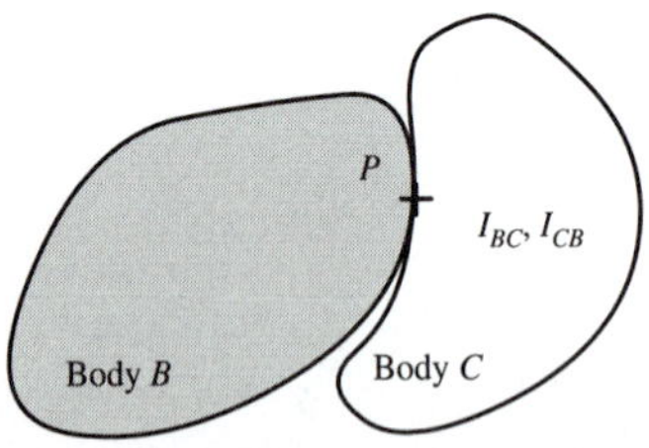

Figure 2.58 Instant center of a rolling contact.

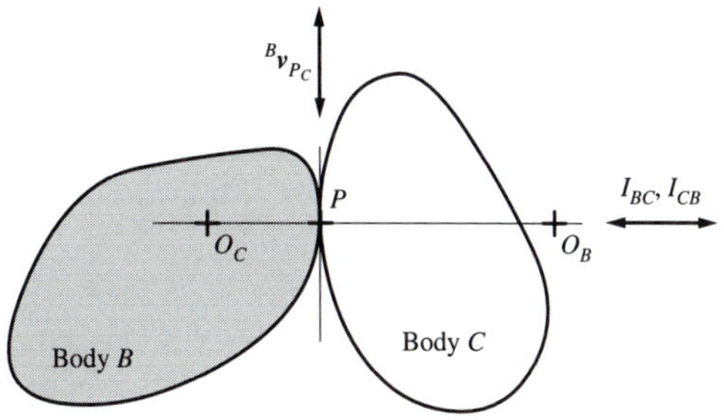

Figure 2.59 Relationship between instant centers and sliding velocity in cam contact.

center where there is zero relative velocity between the two bodies and therefore between the two centrodes. That is, the instant center acts as a point of rolling contact between the two centrodes. This means that as the two bodies move, the two centrodes will roll on each other. Conversely, the relative motion of the two bodies can be faithfully reproduced by rolling one centrode on the other no matter how the original motion was produced. Therefore, the analysis of the relative motion of two bodies moving with planar motion can always be transformed to the study of two bodies rolling on each other. Note that as the two centrodes for links 2 and 3 roll on each other, the contact point (I_{23}) and the instant centers I_{12} and I_{13} will be collinear as illustrated in Fig. 2.60.

These concepts can be extended into spatial motion in which the instant center becomes an instantaneous screw axis (ISA), and the loci of the ISAs in the two bodies are ruled surfaces called axodes. These axodes roll on each other in a direction perpendicular to their generating instantaneous screw axes as well as sliding relative to each other along their instantaneous screw axes.

An example of the fixed and moving centrodes associated with the coupler of a four-bar linkage is shown in Fig. 2.61. This shows the centrodes generated by instant center I_{13}. The centrodes in Fig. 2.61 are very simple, but this is not the typical case. For crank-rocker mechanisms, the centrodes will extend to infinity (when the crank and rocker are parallel) in two directions, and for drag-link mechanisms, the centrodes can form multiple loops. Typical examples are shown in Figs. 2.62 and 2.63. These centrodes were generated with the MATLAB program *centrodes.m* included on the disk with this book.

Another example of centrodes is shown in the model in Fig. 2.64. The mechanism model is a six-bar linkage, and the noncircular gears attached to the two frame-mounted links correspond to the centrodes of relative motion of those links. The motion of the two links attached to the gears is the same relative to the frame and to each other if the linkage is present without the gears or if the gears are present without the linkage.

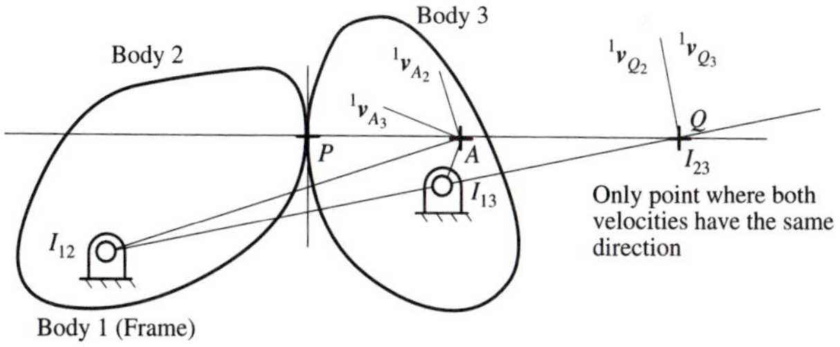

Figure 2.60 The instant center location between two frame-mounted cams.

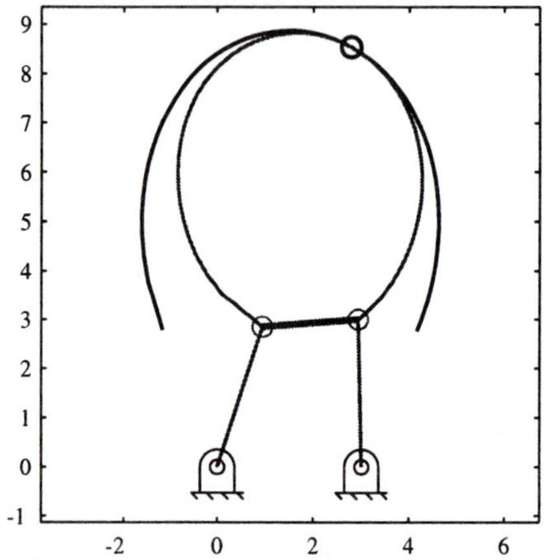

Figure 2.61 Centrodes associated with instant center I_{13} for a simple four-bar linkage.

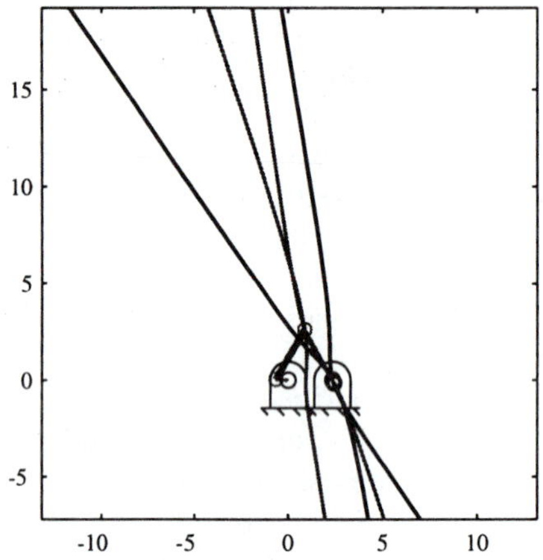

Figure 2.62 Centrodes associated with instant center I_{13} for a crank-rocker four-bar linkage.

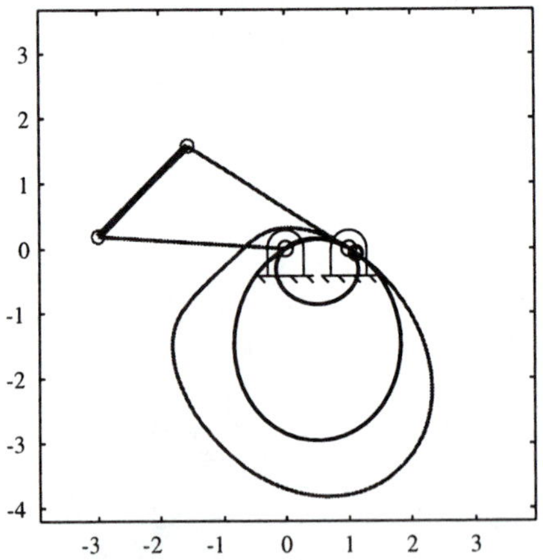

Figure 2.63 Centrodes associated with instant center I_{13} for a drag-link four-bar linkage.

Figure 2.64 Six-bar linkage model with centrodes represented by noncircular gears. The gears roll on each other at the pitch points, and the pitch point is the instant center between the two frame-mounted cranks.

2.17.11 The Kennedy-Aronholdt Theorem

If we have n bodies and we take them two at a time such that $I_{AB} = I_{BA}$, then the total number of instant centers is given by

$$N_{IC} = \frac{n(n-1)}{2} \tag{2.66}$$

Because of the number of instant centers (ICs) occurring in a mechanism with a large number of links, it is desirable to develop a procedure that helps to identify the locations of the instant centers in a systemic manner. This can be done using the results of the Kennedy-Aronholdt theorem.

In the late nineteenth century, Kennedy (England) and Aronholdt (Germany) greatly extended the usefulness of instant centers by independently discovering the theorem of three centers. The theorem is stated as follows:

> *If three bodies are in relative planar motion (or two bodies moving relative to each other and to the fixed reference frame), there are three instant centers pertaining to the relative motion of pairs of those bodies. Those three instant centers are collinear.*

Thus, in Fig. 2.65, given three bodies A, B, and C moving with planar motion in reference frame R, the three instant centers I_{AB}, I_{AC}, and I_{BC} all lie on the same straight line in the plane. To prove the theorem, it is necessary to recognize that the instant center is really two coincident points. One of these two points is embedded in each of the two laminae

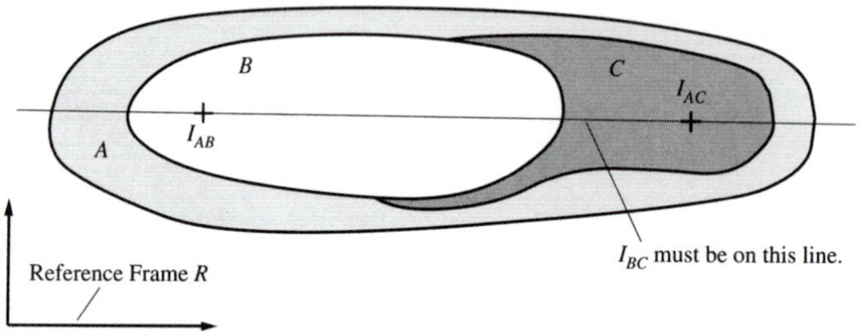

Figure 2.65 The Kennedy-Aronholdt theorem.

for which the instant center describes the relative motion. Hence, in Fig. 2.65:

I_{AB} is two points common to A and B
I_{AC} is two points common to A and C
I_{BC} is two points common to B and C.

Also,

$$
\begin{aligned}
{}^R\mathbf{v}_{(I_{AB})_A} &= {}^R\mathbf{v}_{(I_{AB})_B} \\
{}^R\mathbf{v}_{(I_{AC})_A} &= {}^R\mathbf{v}_{(I_{AC})_C} \\
{}^R\mathbf{v}_{(I_{BC})_B} &= {}^R\mathbf{v}_{(I_{BC})_C}
\end{aligned}
$$

Assume that we know the locations of I_{AB}, I_{AC} and we want to find I_{BC}. We can first write

$$ {}^A\mathbf{v}_{(I_{BC})_B} = {}^A\mathbf{v}_{(I_{BC})_C} $$

or

$$ {}^A\mathbf{v}_{(I_{BC})_B} = {}^A\mathbf{v}_{(I_{BC})_B/(I_{AB})_A} = {}^A\mathbf{v}_{(I_{BC})_B/(I_{AB})_B} = {}^A\boldsymbol{\omega}_B \times \boldsymbol{r}_{(I_{BC})/(I_{AB})} $$

Also,

$$ {}^A\mathbf{v}_{(I_{BC})_C} = {}^A\mathbf{v}_{(I_{BC})_C/(I_{AC})_A} = {}^A\mathbf{v}_{(I_{BC})_C/(I_{AC})_C} = {}^A\boldsymbol{\omega}_C \times \boldsymbol{r}_{(I_{BC})/(I_{AC})} $$

or equating the two relationships,

$$ {}^A\boldsymbol{\omega}_C \times \boldsymbol{r}_{(I_{BC})/(I_{AC})} = {}^A\boldsymbol{\omega}_B \times \boldsymbol{r}_{(I_{BC})/(I_{AB})} $$

Since ${}^B\boldsymbol{\omega}_C$ is parallel to ${}^A\boldsymbol{\omega}_B$, the $\boldsymbol{r}$'s must also be parallel to make the cross-products equal. Because both of the $\boldsymbol{r}$'s pass through I_{BC}, they must be collinear. This can happen only if I_{AB}, I_{AC}, and I_{BC} all lie on the same line.

The Kennedy-Aronholdt theorem can be used in the following way to find instant centers. Assume that we have two groups of three links such that two links are common to both groups. For example, as shown in Fig. 2.66, if we have I_{45} and I_{47} and I_{35} and I_{37}, links 5 and 7 are common to both groups. We know that I_{57} must lie on a line through

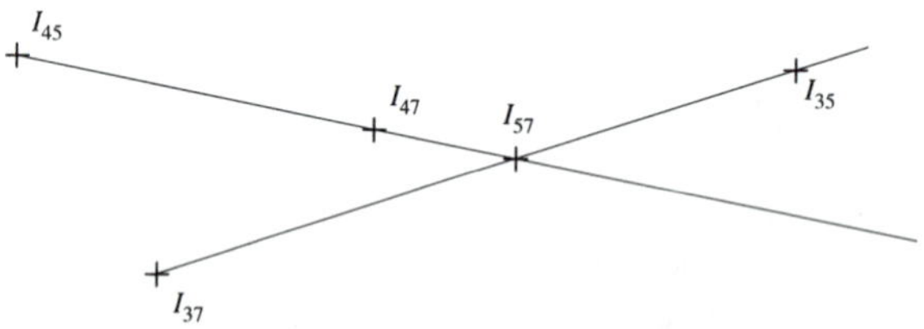

Figure 2.66 Triplets of instant centers.

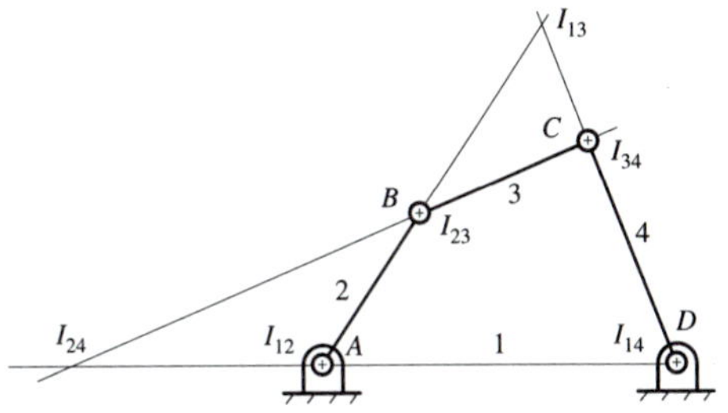

Figure 2.67 Application of the Kennedy-Aronholdt theorem to location of all instant centers of a four-bar linkage.

I_{45}, and I_{47}, and it must also lie on the line through I_{35} and I_{37}. The location is defined by the intersection of the two lines.

Therefore, by selecting two pairs of appropriate instant centers, we can locate the instant center that is common to the two groups of links. A way in which the Kennedy-Aronholdt theorem can be used is illustrated by the following example.

EXAMPLE 2.14 ***(Locating Instant Centers for a Four-Bar Linkage)***

PROBLEM Locate all instant centers of the four-bar linkage in the position shown in Fig. 2.67.

SOLUTION By inspection, I_{12} is at A, I_{23} at B, I_{34} at C, and I_{14} at D. Thus, four of the six instant centers are already identified. To locate I_{13}, note that it is collinear with I_{12} and I_{23} and also with I_{14} and I_{34}. Thus it is at the intersection of BC and AD.

A set of three collinear instant centers always shares the same three subscripts, each subscript appearing on two instant centers. Given two instant centers with a common subscript, the third center, which completes the collinear set, has the two subscripts that are not common to the other two centers. ■

2.17.12 Circle Diagram as a Strategy for Finding Instant Centers

When the number of bodies is large, it is helpful to use some kind of bookkeeping method to help find all of the instant centers. One such method is the circle method, which is based directly on the Kennedy-Aronholdt theorem. The procedure is illustrated on the four-bar linkage in Fig. 2.68 as follows:

1. Draw the kinematic diagram for the mechanism to be analyzed.
2. Draw a circle of arbitrary radius and place tick marks representing all the mechanism member symbols approximately equally spaced around the perimeter of the circle.
3. By inspection, determine as many instant centers as possible, and draw a straight line

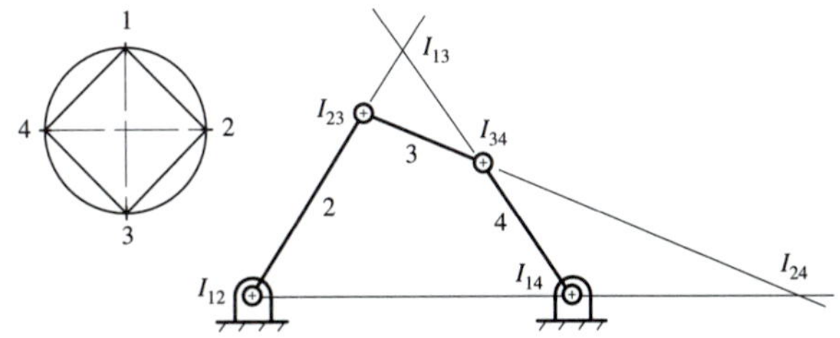

Figure 2.68 Use of the circle diagram when locating instant centers.

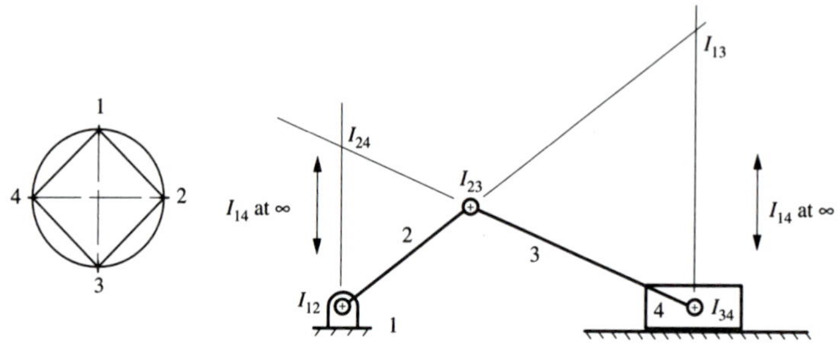

Figure 2.69 The instant centers of a slider-crank linkage.

between the corresponding numbers on the circle. For example, if I_{12} is known, then a line is drawn between symbols 1 and 2.

4. If a line can be drawn between two points on the circle such that the line is the only unknown side of *two* triangles, the instant center represented by that line lies at the intersection of the two lines drawn through the instant center pairs that are identified by the two known sides of each triangle. In other words, the instant center corresponding to any diagonal of a quadrilateral on the circle diagram can be found. Once the instant center is located, the appropriate two points on the circle diagram are connected.
5. Repeat the procedure in step 4 until all of the instant centers of interest are found.

As a second example, consider the slider-crank mechanism shown in Fig. 2.69. Again, the instant centers to be found are I_{24} and I_{13}. These can be found directly; however, it is necessary to note that I_{14} is located at infinity along a line perpendicular to the slider direction given.

2.17.13 Using Instant Centers, the Rotating-Radius Method

Once the proper instant centers are found, these can be used to find the velocity of selected points in a rigid body. This can be done analytically; however, graphical methods are generally much faster to use. An especially useful method for finding velocities is the rotating radius method. To develop the method, assume we have an arbitrary link moving relative to the reference system. For the sake of illustration, assume that the link is 3 and the reference link is the frame (link 1). Let points P and Q be any points fixed to link 3 as shown in Fig. 2.70. Then, we can write

$$^1\boldsymbol{v}_{P_3/Q_3} = {}^1\boldsymbol{\omega}_3 \times \boldsymbol{r}_{P_3/Q_3} = {}^1\boldsymbol{v}_{P_3} - {}^1\boldsymbol{v}_{Q_3}$$

and is perpendicular to the line from P to Q. If point Q_3 has zero velocity relative to link 1, then

$$^1\boldsymbol{v}_{P_3/Q_3} = {}^1\boldsymbol{v}_{P_3}$$

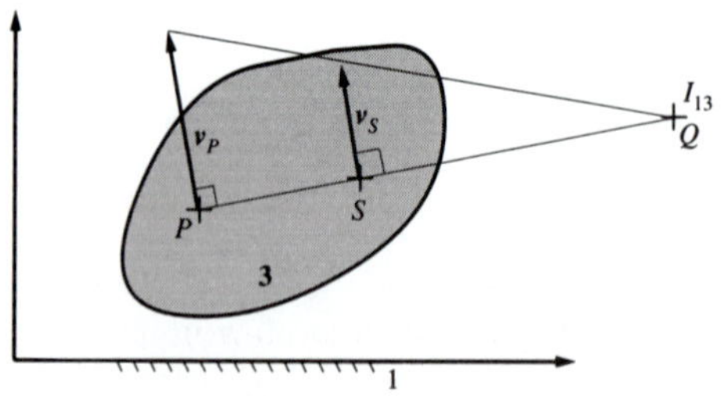

Figure 2.70 The rotating radius method.

However, the only point in link 3 that has zero velocity relative to the frame is I_{13}. Therefore,

$$ {}^1\boldsymbol{v}_{P_3} = {}^1\boldsymbol{\omega}_3 \times \boldsymbol{r}_{P_3/I_{13}} $$

Because point P was *any* arbitrary point in link 3, this equation holds for *all* points in link 3. Therefore, if we know the angular velocity of the link and the instant center relative to the frame, we can compute the absolute velocity of the point. Furthermore, the direction of the absolute velocity is perpendicular to the line from the point to the instant center.

For other points, only the vector $\boldsymbol{r}_{P_3/I_{13}}$ will change as P changes. Considering the magnitude of the velocity,

$$ |{}^1\boldsymbol{v}_{P_3}| = |{}^1\boldsymbol{\omega}_3|\,|\boldsymbol{r}_{P_3/I_{13}}| $$

Because ${}^1\boldsymbol{\omega}_3$ is the same for all points in the link, the magnitude of the velocity for any other point S is given by

$$ |{}^1\boldsymbol{v}_{S_3}| = |{}^1\boldsymbol{\omega}_3|\,|\boldsymbol{r}_{S_3/I_{13}}| $$

Therefore, dividing the two equations

$$ \frac{|{}^1\boldsymbol{v}_{P_3}|}{|{}^1\boldsymbol{v}_{S_3}|} = \frac{|\boldsymbol{r}_{P_3/I_{13}}|}{|\boldsymbol{r}_{S_3/I_{13}}|} $$

or

$$ |{}^1\boldsymbol{v}_{S_3}| = |{}^1\boldsymbol{v}_{P_3}| \frac{|\boldsymbol{r}_{S_3/I_{13}}|}{|\boldsymbol{r}_{P_3/I_{13}}|} $$

This magnitude applies to any point that is the same distance from the instant center. The magnitude of the velocity is directly proportional to its distance from the instant center. Hence if two points in the rigid body have the same $|\boldsymbol{r}_{S_3/I_{13}}|$, they will have the same magnitude of velocity $|{}^1\boldsymbol{v}_{S_3}|$; however, the direction of their velocities will differ because the velocity is perpendicular to the line from the point to the instant center. This is illustrated by S and S' in Fig. 2.71. The actual direction of the velocity is obtained by recognizing that all points will appear to rotate relative to the frame about the instant center.

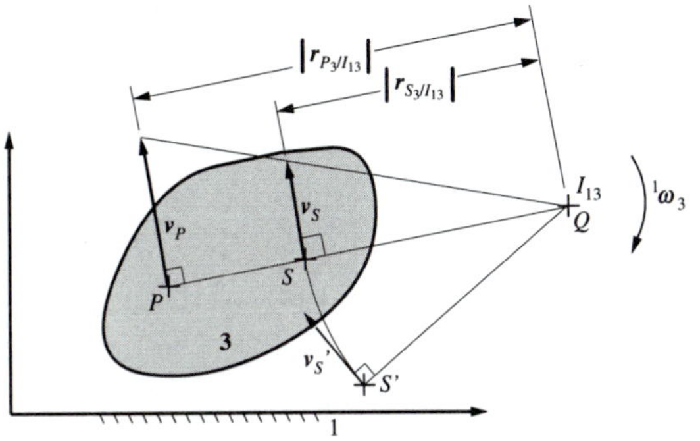

Figure 2.71 The rotating radius method of obtaining the velocity of a point in a body relative to a reference frame (or another body) given the location of the instant center of the body and the velocity of some other point in the body relative to that frame.

This theory is the basis for the rotating radius method. The basic procedure is to find the magnitude of the velocity of one point in the rigid body and draw that velocity vector to scale on the link. The velocity of any other point on the body can then be found by recognizing that the magnitude of the velocity is proportional to the distance from the instant center relative to the frame. Proportional triangles can be drawn by using the line from the original point to the instant center as a base line. Alternately, the line from the new point to the instant center can be used as a base line.

EXAMPLE 2.15 (*Using the Rotating Radius Method for Velocities*)

PROBLEM

For the compound linkage shown in Fig. 2.72, the velocity of point A is given as shown. Find the velocity of point B.

SOLUTION

The first step in the procedure is to determine the instant centers that are needed. This can be done by rewriting the given and desired information in terms of the link numbers and frame number. That is, we are given the velocity ${}^1\boldsymbol{v}_{A_2}$, and we want to find ${}^1\boldsymbol{v}_{B_5}$. Here we see that the reference system is 1 and the two links involved are 2 and 5. In this problem and in general problems using instant centers, we will need to locate three instant centers:

1. I_{12}, which is the instant center between the reference frame and the link where the input information is given.
2. I_{15}, which is the instant center between the reference frame and the link where the velocity is to be found.
3. I_{25}, which is the instant center between the link where a velocity is specified and the link where the velocity is to be found.

When the linkage is analyzed, it is apparent that I_{12} can be found by inspection. Therefore, only I_{25} and I_{15} need to be constructed. This is done by first locating I_{13} using I_{12} and I_{23} and I_{14} and I_{34}. Next, I_{15} is found using I_{13} and I_{35} and I_{16} and I_{56}. Finally, I_{25} is found using I_{15} and I_{12} and I_{23} and I_{35}. The construction lines are shown in Fig. 2.73.

After I_{25} is located, the velocity of I_{25} is found by rotating the triangle formed by the sides $\boldsymbol{v}_{A_2}$ and I_{12}A about I_{12} onto the baseline through I_{12} and I_{25}. The velocity of I_{25} is then found using proportional triangles. Next, the triangle defined by sides $\boldsymbol{v}_{I_{25}}$ and $I_{15}I_{25}$ is rotated about I_{15} onto the baseline through I_{15} and B. The velocity of B is then determined using proportional triangles. Note that when the velocity of I_{25} is found, the instant center is treated as a point in link 2, that is, $(I_{25})_2$. However, when the velocity of B_5 is to be found, the instant center is treated as a point in link 5. This illustrates the fact that the instant center location defines the location of two points, one in link 2 and the other in link 5; however, both points have the same velocity. ■

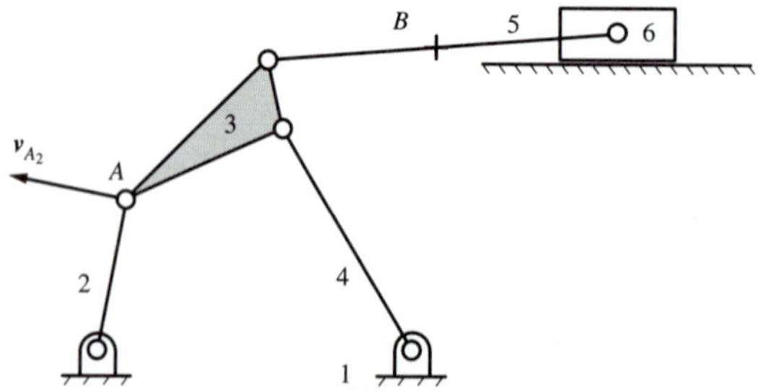

Figure 2.72 Compound linkage for Example 2.15.

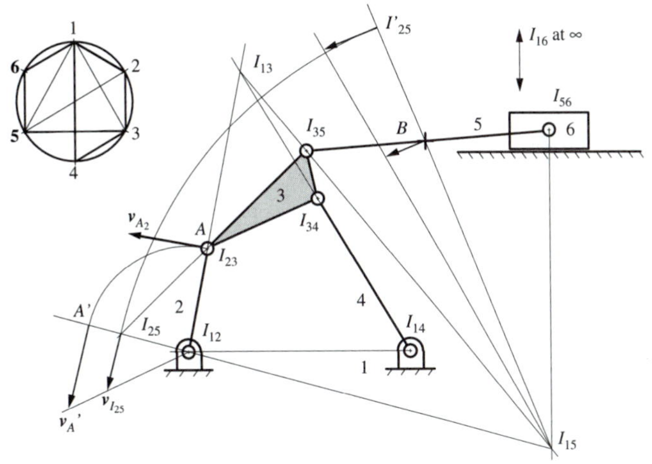

Figure 2.73 Use of the rotating radius method in a compound linkage.

EXAMPLE 2.16 ***(Using Instant Centers to Analyze a Stephenson-II Six-Bar Linkage)***

Problem

Consider the Stephenson-II six-bar linkage in Fig. 2.74. Assume that ${}^1\boldsymbol{\omega}_2$ is given and we want to find ${}^1\boldsymbol{\omega}_5$. This linkage has the characteristics of those described in Section 2.9; that is, the driving link is not included in any four-link loop. The following solution method should be compared to the inversion method in Section 2.9.

SOLUTION

The use of instant centers to solve this problem is especially interesting because the linkage cannot be analyzed using the usual vector polygon approach described earlier.

Again, we need to determine which instant centers are required to solve the problem. Looking at the information that is given and that which is to be found, we see that three links (1, 2, 5) are identified. Therefore, we need to find I_{12}, I_{15}, and I_{25}. Of this set, only I_{25} cannot be determined by inspection. However, it can be found relatively easily from the instant centers that are available by inspection. First using I_{36} and I_{56} and I_{34} and I_{45}, I_{35} can be located. Then using I_{35} and I_{23} and I_{12} and I_{15}, I_{25} can be located. The resulting instant centers are shown in Fig. 2.75. The velocity of the coincident points at I_{25} is given by

$${}^1\boldsymbol{v}_{(I_{25})_2} = {}^1\boldsymbol{\omega}_2 \times \boldsymbol{r}_{I_{25}/I_{12}} = {}^1\boldsymbol{v}_{(I_{25})_5} = {}^1\boldsymbol{\omega}_5 \times \boldsymbol{r}_{I_{25}/I_{15}}$$

Therefore, the magnitudes of the vectors are related by

$$|{}^1\boldsymbol{v}_{(I_{25})_2}| = |{}^1\boldsymbol{\omega}_2||\boldsymbol{r}_{I_{25}/I_{12}}| = |{}^1\boldsymbol{v}_{(I_{25})_5}| = |{}^1\boldsymbol{\omega}_5||\boldsymbol{r}_{I_{25}/I_{15}}|$$

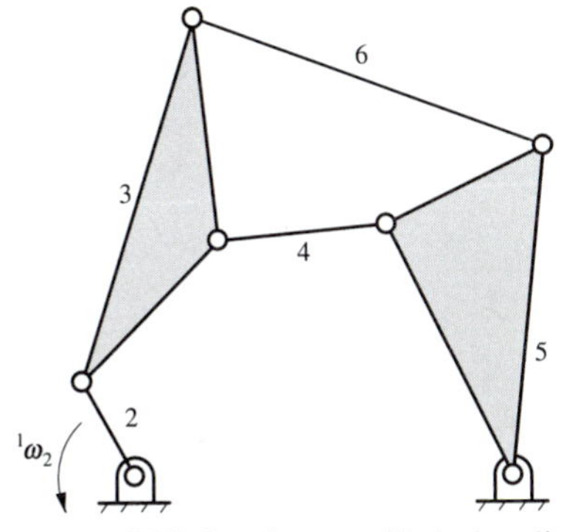

Figure 2.74 Stephenson-II six-bar linkage for Example 2.16.

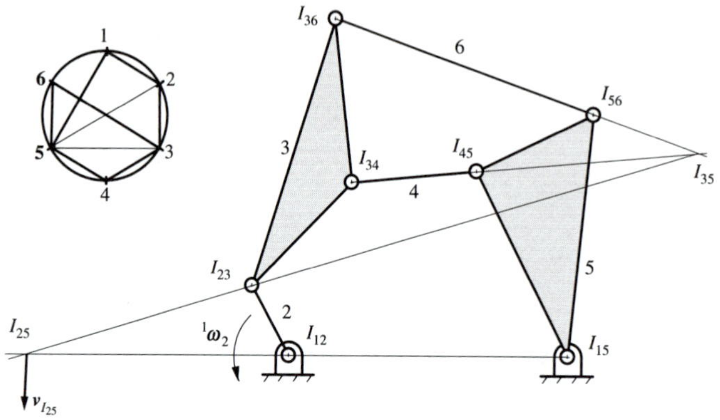

Figure 2.75 The instant center method applied to a Stephenson-II six-bar linkage.

and

$$|{}^1\boldsymbol{\omega}_5| = |{}^1\boldsymbol{\omega}_2| \frac{|\boldsymbol{r}_{I_{25}/I_{12}}|}{|\boldsymbol{r}_{I_{25}/I_{15}}|}$$

This gives the magnitude of ${}^1\boldsymbol{\omega}_5$. We can get the direction by determining the sense of the velocity of ${}^1\boldsymbol{v}_{(I_{25})_5}$. Because the vector is generally downward, the angular velocity must be CCW to satisfy the cross-product sign convention. ■

EXAMPLE 2.17 ***(Finding Instant Centers for a Quick-Return Mechanism)***

PROBLEM

Find all the instant centers for the quick-return linkage shown in Fig. 2.76. The linkage is driven by the crank 2 rotating about the fixed revolute at point O_A. A pin fixed to link 2 at A slides

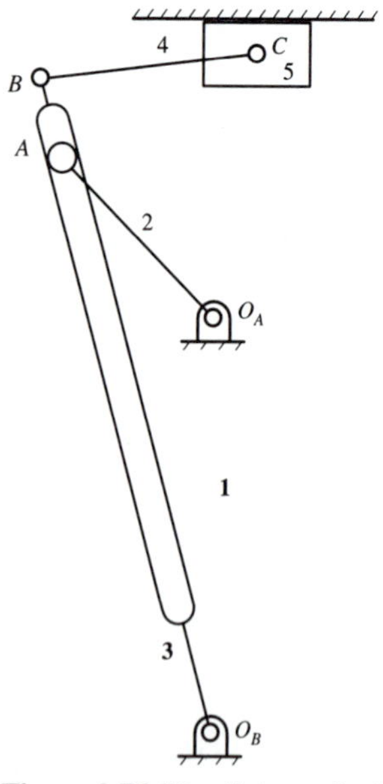

Figure 2.76 The linkage for Example 2.17

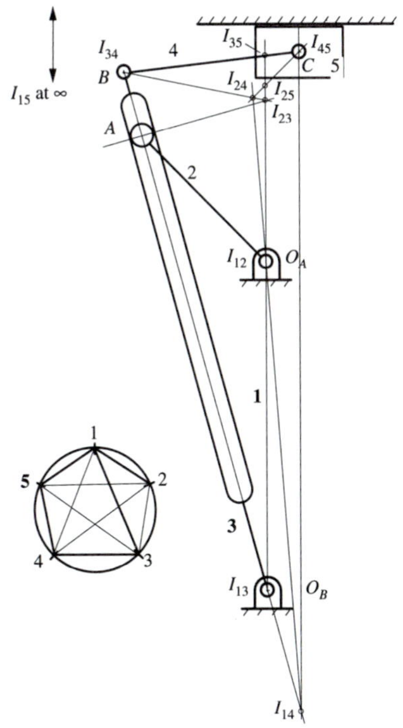

Figure 2.77 Location of instant centers for Example 2.17.

in a slot in link 3. Link 3 rotates about a fixed revolute at point O_B. It is hinged at point B to the connecting link, 4. Link 4 connects to the horizontally sliding block, 5, via a revolute at point C. This type of linkage is used extensively in the machine tools called planers and shapers because it generates a relatively slow and uniform forward, or cutting, stroke and a considerably quicker return stroke. The ratio of the durations of the two strokes can be determined by considering the angles through which the drive crank 2 rotates between the extreme positions of the rocker arm 3. The extreme positions are those in which O_AA is normal to O_BB.

SOLUTION

The instant centers are shown in Fig. 2.77. In practice, it is seldom necessary to locate all instant centers. The great advantage of the instant center technique is its ease of use for complicated linkages, particularly when only the angular velocity of one member or the velocity of one point is to be found. For this problem only three instant centers are needed, although others may be needed in the process of locating them. The three instant centers needed are the set for the input link, output link, and base link. Here the input link is the link whose angular velocity is given or that contains a point whose velocity is given. The output link is the link whose angular velocity is sought or that contains the point whose linear velocity is sought. ■

EXAMPLE 2.18 ***(Finding Instant Centers of a Quick-Return Mechanism in a Singular Position)***

PROBLEM

Find all the instant centers of the quick-return linkage in Example 2.17 when point C is collinear with O_A and O_B. This is shown in Fig. 2.78.

SOLUTION

If an attempt is made to find the instant centers with the procedure used in Example 2.17, it will be possible to find I_{23} and I_{35} directly as shown in Fig. 2.79. However, it is not possible to find the locations of the remaining instant centers by simple construction because all of the remaining instant centers are located on the line defined by I_{12} and I_{13}. To determine the

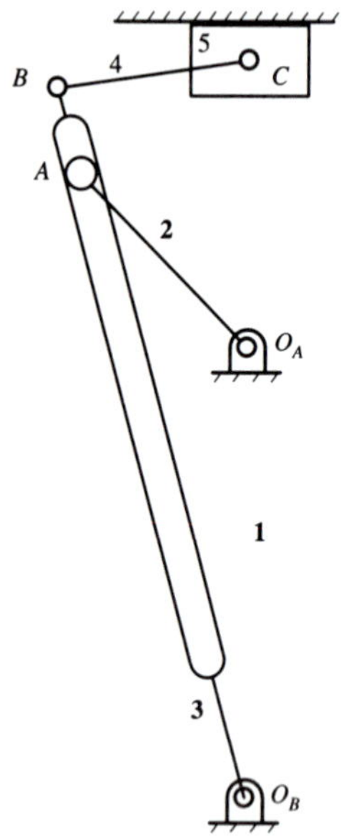

Figure 2.78 The linkage for Example 2.18.

location of the remaining instant centers, let point C be moved slightly off of the line defined by I_{12} and I_{13} and locate the instant centers. The location of the instant centers in the true position can then be determined by visualizing their movement as C approaches its actual position. This is shown in Figs. 2.80 and 2.81. Note that as C moves toward the vertical position, I_{35} becomes coincident with I_{45}, I_{14} becomes coincident with I_{13}, and I_{25} and I_{24} become coincident with I_{23}. ■

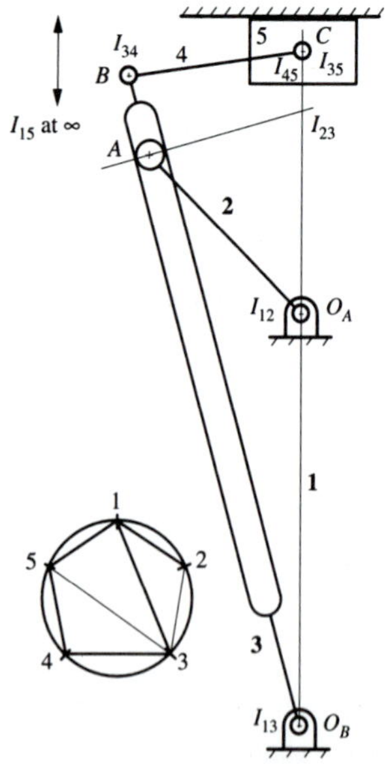

Figure 2.79 Location of I_{23} and I_{35} in Example 2.18.

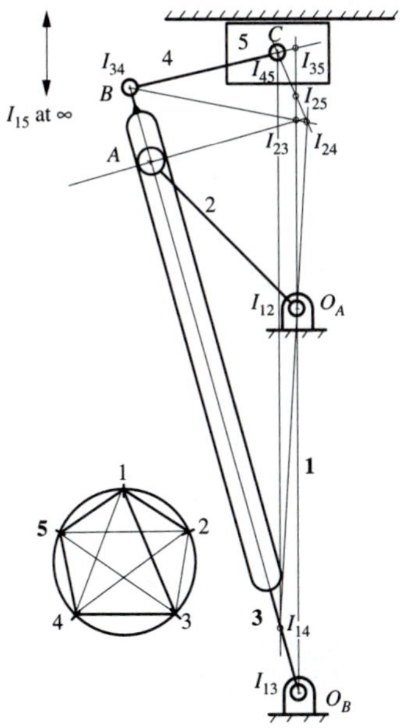

Figure 2.80 Instant centers when C is not in line with Example 2.18.

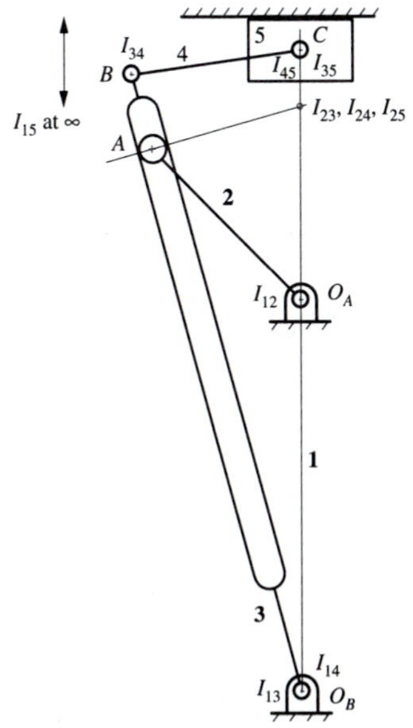

Figure 2.81 Actual location of I_{24}, I_{25}, I_{35}, and I_{14} in Example 2.18.

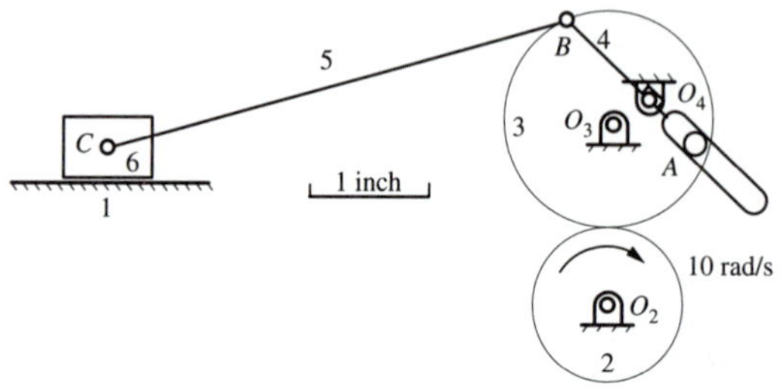

Figure 2.82 The linkage of Example 2.19.

EXAMPLE 2.19 ***(Using Instant Centers to Analyze a Gear Mechanism)***

PROBLEM

Find the velocity of point C in Fig. 2.82 given that the angular velocity of gear 2 is 10 rad/s CW. B is a hinge connecting links 4 and 5 and does not connect to gear 3. Point A is a pin in link 3 that engages a slot in link 4.

SOLUTION

To find the velocity of point C, considered as a point in link 5, from the angular velocity of link 2 relative to link 1, the instant centers I_{12}, I_{15}, and I_{25} are needed. These may be located as shown in Fig. 2.83.

Then,

$$^1\boldsymbol{\omega}_5 = {}^1\boldsymbol{\omega}_2 \times (I_{12}I_{25})/(I_{15}I_{25}) = 10 \times 0.940/7.261 = 1.29 \text{ rad/s CW}$$

$$\boldsymbol{v}_{C_5} = {}^1\boldsymbol{\omega}_5 \times (I_{15}C) = 1.29 \times 4.653 = 6.00 \text{ in/s to the left}$$

Notice that the instant center method is extremely efficient for simple input-output problems, such as this one, in which only two links are of interest. ■

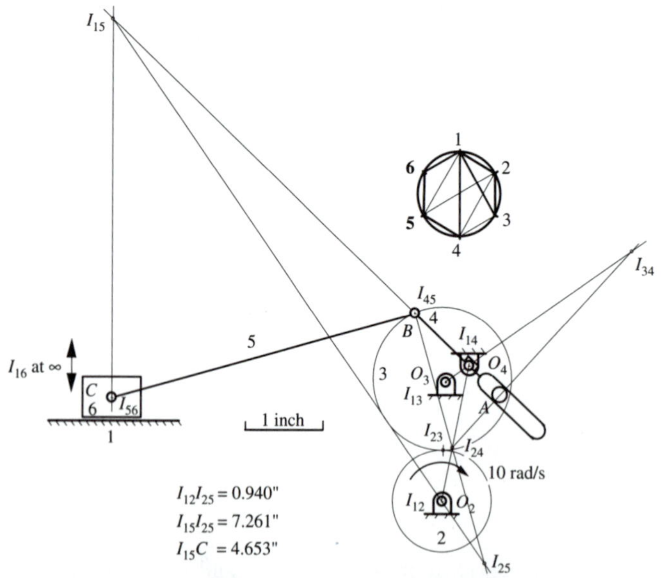

Figure 2.83 Location of instant centers for Example 2.19.

2.18 FINDING INSTANT CENTERS USING DRAFTING PROGRAMS

In this chapter, we have implied that the instant center approach to velocity analysis is a purely graphical approach. However, this does not mean that the actual drawings must be done on a drawing board. A better approach is to use one of the many drafting programs available on computers. The drafting package can be used to:

1. Draw the basic linkage to scale
2. Find the instant centers
3. Use available dimension routines to find appropriate distances
4. Use calculators available on computers to determine the desired velocities

This procedure is relatively fast and accurate, especially if the drafting package will allow the user to draw parallel and perpendicular lines accurately. The results can be easily imported into reports and other documents. This environment also allows the user to explore other design alternatives to obtain desired velocity results. From the examples in this chapter, it is clear that significant changes in the velocity can be made by small alterations in the link lengths or pivot locations.

2.19 CHAPTER 2 Exercise Problems

PROBLEM 2.1

In the mechanism shown below, link 4 moves to the right with a velocity of 75 ft/s. Construct the complete velocity polygon and determine $\boldsymbol{v}_{B_2}$, $\boldsymbol{v}_{G_3}$, ${}^1\boldsymbol{\omega}_2$, and ${}^1\boldsymbol{\omega}_3$. List all equations used, and show necessary computations.

Link Lengths

$AB = 4.8$ in, $BC = 16.0$ in, $BG = 6.0$ in

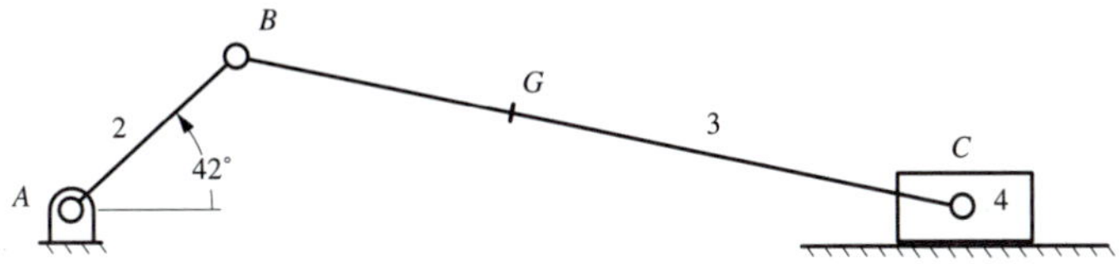

PROBLEM 2.2

In the mechanism shown below, link 2 is rotating CW at the rate of 180 rad/s. Construct the complete velocity polygon, and determine ${}^1\boldsymbol{v}_{B_2}$, ${}^1\boldsymbol{v}_{C_3}$, ${}^1\boldsymbol{v}_{E_3}$, ${}^1\boldsymbol{\omega}_3$, ${}^1\boldsymbol{\omega}_4$.

Link Lengths

$AB = 4.6$ in, $BC = 12.0$ in, $AD = 15.2$ in, $CD = 9.2$ in, $EB = 8.0$ in, $CE = 5.48$ in

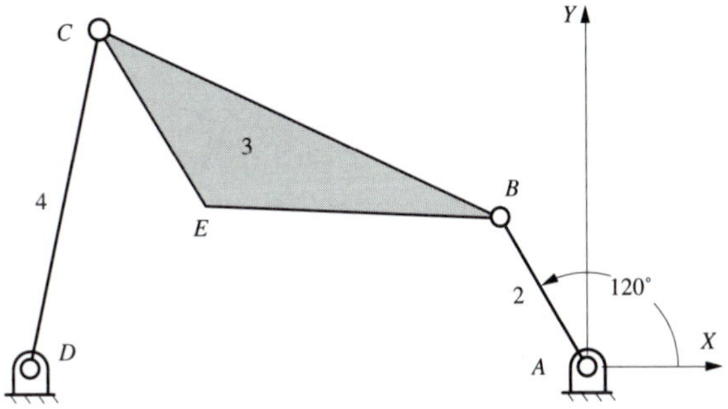

(Problems 2.1–2.11 emphasize velocity analyses.)

PROBLEM 2.3

In the mechanism shown below, ${}^1\mathbf{v}_{B_2} = 120$ in/s. Point C is located 1.5 in above point A and 4.75 in to the right of A. Find ${}^1\mathbf{v}_{C_3}$, ${}^1\mathbf{v}_{D_3}$, ${}^1\boldsymbol{\omega}_3$, and the velocity image of link 3.

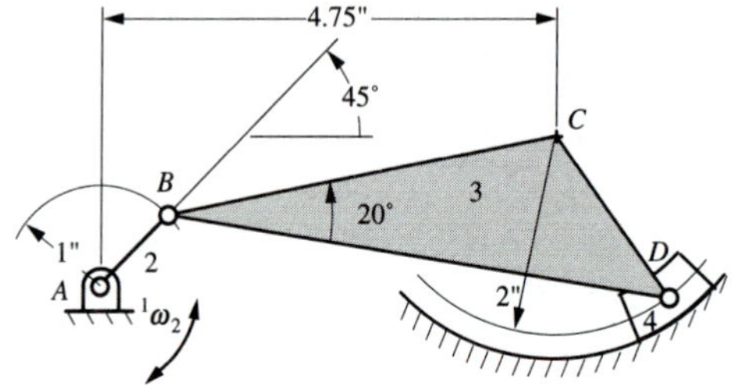

PROBLEM 2.4

In the mechanism shown below, link 2 is turning CW at the rate of 200 rpm. Draw the velocity polygon for the mechanism, and record values for ${}^1\mathbf{v}_{C_3}$ and ${}^1\boldsymbol{\omega}_3$.

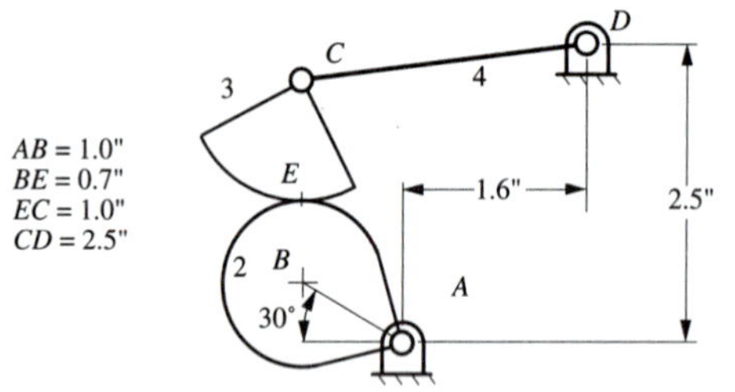

PROBLEM 2.5

Draw the velocity polygon to determine the velocity of link 6. Points A, C, and E have the same vertical coordinate.

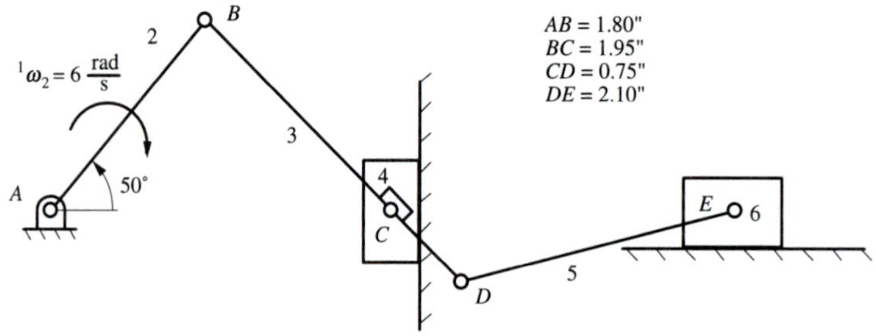

PROBLEM 2.6

In the mechanism shown, ${}^1\mathbf{v}_{A_2} = 20$ in/s. Find ${}^1\boldsymbol{\omega}_5$ and ${}^3\boldsymbol{\omega}_4$. Indicate on link 4 the point that has zero velocity. In the drawing, H and G are the centers of curvature of links 4 and 5, respectively, corresponding to location D. F is the center of curvature of link 3 corresponding to C. Also, point G lies exactly above point E.

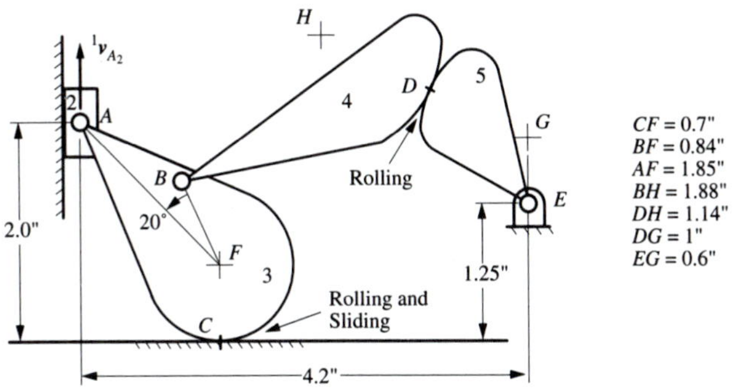

PROBLEM 2.7 Link 2 of the linkage shown in the figure has an angular velocity of 10 rad/s CCW. Find the angular velocity of link 6 and the velocities of points B, C, and D.

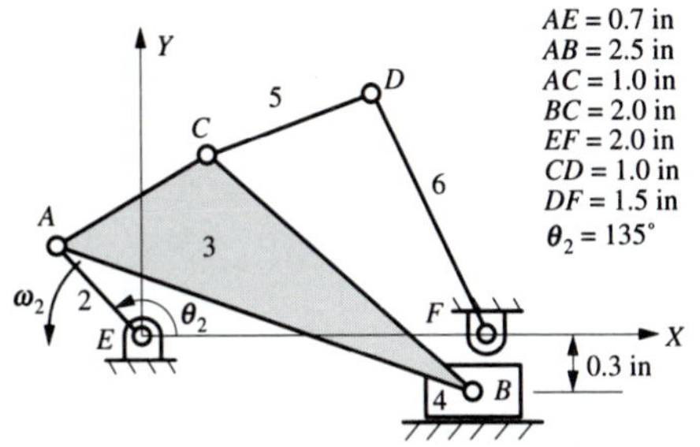

PROBLEM 2.8 In the position shown AB is horizontal. Draw the velocity diagram to determine the sliding velocity of link 6. Determine a new position for point C (between B and D) so that the velocity of link 6 would be equal and opposite to the one calculated for the original position of point C.

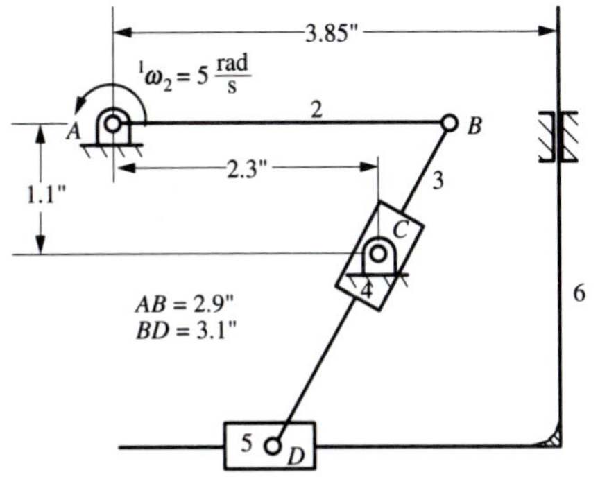

PROBLEM 2.9 The scotch-yoke mechanism is driven by crank 2 at $^1\omega_2$ = 36 rad/s (CCW). Link 4 slides horizontally. Find the velocity of point B on link 4.

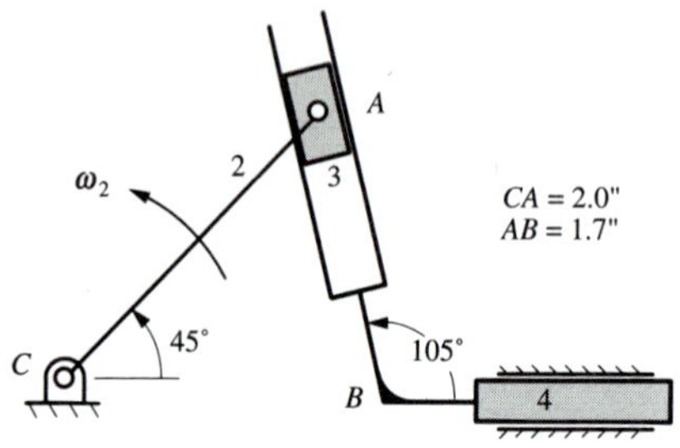

PROBLEM 2.10

In the figure below, points A and C have the same horizontal coordinate, and ${}^1\boldsymbol{\omega}_2 = 30$ rad/s. Draw and dimension the velocity polygon. Identify the sliding velocity and find ${}^1\boldsymbol{\omega}_3$.

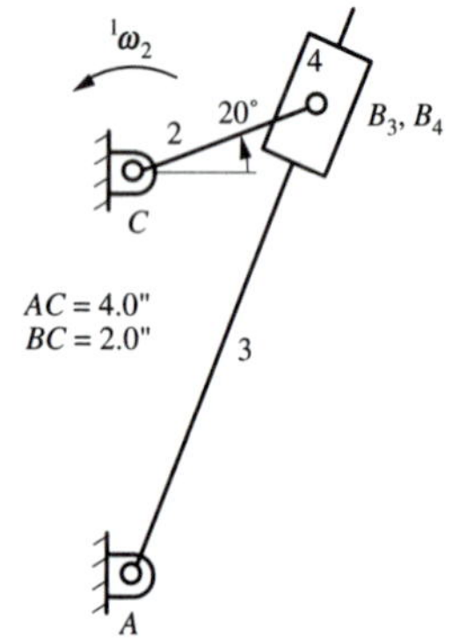

PROBLEM 2.11

The linkage shown is used to raise the fabric roof on a convertible automobile. The dimensions at given as shown. Link 2 is driven by a DC motor through a gear reduction. If the angular velocity, ${}^1\boldsymbol{\omega}_2 = 2$ rad/s, CCW, determine the linear velocity of point J, which is the point where the linkage connects to the automobile near the windshield.

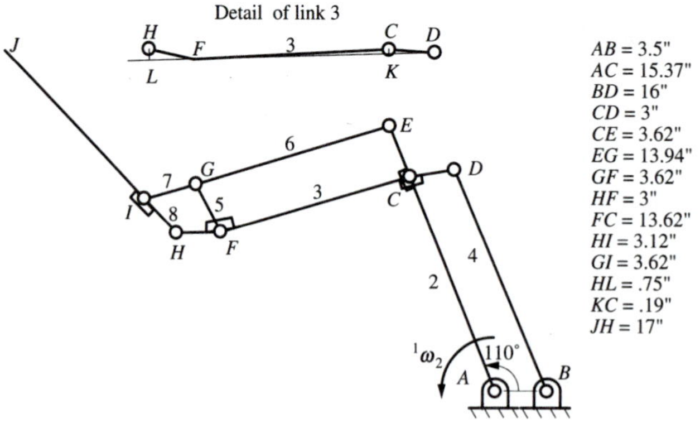

PROBLEM 2.12

In the mechanism shown, ${}^1\boldsymbol{\omega}_2 = 4$ rad/s CCW (constant). Construct the velocity and acceleration polygons and find ${}^1\boldsymbol{\omega}_3$, ${}^1\mathbf{v}_{E_3}$, ${}^1\mathbf{v}_{E_4}$, ${}^1\boldsymbol{\alpha}_3$, ${}^1\mathbf{a}_{E_3}$, and ${}^1\mathbf{a}_{E_4}$ using the image technique when it is applicable. Also find the point in link 3 that has zero acceleration for the position given.

(Problems 2.12–2.13 emphasize velocity and acceleration image.)

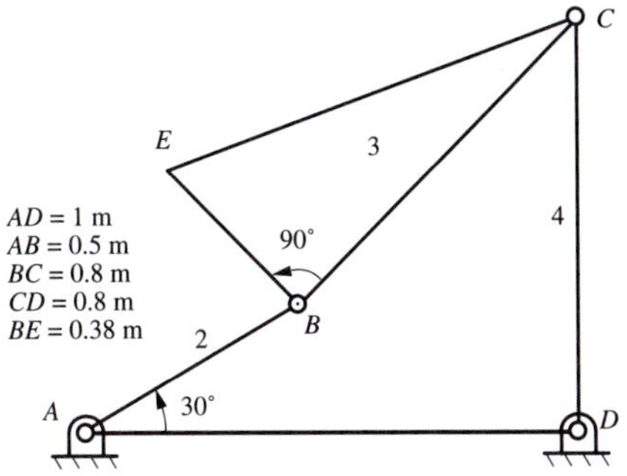

PROBLEM 2.13

In the mechanism shown, point A lies on the X axis. Draw the basic velocity and acceleration polygons and use the image technique to determine the velocity and acceleration of point D_4. Then determine the velocity and acceleration images of link 4. Draw the images on the velocity and acceleration polygons.

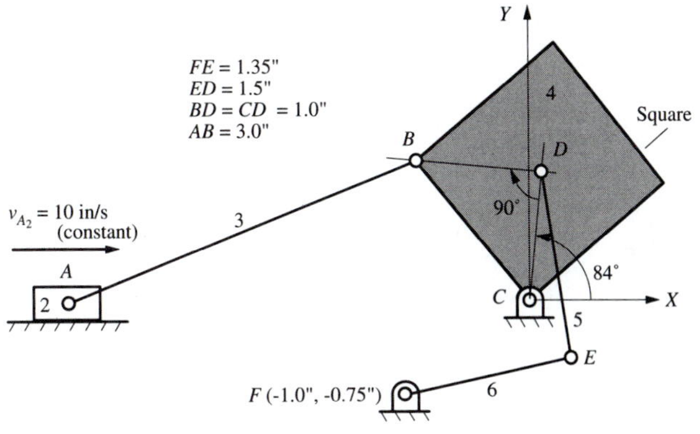

PROBLEM 2.14

The angular velocity of link 2 of the mechanism shown is 20 rad/s, and the angular acceleration is 100 rad/s^2 at the instant being considered. Determine the linear velocity and acceleration of point F_6. Compute the quantities graphically, and properly label all terms.

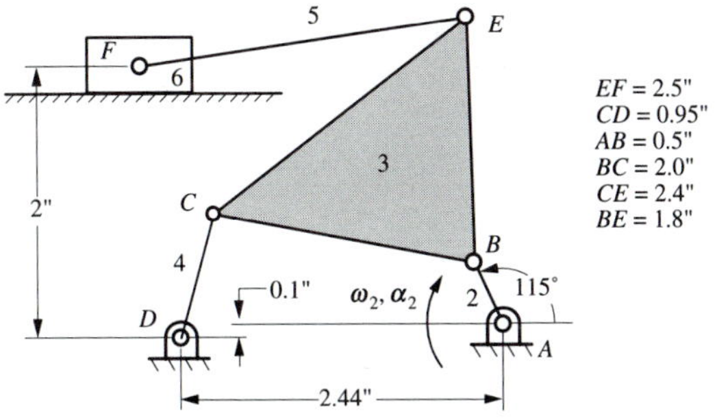

(Problems 2.14–2.28 emphasize velocities and accelerations.)

PROBLEM 2.15

In the drag-link mechanism shown, link 2 is turning CW at the rate of 130 rpm. Construct the velocity and acceleration polygons and compute the following: ${}^1\boldsymbol{a}_{E_5}$, ${}^1\boldsymbol{a}_{F_6}$, and the angular acceleration of link 5.

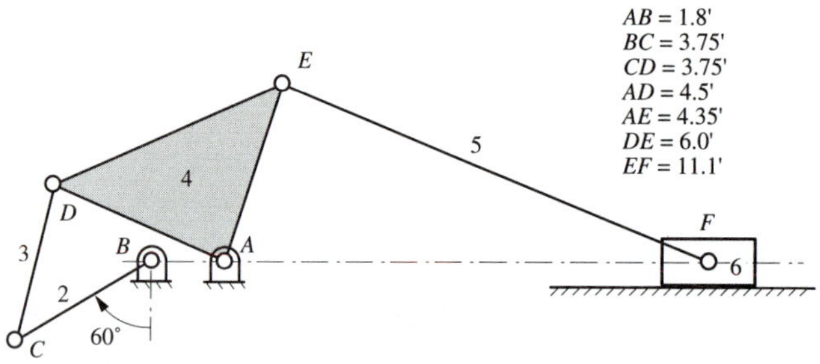

PROBLEM 2.16

In the mechanism shown below, link 2 is turning CW at the rate of 20 rad/s, and link 3 rolls on link 2. Draw the velocity and acceleration polygons for the mechanism, and record values for ${}^1\boldsymbol{a}_{C_3}$ and ${}^1\boldsymbol{\alpha}_3$.

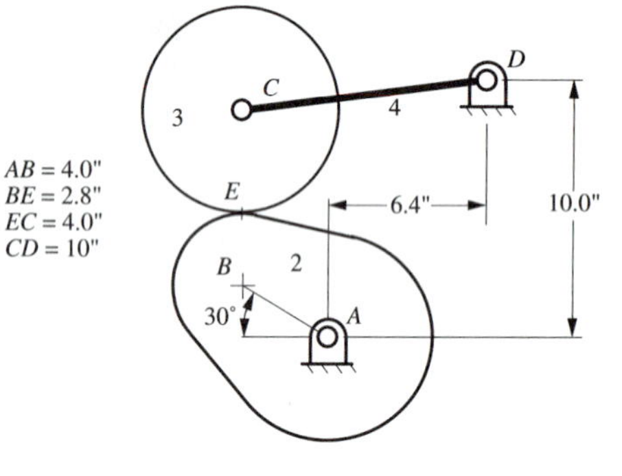

PROBLEM 2.17

In the mechanism shown below, the velocity of A_2 is 10 in/s to the right and is constant. Draw the velocity and acceleration polygons for the mechanism, and record values for angular velocity and acceleration of link 6. Use the image technique to determine the velocity of points D_3 and E_3, and locate the point in link 3 that has zero velocity. Note the links to which the points belong.

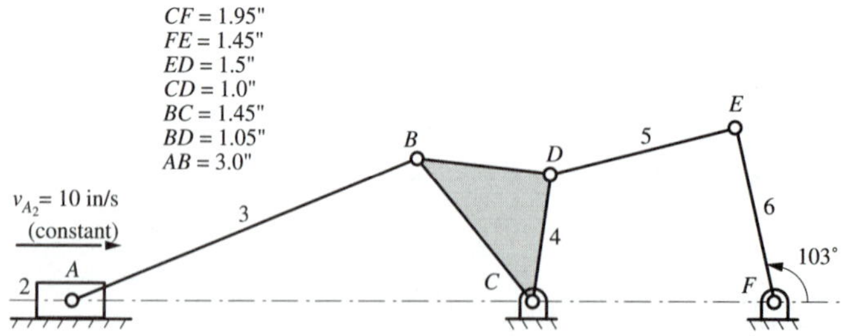

PROBLEM 2.18

For the four-bar linkage, assume that ${}^1\boldsymbol{\omega}_2 = 50$ rad/s CW and ${}^1\boldsymbol{\alpha}_2 = 1600$ rad/s^2 CW. Construct the velocity and acceleration polygons and compute the following: ${}^1\boldsymbol{v}_{B_3}$, ${}^1\boldsymbol{v}_{C_3}$, ${}^1\boldsymbol{v}_{E_3}$, ${}^1\boldsymbol{\omega}_3$, ${}^1\boldsymbol{\omega}_4$, ${}^1\boldsymbol{a}_{B_3}$, ${}^1\boldsymbol{a}_{C_3}$, ${}^1\boldsymbol{a}_{E_3}$, ${}^1\boldsymbol{\alpha}_3$, and ${}^1\boldsymbol{\alpha}_4$.

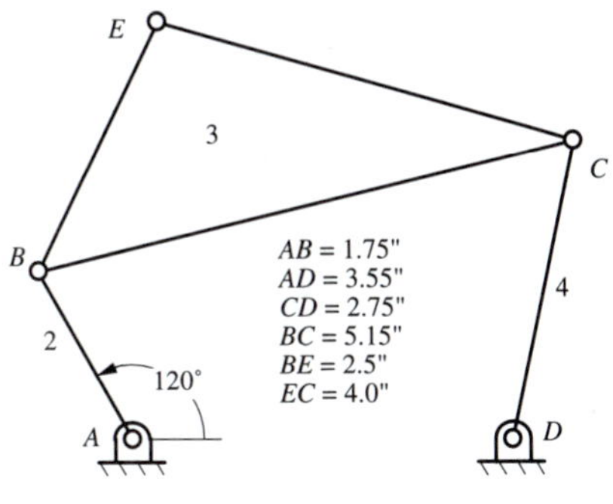

PROBLEM 2.19

Resolve Problem 2.18 if ${}^1\boldsymbol{\omega}_2 = 50$ rad/s CCW and ${}^1\boldsymbol{\alpha}_2 = 0$.

PROBLEM 2.20

The accelerations of points *A* and *B* in the coupler below are as given. Determine the acceleration of the center of mass *G* and the angular acceleration of the body. Draw the vector representing ${}^1\boldsymbol{a}_G$ from *G*.

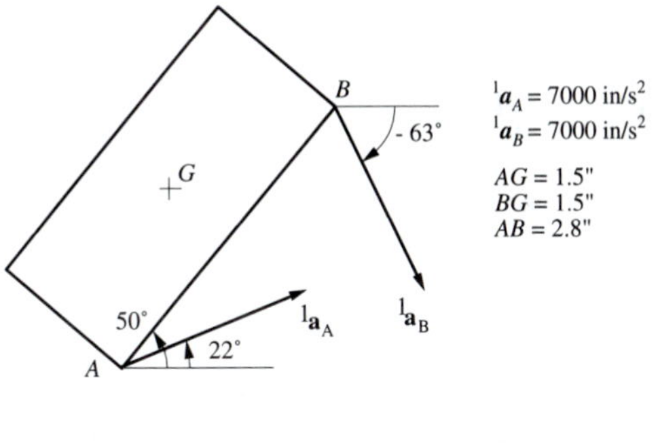

PROBLEM 2.21

The instant center of acceleration of a link can be defined as that point in the link that has zero acceleration. If the accelerations of points *A* and *B* are as given in the rigid body shown below, find the point *C* in that link at which the acceleration is zero.

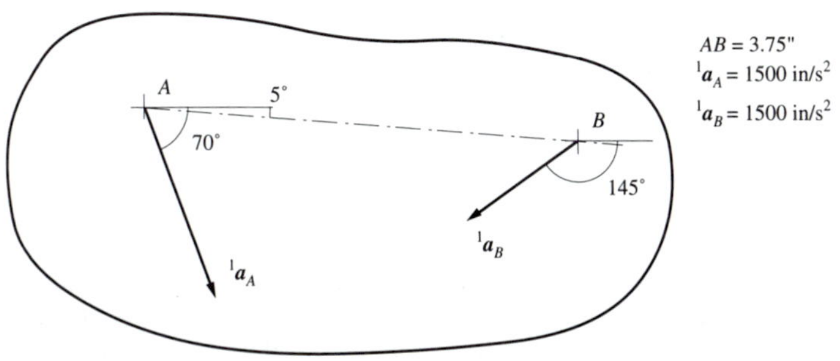

PROBLEM 2.22 Crank 2 of the push-link mechanism shown in the figure is driven at a constant angular velocity ${}^1\omega_2 = 60$ rad/s (CW). Find the velocity and acceleration of point F and the angular velocity and acceleration of links 3 and 4.

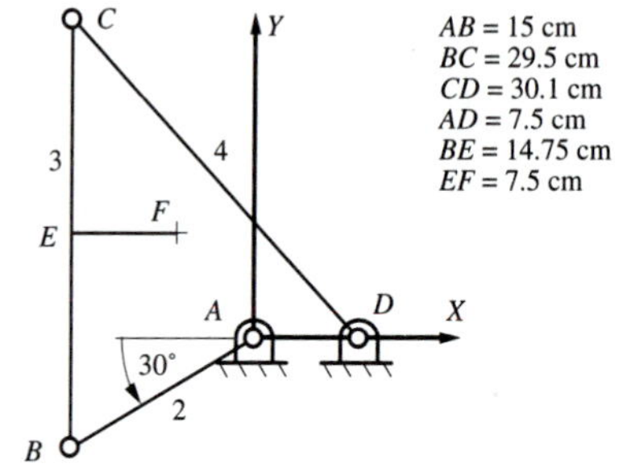

PROBLEM 2.23 For the linkage shown, ${}^1\omega_2 = 10$ rad/s CCW and ${}^1\alpha_2 = 100$ rad/s^2 CCW. Determine ${}^1\omega_3$ and ${}^1\alpha_3$.

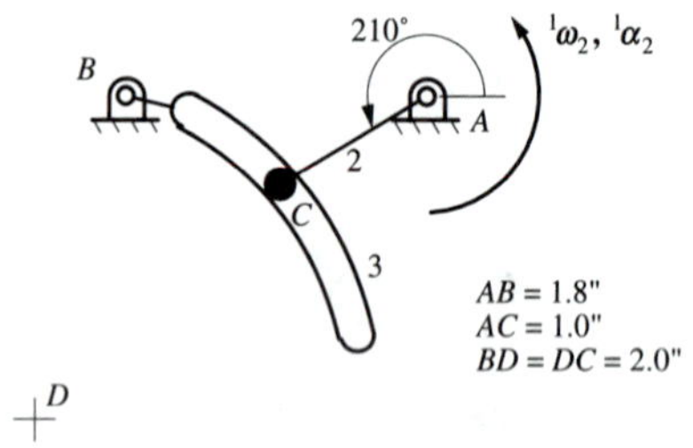

PROBLEM 2.24 For the straight-line mechanism shown in the figure, ${}^1\omega_2 = 20$ rad/s (CW) and ${}^1\alpha_2 = 140$ rad/s^2 (CW). Determine the velocity and acceleration of point B and the angular acceleration of link 3.

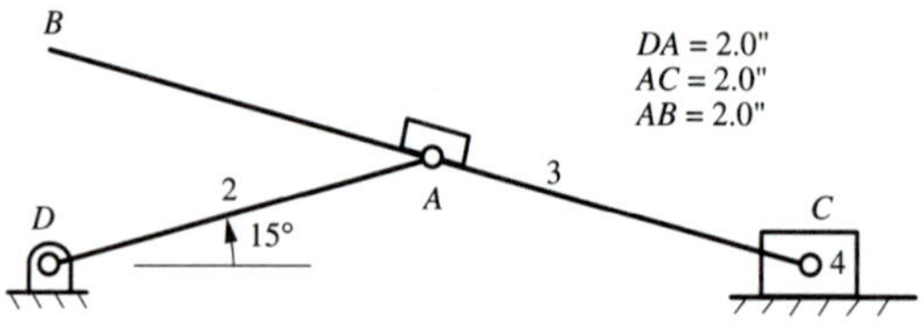

PROBLEM 2.25 The figure shows the mechanism used in two-cylinder 60-degree V-engine consisting, in part, of an articulated connecting rod. Crank 2 rotates at a constant 2000 rpm CW. Find the velocities and acceleration of points B, C, and D and the angular acceleration of links 3 and 5.

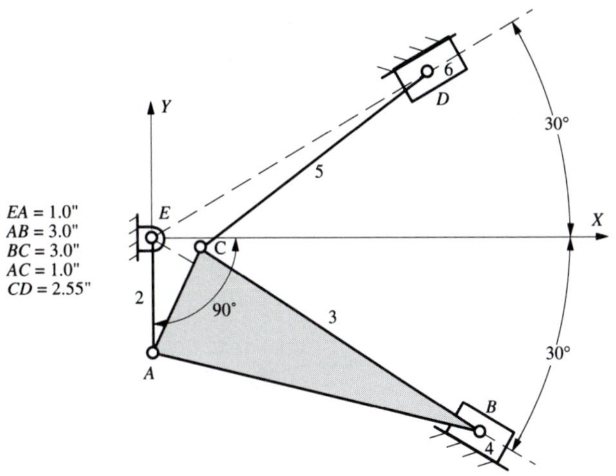

PROBLEM 2.26

In the figure below, ${}^1\boldsymbol{\omega}_2 = 500$ rad/s CCW (constant). Find ${}^1\boldsymbol{\omega}_4$, ${}^2\boldsymbol{\omega}_4$, ${}^1\boldsymbol{\omega}_3$, ${}^6\boldsymbol{\omega}_5$, ${}^3\boldsymbol{\omega}_5$, ${}^1\boldsymbol{v}_D$, ${}^1\boldsymbol{\alpha}_4$, ${}^2\boldsymbol{\alpha}_4$, ${}^1\boldsymbol{\alpha}_3$, ${}^6\boldsymbol{\alpha}_5$, and ${}^1\boldsymbol{a}_D$.

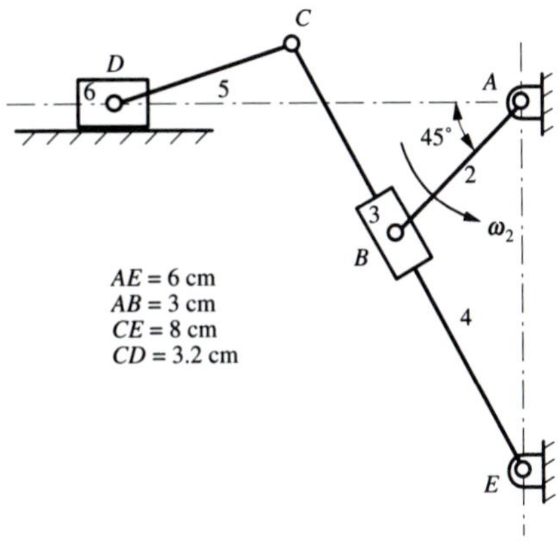

PROBLEM 2.27

The circular cam shown is driven at an angular velocity ${}^1\omega_2 = 15$ rad/s (CW) and ${}^1\alpha_2 = 100$ rad/s^2 (CW). There is rolling contact between the cam and the roller, link 3. Find the angular velocity and angular acceleration of the oscillating follower, link 4.

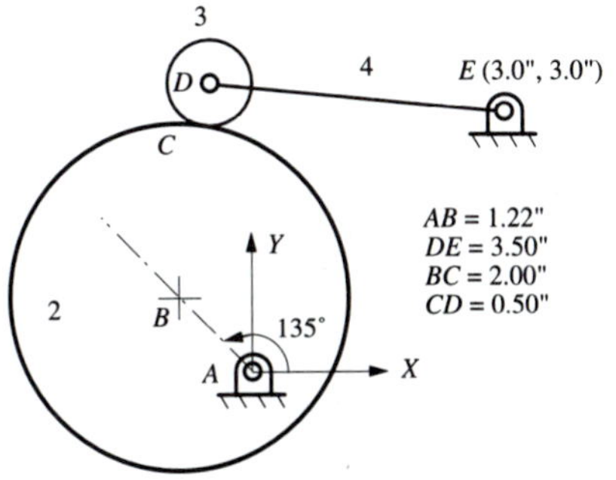

PROBLEM 2.28

Part of a 10-link mechanism is shown in the figure. Links 7 and 8 are drawn to scale, and the velocity and acceleration of points D_7 and F_8 are given. Find ${}^1\boldsymbol{\omega}_8$ and ${}^1\boldsymbol{\alpha}_7$ for the position given. Also find the velocity of G_7 by image.

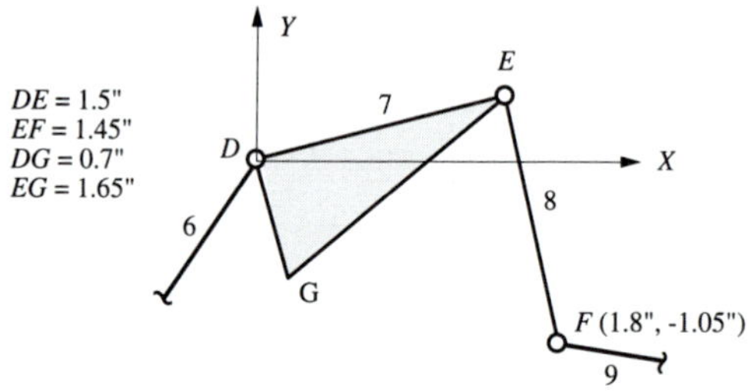

$${}^1\mathbf{v}_{D_7} = 6.0\ \underline{/353^\circ}\ \text{in/s} \qquad {}^1\mathbf{a}_{D_7} = 40\ \underline{/235^\circ}\ \text{in/s}^2$$

$${}^1\mathbf{v}_{F_8} = 7.5\ \underline{/54^\circ}\ \text{in/s} \qquad {}^1\mathbf{a}_{F_8} = 30\ \underline{/305^\circ}\ \text{in/s}^2$$

PROBLEM 2.29

In the two-degree-of-freedom mechanism shown, ${}^1\boldsymbol{\omega}_2$ is given as 10 rad/s CCW. What should the linear velocity of link 6 be so that ${}^1\boldsymbol{\omega}_4 = 5$ rad/s CCW? Draw the complete velocity polygon and show all equations and calculations.

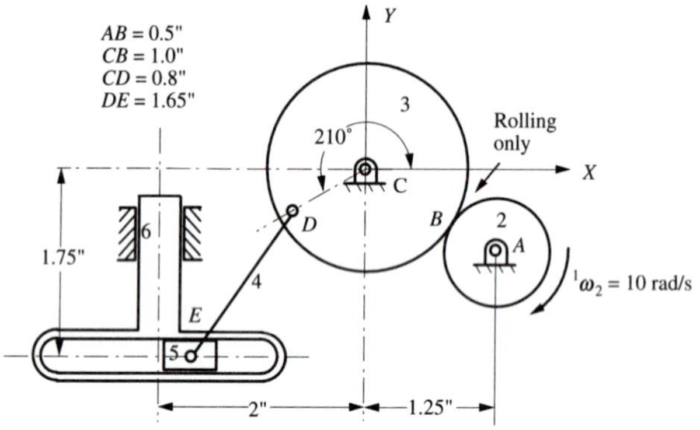

PROBLEM 2.30

For the mechanism shown, assume that link 2 rolls on the frame (link 1) and link 4 rolls on link 3. Assume that link 2 is rotating CW with a constant angular velocity of 100 rad/s. Determine the angular acceleration of link 3 and link 4.

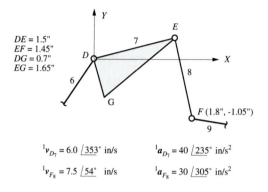

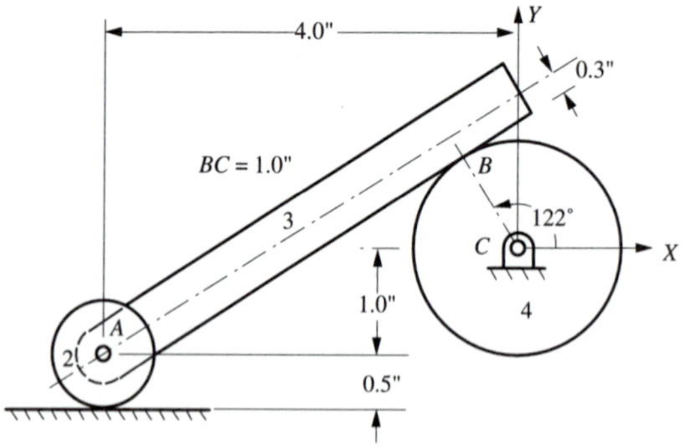

(Problems 2.29–2.32 emphasize rolling contact analyses.)

PROBLEM 2.31

For the mechanism shown, assume that link 4 rolls on the frame (link 1). If link 2 is rotating CW with a constant angular velocity of 10 rad/s, determine the angular accelerations of links 3 and 4 and the acceleration of point E on link 3.

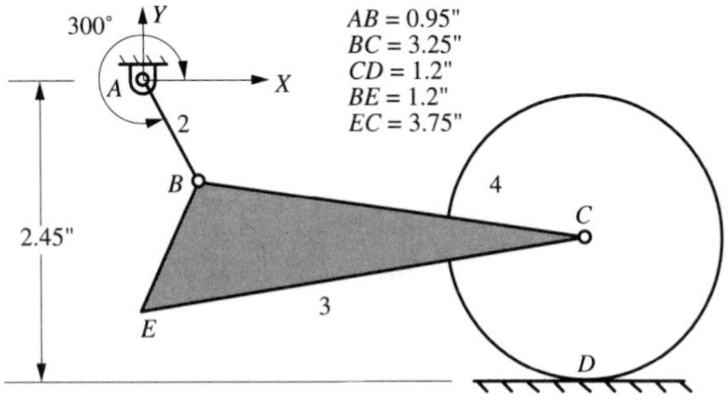

PROBLEM 2.32

If ${}^1\mathbf{v}_{A_2} = 10$ in/s (constant) downward, find ${}^1\boldsymbol{\omega}_3$, ${}^1\boldsymbol{\alpha}_3$, ${}^1\mathbf{v}_{C_3}$, and ${}^1\mathbf{a}_{C_3}$.

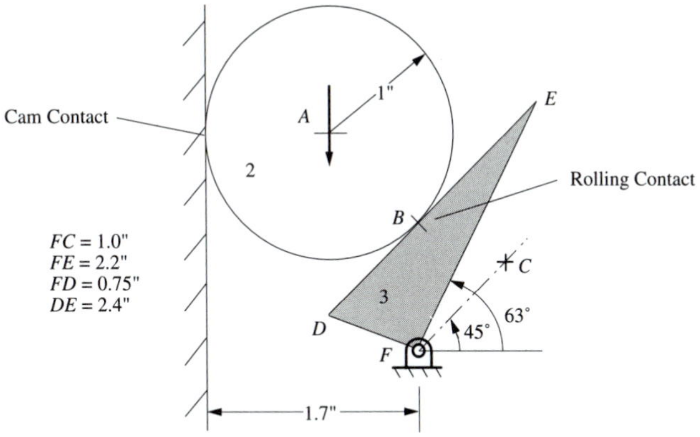

PROBLEM 2.33

In the mechanism shown, ${}^1\omega_2 = 20$ rad/s. At the instant shown, point D, the center of curvature of link 3, lies directly above point E, and point B lies directly above point A. Determine ${}^1\mathbf{v}_{C_3/C_2}$ and ${}^1\boldsymbol{\omega}_3$ using (1) equivalent linkages and (2) coincident points at C.

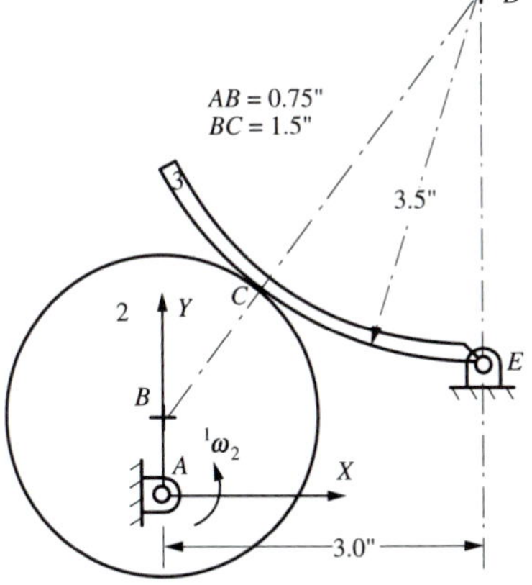

(Problems 2.33–2.34 emphasize cam contact.)

PROBLEM 2.34 In the mechanism shown, ${}^1\boldsymbol{\omega}_2 = 10$ rad/s. Determine ${}^1\boldsymbol{v}_{C_3/C_2}$ and ${}^1\boldsymbol{v}_{C_3}$ using two approaches:

1. Equivalent linkages
2. Coincident points at C

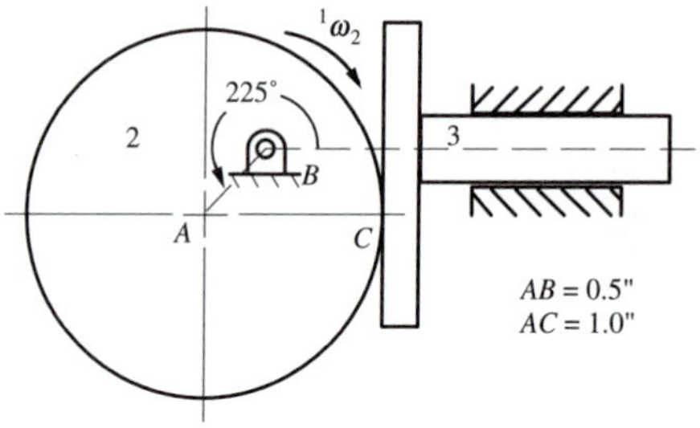

PROBLEM 2.35 On the mechanism shown, link 4 slides on link 1, and link 3 slides on link 4 around the circle arc. Link 2 is pinned to links 1 and 3 as shown. Determine the location of the center of curvature of the path that point P_4 traces on link 2.

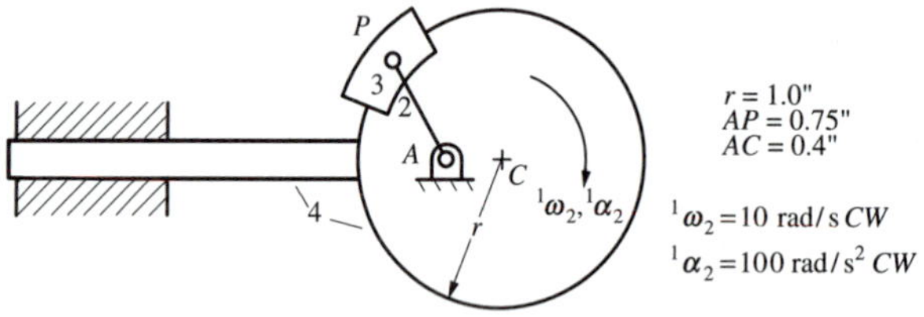

PROBLEM 2.36 For the mechanism shown, find ${}^1\boldsymbol{\omega}_2$, ${}^1\boldsymbol{\alpha}_2$, ${}^1\boldsymbol{v}_{B_2}$, ${}^1\boldsymbol{a}_{B_2}$, ${}^1\boldsymbol{v}_{D_3}$, ${}^1\boldsymbol{a}_{D_3}$, and the location of the center of curvature of the path that point B_3 traces on link 2.

$AB = AC = 10$ cm $\quad CD = 14$ cm
${}^1\boldsymbol{\omega}_3 = 1$ rad/s CCW $\quad {}^1\boldsymbol{\alpha}_3 = 1$ rad/s^2 CW

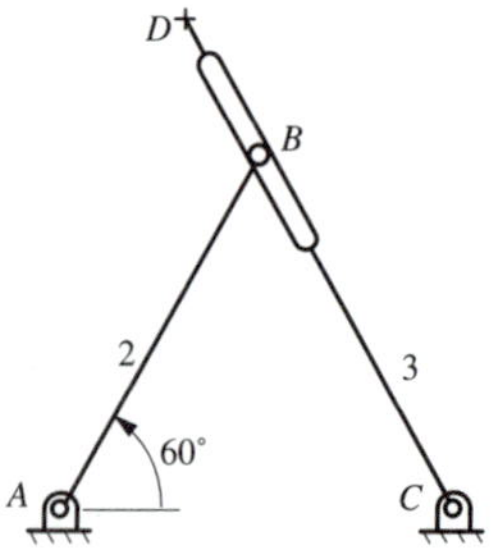

(Problems 2.35–2.42 emphasize coincident points.)

PROBLEM 2.37

In the mechanism below, the velocity and acceleration of point B are given. Determine the angular velocity and acceleration of links 3 and 4. On the velocity and acceleration diagrams, locate the velocity and acceleration of point E on link 3.

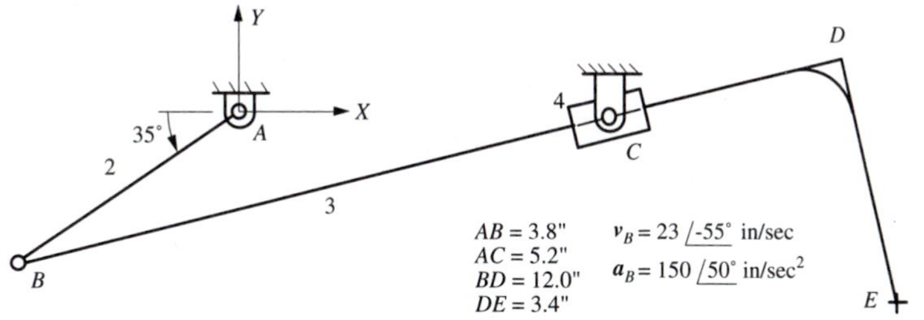

PROBLEM 2.38

In the mechanism below, the angular velocity of link 2 is 60 rpm CCW (constant). Determine the acceleration of point C_6 and the angular velocity of link 6.

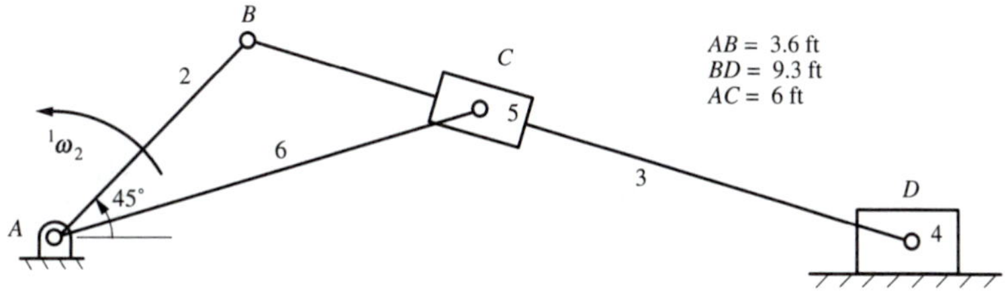

PROBLEM 2.39

In the clamping device shown, links 3 and 4 are an air cylinder. If the opening rate of the air cylinder is 5 cm/s and the opening acceleration of the cylinder is 2 cm/s^2, find the angular velocity and acceleration of link 2 and the linear velocity and acceleration of point D on link 2.

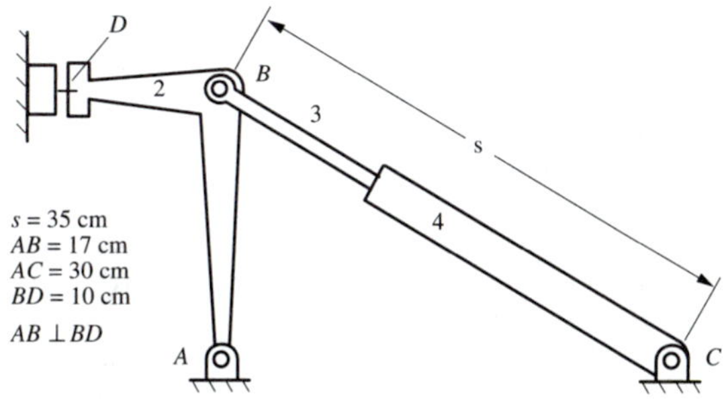

PROBLEM 2.40 If ${}^1\boldsymbol{\omega}_2 = 10$ rad/s (constant), find ${}^1\boldsymbol{v}_{B_2}$, ${}^1\boldsymbol{v}_{B_3}$, ${}^1\boldsymbol{a}_{B_3}$, and ${}^1\boldsymbol{a}_{C_4}$.

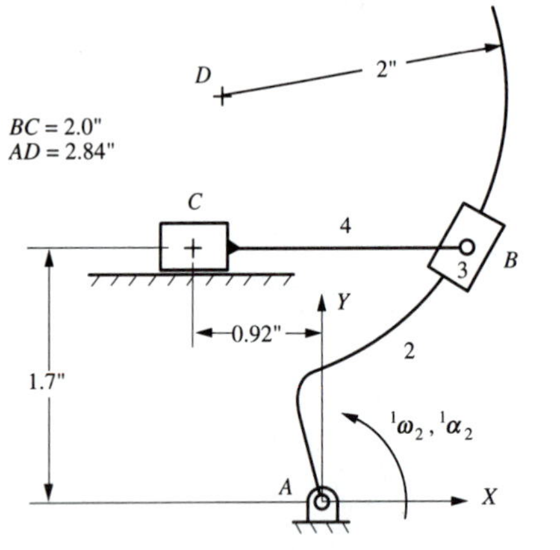

PROBLEM 2.41 In the mechanism shown, ${}^1\boldsymbol{\omega}_2 = 10$ rad/s CW (constant). Determine the angular acceleration of link 3.

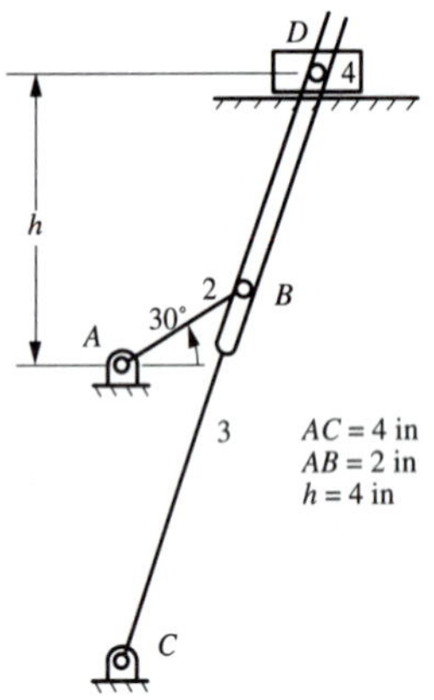

PROBLEM 2.42 In the mechanism shown, slotted links 2 and 3 are independently driven at angular velocities of 30 and 20 rad/s CW and have angular accelerations of 900 and 400 rad/s^2 CW, respectively. Determine the acceleration of point *B*, the center of the pin carried at the intersection of the two slots.

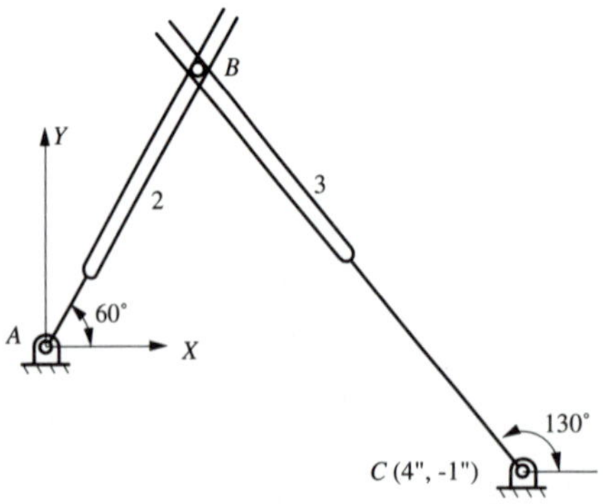

PROBLEM 2.43 Locate all of the instant centers in the mechanism shown below.

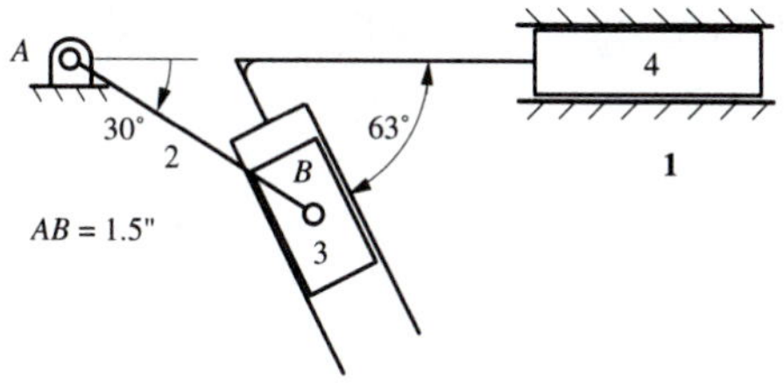

PROBLEM 2.44 Locate all of the instant centers in the mechanism shown below. If link 2 is turning CW at the rate of 60 rad/s, determine the linear velocity of points C and E using instant centers.

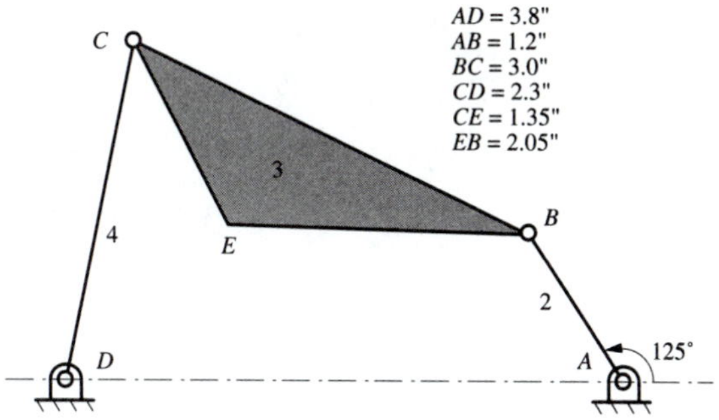

PROBLEM 2.45 Locate all of the instant centers in the mechanism shown below. If the cam (link 2) is turning CW at the rate of 900 rpm, determine the linear velocity of the follower using instant centers.

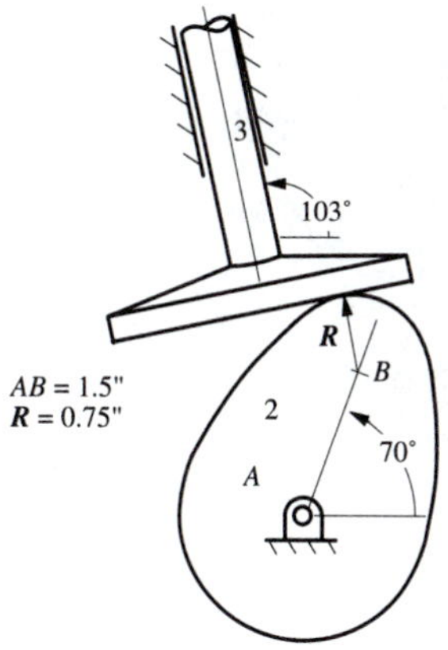

(Problems 2.43–2.63 emphasize instant centers of velocity.)

PROBLEM 2.46 Locate all of the instant centers in the mechanism shown below. If link 2 is turning CW at the rate of 36 rad/s, determine the linear velocity of point B_4 by use of instant centers. Determine the angular velocity of link 4 in rad/s and indicate the direction. Points C and E have the same vertical coordinate, and points A and C have the same horizontal coordinate.

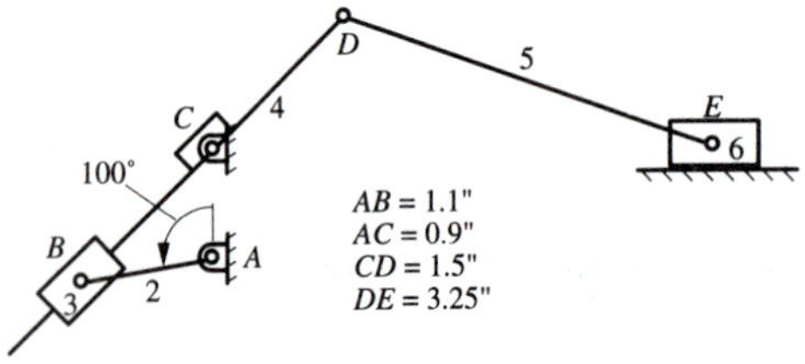

PROBLEM 2.47 Using the instant-center method, find angular velocity of link 6 if link 2 is rotating at 50 rpm CCW.

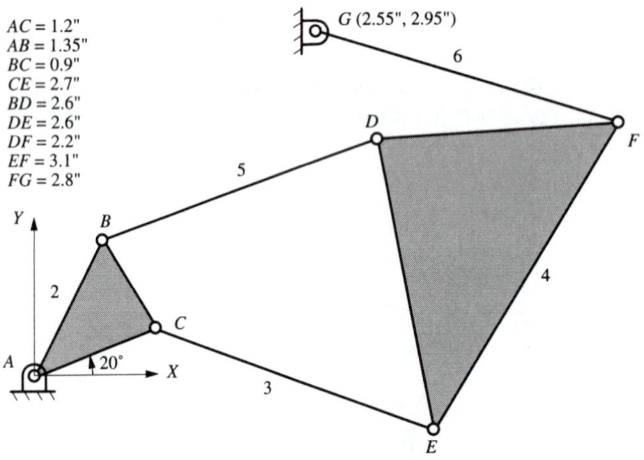

PROBLEM 2.48 In the operation of this mechanism, link 3 strikes and trips link 5, which is initially at rest. High wear has been observed at the point of contact between links 3 and 5. As an engineer, you are asked to correct this situation. Therefore, you decide to do the following:

1. Determine the direction of the velocity of point C on link 3 at the moment of contact.
2. Relocate the ground pivot of link 4 to make the direction of the velocity of point C perpendicular to link 5 (hence less rubbing at the point of contact) when contact occurs.

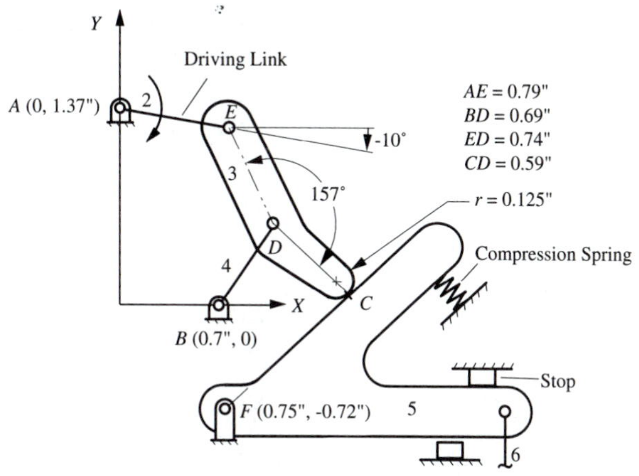

PROBLEM 2.49 For the linkage given, ${}^1\boldsymbol{\omega}_2 = 1$ rad/s CCW. Find I_{26} using the circle diagram method. Using ${}^1\mathbf{v}_{A_2}$ and I_{26}, determine the magnitude and direction of ${}^1\mathbf{v}_{B_6}$ using the rotating radius method.

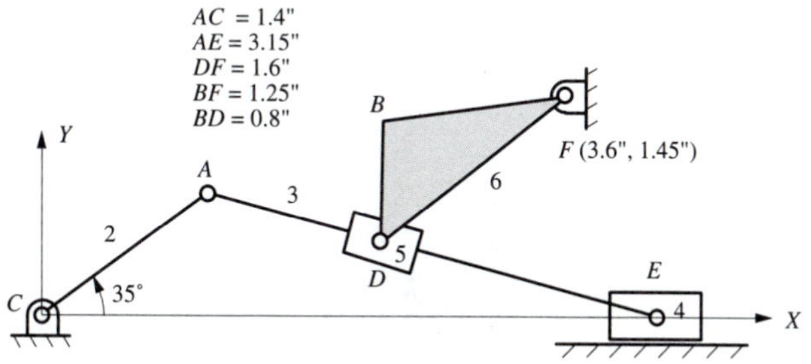

PROBLEM 2.50 Find all of the instant centers of velocity for the mechanism shown below.

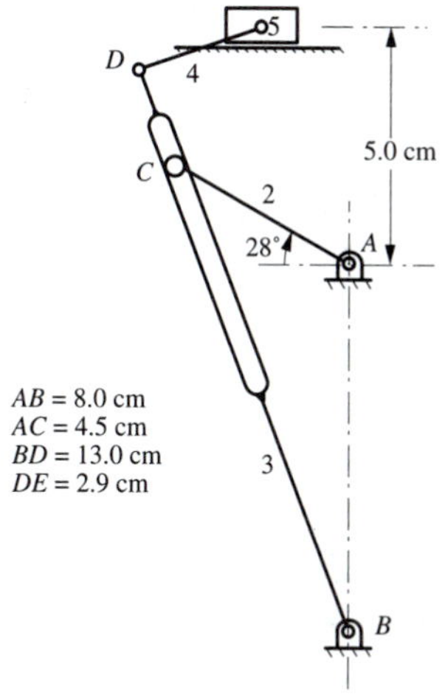

PROBLEM 2.51 Find the velocity of point C given that the angular velocity of gear 2 is 10 rad/s CW. B is a pin joint connecting links 4 and 5. Point A is a pin in link 3 that engages a slot in link 4.

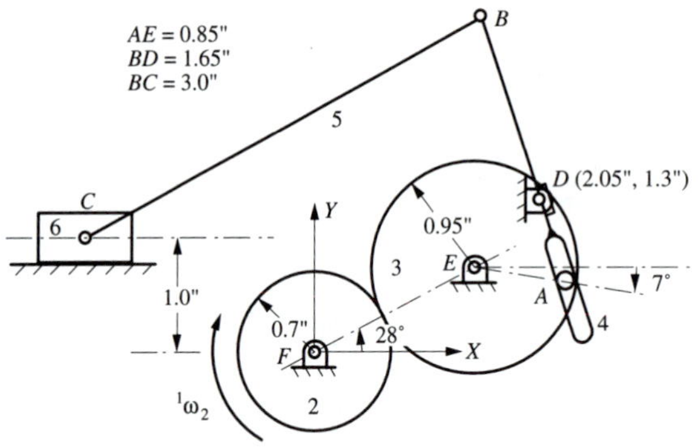

PROBLEM 2.52 If ${}^1\boldsymbol{\omega}_2 = 5$ rad/s CCW, find ${}^1\boldsymbol{\omega}_5$ using instant centers.

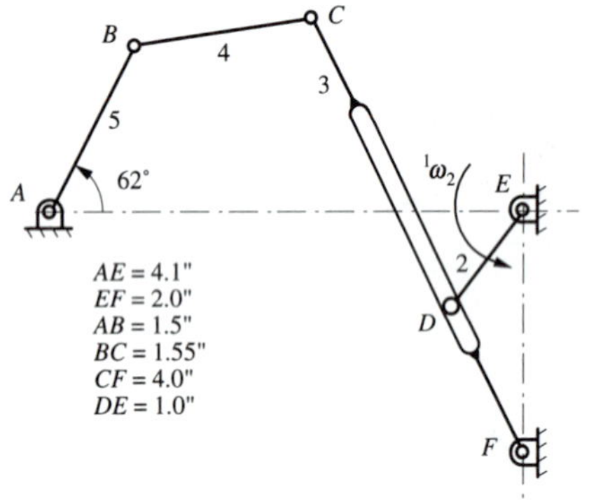

PROBLEM 2.53 If ${}^1\boldsymbol{\omega}_2 = 1$ rad/s CCW, find the velocity of point A on link 6 using the instant center method. Show ${}^1\boldsymbol{v}_{A_6}$ on the drawing.

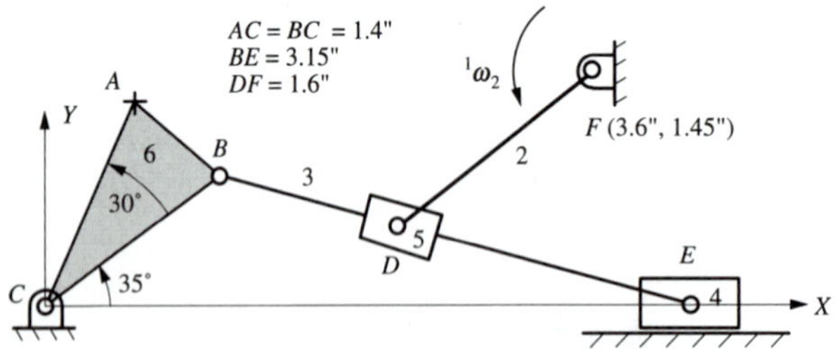

PROBLEM 2.54 If ${}^1\mathbf{v}_{A_2} = 10$ cm/s as shown, find ${}^1\mathbf{v}_{B_4}$ using the instant-center method.

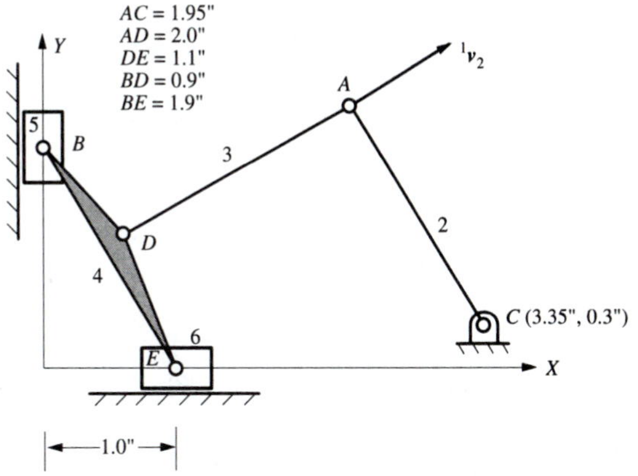

PROBLEM 2.55 If ${}^1\mathbf{v}_{A_2} = 10$ cm/s as shown, find ${}^1\mathbf{v}_{B_4}$ using the instant center method.

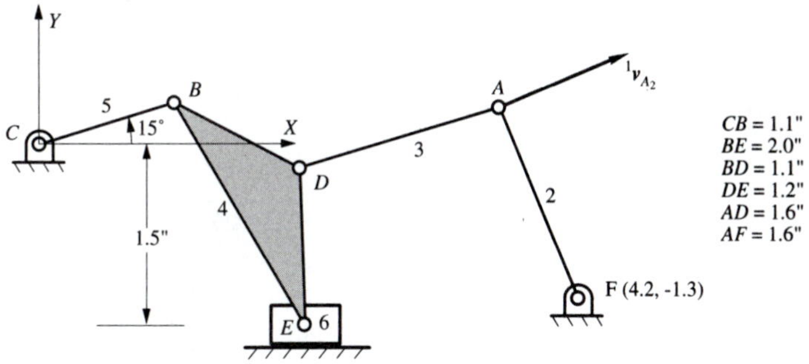

PROBLEM 2.56 If ${}^1\mathbf{v}_{A_6} = 10$ in/s as shown, determine the velocity vector (direction and magnitude) for point B on link 3 using the instant-center method.

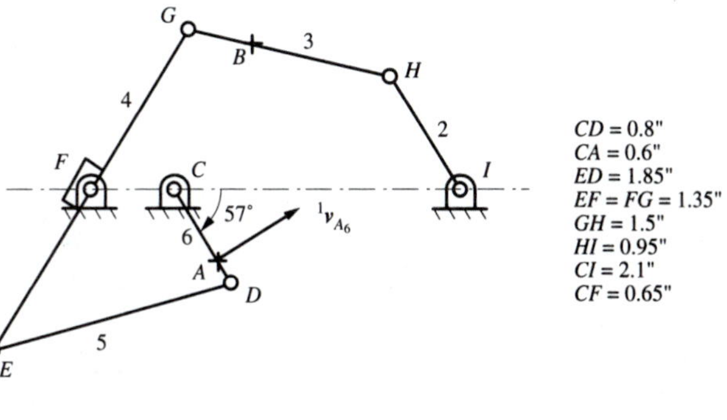

PROBLEM 2.57 In the mechanism below, ${}^1\boldsymbol{\omega}_2$ is 20 rad/s CCW. Find I_{26} and use it to find the angular velocity of link 6.

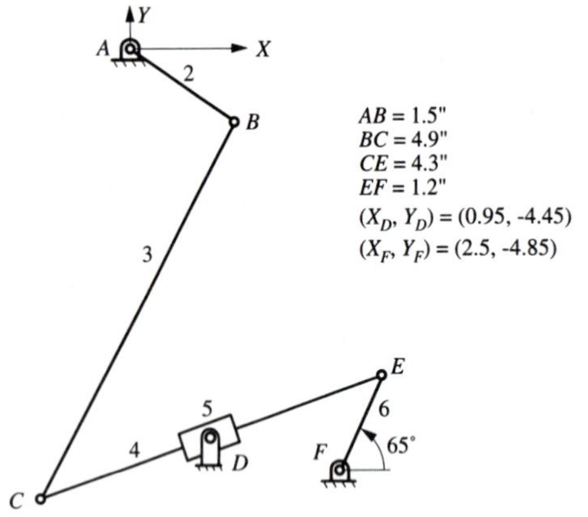

PROBLEM 2.58 If ${}^1\mathbf{v}_{B_2}$ = 10 in/s as shown, determine the velocity vector (direction and magnitude) of point C_4 using the instant center method.

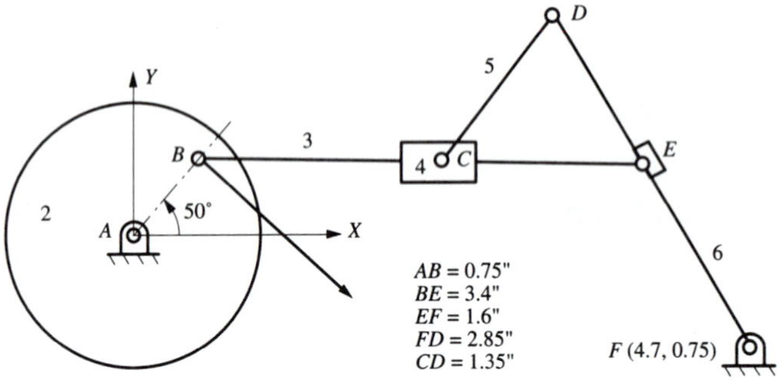

PROBLEM 2.59 If the velocity of A_2 is 10 in/s to the right, find ${}^1\boldsymbol{\omega}_6$ using instant centers.

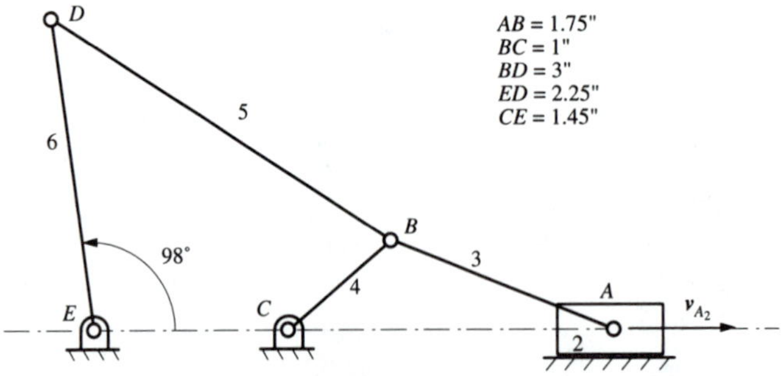

PROBLEM 2.60

Crank 2 of the push-link mechanism shown in the figure is driven at ${}^1\boldsymbol{\omega}_2 = 60$ rad/s (CW). Find the velocity of points B and C and the angular velocity of links 3 and 4 using the instant center method.

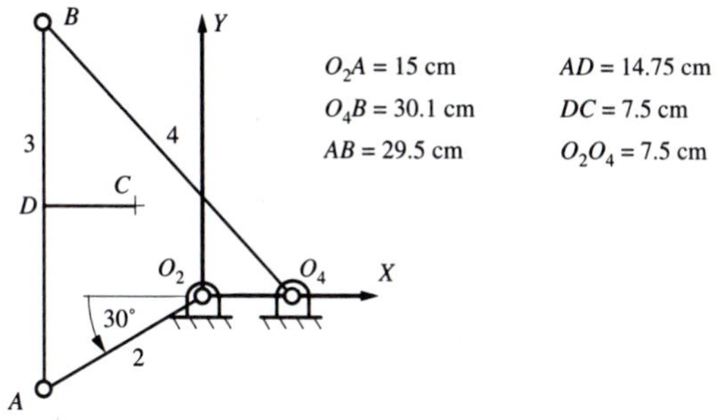

PROBLEM 2.61

The circular cam shown is driven at an angular velocity ${}^1\boldsymbol{\omega}_2 = 15$ rad/s (CW). There is rolling contact between the cam and roller, link 3. Using the instant-center method, find the angular velocity of the oscillating follower, link 4.

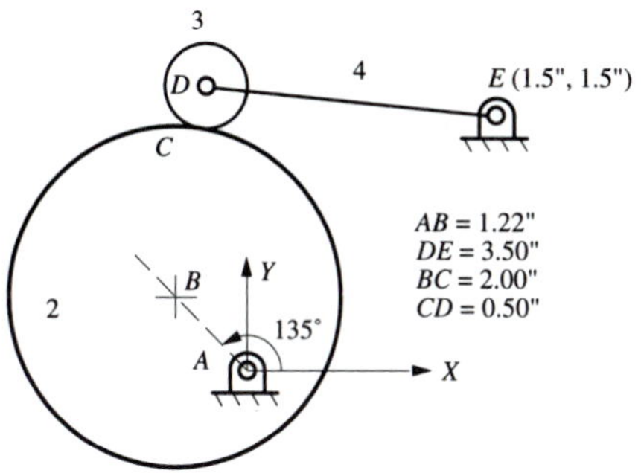

PROBLEM 2.62

If ${}^1\boldsymbol{\omega}_3 = 1$ rad/s CCW, find the velocity of points E and F using the instant center method. Show the velocity vectors ${}^1\boldsymbol{v}_{F_3}$ and ${}^1\boldsymbol{v}_{E_4}$ on the figure.

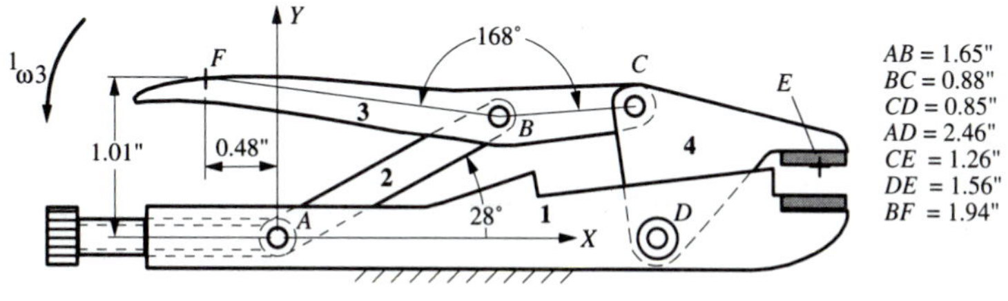

PROBLEM 2.63 In the eight-link mechanism, most of the linkage is contained in the black box and some of the instant centers are located as shown. The velocity of point B is 100 in/s in the direction shown. Compute the velocity of point D_8 and determine the angular velocity of link 2.

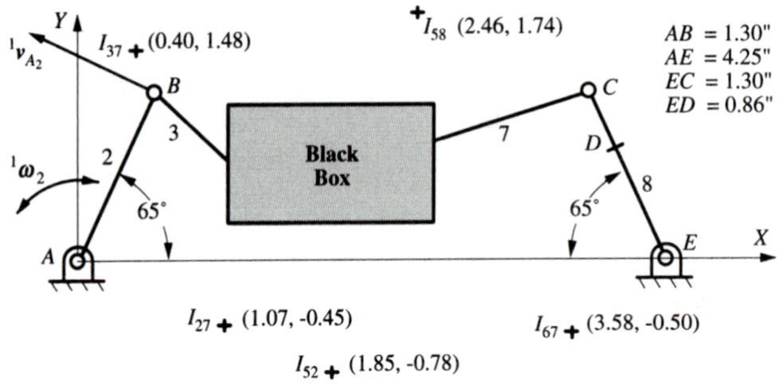

Chapter 3

Analytical Linkage Analysis

3.1 INTRODUCTION

In Chapter 2, graphical techniques for position, velocity, and acceleration analysis of linkages were presented. However, as was pointed out, there are circumstances in which it is preferable to use analytical solution techniques that can be conveniently programmed on a digital computer. In any circumstance in which repetitive or extensive analyses are required, the use of computer software is highly desirable. In the present chapter, the equations used to construct analysis software are developed in detail.

The geometric constraints associated with mechanisms can be formulated using vector displacement, velocity, and acceleration closure equations. The displacement closure equations are based on the observation that there are two different but equivalent distinct paths connecting points on the same vector loop. For example, in the four-bar linkage shown in Fig. 3.1, one can reach point C from point A either by way of point B or point D.

It is convenient to represent the terms in the closure equations by vectors, and the procedures developed in this section work especially well for planar problems. It is also possible to apply the same general approach to spatial linkages. Another popular method for planar mechanisms, which involves slightly more computational work, is the complex number approach, in which the Cartesian vector components are expressed in terms of the real and imaginary parts of a complex number. The use of complex numbers is advantageous in some types of problem; however, the direct vector approach is preferred here. The complex number approach is outlined briefly at the end of this chapter.

There are also specialized techniques for forming closure equations for spatial mechanisms. The general trend is to work with coordinate transformation operators. For this a set of body-fixed coordinates is established at each joint, and the product of a series of joint-to-joint coordinate transformation operators is taken. When this product is continued around the entire mechanism loop, it must be equal to the identity operator. The resulting operator equation can then be manipulated, if required, and corresponding elements equated. Types of operators that have been used include dual complex number 2×2

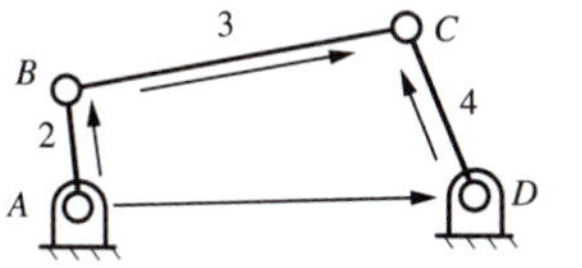

Figure 3.1 Closure of a four-bar linkage.

matrices, dual quaternions, real number 4×4 matrices, and dual number 3×3 matrices. A discussion of the mathematics of these operators is beyond the scope of this text. The description of spatial linkages using matrix transformations is discussed in Ch. 7.

3.2 POSITION, VELOCITY, AND ACCELERATION REPRESENTATIONS

3.2.1 Position Representation

For the purpose of developing an analytical model, we can define the relative locations of a chain of points by a chain of vectors. The points will be associated with the links of a mechanism in some manner, but they do not have to be attached to specific links. An example is given in Fig. 3.2.

The position of point Q in the fixed reference frame is

$$\boldsymbol{r}_Q = \boldsymbol{r}_1 + \boldsymbol{r}_2 + \boldsymbol{r}_3 \tag{3.1}$$

Here, we will represent each vector by a length r_i and an angle θ_i as shown in Fig. 3.3. All angles are measured counterclockwise from a line that remains parallel to the fixed x axis attached to the reference frame.

With this notation, we can resolve each of the vectors in Eq. (3.1) into x and y components making use of the unit vectors $\boldsymbol{i}$ and $\boldsymbol{j}$ as follows:

$$\left.\begin{aligned} \boldsymbol{r}_1 &= r_1(\cos\theta_1\,\boldsymbol{i} + \sin\theta_1\boldsymbol{j}) \\ \boldsymbol{r}_2 &= r_2(\cos\theta_2\,\boldsymbol{i} + \sin\theta_2\boldsymbol{j}) \\ \boldsymbol{r}_3 &= r_3(\cos\theta_3\,\boldsymbol{i} + \sin\theta_3\boldsymbol{j}) \end{aligned}\right\} \tag{3.2}$$

or

$$\boldsymbol{r}_k = r_k(\cos\theta_k\,\boldsymbol{i} + \sin\theta_k\,\boldsymbol{j}), \qquad k = 1, 2, 3 \tag{3.3}$$

3.2.2 Velocity Representation

To determine the velocity of point Q, $\boldsymbol{r}_Q$ can be differentiated. Then

$$\boldsymbol{v}_Q = \dot{\boldsymbol{r}}_Q = \dot{\boldsymbol{r}}_1 + \dot{\boldsymbol{r}}_2 + \dot{\boldsymbol{r}}_3 \tag{3.4}$$

where

$$\dot{\boldsymbol{r}}_k = \frac{d\boldsymbol{r}_k}{dt} \tag{3.5}$$

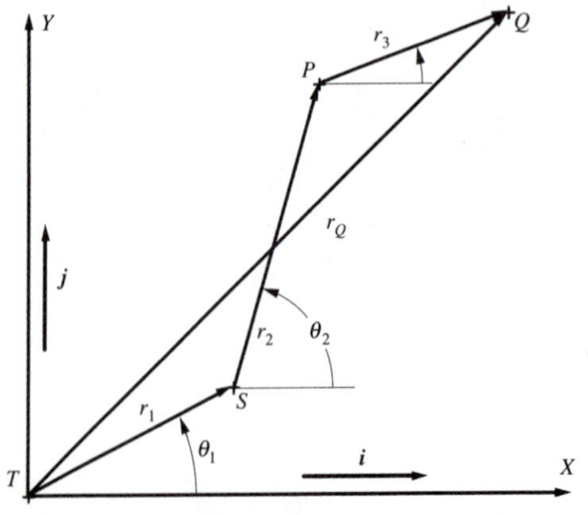

Figure 3.2 Representation of a chain of points by a set of vectors.

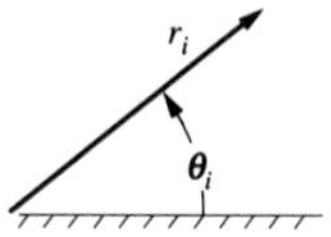

Figure 3.3 Notation used for vectors.

Note that, in general, both the magnitude and direction of $\boldsymbol{r}_k$ can change. When we differentiate Eq. (3.3) using the chain rule of calculus, we obtain

$$\dot{\boldsymbol{r}}_k = \dot{r}_k(\cos\theta_k\,\boldsymbol{i} + \sin\theta_k\,\boldsymbol{j}) + r_k(-\dot{\theta}_k\sin\theta_k\,\boldsymbol{i} + \dot{\theta}_k\cos\theta_k\,\boldsymbol{j}) \tag{3.6}$$

or

$$\dot{\boldsymbol{r}}_k = (\dot{r}_k\cos\theta_k - r_k\dot{\theta}_k\sin\theta_k)\boldsymbol{i} + (\dot{r}_k\sin\theta_k + r_k\dot{\theta}_k\cos\theta_k)\boldsymbol{j} \tag{3.7}$$

If we compare the vector components indicated in Eq. (3.6) with the equations developed in Section 2.3, we will notice a similarity between corresponding terms. In particular, if $\boldsymbol{r}_k$ is the vector defining the relative position between two points P and Q, and body B is moving relative to the reference frame R as shown in Fig. 3.4, then

$$\dot{r}_k(\cos\theta_k\,\boldsymbol{i} + \sin\theta_k\,\boldsymbol{j}) = {}^B\boldsymbol{v}_{Q/P} \tag{3.8}$$

and

$$r_k\dot{\theta}_k(-\sin\theta_k\,\boldsymbol{i} + \cos\theta_k\,\boldsymbol{j}) = {}^R\boldsymbol{\omega}_B \times \boldsymbol{r}_{Q/P} \tag{3.9}$$

Equation (3.8) can be verified by recognizing that it gives the component of the velocity associated with changing the magnitude of the vector between the two points. This component is clearly in the direction of the vector $\boldsymbol{r}_k$. The second term can be verified by computing the cross-product. Recognizing that

$${}^R\boldsymbol{\omega}_B = \dot{\theta}_k\boldsymbol{k}$$

and

$$\boldsymbol{r}_{Q/P} = \boldsymbol{r}_k = r_k(\cos\theta_k\,\boldsymbol{i} + \sin\theta_k\,\boldsymbol{j})$$

Then

$$\begin{aligned}{}^R\boldsymbol{\omega}_B \times \boldsymbol{r}_{Q/P} &= \dot{\theta}_k\boldsymbol{k} \times r_k(\cos\theta_k\,\boldsymbol{i} + \sin\theta_k\,\boldsymbol{j}) = \dot{\theta}_k r_k(\cos\theta_k\,\boldsymbol{k}\times\boldsymbol{i} + \sin\theta_k\,\boldsymbol{k}\times\boldsymbol{j})\\ &= r_k\dot{\theta}_k(-\sin\theta_k\,\boldsymbol{i} + \cos\theta_k\,\boldsymbol{j})\end{aligned}$$

Equation (3.4) can also be expressed as

$$\boldsymbol{v}_Q = \sum_{k=1}^{3} \dot{r}_k(\cos\theta_k\,\boldsymbol{i} + \sin\theta_k\,\boldsymbol{j}) + r_k(-\dot{\theta}_k\sin\theta_k\,\boldsymbol{i} + \dot{\theta}_k\cos\theta_k\,\boldsymbol{j}) \tag{3.10}$$

or

$$\boldsymbol{v}_Q = \sum_{k=1}^{3} (\dot{r}_k\cos\theta_k - r_k\dot{\theta}_k\sin\theta_k)\boldsymbol{i} + (\dot{r}_k\sin\theta_k + r_k\dot{\theta}_k\cos\theta_k)\boldsymbol{j} \tag{3.11}$$

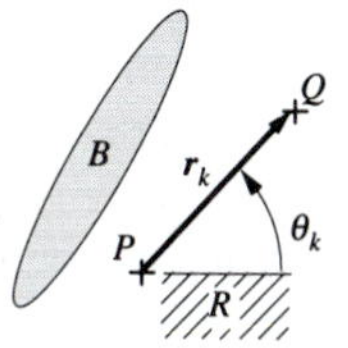

Figure 3.4 Position vector between two points.

3.2.3 Acceleration Representation

To obtain the acceleration expression, we need only to differentiate the velocity expression [Eq. (3.4)]. Symbolically, this is

$$\boldsymbol{a}_Q = \ddot{\boldsymbol{r}}_Q = \ddot{\boldsymbol{r}}_1 + \ddot{\boldsymbol{r}}_2 + \ddot{\boldsymbol{r}}_3 \tag{3.12}$$

where

$$\ddot{\boldsymbol{r}}_k = \frac{d^2\boldsymbol{r}_k}{dt^2}$$

Because the vectors have been defined in a consistent manner (Fig. 3.3), the form for the derivatives for all of the vectors will be the same. Therefore, we can develop the expression with a general vector $\boldsymbol{r}_k$.

Note again that, in general, both the magnitude and direction of $\boldsymbol{r}_k$ can change. When we differentiate Eq. (3.6) using the chain rule of calculus, we obtain

$$\begin{aligned}\ddot{\boldsymbol{r}}_k = {} & \ddot{r}_k(\cos\theta_k\,\boldsymbol{i} + \sin\theta_k\,\boldsymbol{j}) + r_k\ddot{\theta}_k(-\sin\theta_k\,\boldsymbol{i} + \cos\theta_k\,\boldsymbol{j}) - r_k\dot{\theta}_k^2(\cos\theta_k\,\boldsymbol{i} + \sin\theta_k\,\boldsymbol{j}) \\ & + 2\dot{r}_k\dot{\theta}_k(-\sin\theta_k\,\boldsymbol{i} + \cos\theta_k\,\boldsymbol{j})\end{aligned} \tag{3.13}$$

or

$$\begin{aligned}\ddot{\boldsymbol{r}}_k = {} & [(\ddot{r}_k - r_k\dot{\theta}_k^2)\cos\theta_k - (r_k\ddot{\theta}_k + 2\dot{r}_k\dot{\theta}_k)\sin\theta_k]\boldsymbol{i} \\ & + [(\ddot{r}_k - r_k\dot{\theta}_k^2)\sin\theta_k + (r_k\ddot{\theta}_k + 2\dot{r}_k\dot{\theta}_k)\cos\theta_k]\boldsymbol{j}\end{aligned} \tag{3.14}$$

As in the case of the velocity equations, we can compare the vector components indicated in Eq. (3.13) with the acceleration equations developed in Section 2.4. Using the same nomenclature as before (Fig. 3.4),

$$\ddot{r}_k(\cos\theta_k\,\boldsymbol{i} + \sin\theta_k\boldsymbol{j}) = {}^B\boldsymbol{a}_{Q/P} \tag{3.15}$$

$$r_k\ddot{\theta}_k(-\sin\theta_k\,\boldsymbol{i} + \cos\theta_k\boldsymbol{j}) = {}^R\boldsymbol{\alpha}_B \times \boldsymbol{r}_{Q/P} \tag{3.16}$$

$$-r_k\dot{\theta}_k^2(\cos\theta_k\,\boldsymbol{i} + \sin\theta_k\boldsymbol{j}) = {}^R\boldsymbol{\omega}_B \times ({}^R\boldsymbol{\omega}_B \times \boldsymbol{r}_{Q/P}) \tag{3.17}$$

and

$$2\dot{r}_k\dot{\theta}_k(-\sin\theta_k\,\boldsymbol{i} + \cos\theta_k\,\boldsymbol{j}) = 2\,{}^R\boldsymbol{\omega}_B \times {}^B\boldsymbol{v}_{Q/P} \tag{3.18}$$

These can be verified by direct calculation.

If we add the individual components, we can obtain the acceleration of point Q. Then Eq. (3.12) can be expressed as

$$\boldsymbol{a}_Q = \sum_{k=1}^{3}\left\{\begin{aligned}&\ddot{r}_k(\cos\theta_k\,\boldsymbol{i} + \sin\theta_k\,\boldsymbol{j}) + r_k\ddot{\theta}_k(-\sin\theta_k\,\boldsymbol{i} + \cos\theta_k\,\boldsymbol{j}) \\ &- r_k\dot{\theta}_k^2(\cos\theta_k\,\boldsymbol{i} + \sin\theta_k\,\boldsymbol{j}) + 2\dot{r}_k\dot{\theta}_k(-\sin\theta_k\,\boldsymbol{i} + \cos\theta_k\,\boldsymbol{j})\end{aligned}\right\} \tag{3.19}$$

or

$$\boldsymbol{a}_Q = \sum_{k=1}^{3}\left\{\begin{aligned}&[(\ddot{r}_k - r_k\dot{\theta}_k^2)\cos\theta_k - (r_k\ddot{\theta}_k + 2\dot{r}_k\dot{\theta}_k)\sin\theta_k]\boldsymbol{i} \\ &+ [(\ddot{r}_k - r_k\dot{\theta}_k^2)\sin\theta_k + (r_k\ddot{\theta}_k + 2\dot{r}_k\dot{\theta}_k)\cos\theta_k]\boldsymbol{j}\end{aligned}\right\} \tag{3.20}$$

3.2.4 Special Cases

Equations (3.6) and (3.13) or (3.7) and (3.14) are the most general forms of the velocity and acceleration equations. However, in most mechanisms some of the terms will usually be zero because of the special conditions associated with the way in which the vectors are defined. It is possible for any of the terms involved in the velocity and acceleration equations to be zero; however, a common case is to have the magnitude of a given position vector be constant. This is the case when the vector defines the relative positions of two points on a rigid link. When this happens, $\dot{r}$ and $\ddot{r}$ are zero. Then the velocity and acceleration expressions become

$$\dot{\boldsymbol{r}}_k = r_k\dot{\theta}_k(-\sin\theta_k\,\boldsymbol{i} + \cos\theta_k\,\boldsymbol{j}) \tag{3.21}$$

$$\ddot{\boldsymbol{r}}_k = r_k\ddot{\theta}_k(-\sin\theta_k\,\boldsymbol{i} + \cos\theta_k\,\boldsymbol{j}) - r_k\dot{\theta}_k^2(\cos\theta_k\,\boldsymbol{i} + \sin\theta_k\,\boldsymbol{j}) \tag{3.22}$$

or

$$\ddot{\boldsymbol{r}}_k = [-r_k\dot{\theta}_k^2\cos\theta_k - r_k\ddot{\theta}_k\sin\theta_k]\boldsymbol{i} + [-r_k\dot{\theta}_k^2\sin\theta_k + r_k\ddot{\theta}_k\cos\theta_k]\boldsymbol{j} \tag{3.23}$$

3.2.5 Mechanisms to Be Considered

There are six commonly used single-loop chains with revolute and slider joints. We will look at two of these in detail to illustrate how the equations can be developed in each case. Then we will present the results for the remaining four cases. We will then discuss more complex mechanisms that require several vector loops and mechanisms that contain higher pairs.

3.3 ANALYTICAL CLOSURE EQUATIONS FOR FOUR-BAR LINKAGES

We will first give an overview of the development of the equations for the four-bar linkage using the general nomenclature discussed above. The procedures used to solve the equations for the four-bar linkage are similar to the procedures required for solving the equations associated with most other simple mechanisms.

The closure condition simply expresses the condition that a loop of a linkage closes on itself. For the four-bar linkage shown in Fig. 3.5, the closure equations would be

$$\boldsymbol{r}_P = \boldsymbol{r}_2 + \boldsymbol{r}_3 = \boldsymbol{r}_1 + \boldsymbol{r}_4 \tag{3.24}$$

or

$$r_2(\cos\theta_2\,\boldsymbol{i} + \sin\theta_2\,\boldsymbol{j}) + r_3(\cos\theta_3\,\boldsymbol{i} + \sin\theta_3\,\boldsymbol{j}) = r_1(\cos\theta_1\,\boldsymbol{i} + \sin\theta_1\,\boldsymbol{j}) + r_4(\cos\theta_4\,\boldsymbol{i} + \sin\theta_4\,\boldsymbol{j}) \tag{3.25}$$

Rewriting Eq. (3.25) in its component equations, one gets

$$r_2\cos\theta_2 + r_3\cos\theta_3 = r_1\cos\theta_1 + r_4\cos\theta_4 \tag{3.26}$$

$$r_2\sin\theta_2 + r_3\sin\theta_3 = r_1\sin\theta_1 + r_4\sin\theta_4 \tag{3.27}$$

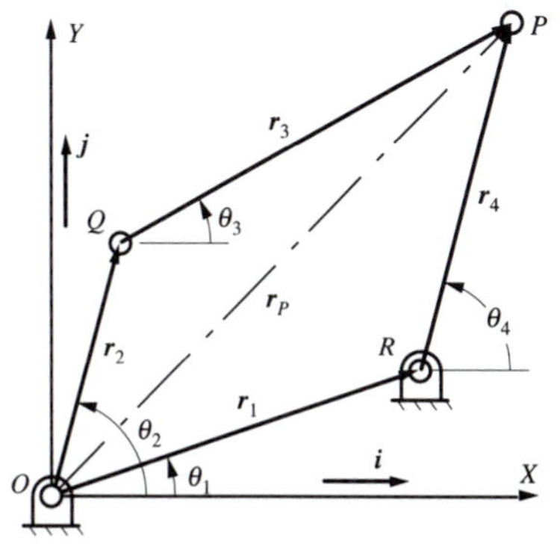

Figure 3.5 Vector closure condition for a four-bar loop. The position of point P obtained by adding the vectors $\boldsymbol{r}_2$ and $\boldsymbol{r}_3$ must always be the same as that obtained by adding vectors $\boldsymbol{r}_1$ and $\boldsymbol{r}_4$. Note that $\boldsymbol{r}_1$ is a constant vector that describes the base member of the linkage. Correspondingly, θ_1 is a constant angle.

Equations (3.26) and (3.27) are the closure equations, and they must be satisfied throughout the motion of the linkage. The base vector will be constant, so r_1 and θ_1 are constants. If θ_2 is given, that is, if crank OQ is a driving crank, it is necessary to solve Eqs. (3.26) and (3.27) for θ_3 and θ_4 in terms of θ_2. Once these expressions are obtained $\dot{\theta}_3$, $\dot{\theta}_4$, $\ddot{\theta}_3$, and $\ddot{\theta}_4$ can be obtained in terms of θ_2, $\dot{\theta}_2$ and $\ddot{\theta}_2$ by differentiation. Velocities and accelerations of points in the mechanism can then be obtained from equations of the type of Eqs. (3.11) and (3.19) recognizing that all of the vector magnitudes are constant ($\dot{r} = \ddot{r} = 0$).

When θ_3 is given, the coupler is the driver, and we must solve (3.26) and (3.27) for θ_2 and θ_4 in terms of θ_3. The procedure for doing this is very similar to that used when θ_2 is the input. Therefore, we will first reconsider briefly the case in which θ_2 is the input.

3.3.1 Solution of Closure Equations for Four-Bar Linkages When Link 2 Is the Driver

The analytical solution procedure follows the same major steps as in the graphical solution. That is, a position analysis must first be performed, then a velocity analysis, and finally the acceleration analysis. The position analysis, for a closed-loop linkage, comprises the solution of the closure equations for the joint angles or link orientations. Once this solution is obtained, the velocity and acceleration states are quickly obtainable using the differentiated equations. It will be seen, however, that the position analysis, which is so easily performed graphically by construction of a drawing to scale, is a complex matter when performed analytically.

For all of the simple mechanisms that we will consider initially, the first step in solving the position equations is to identify the variable to be determined first. When the position equations involve two angles as unknowns, the solution procedure is to isolate the trigonometric function involving the angle to be eliminated on the left-hand side of the equation. In order to eliminate θ_3, first isolate it on one side of Eqs. (3.26) and (3.27) as follows:

$$r_3 \cos \theta_3 = r_1 \cos \theta_1 + r_4 \cos \theta_4 - r_2 \cos \theta_2 \tag{3.28}$$

$$r_3 \sin \theta_3 = r_1 \sin \theta_1 + r_4 \sin \theta_4 - r_2 \sin \theta_2 \tag{3.29}$$

Notice that the angle θ_1 is a known constant. Now square both sides of both equations, add, and simplify the result using the trigonometric identity $sin^2 \theta + cos^2 \theta = 1$. This gives

$$\begin{aligned} r_3^2 = r_1^2 + r_2^2 + r_4^2 + 2r_1r_4(\cos \theta_1 \cos \theta_4 + \sin \theta_1 \sin \theta_4) \\ -2r_1r_2(\cos \theta_1 \cos \theta_2 + \sin \theta_1 \sin \theta_2) - 2r_2r_4(\cos \theta_2 \cos \theta_4 + \sin \theta_2 \sin \theta_4) \end{aligned} \tag{3.30}$$

Equation (3.30) gives θ_4 in terms of the given angle θ_2 (and the constant angle θ_1) but not explicitly. To obtain an explicit expression, simplify Eq. (3.30) by combining the coefficients of $\cos \theta_4$ and $\sin \theta_4$ as follows:

$$A \cos \theta_4 + B \sin \theta_4 + C = 0 \tag{3.31}$$

where

$$\left.\begin{aligned} A &= 2r_1r_4 \cos \theta_1 - 2r_2r_4 \cos \theta_2 \\ B &= 2r_1r_4 \sin \theta_1 - 2r_2r_4 \sin \theta_2 \\ C &= r_1^2 + r_2^2 + r_4^2 - r_3^2 - 2r_1r_2(\cos \theta_1 \cos \theta_2 + \sin \theta_1 \sin \theta_2) \end{aligned}\right\} \tag{3.32}$$

To solve Eq. (3.31), use the standard trigonometric identities for half-angles given in the following:

$$\sin \theta_4 = \frac{2 \tan(\theta_4/2)}{1 + \tan^2(\theta_4/2)} \tag{3.33}$$

$$\cos \theta_4 = \frac{1 - \tan^2(\theta_4/2)}{1 + \tan^2(\theta_4/2)} \tag{3.34}$$

After substitution and simplification, we get

$$(C - A)t^2 + 2Bt + (A + C) = 0$$

where

$$t = \tan\left(\frac{\theta_4}{2}\right)$$

Solving for t gives

$$t = \frac{-2B + \sigma\sqrt{4B^2 - 4(C - A)(C + A)}}{2(C - A)} = \frac{-B + \sigma\sqrt{B^2 - C^2 + A^2}}{C - A} \tag{3.35}$$

and

$$\theta_4 = 2\tan^{-1} t \tag{3.36}$$

where $\sigma = \pm 1$ is a sign variable identifying the assembly mode. Note that $\tan^{-1} t$ has a valid range of $-\pi/2 \leq \tan^{-1} t \leq \pi/2$. Therefore, θ_4 will have the range $-\pi \leq \theta_4 \leq \pi$. Unless the linkage is a Grashof type II linkage in one of the extreme positions of its motion range, there are two solutions for θ_4 corresponding to the two values of σ, and they are both valid. These correspond to two assembly modes or branches for the linkage. Once we pick the value for σ corresponding to the desired mode, the sign in an actual linkage stays the same for any value of θ_2.

Because of the square root in Eq. (3.35), the variable t can be complex $(A^2 + B^2) < C^2$. If this happens, the mechanism cannot be assembled in the position specified. The assembly configurations would then appear as shown in Fig. 3.6.

Equations (3.28) and (3.29) can now be solved for θ_3. Dividing Eq. (3.29) by Eq. (3.28) and solving for θ_3 gives

$$\theta_3 = \tan^{-1}\left[\frac{r_1 \sin\theta_1 + r_4 \sin\theta_4 - r_2 \sin\theta_2}{r_1 \cos\theta_1 + r_4 \cos\theta_4 - r_2 \cos\theta_2}\right] \tag{3.37}$$

Note that in Eq. (3.37), it is essential that the sign of the numerator and denominator be maintained to determine the quadrant in which the angle θ_3 lies. This can be done directly by using the ATAN2 function. The form of this function is

$$\text{ATAN2}(\sin\theta_3, \cos\theta_3) = \tan^{-1}\left[\frac{\sin\theta_3}{\cos\theta_3}\right] \tag{3.38}$$

Equations (3.35)–(3.37) give a complete and consistent solution to the position problem. As indicated before, for any value of θ_2, there are typically two values of θ_3 and θ_4, given by substituting $\sigma = +1$ and -1, respectively, in Eq. (3.35). These two different solutions are shown in Fig. 3.7. The two solutions correspond to an assembly ambiguity that also appears in the graphical construction.

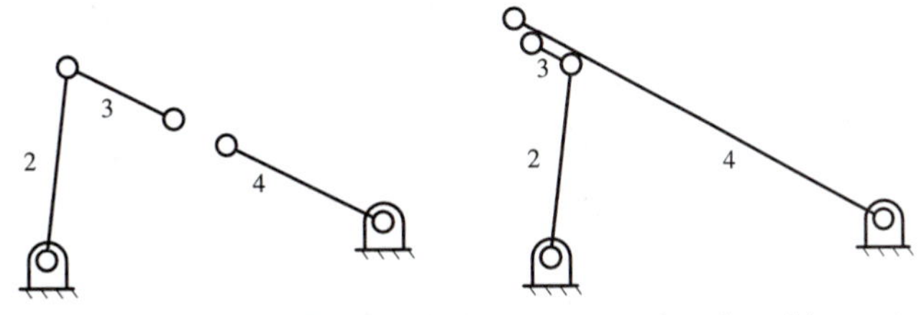

Figure 3.6 Grashof type II linkages cannot be placed in positions that are transitions between solution branches. The variable t would be complex in these cases.

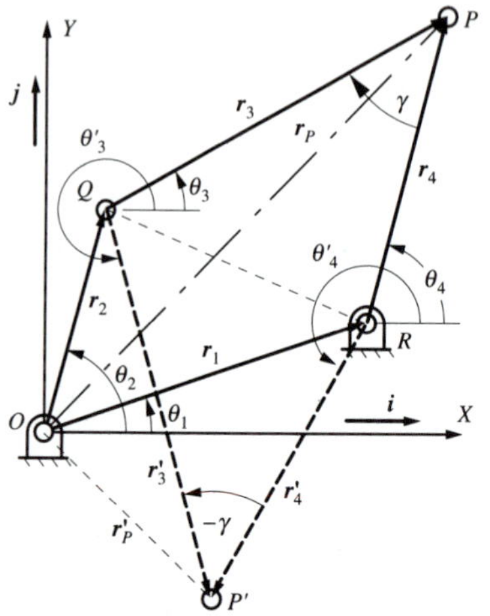

Figure 3.7 The two possible positions (P and P') of the point P for a given value of θ_2. Note that $QP'R$ is the mirror image of QPR about the line QR. Notice that there are two different possible values of θ_3 and two different values of θ_4 corresponding to the two possible positions of point P. The sign of the angle RPQ (γ) is reversed in the second solution, although the magnitude of $RP'Q$ is the same. The sign of γ is a useful graphical indicator of which solution is being examined.

Note that the positions of $\boldsymbol{r}_3$ and $\boldsymbol{r}_4$ are symmetric about the line QR. Therefore, the angle $\gamma = \theta_4 - \theta_3$ has the same magnitude, but opposite sign, in each of the two positions. The sign of γ provides a useful indicator of which of the solution branches has been drawn, from the graphical point of view.

Once all of the angular quantities are known, it is relatively straightforward to compute the coordinates on any point on the vector loops used in the closure equations. In particular, the coordinates of Q, P, and R are given by

$$\boldsymbol{r}_Q = \boldsymbol{r}_2 = r_2(\cos\theta_2\,\boldsymbol{i} + \sin\theta_2\,\boldsymbol{j}) \tag{3.39}$$

$$\begin{aligned}\boldsymbol{r}_P &= \boldsymbol{r}_2 + \boldsymbol{r}_3 = r_2(\cos\theta_2\,\boldsymbol{i} + \sin\theta_2\,\boldsymbol{j}) + r_3(\cos\theta_3\,\boldsymbol{i} + \sin\theta_3\,\boldsymbol{j})\\ &= \boldsymbol{r}_1 + \boldsymbol{r}_4 = r_1(\cos\theta_1\,\boldsymbol{i} + \sin\theta_1\,\boldsymbol{j}) + r_4(\cos\theta_4\,\boldsymbol{i} + \sin\theta_4\,\boldsymbol{j})\end{aligned} \tag{3.40}$$

and

$$\boldsymbol{r}_R = \boldsymbol{r}_1 = r_1(\cos\theta_1\,\boldsymbol{i} + \sin\theta_1\,\boldsymbol{j}) \tag{3.41}$$

3.3.2 Analysis When the Coupler (Link 3) Is the Driving Link

The analytical procedure given above when one of the frame-mounted links (link 2) is the driver is very similar to the graphical procedure. However, if the coupler is the driver, it is difficult to analyze the linkage graphically. The analytical procedure, on the other hand, is very straightforward and no more difficult to conduct than when one of the frame-mounted links is the driver. The details follow exactly the same procedure as that given in Sections 3.3.1 through 3.3.3. Therefore, we will simply outline the procedure and tabulate the results.

In the procedure, we can assume that θ_1 and θ_3, $\dot{\theta}_3$, and $\ddot{\theta}_3$ are known, and θ_2, $\dot{\theta}_2$, $\ddot{\theta}_2$, θ_4, $\dot{\theta}_4$, and $\ddot{\theta}_4$ are to be found. All of the link lengths and θ_1 are constants. For the position analysis, again begin with Eqs. (3.26) and (3.27) and isolate the terms with either θ_2 or θ_4. It is advantageous to select θ_2 for reasons that will become apparent. The resulting equations are

$$r_2 \cos \theta_2 = r_1 \cos \theta_1 + r_4 \cos \theta_4 - r_3 \cos \theta_3 \tag{3.42}$$

$$r_2 \sin \theta_2 = r_1 \sin \theta_1 + r_4 \sin \theta_4 - r_3 \sin \theta_3 \tag{3.43}$$

A comparison of Eqs. (3.42) and (3.43) with Eqs. (3.28) and (3.29) indicates that they are of exactly the same form except that the indices 2 and 3 are interchanged. Therefore, we can use directly the position solution derived in Section 3.4.1 if we interchange the indices 2 and 3.

When the coupler is the driver, there is an assembly-mode ambiguity similar to that when link 2 is the driver. This is illustrated in Fig. 3.8. It is necessary to know the appropriate mode before the analysis is begun; however, once the assembly mode is selected, it is the same for any position of the input link.

The motion of the coupler in a coupler-driven four-bar linkage will be less than 360° unless the four-bar is a type I linkage with the coupler or base as the shortest link. When the linkage reaches its motion limits, links 2 and 4 will be parallel.

3.3.3 Velocity Equations for Four-Bar Linkages

The analytical form of the velocity equations can be developed by differentiating Eq. (3.24). The result is

$$\dot{\boldsymbol{r}}_P = \dot{\boldsymbol{r}}_2 + \dot{\boldsymbol{r}}_3 = \dot{\boldsymbol{r}}_1 + \dot{\boldsymbol{r}}_4 \tag{3.44}$$

When this equation is written in component form, the result is the same as that of differentiating Eqs. (3.26) and (3.27). Recognizing that all of the link lengths are constant, as is θ_1, the resulting component equations are

$$r_2\dot{\theta}_2 \sin \theta_2 + r_3\dot{\theta}_3 \sin \theta_3 = r_4\dot{\theta}_4 \sin \theta_4 \tag{3.45}$$

$$r_2\dot{\theta}_2 \cos \theta_2 + r_3\dot{\theta}_3 \cos \theta_3 = r_4\dot{\theta}_4 \cos \theta_4 \tag{3.46}$$

If $\dot{\theta}_2$ is known, the only new unknowns are $\dot{\theta}_3$ and $\dot{\theta}_4$, and if $\dot{\theta}_3$ is known, the only new unknowns are $\dot{\theta}_2$ and $\dot{\theta}_4$. In either case, the equations can be solved most easily using a linear equation solver. In matrix form, Eqs. (3.45) and (3.46) can be rearranged and rewritten as

$$\begin{bmatrix} -r_J \sin \theta_J & r_4 \sin \theta_4 \\ -r_J \cos \theta_J & r_4 \cos \theta_4 \end{bmatrix} \begin{Bmatrix} \dot{\theta}_J \\ \dot{\theta}_4 \end{Bmatrix} = \begin{Bmatrix} r_M\dot{\theta}_M \sin \theta_M \\ r_M\dot{\theta}_M \cos \theta_M \end{Bmatrix} \tag{3.47}$$

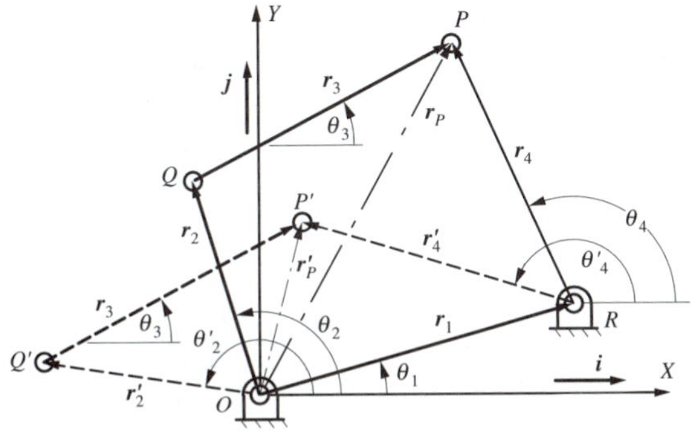

Figure 3.8 The two possible positions (P and P') of the point P for a given value of θ_3. There are two different possible values of θ_2 and two different values of θ_4 corresponding to the two possible positions of point P.

where $M = 2$ and $J = 3$ for $\dot{\theta}_2$ as the input, and $M = 3$ and $J = 2$ for $\dot{\theta}_3$ as the input. The terms in the matrix and vector on the right-hand side of the equation will be known. The equation can therefore be solved manually, on a programmable calculator, or with the matrix solvers in programs such as MATLAB.

Once the angular velocities are known, it is a simple matter to compute the linear velocities of any of the points on the vector loop. The velocities of points Q and P are given by

$$\dot{\boldsymbol{r}}_Q = \dot{\boldsymbol{r}}_2 = r_2\dot{\theta}_2(-\sin\theta_2\,\boldsymbol{i} + \cos\theta_2\,\boldsymbol{j}) \tag{3.48}$$

and

$$\begin{aligned}\dot{\boldsymbol{r}}_P = \dot{\boldsymbol{r}}_2 + \dot{\boldsymbol{r}}_3 &= (-r_2\dot{\theta}_2\sin\theta_2 - r_3\dot{\theta}_3\sin\theta_3)\boldsymbol{i} + (r_2\dot{\theta}_2\cos\theta_2 + r_3\dot{\theta}_3\cos\theta_3)\boldsymbol{j}\\ &= \dot{\boldsymbol{r}}_1 + \dot{\boldsymbol{r}}_4 = (-r_4\dot{\theta}_4\sin\theta_4)\boldsymbol{i} + (r_4\dot{\theta}_4\cos\theta_4)\boldsymbol{j}\end{aligned} \tag{3.49}$$

3.3.4 Acceleration Equations for Four-Bar Linkages

The analytical form of the acceleration equations can be developed by differentiating Eq. (3.44). The result is

$$\ddot{\boldsymbol{r}}_P = \ddot{\boldsymbol{r}}_2 + \ddot{\boldsymbol{r}}_3 = \ddot{\boldsymbol{r}}_1 + \ddot{\boldsymbol{r}}_4 \tag{3.50}$$

When this equation is written in component form, the result is the same as differentiating Eqs. (3.45) and (3.46). The resulting component equations are

$$r_2\ddot{\theta}_2\sin\theta_2 + r_2\dot{\theta}_2^2\cos\theta_2 + r_3\ddot{\theta}_3\sin\theta_3 + r_3\dot{\theta}_3^2\cos\theta_3 = r_4\ddot{\theta}_4\sin\theta_4 + r_4\dot{\theta}_4^2\cos\theta_4 \tag{3.51}$$

$$r_2\ddot{\theta}_2\cos\theta_2 - r_2\dot{\theta}_2^2\sin\theta_2 + r_3\ddot{\theta}_3\cos\theta_3 - r_3\dot{\theta}_3^2\sin\theta_3 = r_4\ddot{\theta}_4\cos\theta_4 - r_4\dot{\theta}_4^2\sin\theta_4 \tag{3.52}$$

When $\ddot{\theta}_2$ is known along with all of the position and velocity terms, the only new unknowns are $\ddot{\theta}_3$ and $\ddot{\theta}_4$, and when $\ddot{\theta}_3$ is known along with all of the position and velocity terms, the only new unknowns are $\ddot{\theta}_2$ and $\ddot{\theta}_4$. Again, because a linear problem is involved, these can be solved for most easily using a linear equation solver. In matrix form, Eqs. (3.51) and (3.52) can be rearranged and rewritten as

$$\begin{aligned}&\begin{bmatrix}-r_J\sin\theta_J & r_4\sin\theta_4\\ -r_J\cos\theta_J & r_4\cos\theta_4\end{bmatrix}\begin{Bmatrix}\ddot{\theta}_J\\ \ddot{\theta}_4\end{Bmatrix}\\ &\quad= \begin{Bmatrix}r_M\ddot{\theta}_M\sin\theta_M + r_M\dot{\theta}_M^2\cos\theta_M + r_J\dot{\theta}_J^2\cos\theta_J - r_4\dot{\theta}_4^2\cos\theta_4\\ r_M\ddot{\theta}_M\cos\theta_M - r_M\dot{\theta}_M^2\sin\theta_M - r_J\dot{\theta}_J^2\sin\theta_J + r_4\dot{\theta}_4^2\sin\theta_4\end{Bmatrix}\end{aligned} \tag{3.53}$$

where $M = 2$ and $J = 3$ for $\ddot{\theta}_2$ as the input, and $M = 3$ and $J = 2$ for $\ddot{\theta}_3$ as the input. The terms in the matrix and vector on the right-hand side of the equation will be known. The equation can therefore be solved manually, on a programmable calculator, or with the matrix solvers in programs such as MATLAB. Notice that the coefficient matrix is the same for both the velocities (Eq. 3.47) and the accelerations (Eq. 3.53).

Once the angular accelerations are known, it is a simple matter to compute the linear accelerations of any of the points in the linkage. The accelerations of points Q and P are given by

$$\ddot{\boldsymbol{r}}_Q = \ddot{\boldsymbol{r}}_2 = (-r_2\ddot{\theta}_2\sin\theta_2 - r_2\dot{\theta}_2^2\cos\theta_2)\boldsymbol{i} + (r_2\ddot{\theta}_2\cos\theta_2 - r_2\dot{\theta}_2^2\sin\theta_2)\boldsymbol{j} \tag{3.54}$$

and

$$\begin{aligned}\ddot{\boldsymbol{r}}_P = \ddot{\boldsymbol{r}}_2 + \ddot{\boldsymbol{r}}_3 &= -(r_2\ddot{\theta}_2\sin\theta_2 + r_2\dot{\theta}_2^2\cos\theta_2 + r_3\ddot{\theta}_3\sin\theta_3 + r_3\dot{\theta}_3^2\cos\theta_3)\boldsymbol{i}\\ &\quad+ (r_2\ddot{\theta}_2\cos\theta_2 - r_2\dot{\theta}_2^2\sin\theta_2 + r_3\ddot{\theta}_3\cos\theta_3 - r_3\dot{\theta}_3^2\sin\theta_3)\boldsymbol{j}\\ &= \ddot{\boldsymbol{r}}_1 + \ddot{\boldsymbol{r}}_4 = -(r_4\ddot{\theta}_4\sin\theta_4 + r_4\dot{\theta}_4^2\cos\theta_4)\boldsymbol{i} + (r_4\ddot{\theta}_4\cos\theta_4 - r_4\dot{\theta}_4^2\sin\theta_4)\boldsymbol{j}\end{aligned} \tag{3.55}$$

Table 3.1 Summary of Position, Velocity, and Acceleration Equations for a Four-Bar Linkage. Link 2 is the Input Link When $M = 2$ and $J = 3$. Link 3 is the Input Link When $M = 3$ and $J = 2$. Link 1 is Assumed to be the Frame. The Link Numbers and Points are Defined in Fig. 3.5

Position

$$A = 2r_1r_4 \cos\theta_1 - 2r_Mr_4\cos\theta_M$$

$$B = 2r_1r_4 \sin\theta_1 - 2r_Mr_4\sin\theta_M$$

$$C = r_1^2 + r_M^2 + r_4^2 - r_J^2 - 2r_1r_M(\cos\theta_1\cos\theta_M + \sin\theta_1\sin\theta_M)$$

$$\theta_4 = 2\tan^{-1}\left[\frac{-B + \sigma\sqrt{B^2 - C^2 + A^2}}{C - A}\right], \qquad \sigma = \pm 1$$

$$\theta_J = \tan^{-1}\left[\frac{r_1\sin\theta_1 + r_4\sin\theta_4 - r_M\sin\theta_M}{r_1\cos\theta_1 + r_4\cos\theta_4 - r_M\cos\theta_M}\right]$$

$$\boldsymbol{r}_Q = \boldsymbol{r}_2 = r_2(\cos\theta_2\,\boldsymbol{i} + \sin\theta_2\,\boldsymbol{j})$$

$$\begin{aligned}\boldsymbol{r}_P &= \boldsymbol{r}_2 + \boldsymbol{r}_3 = r_2(\cos\theta_2\,\boldsymbol{i} + \sin\theta_2\,\boldsymbol{j}) + r_3(\cos\theta_3\,\boldsymbol{i} + \sin\theta_3\,\boldsymbol{j})\\ &= \boldsymbol{r}_1 + \boldsymbol{r}_4 = r_1(\cos\theta_1\,\boldsymbol{i} + \sin\theta_1\,\boldsymbol{j}) + r_4(\cos\theta_4\,\boldsymbol{i} + \sin\theta_4\,\boldsymbol{j})\end{aligned}$$

$$\boldsymbol{r}_R = \boldsymbol{r}_1 = r_1(\cos\theta_1\,\boldsymbol{i} + \sin\theta_1\,\boldsymbol{j})$$

Velocity

$$\begin{bmatrix} -r_J\sin\theta_j & r_4\sin\theta_4 \\ -r_J\cos\theta_j & r_4\cos\theta_4 \end{bmatrix}\begin{Bmatrix}\dot\theta_J \\ \dot\theta_4\end{Bmatrix} = \begin{Bmatrix} r_M\dot\theta_M\sin\theta_M \\ r_M\dot\theta_M\cos\theta_M\end{Bmatrix}$$

$$\dot{\boldsymbol{r}}_Q = \dot{\boldsymbol{r}}_2 = r_2\dot\theta_2(-\sin\theta_2\,\boldsymbol{i} + \cos\theta_2\,\boldsymbol{j})$$

$$\dot{\boldsymbol{r}}_P = (-r_4\dot\theta_4\sin\theta_4)\boldsymbol{i} + (r_4\dot\theta_4\cos\theta_4)\boldsymbol{j}$$

Accelerations

$$\begin{bmatrix} -r_J\sin\theta_j & r_4\sin\theta_4 \\ -r_J\cos\theta_j & r_4\cos\theta_4 \end{bmatrix}\begin{Bmatrix}\ddot\theta_J \\ \ddot\theta_4\end{Bmatrix} = \begin{Bmatrix} r_M\ddot\theta_M\sin\theta_M + r_M\dot\theta_M^2\cos\theta_M + r_J\dot\theta_J^2\cos\theta_J - r_4\dot\theta_4^2\cos\theta_4 \\ r_M\ddot\theta_M\cos\theta_M - r_M\dot\theta_M^2\sin\theta_M - r_J\dot\theta_J^2\sin\theta_J + r_4\dot\theta_4^2\sin\theta_4\end{Bmatrix}$$

$$\ddot{\boldsymbol{r}}_Q = \ddot{\boldsymbol{r}}_2 = (-r_2\ddot\theta_2\sin\theta_2 - r_2\dot\theta_2^2\cos\theta_2)\boldsymbol{i} + (r_2\ddot\theta_2\cos\theta_2 - r_2\dot\theta_2^2\sin\theta_2)\boldsymbol{j}$$

$$\ddot{\boldsymbol{r}}_P = -(r_4\ddot\theta_4\sin\theta_4 + r_4\dot\theta_4^2\cos\theta_4)\boldsymbol{i} + (r_4\ddot\theta_4\cos\theta_4 - r_4\dot\theta_4^2\sin\theta_4)\boldsymbol{j}$$

Now that the equations have been developed, it is relatively simple to write a computer program for the analysis of a four-bar linkage. To aid in this, the equations required are summarized in Table 3.1. The authors have found that MATLAB is a very convenient language for solving simple kinematic equations, and this program runs on a variety of platforms. MATLAB routines for analyzing four-bar linkages are contained on the disk provided with this book.

EXAMPLE 3.1 ***(Position Analysis of a Four-Bar Linkage)***

PROBLEM

For a linkage with $r_1 = 1$, $r_2 = 2$, $r_3 = 3.5$, $r_4 = 4$, and $\theta_1 = 0$, compute the corresponding values of θ_3 and θ_4 for each of the solution branches when the driving crank is in the positions $\theta_2 = 0, \pi/2, \pi, -\pi/2$. Units for the lengths are not explicitly given in this example because the angular results are independent of the units for the lengths.

SOLUTION

The solution procedure is to use the equations in Table 3.1. First compute A, B, C for each value of θ_2 and then select σ. Next compute θ_4 and then θ_3. The calculations for $\theta_2 = 0$ are

Table 3.2 Summary of Results for Example 3.1

θ_2	σ	A	B	C	θ_4	θ_3
0	1	−8	0	4.75	53.58°	66.87°
	−1				−53.58°	−66.87°
$\pi/2$	1	87	−16	8.75	177.28°	−143.85°
	−1				55.85°	21.98°
p	1	24	0	12.75	−122.09°	−75.52°
	−1				122.09°	75.52°
$\pi/2$	1	8	16	8.75	−55.85°	−21.98°
	−1				−177.28°	148.85°

as follows:

$$A = 2r_1r_4\cos\theta_1 - 2r_2r_4\cos\theta_2 = 2(1)(4) - 2(2)(4) = -8$$

$$B = 2r_1r_4\sin\theta_1 - 2r_2r_4\sin\theta_2 = 0$$

$$C = r_1^2 + r_2^2 + r_4^2 - r_3^2 - 2r_1r_2(\cos\theta_1\cos\theta_2 + \sin\theta_1\sin\theta_2) = 1^2 + 2^2 + 4^2 - 3.5^2 - 2(1)(2) = 4.75$$

$$\theta_4 = 2\tan^{-1}\left[\frac{-B + \sigma\sqrt{B^2 - C^2 + A^2}}{C - A}\right]$$

$$= 2\tan^{-1}\left[\frac{-0 + \sqrt{0^2 - 4.75^2 + (-8)^2}}{4.75 + 8}\right] = 2\tan^{-1}(0.5049) = 53.58°$$

$$\theta_3 = \tan^{-1}\left[\frac{r_1\sin\theta_1 + r_4\sin\theta_4 - r_2\sin\theta_2}{r_1\cos\theta_1 + r_4\cos\theta_4 - r_2\cos\theta_2}\right]$$

$$= \tan^{-1}\left[\frac{4\sin(53.58)}{1 + 4\cos(53.58) - 2}\right] = \tan^{-1}\left[\frac{3.2187}{1.3748}\right] = \tan^{-1}(2.3412) = 66.87°$$

The remainder of the solution is summarized in Table 3.2.

The arithmetic may also be checked by comparing $\gamma = \theta_4 - \theta_3$ for $\sigma = \pm 1$. One value should be minus the other if both values are in the range $-\pi < \gamma < \pi$. It may be necessary to add or subtract 2π to either value to bring γ into the range $-\pi < \gamma \le \pi$. ■

EXAMPLE 3.2 ***(Velocity and Acceleration Analysis of a Four-Bar Linkage with Crank Input)***

PROBLEM

If, for the linkage in Example 3.1, $\dot{\theta}_2 = 10$ rad/s and $\ddot{\theta}_2 = 0$, compute $\dot{\theta}_3$, $\dot{\theta}_4$, $\ddot{\theta}_3$, and $\ddot{\theta}_4$ in the first of the four positions ($\theta_2 = 0$).

SOLUTION

Using the equations in Table 3.1, the velocity expression is

$$\begin{bmatrix} -r_3\sin\theta_3 & r_4\sin\theta_4 \\ -r_3\cos\theta_3 & r_4\cos\theta_4 \end{bmatrix}\begin{Bmatrix}\dot{\theta}_3 \\ \dot{\theta}_4\end{Bmatrix} = \begin{Bmatrix} r_2\dot{\theta}_2\sin\theta_2 \\ r_2\dot{\theta}_2\cos\theta_2\end{Bmatrix}$$

$$= \begin{bmatrix} -3.5\sin(66.87) & 4\sin(53.58) \\ -3.5\cos(66.87) & 4\cos(53.58) \end{bmatrix}\begin{Bmatrix}\dot{\theta}_3 \\ \dot{\theta}_4\end{Bmatrix} = \begin{Bmatrix} 2(10)\sin(0) \\ 2(10)\cos(0)\end{Bmatrix}$$

$$= \begin{bmatrix} -3.2187 & 3.2187 \\ -1.3749 & 2.3748 \end{bmatrix}\begin{Bmatrix}\dot{\theta}_3 \\ \dot{\theta}_4\end{Bmatrix} = \begin{Bmatrix} 0 \\ 20\end{Bmatrix}$$

Solving the linear set of equations gives $\dot{\theta}_3 = 20$ rad/s and $\dot{\theta}_4 = 20.0$ rad/s. Both values are positive, so the corresponding angular velocities are counterclockwise. The acceleration expression is

$$\begin{bmatrix} -r_3 \sin\theta_3 & r_4 \sin\theta_4 \\ -r_3 \cos\theta_3 & r_4 \cos\theta_4 \end{bmatrix} \begin{Bmatrix} \ddot{\theta}_3 \\ \ddot{\theta}_4 \end{Bmatrix} = \begin{Bmatrix} r_2\ddot{\theta}_2 \sin\theta_2 + r_2\dot{\theta}_2^2 \cos\theta_2 + r_3\dot{\theta}_3^2 \cos\theta_3 - r_4\dot{\theta}_4^2 \cos\theta_4 \\ r_2\ddot{\theta}_2 \cos\theta_2 - r_2\dot{\theta}_2^2 \sin\theta_2 - r_3\dot{\theta}_3^2 \sin\theta_3 + r_4\dot{\theta}_4^2 \sin\theta_4 \end{Bmatrix}$$

$$\begin{bmatrix} -3.2187 & 3.2187 \\ -1.3749 & 2.3748 \end{bmatrix} \begin{Bmatrix} \ddot{\theta}_3 \\ \ddot{\theta}_4 \end{Bmatrix} = \begin{Bmatrix} 0 + 2(10)^2 + 3.5(20)^2 \cos(66.87) - 4(20)^2 \cos(53.58) \\ 0 - 0 - 3.5(20)^2 \sin(66.87) + 4(20)^2 \sin(53.58) \end{Bmatrix}$$

$$= \begin{Bmatrix} -200.0265 \\ -0.0363 \end{Bmatrix}$$

Solving the linear set of equations gives $\ddot{\theta}_3 = 147.5634 \text{ rad/s}^2$ and $\ddot{\theta}_4 = 85.4150 \text{ rad/s}^2$. Again, both values are positive, so the corresponding angular accelerations are counterclockwise. ■

3.4 ANALYTICAL EQUATIONS FOR A RIGID BODY AFTER THE KINEMATIC PROPERTIES OF TWO POINTS ARE KNOWN

The equations presented so far will permit the kinematic properties of the points on the vector loop to be computed directly. However, often we need to compute the position, velocity, and acceleration of points that are not directly on the vector loops. In general, given the kinematic properties of *one* point on a rigid body and the angular position, angular velocity, and angular acceleration of the body, we can compute the position, velocity, and acceleration of *any* defined point on the rigid body.

Consider the rigid body represented in Fig. 3.9. Assume that A and B are two points attached to an arbitrary link, say link 5, and a third point is defined relative to the line between points A and B by the angle β and the distance $r_{C/A}$, which is represented in Fig. 3.9 as r_6. Then the linear position, velocity, and acceleration of point C_5 can be computed directly if the following are known: $\boldsymbol{r}_A$, $\dot{\boldsymbol{r}}_A$, $\ddot{\boldsymbol{r}}_A$, θ_5, $\dot{\theta}_5$, and $\ddot{\theta}_5$. The position of point C is given as

$$\boldsymbol{r}_C = \boldsymbol{r}_A + \boldsymbol{r}_6$$

or

$$\boldsymbol{r}_C = \boldsymbol{r}_A + r_6(\cos\theta_6\,\boldsymbol{i} + \sin\theta_6\,\boldsymbol{j}) \tag{3.56}$$

where

$$\theta_6 = \beta + \theta_5 \tag{3.57}$$

Recognizing that β is a constant, the velocity of point C is given by

$$\dot{\boldsymbol{r}}_C = \dot{\boldsymbol{r}}_A + r_6\dot{\theta}_5(-\sin\theta_6\,\boldsymbol{i} + \cos\theta_6\,\boldsymbol{j}) \tag{3.58}$$

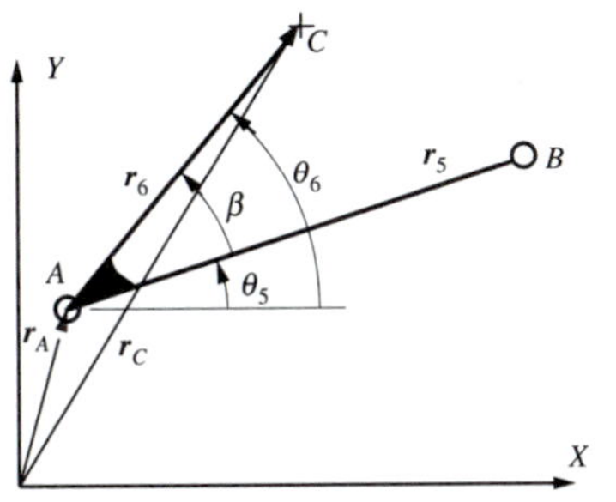

Figure 3.9 Calculation of the kinematic properties of a point on a link after the kinematic properties of one point and the angular velocity and acceleration of the link are known.

and the acceleration is given by

$$\ddot{\boldsymbol{r}}_C = \ddot{\boldsymbol{r}}_A + r_6\ddot{\theta}_5(-\sin\theta_6\,\boldsymbol{i} + \cos\theta_6\,\boldsymbol{j}) - r_6\dot{\theta}_5^2(\cos\theta_6\,\boldsymbol{i} + \sin\theta_6\,\boldsymbol{j}) \tag{3.59}$$

Note that we have assumed here that θ_5, $\dot{\theta}_5$, and $\ddot{\theta}_5$ are known. Often, we will know the kinematic information for two points on a rigid link instead of these angular quantities. If we know the position, velocity, and acceleration of two points (say A and B), we can compute θ_5, $\dot{\theta}_5$, and $\ddot{\theta}_5$ and proceed as before. The angle can be computed from the x and y components of the position vectors for A and B using

$$\theta_5 = \tan^{-1}\left[\frac{r_{B_y} - r_{A_y}}{r_{B_x} - r_{A_x}}\right]$$

The angular velocity can be computed by rewriting Eq. (3.58) in terms of points A and B. That is,

$$\dot{\boldsymbol{r}}_B = \dot{\boldsymbol{r}}_A + r_5\dot{\theta}_5(-\sin\theta_5\,\boldsymbol{i} + \cos\theta_5\,\boldsymbol{j})$$

Therefore,

$$\dot{\theta}_5 = -\frac{\dot{r}_{B_x} - \dot{r}_{A_x}}{r_5\sin\theta_5} = \frac{\dot{r}_{B_y} - \dot{r}_{A_y}}{r_5\cos\theta_5}$$

Similarly, the angular acceleration can be computed by rewriting Eq. (3.59) in terms of A and B. That is,

$$\ddot{\boldsymbol{r}}_B = \ddot{\boldsymbol{r}}_A + r_5\ddot{\theta}_5(-\sin\theta_5\,\boldsymbol{i} + \cos\theta_5\,\boldsymbol{j}) - r_5\dot{\theta}_5^2(\cos\theta_5\,\boldsymbol{i} + \sin\theta_5\,\boldsymbol{j})$$

Therefore,

$$\ddot{\theta}_5 = -\frac{(\ddot{r}_{B_x} - \ddot{r}_{A_x}) + r_5\dot{\theta}_5^2\cos\theta_5}{r_5\sin\theta_5} = \frac{(\ddot{r}_{B_y} - \ddot{r}_{A_y}) + r_5\dot{\theta}_5^2\sin\theta_5}{r_5\cos\theta_5}$$

These equations are summarized in Table 3.3, and a MATLAB function routine for the calculations is included on the disk with this book.

Table 3.3 Summary of Position, Velocity, and Acceleration Equations for an Arbitrary Point on a Rigid Body. The Vectors and Points are Defined in Fig. 3.10

If $\boldsymbol{r}_A$ and $\boldsymbol{r}_B$ are given instead of θ_5, $\dot{\theta}_5$, and $\ddot{\theta}_5$, first compute θ_5, $\dot{\theta}_5$, and $\ddot{\theta}_5$ using the following:

$$\theta_5 = \tan^{-1}\left[\frac{r_{B_y} - r_{A_y}}{r_{B_x} - r_{A_x}}\right]$$

$$\dot{\theta}_5 = -\frac{\dot{r}_{B_x} - \dot{r}_{A_x}}{r_5\sin\theta_5} = \frac{\dot{r}_{B_y} - \dot{r}_{A_y}}{r_5\cos\theta_5}$$

$$\ddot{\theta}_5 = -\frac{(\ddot{r}_{B_x} - \ddot{r}_{A_x}) + r_5\dot{\theta}_5^2\cos\theta_5}{r_5\sin\theta_5} = \frac{(\ddot{r}_{B_y} - \ddot{r}_{A_y}) + r_5\dot{\theta}_5^2\sin\theta_5}{r_5\cos\theta_5}$$

Position

$$\boldsymbol{r}_C = \boldsymbol{r}_A + r_6(\cos\theta_6\,\boldsymbol{i} + \sin\theta_6\,\boldsymbol{j})$$

$$\theta_6 = \beta + \theta_5$$

Velocity

$$\dot{\boldsymbol{r}}_C = \dot{\boldsymbol{r}}_A + r_6\dot{\theta}_5(-\sin\theta_6\,\boldsymbol{i} + \cos\theta_6\,\boldsymbol{j})$$

Accelerations

$$\ddot{\boldsymbol{r}}_C = \ddot{\boldsymbol{r}}_A + r_6\ddot{\theta}_5(-\sin\theta_6\,\boldsymbol{i} + \cos\theta_6\,\boldsymbol{j}) - r_6\dot{\theta}_5^2(\cos\theta_6\,\boldsymbol{i} + \sin\theta_6\,\boldsymbol{j})$$

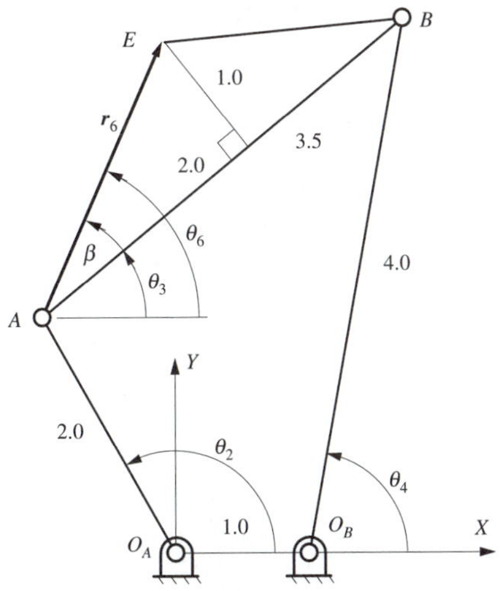

Figure 3.10 The linkage for Example 3.3.

EXAMPLE 3.3 ***(Velocity and Acceleration Analysis of Coupler Point)***

PROBLEM

For the linkage in Examples 3.1 and 3.2 shown in Fig. 3.10, compute the velocity and acceleration of point E_3 when $\theta_2 = 0$, $\dot{\theta}_2 = 10$ rad/s, $\ddot{\theta}_2 = 0$, and $\sigma = 1$. Assume that the lengths are given in centimeters.

SOLUTION

First compute the angle β between the line AB and the line AE. The angle is given by

$$\beta = \tan^{-1}\left(\frac{1}{2}\right) = 26.565°$$

and the length AE is given by

$$r_6 = AE = 2.0/\cos(26.565) = 2.236 \text{ cm}$$

Then the velocity of E_3 is given by

$$\dot{\boldsymbol{r}}_{E_3} = r_2\dot{\theta}_2(-\sin\theta_2\,\boldsymbol{i} + \cos\theta_2\boldsymbol{j}) + r_6\dot{\theta}_3(-\sin\theta_6\,\boldsymbol{i} + \cos\theta_6\boldsymbol{j})$$

Substitution of $\theta_2 = 0$, $\theta_3 = 66.87°$, $\dot{\theta}_2 = 10.0$ rad/s, and $\dot{\theta}_3 = 20.0$ rad/s from Example 3.2 gives

$$\theta_6 = \theta_3 + \beta = 66.87 + 26.565 = 93.435°$$

and

$$\dot{\boldsymbol{r}}_{E_3} = 2(10)(0\boldsymbol{i} + \boldsymbol{j}) + 2.236(20.0)[-\sin(93.435)\boldsymbol{i} + \cos(93.435)\boldsymbol{j}]$$
$$= 20\boldsymbol{j} - 44.640\boldsymbol{i} - 2.679\boldsymbol{j} = -44.64\boldsymbol{i} + 17.32\boldsymbol{j} \text{ cm/s}$$

The acceleration of E is given by

$$\ddot{\boldsymbol{r}}_{E_3} = r_2\ddot{\theta}_2(-\sin\theta_2\,\boldsymbol{i} + \cos\theta_2\,\boldsymbol{j}) - r_2\dot{\theta}_2^2(\cos\theta_2\,\boldsymbol{i} + \sin\theta_2\,\boldsymbol{j})$$
$$+ r_6\ddot{\theta}_3(-\sin\theta_6\,\boldsymbol{i} + \cos\theta_6\,\boldsymbol{j}) - r_6\dot{\theta}_3^2(\cos\theta_6\,\boldsymbol{i} + \sin\theta_6\,\boldsymbol{j})$$

Substitution of $\theta_2 = 0$, $\theta_3 = 66.87°$, $\dot{\theta}_2 = 10.0$ rad/s, $\ddot{\theta}_2 = 0$ rad/s^2, $\dot{\theta}_3 = 20.0$ rad/s, and $\ddot{\theta}_3 = 147.56$ rad/s^2 gives

$$\ddot{\boldsymbol{r}}_{E_3} = 0 - 2(10)^2(\boldsymbol{i} + 0\boldsymbol{j}) + 2.236(147.56)[-\sin(93.435)\boldsymbol{i} + \cos(93.435)\boldsymbol{j}]$$
$$-2.236(20.0)^2[\cos(93.435)\boldsymbol{i} + \sin(93.435)\boldsymbol{j}] = -475.76\boldsymbol{i} - 912.56\boldsymbol{j} \text{ cm/s}^2$$

■

3.5 ANALYTICAL EQUATIONS FOR SLIDER-CRANK MECHANISMS

Next to the four-bar linkage, the slider-crank is probably the most commonly used mechanism. It appears in all internal combustion engines (Fig. 3.11) and in numerous industrial (Fig. 3.12) and household devices (Fig. 3.13). A general slider-crank mechanism is represented in Fig. 3.14. To develop the closure equations, locate vectors $\boldsymbol{r}_2$ and $\boldsymbol{r}_3$ as was done in the regular four-bar linkage. To form the other part of the closure equation, draw two vectors, one in the direction of the slider velocity and one perpendicular to the velocity

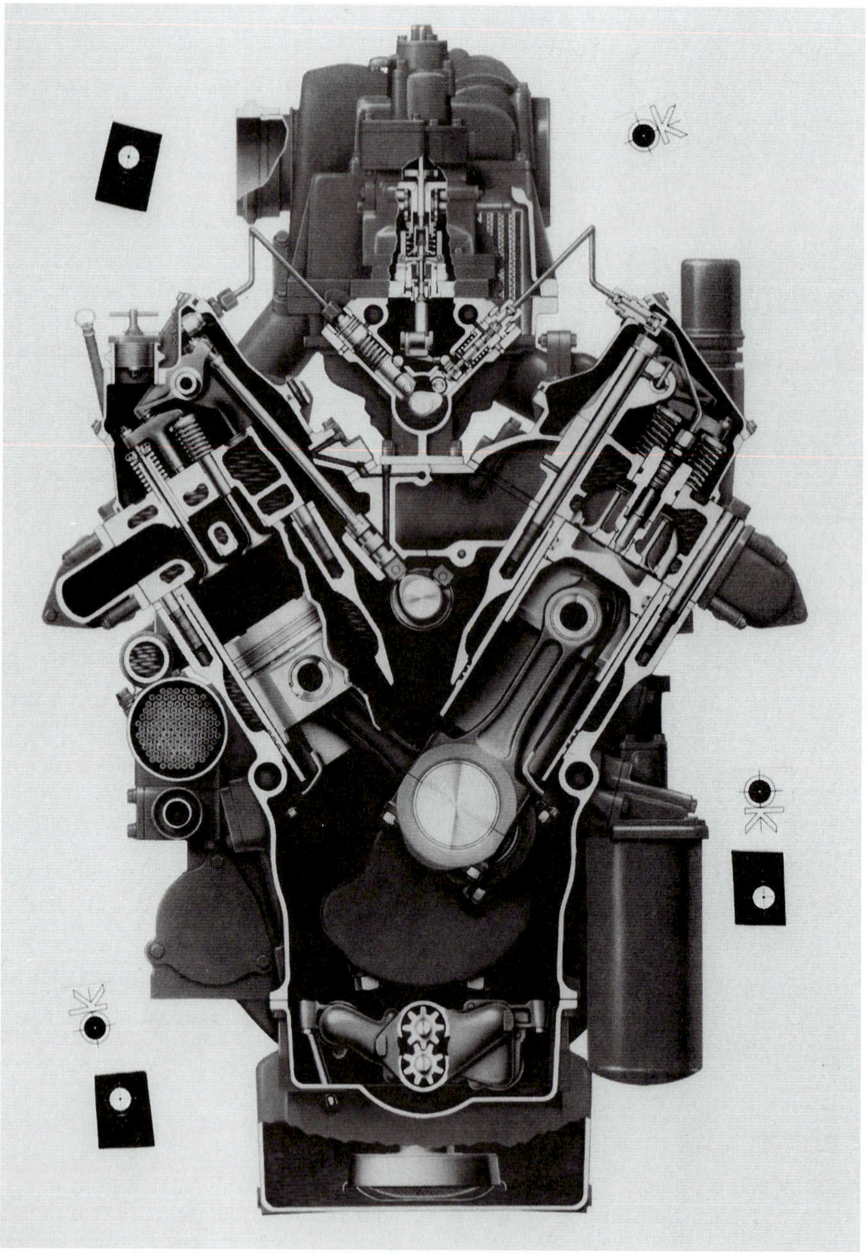

Figure 3.11 Internal combustion engine. An example of a slider-crank mechanism where the crank is the output link. (Courtesy of Caterpillar, Inc., Peoria, IL.)

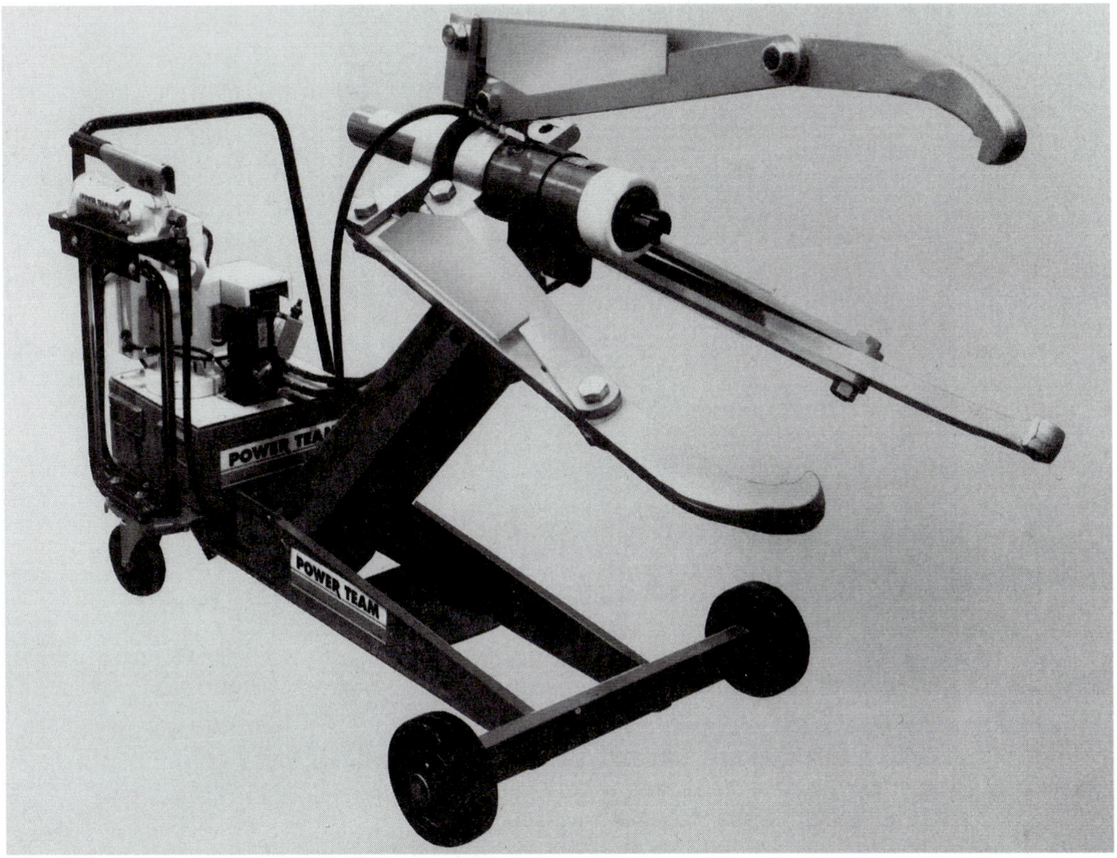

Figure 3.12 Hydraulic shaft puller. An example of a slider-crank mechanism where the slider is the input link. (Courtesy of Power Team, Owatonna, MN.)

direction. The variables associated with the problem are then located as shown in Fig. 3.14. The loop closure equation is then the same as that for the regular four-bar linkage:

$$\boldsymbol{r}_P = \boldsymbol{r}_2 + \boldsymbol{r}_3 = \boldsymbol{r}_1 + \boldsymbol{r}_4 \tag{3.60}$$

or

$$\begin{aligned} r_2(\cos\theta_2\,\boldsymbol{i} + \sin\theta_2\,\boldsymbol{j}) + r_3(\cos\theta_3\,\boldsymbol{i} + \sin\theta_3\,\boldsymbol{j}) \\ = r_1(\cos\theta_1\,\boldsymbol{i} + \sin\theta_1\,\boldsymbol{j}) + r_4(\cos\theta_4\,\boldsymbol{i} + \sin\theta_4\,\boldsymbol{j}) \end{aligned} \tag{3.61}$$

where

$$\theta_4 = \theta_1 + \pi/2 \tag{3.62}$$

Rewriting Eq. (3.61) in its component equations gives

$$r_2\cos\theta_2 + r_3\cos\theta_3 = r_1\cos\theta_1 + r_4\cos\theta_4 \tag{3.63}$$

$$r_2\sin\theta_2 + r_3\sin\theta_3 = r_1\sin\theta_1 + r_4\sin\theta_4 \tag{3.64}$$

Equations (3.62)–(3.64) must be satisfied throughout the motion of the linkage. The base vector, $\boldsymbol{r}_1$, will vary in magnitude but be constant in direction. The vector $\boldsymbol{r}_4$ will be constant. Therefore, r_2, r_3, r_4, θ_1, and θ_4 are constants. If θ_2 is given, it is necessary to solve Eqs. (3.63) and (3.64) for θ_3 and r_1 in terms of θ_2. If r_1 is given, it is necessary to solve the same equations for θ_2 and θ_3. Finally, if θ_3 is given, it is necessary to solve the equations

Figure 3.13 Ping-Pong table linkage. An example of a slider-crank mechanism where the coupler is the input link.

for θ_2 and r_1. Once these expressions are obtained, the unknown velocities and accelerations can be computed in terms of the knowns by differentiation.

3.5.1 Solution to Position Equations When θ_2 Is Input

The analytical solution procedure follows the same major steps as in the four-bar linkage case. To eliminate θ_3, first isolate it in using Eqs. (3.63) and (3.64) as follows

$$r_3 \cos \theta_3 = r_1 \cos \theta_1 + r_4 \cos \theta_4 - r_2 \cos \theta_2 \quad \textbf{(3.65)}$$

$$r_3 \sin \theta_3 = r_1 \sin \theta_1 + r_4 \sin \theta_4 - r_2 \sin \theta_2 \quad \textbf{(3.66)}$$

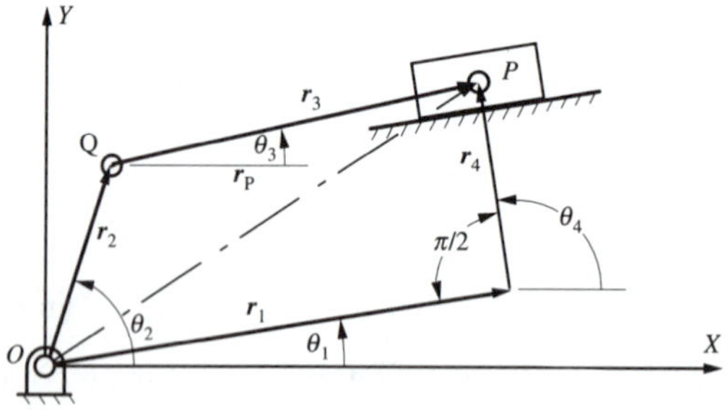

Figure 3.14 Vector closure condition for a slider-crank mechanism. The position of point P obtained by adding the vectors $\boldsymbol{r}_2$ and $\boldsymbol{r}_3$ is the same as that obtained by adding vectors $\boldsymbol{r}_1$ and $\boldsymbol{r}_4$.

Notice that the angle θ_1 is a known constant, but r_1 varies and is unknown. Now square both sides of both equations and add. This gives

$$r_3^2(\cos^2\theta_3 + \sin^2\theta_3) = (r_1\cos\theta_1 + r_4\cos\theta_4 - r_2\cos\theta_2)^2 + (r_1\sin\theta_1 + r_4\sin\theta_4 - r_2\sin\theta_2)^2$$

Expansion and simplification using the trigonometric identity in Eq. (3.30) give

$$\begin{aligned} r_3^2 = {} & r_1^2 + r_2^2 + r_4^2 + 2r_1r_4(\cos\theta_1\cos\theta_4 + \sin\theta_1\sin\theta_4) \\ & -2r_1r_2(\cos\theta_1\cos\theta_2 + \sin\theta_1\sin\theta_2) - 2r_2r_4(\cos\theta_2\cos\theta_4 + \sin\theta_2\sin\theta_4) \end{aligned} \tag{3.67}$$

Equation (3.67) gives r_1 in a quadratic expression involving θ_2 and the other known variables. To obtain a solution, collect together the coefficients of the different powers of r_1 as follows:

$$r_1^2 + Ar_1 + B = 0 \tag{3.68}$$

where

$$\begin{aligned} A &= 2r_4(\cos\theta_1\cos\theta_4 + \sin\theta_1\sin\theta_4) - 2r_2(\cos\theta_1\cos\theta_2 + \sin\theta_1\sin\theta_2) \\ B &= r_2^2 + r_4^2 - r_3^2 - 2r_2r_4(\cos\theta_2\cos\theta_4 + \sin\theta_2\sin\theta_4) \end{aligned} \tag{3.69}$$

Solving for r_1 gives

$$r_1 = \frac{-A + \sigma\sqrt{A^2 - 4B}}{2} \tag{3.70}$$

where $\sigma = \pm 1$ is a sign variable identifying the assembly mode. There are two assembly modes corresponding to the two configurations shown in Fig. 3.15.

As in the case of the four-bar linkage, once we pick the value for σ corresponding to the desired mode, the sign in an actual linkage stays the same for any value of θ_2.

Because of the square root in Eq. (3.70), the variable r_1 can be complex ($A^2 < 4B$). If this happens, the mechanism cannot be assembled in the position specified. The assembly would then appear as one of the configurations shown in Fig. 3.16.

Once a value for r_1 is determined, Eqs. (3.65) and (3.66) can be solved for θ_3. Dividing Eq. (3.66) by Eq. (3.65) and solving for θ_3 gives

$$\theta_3 = \tan^{-1}\left[\frac{r_1\sin\theta_1 + r_4\sin\theta_4 - r_2\sin\theta_2}{r_1\cos\theta_1 + r_4\cos\theta_4 - r_2\cos\theta_2}\right] \tag{3.71}$$

As in the case of the four-bar linkage, it is essential that the sign of the numerator and denominator in Eq. (3.71) be maintained to determine the quadrant in which the angle θ_3 lies.

Once all of the angular quantities are known, it is relatively straightforward to compute the coordinates of any point on the vector loops used in the closure equations. In particular,

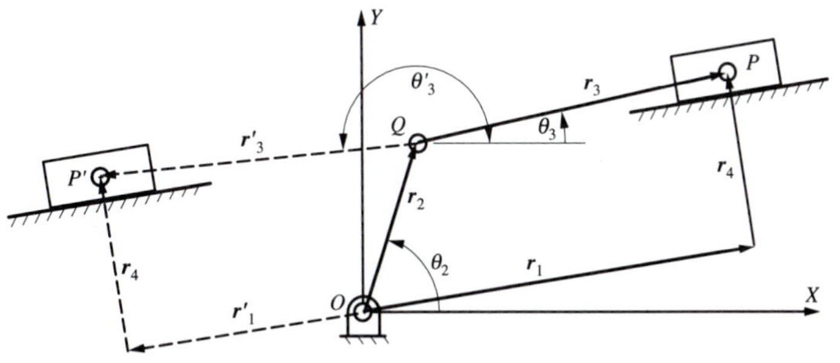

Figure 3.15 The two possible positions (P and P') of the point P for a given value of θ_2 in a slider-crank mechanism.

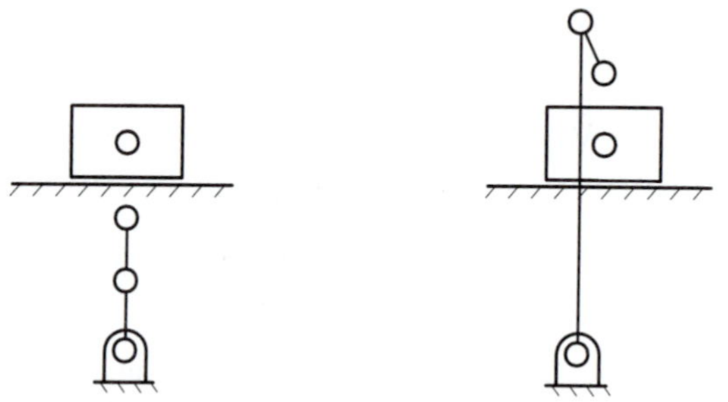

Figure 3.16 Configurations giving complex solutions for slider-crank position problem.

the coordinates of Q and P are given by

$$\boldsymbol{r}_Q = \boldsymbol{r}_2 = r_2(\cos\theta_2\,\boldsymbol{i} + \sin\theta_2\,\boldsymbol{j}) \tag{3.72}$$

and

$$\boldsymbol{r}_P = \boldsymbol{r}_2 + \boldsymbol{r}_3 = r_2(\cos\theta_2\,\boldsymbol{i} + \sin\theta_2\,\boldsymbol{j}) + r_3(\cos\theta_3\,\boldsymbol{i} + \sin\theta_3\,\boldsymbol{j}) \tag{3.73}$$

3.5.2 Solution to Position Equations When r_1 Is Input

The analytical solution procedure follows the same major steps as in the previous case. Again we start by eliminating θ_3 from Eqs. (3.63) and (3.64) to get Eq. (3.67). Then we simplify Eq. (3.67) as follows:

$$A\cos\theta_2 + B\sin\theta_2 + C = 0 \tag{3.74}$$

where:

$$\left.\begin{aligned} A &= -2r_1r_2\cos\theta_1 - 2r_2r_4\cos\theta_4 \\ B &= -2r_1r_2\sin\theta_1 - 2r_2r_4\sin\theta_4 \\ C &= r_1^2 + r_2^2 + r_4^2 - r_3^2 + 2r_1r_4(\cos\theta_1\cos\theta_4 + \sin\theta_1\sin\theta_4) \end{aligned}\right\} \tag{3.75}$$

To solve Eq. (3.74), the trigonometric half-angle identities given in Eqs. (3.33–3.34) can be used. Using these identities in Eq. (3.74) and simplifying gives

$$A(1 - t^2) + B(2t) + C(1 + t^2) = 0$$

where

$$t = \tan\left(\frac{\theta_2}{2}\right)$$

Simplifying gives

$$(C - A)t^2 + 2Bt + (A + C) = 0$$

Solving for t gives

$$t = \frac{-2B + \sigma\sqrt{4B^2 - 4(C - A)(C + A)}}{2(C - A)} = \frac{-B + \sigma\sqrt{B^2 - C^2 + A^2}}{C - A} \tag{3.76}$$

and

$$\theta_2 = 2\tan^{-1} t \tag{3.77}$$

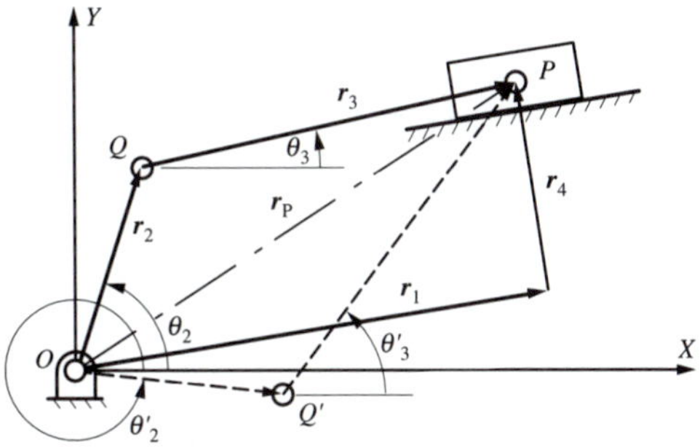

Figure 3.17 Two possible assembly modes when the position, $\mathbf{r}_1$, of the slider is given as an input.

where $\sigma = \pm 1$ is a sign variable identifying the assembly mode. Once again, because $\tan^{-1} t$ has a valid range of values $-\pi/2 \le \tan^{-1} t \le \pi/2$, θ_2 will have the range $-\pi \le \theta_2 \le \pi$. Typically, there are two solutions for θ_2 corresponding to the two values of σ, and they are both valid. These correspond to the two assembly modes shown in Fig. 3.17. Once we pick the value for σ corresponding to the desired mode, the sign in an actual linkage stays the same for any value of r_1.

Because of the square root in Eq. (3.76), the variable t can be complex $(A^2 + B^2) < C^2$. If this happens, the mechanism cannot be assembled for the specified value of r_1. The assembly configurations would then appear as shown in Fig. 3.18.

Equations (3.65) and (3.66) can now be solved for θ_3. The resulting equation is Eq. (3.71). As in the previous cases, it is essential that the sign of the numerator and denominator in Eq. (3.71) be maintained to determine the quadrant in which the angle θ_3 lies. Note that the positions of $\mathbf{r}_2$ and $\mathbf{r}_3$ are symmetric about the line OP.

Once all of the angular quantities are known, it is relatively straightforward to compute the coordinates of any point on the vector loops used in the closure equations. The coordinates of Q and P are again given by Eqs. (3.63) and (3.64).

3.5.3 Solution to Position Equations When θ_3 Is Input

When the coupler is the input link, values for θ_3 and its derivatives will be known. The analytical procedure for solving the position equations follows the same major steps as when θ_2 is the input. Therefore, we will simply outline the procedure and tabulate the results.

In the procedure, we can assume that θ_1, θ_3, $\dot{\theta}_3$, and $\ddot{\theta}_3$ are known and θ_2, $\dot{\theta}_2$, $\ddot{\theta}_2$, r_1, $\dot{r}_1$, and $\ddot{r}_1$ are to be found. The link lengths r_2 and r_3 and θ_1 are constants. For the position

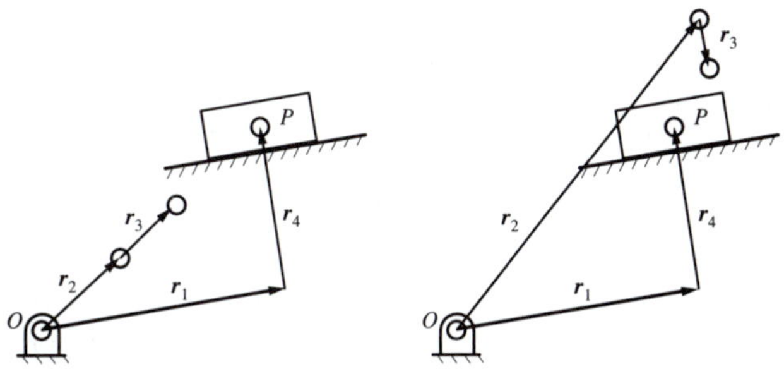

Figure 3.18 Slider-crank mechanisms that cannot be assembled in the position chosen for $\mathbf{r}_1$. The variable t would be complex in these cases.

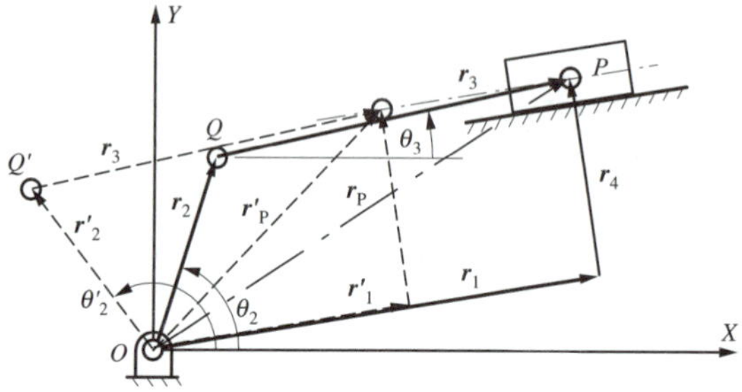

Figure 3.19 Two possible assembly modes when the coupler is the input link.

analysis, again begin with Eqs. (3.63) and (3.64) and isolate the terms with either θ_2 or θ_4. It is advantageous to select θ_2 for the reasons given below. The resulting equations are

$$r_2 \cos\theta_2 = r_1 \cos\theta_1 + r_4 \cos\theta_4 - r_3 \cos\theta_3 \tag{3.78}$$

$$r_2 \sin\theta_2 = r_1 \sin\theta_1 + r_4 \sin\theta_4 - r_3 \sin\theta_3 \tag{3.79}$$

A comparison of Eqs. (3.78) and (3.79) with Eqs. (3.65) and (3.66) indicates that they are of the same form except that the indices 2 and 3 are interchanged. Therefore, we can use directly the position solution derived in Section 3.5.1 if we interchange the indices 2 and 3.

When the coupler is the driver, there is an assembly-mode ambiguity similar to that observed when link 2 is the driver. This is illustrated in Fig. 3.19. It is necessary to know the appropriate mode before the analysis can be completed; however, once the assembly mode is selected, it is the same for all positions of the input.

3.5.4 Velocity Equations for Slider-Crank Mechanism

The analytical form of the velocity equations can be developed by differentiating Eq. (3.60). The result is

$$\dot{\boldsymbol{r}}_P = \dot{\boldsymbol{r}}_2 + \dot{\boldsymbol{r}}_3 = \dot{\boldsymbol{r}}_1 + \dot{\boldsymbol{r}}_4 \tag{3.80}$$

When this equation is written in component form, the result is the same as differentiating Eqs. (3.63) and (3.64). Recognizing that r_2, r_3, r_4, θ_1, and θ_4 are constants, the resulting component equations are

$$-r_2\dot{\theta}_2 \sin\theta_2 - r_3\dot{\theta}_3 \sin\theta_3 = \dot{r}_1 \cos\theta_1 \tag{3.81}$$

$$r_2\dot{\theta}_2 \cos\theta_2 + r_3\dot{\theta}_3 \cos\theta_3 = \dot{r}_1 \sin\theta_1 \tag{3.82}$$

The solution procedure depends on whether $\dot{r}_1$, $\dot{\theta}_2$, or $\dot{\theta}_3$ is known. If $\dot{\theta}_2$ (or $\dot{\theta}_3$) is input, then $\dot{r}_1$ and $\dot{\theta}_3$ (or $\dot{\theta}_2$) will be unknown. Therefore, the matrix equation to be solved is

$$\begin{bmatrix} \cos\theta_1 & r_J \sin\theta_J \\ \sin\theta_1 & -r_J \cos\theta_J \end{bmatrix} \begin{Bmatrix} \dot{r}_1 \\ \dot{\theta}_J \end{Bmatrix} = \begin{Bmatrix} -r_M\dot{\theta}_M \sin\theta_M \\ r_M\dot{\theta}_M \cos\theta_M \end{Bmatrix} \tag{3.83}$$

where $M = 2$ and $J = 3$ for a link 2 as the input, and $M = 3$ and $J = 2$ for link 3 as the input. If $\dot{r}_1$ is input, then $\dot{\theta}_2$ and $\dot{\theta}_3$ will be unknown. The matrix equation to be solved then is

$$\begin{bmatrix} -r_2 \sin\theta_2 & -r_3 \sin\theta_3 \\ r_2 \cos\theta_2 & r_3 \cos\theta_3 \end{bmatrix} \begin{Bmatrix} \dot{\theta}_2 \\ \dot{\theta}_3 \end{Bmatrix} = \begin{Bmatrix} \dot{r}_1 \cos\theta_1 \\ \dot{r}_1 \sin\theta_1 \end{Bmatrix} \tag{3.84}$$

The terms in the matrix and in the vector on the right-hand sides of Eqs. (3.83) and (3.84) will be known. The equation can therefore be solved manually, on a programmable calculator, or with the matrix solvers in programs such as MATLAB.

Once the angular velocities are known, it is a simple matter to compute the linear velocities of any of the points on the vector loop. The velocities of points Q and P are given by

$$\dot{\boldsymbol{r}}_Q = \dot{\boldsymbol{r}}_2 = r_2\dot{\theta}_2(-\sin\theta_2\,\boldsymbol{i} + \cos\theta_2\,\boldsymbol{j}) \tag{3.85}$$

and

$$\dot{\boldsymbol{r}}_P = \dot{\boldsymbol{r}}_2 + \dot{\boldsymbol{r}}_3 = (-r_2\dot{\theta}_2\sin\theta_2 - r_3\dot{\theta}_3\sin\theta_3)\boldsymbol{i} + (r_2\dot{\theta}_2\cos\theta_2 + r_3\dot{\theta}_3\cos\theta_3)\boldsymbol{j} \tag{3.86}$$

3.5.5 Acceleration Equations for Slider-Crank Mechanism

The analytical form of the acceleration equations can be developed by differentiating Eq. (3.80). The result is

$$\ddot{\boldsymbol{r}}_P = \ddot{\boldsymbol{r}}_2 + \ddot{\boldsymbol{r}}_3 = \ddot{\boldsymbol{r}}_1 + \ddot{\boldsymbol{r}}_4$$

When this equation is written in component form, the result is the same as differentiating Eqs. (3.81) and (3.82). The resulting component equations are

$$-r_2\ddot{\theta}_2\sin\theta_2 - r_2\dot{\theta}_2^2\cos\theta_2 - r_3\ddot{\theta}_3\sin\theta_3 - r_3\dot{\theta}_3^2\cos\theta_3 = \ddot{r}_1\cos\theta_1 \tag{3.87}$$

$$r_2\ddot{\theta}_2\cos\theta_2 - r_2\dot{\theta}_2^2\sin\theta_2 + r_3\ddot{\theta}_3\cos\theta_3 - r_3\dot{\theta}_3^2\sin\theta_3 = \ddot{r}_1\sin\theta_1 \tag{3.88}$$

As was the case for velocities, the solution procedure depends on whether $\ddot{\theta}_2$, $\ddot{\theta}_3$, or $\ddot{r}_1$ is known. If $\ddot{\theta}_2$ (or $\ddot{\theta}_3$) is input, then $\ddot{r}_1$ and $\ddot{\theta}_3$ (or $\ddot{\theta}_2$) will be unknown, and the matrix equation to be solved is

$$\begin{bmatrix} \cos\theta_1 & r_J\sin\theta_J \\ \sin\theta_1 & -r_J\cos\theta_J \end{bmatrix} \begin{Bmatrix} \ddot{r}_1 \\ \ddot{\theta}_J \end{Bmatrix} = \begin{Bmatrix} -r_M\ddot{\theta}_M\sin\theta_M - r_M\dot{\theta}_M^2\cos\theta_M - r_J\dot{\theta}_J^2\cos\theta_J \\ r_M\ddot{\theta}_M\cos\theta_M - r_M\dot{\theta}_M^2\sin\theta_M - r_J\dot{\theta}_J^2\sin\theta_J \end{Bmatrix} \tag{3.89}$$

If $\ddot{r}_1$ is input, then $\ddot{\theta}_2$ and $\ddot{\theta}_3$ will be unknown, and the matrix equation to be solved is then

$$\begin{bmatrix} -r_2\sin\theta_2 & -r_3\sin\theta_3 \\ r_2\cos\theta_2 & r_3\cos\theta_3 \end{bmatrix} \begin{Bmatrix} \ddot{\theta}_2 \\ \ddot{\theta}_3 \end{Bmatrix} = \begin{Bmatrix} r_2\dot{\theta}_2^2\cos\theta_2 + r_3\dot{\theta}_3^2\cos\theta_3 + \ddot{r}_1\cos\theta_1 \\ r_2\dot{\theta}_2^2\sin\theta_2 + r_3\dot{\theta}_3^2\sin\theta_3 + \ddot{r}_1\sin\theta_1 \end{Bmatrix} \tag{3.90}$$

The terms in the matrix and in the vector on the right-hand sides of Eqs. (3.89) and (3.90) will be known. The equation can therefore be solved manually, on a programmable calculator, or with the matrix solvers in programs such as MATLAB. Notice again that the coefficient matrix is the same for both the velocities (Eqs. 3.83 and 3.84) and for the accelerations (Eqs. 3.89 and 3.90).

Once the angular accelerations are known, it is a simple matter to compute the linear acceleration of any point on the vector loop. The accelerations of points Q and P are given by

$$\ddot{\boldsymbol{r}}_Q = \ddot{\boldsymbol{r}}_2 = (-r_2\ddot{\theta}_2\sin\theta_2 - r_2\dot{\theta}_2^2\cos\theta_2)\boldsymbol{i} + (r_2\ddot{\theta}_2\cos\theta_2 - r_2\dot{\theta}_2^2\sin\theta_2)\boldsymbol{j} \tag{3.91}$$

and

$$\begin{aligned}\ddot{\boldsymbol{r}}_P = \ddot{\boldsymbol{r}}_2 + \ddot{\boldsymbol{r}}_3 &= -(r_2\ddot{\theta}_2\sin\theta_2 + r_2\dot{\theta}_2^2\cos\theta_2 + r_3\ddot{\theta}_3\sin\theta_3 + r_3\dot{\theta}_3^2\cos\theta_3)\boldsymbol{i} \\ &+ (r_2\ddot{\theta}_2\cos\theta_2 - r_2\dot{\theta}_2^2\sin\theta_2 + r_3\ddot{\theta}_3\cos\theta_3 - r_3\dot{\theta}_3^2\sin\theta_3)\boldsymbol{j}\end{aligned} \tag{3.92}$$

Now that the equations have been developed, it is relatively simple to write a computer program for the analysis of a slider-crank linkage. To aid in this, the equations required are summarized in Tables 3.4 and 3.5. MATLAB programs for analyzing slider-crank linkages are included on the disk with this book.

Table 3.4 Summary of Position, Velocity, and Acceleration Equations for a Slider-Crank Mechanism When Either the Crank or the Coupler is the Input. Link 2 is the Input Link When $M = 2$ and $J = 3$. Link 3 is the Input Link When $M = 3$ and $J = 2$. The Link Numbers and Points are Defined in Fig. 3.14

Position

$$A = 2r_4(\cos\theta_1 \cos\theta_4 + \sin\theta_1 \sin\theta_4) - 2r_M(\cos\theta_1 \cos\theta_M + \sin\theta_1 \sin\theta_M)$$

$$B = r_M^2 + r_4^2 - r_J^2 - 2r_M r_4(\cos\theta_M \cos\theta_4 + \sin\theta_M \sin\theta_4)$$

$$r_1 = \frac{-A + \sigma\sqrt{A^2 - 4B}}{2}, \qquad \sigma = \pm 1$$

$$\theta_J = \tan^{-1}\left[\frac{r_1 \sin\theta_1 + r_4 \sin\theta_4 - r_M \sin\theta_M}{r_1 \cos\theta_1 + r_4 \cos\theta_4 - r_M \cos\theta_M}\right]$$

$$\boldsymbol{r}_Q = \boldsymbol{r}_2 = r_2(\cos\theta_2\,\boldsymbol{i} + \sin\theta_2\boldsymbol{j})$$

$$\boldsymbol{r}_P = \boldsymbol{r}_2 + \boldsymbol{r}_3 = r_2(\cos\theta_2\,\boldsymbol{i} + \sin\theta_2\boldsymbol{j}) + r_3(\cos\theta_3\,\boldsymbol{i} + \sin\theta_3\boldsymbol{j})$$

Velocity

$$\begin{bmatrix} \cos\theta_1 & r_J \sin\theta_J \\ \sin\theta_1 & -r_J \cos\theta_J \end{bmatrix} \begin{Bmatrix} \dot{r}_1 \\ \dot{\theta}_J \end{Bmatrix} = \begin{Bmatrix} -r_M\dot{\theta}_M \sin\theta_M \\ r_M\dot{\theta}_M \cos\theta_M \end{Bmatrix}$$

$$\dot{\boldsymbol{r}}_Q = r_2\dot{\theta}_2(-\sin\theta_2\,\boldsymbol{i} + \cos\theta_2\,\boldsymbol{j})$$

$$\dot{\boldsymbol{r}}_P = (-r_2\dot{\theta}_2 \sin\theta_2 - r_3\dot{\theta}_3 \sin\theta_3)\boldsymbol{i} + (r_2\dot{\theta}_2 \cos\theta_2 + r_3\dot{\theta}_3 \cos\theta_3)\boldsymbol{j}$$

Accelerations

$$\begin{bmatrix} \cos\theta_1 & r_J \sin\theta_J \\ \sin\theta_1 & -r_J \cos\theta_J \end{bmatrix} \begin{Bmatrix} \ddot{r}_1 \\ \ddot{\theta}_J \end{Bmatrix} = \begin{Bmatrix} -r_M\ddot{\theta}_M \sin\theta_M - r_M\dot{\theta}_M^2 \cos\theta_M - r_J\dot{\theta}_J^2 \cos\theta_J \\ r_M\ddot{\theta}_M \cos\theta_M - r_M\dot{\theta}_M^2 \sin\theta_M - r_J\dot{\theta}_J^2 \sin\theta_J \end{Bmatrix}$$

$$\ddot{\boldsymbol{r}}_Q = (-r_2\ddot{\theta}_2 \sin\theta_2 - r_2\dot{\theta}_2^2 \cos\theta_2)\boldsymbol{i} + (r_2\ddot{\theta}_2 \cos\theta_2 - r_2\dot{\theta}_2^2 \sin\theta_2)\boldsymbol{j}$$

$$\ddot{\boldsymbol{r}}_P = -(r_2\ddot{\theta}_2 \sin\theta_2 + r_2\dot{\theta}_2^2 \cos\theta_2 + r_3\ddot{\theta}_3 \sin\theta_3 + r_3\dot{\theta}_3^2 \cos\theta_3)\boldsymbol{i} + (r_2\ddot{\theta}_2 \cos\theta_2 - r_2\dot{\theta}_2^2 \sin\theta_2 + r_3\ddot{\theta}_3 \cos\theta_3 - r_3\dot{\theta}_3^2 \sin\theta_3)\boldsymbol{j}$$

EXAMPLE 3.4 ***(Kinematic Analysis of Slider-Crank Mechanism with Crank Input)***

PROBLEM

In the slider-crank mechanism shown in Fig. 3.20, $\theta_2 = 45°$, $\dot{\theta}_2 = 10$ rad/s, and $\ddot{\theta}_2 = 0$. The link lengths, r_2 and r_3, are as shown, and the line of motion of point C_4 is along the line AC. Find the position, velocity, and acceleration of C_4 and the angular velocity and acceleration of link 3.

SOLUTION

For this problem, the crank is the input, and the analysis can be conducted using the equations in Table 3.4 with $M = 2$ and $J = 3$. The known input information is:

$$\theta_1 = 0° \quad \theta_2 = 45° \quad \dot{\theta}_2 = 10 \text{ rad/s} \quad \ddot{\theta}_2 = 0$$

$$r_2 = 5 \text{ in} \quad r_3 = 8 \text{ in} \quad r_4 = 0 \text{ in}$$

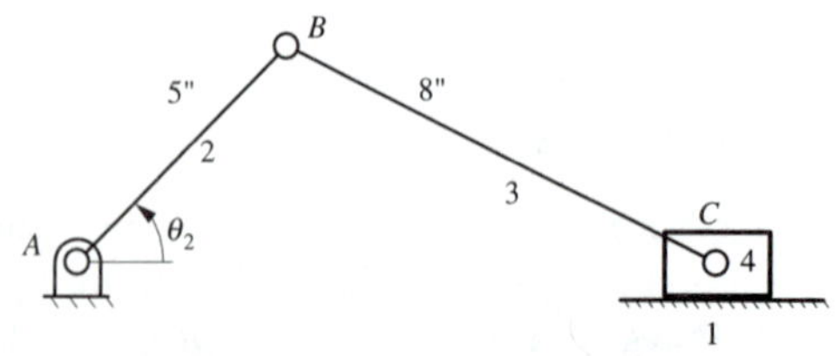

Figure 3.20 The slider-crank linkage to be analyzed in Example 3.4.

Table 3.5 Summary of Position, Velocity, and Acceleration Equations for a Slider-Crank Mechanism When the Slider (Link 4) is the Input Link. The Link Numbers and Points are Defined in Fig. 3.14

Position

$$A = -2r_1r_2\cos\theta_1 - 2r_2r_4\cos\theta_4$$
$$B = -2r_1r_2\sin\theta_1 - 2r_2r_4\sin\theta_4$$
$$C = r_1^2 + r_2^2 + r_4^2 - r_3^2 + 2r_1r_4(\cos\theta_1\cos\theta_4 + \sin\theta_1\sin\theta_4)$$
$$\theta_2 = 2\tan^{-1}\left[\frac{-B + \sigma\sqrt{B^2 - C^2 + A^2}}{C - A}\right], \qquad \sigma = \pm 1$$
$$\theta_3 = \tan^{-1}\left[\frac{r_1\sin\theta_1 + r_4\sin\theta_4 - r_2\sin\theta_2}{r_1\cos\theta_1 + r_4\cos\theta_4 - r_2\cos\theta_2}\right]$$
$$\boldsymbol{r}_Q = \boldsymbol{r}_2 = r_2(\cos\theta_2\,\boldsymbol{i} + \sin\theta_2\,\boldsymbol{j})$$
$$\boldsymbol{r}_P = \boldsymbol{r}_2 + \boldsymbol{r}_3 = r_2(\cos\theta_2\,\boldsymbol{i} + \sin\theta_2\,\boldsymbol{j}) + r_3(\cos\theta_3\,\boldsymbol{i} + \sin\theta_3\,\boldsymbol{j})$$

Velocity

$$\begin{bmatrix} -r_2\sin\theta_2 & -r_3\sin\theta_3 \\ r_2\cos\theta_2 & r_3\cos\theta_3 \end{bmatrix}\begin{Bmatrix} \dot\theta_2 \\ \dot\theta_3 \end{Bmatrix} = \begin{Bmatrix} \dot r_1\cos\theta_1 \\ \dot r_1\sin\theta_1 \end{Bmatrix}$$
$$\dot{\boldsymbol{r}}_Q = r_2\dot\theta_2(-\sin\theta_2\,\boldsymbol{i} + \cos\theta_2\,\boldsymbol{j})$$
$$\dot{\boldsymbol{r}}_P = (-r_2\dot\theta_2\sin\theta_2 - r_3\dot\theta_3\sin\theta_3)\boldsymbol{i} + (r_2\dot\theta_2\cos\theta_2 + r_3\dot\theta_3\cos\theta_3)\boldsymbol{j}$$

Accelerations

$$\begin{bmatrix} -r_2\sin\theta_2 & -r_3\sin\theta_3 \\ r_2\cos\theta_2 & r_3\cos\theta_3 \end{bmatrix}\begin{Bmatrix} \ddot\theta_2 \\ \ddot\theta_3 \end{Bmatrix} = \begin{Bmatrix} r_2\dot\theta_2^2\cos\theta_2 + r_3\dot\theta_3^2\cos\theta_3 + \ddot r_1\cos\theta_1 \\ r_2\dot\theta_2^2\sin\theta_2 + r_3\dot\theta_3^2\sin\theta_3 + \ddot r_1\sin\theta_1 \end{Bmatrix}$$
$$\ddot{\boldsymbol{r}}_Q = (-r_2\ddot\theta_2\sin\theta_2 - r_2\dot\theta_2^2\cos\theta_2)\boldsymbol{i} + (r_2\ddot\theta_2\cos\theta_2 - r_2\dot\theta_2^2\sin\theta_2)\boldsymbol{j}$$
$$\ddot{\boldsymbol{r}}_P = -(r_2\ddot\theta_2\sin\theta_2 + r_2\dot\theta_2^2\cos\theta_2 + r_3\ddot\theta_3\sin\theta_3 + r_3\dot\theta_3^2\cos\theta_3)\boldsymbol{i}$$
$$+ (r_2\ddot\theta_2\cos\theta_2 - r_2\dot\theta_2^2\sin\theta_2 + r_3\ddot\theta_3\cos\theta_3 - r_3\dot\theta_3^2\sin\theta_3)\boldsymbol{j}$$

Start with the position analysis, and first compute the constants A and B from Eq. (3.69):

$$A = 2r_4(\cos\theta_1\cos\theta_4 + \sin\theta_1\sin\theta_4) - 2r_2(\cos\theta_1\cos\theta_2 + \sin\theta_1\sin\theta_2)$$
$$= -2(5)(\cos(0)\cos(45) + \sin(0)\sin(45)) = -7.70711$$
$$B = r_2^2 + r_4^2 - r_3^2 - 2r_2r_4(\cos\theta_2\cos\theta_4 + \sin\theta_2\sin\theta_4)$$
$$= (5)^2 - (8)^2 = -39$$

The desired configuration of the linkage corresponds to the position of the slider with the larger x coordinate. Therefore, $\sigma = +1$. Then,

$$r_1 = \frac{-A + \sigma\sqrt{A^2 - 4B}}{2} = \frac{-(-7.70711) + \sqrt{(-7.70711)^2 - 4(-39)}}{2} = 10.712$$

Then θ_3 is given by

$$\theta_3 = \tan^{-1}\left[\frac{r_1\sin\theta_1 + r_4\sin\theta_4 - r_2\sin\theta_2}{r_1\cos\theta_1 + r_4\cos\theta_4 - r_2\cos\theta_2}\right] = \tan^{-1}\left[\frac{-5\sin(45)}{10.712 - 5\cos(45)}\right] = -26.228°$$

For the velocities, solve the linear set of velocity equations

$$\begin{bmatrix} \cos\theta_1 & r_3\sin\theta_3 \\ \sin\theta_1 & -r_3\cos\theta_3 \end{bmatrix}\begin{Bmatrix} \dot r_1 \\ \dot\theta_3 \end{Bmatrix} = \begin{Bmatrix} -r_2\dot\theta_2\sin\theta_2 \\ r_2\dot\theta_2\cos\theta_2 \end{Bmatrix} \quad \text{or} \quad \begin{bmatrix} 1 & 8\sin(-26.228) \\ 0 & -8\cos(-26.228) \end{bmatrix}\begin{Bmatrix} \dot r_1 \\ \dot\theta_3 \end{Bmatrix}$$
$$= \begin{Bmatrix} -5(10)\sin(45) \\ 5(10)\cos(45) \end{Bmatrix}$$

then

$$\begin{bmatrix} 1 & -3.5355 \\ 0 & -7.1764 \end{bmatrix} \begin{Bmatrix} \dot{r}_1 \\ \dot{\theta}_3 \end{Bmatrix} = \begin{Bmatrix} -35.3553 \\ 35.3553 \end{Bmatrix} \quad \text{or} \quad \begin{Bmatrix} \dot{r}_1 \\ \dot{\theta}_3 \end{Bmatrix} = \begin{Bmatrix} -52.774 \\ -4.927 \end{Bmatrix}$$

Therefore, $\dot{r}_1 = -52.774$ in/s and $\dot{\theta}_3 = -4.927$ rad/s CCW or 4.927 rad/s CW.
For the accelerations, solve the linear set of acceleration equations,

$$\begin{bmatrix} \cos\theta_1 & r_3 \sin\theta_3 \\ \sin\theta_1 & -r_3\cos\theta_3 \end{bmatrix} \begin{Bmatrix} \ddot{r}_1 \\ \ddot{\theta}_3 \end{Bmatrix} = \begin{Bmatrix} -r_2\ddot{\theta}_2 \sin\theta_2 - r_2\dot{\theta}_2^2\cos\theta_2 - r_3\dot{\theta}_3^2\cos\theta_3 \\ r_2\ddot{\theta}_2\cos\theta_2 - r_2\dot{\theta}_2^2\sin\theta_2 - r_3\dot{\theta}_3^2\sin\theta_3 \end{Bmatrix}$$

or

$$\begin{bmatrix} 1 & -3.5355 \\ 0 & -7.1764 \end{bmatrix} \begin{Bmatrix} \ddot{r}_1 \\ \ddot{\theta}_3 \end{Bmatrix} = \begin{Bmatrix} -5(10)^2\cos(45) - 8(-4.9266)^2\cos(-26.228) \\ -5(10)^2\sin(45) - 8(-4.9266)^2\sin(-26.228) \end{Bmatrix} = \begin{Bmatrix} -527.7366 \\ -267.7395 \end{Bmatrix}$$

then

$$\begin{Bmatrix} \ddot{r}_1 \\ \ddot{\theta}_3 \end{Bmatrix} = \begin{Bmatrix} -395.83 \\ 37.309 \end{Bmatrix}$$

Therefore, $\ddot{r}_1 = -395.83$ in/s^2, and $\ddot{\theta}_3 = 37.30$ rad/s^2 CCW. The results can be checked with the graphical analysis in Example 3.3. ■

EXAMPLE 3.5 (*Kinematic Analysis of a Slider-Crank Mechanism with a Slider Input*)

PROBLEM

Reanalyze the slider-crank mechanism shown in Fig. 3.20 when $r_1 = 10.75$ in, $\dot{r}_1 = 50$ in/s, and $\ddot{r}_1 = 400$ in/s^2. The link lengths, r_2 and r_3, are the same as in Example 3.4, and again the line of action of point C_4 is along the line AC. Find the position, angular velocity, and angular acceleration of link 2 and of link 3.

SOLUTION

This is essentially the same problem as in Example 3.4 except that now the slider is the input link and link 2 is the output. The analysis can be conducted using the equations in Table 3.5. The known input information is

$\theta_1 = 0°$ $\quad r_1 = 10.75$ $\quad \dot{r}_1 = 50$ in/s $\quad \ddot{r}_1 = 400$ in/s^2

$r_2 = 5$ in $\quad r_3 = 8$ in $\quad r_4 = 0$ in

Start with the position analysis, and first compute constants A, B, and C:

$$A = -2r_1r_2\cos\theta_1 - 2r_2r_4\cos\theta_4 = -2(10.75)(5) = -107.5$$

$$B = -2r_1r_2\sin\theta_1 - 2r_2r_4\sin\theta_4 = 0$$

$$C = r_1^2 + r_2^2 + r_4^2 - r_3^2 + 2r_1r_4(\cos\theta_1\cos\theta_4 + \sin\theta_1\sin\theta_4) = (10.75)^2 + (5)^2 - (8)^2 = 76.56$$

For the configuration in Fig. 3.20, $\sigma = 1$. Then,

$$\theta_2 = 2\tan^{-1}\left[\frac{-B + \sigma\sqrt{B^2 - C^2 + A^2}}{C - A}\right] = 2\tan^{-1}\left[\frac{+1\sqrt{-(76.56)^2 + (-107.5)^2}}{76.56 - (-107.5)}\right] = 44.5850°$$

and

$$\theta_3 = \tan^{-1}\left[\frac{r_1\sin\theta_1 + r_4\sin\theta_4 - r_2\sin\theta_2}{r_1\cos\theta_1 + r_4\cos\theta_4 - r_2\cos\theta_2}\right] = \tan^{-1}\left[\frac{-5\sin(44.585)}{10.75 - 5\cos(44.585)}\right] = -26.02°$$

For the velocities, solve the linear set of velocity equations

$$\begin{bmatrix} -r_2\sin\theta_2 & -r_3\sin\theta_3 \\ r_2\cos\theta_2 & r_3\cos\theta_3 \end{bmatrix}\begin{Bmatrix} \dot\theta_2 \\ \dot\theta_3 \end{Bmatrix} = \begin{Bmatrix} \dot r_1\cos\theta_1 \\ \dot r_1\sin\theta_1 \end{Bmatrix} \quad \text{or} \quad \begin{bmatrix} -5\sin(44.585) & -8\sin(-26.02) \\ 5\cos(44.585) & 8\cos(-26.02) \end{bmatrix}\begin{Bmatrix} \dot\theta_2 \\ \dot\theta_3 \end{Bmatrix} = \begin{Bmatrix} 50 \\ 0 \end{Bmatrix}$$

then

$$\begin{bmatrix} -3.5098 & 3.5098 \\ 3.5610 & 7.189 \end{bmatrix}\begin{Bmatrix} \dot\theta_2 \\ \dot\theta_3 \end{Bmatrix} = \begin{Bmatrix} 50 \\ 0 \end{Bmatrix} \quad \text{or} \quad \begin{Bmatrix} \dot\theta_2 \\ \dot\theta_3 \end{Bmatrix} = \begin{Bmatrix} -9.527 \\ 4.719 \end{Bmatrix}$$

Therefore, $\dot\theta_2 = -9.527$ rad/s CCW or $\dot\theta_2 = 9.527$ rad/s CW, and $\dot\theta_3 = 4.719$ rad/s CCW.

For the accelerations, solve the linear set of acceleration equations

$$\begin{bmatrix} -r_2\sin\theta_2 & -r_3\sin\theta_3 \\ r_2\cos\theta_2 & r_3\cos\theta_3 \end{bmatrix}\begin{Bmatrix} \ddot\theta_2 \\ \ddot\theta_3 \end{Bmatrix} = \begin{Bmatrix} r_2\dot\theta_2^2\cos\theta_2 + r_3\dot\theta_3^2\cos\theta_3 + \ddot r_1\cos\theta_1 \\ r_2\dot\theta_2^2\sin\theta_2 + r_3\dot\theta_3^2\sin\theta_3 + \ddot r_1\sin\theta_1 \end{Bmatrix}$$

or

$$\begin{bmatrix} -3.5098 & 3.5098 \\ 3.5610 & 7.189 \end{bmatrix}\begin{Bmatrix} \ddot\theta_2 \\ \ddot\theta_3 \end{Bmatrix} = \begin{Bmatrix} 5(9.527)^2\cos(44.585) + 8(4.719)^2\cos(-26.02) + 400 \\ 5(9.527)^2\sin(44.585) + 8(4.719)^2\sin(-26.02) \end{Bmatrix} = \begin{Bmatrix} 883.309 \\ 240.381 \end{Bmatrix}$$

Then

$$\begin{Bmatrix} \ddot\theta_2 \\ \ddot\theta_3 \end{Bmatrix} = \begin{Bmatrix} -145.933 \\ 105.726 \end{Bmatrix}$$

Therefore, $\ddot\theta_2 = -145.933$ rad/s^2 CCW or $\ddot\theta_2 = 145.933$ rad/s^2 CW, and $\ddot\theta_3 = 105.726$ rad/s^2 CCW. ■

3.6 ANALYTICAL EQUATIONS FOR THE SLIDER-CRANK INVERSION

The slider-crank inversion is a common mechanism when linear actuators are involved (e.g., Figs. 1.35 and 3.21). It is also used in various pump mechanisms. As discussed in Chapter 1, for low-load conditions, the slider is often replaced by a pin-in-a-slot joint. The resulting mechanisms can be analyzed using the equations developed in this section by modeling the pin-in-a-slot joint as a revolute joint and slider joint connected by a link. A device that could be analyzed using this procedure is the walking toy shown in Fig. 3.22. To develop the closure equations, first locate vectors $\boldsymbol{r}_2$ and $\boldsymbol{r}_1$ as was done in the previous linkage. To form the other part of the closure equation, draw two vectors, one ($\boldsymbol{r}_3$) in the direction of the slider velocity from P to Q and one ($\boldsymbol{r}_4$) perpendicular to the velocity direction. The variables associated with the problem are then located as shown in Fig. 3.23, and the loop closure equation is given by

$$\boldsymbol{r}_P = \boldsymbol{r}_2 = \boldsymbol{r}_1 + \boldsymbol{r}_3 + \boldsymbol{r}_4 \tag{3.93}$$

or

$$r_2(\cos\theta_2\,\boldsymbol{i} + \sin\theta_2\,\boldsymbol{j}) = r_1(\cos\theta_1\,\boldsymbol{i} + \sin\theta_1\,\boldsymbol{j}) + r_3(\cos\theta_3\,\boldsymbol{i} + \sin\theta_3\,\boldsymbol{j}) + r_4(\cos\theta_4\,\boldsymbol{i} + \sin\theta_4\,\boldsymbol{j}) \tag{3.94}$$

where

$$\theta_4 = \theta_3 - \pi/2 \tag{3.95}$$

Note that r_4 can be negative.

Figure 3.21 Backhoe. Each joint is actuated by an inversion of the slider crank-mechanism (Courtesy of Deere & Co., Molene, IL.)

Figure 3.22 Walking toy. The pin-in-a-slot joints can be modeled as a separate revolute and slider joint connected by a link. The resulting mechanism can be analyzed using the equations developed in this section.

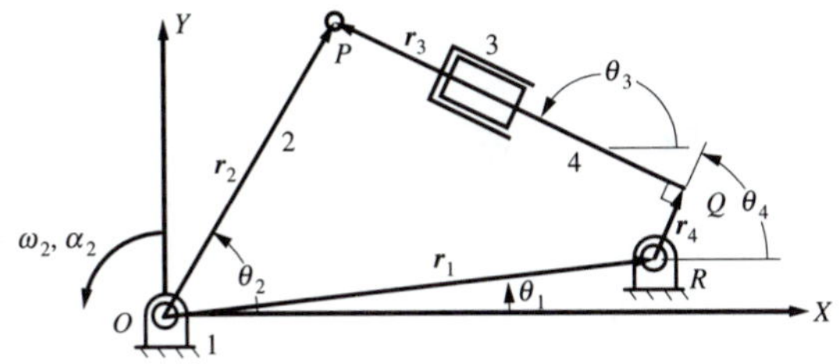

Figure 3.23 Vector closure condition for the slider-crank inversion. The position of point Q indicated by $\boldsymbol{r}_2 + \boldsymbol{r}_3$ is the same as that obtained by adding vectors $\boldsymbol{r}_1 + \boldsymbol{r}_4$.

Rewriting Eq. (3.94) in its component equations gives

$$r_2 \cos \theta_2 = r_1 \cos \theta_1 + r_3 \cos \theta_3 + r_4 \cos \theta_4 \tag{3.96}$$

$$r_2 \sin \theta_2 = r_1 \sin \theta_1 + r_3 \sin \theta_3 + r_4 \sin \theta_4 \tag{3.97}$$

Equations (3.95)–(3.97) must be satisfied throughout the motion of the linkage. The base vector $\boldsymbol{r}_1$ will be constant in direction and magnitude. The vectors $\boldsymbol{r}_2$ and $\boldsymbol{r}_4$ will be constant in magnitude, and $\boldsymbol{r}_3$ will vary in both magnitude and direction. Therefore, r_1, r_2, r_4, θ_1, and θ_4 are constants.

If θ_2 is given, it is necessary to solve Eqs. (3.95)–(3.97) for θ_3 and r_3 in terms of θ_2, and if θ_3 is given, it is necessary to solve Eqs. (3.95)–(3.97) for θ_2 and r_3 in terms of θ_3. If r_3 is given, it is necessary to solve the same equations for θ_2 and θ_3. Once the position equations are solved, the equations for the unknown velocities and accelerations can be established in terms of the knowns by differentiation.

3.6.1 Solution to Position Equations When θ_2 Is Input

The analytical solution procedure is slightly different from that used in the previous cases because θ_4 is a function of θ_3. Therefore, θ_3 cannot be elimінted without first considering θ_4. To proceed, first eliminate θ_4 from Eqs. (3.96) and (3.97) by using Eq. (3.95) and isolate the terms containing θ_3 on the right-hand side of the equations. Then,

$$r_2 \cos \theta_2 - r_1 \cos \theta_1 = r_3 \cos \theta_3 + r_4 \sin \theta_3 \tag{3.98}$$

$$r_2 \sin \theta_2 - r_1 \sin \theta_1 = r_3 \sin \theta_3 - r_4 \cos \theta_3 \tag{3.99}$$

Now square both sides of both equations and add. After simplifying using the trigonometric identity in Eq. (3.30) and solving for r_3, the resulting equation is

$$r_3 = \sqrt{r_2^2 + r_1^2 - r_4^2 - 2r_1r_2(\cos \theta_1 \cos \theta_2 + \sin \theta_1 \sin \theta_2)} \tag{3.100}$$

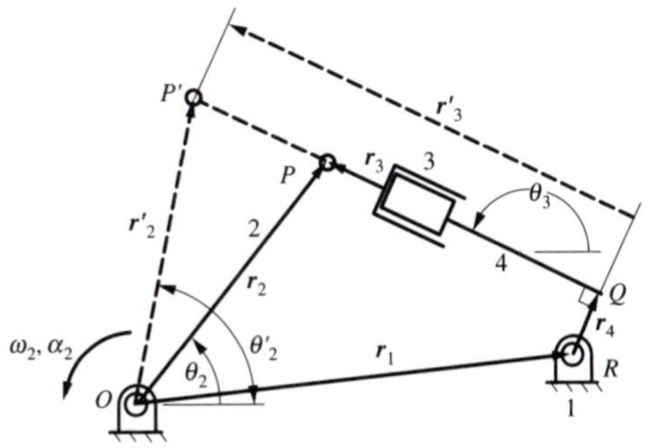

Figure 3.24 Two possible assembly modes for link 2 in Fig. 3.22 if θ_3 is given.

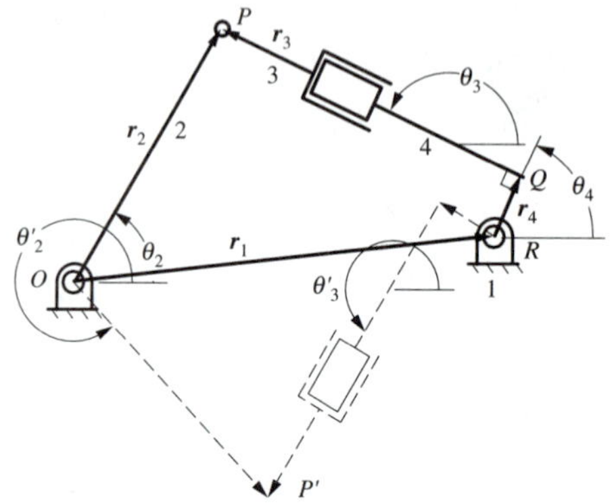

Figure 3.25 Two possible configurations for line PR in Fig. 3.22 if r_3 is given.

To solve for θ_3, replace $\cos\theta_3$ and $\sin\theta_3$ in Eq. (3.98) by the trigonometric half-angle identities given in Eqs. (3.33–3.34).

Equation (3.98) then becomes

$$A(1 + t^2) - r_3(1 - t^2) - r_4(2t) = 0 \tag{3.101}$$

where

$$A = r_2 \cos\theta_2 - r_1 \cos\theta_1$$

and

$$t = \tan\left(\frac{\theta_3}{2}\right)$$

Collecting terms in Eqs. (3.101) gives

$$(A + r_3)t^2 - 2r_4t + (A - r_3) = 0$$

This equation will give two roots for t, but one root is extraneous in this problem. The roots are

$$t = \frac{r_4 + \beta\sqrt{r_4^2 - A^2 + r_3^2}}{(A + r_3)} \tag{3.102}$$

where $\beta = \pm 1$. To determine the correct value of β for the problem, we must first compute a value of θ_3 for each value of t using

$$\theta_3 = 2\tan^{-1} t$$

Next substitute both values of θ_3 into Eq. (3.99). The correct value of β will correspond to the value of θ_3 satisfying Eq. (3.99). The value of β must be computed for each value of θ_2 if more than one position is analyzed.

Because of the square root in Eq. (3.100), the value of r_3 can be complex ($r_2^2 + r_1^2 - r_4^2 - 2r_1r_2(\cos\theta_1 \cos\theta_2 + \sin\theta_1 \sin\theta_2) < 0$). If this happens, the mechanism cannot be assembled for the value of θ_2 specified and for the given values of the link lengths. The results are summarized in Table 3.6.

3.6.2 Solution to Position Equations When θ_3 Is Input

When the coupler angle is the input variable, values for θ_3 and its derivative will be known. The analytical procedure for solving the position equations follows the same steps as when θ_2 is the input, although the form of the equations is slightly different.

Table 3.6 Summary of Position, Velocity, and Acceleration Equations for an Inverted Slider-Crank Mechanism when θ_2 is the Input Variable. The Link Numbers and Points are Defined in Fig. 3.21.

Position

$$r_3 = \sqrt{r_2^2 + r_1^2 - r_4^2 - 2r_1r_2(\cos\theta_1\cos\theta_2 + \sin\theta_1\sin\theta_2)}$$

$$A = r_2\cos\theta_2 - r_1\cos\theta_1$$

$$\theta_3 = 2\tan^{-1}\left[\frac{r_4 + \beta\sqrt{r_4^2 - A^2 + r_3^2}}{(A + r_3)}\right], \qquad \beta = \pm 1 \text{ but only one value is valid}$$

$$\theta_4 = \theta_3 - \pi/2$$

$$\boldsymbol{r}_P = \boldsymbol{r}_2 = r_2(\cos\theta_2\,\boldsymbol{i} + \sin\theta_2\boldsymbol{j})$$

$$\boldsymbol{r}_Q = \boldsymbol{r}_1 + \boldsymbol{r}_4 = r_1(\cos\theta_1\,\boldsymbol{i} + \sin\theta_1\boldsymbol{j}) + r_4(\cos\theta_4\,\boldsymbol{i} + \sin\theta_4\boldsymbol{j})$$

Velocity

$$\begin{bmatrix}\cos\theta_3 & -r_3\sin\theta_3 - r_4\sin\theta_4\\ \sin\theta_3 & r_3\cos\theta_3 + r_4\cos\theta_4\end{bmatrix}\begin{Bmatrix}\dot{r}_3\\ \dot{\theta}_3\end{Bmatrix} = \begin{Bmatrix}-r_2\dot{\theta}_2\sin\theta_2\\ r_2\dot{\theta}_2\cos\theta_2\end{Bmatrix}$$

$$\dot{\boldsymbol{r}}_P = \dot{\boldsymbol{r}}_2 = r_2\ddot{\theta}_2(-\sin\theta_2\,\boldsymbol{i} + \cos\theta_2\boldsymbol{j})$$

$$\dot{\boldsymbol{r}}_Q = \dot{\boldsymbol{r}}_1 + \dot{\boldsymbol{r}}_4 = r_4\dot{\theta}_3(-\sin\theta_4\,\boldsymbol{i} + \cos\theta_4\boldsymbol{j})$$

Accelerations

$$\begin{bmatrix}\cos\theta_3 & -r_3\sin\theta_3 - r_4\sin\theta_4\\ \sin\theta_3 & r_3\cos\theta_3 + r_4\cos\theta_4\end{bmatrix}\begin{Bmatrix}\ddot{r}_3\\ \ddot{\theta}_3\end{Bmatrix}$$

$$= \begin{Bmatrix}-r_2\ddot{\theta}_2\sin\theta_2 - r_2\dot{\theta}_2^2\cos\theta_2 + r_3\dot{\theta}_3^2\cos\theta_3 + 2\dot{r}_3\dot{\theta}_3\sin\theta_3 + r_4\dot{\theta}_3^2\cos\theta_4\\ r_2\ddot{\theta}_2\cos\theta_2 - r_2\dot{\theta}_2^2\sin\theta_2 + r_3\dot{\theta}_3^2\sin\theta_3 - 2\dot{r}_3\dot{\theta}_3\cos\theta_3 + r_4\dot{\theta}_3^2\sin\theta_4\end{Bmatrix}$$

$$\ddot{\boldsymbol{r}}_P = \ddot{\boldsymbol{r}}_2 = -(r_2\ddot{\theta}_2\sin\theta_2 + r_2\dot{\theta}_2^2\cos\theta_2)\,\boldsymbol{i} + (r_2\ddot{\theta}_2\cos\theta_2 - r_2\dot{\theta}_2^2\sin\theta_2)\boldsymbol{j}$$

$$\ddot{\boldsymbol{r}}_Q = \ddot{\boldsymbol{r}}_1 + \ddot{\boldsymbol{r}}_4 = -(r_4\ddot{\theta}_3\sin\theta_4 + r_4\dot{\theta}_3^2\cos\theta_4)\,\boldsymbol{i} + (r_4\ddot{\theta}_3\cos\theta_4 - r_4\dot{\theta}_3^2\sin\theta_4)\boldsymbol{j}$$

For the position analysis, again begin with the vector loop equations, decompose into components and isolate the terms with θ_2. The resulting equations are

$$r_2\cos\theta_2 = r_1\cos\theta_1 + r_3\cos\theta_3 + r_4\sin\theta_3 \tag{3.103}$$

$$r_2\sin\theta_2 = r_1\sin\theta_1 + r_3\sin\theta_3 - r_4\cos\theta_3 \tag{3.104}$$

These equations can be solved first for r_3 and then for θ_2 using the procedures given in previous sections. The results are summarized in Table 3.7.

3.6.3 Solution to Position Equations When r_3 Is Input

When r_3 is input, we can elminate θ_3 from the component equations as was done in the previous section to obtain Eq. (3.91). We can then solve for θ_2 and then for θ_3. The results are summarized in Table 3.8.

3.6.4 Velocity Equations for the Slider-Crank Inversion

The analytical form of the velocity equations can be developed by differentiating Eq. (3.93). The result is

$$\dot{\boldsymbol{r}}_P = \dot{\boldsymbol{r}}_2 = \dot{\boldsymbol{r}}_1 + \dot{\boldsymbol{r}}_4 + \dot{\boldsymbol{r}}_3 \tag{3.105}$$

When this equation is written in component form, the result is the same as differentiating Eqs. (3.96) and (3.97). Recognizing that r_1, r_2, r_4, and θ_1 are constants and that from Eq.

Table 3.7 Summary of Position, Velocity, and Acceleration Equations for an Inverted Slider-Crank Mechanism when θ_3 is the Input Variable. The Link Numbers and Points are Defined in Fig. 3.21.

Position

$$r_3 = \tfrac{1}{2}[-B + \beta\sqrt{B^2 - 4C}], \beta = \pm 1$$

$$B = 2r_1(\cos\theta_1 \cos\theta_3 + \sin\theta_1 \sin\theta_3)$$

$$C = r_1^2 - r_2^2 + r_4^2 + 2r_1r_4(\cos\theta_1 \sin\theta_3 - \sin\theta_1 \cos\theta_3)$$

$$\theta_2 = \tan^{-1}\left[\frac{r_1 \sin\theta_1 + r_3 \sin\theta_3 - r_4 \cos\theta_3}{r_1 \cos\theta_1 + r_3 \cos\theta_3 + r_4 \sin\theta_3}\right]$$

$$\theta_4 = \theta_3 - \pi/2$$

$$\boldsymbol{r}_P = \boldsymbol{r}_2 = r_2(\cos\theta_2\, \boldsymbol{i} + \sin\theta_2\, \boldsymbol{j})$$

$$\boldsymbol{r}_Q = \boldsymbol{r}_1 + \boldsymbol{r}_4 = r_1(\cos\theta_1\, \boldsymbol{i} + \sin\theta_1\, \boldsymbol{j}) + r_4(\cos\theta_4\, \boldsymbol{i} + \sin\theta_4\, \boldsymbol{j})$$

Velocity

$$\begin{bmatrix} -r_2 \sin\theta_2 & -\cos\theta_3 \\ r_2 \cos\theta_2 & -\sin\theta_3 \end{bmatrix} \begin{Bmatrix} \dot\theta_2 \\ \dot r_3 \end{Bmatrix} = \begin{Bmatrix} -r_3\dot\theta_3 \sin\theta_3 - r_4\dot\theta_3 \sin\theta_4 \\ r_3\dot\theta_3 \cos\theta_3 + r_4\dot\theta_3 \cos\theta_4 \end{Bmatrix}$$

$$\dot{\boldsymbol{r}}_P = \dot{\boldsymbol{r}}_2 = r_2\dot\theta_2(-\sin\theta_2\, \boldsymbol{i} + \cos\theta_2\, \boldsymbol{j})$$

$$\dot{\boldsymbol{r}}_Q = \dot{\boldsymbol{r}}_1 + \dot{\boldsymbol{r}}_4 = r_4\dot\theta_3(-\sin\theta_4\, \boldsymbol{i} + \cos\theta_4\, \boldsymbol{j})$$

Accelerations

$$\begin{bmatrix} -r_2 \sin\theta_2 & -\cos\theta_3 \\ r_2 \cos\theta_2 & -\sin\theta_3 \end{bmatrix} \begin{Bmatrix} \ddot\theta_2 \\ \ddot r_3 \end{Bmatrix}$$

$$= \begin{Bmatrix} r_2\dot\theta_2^2 \cos\theta_2 - r_3\ddot\theta_3 \sin\theta_3 - r_3\dot\theta_3^2 \cos\theta_3 - 2\dot r_3\dot\theta_3 \sin\theta_3 - r_4\ddot\theta_3 \sin\theta_4 - r_4\dot\theta_3^2 \cos\theta_4 \\ r_2\dot\theta_2^2 \sin\theta_2 + r_3\ddot\theta_3 \cos\theta_3 - r_3\dot\theta_3^2 \sin\theta_3 + 2\dot r_3\dot\theta_3 \cos\theta_3 + r_4\ddot\theta_3 \cos\theta_4 - r_4\dot\theta_3^2 \sin\theta_4 \end{Bmatrix}$$

$$\ddot{\boldsymbol{r}}_P = \ddot{\boldsymbol{r}}_2 = -(r_2\ddot\theta_2 \sin\theta_2 + r_2\dot\theta_2^2 \cos\theta_2)\, \boldsymbol{i} + (r_2\ddot\theta_2 \cos\theta_2 - r_2\dot\theta_2^2 \sin\theta_2)\boldsymbol{j}$$

$$\ddot{\boldsymbol{r}}_Q = \ddot{\boldsymbol{r}}_1 + \ddot{\boldsymbol{r}}_4 = -(r_4\ddot\theta_3 \sin\theta_4 + r_4\dot\theta_3^2 \cos\theta_4)\, \boldsymbol{i} + (r_4\ddot\theta_3 \cos\theta_4 - r_4\dot\theta_3^2 \sin\theta_4)\boldsymbol{j}$$

(3.95) $\dot\theta_3 = \dot\theta_4$, the resulting component equations are

$$-r_2\dot\theta_2 \sin\theta_2 = -r_3\dot\theta_3 \sin\theta_3 + \dot r_3 \cos\theta_3 - r_4\dot\theta_3 \sin\theta_4 \quad \textbf{(3.106)}$$

$$r_2\dot\theta_2 \cos\theta_2 = r_3\dot\theta_3 \cos\theta_3 + \dot r_3 \sin\theta_3 + r_4\dot\theta_3 \cos\theta_4 \quad \textbf{(3.107)}$$

The solution procedure depends on whether $\dot\theta_2$, $\dot\theta_3$, or $\dot r_3$ is known. If $\dot\theta_2$ is input, then $\dot r_3$ and $\dot\theta_3$ will be unknown. Therefore, the matrix equation to be solved is

$$\begin{bmatrix} \cos\theta_3 & -r_3 \sin\theta_3 - r_4 \sin\theta_4 \\ \sin\theta_3 & r_3 \cos\theta_3 + r_4 \cos\theta_4 \end{bmatrix} \begin{Bmatrix} \dot r_3 \\ \dot\theta_3 \end{Bmatrix} = \begin{Bmatrix} -r_2\dot\theta_2 \sin\theta_2 \\ r_2\dot\theta_2 \cos\theta_2 \end{Bmatrix} \quad \textbf{(3.108)}$$

If $\dot\theta_3$ is input, then $\dot r_3$ and $\dot\theta_2$ will be unknown, and the matrix equation to be solved is

$$\begin{bmatrix} -r_2 \sin\theta_2 & -\cos\theta_3 \\ r_2 \cos\theta_2 & -\sin\theta_3 \end{bmatrix} \begin{Bmatrix} \dot\theta_2 \\ \dot r_3 \end{Bmatrix} = \begin{Bmatrix} -r_3\dot\theta_3 \sin\theta_3 - r_4\dot\theta_3 \sin\theta_4 \\ r_3\dot\theta_3 \cos\theta_3 + r_4\dot\theta_3 \cos\theta_4 \end{Bmatrix} \quad \textbf{(3.109)}$$

If $\dot r_3$ is input, then $\dot\theta_2$ and $\dot\theta_3$ will be unknown. The matrix equation to be solved then is

$$\begin{bmatrix} -r_2 \sin\theta_2 & r_3 \sin\theta_3 + r_4 \sin\theta_4 \\ r_2 \cos\theta_2 & -r_3 \cos\theta_3 - r_4 \cos\theta_4 \end{bmatrix} \begin{Bmatrix} \dot\theta_2 \\ \dot\theta_3 \end{Bmatrix} = \begin{Bmatrix} \dot r_3 \cos\theta_3 \\ \dot r_3 \sin\theta_3 \end{Bmatrix} \quad \textbf{(3.110)}$$

The terms in the matrix and vector on the right-hand sides of Eqs. (3.108)–(3.110) will be known. The equation can therefore be solved manually, on a programmable calculator, or with the matrix solvers in programs such as MATLAB.

Table 3.8 Summary of Position, Velocity, and Acceleration Equations for Slider-Crank Inversion when $\boldsymbol{r}_3$ is the Input Variable. The Link Numbers and Points are Defined in Fig. 3.21.

Position

$$\left.\begin{aligned} A &= -2r_1r_2\cos\theta_1 \\ B &= -2r_1r_2\sin\theta_1 \\ C &= r_2^2 + r_1^2 - r_4^2 - r_3^2 \end{aligned}\right\}$$

$$\theta_2 = 2\tan^{-1}\left[\frac{-B + \beta\sqrt{B^2 - C^2 + A^2}}{C - A}\right], \qquad \beta = \pm 1$$

$$A = r_2\cos\theta_2 - r_1\cos\theta_1$$

$$\theta_3 = 2\tan^{-1}\left[\frac{r_4 + \sigma\sqrt{r_4^2 - A^2 + r_3^2}}{(A + r_3)}\right], \qquad \sigma = \pm 1$$

$$\theta_4 = \theta_3 - \pi/2$$

$$\boldsymbol{r}_P = \boldsymbol{r}_2 = r_2(\cos\theta_2\,\boldsymbol{i} + \sin\theta_2\,\boldsymbol{j})$$

$$\boldsymbol{r}_Q = \boldsymbol{r}_1 + \boldsymbol{r}_4 = r_1(\cos\theta_1\,\boldsymbol{i} + \sin\theta_1\,\boldsymbol{j}) + r_4(\cos\theta_4\,\boldsymbol{i} + \sin\theta_4\,\boldsymbol{j})$$

Velocity

$$\begin{bmatrix} -r_2\sin\theta_2 & r_3\sin\theta_3 + r_4\sin\theta_4 \\ r_2\cos\theta_2 & -r_3\cos\theta_3 - r_4\cos\theta_4 \end{bmatrix}\begin{Bmatrix} \dot\theta_2 \\ \dot\theta_3 \end{Bmatrix} = \begin{Bmatrix} \dot r_3\cos\theta_3 \\ \dot r_3\sin\theta_3 \end{Bmatrix}$$

$$\dot{\boldsymbol{r}}_P = \dot{\boldsymbol{r}}_2 = r_2\dot\theta_2(-\sin\theta_2\,\boldsymbol{i} + \cos\theta_2\,\boldsymbol{j})$$

$$\dot{\boldsymbol{r}}_Q = \dot{\boldsymbol{r}}_1 + \dot{\boldsymbol{r}}_4 = r_4\dot\theta_3(-\sin\theta_4\,\boldsymbol{i} + \cos\theta_4\,\boldsymbol{j})$$

Accelerations

$$\begin{bmatrix} -r_2\sin\theta_2 & r_3\sin\theta_3 + r_4\sin\theta_4 \\ r_2\cos\theta_2 & -r_3\cos\theta_3 - r_4\cos\theta_4 \end{bmatrix}\begin{Bmatrix} \ddot\theta_2 \\ \ddot\theta_3 \end{Bmatrix}$$

$$= \begin{Bmatrix} r_2\dot\theta_2^2\cos\theta_2 - r_3\dot\theta_3^2\cos\theta_3 - 2\dot r_3\dot\theta_3\sin\theta_3 + \ddot r_3\cos\theta_3 - r_4\dot\theta_3^2\cos\theta_4 \\ r_2\dot\theta_2^2\sin\theta_2 - r_3\dot\theta_3^2\sin\theta_3 + 2\dot r_3\dot\theta_3\cos\theta_3 + \ddot r_3\sin\theta_3 - r_4\dot\theta_3^2\sin\theta_4 \end{Bmatrix}$$

$$\ddot{\boldsymbol{r}}_P = \ddot{\boldsymbol{r}}_2 = -(r_2\ddot\theta_2\sin\theta_2 + r_2\dot\theta_2^2\cos\theta_2)\boldsymbol{i} + (r_2\ddot\theta_2\cos\theta_2 - r_2\dot\theta_2^2\sin\theta_2)\boldsymbol{j}$$

$$\ddot{\boldsymbol{r}}_Q = \ddot{\boldsymbol{r}}_1 + \ddot{\boldsymbol{r}}_4 = -(r_4\ddot\theta_3\sin\theta_4 + r_4\dot\theta_3^2\cos\theta_4)\boldsymbol{i} + (r_4\ddot\theta_3\cos\theta_4 - r_4\dot\theta_3^2\sin\theta_4)\boldsymbol{j}$$

Once the angular velocities are known, it is a simple matter to compute the linear velocities of any of the points on the vector loop. The velocities of points Q and P are given by

$$\dot{\boldsymbol{r}}_P = \dot{\boldsymbol{r}}_2 = r_2\dot\theta_2(-\sin\theta_2\,\boldsymbol{i} + \cos\theta_2\boldsymbol{j}) \tag{3.111}$$

and

$$\dot{\boldsymbol{r}}_Q = \dot{\boldsymbol{r}}_1 + \dot{\boldsymbol{r}}_4 = r_4\dot\theta_3(-\sin\dot\theta_4\,\boldsymbol{i} + \cos\theta_4\boldsymbol{j}) \tag{3.112}$$

3.6.5 Acceleration Equations for the Slider-Crank Inversion

The analytical form of the acceleration equations can be developed by differentiating Eq. (3.105). The result is

$$\ddot{\boldsymbol{r}}_p = \ddot{\boldsymbol{r}}_2 = \ddot{\boldsymbol{r}}_1 + \ddot{\boldsymbol{r}}_3 + \ddot{\boldsymbol{r}}_4 \tag{3.113}$$

When this equation is written in component form, the result is the same as differentiating Eqs. (3.106) and (3.107). The resulting component equations are

$$\begin{aligned} &-r_2\ddot\theta_2\sin\theta_2 - r_2\dot\theta_2^2\cos\theta_2 \\ &\quad = -r_3\ddot\theta_3\sin\theta_3 - r_3\dot\theta_3^2\cos\theta_3 + \ddot r_3\cos\theta_3 - 2\dot r_3\dot\theta_3\sin\theta_3 - r_4\ddot\theta_3\sin\theta_4 - r_4\dot\theta_3^2\cos\theta_4 \end{aligned} \tag{3.114}$$

$$r_2\ddot{\theta}_2\cos\theta_2 - r_2\dot{\theta}_2^2\sin\theta_2$$
$$= r_3\ddot{\theta}_3\cos\theta_3 - r_3\dot{\theta}_3^2\sin\theta_3 + \ddot{r}_3\sin\theta_3 + 2\dot{r}_3\dot{\theta}_3\cos\theta_3 + r_4\ddot{\theta}_3\cos\theta_4 - r_4\dot{\theta}_3^2\sin\theta_4 \quad \textbf{(3.115)}$$

In a manner similar to that in the case of velocities, the solution procedure depends on whether $\ddot{r}_3$ or $\ddot{\theta}_2$ is known. If $\ddot{\theta}_2$ is input, then $\ddot{r}_3$ and $\ddot{\theta}_3$ will be unknown. Therefore, the matrix equation to be solved is

$$\begin{bmatrix} \cos\theta_3 & -r_3\sin\theta_3 - r_4\sin\theta_4 \\ \sin\theta_3 & r_3\cos\theta_3 + r_4\cos\theta_4 \end{bmatrix} \begin{Bmatrix} \ddot{r}_3 \\ \ddot{\theta}_3 \end{Bmatrix} = \begin{Bmatrix} -r_2\ddot{\theta}_2\sin\theta_2 - r_2\dot{\theta}_2^2\cos\theta_2 + r_3\dot{\theta}_3^2\cos\theta_3 + 2\dot{r}_3\dot{\theta}_3\sin\theta_3 + r_4\dot{\theta}_3^2\cos\theta_4 \\ r_2\ddot{\theta}_2\cos\theta_2 - r_2\dot{\theta}_2^2\sin\theta_2 + r_3\dot{\theta}_3^2\sin\theta_3 - 2\dot{r}_3\dot{\theta}_3\cos\theta_3 + r_4\dot{\theta}_3^2\sin\theta_4 \end{Bmatrix} \quad \textbf{(3.116)}$$

If $\ddot{\theta}_3$ is input, then $\ddot{r}_3$ and $\ddot{\theta}_2$ will be unknown, and the matrix equation to be solved is

$$\begin{bmatrix} -r_2\sin\theta_2 & -\cos\theta_3 \\ r_2\cos\theta_2 & -\sin\theta_3 \end{bmatrix} \begin{Bmatrix} \ddot{\theta}_2 \\ \ddot{r}_3 \end{Bmatrix} = \begin{Bmatrix} r_2\dot{\theta}_2^2\cos\theta_2 - r_3\ddot{\theta}_3\sin\theta_3 - r_3\dot{\theta}_3^2\cos\theta_3 - 2\dot{r}_3\dot{\theta}_3\sin\theta_3 - r_4\ddot{\theta}_3\sin\theta_4 - r_4\dot{\theta}_3^2\cos\theta_4 \\ r_2\dot{\theta}_2^2\sin\theta_2 + r_3\ddot{\theta}_3\cos\theta_3 - r_3\dot{\theta}_3^2\sin\theta_3 + 2\dot{r}_3\dot{\theta}_3\cos\theta_3 + r_4\ddot{\theta}_3\cos\theta_4 - r_4\dot{\theta}_3^2\sin\theta_4 \end{Bmatrix} \quad \textbf{(3.117)}$$

If $\ddot{r}_3$ is input, then $\ddot{\theta}_2$ and $\ddot{\theta}_3$ will be unknown. The matrix equation to be solved is then

$$\begin{bmatrix} -r_2\sin\theta_2 & r_3\sin\theta_3 + r_4\sin\theta_4 \\ r_2\cos\theta_2 & -r_3\cos\theta_3 - r_4\cos\theta_4 \end{bmatrix} \begin{Bmatrix} \ddot{\theta}_2 \\ \ddot{\theta}_3 \end{Bmatrix} = \begin{Bmatrix} r_2\dot{\theta}_2^2\cos\theta_2 - r_3\dot{\theta}_3^2\cos\theta_3 - 2\dot{r}_3\dot{\theta}_3\sin\theta_3 + \ddot{r}_3\cos\theta_3 - r_4\dot{\theta}_3^2\cos\theta_4 \\ r_2\dot{\theta}_2^2\sin\theta_2 - r_3\dot{\theta}_3^2\sin\theta_3 + 2\dot{r}_3\dot{\theta}_3\cos\theta_3 + \ddot{r}_3\sin\theta_3 - r_4\dot{\theta}_3^2\sin\theta_4 \end{Bmatrix} \quad \textbf{(3.118)}$$

The terms in the matrix and vector on the right-hand sides of Eqs. (3.116)–(3.118) will be known. The equation can therefore be solved manually, on a programmable calculator, or with the matrix solvers in programs such as MATLAB. Notice again that the coefficient matrix is the same for both the velocities (Eqs. 3.108 and 3.110) and the accelerations (Eqs. 3.116–3.118).

Once the angular accelerations are known, it is a simple matter to compute the linear accelerations of any of the points on the vector loop. The accelerations of points P and Q are given by

$$\ddot{\boldsymbol{r}}_P = \ddot{\boldsymbol{r}}_2 = -(r_2\ddot{\theta}_2\sin\theta_2 + r_2\dot{\theta}_2^2\cos\theta_2)\boldsymbol{i} + (r_2\ddot{\theta}_2\cos\theta_2 - r_2\dot{\theta}_2^2\sin\theta_2)\boldsymbol{j} \quad \textbf{(3.119)}$$

and

$$\ddot{\boldsymbol{r}}_Q = \ddot{\boldsymbol{r}}_1 + \ddot{\boldsymbol{r}}_4 = -(r_4\ddot{\theta}_3\sin\theta_4 + r_4\dot{\theta}_3^2\cos\theta_4)\boldsymbol{i} + (r_4\ddot{\theta}_3\cos\theta_4 - r_4\dot{\theta}_3^2\sin\theta_4)\boldsymbol{j} \quad \textbf{(3.120)}$$

Now that the equations have been developed, it is a relatively simple matter to write a computer program for the analysis of an inverted slider-crank linkage. To aid in this, the equations required are summarized in Tables 3.6, 3.7, and 3.8. A MATLAB program for analyzing an inverted slider-crank linkage is included on the disk that accompanies this book.

EXAMPLE 3.6 ***(Kinematic Analysis of Foot-Pump Mechanism)***

PROBLEM

The foot pump shown in Fig. 3.26 is to be analyzed in one position ($\theta_2 = 60°$) as an inverted slider-crank linkage. Dimensions for the mechanism are given in Fig. 3.27, and the vector diagram for the analysis is given in Fig. 3.28. Assume that the angular velocity of the driver

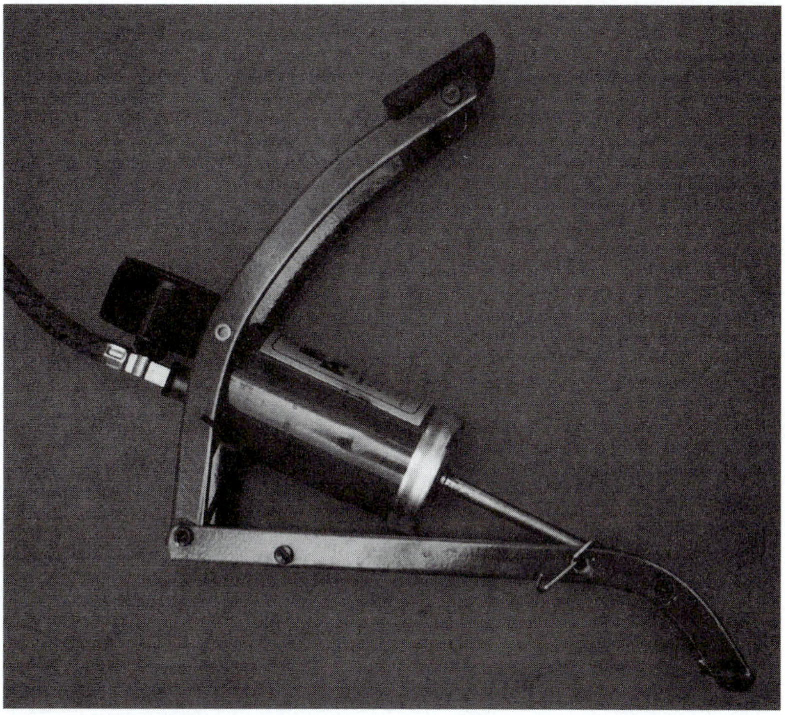

Figure 3.26 Foot-pump mechanism.

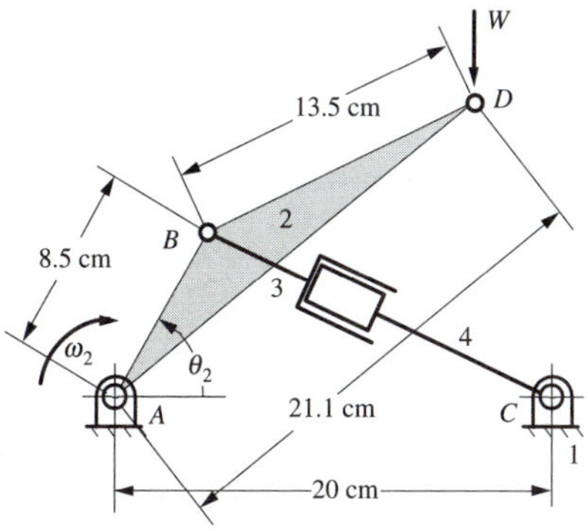

Figure 3.27 Kinematic model of foot-pump mechanism used in Example 3.6.

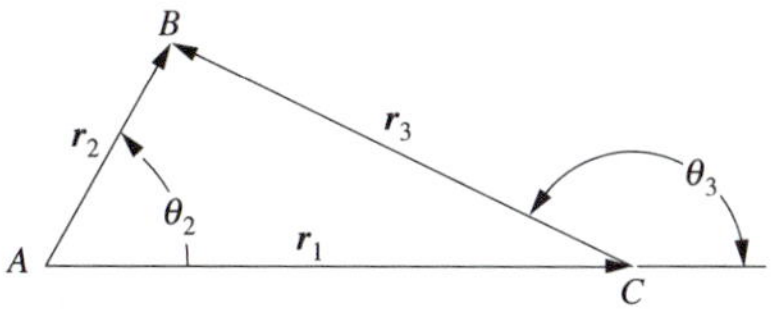

Figure 3.28 Position polygon for Example 3.6.

(link 2) is constant at 2.5 radians per second clockwise. Conduct a position, velocity, and acceleration analysis to determine, respectively, the combined length of links 3 and 4, the angular velocity and acceleration of link 3 (or link 4), and the velocity and acceleration of point B observed from link 3. This information can be used to study the pumping action between links 3 and 4.

SOLUTION The analysis can be conducted using the equations in Table 3.6. The known input information is

$$
\begin{array}{llll}
r_1 = 20 \text{ cm} & \theta_1 = 0 & & \\
r_2 = 8.5 \text{ cm} & \theta_2 = 60^\circ & \dot{\theta}_2 = -2.5 \text{ rad/s} & \ddot{\theta}_2 = 0 \\
r_3 = ? & \theta_3 = ? & \dot{\theta}_3 = ? & \ddot{\theta}_3 = ? \\
r_4 = 0 & \theta_4 = 0 & \dot{\theta}_4 = 0 & \ddot{\theta}_4 = 0
\end{array}
$$

From Table 3.6, r_3 is given by

$$
\begin{aligned}
r_3 &= \sqrt{r_2^2 + r_1^2 - r_4^2 - 2r_1r_2(\cos\theta_1\cos\theta_2 + \sin\theta_1\sin\theta_2)} \\
&= \sqrt{8.5^2 + 20^2 - 2(20)(8.5)\cos(60)} = \sqrt{302.2} = 17.38 \text{ cm}
\end{aligned}
$$

Next compute θ_3. Start with the value of A:

$$A = r_2\cos\theta_2 - r_1\cos\theta_1 = 8.5\cos 60 - 20 = -15.75$$

In order to determine θ_3, we need to specify β. Often, we need to compute both values of θ_3 to determine the proper value for β.

$$\theta_3 = 2\tan^{-1}\left[\frac{r_4 + \beta\sqrt{r_4^2 - A^2 + r_3^2}}{(A + r_3)}\right] = 2\tan^{-1}\left[\frac{\beta\sqrt{-(-15.75)^2 + (17.38)^2}}{(-15.75 + 17.38)}\right]$$

$$= 2\tan^{-1}\left[\frac{\beta 7.3486}{1.63}\right] = \pm 154.98^\circ$$

For this problem, $\beta = +1$.

For the velocity analysis, we can use Eq. (3.108). Then,

$$\begin{bmatrix}\cos\theta_3 & -r_3\sin\theta_3 - r_4\sin\theta_4 \\ \sin\theta_3 & r_3\cos\theta_3 + r_4\cos\theta_4\end{bmatrix}\begin{Bmatrix}\dot{r}_3 \\ \dot{\theta}_3\end{Bmatrix} = \begin{Bmatrix}-r_2\dot{\theta}_2\sin\theta_2 \\ r_2\dot{\theta}_2\cos\theta_2\end{Bmatrix}$$

or

$$\begin{bmatrix}\cos(154.99) & -17.38\sin(154.99) \\ \sin(154.99) & 17.38\cos(154.99)\end{bmatrix}\begin{Bmatrix}\dot{r}_3 \\ \dot{\theta}_3\end{Bmatrix} = \begin{Bmatrix}-8.5(-2.5)\sin(60) \\ 8.5(-2.5)\cos(60)\end{Bmatrix}$$

or

$$\begin{bmatrix}-0.9059 & -7.3612 \\ 0.4234 & -15.750\end{bmatrix}\begin{Bmatrix}\dot{r}_3 \\ \dot{\theta}_3\end{Bmatrix} = \begin{Bmatrix}18.403 \\ -10.625\end{Bmatrix} \Rightarrow \begin{Bmatrix}\dot{r}_3 \\ \dot{\theta}_3\end{Bmatrix} = \begin{Bmatrix}-21.171 \text{ cm/s} \\ 0.1055 \text{ rad/s}^2\end{Bmatrix}$$

For the acceleration analysis, we can use Eq. (3.117). Then,

$$\begin{Bmatrix}\cos\theta_3 & -r_3\sin\theta_3 - r_4\sin\theta_4 \\ \sin\theta_3 & r_3\cos\theta_3 + r_4\cos\theta_4\end{Bmatrix}\begin{Bmatrix}\ddot{r}_3 \\ \ddot{\theta}_3\end{Bmatrix}$$

$$= \begin{Bmatrix}-r_2\ddot{\theta}_2\sin\theta_2 - r_2\dot{\theta}_2^2\cos\theta_2 + r_3\dot{\theta}_3^2\cos\theta_3 + 2\dot{r}_3\dot{\theta}_3\sin\theta_3 + r_4\dot{\theta}_3^2\cos\theta_4 \\ r_2\ddot{\theta}_2\cos\theta_2 - r_2\dot{\theta}_2^2\sin\theta_2 + r_3\dot{\theta}_3^2\sin\theta_3 - 2\dot{r}_3\dot{\theta}_3\cos\theta_3 + r_4\dot{\theta}_3^2\sin\theta_4\end{Bmatrix}$$

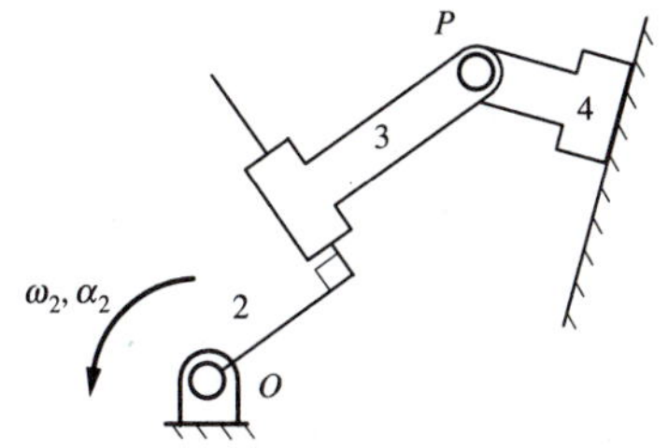

Figure 3.29 Schematic diagram of RPRP mechanism.

or

$$\begin{bmatrix} -0.9059 & -7.3612 \\ 0.4234 & -15.750 \end{bmatrix} \begin{Bmatrix} \ddot{r}_3 \\ \ddot{\theta}_3 \end{Bmatrix}$$
$$= \begin{Bmatrix} -(8.5)(-2.5)^2\cos(60) + 17.38(0.1055)^2\cos(154.99) + 2(-21.171)(0.1055)\sin(154.99) \\ -(8.5)(-2.5)^2\sin(60) + 17.38(0.1055)^2\sin(154.99) + 2(-21.171)(0.1055)\cos(154.99) \end{Bmatrix}$$
$$= \begin{Bmatrix} -28.6283 \\ -49.9710 \end{Bmatrix} \Rightarrow \begin{Bmatrix} \ddot{r}_3 \\ \ddot{\theta}_3 \end{Bmatrix} = \begin{Bmatrix} 4.777 \text{ cm/s}^2 \\ 3.012 \text{ rad/s}^2 \end{Bmatrix}$$

The velocity and acceleration of point B when observed from link 4 are $\dot{r}_3$ and $\ddot{r}_3$, respectively. Note that $\dot{r}_3$ is negative when $\dot{\theta}_2$ is CW. ■

3.7 ANALYTICAL EQUATIONS FOR AN RPRP MECHANISM

A schematic drawing of the RPRP mechanism is shown in Fig. 3.29. In the mechanism shown, link 2 is connected to the frame by a revolute joint and to the coupler by a prismatic joint. Link 4 is connected to the frame through a prismatic joint and to the coupler by a revolute joint. This is a less common mechanism than the various four-bar linkages and slider cranks; however, it does occur in industrial machinery. For example, one variation of it, the Rapson slide, is used in marine steering gear.

When there is a slider joint between two links, the actual location of the slider does not matter from a kinematic standpoint. Therefore, for simplicity, we can analyze the mechanism as if both sliders were at point P. The resulting mechanism then appears as shown in Fig. 3.30. In Fig. 3.30, the angles are not indicated for simplicity. The angles are again always measured counterclockwise from the horizontal as shown in Fig. 3.3. To develop the closure equations, locate vectors $\boldsymbol{r}_1$ through $\boldsymbol{r}_4$ as shown in Fig. 3.30. By locating point P using two different sets of vectors, the loop closure equation is seen to be the same as

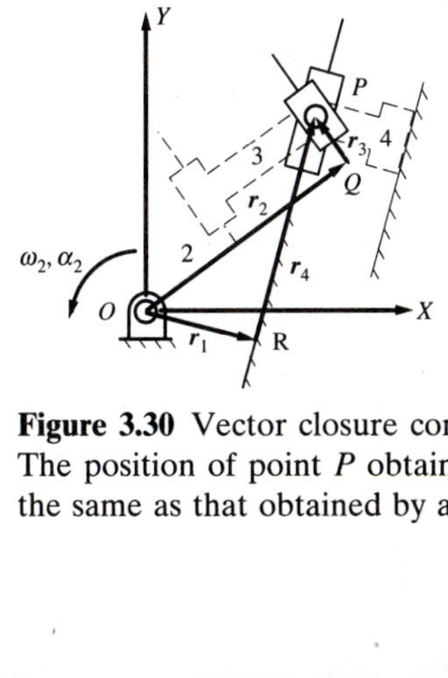

Figure 3.30 Vector closure condition for RPRP mechanism. The position of point P obtained by the vectors $\boldsymbol{r}_2$ and $\boldsymbol{r}_3$ is the same as that obtained by adding vectors $\boldsymbol{r}_1$ and $\boldsymbol{r}_4$.

that for the regular four-bar linkage. Namely,

$$\boldsymbol{r}_P = \boldsymbol{r}_2 + \boldsymbol{r}_3 = \boldsymbol{r}_1 + \boldsymbol{r}_4 \tag{3.121}$$

or

$$r_2(\cos\theta_2\,\boldsymbol{i} + \sin\theta_2\boldsymbol{j}) + r_3(\cos\theta_3\,\boldsymbol{i} + \sin\theta_3\boldsymbol{j}) = r_1(\cos\theta_1\,\boldsymbol{i} + \sin\theta_1\boldsymbol{j}) + r_4(\cos\theta_4\,\boldsymbol{i} + \sin\theta_4\boldsymbol{j}) \tag{3.122}$$

Rewriting Eq. (3.122) in its component equations gives

$$r_2\cos\theta_2 + r_3\cos\theta_3 = r_1\cos\theta_1 + r_4\cos\theta_4 \tag{3.123}$$

$$r_2\sin\theta_2 + r_3\sin\theta_3 = r_1\sin\theta_1 + r_4\sin\theta_4 \tag{3.124}$$

Before solving the equations, it is necessary to identify the magnitudes and directions that are constants. There are eight quantities to identify: r_1 and θ_1, r_2 and θ_2, r_3 and θ_3, and r_4 and θ_4. From the diagram in Fig. 3.30, the following are constants: r_1, θ_1, r_2, and θ_4. Furthermore, we know that

$$\theta_3 = \theta_2 + \pi/2 \tag{3.125}$$

and

$$\theta_4 = \theta_1 + \pi/2 \tag{3.126}$$

The variables are r_3, θ_2, r_4, and θ_3. For a one-degree-of-freedom mechanism, one of the variables must be an input (i.e., known) variable. Therefore, there is a total of three unknowns. Because we have three equations (3.123, 3.124, and 3.125) involving the unknowns, we can solve for them.

Once all of the position variables are obtained, the unknown velocities and accelerations can be obtained by differentiating Eqs. (3.123), (3.124), and (3.125) and solving the resulting set of linear equations for the unknowns.

Before solving the equations for the unknowns, it is necessary to select an input variable. Any of r_3, θ_2, r_4, or θ_3 could be chosen. Because θ_2 and θ_3 are related by Eq. (3.125), there is no practical difference between specifying one or the other as the input. Therefore, the choices for inputs reduce to θ_2 (or θ_3) and r_4 (or r_3). The procedure for developing the equations is the same as that in Sections 3.4–3.6. Therefore, the detailed development of the equations will not be given here. Rather, an overview of each case will be given and the results summarized in a table.

3.7.1 Solution of Closure Equations When θ_2 Is Known

The analytical solution procedure follows the same major steps as in the previous cases. That is, a position analysis must be performed first, then a velocity analysis, and finally the acceleration analysis. The case in which θ_2 is an input is especially simple because θ_3 can be computed from Eq. (3.125), and Eqs. (3.123) and (3.124) then become linear in the unknowns r_3 and r_4. The equations to be solved are

$$r_3\cos\theta_3 - r_4\cos\theta_4 = r_1\cos\theta_1 - r_2\cos\theta_2 \tag{3.127}$$

$$r_3\sin\theta_3 - r_4\sin\theta_4 = r_1\sin\theta_1 - r_2\sin\theta_2 \tag{3.128}$$

or in matrix form

$$\begin{bmatrix}\cos\theta_3 & -\cos\theta_4\\ \sin\theta_3 & -\sin\theta_4\end{bmatrix}\begin{Bmatrix}r_3\\ r_4\end{Bmatrix} = \begin{Bmatrix}r_1\cos\theta_1 - r_2\cos\theta_2\\ r_1\sin\theta_1 - r_2\sin\theta_2\end{Bmatrix} \tag{3.129}$$

The terms in the matrix and in the vector on the right-hand side of the equation will be known. The matrix equation can therefore be solved manually, on a programmable calculator, or with the matrix solvers in programs such as MATLAB.

Table 3.9 Summary of Position, Velocity, and Acceleration Equations for an RPRP Mechanism When θ_2 is the Input Variable. The Link Numbers and Points are Defined in Fig. 3.30.

Position

$$\theta_3 = \theta_2 + \pi/2$$

$$\theta_4 = \theta_1 + \pi/2$$

$$\begin{bmatrix} \cos\theta_3 & -\cos\theta_4 \\ \sin\theta_3 & -\sin\theta_4 \end{bmatrix} \begin{Bmatrix} r_3 \\ r_4 \end{Bmatrix} = \begin{Bmatrix} r_1\cos\theta_1 - r_2\cos\theta_2 \\ r_1\sin\theta_1 - r_2\sin\theta_2 \end{Bmatrix}$$

$$\boldsymbol{r}_P = \boldsymbol{r}_2 + \boldsymbol{r}_3 = (r_2\cos\theta_2 + r_3\cos\theta_3)\boldsymbol{i} + (r_2\sin\theta_2 + r_3\sin\theta_3)\boldsymbol{j}$$
$$= \boldsymbol{r}_1 + \boldsymbol{r}_4 = (r_1\cos\theta_1 + r_4\cos\theta_4)\boldsymbol{i} + (r_1\sin\theta_1 + r_4\sin\theta_4)\boldsymbol{j}$$

$$\boldsymbol{r}_Q = \boldsymbol{r}_2 = (r_2\cos\theta_2)\boldsymbol{i} + (r_2\sin\theta_2)\boldsymbol{j}$$

$$\boldsymbol{r}_R = \boldsymbol{r}_1 = (r_1\cos\theta_1)\boldsymbol{i} + (r_1\sin\theta_1)\boldsymbol{j}$$

Velocity

$$\dot\theta_2 = \dot\theta_2$$

$$\begin{bmatrix} \cos\theta_3 & -\cos\theta_4 \\ \sin\theta_3 & -\sin\theta_4 \end{bmatrix} \begin{Bmatrix} \dot r_3 \\ \dot r_4 \end{Bmatrix} = \begin{Bmatrix} r_2\dot\theta_2\sin\theta_2 + r_3\dot\theta_3\sin\theta_3 \\ -r_2\dot\theta_2\cos\theta_2 - r_3\dot\theta_3\cos\theta_3 \end{Bmatrix}$$

$$\dot{\boldsymbol{r}}_P = \dot{\boldsymbol{r}}_2 + \dot{\boldsymbol{r}}_3 = (-r_2\dot\theta_2\sin\theta_2 - r_3\dot\theta_3\sin\theta_3 + \dot r_3\cos\theta_3)\boldsymbol{i} + (r_2\dot\theta_2\cos\theta_2 + r_3\dot\theta_3\cos\theta_3 + \dot r_3\sin\theta_3)\boldsymbol{j}$$
$$= \dot{\boldsymbol{r}}_4 = (\dot r_4\cos\theta_4)\boldsymbol{i} + (\dot r_4\sin\theta_4)\boldsymbol{j}$$

$$\dot{\boldsymbol{r}}_Q = \dot{\boldsymbol{r}}_2 = (-r_2\dot\theta_2\sin\theta_2)\boldsymbol{i} + (r_2\dot\theta_2\cos\theta_2)\boldsymbol{j}$$

Accelerations

$$\ddot\theta_3 = \ddot\theta_2$$

$$\begin{bmatrix} \cos\theta_3 & -\cos\theta_4 \\ \sin\theta_3 & -\sin\theta_4 \end{bmatrix} \begin{Bmatrix} \ddot r_3 \\ \ddot r_4 \end{Bmatrix}$$
$$= \begin{Bmatrix} r_2\ddot\theta_2\sin\theta_2 + r_2\dot\theta_2^2\cos\theta_2 + 2\dot r_3\dot\theta_3\sin\theta_3 + r_3\ddot\theta_3\sin\theta_3 + r_3\dot\theta_3^2\cos\theta_3 \\ -r_2\ddot\theta_2\cos\theta_2 + r_2\dot\theta_2^2\sin\theta_2 - 2\dot r_3\dot\theta_3\cos\theta_3 - r_3\ddot\theta_3\cos\theta_3 + r_3\dot\theta_3^2\sin\theta_3 \end{Bmatrix}$$

$$\ddot{\boldsymbol{r}}_P = \ddot{\boldsymbol{r}}_4 = (\ddot r_4\cos\theta_4)\boldsymbol{i} + (\ddot r_4\sin\theta_4)\boldsymbol{j}$$

$$\ddot{\boldsymbol{r}}_Q = \ddot{\boldsymbol{r}}_2 = (-r_2\ddot\theta_2\sin\theta_2 - r_2\dot\theta_2^2\cos\theta_2)\boldsymbol{i} + (r_2\ddot\theta_2\cos\theta_2 - r_2\dot\theta_2^2\sin\theta_2)\boldsymbol{j}$$

Once the position equations are solved, the coordinates of points P, Q, and R can be computed directly. The equations are given in Table 3.9.

3.7.2 Solution of Closure Equations When r_4 Is Known

When r_4 is known, we must determine θ_2, θ_3, and r_3. Of these unknowns, θ_2 and θ_3 are related in a trivial manner through Eq. (3.125). Therefore, the principal unknowns are θ_2 and r_3. The equations to be solved are Eqs. (3.123) and (3.124). In general, there are two solutions for r_3, and they are both valid. The two assembly modes are represented in Fig. 3.31. The equations from the position analysis are summarized as part of Table 3.10. The different assembly modes are identified by selecting values of σ in Table 3.10 as either $+1$ or -1.

It is also possible to specify a value for r_4 that will prevent the mechanism from being assembled. This is indicated when the argument of the square root in Table 3.10 is negative.

Once both r_3 and r_4 are known, we can find θ_2 (and θ_3) using simple geometry. First compute the angle ϕ and β shown in Fig. 3.31 using

$$\phi = \tan^{-1}\left(\frac{r_3}{r_2}\right) \tag{3.130}$$

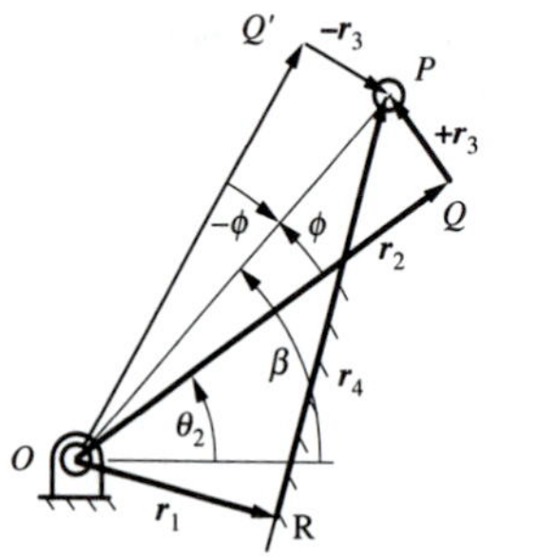

Figure 3.31 Two assembly modes when r_4 is input.

and

$$\beta = \tan^{-1}\left(\frac{r_{P_y}}{r_{P_x}}\right) = \tan^{-1}\left(\frac{r_1 \sin\theta_1 + r_4 \sin\theta_4}{r_1 \cos\theta_1 + r_4 \cos\theta_4}\right) \quad \textbf{(3.131)}$$

Then,

$$\theta_2 = \beta - \phi \quad \textbf{(3.132)}$$

Table 3.10 Summary of Position, Velocity, and Acceleration Equations for an RPRP Mechanism When r_4 is the Input Variable. The Link Numbers and Points are Defined in Fig. 3.30.

Position

$$r_3 = \sigma\sqrt{r_4^2 + r_1^2 - r_2^2}, \qquad \sigma = \pm 1$$

$$\phi = \tan^{-1}\left(\frac{r_3}{r_2}\right)$$

$$\beta = \tan^{-1}\left(\frac{r_1 \sin\theta_1 + r_4 \sin\theta_4}{r_1 \cos\theta_1 + r_4 \cos\theta_4}\right)$$

$$\theta_2 = \beta - \phi, \qquad \theta_3 = \theta_2 + \pi/2, \qquad \theta_4 = \theta_1 + \pi/2$$

$$\boldsymbol{r}_p = \boldsymbol{r}_2 + \boldsymbol{r}_3 = (r_2\cos\theta_2 + r_3\cos\theta_3)\boldsymbol{i} + (r_2\sin\theta_2 + r_3\sin\theta_3)\boldsymbol{j}$$
$$= \boldsymbol{r}_1 + \boldsymbol{r}_4 = (r_1\cos\theta_1 + r_4\cos\theta_4)\boldsymbol{i} + (r_1\sin\theta_1 + r_4\sin\theta_4)\boldsymbol{j}$$

$$\boldsymbol{r}_Q = \boldsymbol{r}_2 = (r_2\cos\theta_2)\boldsymbol{i} + (r_2\sin\theta_2)\boldsymbol{j}$$

Velocity

$$\begin{bmatrix} \cos\theta_3 & -(r_3\sin\theta_3 + r_2\sin\theta_2) \\ \sin\theta_3 & (r_3\cos\theta_3 + r_2\cos\theta_2) \end{bmatrix} \begin{Bmatrix} \dot{r}_3 \\ \dot{\theta}_2 \end{Bmatrix} = \begin{Bmatrix} \dot{r}_4\cos\theta_4 \\ \dot{r}_4\sin\theta_4 \end{Bmatrix}$$

$$\dot{\theta}_3 = \dot{\theta}_2$$

$$\dot{\boldsymbol{r}}_P = \dot{\boldsymbol{r}}_2 + \dot{\boldsymbol{r}}_3 = (-r_2\dot{\theta}_2\sin\theta_2 - r_3\dot{\theta}_3\sin\theta_3 + \dot{r}_3\cos\theta_3)\boldsymbol{i} + (r_2\dot{\theta}_2\cos\theta_2 + r_3\dot{\theta}_3\cos\theta_3 + \dot{r}_3\sin\theta_3)\boldsymbol{j}$$
$$= \dot{\boldsymbol{r}}_4 = (\dot{r}_4\cos\theta_4)\boldsymbol{i} + (\dot{r}_4\sin\theta_4)\boldsymbol{j}$$

$$\dot{\boldsymbol{r}}_Q = \dot{\boldsymbol{r}}_2 = (-r_2\dot{\theta}_2\sin\theta_2)\boldsymbol{i} + (r_2\dot{\theta}_2\cos\theta_2)\boldsymbol{j}$$

Accelerations

$$\begin{bmatrix} \cos\theta_3 & -(r_3\sin\theta_3 + r_2\sin\theta_2) \\ \sin\theta_3 & (r_3\cos\theta_3 + r_2\cos\theta_2) \end{bmatrix} \begin{Bmatrix} \ddot{r}_3 \\ \ddot{\theta}_2 \end{Bmatrix} = \begin{Bmatrix} \ddot{r}_4\cos\theta_4 + r_2\dot{\theta}_2^2\cos\theta_2 + 2\dot{r}_3\dot{\theta}_3\sin\theta_3 + r_3\dot{\theta}_3^2\cos\theta_3 \\ \ddot{r}_4\sin\theta_4 + r_2\dot{\theta}_2^2\sin\theta_2 - 2\dot{r}_3\dot{\theta}_3\cos\theta_3 + r_3\dot{\theta}_3^2\sin\theta_3 \end{Bmatrix}$$

$$\ddot{\theta}_3 = \ddot{\theta}_2$$

$$\ddot{\boldsymbol{r}}_P = \ddot{\boldsymbol{r}}_4 = (\ddot{r}_4\cos\theta_4)\boldsymbol{i} + (\ddot{r}_4\sin\theta_4)\boldsymbol{j}$$

$$\ddot{\boldsymbol{r}}_Q = \ddot{\boldsymbol{r}}_2 = (-r_2\ddot{\theta}_2\sin\theta_2 - r_2\dot{\theta}_2^2\cos\theta_2)\boldsymbol{i} + (r_2\ddot{\theta}_2\cos\theta_2 - r_2\dot{\theta}_2^2\sin\theta_2)\boldsymbol{j}$$

and

$$\theta_3 = \theta_2 + \pi/2 \tag{3.133}$$

Equation (3.132) is valid for both the plus and minus values for r_3 because the sign of ϕ will be positive when r_3 is positive and negative when r_3 is negative.

3.7.3 Solution of Closure Equations When r_3 Is Known

When r_3 is known, we must determine θ_2, θ_3, and r_4. Of these unknowns, θ_2 and θ_3 are related in a trivial manner through Eq. (3.125). Therefore, the principal unknowns are θ_2 and r_4. The procedure for solving the position equations is very similar to that when r_4 was input. In general, there are two solutions for r_4, and they are both valid. The equations from the position analysis are summarized as part of Table 3.11. Note that it is possible to specify a value for r_3 that will prevent the mechanism from being assembled. This is indicated when the argument of the square root in Table 3.11 is negative.

Table 3.11 Summary of Position, Velocity, and Acceleration Equations for an RPRP Mechanism When $\boldsymbol{r}_3$ is the Input Variable. The Link Numbers and Points are Defined in Fig. 3.30.

Position

$$r_4 = \sigma\sqrt{r_3^2 - r_1^2 + r_2^2}, \qquad \sigma = \pm 1$$

$$\phi = \tan^{-1}\left(\frac{r_3}{r_2}\right)$$

$$\beta = \tan^{-1}\left(\frac{r_1 \sin\theta_1 + r_4 \sin\theta_4}{r_1 \cos\theta_1 + r_4 \cos\theta_4}\right)$$

$$\theta_2 = \beta - \phi, \qquad \theta_3 = \theta_2 + \pi/2, \qquad \theta_4 = \theta_1 + \pi/2$$

$$\boldsymbol{r}_P = \boldsymbol{r}_2 + \boldsymbol{r}_3 = (r_2\cos\theta_2 + r_3\cos\theta_3)\boldsymbol{i} + (r_2\sin\theta_2 + r_3\sin\theta_3)\boldsymbol{j}$$
$$= \boldsymbol{r}_1 + \boldsymbol{r}_4 = (r_1\cos\theta_1 + r_4\cos\theta_4)\boldsymbol{i} + (r_1\sin\theta_1 + r_4\sin\theta_4)\boldsymbol{j}$$

$$\boldsymbol{r}_Q = \boldsymbol{r}_2 = (r_2\cos\theta_2)\boldsymbol{i} + (r_2\sin\theta_2)\boldsymbol{j}$$

Velocity

$$\begin{bmatrix} \cos\theta_4 & (r_3\sin\theta_3 + r_2\sin\theta_2) \\ \sin\theta_4 & -(r_3\cos\theta_3 + r_2\cos\theta_2) \end{bmatrix} \begin{Bmatrix} \dot{r}_4 \\ \dot{\theta}_2 \end{Bmatrix} = \begin{Bmatrix} \dot{r}_3\cos\theta_3 \\ \dot{r}_3\sin\theta_3 \end{Bmatrix}$$

$$\dot{\theta}_3 = \dot{\theta}_2$$

$$\dot{\boldsymbol{r}}_P = \dot{\boldsymbol{r}}_2 + \dot{\boldsymbol{r}}_3 = (-r_2\dot{\theta}_2\sin\theta_2 - r_3\dot{\theta}_3\sin\theta_3 + \dot{r}_3\cos\theta_3)\boldsymbol{i} + (r_2\dot{\theta}_2\cos\theta_2 + r_3\dot{\theta}_3\cos\theta_3 + \dot{r}_3\sin\theta_3)\boldsymbol{j}$$
$$= \dot{\boldsymbol{r}}_4 = (\dot{r}_4\cos\theta_4)\boldsymbol{i} + (\dot{r}_4\sin\theta_4)\boldsymbol{j}$$

$$\dot{\boldsymbol{r}}_Q = \dot{\boldsymbol{r}}_2 = (-r_2\dot{\theta}_2\sin\theta_2)\boldsymbol{i} + (r_2\dot{\theta}_2\cos\theta_2)\boldsymbol{j}$$

Accelerations

$$\begin{bmatrix} \cos\theta_4 & (r_3\sin\theta_3 + r_2\sin\theta_2) \\ \sin\theta_4 & -(r_3\cos\theta_3 + r_2\cos\theta_2) \end{bmatrix} \begin{Bmatrix} \ddot{r}_4 \\ \ddot{\theta}_2 \end{Bmatrix}$$
$$= \begin{Bmatrix} -r_2\dot{\theta}_2^2\cos\theta_2 - 2\dot{r}_3\dot{\theta}_3\sin\theta_3 - r_3\dot{\theta}_3^2\cos\theta_3 + \ddot{r}_3\cos\theta_3 \\ -r_2\dot{\theta}_2^2\sin\theta_2 + 2\dot{r}_3\dot{\theta}_3\cos\theta_3 - r_3\dot{\theta}_3^2\sin\theta_3 + \ddot{r}_3\sin\theta_3 \end{Bmatrix}$$

$$\ddot{\theta}_3 = \ddot{\theta}_2$$

$$\ddot{\boldsymbol{r}}_P = \ddot{\boldsymbol{r}}_4 = (\ddot{r}_4\cos\theta_4)\boldsymbol{i} + (\ddot{r}_4\sin\theta_4)\boldsymbol{j}$$

$$\ddot{\boldsymbol{r}}_Q = \ddot{\boldsymbol{r}}_2 = (-r_2\ddot{\theta}_2\sin\theta_2 - r_2\dot{\theta}_2^2\cos\theta_2)\boldsymbol{i} + (r_2\ddot{\theta}_2\cos\theta_2 - r_2\dot{\theta}_2^2\sin\theta_2)\boldsymbol{j}$$

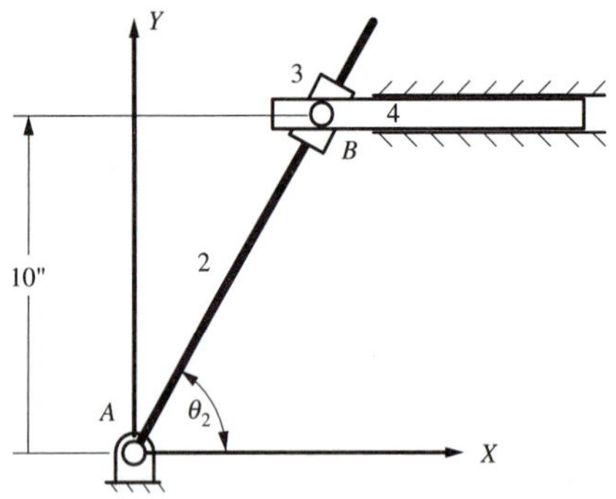

Figure 3.32 Mechanism for Example 3.7.

3.7.4 Velocity and Acceleration Equations for RPRP Mechanism

The analytical form of the velocity equations can be developed by differentiating the position equations and solving the resulting linear set of equations for the unknowns as done in the previous examples. The acceleration equations are developed by differentiating the velocity equations. The acceleration equations are also linear and can easily be solved. The results for the RPRP mechanism are given in Tables 3.9, 3.10, and 3.11. These equations can easily be programmed, and MATLAB programs for solving the equations in the three tables are given on the disk with this book.

EXAMPLE 3.7 ***(Kinematic Analysis of an RPRP Mechanism)***

PROBLEM

In the mechanism shown in Fig. 3.32, link 3 slides on link 2, and link 4 is pinned to link 3. Link 4 also slides on the frame (link 1). If ${}^1\omega_2 = 10$ rad/s CCW and is constant, determine the velocity and acceleration of link 4 for the position defined by $\theta_2 = 60$ degrees.

SOLUTION

Using the nomenclature developed earlier, the basic vector closure diagram is as shown in Fig. 3.33. Note that in this example, $r_2 = 0$. From Figs. 3.32 and 3.33, the following geometric quantities can be determined.

$$r_2 = 0 \qquad r_1 = 10$$

$$\theta_1 = 90° \qquad \theta_2 = 60° \qquad \theta_3 = 60° \qquad \theta_4 = 180°$$

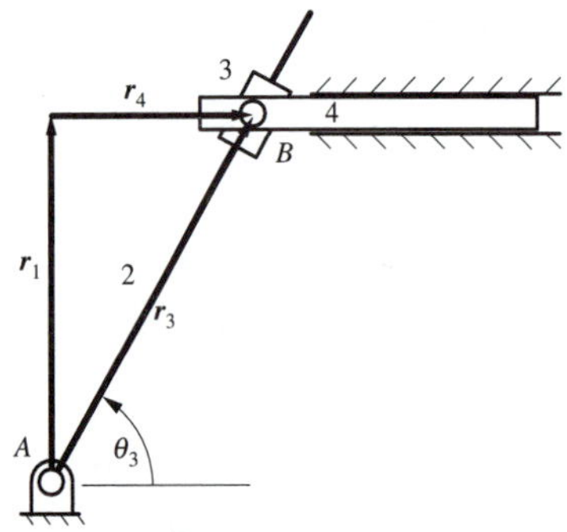

Figure 3.33 Vector closure diagram for Example 3.7.

The geometric unknowns are r_3 and r_4. Because $\theta_2 = \theta_3$ is the input variable, we can compute the results using the equations in Table 3.9. To solve for r_3 and r_4, solve

$$\begin{bmatrix} \cos\theta_3 & -\cos\theta_4 \\ \sin\theta_3 & -\sin\theta_4 \end{bmatrix} \begin{Bmatrix} r_3 \\ r_4 \end{Bmatrix} = \begin{Bmatrix} r_1\cos\theta_1 - r_2\cos\theta_2 \\ r_1\sin\theta_1 - r_2\sin\theta_2 \end{Bmatrix}$$

or

$$\begin{bmatrix} \cos 60^\circ & -\cos 180^\circ \\ \sin 60^\circ & -\sin 180^\circ \end{bmatrix} \begin{Bmatrix} r_3 \\ r_4 \end{Bmatrix} = \begin{Bmatrix} 10\cos 90^\circ - 0 \\ 10\sin 90^\circ - 0 \end{Bmatrix}$$

Simplifying,

$$\begin{bmatrix} 0.5 & 1 \\ 0.8660 & 0 \end{bmatrix} \begin{Bmatrix} r_3 \\ r_4 \end{Bmatrix} = \begin{Bmatrix} 0 \\ 10 \end{Bmatrix}$$

and solving,

$$\begin{Bmatrix} r_3 \\ r_4 \end{Bmatrix} = \begin{Bmatrix} 11.547 \\ -5.774 \end{Bmatrix}$$

The negative sign for r_4 means that it is pointing in the opposite direction to that given by $\theta_4 = 180^\circ$. Clearly, the geometry in Fig. 3.33 could have been solved directly using geometry; however, the more general equations are used here to illustrate the procedure.

For the velocities, the unknowns are $\dot{r}_3$ and $\dot{r}_4$. From the problem statement, $\dot{\theta}_2 = \dot{\theta}_3 = 10$ rad/s CCW. Therefore, from Table 3.9,

$$\begin{bmatrix} \cos\theta_3 & -\cos\theta_4 \\ \sin\theta_3 & -\sin\theta_4 \end{bmatrix} \begin{Bmatrix} \dot{r}_3 \\ \dot{r}_4 \end{Bmatrix} = \begin{Bmatrix} r_2\dot{\theta}_2\sin\theta_2 + r_3\dot{\theta}_3\sin\theta_3 \\ -r_2\dot{\theta}_2\cos\theta_2 - r_3\dot{\theta}_3\cos\theta_3 \end{Bmatrix}$$

Note that the coefficient matrix on the left-hand side of the velocity equation is the same as the corresponding position matrix.

Substituting the known quantities into the matrix equation gives

$$\begin{Bmatrix} 0.5 & 1 \\ 0.8660 & 0 \end{Bmatrix} \begin{Bmatrix} \dot{r}_3 \\ \dot{r}_4 \end{Bmatrix} = \begin{Bmatrix} 0 + 1.547(10)\sin 60^\circ \\ 0 - 11.547(10)\cos 60^\circ \end{Bmatrix} = \begin{Bmatrix} 100 \\ -57.74 \end{Bmatrix}$$

or

$$\begin{Bmatrix} \dot{r}_3 \\ \dot{r}_4 \end{Bmatrix} = \begin{Bmatrix} -66.67 \\ 133.33 \end{Bmatrix}$$

Note that the positive value for $\dot{r}_4$ means that r_4 is increasing with time. Because r_4 is negative when it points in the positive x direction ($\theta_4 = 180^\circ$), a positive sign for $\dot{r}_4$ means that r_4 is increasing or becoming less negative with time. This means that link 4 is moving in the $-x$ direction.

For the acceleration analysis, the unknowns are $\ddot{r}_3$ and $\ddot{r}_4$. From the problem statement, $\ddot{\theta}_2 = \ddot{\theta}_3 = 0$, and from Table 3.9,

$$\begin{bmatrix} \cos\theta_3 & -\cos\theta_4 \\ \sin\theta_3 & -\sin\theta_4 \end{bmatrix} \begin{Bmatrix} \ddot{r}_3 \\ \ddot{r}_4 \end{Bmatrix} = \begin{Bmatrix} r_2\ddot{\theta}_2\sin\theta_2 + r_2\dot{\theta}_2^2\cos\theta_2 + 2\dot{r}_3\dot{\theta}_3\sin\theta_3 + r_3\ddot{\theta}_3\sin\theta_3 + r_3\dot{\theta}_3^2\cos\theta_3 \\ -r_2\ddot{\theta}_2\cos\theta_2 + r_2\dot{\theta}_2^2\sin\theta_2 - 2\dot{r}_3\dot{\theta}_3\cos\theta_3 - r_3\ddot{\theta}_3\cos\theta_3 + r_3\dot{\theta}_3^2\sin\theta_3 \end{Bmatrix}$$

Note again that the coefficient matrix on the left-hand side of the acceleration equation is the same as the corresponding position matrix. Substituting the known quantities into

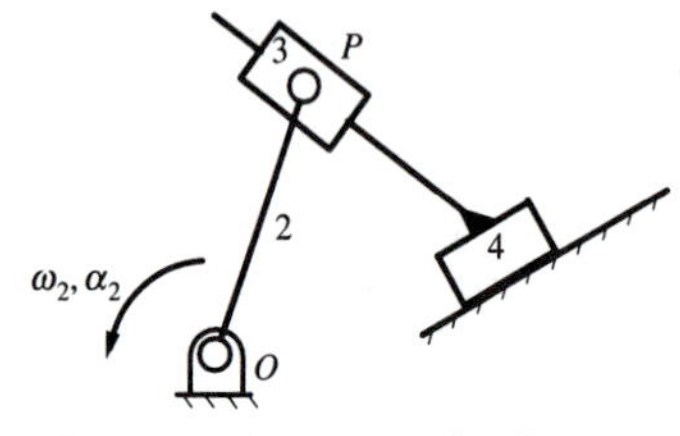

Figure 3.34 Schematic diagram of RRPP mechanism.

the matrix equation gives

$$\begin{bmatrix} 0.5 & 1 \\ 0.8660 & 0 \end{bmatrix} \begin{Bmatrix} \ddot{r}_3 \\ \ddot{r}_4 \end{Bmatrix} = \begin{Bmatrix} -577.4 \\ 1666.7 \end{Bmatrix}$$

or

$$\begin{Bmatrix} \ddot{r}_3 \\ \ddot{r}_4 \end{Bmatrix} = \begin{Bmatrix} 1924.5 \\ -1539.6 \end{Bmatrix}$$

Again, the positive value for $\ddot{r}_4$ means the slider is accelerating in the positive r_4 direction, that is, to the left. ■

3.8 ANALYTICAL EQUATIONS FOR AN RRPP MECHANISM

A schematic drawing of the RRPP mechanism is shown in Fig. 3.34. In the mechanism shown, link 2 is connected to the frame and to the coupler (link 3) by revolute joints. The coupler is connected to link 4 by a prismatic joint, and link 4 is connected to the frame through a prismatic joint. This mechanism occurs frequently in industrial machinery and household appliances. A common version of it is the Scotch yoke, which is a compact mechanism for converting rotary motion to reciprocating motion.

To analyze the mechanism using vector closure equations, we must align vectors in the directions of the slider motions as shown in Fig. 3.35. Only three vectors are required to model the motion. Vector $\mathbf{r}_1$ is fixed at an angle θ_1 and is of variable length. This vector begins at point O and ends at point Q, where Q is the intersection of a line through O and in the direction of the velocity of link 4 relative to link 1 and a second line through P in the direction of the velocity of link 3 relative to link 4. The two lines intersect at an angle β. Vector $\mathbf{r}_2$ is the crank and of fixed length but variable orientation. Vector $\mathbf{r}_3$ is measured from point Q and gives the displacement of slider 3 relative to link 4.

To develop the closure equations, locate point P with vector $\mathbf{r}_2$ and with vectors $\mathbf{r}_1 + \mathbf{r}_3$ as shown in Fig. 3.35. Then the vector closure equation is

$$\mathbf{r}_P = \mathbf{r}_2 = \mathbf{r}_1 + \mathbf{r}_3 \tag{3.134}$$

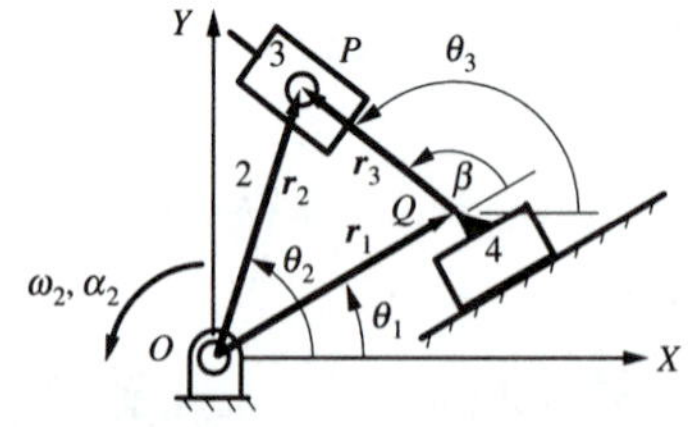

Figure 3.35 Vector closure diagram of RRPP mechanism.

or

$$r_2(\cos\theta_2\,\boldsymbol{i} + \sin\theta_2\boldsymbol{j}) = r_1(\cos\theta_1\,\boldsymbol{i} + \sin\theta_1\boldsymbol{j}) + r_3(\cos\theta_3\,\boldsymbol{i} + \sin\theta_3\boldsymbol{j}) \tag{3.135}$$

Rewriting the components in Eq. (3.134) as separate equations gives

$$r_2\cos\theta_2 = r_1\cos\theta_1 + r_3\cos\theta_3 \tag{3.136}$$

$$r_2\sin\theta_2 = r_1\sin\theta_1 + r_3\sin\theta_3 \tag{3.137}$$

Equations (3.136) and (3.137) must be satisfied throughout the motion of the linkage. There are six quantities to identify: r_1 and θ_1, r_2 and θ_2, and r_3 and θ_3. From the diagram in Fig. 3.35, θ_1, r_2, and θ_3 are constants. Furthermore, we know that

$$\theta_3 = \theta_1 + \beta \tag{3.138}$$

where β is a constant.

The variables are r_1, θ_2, and r_3. For a one-degree-of-freedom mechanism, one of the variables must be an input (i.e., a known) variable. Therefore, there is a total of two unknowns, and we can solve for them using the two equations (3.151 and 3.152) involving the unknowns.

Once all of the position variables are obtained, the unknown velocities and accelerations can be obtained by differentiating Eqs. (3.151) and (3.152) and solving the resulting sets of linear equations for the unknowns.

Before solving the equations for the unknowns, it is necessary to select an input variable. Any of the variables r_1, θ_2, or r_3 could be chosen, and the equations have been developed for each case. Again, we will not give a detailed development of the solution procedure because it is similar to the examples discussed before. An overview of each case is given in the following, and the results are summarized in tables.

3.8.1 Solution When θ_2 Is Known

The case in which θ_2 is an input is especially simple because Eqs. (3.151) and (3.152) then become linear in the unknowns r_1 and r_3. Only one assembly mode is possible, and if r_2 is nonzero, there are no positions in which the mechanism cannot be assembled. The velocity and acceleration equations are obtained by differentiation. The solution for this case is given in Table 3.12, and a MATLAB program for solving the equations is given on the disk with this book.

3.8.2 Solution When r_1 Is Known

When r_1 is known, we must solve Eqs. (3.151) and (3.152) for θ_2 and r_3. The solution is similar to that for the slider-crank inversion. In general, there are two solutions for r_3, corresponding to the two assembly modes or branches for the linkage represented in Fig. 3.36. Also, it is possible to specify a value for r_1 that will prevent the mechanism from

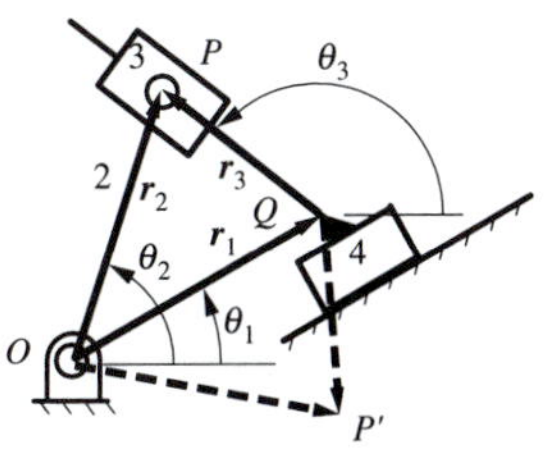

Figure 3.36 Two configurations possible when r_1 is input.

Table 3.12 Summary of Position, Velocity, and Acceleration Equations for an RRPP Mechanism When θ_2 is the Input Variable. The Link Numbers and Points are Defined in Fig. 3.34.

Position

$$\theta_3 = \theta_1 + \beta$$

$$\begin{bmatrix} \cos\theta_1 & \cos\theta_3 \\ \sin\theta_1 & \sin\theta_3 \end{bmatrix} \begin{Bmatrix} r_1 \\ r_3 \end{Bmatrix} = \begin{Bmatrix} r_2\cos\theta_2 \\ r_2\sin\theta_2 \end{Bmatrix}$$

$$\boldsymbol{r}_P = \boldsymbol{r}_2 = (r_2\cos\theta_2)\boldsymbol{i} + (r_2\sin\theta_2)\boldsymbol{j}$$

$$\boldsymbol{r}_Q = \boldsymbol{r}_1 = (r_1\cos\theta_1)\boldsymbol{i} + (r_1\sin\theta_1)\boldsymbol{j}$$

Velocity

$$\begin{bmatrix} \cos\theta_1 & \cos\theta_3 \\ \sin\theta_1 & \sin\theta_3 \end{bmatrix} \begin{Bmatrix} \dot{r}_1 \\ \dot{r}_3 \end{Bmatrix} = \begin{Bmatrix} -r_2\dot{\theta}_2\sin\theta_2 \\ r_2\dot{\theta}_2\cos\theta_2 \end{Bmatrix}$$

$$\dot{\boldsymbol{r}}_P = \dot{\boldsymbol{r}}_2 = (-r_2\dot{\theta}_2\sin\theta_2)\boldsymbol{i} + (r_2\dot{\theta}_2\cos\theta_2)\boldsymbol{j}$$

$$\dot{\boldsymbol{r}}_Q = \dot{\boldsymbol{r}}_1 = (\dot{r}_1\cos\theta_1)\boldsymbol{i} + (\dot{r}_1\sin\theta_1)\boldsymbol{j}$$

Accelerations

$$\begin{bmatrix} \cos\theta_1 & \cos\theta_3 \\ \sin\theta_1 & \sin\theta_3 \end{bmatrix} \begin{Bmatrix} \ddot{r}_1 \\ \ddot{r}_3 \end{Bmatrix} = \begin{Bmatrix} -r_2\ddot{\theta}_2\sin\theta_2 - r_2\dot{\theta}_2^2\cos\theta_2 \\ r_2\ddot{\theta}_2\cos\theta_2 - r_2\dot{\theta}_2^2\sin\theta_2 \end{Bmatrix}$$

$$\ddot{\boldsymbol{r}}_P = \ddot{\boldsymbol{r}}_2 = (-r_2\ddot{\theta}_2\sin\theta_2 - r_2\dot{\theta}_2^2\cos\theta_2)\boldsymbol{i} + (r_2\ddot{\theta}_2\cos\theta_2 - r_2\dot{\theta}_2^2\sin\theta_2)\boldsymbol{j}$$

$$\ddot{\boldsymbol{r}}_Q = \ddot{\boldsymbol{r}}_1 = (\ddot{r}_1\cos\theta_1)\boldsymbol{i} + (\ddot{r}_1\sin\theta_1)\boldsymbol{j}$$

Table 3.13 Summary of Position, Velocity, and Acceleration Equations for an RRPP Mechanism When Either r_1 or r_3 is the Input Variable. When r_1 is the Input, $M = 1$ and $J = 3$. When r_3 is the Input, $M = 3$ and $J = 1$. The Link Numbers and Points are Defined in Fig. 3.34.

Position

$$\theta_3 = \theta_1 + \beta$$

$$B = r_M(\cos\theta_M\cos\theta_J + \sin\theta_M\sin\theta_J)$$

$$C = (r_M^2 - r_2^2)$$

$$r_J = -B + \sigma\sqrt{B^2 - C}, \qquad \sigma = \pm 1$$

$$\theta_2 = \tan^{-1}\frac{r_1\sin\theta_1 + r_3\sin\theta_3}{r_1\cos\theta_1 + r_3\cos\theta_3}$$

$$\boldsymbol{r}_P = \boldsymbol{r}_2 = (r_2\cos\theta_2)\boldsymbol{i} + (r_2\sin\theta_2)\boldsymbol{j}$$

$$\boldsymbol{r}_Q = \boldsymbol{r}_1 = (r_1\cos\theta_1)\boldsymbol{i} + (r_1\sin\theta_1)\boldsymbol{j}$$

Velocity

$$\begin{bmatrix} -r_2\sin\theta_2 & -\cos\theta_J \\ r_2\cos\theta_2 & -\sin\theta_J \end{bmatrix} \begin{Bmatrix} \dot{\theta}_2 \\ \dot{r}_J \end{Bmatrix} = \begin{Bmatrix} \dot{r}_M\cos\theta_M \\ \dot{r}_M\sin\theta_M \end{Bmatrix}$$

$$\dot{\boldsymbol{r}}_P = \dot{\boldsymbol{r}}_2 = (-r_2\dot{\theta}_2\sin\theta_2)\boldsymbol{i} + (r_2\dot{\theta}_2\cos\theta_2)\boldsymbol{j}$$

$$\dot{\boldsymbol{r}}_Q = \dot{\boldsymbol{r}}_1 = (\dot{r}_1\cos\theta_1)\boldsymbol{i} + (\dot{r}_1\sin\theta_1)\boldsymbol{j}$$

Accelerations

$$\begin{bmatrix} -r_2\sin\theta_2 & -\cos\theta_J \\ r_2\cos\theta_2 & -\sin\theta_J \end{bmatrix} \begin{Bmatrix} \ddot{\theta}_2 \\ \ddot{r}_J \end{Bmatrix} = \begin{Bmatrix} \ddot{r}_M\cos\theta_M + r_2\dot{\theta}_2^2\cos\theta_2 \\ \ddot{r}_M\sin\theta_M + r_2\dot{\theta}_2^2\sin\theta_2 \end{Bmatrix}$$

$$\ddot{\boldsymbol{r}}_P = \ddot{\boldsymbol{r}}_2 = (-r_2\ddot{\theta}_2\sin\theta_2 - r_2\dot{\theta}_2^2\cos\theta_2)\boldsymbol{i} + (r_2\ddot{\theta}_2\cos\theta_2 - r_2\dot{\theta}_2^2\sin\theta_2)\boldsymbol{j}$$

$$\ddot{\boldsymbol{r}}_Q = \ddot{\boldsymbol{r}}_1 = (\ddot{r}_1\cos\theta_1)\boldsymbol{i} + (\ddot{r}_1\sin\theta_1)\boldsymbol{j}$$

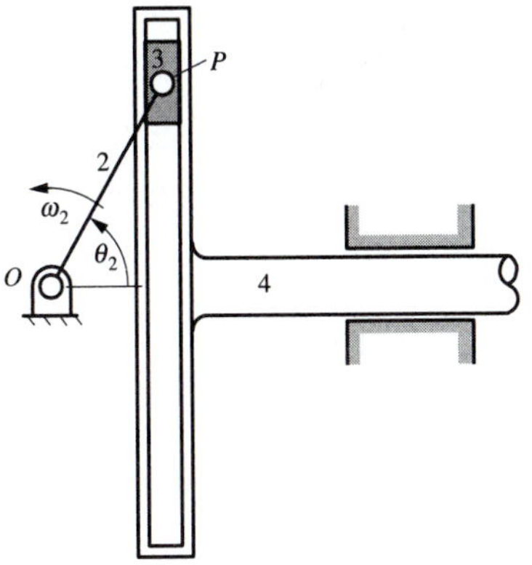

Figure 3.37 Scotch yoke mechanism.

being assembled. After the position equations are solved, the velocity and acceleration equations are obtained by differentiation. The results are summarized in Table 3.13 and a MATLAB program for solving the equations is given on the disk with this book.

3.8.3 Solution When r_3 Is Known

When r_3 is known, we must solve Eqs. (3.151) and (3.152) for θ_2 and r_1. The solution is almost identical to that for the case in which r_1 is known. We need only switch the indices for 1 and 3, and the results are otherwise the same. This is indicated in Table 3.13. This case is also included in a MATLAB routine on the disk with this book.

EXAMPLE 3.8 ***(Kinematic Analysis of Scotch Yoke)***

PROBLEM

In the Scotch yoke mechanism shown in Fig. 3.37, the angular velocity of link 2 relative to the frame is 1 rad/s CCW (constant) when the angle θ_2 is 60°. Also, the length OP = 2 inches. Determine the velocity and acceleration of link 4.

SOLUTION

Using the nomenclature developed earlier, the basic vector closure diagram is as shown in Fig. 3.38. From Figs. 3.37 and 3.38, the following geometry quantities can be determined.

$$r_2 = 2$$

$$\theta_1 = 0°, \quad \theta_2 = 60°, \quad \beta = 90°, \quad \theta_3 = 90°$$

$$\dot{\theta}_2 = 1, \quad \ddot{\theta}_2 = 0$$

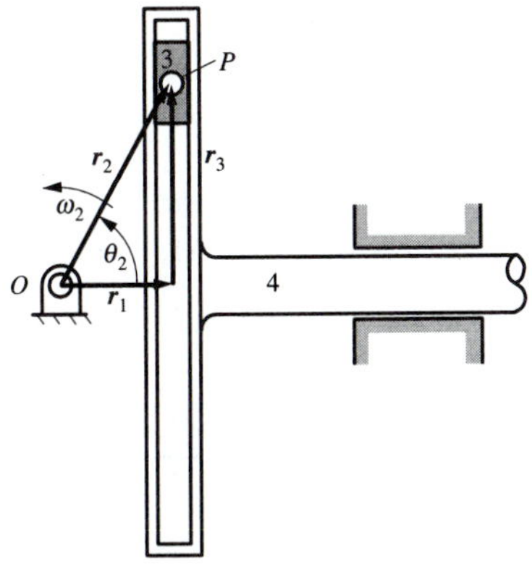

Figure 3.38 Vector diagram for Scotch yoke mechanism.

Because the crank (link 2) is the driving link, we can use the equations in Table 3.12 to solve the problem. The geometric unknowns are r_1 and r_3. To determine r_1 and r_3, solve

$$\begin{bmatrix} \cos\theta_1 & \cos\theta_3 \\ \sin\theta_1 & \sin\theta_3 \end{bmatrix} \begin{Bmatrix} r_1 \\ r_3 \end{Bmatrix} = \begin{Bmatrix} r_2\cos\theta_2 \\ r_2\sin\theta_2 \end{Bmatrix}$$

or

$$\begin{bmatrix} 1 & 0 \\ 0 & 1 \end{bmatrix} \begin{Bmatrix} r_1 \\ r_3 \end{Bmatrix} = \begin{Bmatrix} r_2\cos\theta_2 \\ r_2\sin\theta_2 \end{Bmatrix}$$

Then,

$$r_1 = r_2\cos\theta_2$$

and

$$r_3 = r_2\sin\theta_2$$

Notice that the motion of the slider is a sinusoidal function of the input rotation. This is one of the benefits of the Scotch yoke and is one of the reasons that it is used. For the given input values ($r_2 = 2$ and $\theta_2 = 60°$) it is clear that $r_1 = 1.0$ and $r_3 = 1.732$.

For the velocities, the unknowns are $\dot{r}_1$ and $\dot{r}_3$. These can be determined from the matrix equation

$$\begin{bmatrix} \cos\theta_1 & \cos\theta_3 \\ \sin\theta_1 & \sin\theta_3 \end{bmatrix} \begin{Bmatrix} \dot{r}_1 \\ \dot{r}_3 \end{Bmatrix} = \begin{Bmatrix} -r_2\dot{\theta}_2\sin\theta_2 \\ r_2\dot{\theta}_2\cos\theta_2 \end{Bmatrix}$$

or

$$\begin{bmatrix} 1 & 0 \\ 0 & 1 \end{bmatrix} \begin{Bmatrix} \dot{r}_1 \\ \dot{r}_3 \end{Bmatrix} = \begin{Bmatrix} -2(1)\sin 60 \\ 2(1)\cos 60 \end{Bmatrix} = \begin{Bmatrix} -1.732 \\ 1 \end{Bmatrix}$$

or

$$\dot{r}_1 = -1.732 \text{ in/s}$$

and

$$\dot{r}_3 = 1 \text{ in/s}$$

The negative sign means that r_1 is decreasing in length with increasing time.

For the accelerations, the unknowns are $\ddot{r}_1$ and $\ddot{r}_3$. These can be determined by solving

$$\begin{bmatrix} \cos\theta_1 & \cos\theta_3 \\ \sin\theta_1 & \sin\theta_3 \end{bmatrix} \begin{Bmatrix} \ddot{r}_1 \\ \ddot{r}_3 \end{Bmatrix} = \begin{Bmatrix} -r_2\ddot{\theta}_2\sin\theta_2 - r_2\dot{\theta}_2^2\cos\theta_2 \\ r_2\ddot{\theta}_2\cos\theta_2 - r_2\dot{\theta}_2^2\sin\theta_2 \end{Bmatrix}$$

Substituting into the equation the known values,

$$\begin{bmatrix} 1 & 0 \\ 0 & 1 \end{bmatrix} \begin{Bmatrix} \ddot{r}_1 \\ \ddot{r}_3 \end{Bmatrix} = \begin{Bmatrix} 0 - 2(1)^2\cos 60 \\ 0 - 2(1)^2\sin 60 \end{Bmatrix} = \begin{Bmatrix} -1 \\ -1.732 \end{Bmatrix}$$

or

$$\ddot{r}_1 = -1 \text{ in/s}^2$$

and

$$\ddot{r}_3 = -1.732 \text{ in/s}^2$$

■

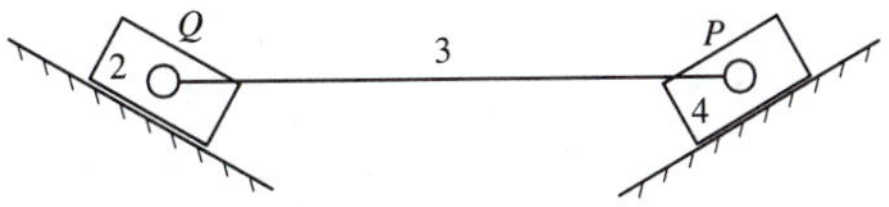

Figure 3.39 Schematic diagram of elliptic trammel.

3.9 ANALYTICAL EQUATIONS FOR ELLIPTIC TRAMMEL

The elliptic trammel is an inversion of the RRPP mechanism, and a schematic drawing of this mechanism is shown in Fig. 3.39. In the mechanism, links 2 and 4 are connected to the frame by prismatic joints and to the coupler by revolute joints. A significant feature of this mechanism is that coupler points trace ellipses on the frame, and it is used in machine tools for this purpose.

To analyze the mechanism using vector closure equations, we must again align vectors in the directions of the slider motions as shown in Fig. 3.40. As in the case of the RRPP mechanism, only three vectors are required to model the motion. Vector $\boldsymbol{r}_1$ is fixed at an angle θ_1 and of variable length. This vector begins at point O and ends at point Q, where O is the intersection of a line through Q and in the direction of the velocity of link 2 relative to link 1 and a second line through P in the direction of the velocity of link 4 relative to link 1. Point O is the origin of the frame coordinate system. The two lines intersect at an angle β, measured from $\boldsymbol{r}_2$ to $\boldsymbol{r}_1$. Vector $\boldsymbol{r}_2$ is fixed at an angle θ_2 and of variable length. Vector $\boldsymbol{r}_3$ is measured from point P to point Q.

As shown in Fig. 3.40, to develop the closure equations, locate point Q with vector $\boldsymbol{r}_1$ and with vectors $\boldsymbol{r}_2 + \boldsymbol{r}_3$. Then the vector closure equation is

$$\boldsymbol{r}_Q = \boldsymbol{r}_1 = \boldsymbol{r}_2 + \boldsymbol{r}_3 \tag{3.139}$$

or

$$r_1(\cos\theta_1\,\boldsymbol{i} + \sin\theta_1\boldsymbol{j}) = r_2(\cos\theta_2\,\boldsymbol{i} + \sin\theta_2\boldsymbol{j}) + r_3(\cos\theta_3\,\boldsymbol{i} + \sin\theta_3\boldsymbol{j}) \tag{3.140}$$

Rewriting the components in Eq. (3.139) as separate equations gives

$$r_1\cos\theta_1 = r_2\cos\theta_2 + r_3\cos\theta_3 \tag{3.141}$$

$$r_1\sin\theta_1 = r_2\sin\theta_2 + r_3\sin\theta_3 \tag{3.142}$$

Equations (3.141) and (3.142) must be satisfied throughout the motion of the linkage. As in the case of the RRPP mechanism, there are six quantities to identify: r_1 and θ_1, r_2 and θ_2, and r_3 and θ_3. From the diagram in Fig. 3.40, the following are constants: θ_1, r_3, and θ_2. Furthermore, we know that

$$\theta_1 = \theta_2 + \beta \tag{3.143}$$

where β is a constant.

The variables are r_1, θ_3, and r_2. For a one-degree-of-freedom mechanism, one of the variables must be an input (i.e., known) variable. Therefore, there is a total of two un-

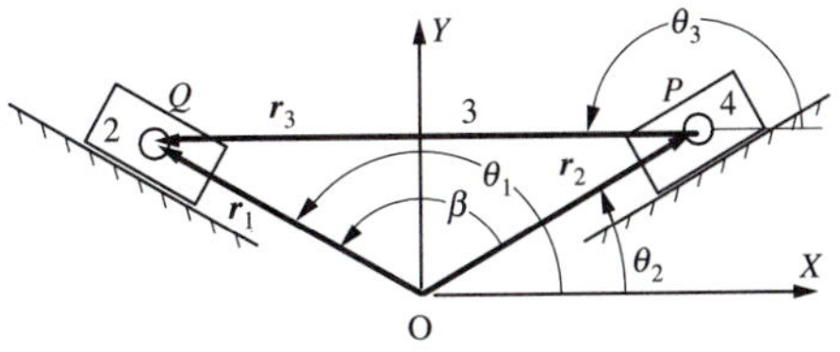

Figure 3.40 Vector closure diagram for elliptic trammel mechanism.

knowns, and we can solve for them using the two equations (3.172 and 3.173) involving the unknowns. Once all of the position variables are obtained, the unknown velocities and accelerations can be found by differentiating Eqs. (3.172) and (3.173) and solving the resulting set of linear equations for the unknowns.

Before solving the equations for the unknowns it is necessary to select an input variable from r_1, θ_3, and r_2. Because of the symmetry of the mechanism, choosing r_1 or r_2 will give a similar set of equations. That is, if we establish the input-output relationships for r_1 as the input, we can derive the relationships for r_2 as the input by simply interchanging the subscripts 1 and 2 in the relationships derived for r_1 as the driver. Therefore, we need to consider only the cases for θ_3 and r_1 as input variables.

3.9.1 Analysis When θ_3 Is Known

The analytical solution procedure follows the same major steps as were followed in the case of the RRPP mechanism. When θ_3 is an input, Eqs. (3.172) and (3.173) become linear in the unknowns r_1 and r_2, and the equations for position, velocity, and acceleration can easily be solved. The resulting equations are summarized in Table 3.14.

3.9.2 Analysis When r_1 Is Known

When r_1 is known, we must determine θ_3 and r_2. Two solutions result corrresponding to the assembly modes shown in Fig. 3.41. It is also possible to specify values of r_1 for which the mechanism cannot be assembled. The analytical form of the velocity and acceleration equations can be developed by differentiating the position equations and solving the resulting linear equations for the unknowns. The results for r_1 as the input are summarized in Table 3.15.

Table 3.14 Summary of Position, Velocity, and Acceleration Equations for an Elliptic Trammel Mechanism When θ_3 is the Input Variable. The Link Numbers and Points are Defined in Fig. 3.40.

Position

$$\theta_1 = \theta_2 + \beta$$

$$\begin{bmatrix} \cos\theta_1 & -\cos\theta_2 \\ \sin\theta_1 & -\sin\theta_2 \end{bmatrix} \begin{Bmatrix} r_1 \\ r_2 \end{Bmatrix} = \begin{Bmatrix} r_3\cos\theta_3 \\ r_3\sin\theta_3 \end{Bmatrix}$$

$$\boldsymbol{r}_P = \boldsymbol{r}_2 = (r_2\cos\theta_2)\boldsymbol{i} + (r_2\sin\theta_2)\boldsymbol{j}$$

$$\boldsymbol{r}_Q = \boldsymbol{r}_1 = (r_1\cos\theta_1)\boldsymbol{i} + (r_1\sin\theta_1)\boldsymbol{j}$$

Velocity

$$\begin{bmatrix} \cos\theta_1 & -\cos\theta_2 \\ \sin\theta_1 & -\sin\theta_2 \end{bmatrix} \begin{Bmatrix} \dot{r}_1 \\ \dot{r}_2 \end{Bmatrix} = \begin{Bmatrix} -r_3\dot{\theta}_3\sin\theta_3 \\ r_3\dot{\theta}_3\cos\theta_3 \end{Bmatrix}$$

$$\dot{\boldsymbol{r}}_P = \dot{\boldsymbol{r}}_2 = (\dot{r}_2\cos\theta_2)\boldsymbol{i} + (\dot{r}_2\sin\theta_2)\boldsymbol{j}$$

$$\dot{\boldsymbol{r}}_Q = \dot{\boldsymbol{r}}_1 = (\dot{r}_1\cos\theta_1)\boldsymbol{i} + (\dot{r}_1\sin\theta_1)\boldsymbol{j}$$

Accelerations:

$$\begin{bmatrix} \cos\theta_1 & -\cos\theta_2 \\ \sin\theta_1 & -\sin\theta_2 \end{bmatrix} \begin{Bmatrix} \ddot{r}_1 \\ \ddot{r}_2 \end{Bmatrix} = \begin{Bmatrix} -r_3\ddot{\theta}_3\sin\theta_3 - r_3\dot{\theta}_3^2\cos\theta_3 \\ r_3\ddot{\theta}_3\cos\theta_3 - r_3\dot{\theta}_3^2\sin\theta_3 \end{Bmatrix}$$

$$\ddot{\boldsymbol{r}}_P = \ddot{\boldsymbol{r}}_2 = (\ddot{r}_2\cos\theta_2)\boldsymbol{i} + (\ddot{r}_2\sin\theta_2)\boldsymbol{j}$$

$$\ddot{\boldsymbol{r}}_Q = \ddot{\boldsymbol{r}}_1 = (\ddot{r}_1\cos\theta_1)\boldsymbol{i} + (\ddot{r}_1\sin\theta_1)\boldsymbol{j}$$

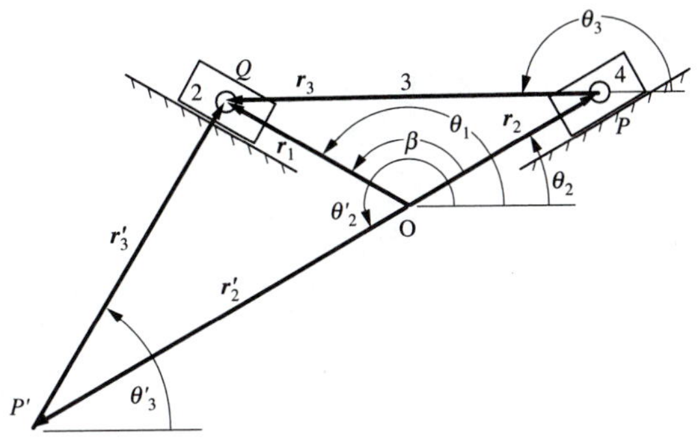

Figure 3.41 Two configurations possible for elliptic trammel when r_1 is input.

EXAMPLE 3.9 ***(Kinematic Analysis of Elliptic Trammel)***

PROBLEM

In the elliptic trammel mechanism shown in Fig. 3.41, the angular velocity of link 3 relative to the frame is 10 rad/s CCW (constant). Also, the legnth $QP = 10$ cm and QR is 20 cm. Determine the position of point R and the velocity and acceleration of point P for a full rotation of the coupler.

SOLUTION

Using the nomenclature developed earlier, the basic vector closure diagram for the linkage is as shown in Fig. 3.42a. From Figs. 3.41 and 3.42, the following geometric quantities can

Table 3.15 Summary of Position, Velocity, and Acceleration Equations for an Elliptic Trammel Mechanism When $\boldsymbol{r}_1$ is the Input. The Link Numbers and Points are Defined in Fig. 3.40.

Position

$$\theta_1 = \theta_2 + \beta$$

$$B = -r_1(\cos\theta_1 \cos\theta_2 + \sin\theta_1 \sin\theta_2)$$

$$C = (r_1^2 - r_3^2)$$

$$r_2 = -B + \sigma\sqrt{B^2 - C}$$

$$\theta_3 = \tan^{-1}\frac{r_1 \sin\theta_1 - r_2 \sin\theta_2}{r_1 \cos\theta_1 - r_2 \cos\theta_2}$$

$$\boldsymbol{r}_P = \boldsymbol{r}_2 = (r_2 \cos\theta_2)\boldsymbol{i} + (r_2 \sin\theta_2)\boldsymbol{j}$$

$$\boldsymbol{r}_Q = \boldsymbol{r}_1 = (r_1 \cos\theta_1)\boldsymbol{i} + (r_1 \sin\theta_1)\boldsymbol{j}$$

Velocity

$$\begin{bmatrix} \cos\theta_2 & -r_3 \sin\theta_3 \\ \sin\theta_2 & r_3 \cos\theta_3 \end{bmatrix} \begin{Bmatrix} \dot{r}_2 \\ \dot{\theta}_3 \end{Bmatrix} = \begin{Bmatrix} \dot{r}_1 \cos\theta_1 \\ \dot{r}_1 \sin\theta_1 \end{Bmatrix}$$

$$\dot{\boldsymbol{r}}_P = \dot{\boldsymbol{r}}_2 = (\dot{r}_2 \cos\theta_2)\boldsymbol{i} + (\dot{r}_2 \sin\theta_2)\boldsymbol{j}$$

$$\dot{\boldsymbol{r}}_Q = \dot{\boldsymbol{r}}_1 = (\dot{r}_1 \cos\theta_1)\boldsymbol{i} + (\dot{r}_1 \sin\theta_1)\boldsymbol{j}$$

Accelerations

$$\begin{bmatrix} \cos\theta_2 & -r_3 \sin\theta_3 \\ \sin\theta_2 & r_3 \cos\theta_3 \end{bmatrix} \begin{Bmatrix} \ddot{r}_2 \\ \ddot{\theta}_3 \end{Bmatrix} = \begin{Bmatrix} \ddot{r}_1 \cos\theta_1 + r_3\dot{\theta}_3^2 \cos\theta_3 \\ \ddot{r}_1 \sin\theta_1 + r_3\dot{\theta}_3^2 \sin\theta_3 \end{Bmatrix}$$

$$\ddot{\boldsymbol{r}}_p = \ddot{\boldsymbol{r}}_2 = (\ddot{r}_2 \cos\theta_2)\boldsymbol{i} + (\ddot{r}_2 \sin\theta_2)\boldsymbol{j}$$

$$\ddot{\boldsymbol{r}}_Q = \ddot{\boldsymbol{r}}_1 = (\ddot{r}_1 \cos\theta_1)\boldsymbol{i} + (\ddot{r}_1 \sin\theta_1)\boldsymbol{j}$$

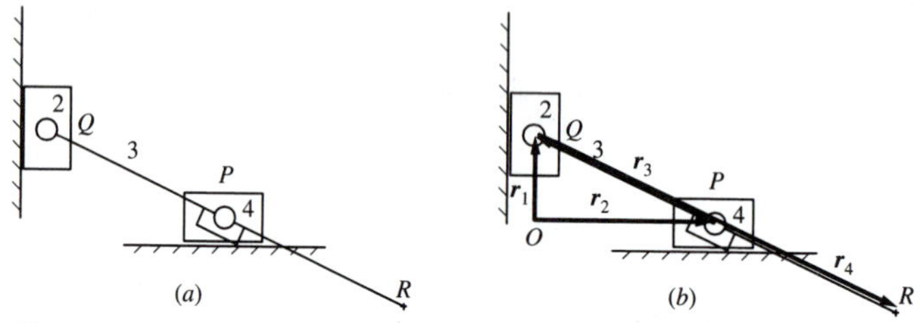

Figure 3.42 Elliptic trammel mechanism (a) and vector diagram (b) for Example 3.9.

be determined.

$$r_3 = 10$$
$$\theta_1 = 90°, \qquad \theta_2 = 0°, \qquad \beta = 90°$$
$$\dot{\theta}_3 = 10, \qquad \ddot{\theta}_3 = 0$$

Because we are interested in the behavior of the mechanism for its full range of motion, we must solve the position, velocity, and acceleration equation in terms of θ_3. The vector diagram establishing the quantities involved is shown in Fig. 3.42b.

The position of R_3 is given by

$$\boldsymbol{r}_{R_3} = \boldsymbol{r}_1 + \boldsymbol{r}_4$$

or in component form,

$$\boldsymbol{r}_{R_3} = (r_1 \cos\theta_1 + r_4 \cos\theta_4)\boldsymbol{i} + (r_1 \sin\theta_1 + r_4 \sin\theta_4)\boldsymbol{j} \tag{3.143}$$

From Fig. 3.42, $\theta_4 = \theta_3 + \pi$. Therefore, Eq. (3.143) becomes

$$\boldsymbol{r}_{R_3} = (r_1 \cos\theta_1 - r_4 \cos\theta_3)\boldsymbol{i} + (r_1 \sin\theta_1 - r_4 \sin\theta_3)\boldsymbol{j} \tag{3.144}$$

To solve for the position, velocity, and acceleration of P_3, we must determine the corresponding values for $\boldsymbol{r}_1$. For this problem, the equations to be solved are:

$$\begin{bmatrix} 0 & -1 \\ 1 & 0 \end{bmatrix} \begin{Bmatrix} r_1 \\ r_2 \end{Bmatrix} = \begin{Bmatrix} r_3 \cos\theta_3 \\ r_3 \sin\theta_3 \end{Bmatrix} \tag{3.145}$$

$$\begin{bmatrix} 0 & -1 \\ 1 & 0 \end{bmatrix} \begin{Bmatrix} \dot{r}_1 \\ \dot{r}_2 \end{Bmatrix} = \begin{Bmatrix} -r_3\dot{\theta}_3 \sin\theta_3 \\ r_3\dot{\theta}_3 \cos\theta_3 \end{Bmatrix} \tag{3.146}$$

and

$$\begin{bmatrix} 0 & -1 \\ 1 & 0 \end{bmatrix} \begin{Bmatrix} \ddot{r}_1 \\ \ddot{r}_2 \end{Bmatrix} = \begin{Bmatrix} -r_3\ddot{\theta}_3 \sin\theta_3 - r_3\dot{\theta}_3^2 \cos\theta_3 \\ r_3\ddot{\theta}_3 \cos\theta_3 - r_3\dot{\theta}_3^2 \sin\theta_3 \end{Bmatrix} \tag{3.147}$$

From Eqs. (3.145)–(3.147),

$$r_1 = r_3 \sin\theta_3 \tag{3.148}$$
$$\dot{r}_1 = \dot{r}_3\dot{\theta}_3 \cos\theta_3$$

and

$$\ddot{r}_1 = r_3\ddot{\theta}_3 \cos\theta_3 - r_3\dot{\theta}_3^2 \sin\theta_3 \tag{3.150}$$

Equations (3.148)–(3.150) can be combined with Eq. (3.144) to solve for the position of R_3 and for the velocity and acceleration of P_3 as a function of θ_3. The equations can easily be computed, and the results are plotted in Fig. 3.43. The MATLAB program used to generate the curves is included on the disk with this book. Notice that the path of R_3 is an ellipse. ■

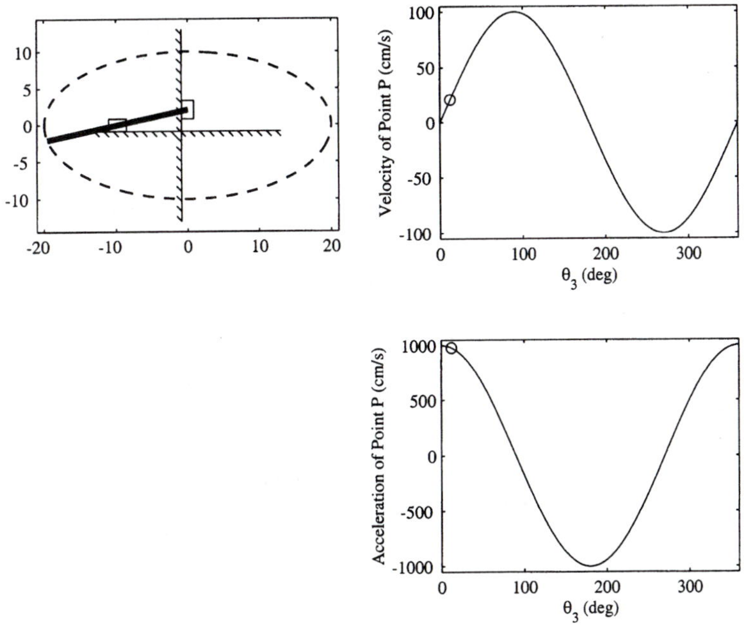

Figure 3.43 Results for Example 3.9.

3.10 ANALYTICAL EQUATIONS FOR OLDHAM MECHANISM

The Oldham (RPPR) mechanism is another inversion of the RRPP mechanism, and a schematic drawing of this mechanism is shown in Fig. 3.44. In the mechanism, links 2 and 4 are connected to the frame by revolute joints and to the coupler by prismatic joints. Therefore, the angle, β, between links 2 and 4 is fixed.

In order to analyze the mechanism using vector closure equations, we must again align vectors in the directions of the slider motions as shown in Fig. 3.45. As in the case of the elliptic trammel, only three vectors are required to model the motion. Vector $\boldsymbol{r}_1$ is fixed at an angle θ_1 and of constant length. Point O is the origin of the frame coordinate system. Links 2 and 4 intersect at an angle β, where β is measured from $\boldsymbol{r}_2$ to $\boldsymbol{r}_4$. Vector $\boldsymbol{r}_2$ is at an angle θ_2 with respect to a horizontal line and of variable length. Vector $\boldsymbol{r}_4$ is at an angle θ_4 with respect to a horizontal line and of variable length.

As shown in Fig. 3.44, to develop the closure equations, locate point P with vector $\boldsymbol{r}_2$ and with vectors $\boldsymbol{r}_1 + \boldsymbol{r}_4$. Then the vector closure equation is

$$\boldsymbol{r}_P = \boldsymbol{r}_2 = \boldsymbol{r}_1 + \boldsymbol{r}_4 \quad \textbf{(3.151)}$$

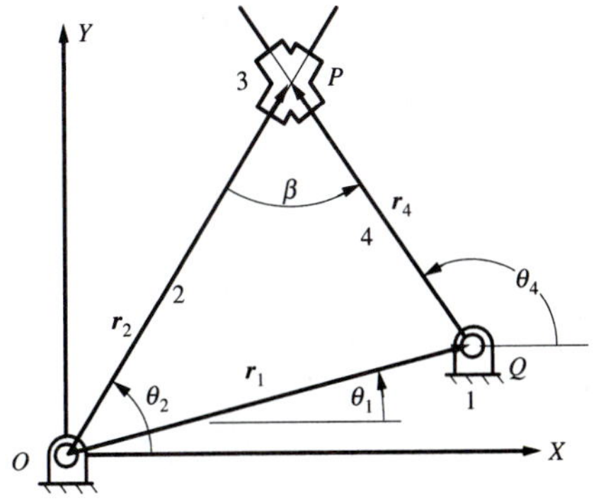

Figure 3.44 Schematic diagram of Oldham mechanism.

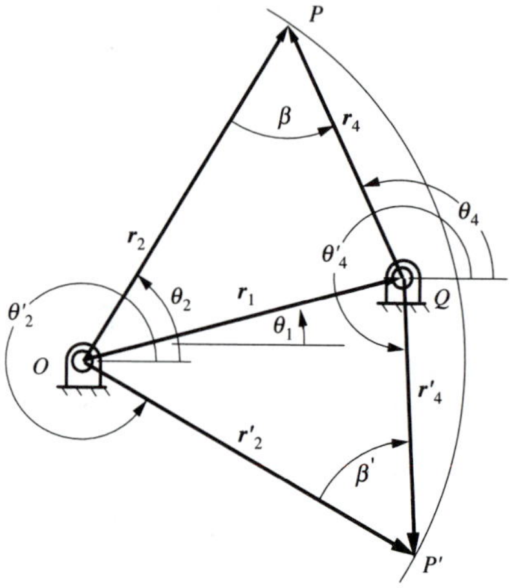

Figure 3.45 Two assembly modes for Oldham mechanism when r_2 is the input variable.

or

$$r_2(\cos\theta_2\,\boldsymbol{i} + \sin\theta_2\boldsymbol{j}) = r_1(\cos\theta_1\,\boldsymbol{i} + \sin\theta_1\boldsymbol{j}) + r_4(\cos\theta_4\,\boldsymbol{i} + \sin\theta_4\boldsymbol{j}) \tag{3.152}$$

Rewriting the components in Eq. (3.151) as separate equations gives

$$r_2\cos\theta_2 = r_1\cos\theta_1 + r_4\cos\theta_4 \tag{3.153}$$

$$r_2\sin\theta_2 = r_1\sin\theta_1 + r_4\sin\theta_4 \tag{3.154}$$

As was the case for the other RRPP mechanisms, there are six quantities to identify: r_1 and θ_1, r_2 and θ_2, and r_4 and θ_4. From the diagram in Fig. 3.44, θ_1, and r_1 are constants. Furthermore, we know that

$$\theta_4 = \theta_2 + \beta \tag{3.155}$$

where β is a constant.

The variables are θ_2, θ_4, r_2, and r_4. For a one-degree-of-freedom mechanism, one of the variables must be an input (i.e., known) variable, and there is a total of three unknowns. We can solve for them using the three equations (3.153–3.155) involving the unknowns. Once all of the position variables are obtained, the unknown velocities and accelerations can be found by differentiating Eqs. (3.153)–(3.155) and solving the resulting set of linear equations for the unknowns.

Before solving the equations for the unknowns, it is necessary to select an input variable from among r_2, r_4, θ_2, and θ_4. The equations will be similar when either r_2 or r_4 is selected and when either θ_2 or θ_4 is selected. Therefore, we will consider only the two cases in which r_2 is a variable and θ_2 is a variable.

3.10.1 Analysis When θ_2 Is Known

When θ_2 is the input variable, Eqs. (3.153) and (3.154) become linear in the unknowns r_2 and r_4, and the equations for position, velocity, and acceleration can be easily solved. The resulting equations are summarized in Table 3.16. The equations can easily be programmed, and a MATLAB program for analyzing the mechanism when θ_2 is the driver is included on the disk with this book.

Table 3.16 Summary of Position, Velocity, and Acceleration Equations for an Oldham Mechanism When θ_2 is the Input Variable. The Link Numbers and Points are Defined in Fig. 3.44.

Position

$$\theta_4 = \theta_2 + \beta$$

$$\begin{bmatrix} \cos\theta_2 & -\cos\theta_4 \\ \sin\theta_2 & -\sin\theta_4 \end{bmatrix} \begin{Bmatrix} r_2 \\ r_4 \end{Bmatrix} = \begin{Bmatrix} r_1\cos\theta_1 \\ r_1\sin\theta_1 \end{Bmatrix}$$

$$\boldsymbol{r}_P = \boldsymbol{r}_2 = (r_2\cos\theta_2)\boldsymbol{i} + (r_2\sin\theta_2)\boldsymbol{j}$$
$$\boldsymbol{r}_Q = \boldsymbol{r}_1 = (r_1\cos\theta_1)\boldsymbol{i} + (r_1\sin\theta_1)\boldsymbol{j}$$

Velocity

$$\dot\theta_4 = \dot\theta_2$$

$$\begin{bmatrix} \cos\theta_2 & -\cos\theta_4 \\ \sin\theta_2 & -\sin\theta_4 \end{bmatrix} \begin{Bmatrix} \dot r_2 \\ \dot r_4 \end{Bmatrix} = \begin{Bmatrix} r_2\dot\theta_2\sin\theta_2 - r_4\dot\theta_2\sin\theta_4 \\ -r_2\dot\theta_2\cos\theta_2 + r_4\dot\theta_2\cos\theta_4 \end{Bmatrix}$$

$$\dot{\boldsymbol{r}}_{P_2} = (-r_2\dot\theta_2\sin\theta_2)\boldsymbol{i} + (r_2\dot\theta_2\cos\theta_2)\boldsymbol{j}$$
$$\dot{\boldsymbol{r}}_{P_3} = (\dot r_2\cos\theta_2 - r_2\dot\theta_2\sin\theta_2)\boldsymbol{i} + (\dot r_2\sin\theta_2 + r_2\dot\theta_2\cos\theta_2)\boldsymbol{j}$$

Accelerations

$$\ddot\theta_4 = \ddot\theta_2$$

$$\begin{bmatrix} \cos\theta_2 & -\cos\theta_4 \\ \sin\theta_2 & -\sin\theta_4 \end{bmatrix} \begin{Bmatrix} \ddot r_2 \\ \ddot r_4 \end{Bmatrix}$$
$$= \begin{Bmatrix} (r_2\ddot\theta_2 + 2\dot r_2\dot\theta_2)\sin\theta_2 - (r_4\ddot\theta_2 + 2\dot r_4\dot\theta_2)\sin\theta_4 + r_2\dot\theta_2^2\cos\theta_2 - r_4\dot\theta_2^2\cos\theta_4 \\ -(r_2\ddot\theta_2 + 2\dot r_2\dot\theta_2)\cos\theta_2 + (r_4\ddot\theta_2 + 2\dot r_4\dot\theta_2)\cos\theta_4 + r_2\dot\theta_2^2\sin\theta_2 - r_4\dot\theta_2^2\sin\theta_4 \end{Bmatrix}$$

$$\ddot{\boldsymbol{r}}_{P_2} = (-r_2\ddot\theta_2\sin\theta_2 - r_2\dot\theta_2^2\cos\theta_2)\boldsymbol{i} + (r_2\ddot\theta_2\cos\theta_2 - r_2\dot\theta_2^2\sin\theta_2)\boldsymbol{j}$$

$$\ddot{\boldsymbol{r}}_{P_3} = (\ddot r_2\cos\theta_2 - 2\dot r_2\dot\theta_2\sin\theta_2 - r_2\ddot\theta_2\sin\theta_2 - r_2\dot\theta_2^2\cos\theta_2)\boldsymbol{i}$$
$$+ (\ddot r_2\sin\theta_2 + 2\dot r_2\dot\theta_2\cos\theta_2 + r_2\ddot\theta_2\cos\theta_2 - r_2\dot\theta_2^2\sin\theta_2)\boldsymbol{j}$$

3.10.2 Analysis When r_2 Is Known

When r_2 is known, we must determine θ_2, θ_4, and r_4, and the equations required for the analysis are given in Table 3.17. Two solutions corresponding to the assembly modes shown in Fig. 3.45 result. Note that in one assembly mode, β is positive, and in the other, β is negative. If the sign of β must be positive, the solution corresponding to a minus angle would be discarded. The value of γ that is valid is the one satisfying Eqs. (3.153) and (3.154). It is also possible to specify values for r_2 for which the mechanism cannot be assembled. The analytical form of the velocity and acceleration equations can be developed by differentiating the position equations and solving the resulting linear equations for the unknowns. Again, the equations can easily be programmed, and a MATLAB program for analyzing the mechanism when r_2 is the driver is included on the disk with this book.

EXAMPLE 3.10 *(Kinematic Analysis of Oldham Mechanism)*

PROBLEM

In the Oldham mechanism shown in Fig. 3.46a, the angular velocity of link 2 relative to the frame is 10 rad/s CCW and the angular acceleration is 100 rad/s^2 CCW. Also, the length OQ = 10 cm and the angle β is 45°. Determine the position of point P and the velocity and acceleration of points P_2 and P_3 in the position given.

Table 3.17 Summary of Position, Velocity, and Acceleration Equations for an Oldham Mechanism When $\boldsymbol{r}_2$ is the Input Variable. The Link Numbers and Points are Defined in Fig. 3.44.

Position

$$r_4 = r_2\cos\beta + \sigma\sqrt{r_2^2(\cos^2\beta - 1) + r_1^2}, \qquad \sigma = \pm 1$$

$$A = r_1\sin\theta_1 - r_4\sin\beta$$

$$B = -2(r_2 - r_4\cos\beta)$$

$$C = r_4\sin\beta + r_1\sin\theta_1$$

$$\theta_2 = 2\tan^{-1}\left[\frac{-B + \gamma\sqrt{B^2 - C^2 + A^2}}{(C - A)}\right], \qquad \gamma = +1 \text{ or } -1; \text{valid value satisfies Eq. (3.153)}$$

$$\theta_4 = \theta_2 + \beta$$

$$\boldsymbol{r}_P = \boldsymbol{r}_2 = (r_2\cos\theta_2)\boldsymbol{i} + (r_2\sin\theta_2)\boldsymbol{j}$$

$$\boldsymbol{r}_Q = \boldsymbol{r}_1 = (r_1\cos\theta_1)\boldsymbol{i} + (r_1\sin\theta_1)\boldsymbol{j}$$

Velocity

$$\dot\theta_4 = \dot\theta_2$$

$$\begin{bmatrix} -r_2\sin\theta_2 + r_4\sin\theta_4 & -\cos\theta_4 \\ r_2\cos\theta_2 - r_4\cos\theta_4 & -\sin\theta_4 \end{bmatrix}\begin{Bmatrix} \dot\theta_2 \\ \dot r_4 \end{Bmatrix} = \begin{Bmatrix} -\dot r_2\cos\theta_2 \\ -\dot r_2\sin\theta_2 \end{Bmatrix}$$

$$\boldsymbol{r}_{P_2} = (-r_2\dot\theta_2\sin\theta_2)\boldsymbol{i} + (r_2\dot\theta_2\cos\theta_2)\boldsymbol{j}$$

$$\boldsymbol{r}_{P_3} = (\dot r_2\cos\theta_2 - r_2\dot\theta_2\sin\theta_2)\boldsymbol{i} + (\dot r_2\sin\theta_2 + r_2\dot\theta_2\cos\theta_2)\boldsymbol{j}$$

Accelerations

$$\ddot\theta_4 = \ddot\theta_2$$

$$\begin{bmatrix} -r_2\sin\theta_2 + r_4\sin\theta_4 & -\cos\theta_4 \\ r_2\cos\theta_2 - r_4\cos\theta_4 & -\sin\theta_4 \end{bmatrix}\begin{Bmatrix} \ddot\theta_2 \\ \ddot r_4 \end{Bmatrix}$$

$$= \begin{Bmatrix} 2\dot r_2\dot\theta_2\sin\theta_2 - 2\dot r_4\dot\theta_2\sin\theta_4 + r_2\dot\theta_2^2\cos\theta_2 - \ddot r_2\cos\theta_2 - r_4\dot\theta_2^2\cos\theta_4 \\ -2\dot r_2\dot\theta_2\cos\theta_2 + 2\dot r_4\dot\theta_2\cos\theta_4 + r_2\dot\theta_2^2\sin\theta_2 - \ddot r_2\sin\theta_2 - r_4\dot\theta_2^2\sin\theta_4 \end{Bmatrix}$$

$$\ddot{\boldsymbol{r}}_{P_2} = (-r_2\ddot\theta_2\sin\theta_2 - r_2\dot\theta_2^2\cos\theta_2)\boldsymbol{i} + (r_2\ddot\theta_2\cos\theta_2 - r_2\dot\theta_2^2\sin\theta_2)\boldsymbol{j}$$

$$\ddot{\boldsymbol{r}}_{P_3} = (\ddot r_2\cos\theta_2 - 2\dot r_2\dot\theta_2\sin\theta_2 - r_2\ddot\theta_2\sin\theta_2 - r_2\dot\theta_2^2\cos\theta_2)\boldsymbol{i}$$
$$+ (\ddot r_2\sin\theta_2 + 2\dot r_2\dot\theta_2\cos\theta_2 + r_2\ddot\theta_2\cos\theta_2 - r_2\dot\theta_2^2\sin\theta_2)\boldsymbol{j}$$

SOLUTION

Using the nomenclature developed earlier, the basic vector closure diagram for the linkage is as shown in Fig. 3.46b. From Fig. 3.46, the following geometry quantities can be determined.

$$\theta_1 = 0°, \qquad \theta_2 = 60°, \qquad \theta_4 = 105°$$

$$r_1 = 10 \text{ cm}$$

To perform the analysis, we can use the equations in Table 3.16. For the position analysis, we need to solve

$$\begin{bmatrix} \cos\theta_2 & -\cos\theta_4 \\ \sin\theta_2 & -\sin\theta_4 \end{bmatrix}\begin{Bmatrix} r_2 \\ r_4 \end{Bmatrix} = \begin{Bmatrix} r_1\cos\theta_1 \\ r_1\sin\theta_1 \end{Bmatrix} \quad \text{or} \quad \begin{bmatrix} \cos(60) & -\cos(105) \\ \sin(60) & -\sin(105) \end{bmatrix}\begin{Bmatrix} r_2 \\ r_4 \end{Bmatrix} = \begin{Bmatrix} 10\cos 0 \\ 10\sin 0 \end{Bmatrix}$$

or

$$\begin{bmatrix} 0.500 & 0.2588 \\ 0.8660 & -0.9659 \end{bmatrix}\begin{Bmatrix} r_2 \\ r_4 \end{Bmatrix} = \begin{Bmatrix} 10 \\ 0 \end{Bmatrix} \Rightarrow \begin{Bmatrix} r_2 \\ r_4 \end{Bmatrix} = \begin{Bmatrix} 13.6603 \\ 12.2474 \end{Bmatrix}$$

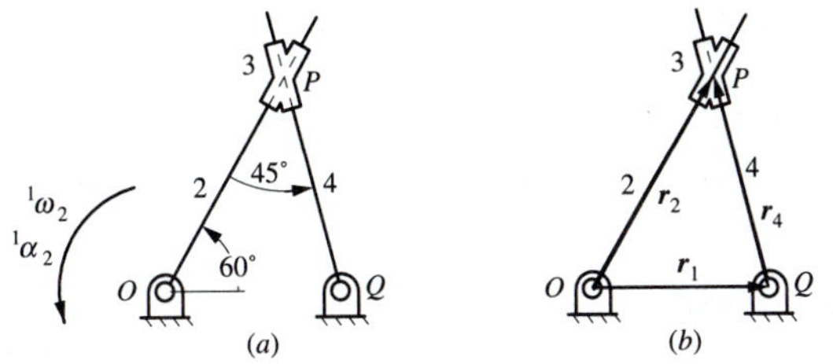

Figure 3.46 Oldham mechanism (a) and vector diagram (b) for Example 3.10.

The positions of P and Q are given by

$$\boldsymbol{r}_p = (r_2 \cos \theta_2)\boldsymbol{i} + (r_2 \sin \theta_2)\mathbf{j} = 6.8301\boldsymbol{i} + 11.8301\boldsymbol{j}$$

and

$$\boldsymbol{r}_Q = (r_1 \cos \theta_1)\boldsymbol{i} + (r_1 \sin \theta_1)\boldsymbol{j} = 10\boldsymbol{i}$$

For the velocity analysis, solve

$$\begin{bmatrix} \cos \theta_2 & -\cos \theta_4 \\ \sin \theta_2 & -\sin \theta_4 \end{bmatrix} \begin{Bmatrix} \dot{r}_2 \\ \dot{r}_4 \end{Bmatrix} = \begin{Bmatrix} r_2\dot{\theta}_2 \sin \theta_2 - r_4\dot{\theta}_2 \sin \theta_4 \\ -r_2\dot{\theta}_2 \cos \theta_2 + r_4\dot{\theta}_2 \cos \theta_4 \end{Bmatrix} \Rightarrow \begin{bmatrix} 0.500 & 0.2588 \\ 0.8660 & -0.9659 \end{bmatrix} \begin{Bmatrix} \dot{r}_2 \\ \dot{r}_4 \end{Bmatrix} = \begin{Bmatrix} 0 \\ -100 \end{Bmatrix}$$

or

$$\begin{Bmatrix} \dot{r}_2 \\ \dot{r}_4 \end{Bmatrix} = \begin{Bmatrix} -36.6025 \\ 70.7107 \end{Bmatrix}$$

The velocities of points P_2 and P_3 are

$$\dot{\boldsymbol{r}}_{P_2} = (-r_2\dot{\theta} \sin \theta_2)\boldsymbol{i} + (r_2\dot{\theta}_2 \cos \theta_2)\boldsymbol{j} = -118.3031\boldsymbol{i} + 68.3013\boldsymbol{j}$$

and

$$\dot{\boldsymbol{r}}_{P_3} = (\dot{r}_2 \cos \theta_2 - r_2\dot{\theta}_2 \sin \theta_2)\boldsymbol{i} + (\dot{r}_2 \sin \theta_2 + r_2\dot{\theta}_2 \cos \theta_2)\boldsymbol{j} = -136.6025\boldsymbol{i} + 36.6025\boldsymbol{j}$$

For the acceleration analysis, solve

$$\begin{bmatrix} \cos \theta_2 & -\cos\theta_4 \\ \sin \theta_2 & -\sin\theta_4 \end{bmatrix} \begin{Bmatrix} \ddot{r}_2 \\ \ddot{r}_4 \end{Bmatrix}$$

$$= \begin{Bmatrix} (r_2\ddot{\theta}_2 + 2\dot{r}_2\dot{\theta}_2) \sin \theta_2 - (r_4\ddot{\theta}_2 + 2\dot{r}_4\dot{\theta}_2) \sin \theta_4 + r_2\dot{\theta}_2^2 \cos \theta_2 - r_4\dot{\theta}_2^2 \cos \theta_4 \\ (-r_2\ddot{\theta}_2 + 2\dot{r}_2\dot{\theta}_2) \cos \theta_2 + (r_4\ddot{\theta}_2 + 2\dot{r}_4\dot{\theta}_2) \cos \theta_4 + r_2\dot{\theta}_2^2 \sin \theta_2 - r_4\dot{\theta}_2^2 \sin \theta_4 \end{Bmatrix}$$

or

$$\begin{bmatrix} 0.500 & 0.2588 \\ 0.8660 & -0.9659 \end{bmatrix} \begin{Bmatrix} \ddot{r}_2 \\ \ddot{r}_4 \end{Bmatrix} = \begin{Bmatrix} -1000 \\ -1000 \end{Bmatrix}$$

Then,

$$\begin{Bmatrix} \ddot{r}_2 \\ \ddot{r}_4 \end{Bmatrix} = \begin{Bmatrix} -1732.1 \\ -517.6 \end{Bmatrix}$$

Also,

$$\ddot{\boldsymbol{r}}_{P_2} = (-r_2\ddot{\theta}_2 \sin \theta_2 - r_2\dot{\theta}_2^2 \cos \theta_2)\boldsymbol{i} + (r_2\ddot{\theta}_2 \cos \theta_2 - r_2\dot{\theta}_2^2 \sin \theta_2)\boldsymbol{j} = -1866\boldsymbol{i} - 500\boldsymbol{j}$$

and

$$\ddot{\boldsymbol{r}}_{P_3} = (\ddot{r}_2 \cos\theta_2 - 2\dot{r}_2\dot{\theta}_2 \sin\theta_2 - r_2\ddot{\theta}_2 \sin\theta_2 - r_2\dot{\theta}_2^2 \cos\theta_2)\boldsymbol{i} + (\ddot{r}_2 \sin\theta_2 + 2\dot{r}_2\dot{\theta}_2 \cos\theta_2 + r_2\ddot{\theta}_2 \cos\theta_2 - r_2\dot{\theta}_2^2 \sin\theta_2)\boldsymbol{j} = -2098.1\boldsymbol{i} - 2366\boldsymbol{j}$$ ■

3.11 CLOSURE OR LOOP EQUATION APPROACH FOR COMPOUND MECHANISMS

As in the case of simple, single-loop mechanisms, each vector is represented by a length r_i and an angle θ_i. All angles are measured counterclockwise from a fixed line parallel to the x axis attached to the frame as shown in Fig. 3.47.

To illustrate the method for compound mechanisms, consider the kinematic diagram of the mechanism given in Fig. 3.47. Each member is represented by a directed length and an angle. The formulation of the analytical procedure based on vector loops for compound mechanisms is straightforward, but it requires a system if results are to be meaningful. A procedure will be outlined in the following and illustrated on the mechanism in Fig. 3.47. It will be noted that the procedure presented is a generalization of that used for the single-loop mechanisms. In the mechanism shown, assume that θ_2, ${}^1\omega_2$, and ${}^1\alpha_2$ are known values.

Procedure

1. Draw a kinematic sketch of the mechanism. The sketch need not be to scale; however, it must be accurate enough that the assembly mode can be determined by inspection.

2. Establish a global coordinate system for the mechanism. This will establish the horizontal axis from which all angles will be measured and identify the system from which all global coordinates will be determined.

3. Represent the link between adjacent joints by a vector $\boldsymbol{r}_i$ defined by a directed line and an angle measured positive CCW from the x axis (or a line parallel to the x axis).

$$\boldsymbol{r}_i = r_i \angle\theta_i$$

4. If sliders are involved, locate the slider by two vectors, one in the direction of the relative velocity between the slider and the slide and the second in a direction perpendicular to the direction of the velocity (see r_7 and r_8 in Fig. 3.47).

5. Note which lengths and angles are fixed and which are variable. In the mechanism above, r_7 is the only *variable* length and θ_1, θ_7, θ_8, and β are the only *fixed* angles.

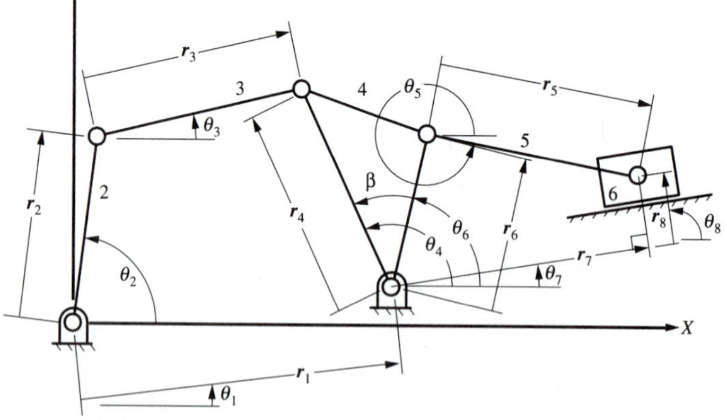

Figure 3.47 Example of formulation of solution procedure using vector loops.

If the vectors are properly defined, they will all be expressible as

$$\boldsymbol{r}_i = r_i(\cos\theta_i \boldsymbol{i} + \sin\theta_i \boldsymbol{j})$$

The cosine term will always go with $\boldsymbol{i}$ and the sine term will go with $\boldsymbol{j}$. Some angles may be functions of others. For example, $\theta_6 = \theta_4 - \beta$.

6. Identify all of the joints on the linkage, and be sure that each one is located at the end of one of the vectors. Then identify all of the independent vector loops in the linkage, and write a vector equation for each loop. For the mechanism in Fig. 3.47, there are two obvious vector loops represented by the following equations:

$$\boldsymbol{r}_1 + \boldsymbol{r}_4 = \boldsymbol{r}_2 + \boldsymbol{r}_3 \tag{3.156}$$

$$\boldsymbol{r}_5 + \boldsymbol{r}_6 = \boldsymbol{r}_7 + \boldsymbol{r}_8 \tag{3.157}$$

7. Write the x, y scalar equations for each vector equation. Notice that the form of the equations is consistent, and once the basic vector equations are given (e.g., Eqs. 3.156 and 3.157), it is not even necessary to look at the mechanism to be able to write the component equations.

$$x \Rightarrow r_1\cos\theta_1 + r_4\cos\theta_4 = r_2\cos\theta_2 + r_3\cos\theta_3 \tag{3.158}$$

$$y \Rightarrow r_1\sin\theta_1 + r_4\sin\theta_4 = r_2\sin\theta_2 + r_3\sin\theta_3 \tag{3.159}$$

$$x \Rightarrow r_5\cos\theta_5 + r_6\cos\theta_6 = r_7\cos\theta_7 + r_8\cos\theta_8 \tag{3.160}$$

$$y \Rightarrow r_5\sin\theta_5 + r_6\sin\theta_6 = r_7\sin\theta_7 + r_8\sin\theta_8 \tag{3.161}$$

8. Identify any constraints among the lengths or angles that are not identified by the vector loops. In the mechanism in Fig. 3.47, θ_7 is related to θ_8 by $\pi/2$, and θ_6 is related to θ_4 through β. Therefore, the extra constraint equations are

$$\theta_8 = \theta_7 + \pi/2 \tag{3.162}$$

and

$$\theta_6 = \theta_4 - \beta \tag{3.163}$$

9. Count the total number of variables in the component equations and the extra constraint equations. If n is the total number of equations and f is the number of degrees of freedom for the mechanism, the total number of unknowns should be $n + f$. If the number of unknowns is larger than this, it is necessary to identify additional constraints or to reformulate the loop closure equations. In the mechanism in Fig. 3.47, the total number of unknowns is seven (θ_2, θ_3, θ_4, θ_5, θ_6, θ_8, and r_7), and the mechanism has only one degree of freedom. Therefore, $n + f = 7$, which indicates that the problem should be solvable. Note that the number of unknowns and the number of variables are not necessarily the same. In this mechanism, θ_8 is a constant but initially unknown. It must be computed using Eq. (3.162). Equations (3.158)–(3.163) are nonlinear in the unknowns (θ_2, θ_3, θ_4, θ_5, θ_6, θ_8, and r_7), and most of the analysis difficulties are concentrated here.

10. For velocities, differentiate the position equations (x and y components) term by term. In the example case, the velocity equations are

$$\dot{\mathbf{r}}_1 + \dot{\mathbf{r}}_4 = \dot{\mathbf{r}}_2 + \dot{\mathbf{r}}_3 \tag{3.164}$$

$$\dot{\mathbf{r}}_5 + \dot{\mathbf{r}}_6 = \dot{\mathbf{r}}_7 + \dot{\mathbf{r}}_8 \tag{3.165}$$

and recognizing which terms are constants and which are variables,

$$x \Rightarrow -r_4\dot{\theta}_4\sin\theta_4 = -r_2\dot{\theta}_2\sin\theta_2 - r_3\dot{\theta}_3\sin\theta_3 \tag{3.166}$$

$$y \Rightarrow r_4\dot{\theta}_4\cos\theta_4 = r_2\dot{\theta}_2\cos\theta_2 + r_3\dot{\theta}_3\cos\theta_3 \tag{3.167}$$

$$\dot{\theta}_6 = \dot{\theta}_4 \tag{3.168}$$

$$x \Rightarrow -r_5\dot{\theta}_5\sin\theta_5 - r_6\dot{\theta}_6\sin\theta_6 = \dot{r}_7\cos\theta_7 \tag{3.169}$$

$$y \Rightarrow r_5\dot{\theta}_5\cos\theta_5 + r_6\dot{\theta}_6\cos\theta_6 = \dot{r}_7\sin\theta_7 \tag{3.170}$$

Note that once we have solved the position equations, only the angle and length derivatives will be unknown. Hence the equations are linear in the unknowns and can be easily solved. There are five linear equations in five unknowns ($\dot{\theta}_3$, $\dot{\theta}_4$, $\dot{\theta}_5$, $\dot{\theta}_6$, $\dot{r}_7$). These can be solved directly by Gaussian elimination, by using a programmable calculator, or by using a matrix solver such as MATLAB.

11. For accelerations, differentiate the velocity equations (x and y components) term by term. In the example case, the acceleration equations are

$$\ddot{\boldsymbol{r}}_1 + \ddot{\boldsymbol{r}}_4 = \ddot{\boldsymbol{r}}_2 + \ddot{\boldsymbol{r}}_3 \tag{3.171}$$

$$\ddot{\boldsymbol{r}}_5 + \ddot{\boldsymbol{r}}_6 = \ddot{\boldsymbol{r}}_7 + \ddot{\boldsymbol{r}}_8 \tag{3.172}$$

and in terms of components,

$$x \Rightarrow -r_4\ddot{\theta}_4 \sin\theta_4 - r_4\dot{\theta}_4^2 \cos\theta_4 = -r_2\ddot{\theta}_2 \sin\theta_2 - r_2\dot{\theta}_2^2 \cos\theta_2 - r_3\ddot{\theta}_3 \sin\theta_3 - r_3\dot{\theta}_3^2 \cos\theta_3 \tag{3.173}$$

$$y \Rightarrow r_4\ddot{\theta}_4 \cos\theta_4 - r_4\dot{\theta}_4^2 \sin\theta_4 = r_2\ddot{\theta}_2 \cos\theta_2 - r_2\dot{\theta}_2^2 \sin\theta_2 + r_3\ddot{\theta}_3 \cos\theta_3 - r_3\dot{\theta}_3^2 \sin\theta_3 \tag{3.174}$$

$$\ddot{\theta}_6 = \ddot{\theta}_4 \tag{3.175}$$

$$x \Rightarrow -r_5\ddot{\theta}_5 \sin\theta_5 - r_5\dot{\theta}_5^2 \cos\theta_5 - r_6\ddot{\theta}_6 \sin\theta_6 - r_6\dot{\theta}_6^2 \cos\theta_6 = \ddot{r}_7 \cos\theta_7 \tag{3.176}$$

$$y \Rightarrow r_5\ddot{\theta}_5 \cos\theta_5 - r_5\dot{\theta}_5^2 \sin\theta_5 + r_6\ddot{\theta}_6 \cos\theta_6 - r_6\dot{\theta}_6^2 \sin\theta_6 = \ddot{r}_7 \sin\theta_7 \tag{3.177}$$

Note that once we have solved the position and velocity equations, only the derivatives of velocity will be unknown. Hence, the equations are linear in the unknowns and can easily be solved. There are five linear equations in five unknowns ($\ddot{\theta}_3$, $\ddot{\theta}_4$, $\ddot{\theta}_5$, $\ddot{\theta}_6$, $\ddot{r}_7$). Once again, these can be solved directly by Gaussian elimination, by using a programmable calculator, or by using a matrix solver such as MATLAB.

3.11.1 Handling Points Not on the Vector Loops

The solution procedure outlined above will give the position, velocity, and acceleration of each point at a vertex of a vector loop in addition to the angular velocity and acceleration of each link. The angular velocities and accelerations are the $\dot{\theta}_i$ and $\ddot{\theta}_i$ terms, respectively. In general,

$$^1\omega_i = \dot{\theta}_i \boldsymbol{k}$$

and

$$^1\alpha_i = \ddot{\theta}_i \boldsymbol{k}$$

Once the basic analysis is completed by solving the vector loop equations, we will be able to locate at least one point on each rigid body (link) as a function of time. We will also be able to determine the orientation of each rigid body as a function of time, that is, θ_i, $^1\omega_i$, and $^1\alpha_i$ will be known or can be determined for each link.

Points that are not vertices of the vector loops must be associated with one of the rigid bodies in the mechanism. To determine the kinematic properties of a given point, we simply identify the point by a vector in terms of the known quantities, determine the x, y components of the vector, and differentiate. For example, assume that we want to know the kinematic properties of a point Q on link 3 as shown in Fig. 3.48. Then,

$$\boldsymbol{r}_{Q_3/A_2} = \boldsymbol{r}_2 + \boldsymbol{r}_9$$

or

$$\begin{aligned} \boldsymbol{r}_{Q_3/A_2} &= r_2(\cos\theta_2 \boldsymbol{i} + \sin\theta_2 \boldsymbol{j}) + r_9[\cos(\Delta + \theta_3)\boldsymbol{i} + \sin(\Delta + \theta_3)\boldsymbol{j}] \\ &= [r_2 \cos\theta_2 + r_9 \cos(\Delta + \theta_3)]\boldsymbol{i} + [r_2 \sin\theta_2 + r_9 \sin(\Delta + \theta_3)]\boldsymbol{j} \end{aligned} \tag{3.178}$$

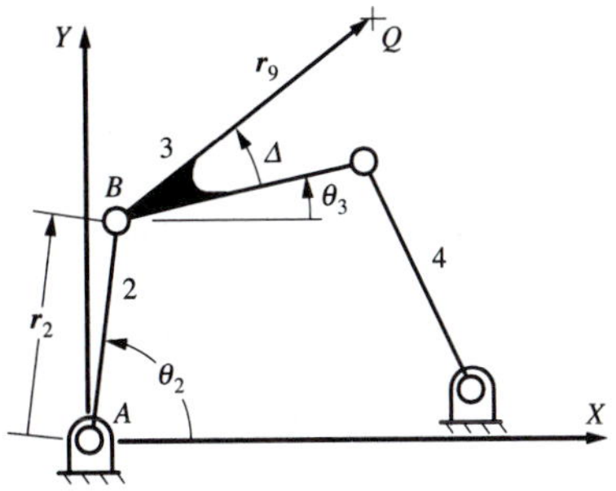

Figure 3.48 Determination of velocity and acceleration of point Q which is not a vertex of a vector loop.

All terms on the right-hand side of Eq. (3.178) will be known. Therefore, the position vector can be computed directly. The velocity is given by

$$ {}^1\boldsymbol{v}_{Q_3/A_2} = \dot{\boldsymbol{r}}_2 + \dot{\boldsymbol{r}}_9 $$

and

$$ \begin{aligned} {}^1\boldsymbol{v}_{Q_3/A_2} &= r_2\dot{\theta}_2(-\sin\theta_2\boldsymbol{i} + \cos\theta_2\boldsymbol{j}) + r_9\dot{\theta}_3[-\sin(\Delta+\theta_3)\boldsymbol{i} + \cos(\Delta+\theta_3)\boldsymbol{j}] \\ &= -[r_2\dot{\theta}_2\sin\theta_2 + r_9\dot{\theta}_3\sin(\Delta+\theta_3)]\boldsymbol{i} + [r_2\dot{\theta}_2\cos\theta_2 + r_9\dot{\theta}_3\cos(\Delta+\theta_3)]\boldsymbol{j} \end{aligned} \tag{3.179} $$

Again, all quantities on the right-hand side of Eq. (3.179) are known and so the velocity of point Q can be computed without difficulty. For the acceleration, differentiate the velocity expression. Then,

$$ {}^1\boldsymbol{a}_{Q_3/A_2} = \ddot{\boldsymbol{r}}_2 + \ddot{\boldsymbol{r}}_9 $$

and

$$ \begin{aligned} {}^1\boldsymbol{a}_{Q_3/A_2} &= r_2\ddot{\theta}_2(-\sin\theta_2\boldsymbol{i} + \cos\theta_2\boldsymbol{j}) - r_2(\dot{\theta}_2)^2(\cos\theta_2\boldsymbol{i} + \sin\theta_2\boldsymbol{j}) \\ &\quad + r_9\ddot{\theta}_3[-\sin(\Delta+\theta_3)\boldsymbol{i} + \cos(\Delta+\theta_3)\boldsymbol{j}] - r_9(\dot{\theta}_3)^2[\cos(\Delta+\theta_3)\boldsymbol{i} + \sin(\Delta+\theta_3)\boldsymbol{j}] \\ &= [-r_2\ddot{\theta}_2\sin\theta_2 - r_9\ddot{\theta}_3\sin(\Delta+\theta_3) - r_2(\dot{\theta}_2)^2\cos\theta_2 - r_9(\dot{\theta}_3)^2\cos(\Delta+\theta_3)]\boldsymbol{i} \\ &\quad + [r_2\ddot{\theta}_2\cos\theta_2 + r_9\ddot{\theta}_3\cos(\Delta+\theta_3) - r_2(\dot{\theta}_2)^2\sin\theta_2 - r_9(\dot{\theta}_3)^2\sin(\Delta+\theta_3)]\boldsymbol{j} \end{aligned} \tag{3.180} $$

Again, all quantities on the right-hand side of Eq. (3.180) are known, and so the acceleration of point Q can be computed without difficulty. Note that this procedure is simply a variation on the rigid body analysis given in Section 3.5

3.11.2 Solving the Position Equations

A review of the analysis just developed shows that only the position equations are nonlinear in the unknowns. Therefore specialized techniques are required to solve them. If a numerical solution is chosen, then an initial guess for the variables is required. This is best obtained by sketching the mechanism to scale. A numerical iteration method such as the Newton-Raphson method can be used to obtain refined values. If a series of input angles is to be investigated, then the final variable values for the previous input value can be used as the initial estimates of the variables for the next input value provided that the input angle increments are relatively small (i.e., within about 10 degrees of each other).

Another numerical approach which is computationally more efficient than using Newton's method, but sometimes has convergence problems at end-of-travel positions, is to numerically integrate the velocity equations after a precise set of values for the variables is obtained by Newton's method. The input step size for this integration should not exceed 2 degrees. This method is very convenient if a numerical integration is already needed for dynamic problems in which the equations of motion are required.

In cases in which it is possible to obtain the displacement variable values analytically, that method is generally preferred due to the elimination of numerical instability problems. In general, it is always possible to solve the equation analytically if the mechanism can be analyzed by hand using traditional graphical methods with vector polygons as presented in Chapter 2. When it is possible do to this, the position equations can be solved in sets of two equations in two unknowns. If it is not possible to reduce the equations to a series of two equations in two unknowns, the equations must be solved iteratively using a numerical procedure such as Newton's method.

When it is possible to solve the position equations algebraically, one of two situations will usually occur. In the first situation, the compound mechanism can be treated as a series of simple mechanisms. In the second case, the compound mechanism cannot be represented as a series of simple mechanisms; however, the equations can be partitioned into a sequential set of two equations in two unknowns. These two situations will be presented separately.

3.11.2.1 Compound Linkage as a Series of Simple Mechanisms

When the compound linkage is a series of simple mechanisms, we can analyze each mechanism in sequence. The output for one mechanism is the input to the next mechanism. If we have computer routines to analyze the single-loop mechanisms, the routines can be concatenated to analyze the entire linkage. This is the case that exists in the mechanism of Fig. 3.47. When we examine the mechanism, we find that the first linkage is a four-bar linkage defined by vectors $\boldsymbol{r}_1$, $\boldsymbol{r}_2$, $\boldsymbol{r}_3$, and $\boldsymbol{r}_4$. This mechanism can be analyzed using the equations developed in Section 3.4. The four-bar loop drives link 4. Therefore, once the position, velocity, and acceleration for $\boldsymbol{r}_4$ are known, the corresponding values for $\boldsymbol{r}_6$ can be found using rigid body conditions (Section 3.5). Link 4 is the input for the slider-crank mechanism defined by $\boldsymbol{r}_5$, $\boldsymbol{r}_6$, $\boldsymbol{r}_7$, and $\boldsymbol{r}_8$. This mechanism can be analyzed using the equations in Section 3.6 to determine the kinematic properties of the slider.

EXAMPLE 3.11 ***(Kinematic Analysis of Compound Linkage)***

PROBLEM

Determine the angular position, velocity, and acceleration of link 6 in the mechanism in Fig. 3.49 if the slider is moving at 10 cm/s (constant) to the right. The following dimensions are known:

$AB = 22.7$ cm $\quad CD = 7.5$ cm $\quad EF = 10.6$ cm

$BC = 10.6$ cm $\quad CF = 14.6$ cm

$AC = 28$ cm $\quad DE = 11.4$ cm

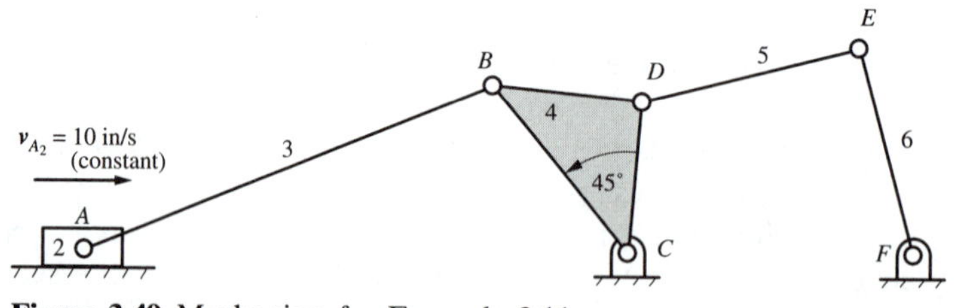

Figure 3.49 Mechanism for Example 3.11.

SOLUTION

We will analyze the mechanism as three linkage systems in series. First we will analyze the slider-crank mechanism using the equations in Table 3.5. Next we will compute the position of *CD* from rigid body conditions (Table 3.3). Last, the four-bar linkage (*CDEF*) can be analyzed using Table 3.1. The actual numerical calculations can be made using the MATLAB routines included on the disk with this book.

To facilitate the analysis, the mechanism in Fig. 3.49 is represented by the vectors indicated in Fig. 3.50.

For the slider-crank part of the mechanism, the following magnitudes and directions are known:

$$r_1 = 28 \qquad \theta_1 = 180° \qquad \dot{r}_1 = 10 \qquad \ddot{r}_1 = 0$$

$$r_2 = 10.6 \qquad \theta_2, \dot{\theta}_2, \ddot{\theta}_2 = ?$$

$$r_3 = 22.7 \qquad \theta_3, \dot{\theta}_3, \ddot{\theta}_3 = ?$$

For this set of values, the equations in Table 3.5 can be used. The value of $\boldsymbol{\sigma}$ is -1 for the geometry given, and the results are

$$\theta_1 = 180° \qquad \dot{\theta}_1 = 0 \text{ rad/s} \qquad \ddot{\theta}_1 = 0 \text{ rad/s}^2$$

$$\theta_2 = 129.94° \qquad \dot{\theta}_2 = -0.9314 \text{ rad/s} \qquad \ddot{\theta}_2 = 0.5005 \text{ rad/s}^2$$

$$\theta_3 = -159.02° \qquad \dot{\theta}_3 = 0.299 \text{ rad/s} \qquad \ddot{\theta}_3 = -0.459 \text{ rad/s}^2$$

The orientation of the vector r_4 will be related to that of the vector r_2 through the equation

$$\theta_4 = \theta_2 - 45°$$

Therefore, the magnitudes and directions for the vectors defining the four-bar linkage are

$$r_4 = 7.5 \qquad \theta_4 = 84.94° \qquad \dot{\theta}_4 = -0.9314 \text{ rad/s} \qquad \ddot{\theta}_4 = 0.5005 \text{ rad/s}^2$$

$$r_5 = 11.4 \qquad \theta_5, \dot{\theta}_5, \ddot{\theta}_5 = ?$$

$$r_6 = 10.6 \qquad \theta_6, \dot{\theta}_6, \ddot{\theta}_6 = ?$$

$$r_7 = 14.6 \qquad \theta_7 = 0°$$

For this set of values, the equations in Table 3.1 can be used. The value of σ is again -1 for the geometry given, and the results are

$$\theta_4 = 84.94° \qquad \dot{\theta}_4 = -0.9314 \text{ rad/s} \qquad \ddot{\theta}_4 = 0.5005 \text{ rad/s}^2$$

$$\theta_5 = 13.87° \qquad \dot{\theta}_5 = 0.2175 \text{ rad/s} \qquad \ddot{\theta}_5 = 0.0536 \text{ rad/s}^2$$

$$\theta_6 = 105.7146° \qquad \dot{\theta}_6 = -0.6237 \text{ rad/s} \qquad \ddot{\theta}_6 = 0.5978 \text{ rad/s}^2$$

$$\theta_7 = 0° \qquad \dot{\theta}_7 = 0 \text{ rad/s} \qquad \ddot{\theta}_7 = 0 \text{ rad/s}^2$$

■

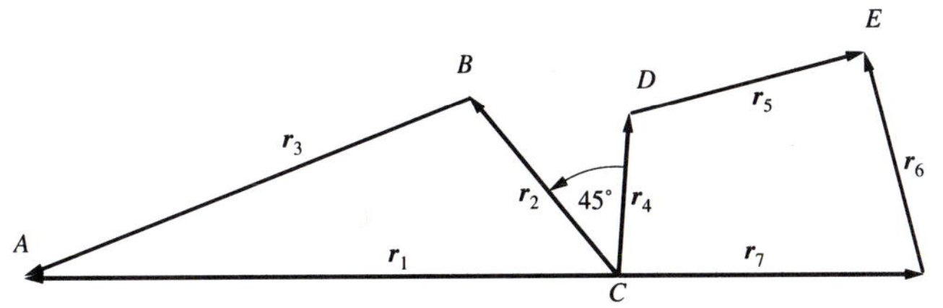

Figure 3.50 Vectors representing mechanism in Fig. 3.49.

3.11.2.2 General Cases in Which Two Equations in Two Unknowns Result

For simple lower-pair mechanisms with one loop equation, the position analysis will reduce to two scalar equations in two unknowns, and it is relatively easy to develop closed-form equations for the unknown variables. However, when analyzing more complex lower-pair mechanisms with n loop equations, the number of equations and the number of variable unknowns are both $2n$, and the solution can become much more complicated. However, not all the pair variables appear in each of the equations. Fortunately, it is often possible to group the equations into smaller sets that can be solved independently in a serial fashion.

If a given lower-pair mechanism can be analyzed using the traditional vector-polygon approach, it is always possible to group the position equations in such a way that no more than two equations in two unknowns must be solved at any one time. For such mechanisms, the position equations can always be solved in closed form, and these types of mechanisms form the vast majority of the linkages that an engineer might design. For complex mechanisms that cannot be analyzed entirely using closed-form equations, it is often possible to analyze a part of the mechanism with closed-form equations after other parts are analyzed numerically.

For simplicity, the vector form of the loop closure equation for each loop will be represented in homogeneous form as

$$\sum_{i=1}^{k} \sigma_i \boldsymbol{r}_i = 0$$

where $\sigma_i = \pm 1$. This equation can be divided into x and y components, and the corresponding component equations are

$$\sum_{i=1}^{k} \sigma_i r_i \cos\theta_i = 0 \qquad \textbf{(3.181)}$$

$$\sum_{i=1}^{k} \sigma_i r_i \sin\theta_i = 0 \qquad \textbf{(3.182)}$$

where r_i is the length of vector i, θ_i is the angle (measured CCW) between link i and a horizontal line, and k is the number of vectors in a given loop. Equations (3.181) and (3.182) are written for each closure loop of the mechanism, and for n loops there will be $2n$ equations and $2n$ unknowns. When the links contain more than two joints, there will also be auxiliary equations that relate joint angles. As shown in Fig. 3.51, these auxiliary equations can generally be written as

$$\theta_q = \theta_r - \Delta \qquad \textbf{(3.183)}$$

where Δ is a constant. When such equations are necessary, the equations can be written such that θ_r is solved first so that it is a trivial matter to solve for θ_q. Note that Eq. (3.183) is linear.

When all of the equations of the forms given by Eqs. (3.181)–(3.183) are considered as a set, it is usually possible to separate the $2n$ nonlinear equations into smaller groups, which can be solved serially, and in most cases the nonlinear equations can be grouped

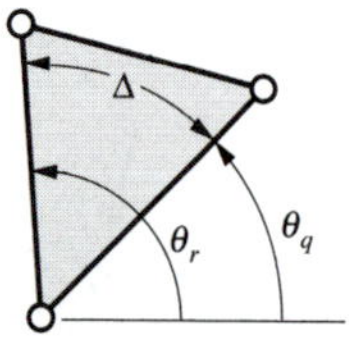

Figure 3.51 Geometry described by auxiliary equations.

into sets of two equations and two unknowns of the form:

$$\sigma_p r_p \cos\theta_p + \sigma_m r_m \cos\theta_m + \sum_{i=1}^{p-1} \sigma_i r_i \cos\theta_i + \sum_{i=p+1}^{m-1} \sigma_i r_i \cos\theta_i + \sum_{i=m+1}^{k} \sigma_i r_i \cos\theta_i = 0 \quad \textbf{(3.184)}$$

$$\sigma_p r_p \sin\theta_p + \sigma_m r_m \sin\theta_m + \sum_{i=1}^{p-1} \sigma_i r_i \sin\theta_i + \sum_{i=p+1}^{m-1} \sigma_i r_i \sin\theta_i + \sum_{i=m+1}^{k} \sigma_i r_i \sin\theta_i = 0 \quad \textbf{(3.185)}$$

where $\sigma_p = \sigma_m$.

Then,

$$r_p \cos\theta_p + r_m \cos\theta_m = C_1 \quad \textbf{(3.186)}$$

$$r_p \sin\theta_p + r_m \sin\theta_m = C_2 \quad \textbf{(3.187)}$$

where

$$C_1 = -\frac{1}{\sigma_m}\left(\sum_{i=1}^{p-1} \sigma_i r_i \cos\theta_i + \sum_{i=p+1}^{m-1} \sigma_i r_i \cos\theta_i + \sum_{i=m+1}^{k} \sigma_i r_i \cos\theta_i\right)$$

and

$$C_2 = -\frac{1}{\sigma_m}\left(\sum_{i=1}^{p-1} \sigma_i r_i \sin\theta_i + \sum_{i=p+1}^{m-1} \sigma_i r_i \sin\theta_i + \sum_{i=m+1}^{k} \sigma_i r_i \sin\theta_i\right)$$

In Eqs. (3.186) and (3.187), two of the four variables r_p, θ_p, r_m, and θ_m can be unknown, resulting in six possible combinations; however, only four of these six combinations are unique. The four cases that must be considered are case 1, r_p and θ_p or r_m and θ_m are the unknowns; case 2, r_p and θ_m or r_m and θ_p are the unknowns; case 3, r_p and r_m are the unknowns; case 4, θ_p and θ_m are the unknowns. The rest of the variables in the four cases are known for each position of the mechanism. When solving each of the cases, the three trigonometric identities discussed earlier for sin θ, cos θ, and tan ($\theta/2$) are used:

$$\cos^2\theta + \sin^2\theta = 1 \quad \textbf{(3.188)}$$

$$\cos\theta = \frac{1-\tan^2(\theta/2)}{1+\tan^2(\theta/2)} \quad \textbf{(3.189)}$$

and

$$\sin\theta = \frac{2\tan(\theta/2)}{1+\tan^2(\theta/2)} \quad \textbf{(3.190)}$$

The equations for calculating the unknown variables in each of these cases are developed in the following.

3.11.2.2.1 *Case 1, r_p and θ_p Unknown*

To solve this case, the terms $r_p \cos\theta_p$ and $r_p \sin\theta_p$ are first isolated on the left-hand side of Eqs. (3.186) and (3.187).

$$r_p \cos\theta_p = C_1 - r_m \cos\theta_m \quad \textbf{(3.191)}$$

$$r_p \sin\theta_p = C_2 - r_m \sin\theta_m \quad \textbf{(3.192)}$$

Equation (3.192) is then divided by Eq. (3.191) to provide the solution for θ_p. Next the two equations are squared, added together, and simplified using Eq. (3.188) to obtain a solution for r_p. The resulting expressions for r_p and θ_p are

$$r_p = \sqrt{C_1^2 + C_2^2 + r_m^2 - 2C_1 r_m \cos\theta_m - 2C_2 r_m \sin\theta_m} \quad \textbf{(3.193)}$$

$$\theta_p = \tan^{-1}\left[\frac{C_2 - r_m \sin\theta_m}{C_1 - r_m \cos\theta_m}\right] \quad \textbf{(3.194)}$$

3.11.2.2.2 Case 2, r_p and θ_m Unknown

To solve for the two unknown variables, r_p and θ_m, the terms $r_m \cos \theta_m$ and $r_m \sin \theta_m$ are first isolated on the left-hand side of Eqs. (3.186) and (3.187).

$$r_m \cos \theta_m = C_1 - r_p \cos \theta_p \quad \textbf{(3.195)}$$

$$r_m \sin \theta_m = C_2 - r_p \sin \theta_p \quad \textbf{(3.196)}$$

Equations (3.195) and (3.196) are then squared, added together, and simplified using Eq. (3.188) to give a quadratic equation in the variable r_p. The solution to the resulting equation is

$$r_p = \frac{-b \pm \sqrt{b^2 - 4c}}{2} \quad \textbf{(3.197)}$$

where

$$b = -2C_1 \cos \theta_p - 2C_2 \sin \theta_p$$

and

$$c = C_1^2 + C_2^2 - r_m^2$$

The angle θ_m is found by dividing Eq. (3.186) by Eq. (3.239) and solving for θ_m:

$$\theta_m = \tan^{-1}\left[\frac{-r_p \sin \theta_p + C_2}{-r_p \cos \theta_p + C_1}\right] \quad \textbf{(3.198)}$$

Equations (3.197) and (3.198) each have two solutions corresponding to the two assembly modes of this part of the linkage. The proper mode must be specified at the time of the analysis. This can be done directly or by providing an initial estimate of the position of the mechanism and determining which solution is closest to that indicated by the initial estimate. In practice, the initial estimate of the position of the mechanism could be provided by an approximate sketch drawn on a computer screen.

3.11.2.2.3 Case 3, r_p and r_m Unknown

To solve for r_p, Eq. (3.186) is first multiplied by tan θ_m and the result simplified. Then Eq. (3.187) is multiplied by cos θ_m and the resulting equations from the two operations are subtracted. After simplification, the expression for r_p is

$$r_p = \frac{C_1 \sin \theta_m - C_2 \cos \theta_m}{\sin(\theta_m - \theta_p)} \quad \textbf{(3.199)}$$

After r_p is known, Eqs. (3.186) and (3.187) can be solved directly for r_m:

$$r_m = \frac{C_1 - r_p \cos \theta_p}{\cos \theta_m} = \frac{C_2 - r_p \sin \theta_p}{\sin \theta_m} \quad \textbf{(3.200)}$$

3.11.2.2.4 Case 4, θ_p and θ_m Unknown

To solve for θ_m, Eqs. (3.191) and (3.192) are squared, added together, and simplified with the aid of Eqs. (3.188)–(3.190) to give a quadratic equation in the variable tan(θ_m/2). The resulting solution is

$$\theta_m = 2\tan^{-1}\left(\frac{-C_4 \pm \sqrt{C_4^2 - 4C_3C_5}}{2C_3}\right) \quad \textbf{(3.201)}$$

where

$$C_3 = r_p^2 - C_1^2 - C_2^2 - r_m^2 - 2C_1 r_m$$

$$C_4 = 4C_2 r_m$$

and

$$C_5 = r_p^2 - C_1^2 - C_2^2 - r_m^2 + 2C_1 r$$

Given θ_m, an expression for θ_p can be found by dividing Eq. (3.202) by Eq. (3.201) and simplifying

$$\theta_p = \tan^{-1}\left(\frac{C_2 - r_m \sin\theta_m}{C_1 - r_m \cos\theta_m}\right) \qquad \textbf{(3.202)}$$

Equations (3.201) and (3.202) each have two solutions corresponding to the two assembly modes of this part of the linkage. As with case 2, the proper assembly mode must be identified before the analysis can be conducted.

3.12 CLOSURE EQUATIONS FOR MECHANISMS WITH HIGHER PAIRS

The closure equation approach can also be used for mechanisms with higher pairs if we use the centers of curvature of the contact surfaces corresponding to the contact points. This is exactly the approach employed when equivalent mechanisms are used, and, in fact, we could represent the higher pair mechanisms by their equivalent lower-pair mechanism and determine the kinematic properties by analyzing the corresponding lower-pair mechanism. By using the centers of curvature, however, we can approach the problem without using equivalent mechanisms directly.

The approach using centers of curvature can be applied directly to mechanisms with cam joints and to mechanisms with rolling joints if the contact points (and the corresponding centers of curvature) are known. With rolling contact, locating the contact point as a function of the input motion requires that we know the initial contact point when the mechanism begins to move. Subsequent contact points are then located by enforcing the constraint that there is no slipping at the contacting surfaces. If circle arcs are involved, the resulting constraint equations are simple; however, if general surfaces are involved, the constraint equations require that the arc length on each contacting surface be determined by integration. For simplicity, we will limit the discussions here to cases in which the contact point either is known or can be determined simply.

For higher pair mechanisms, the vector closure diagrams are set up using the same procedure as would be used when the mechanism is drawn. In general, the same points and vectors will be used. The procedure will be illustrated with three examples.

EXAMPLE 3.12 ***(Kinematic Analysis of Mechanism with Cam Contact)***

PROBLEM

In the mechanism shown in Fig. 3.52, ${}^1\boldsymbol{\omega}_2 = 10$ rad/s and is constant. Determine ${}^1\boldsymbol{v}_{C_3/C_2}$, ${}^1\boldsymbol{v}_{C_3}$, ${}^1\boldsymbol{a}_{C_3/C_2}$, and ${}^1\boldsymbol{a}_{C_3}$ using vector closure equations.

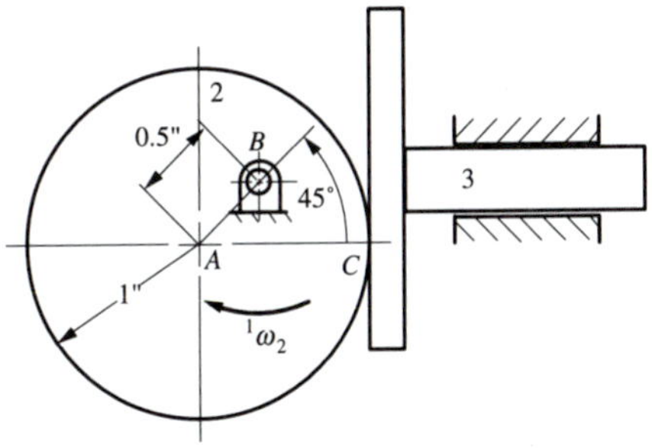

Figure 3.52 Figure for Example 3.12.

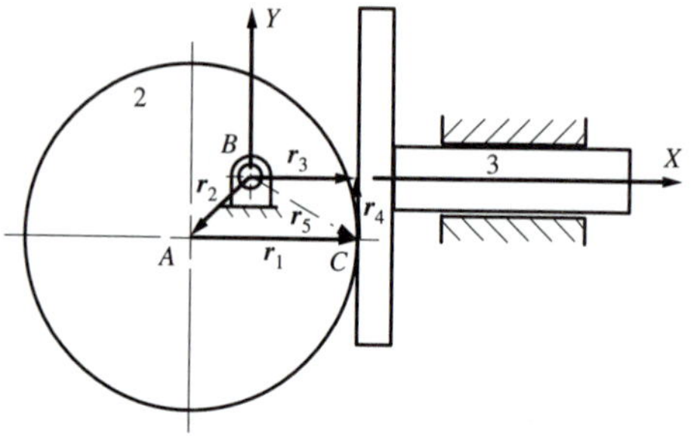

Figure 3.53 Vector closure for Example 3.12.

To solve the problem, set up four vectors as shown in Fig. 3.53. The vector $\boldsymbol{r}_2$ is from point B to point A, the center of curvature of link 2 corresponding to the contact point at C. Vector $\boldsymbol{r}_1$ is from point A to point C, the contact location. The vector $\boldsymbol{r}_1$ is constant in both direction and magnitude. Vector $\boldsymbol{r}_3$ is from point B to the face of the cam follower. The direction of $\boldsymbol{r}_3$ is constant, but the magnitude varies. Because $\boldsymbol{r}_3$ is measured from a fixed point on the frame (point B) to the face of the cam follower, the first and second derivatives of $\boldsymbol{r}_3$ correspond to the velocity and acceleration, respectively, of the cam follower. Vector $\boldsymbol{r}_4$ is measured from the contact point to a line through B and in the direction of travel of the cam.

The vector closure equation for the mechanism is

$$\boldsymbol{r}_3 = \boldsymbol{r}_2 + \boldsymbol{r}_1 + \boldsymbol{r}_4 \tag{3.203}$$

and the corresponding velocity and acceleration expressions are given by

$$\dot{\boldsymbol{r}}_3 = \dot{\boldsymbol{r}}_2 + \dot{\boldsymbol{r}}_1 + \dot{\boldsymbol{r}}_4 \tag{3.204}$$

and

$$\ddot{\boldsymbol{r}}_3 = \ddot{\boldsymbol{r}}_2 + \ddot{\boldsymbol{r}}_1 + \ddot{\boldsymbol{r}}_4 \tag{3.205}$$

Before actually solving the equations, we can summarize the variables that are known and unknown. These are

$r_1 = 1.0$ in	$\theta_1 = 0°$	$\dot{\theta}_1 = 0$ rad/s	$\ddot{\theta}_1 = 0$
$r_2 = 0.5$ in	$\theta_2 = 225°$	$\dot{\theta}_2 = -10$ rad/s	$\ddot{\theta}_2 = 0$
$r_3 = ?$	$\theta_3 = 0°$	$\dot{\theta}_3 = 0$ rad/s	$\ddot{\theta}_3 = 0$
$r_4 = ?$	$\theta_4 = 90°$	$\dot{\theta}_4 = 0$ rad/s	$\ddot{\theta}_4 = 0$

As in the cases of lower-pair mechanisms, the position equation must be solved first. The resulting linear velocity and acceleration equations can then be solved easily. Rewriting the position closure equation in component form gives

$$r_3 \cos\theta_3 = r_2 \cos\theta_2 + r_1 \cos\theta_1 + r_4 \cos\theta_4$$
$$r_3 \sin\theta_3 = r_2 \sin\theta_2 + r_1 \sin\theta_1 + r_4 \sin\theta_4$$

Simplifying based on the input values,

$$\begin{aligned} r_3 &= r_2 \cos\theta_2 + r_1 \\ 0 &= r_2 \sin\theta_2 + r_4 \end{aligned} \tag{3.206}$$

Equations (3.206) are linear in the unknowns (r_3 and r_4) and can easily be solved. The results are

$$r_3 = r_2 \cos\theta_2 + r_1 = 0.5 \cos(225) + 1.0 = 0.646 \text{ in}$$

and

$$r_4 = -r_2 \sin \theta_2 = 0.354 \text{ in}$$

To conduct the velocity analysis, rewrite Eq. (3.204) in component form and simplify or differentiate Eqs. (3.206) and simplify. In either case, the results are

$$\begin{aligned} \dot{r}_3 &= -r_2\dot{\theta}_2 \sin \theta_2 \\ \dot{r}_4 &= -r_2\dot{\theta}_2 \cos \theta_2 \end{aligned} \qquad \textbf{(3.207)}$$

Substituting in the known values,

$$\begin{aligned} \dot{r}_3 &= -r_2\dot{\theta}_2 \sin \theta_2 = -0.5(-10)\sin(225) = -3.535 \text{ in/s} \\ \dot{r}_4 &= -r_2\dot{\theta}_2 \cos \theta_2 = -0.5(-10)\cos(225) = -3.535 \text{ in/s} \end{aligned}$$

The location of both C_2 and C_3 is given by $\boldsymbol{r}_5 = \boldsymbol{r}_1 + \boldsymbol{r}_2$ in Fig. 3.53. Both points momentarily have the same coordinates. However, the velocities of the corresponding points are different. To determine the velocities, we must carefully interpret the vectors. The velocity of all points on the follower is the same. Therefore, the velocity of C_3 is given by $\dot{\boldsymbol{r}}_3$ if $\boldsymbol{r}_3$ remains horizontal. The velocity of C_2 is given by the derivative of a vector fixed to link 2 and directed from point B to C. This is $\dot{\boldsymbol{r}}_5$ if we assume $\boldsymbol{r}_5$ is fixed to link 2. Then the velocity of C_2 is given by $\dot{\boldsymbol{r}}_5 = \dot{\boldsymbol{r}}_1 + \dot{\boldsymbol{r}}_2$ if we assume that both $\boldsymbol{r}_1$ and $\boldsymbol{r}_2$ are fixed to (i.e., rotate with) link 2. Then the components of the velocity of C_2 will be given by

$$\boldsymbol{v}_{C_2} = \dot{\boldsymbol{r}}_5 = \dot{\boldsymbol{r}}_1 + \dot{\boldsymbol{r}}_2 = (-r_1\dot{\theta}_2 \sin \theta_1 - r_2\dot{\theta}_2 \sin \theta_2)\boldsymbol{i} + (r_1\dot{\theta}_2 \cos \theta_1 + r_2\dot{\theta}_2 \cos \theta_2)\boldsymbol{j}$$

The relative velocity is given by

$${}^1\boldsymbol{v}_{C_3/C_2} = {}^1\boldsymbol{v}_{C_3} - {}^1\boldsymbol{v}_{C_2} = (\dot{r}_3 + r_1\dot{\theta}_2 \sin \theta_1 + r_2\dot{\theta}_2 \sin \theta_2)\boldsymbol{i} - (r_1\dot{\theta}_2 \cos \theta_1 + r_2\dot{\theta}_2 \cos \theta_2)\boldsymbol{j}$$

Substituting values for the variables,

$${}^1\boldsymbol{v}_{C_3/C_2} = {}^1\boldsymbol{v}_{C_3} - {}^1\boldsymbol{v}_{C_2} = [0]_i - [1.0(-10)\cos(0) + 0.5(-10)\cos(225)]\boldsymbol{j} = 6.464\boldsymbol{j} \text{ in/s}$$

For the acceleration analysis, differentiate Eqs. (3.207) and simplify. Then,

$$\begin{aligned} \ddot{r}_3 &= -r_2\ddot{\theta}_2 \sin \theta_2 - r_2\dot{\theta}_2^2 \cos \theta_2 \\ \ddot{r}_4 &= -r_2\ddot{\theta}_2 \cos \theta_2 + r_2\dot{\theta}_2^2 \sin \theta_2 \end{aligned} \qquad \textbf{(3.208)}$$

Substituting in the known values,

$$\begin{aligned} \ddot{r}_3 &= -r_2\ddot{\theta}_2 \sin \theta_2 - r_2\dot{\theta}_2^2 \cos \theta_2 = 0 - 0.5(-10)^2 \cos(225) = 35.35 \text{ in/s}^2 \\ \ddot{r}_4 &= -r_2\ddot{\theta}_2 \cos \theta_2 + r_2\dot{\theta}_2^2 \sin \theta_2 = 0 + 0.5(-10)^2 \sin(225) = -35.35 \text{ in/s}^2 \end{aligned}$$

Finally,

$$\begin{aligned} \boldsymbol{a}_{C_2} = \ddot{\boldsymbol{r}}_5 = \ddot{\boldsymbol{r}}_1 + \ddot{\boldsymbol{r}}_2 &= (-r_1\ddot{\theta}_2 \sin \theta_1 - r_2\ddot{\theta}_2 \sin \theta_2 - r_1\dot{\theta}_2^2 \cos \theta_1 - r_2\dot{\theta}_2^2 \cos \theta_2)\boldsymbol{i} \\ &+ (r_1\ddot{\theta}_2 \cos \theta_1 + r_2\ddot{\theta}_2 \cos \theta_2 - r_1\dot{\theta}_2^2 \sin \theta_1 - r_2\dot{\theta}_2^2 \sin \theta_2)\boldsymbol{j} \\ &= [0 - 0 - 1(-10)^2 - 0.5(-10)^2 \cos(225)]\boldsymbol{i} + [0 + 0 - 0 - 0.5(-10)^2 \sin(225)]\boldsymbol{j} \\ &= [-100 + 35.35]\boldsymbol{i} + [35.35]\boldsymbol{j} = -64.65\boldsymbol{i} + 35.35\boldsymbol{j} \end{aligned}$$

and

$${}^1\boldsymbol{a}_{C_3/C_2} = {}^1\boldsymbol{a}_{C_3} - {}^1\boldsymbol{a}_{C_2} = [35.35 + 64.65]\boldsymbol{i} - 35.35\boldsymbol{j} = 100\boldsymbol{i} - 35.35\boldsymbol{j} \text{ in/s}^2$$

■

EXAMPLE 3.13 ***(Kinematic Analysis of a Mechanism with a Pin-in-Slot Joint)***

PROBLEM

For the mechanism shown in Fig. 3.54, find ${}^1\boldsymbol{\omega}_3$ and ${}^1\boldsymbol{\alpha}_3$ if $\theta_2 = 100°$ and ${}^1\boldsymbol{\omega}_2 = 50$ rad/s CCW and is constant.

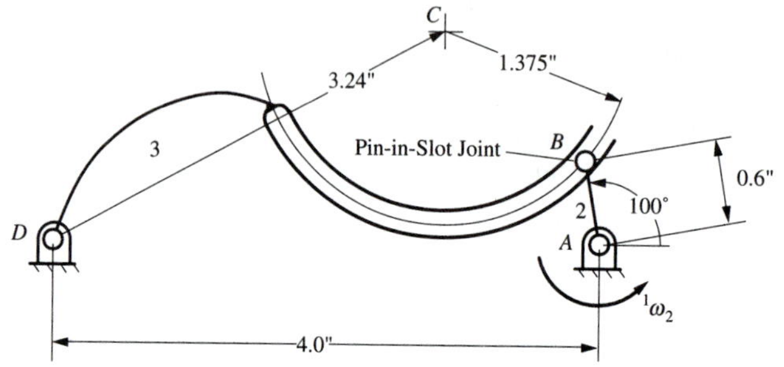

Figure 3.54

SOLUTION

To solve the problem, set up four vectors as shown in Fig. 3.55. The vector $\mathbf{r}_1$ is from point A to D, and $\mathbf{r}_2$ is from point A to B. The other two vectors involve the center of curvature, C, of the path that point B_2 traces on link 3. Vector $\mathbf{r}_3$ is from point B to C, and $\mathbf{r}_4$ is from point D to C. Both points D and C are fixed to link 3; therefore, $\mathbf{r}_4$ is fixed to link 3. All of the vectors have constant lengths. The unknown angles are θ_3 and θ_4 because θ_1 is fixed and θ_2 is the known input angle. The known and unknown information can be summarized as follows:

$r_1 = 4.0$ in	$\theta_1 = 180°$	$\dot{\theta}_1 = 0$ rad/s	$\ddot{\theta}_1 = 0$ rad/s^2
$r_2 = 0.6$ in	$\theta_2 = 100°$	$\dot{\theta}_2 = 50$ rad/s	$\ddot{\theta}_2 = 0$ rad/s^2
$r_3 = 1.375$ in	$\theta_3 = ?$	$\dot{\theta}_3 = ?$	$\ddot{\theta}_3 = ?$
$r_4 = 3.24$ in	$\theta_4 = ?$	$\dot{\theta}_4 = ?$	$\ddot{\theta}_4 = ?$

Based on Fig. 3.55, the vector closure equation for this mechanism is

$$\mathbf{r}_2 + \mathbf{r}_3 = \mathbf{r}_1 + \mathbf{r}_4 \tag{3.209}$$

This equation is exactly the same as that for a four-bar linkage (Eq. 3.24). Therefore, the equations developed for a four-bar linkage and summarized in Table 3.1 can be applied directly to this example. The results are

$\theta_3 = 138.31°$	$\dot{\theta}_3 = -22.21$ rad/s	$\ddot{\theta}_3 = -73.77$ rad/s^2
$\theta_4 = 27.69°$	$\dot{\theta}_4 = 6.133$ rad/s	$\ddot{\theta}_4 = -625.98$ rad/s^2

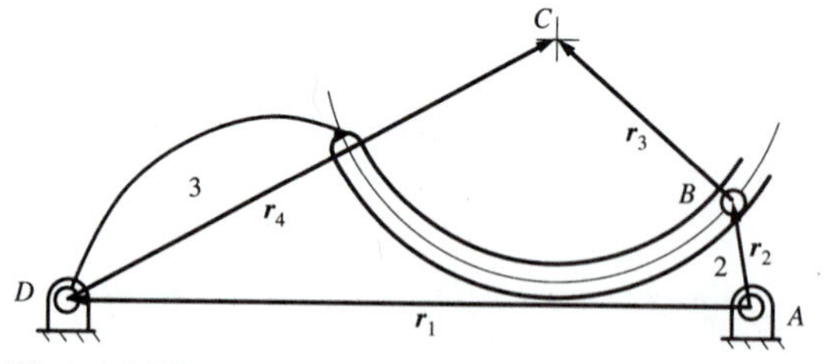

Figure 3.55

In the mechanism, vector $\boldsymbol{r}_4$ is fixed to link 3. Therefore,

$$^1\omega_3 = 6.133 \text{ rad/s (CCW)} \quad \text{and} \quad ^1\alpha_3 = -625.98 \text{ rad/s}^2 \text{ (CW)}$$ ■

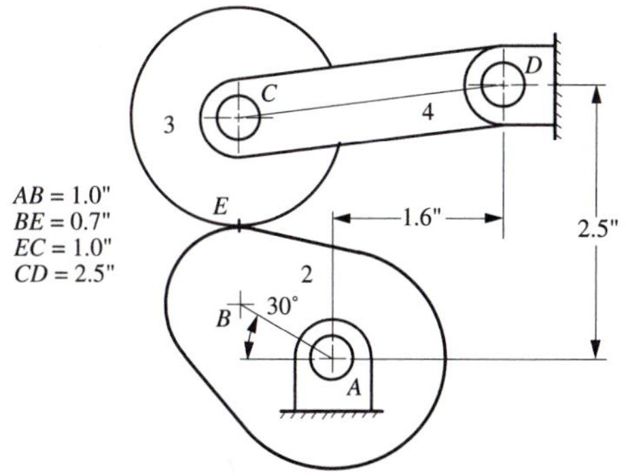

Figure 3.56

EXAMPLE 3.14 ***(Kinematic Analysis of Mechanism with Rolling Contact)***

PROBLEM

In the mechanism shown in Fig. 3.56, link 2 is turning with a constant angular velocity of 200 rpm CCW. Determine the angular velocity and acceleration of link 4.

SOLUTION

This mechanism involves rolling contact at point *E*. It is relatively straightforward to determine the angular quantities associated with link 4 if we locate the vectors for the closure equations using the centers of curvature of links 2 and 3 corresponding to the contact location *E*. This approach will not yield any angular information for link 3, however. In fact, the velocity and acceleration of link 4 are the same whether there is rolling or slipping at *E*.

The vector closure diagram is given in Fig. 3.57. The vector $\boldsymbol{r}_1$ is from point *A* to *D*, and $\boldsymbol{r}_2$ is from point *A* to *B*. The other two vectors involve the center of curvature, *C*, of the path that point B_2 traces on link 3. Vector $\boldsymbol{r}_3$ is from point *B* to *C*, the centers of curvature corresponding to E_2 and E_3, respectively, and $\boldsymbol{r}_4$ is from point *D* to *C*. Note that the points *D* and *C* of interest are those fixed to link 4. Therefore, $\boldsymbol{r}_4$ can be treated as a vector fixed to link 4.

All of the vectors have constant lengths. The unknown angles are θ_3 and θ_4 because θ_1 is fixed and θ_2 is the known input angle. The angle θ_1 can be computed from

$$\theta_1 = \tan^{-1}\left[\frac{2.5}{1.6}\right] = 57.38°$$

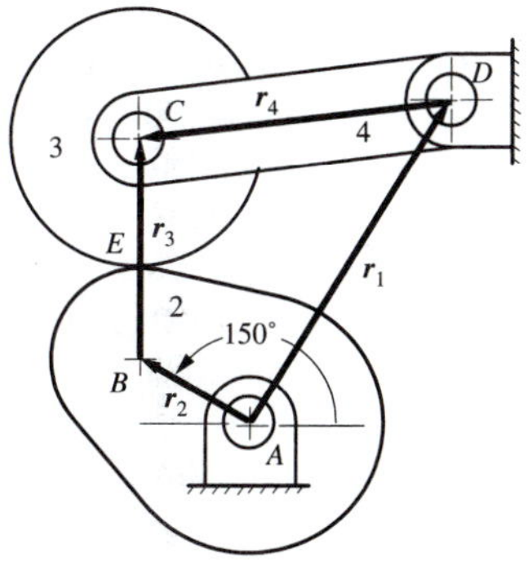

Figure 3.57

The known and unknown information can be summarized as follows:

$r_1 = \sqrt{2.5^2 + 1.6^2} = 2.968$ in $\quad \theta_1 = 57.38°\quad \dot{\theta}_1 = 0$ rad/s $\quad \ddot{\theta}_1 = 0$

$r_2 = 1.0$ in $\quad \theta_2 = 150° \quad \dot{\theta}_2 = 200\left(\frac{2\pi}{60}\right) = 20.94$ rad/s $\quad \ddot{\theta} = 0$

$r_3 = (0.7 + 1.0) = 1.7$ in $\quad \theta_3 = ? \quad \dot{\theta}_3 = ? \quad \ddot{\theta}_3 = ?$

$r_4 = 2.5$ in $\quad \theta_4 = ? \quad \dot{\theta}_4 = ? \quad \ddot{\theta}_4 = ?$

Based on Fig. 3.57, the vector closure equation for this mechanism is

$$\boldsymbol{r}_2 + \boldsymbol{r}_3 = \boldsymbol{r}_1 + \boldsymbol{r}_4$$

This equation is again exactly the same as for a four-bar linkage (Eq. 3.24). Therefore, the equations developed for a four-bar linkage and summarized in Table 3.1 can again be applied directly to this example. The results are

$$\theta_3 = 138.31° \quad \dot{\theta}_3 = -22.21 \text{ rad/s} \quad \ddot{\theta}_3 = -73.77 \text{ rad/s}^2$$

$$\theta_4 = 27.69° \quad \dot{\theta}_4 = 6.133 \text{ rad/s} \quad \ddot{\theta}_4 = -625.98 \text{ rad/s}^2$$

In the mechanism, vector $\boldsymbol{r}_4$ is fixed to link 3. Therefore, ${}^1\boldsymbol{\omega}_4 = 6.133$ rad/s (CCW) and ${}^1\boldsymbol{\alpha}_4 = -625.98$ rad/s^2 (CW).

3.13 NOTATIONAL DIFFERENCES: VECTORS AND COMPLEX NUMBERS

Several different notations are in widespread use for analytical solution of planar kinematic problems. The two principal notations are based on vectors and complex numbers. It is the purpose of this section to compare these two notations. In principle, they are completely equivalent to one another, with every relationship written in one notation directly translatable to the other. Nevertheless, some relationships are more easily discerned when using one in preference to the other. Broadly speaking, the complex number notation tends to be most compatible with relationships that are most naturally expressed in polar coordinates. This includes most relationships describing the instantaneous motion state of a rigid body. These relationships are usually most compactly expressed in complex notation. Vector notation is, again broadly speaking, most compatible with relationships that are most naturally expressed in Cartesian coordinates. This is usually true whenever there is no single point that dominates the geometry of the system. In the opinion of the authors of this book, this includes the majority of situations to be studied. Also, planar vector notation is fully compatible with the corresponding techniques used for three-dimensional representation. Therefore, if only one notation is to be used, it should be the vector notation. For this reason, this text is based on the use of vector notation. Of course, advanced students of the subject should seek proficiency in both types of notation.

In complex number notation, planar vector quantities are represented by identifying the real and imaginary parts with orthogonal components. Normally, the x component is represented by the real part and the y component by the imaginary part. That is, the complex number

$$z = x + iy$$

represents the vector (x, y). An important alternative form for z is

$$z = re^{i\theta} = r(\cos\theta + i\sin\theta) \tag{3.209}$$

where r is the length of the vector and θ is its direction relative to the x axis. That is,

$$r = \sqrt{x^2 + y^2}$$

and

$$\theta = \tan^{-1}(y/x)$$

It is this form that is effective in expressing polar relationships.

Referring to Fig. 3.58, we can write the basic closure equation for the four-bar linkage with the vectors $\boldsymbol{a}$, $\boldsymbol{b}$, $\boldsymbol{c}$, and $\boldsymbol{d}$ interpreted as a complex numbers. Then

$$\boldsymbol{b} + \boldsymbol{c} = \boldsymbol{a} + \boldsymbol{d}$$

Using the form of Eq. (3.209), this can be written

$$be^{i\theta} + ce^{i\psi} = a + de^{i\phi} \tag{3.210}$$

Decomposition of this expression into its real and imaginary parts, respectively, gives

$$b \cos \theta + c \cos \psi = a + d \cos \phi$$
$$b \sin \theta + c \sin \psi = d \sin \phi$$

which are identical to the component equations developed using the vector formulation. The development of a position solution and of the velocity and acceleration solutions of this chapter is then identical to that given earlier.

Alternatively, the elimination of one of the variables may be pursued in the complex variable form. Equation (3.2.10) may be written in the form

$$\boldsymbol{b} + \boldsymbol{c} - \boldsymbol{a} = \boldsymbol{d} \tag{3.2.11}$$

Now, the conjugates of $\boldsymbol{b}$, $\boldsymbol{c}$, and $\boldsymbol{d}$ are, respectively,

$$\tilde{\boldsymbol{b}} = be^{-i\theta}, \qquad \tilde{\boldsymbol{c}} = ce^{-i\psi}, \qquad \tilde{\boldsymbol{d}} = de^{-i\phi}$$

Also, the conjugate of $\boldsymbol{a}$ is $\boldsymbol{a}$ since $\boldsymbol{a}$ is a real number. The conjugate of Eq. (3.2.10)

$$\tilde{\boldsymbol{b}} + \tilde{\boldsymbol{c}} - \tilde{\boldsymbol{a}} = \tilde{\boldsymbol{d}}$$

is also true because the process of forming the conjugate simply changes the signs on all imaginary parts.

Multiplication of each side of Eq. (3.2.11) by its conjugate gives

$$(\boldsymbol{b} + \boldsymbol{c} - \boldsymbol{a})(\tilde{\boldsymbol{b}} + \tilde{\boldsymbol{c}} - \tilde{\boldsymbol{a}}) = \boldsymbol{d}\tilde{\boldsymbol{d}}$$

Now, referring to Eq. (3.2.10),

$$\boldsymbol{z}\tilde{\boldsymbol{z}} = re^{i\theta}re^{-i\theta} = r^2$$

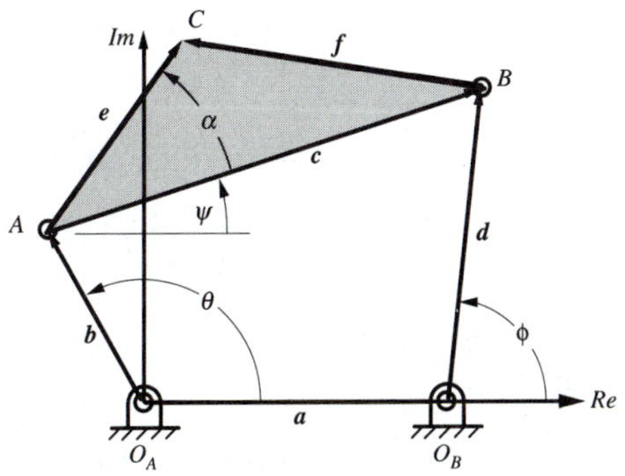

Figure 3.58

Also

$$z + \tilde{z} = re^{i\theta} + re^{-i\theta} = r(\cos\theta + i\sin\theta) + r(\cos\theta - i\sin\theta) = 2r\cos\theta$$

Hence

$$\mathbf{b}\tilde{\mathbf{b}} = b^2, \qquad \mathbf{c}\tilde{\mathbf{c}} = c^2, \qquad \mathbf{d}\tilde{\mathbf{d}} = d^2$$

Thus, expansion of the foregoing expression gives

$$\mathbf{b}\tilde{\mathbf{b}} + \mathbf{c}\tilde{\mathbf{c}} + \mathbf{a}\tilde{\mathbf{a}} + \mathbf{b}\tilde{\mathbf{c}} + \tilde{\mathbf{b}}\mathbf{c} - \mathbf{b}\tilde{\mathbf{a}} - \tilde{\mathbf{b}}\mathbf{a} - \mathbf{c}\tilde{\mathbf{a}} - \tilde{\mathbf{c}}\mathbf{a} = \mathbf{d}\tilde{\mathbf{d}}$$

or

$$b^2 + c^2 + a^2 + be^{i\theta}ce^{-i\psi} + be^{-i\theta}ce^{i\psi} - abe^{i\theta} - abe^{-i\theta} - ace^{i\psi} - ace^{-i\psi} = d^2$$

or

$$b^2 + c^2 + a^2 + bce^{i(\theta-\psi)} + bce^{-i(\theta-\psi)} - abe^{i\theta} - abe^{-i\theta} - ace^{i\psi} - ace^{-i\psi} = d^2$$

or

$$b^2 + c^2 + a^2 + 2\,bc\cos(\theta - \psi) - 2ab\cos\theta - 2ac\cos\psi = d^2$$

which is the same as the form derived earlier.

Similarly, multiplying each side of the equation

$$\mathbf{b} - \mathbf{d} - \mathbf{a} = -\mathbf{c}$$

by its conjugate gives

$$\mathbf{b}\tilde{\mathbf{b}} + \mathbf{d}\tilde{\mathbf{d}} + \mathbf{a}\tilde{\mathbf{a}} - \mathbf{b}\tilde{\mathbf{d}} - \tilde{\mathbf{b}}\mathbf{d} - \mathbf{b}\tilde{\mathbf{a}} - \tilde{\mathbf{b}}\mathbf{a} + \mathbf{d}\tilde{\mathbf{a}} + \tilde{\mathbf{d}}\mathbf{a} = \mathbf{c}\tilde{\mathbf{c}}$$

or

$$b^2 + d^2 + a^2 - bde^{i(\theta-\phi)} - bde^{-i(\theta-\phi)} - abe^{i\theta} - abe^{-i\theta} + ade^{i\phi} + ade^{-i\phi} = c^2$$

giving

$$b^2 + d^2 + a^2 - 2bd\cos(\theta - \phi) - 2ab\cos\theta + 2ad\cos\phi = c^2$$

which is also the same as the form derived earlier

Equation (3.2.10) lends itself to development of velocity and acceleration expressions. Differentiation with respect to time gives

$$ib\dot{\theta}e^{i\theta} + ic\dot{\psi}e^{i\psi} = id\dot{\phi}e^{i\phi}$$

or, removing the common factor i,

$$b\dot{\theta}e^{i\theta} + c\dot{\psi}e^{i\psi} = d\dot{\phi}e^{i\phi} \tag{3.212}$$

Separation into the real and imaginary parts gives, respectively,

$$b\dot{\theta}\cos\theta + c\dot{\psi}\cos\psi = d\dot{\phi}\cos\phi$$

and

$$b\dot{\theta}\sin\theta + c\dot{\psi}\sin\psi = d\dot{\phi}\sin\phi$$

which may be recognized as the same form as given in Table 3.1.

Differentiation of Eq. (3.212) with respect to time gives

$$b\ddot{\theta}e^{i\theta} + ib\dot{\theta}^2e^{i\theta} + c\ddot{\psi}e^{i\psi} + ic\dot{\psi}^2e^{i\psi} = d\ddot{\phi}e^{i\phi} + id\dot{\phi}^2e^{i\phi}$$

Expansion of the $e^{i\theta}$ terms gives

$$\begin{aligned} &b\ddot{\theta}(\cos\theta + i\sin\theta) + b\dot{\theta}^2(i\cos\theta - \sin\theta) + c\ddot{\psi}(\cos\psi + i\sin\psi) + c\dot{\psi}^2(i\cos\psi - \sin\psi) \\ &\quad = d\ddot{\phi}(\cos\phi + i\sin\phi) + d\dot{\phi}^2(i\cos\phi - \sin\phi) \end{aligned}$$

Hence, separation into the real and imaginary parts gives

$$b\ddot{\theta}\cos\theta - b\dot{\theta}^2\sin\theta + c\ddot{\psi}\cos\psi - c\dot{\psi}^2\sin\psi = d\ddot{\phi}\cos\phi - d\dot{\phi}^2\sin\phi$$

and

$$b\ddot{\theta}\sin\theta + b\dot{\theta}^2\cos\theta + c\ddot{\psi}\sin\psi + c\dot{\psi}^2\cos\psi = d\ddot{\phi}\sin\phi + d\dot{\phi}^2\cos\phi$$

which can be recognized as the same form as those given in Table 3.1.

The foregoing illustrates the equivalence of the vector and complex-number representations for simple planar mechanisms.

3.14 CHAPTER 3 *Exercise Problems*

PROBLEM 3.1

For the mechanism shown, do the following:

(a) Write the vector equation of the above linkage.
(b) Write the x and y displacement equations.
(c) Find the velocity component equations.
(d) Find the acceleration component equations.

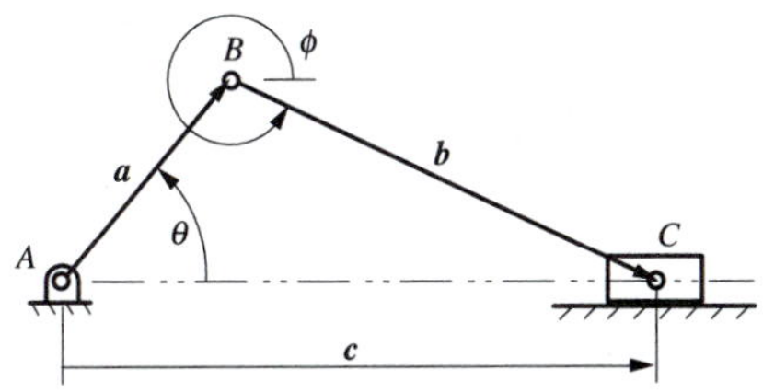

PROBLEM 3.2

In the mechanism given, determine $\dot{\phi}$ *analytically*

AB = 1 cm, BC = 4 cm, $\phi = 12.50°$, $\theta = 60°$, $\dot{\theta} = 10$ rad/s

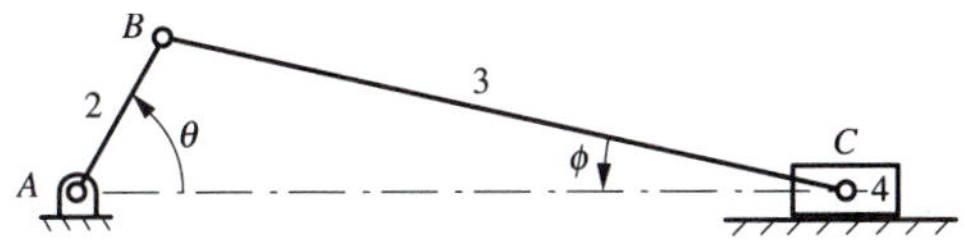

PROBLEM 3.3

In the mechanism shown, $\dot{s} = -10$ in/s and $\ddot{s} = 0$ for the position corresponding to $\phi = 60°$. Find $\dot{\phi}$ and $\ddot{\phi}$ for that position using the loop equation approach.

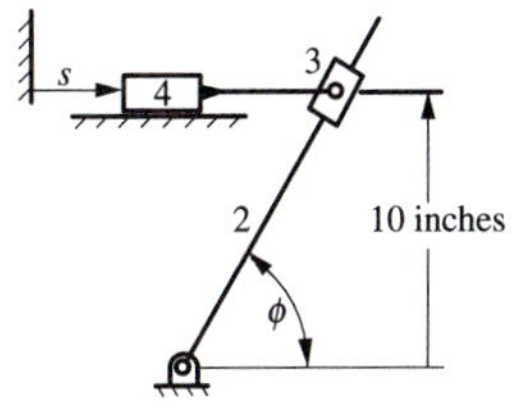

PROBLEM 3.4

In the mechanism given, point A is moving to the right with a velocity of 10 cm/s. Use the loop equation approach to determine the angular velocity of link 3. Link 3 is 10 cm long, and ϕ is 120° in the positions shown.

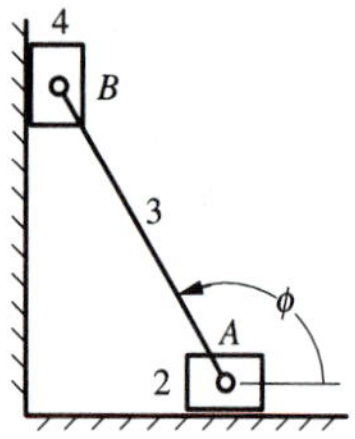

PROBLEM 3.5

The mechanism shown is a marine steering gear called Raphson's slide. *AB* is the tiller, and *CD* is the actuating rod. If the velocity of rod *CD* is a constant 10 inches per minute to the right, use the loop-equation approach to determine the angular acceleration of the tiller.

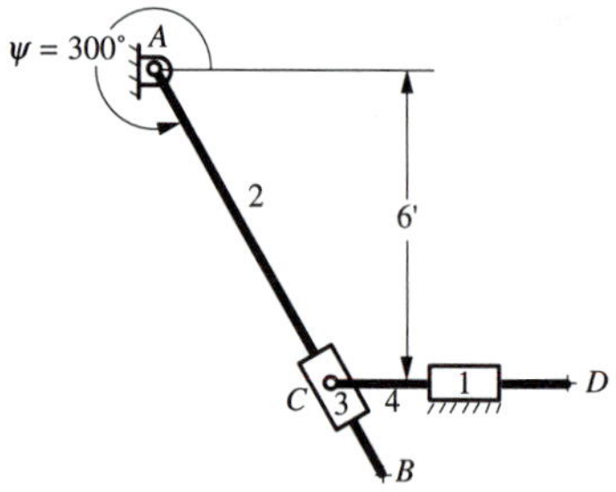

PROBLEM 3.6

Use loop equations to determine the velocity and acceleration of point *B* on link 2 when $\theta_3 = 30°$. Make point *A* the origin of your reference coordinate system.

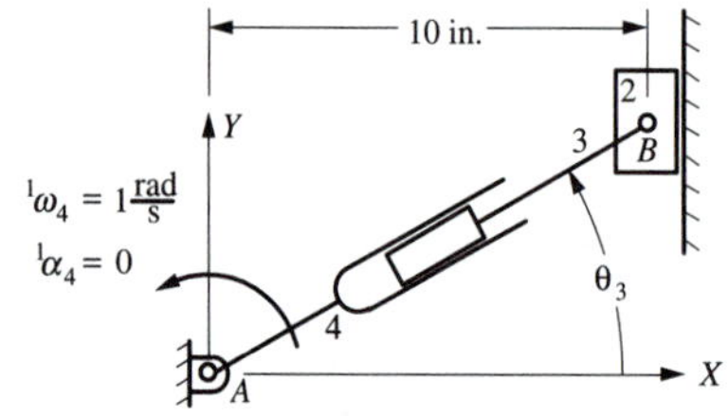

PROBLEM 3.7

For the mechanism in the position shown, link 2 is the driver and rotates with a constant angular velocity of 100 rad/s CCW. Write vector loop equations for position, velocity, and acceleration, and solve for the velocity and acceleration of point *C* on link 4.

$AB = 0.9''$, $AD = 1.7''$, $BC = 2.6''$, $h = 0.8''$, $\theta_1 = 6°$, $\phi = 120°$

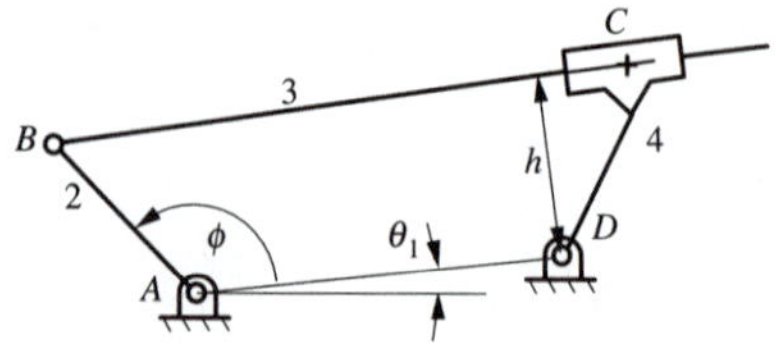

PROBLEM 3.8

For the mechanism in the position shown, link 2 is the driver and rotates with a constant angular velocity of 50 rad/s CCW. Write vector loop equations for position, velocity, and acceleration, and solve for the velocity and acceleration of point C on link 3.

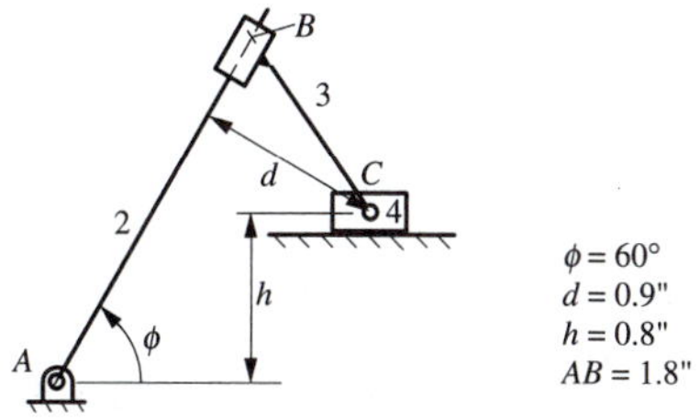

PROBLEM 3.9

In the mechanism in shown, link 3 slides on link 2, and link 4 is pinned to link 3 and slides on the frame. If ${}^1\boldsymbol{\omega}_2$ = 10 rad/s CCW (constant), use loop equations to find the acceleration of link 4 for the position defined by $\phi = 90°$.

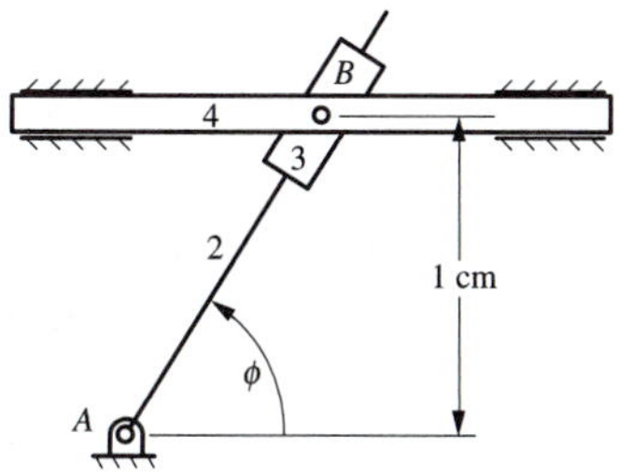

PROBLEM 3.10

For the mechanism in the position shown, the cam (link 2) rotates with an angular velocity of 200 rad/s. Write the vector loop equations for position, velocity, and acceleration and determine the angular velocity and acceleration of the follower (link 3). Use $\phi = 60°$ and neglect the follower thickness (i.e., assume that it is zero).

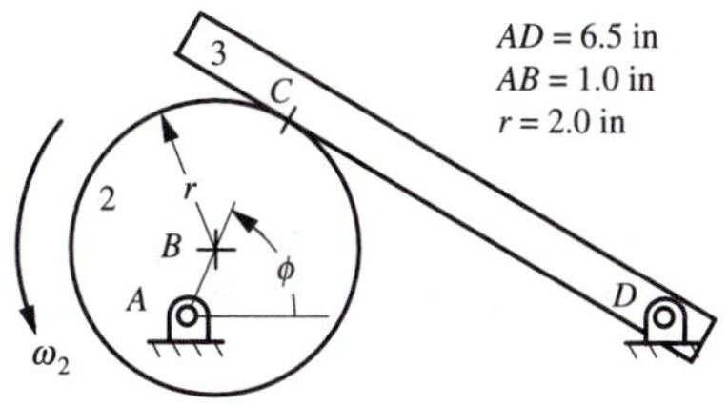

PROBLEM 3.11

In the mechanism shown, link 3 is perpendicular to link 2. Write the vector loop equations for position and velocity. If the angular velocity of link 2 is 100 rad/s CCW, use the vector loop equations to solve for the velocity of point C_4 for the position corresponding to $\phi = 60°$.

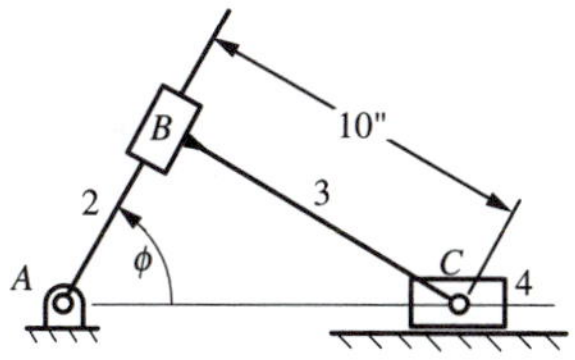

PROBLEM 3.12

In the simple, two-link mechanism given, ${}^1\mathbf{v}_{B_2}$ is 10 in/s to the right. Use the loop equation approach to determine ${}^1\mathbf{v}_{A_2}$ and ${}^1\boldsymbol{\omega}_2$.

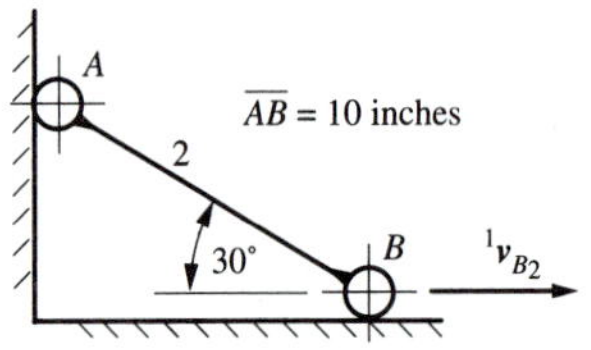

PROBLEM 3.13

In the mechanism below, the angular velocity of link 2 is 100 rad/s CCW and the dimensions of various links are given. Use loop equations to find the position and velocity of point D on link 3 when θ_2 is 90°.

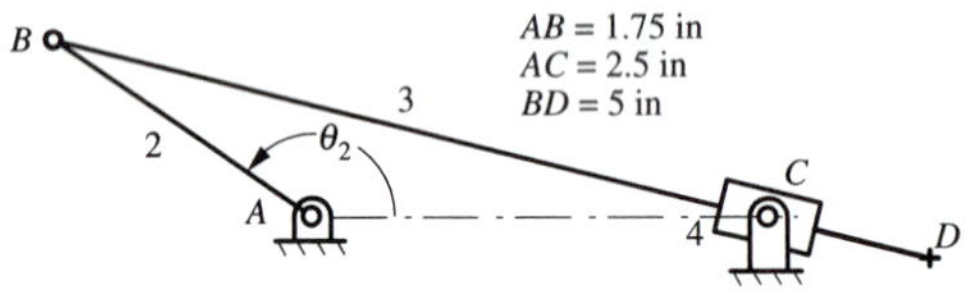

PROBLEM 3.14

In the Scotch yoke mechanism shown, ${}^1\boldsymbol{\omega}_2 = 10$ rad/s, ${}^1\boldsymbol{\alpha}_2 = 100$ rad/s^2, and $\theta_2 = 60°$. Also, length $O_2A = 20$ inches. Determine ${}^1\mathbf{v}_{A_4}$ and ${}^1\mathbf{a}_{A_4}$ using loop equations.

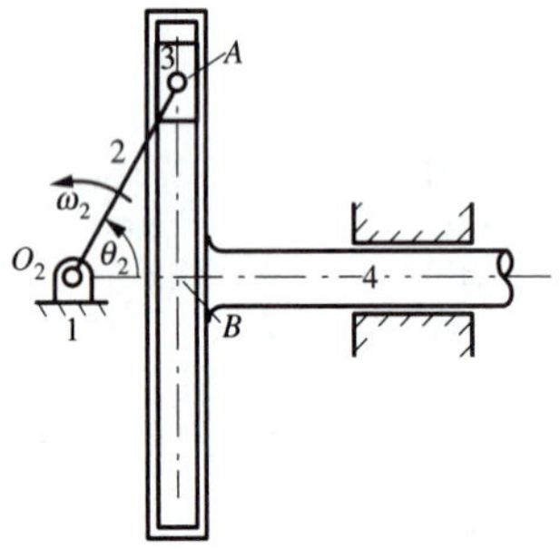

PROBLEM 3.15

Use loop equations to determine the velocity and acceleration of point B on link 4. The angular velocity of link 2 is constant at 10 rad/s counterclockwise.

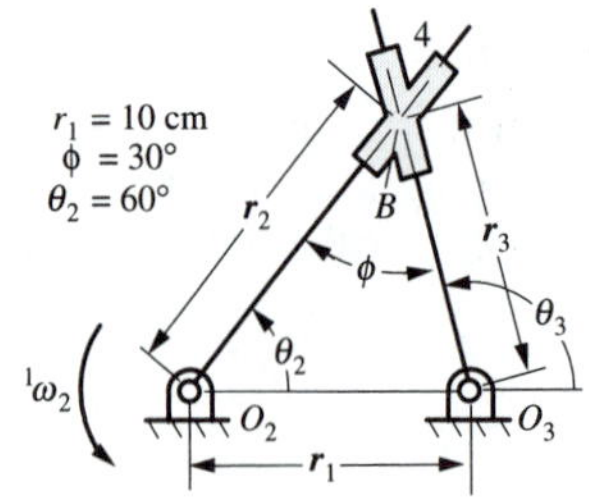

PROBLEM 3.16

The oscillating fan shown below is to be analyzed as a double rocker. The fan is link 2, the motor shaft is connected to link 3, and link 4 is connected from the coupler to the frame. The actual input of the mechanism is the coupler and ${}^2\omega_3$, which is a constant 956 (rad/s) in the counterclockwise direction. Compute the angular velocity and angular acceleration of link 2 if $\theta = 120°$, $AD = 0.75$ in, $AB = DC = 3.0$ in, $BC = 0.50$ in.

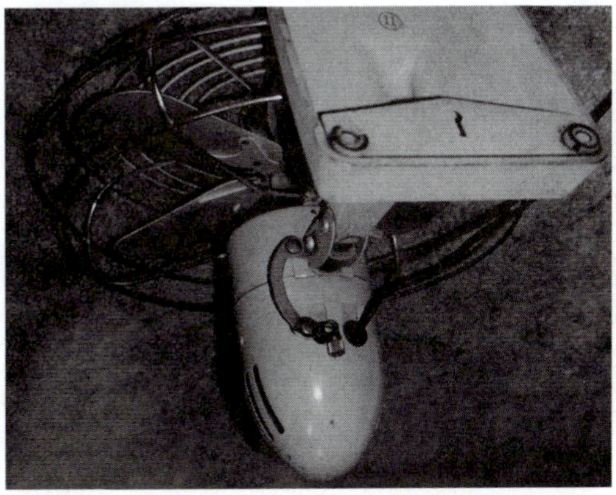

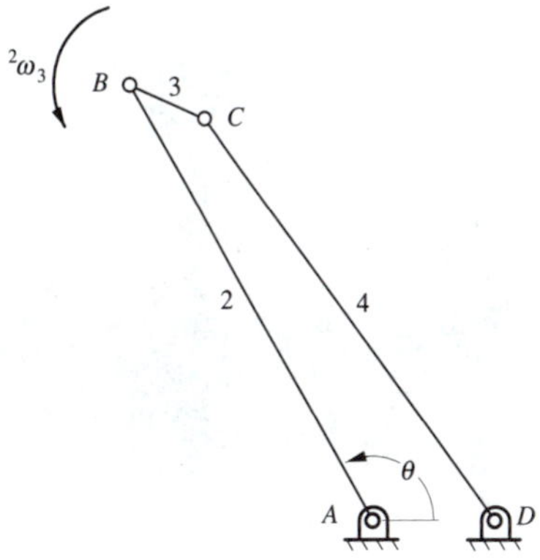

PROBLEM 3.17

The rear motorcycle suspension can be analyzed as an inverted slider-crank mechanism. The frame of the motorcycle is link 1, the tire assembly is attached to link 2 at point C. The shock absorber is links 3 and 4. As the bicycle goes over a bump in the position shown, the angular velocity of link 2 relative to the frame is ${}^1\omega_2$ is 5 (rad/s), and the angular acceleration is ${}^1\alpha_2$ is 45 (rad/s^2), both in the clockwise direction. Compute the angular velocity and angular acceleration of link 3 for the position defined by $\theta = 187°$.

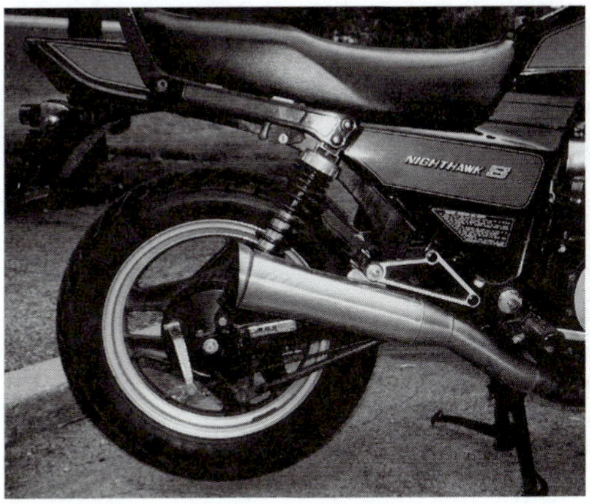

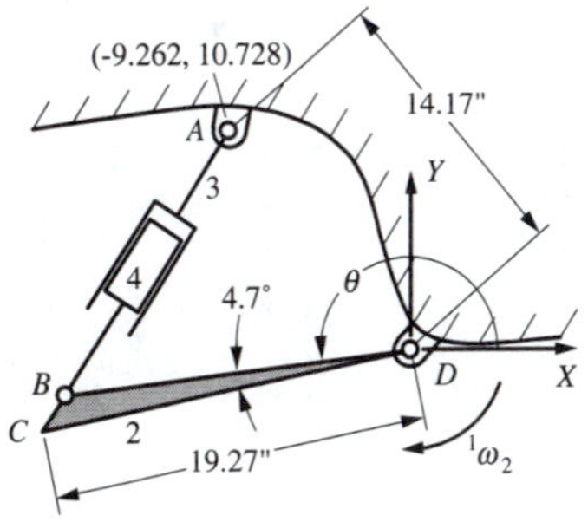

PROBLEM 3.18

The door-closing linkage shown is to be analyzed as a slider-crank linkage. Link 2 is the door, and links 3 and 4 are the two links of the door closer. Assume that the angular velocity of the door (link 2) is a constant at 3.71 radians per second clockwise. Compute the angular velocity and angular acceleration of link 4 if the dimensions are as follows:

Coordinates of D $(-25, -3.0)$
$AB = 17.0$ inches

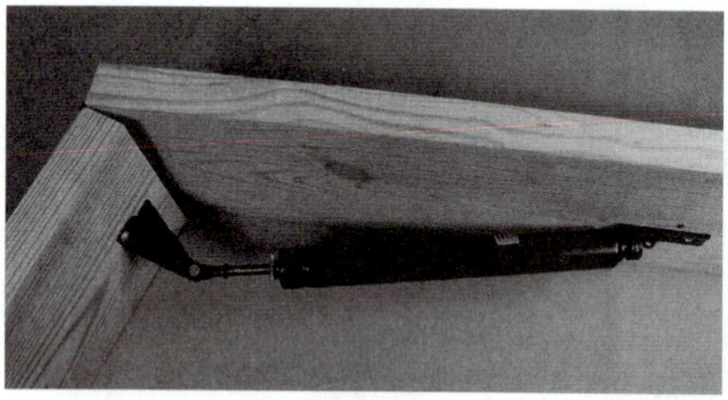

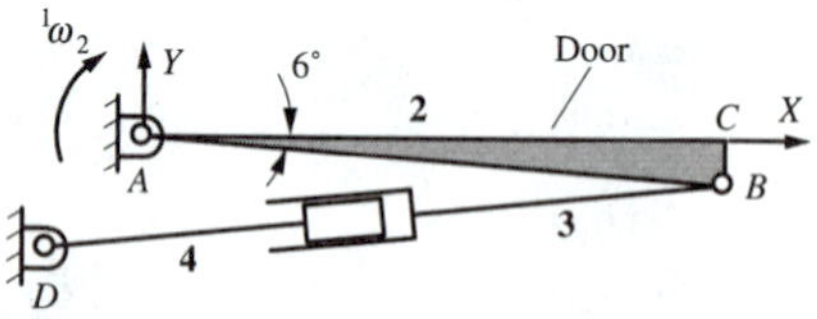

PROBLEM 3.19

The general action of a person who is doing pushups can be modeled as a four-bar linkage as shown below. The floor is the base link, and link 4 is the back and legs. Link 2 is the forearm, and link 3 is the upper arm. For the purposes of analysis, the motion that is controlled is the motion of link 3 relative to link 2 (elbow joint). Assume that $^2\omega_3$ is a constant 6.0 rad/s in the counterclockwise direction. Compute the angular velocity and angular acceleration of link 4 if link 2 is oriented at 45° to the horizontal.

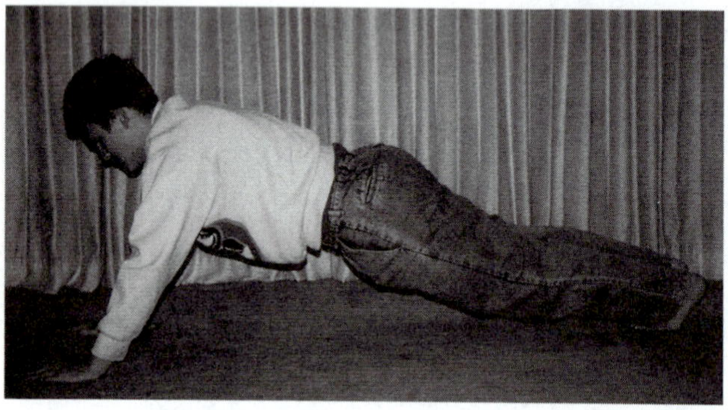

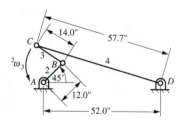

PROBLEM 3.20

A carousel mechanism can be modeled as an inverted slider-crank mechanism as shown. Point D is the location of the saddle on the horse. Assume that the angular velocity of the driver (link 2) is a constant 2 rad/s counterclockwise. Compute the velocity and acceleration of D_3 in the position shown if AB = 8.0 in, BC = 96.0 in, and BD = 54 in.

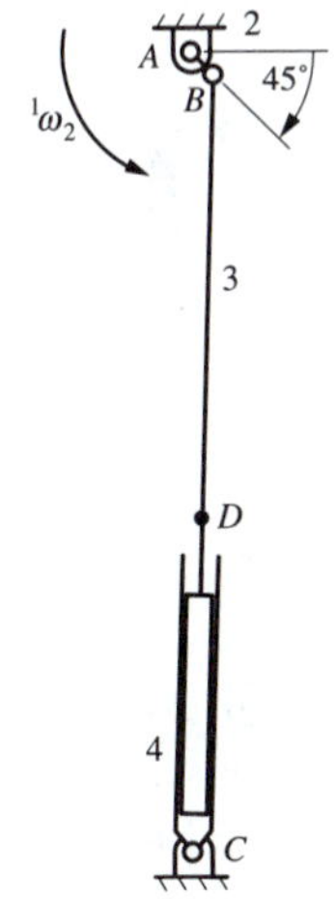

Chapter 4

Planar Linkage Design

4.1 INTRODUCTION

The machine designer is often called upon to provide a means of generating an irregular motion. For our purposes, an irregular motion can be regarded as anything except either uniform rotation about a fixed axis or uniform rectilinear translation. The means of generating irregular motions by means of one-degree-of-freedom mechanisms are two: cams and linkages. As irregular motion generators, they each have advantages and disadvantages. In general, cams are easily designed but are relatively difficult, and therefore expensive, to manufacture. They are also relatively unreliable due to wear problems. Linkages are difficult to design but are inexpensive to manufacture and relatively reliable. The subject of this chapter is linkage design.

One naturally attempts to use the simplest mechanism capable of performing the desired function. For this reason, four-link mechanisms are by far the most widely used. The techniques used for the design of five- and six-bar mechanisms are basically extensions of those used for four-link mechanisms. Thus, the primary emphasis of this chapter will be on four-link mechanism synthesis.

The joints most commonly used in mechanisms are those in which the joint constraints are provided by two surfaces in contact, which, ideally, occurs over an area. This is as opposed to point or line contact as is used in cams and gears. Surface contact is desirable from the point of view of lubrication and wear resistance. The only surface contact, or lower pair, joints that are available for use in planar mechanisms are hinges and prismatic slides. There are, therefore, four possible basic types of four-link mechanisms with surface contact joints.

1. ***The four-bar linkage.*** In this linkage, all four joints are hinges as shown in Fig. 4.1. This is by far the most widely used linkage for irregular motion generation.
2. ***The slider-crank (and its inversions).*** The slider-crank chain is shown schematically in Fig. 4.2. Linkages based on this chain are very commonly used to convert linear to rotary motion and vice versa. It is little used when neither a linear input nor a linear output is needed.
3. ***The elliptic trammel (and its inversions).*** The chain for the elliptic trammel is shown in Fig. 4.3. Except for the Scotch yoke and Oldham inversions, the elliptic trammel is little used because of slip-stick friction problems in the two slides. Analysis equations for two inversions of this linkage are given in Sections 3.9 and 3.10.
4. ***Rapson slide.*** A schematic diagram of the chain for the Rapson slide is shown in Fig. 4.4. There are two sliders that must be carefully designed if mechanisms based on the chain are to work properly. In practice, the Rapson slide is much less used than the four-bar linkage or slider crank because neither rotary joint can be made to rotate 360°

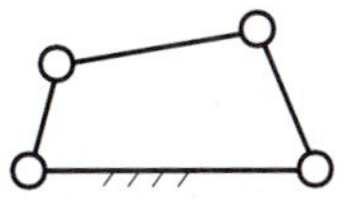

Figure 4.1 The four revolute four-bar linkage. This is one of four basic planar single-loop linkages. It is the most commonly used mechanism for generating irregular motions.

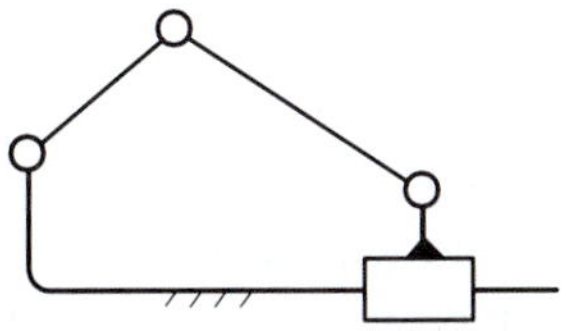

Figure 4.2 The slider-crank linkage is obtained by replacing one revolute joint in a four-bar linkage with a prismatic joint. When inverted onto the crank, or the coupler, so that the slide rotates, the linkage becomes a turning block linkage.

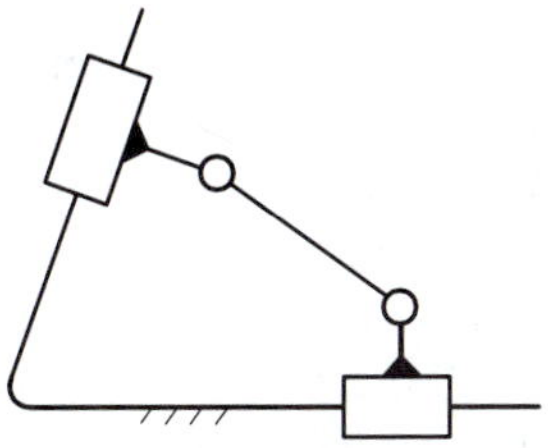

Figure 4.3 Elliptic trammel linkage. The paths of all points in the coupler are ellipses. When inverted onto one of the revolute-prismatic members, this becomes a Scotch yoke linkage. The Scotch yoke is sometimes used as a harmonic motion generator. The other possible inversion, onto the coupler, is used in practice as the Oldham coupling. This is a simple mechanism for accommodating misalignment between shafts.

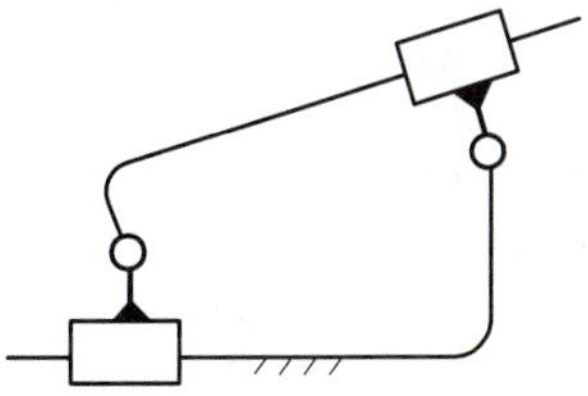

Figure 4.4 The Rapson slide linkage. Its inversions are also Rapson slide linkages.

and because of slip-stick friction in the two slides. The analysis equations for one inversion of this mechanism are given in Section 3.7.

The majority of the techniques discussed in this book are intended for four-bar linkage synthesis. This is primarily because of the large number of dimensions that can be varied, allowing more flexibility in design. Unfortunately, it also results in more complicated design techniques. When the techniques are applied to linkages having one or more slider joints, the results are somewhat simpler.

It is very rare for the desired motion to be exactly producible by a four-bar linkage. Thus, we can typically only approximate the desired motion. One approach is to select a number of positions (precision points) along the desired path and compel the linkage to move exactly through those positions. Using this method, one has no direct control over the behavior of the linkage between the design positions. One works in the (sometimes pious) hope that the linkage movement will not deviate too far from that desired between the design positions. It is, in fact, remarkable how accurate this method can be in favorable circumstances. It is possible to design a four-bar linkage for which the path of a point on the coupler deviates no more than one-thousandth of an inch from a straight line over a 10-inch line length in this way.

The types of problems most usually tackled using the precision position approach permit a graphical solution. This is straightforward for problems with two and three design positions but becomes complex and laborious for four or five design positions. Most precision position problems do not admit more than five design positions. Computer packages, such as KINSYN,[1] RECSYN,[2] and LINCAGES,[3] have been developed to automate the solution of precision position problems. Graphical techniques that are useful for small numbers of design positions will be described in this chapter. They form a basis for understanding the techniques of computer-aided synthesis required for more demanding problems.

The second basic approach is to select a rather larger number of design positions and, instead of requiring the mechanism to pass through them exactly, minimize the sum of the squares of the deviations of the mechanism position from those positions. Thus, the linkage motion approaches a rather larger number of design positions but does not exactly pass through any of them. This method makes use of numerical optimization techniques to produce solution linkages. Consequently, the use of a computer is essential. Used directly, this type of approach requires the user to manipulate the mathematical constraints to obtain control over the type and properties of the solution linkage. Some packages, such as the automatic synthesis module of RECSYN, attempt to provide that control in a more user-friendly form.

In a given problem, either of these approaches may yield the best results. The choice is most often decided by the techniques with which the designer is most familiar and by what aids, such as synthesis programs, he or she may have available.

The range of synthesis problems that arise is infinite. We will restrict our study to a few classes of problems that, because of a combination of practical importance and a well-developed theory, are most usually treated. They are:

1. ***The motion generation problem.*** A linkage is to be synthesized whose coupler, as a whole, is to follow a desired trajectory. That is, the movement of the coupler as a whole is specified, not just that of a point or line lying on it.
2. ***The function generation problem.*** In this case the angles of the two cranks are to be coordinated. The name "function generation" originated in the days in which mechanical analog computers were used to perform complex mathematical calculations in such devices as naval gunsights. Linkages were used to generate angular relationships approximating logarithms, trigonometric functions, and so forth.
3. ***The point path problem.*** A single point on the coupler is to follow a nominated curve. In this form the problem does not admit a direct graphical solution. However, this class

of problem is important from a practical point of view, and design methods will be presented. These are trial-and-error techniques starting with selection of an approximate coupler-point path from an atlas of coupler curves or from curves generated with a computer program. Simple computer programs can be important aids in the trial-and-error process.

A modified type of point path problem in which the progression of the coupler point between design positions is coordinated with the corresponding angular displacements of the driving crank does permit direct graphical solution. This is referred to as the path-angle problem type. The techniques required for solution of this type of problem are beyond the scope of this book. KINSYN, RECSYN, and LINCAGES do provide the capability for its solution.

4. ***The rocker amplitude problem.*** In the rocker amplitude problem, the output link is to oscillate through a specified angular amplitude. Typically, the required linkage is a crank rocker with continuously rotating driving crank. An oscillatory output motion of specified amplitude is required.

4.2 MOTION GENERATION

4.2.1 Introduction

Figure 4.5 shows the path of a moving lamina as described by the paths of three points embedded in it: A, B, and C. That is, A_1 is the first position of point A, A_2 is its second position, and A_3 is its third position, and similarly for points B and C. We will use this notation extensively in the following. As viewed in the moving lamina, there is only one point, A. As seen from the fixed reference frame, this point assumes three different positions, A_1, A_2, A_3, as the moving lamina moves through the three positions shown.

Actually, only the path of one point and the changes in the orientation of a line drawn on the lamina are needed to describe its motion. In order to synthesize a four-bar linkage whose coupler will approximate the given motion, we choose a number of positions on the trajectory, such as $A_1B_1C_1$, $A_2B_2C_2$, and $A_3B_3C_3$ as design positions. This is shown in Fig. 4.5. The coupler will be made to pass through these positions precisely. Depending on the degree of accuracy required, a larger or smaller number of design positions should be chosen. Synthesis of the linkage is easier and the flexibility available to the designer is greater if fewer positions are used. Five is the upper limit to the number of design positions that can be used.

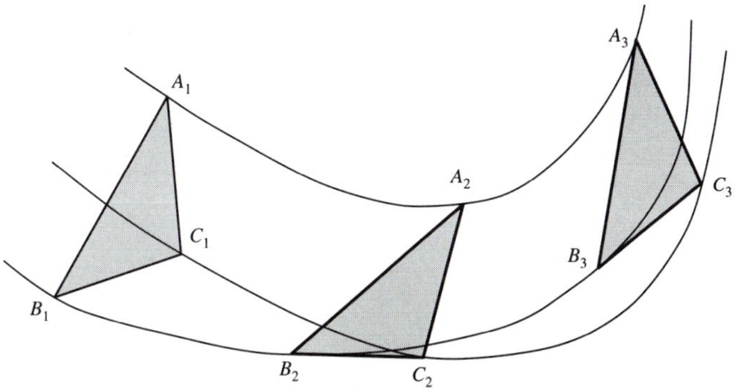

Figure 4.5 Motion of a lamina along a continuous trajectory. Each point in the lamina moves along a continuous curve. The triangle ABC drawn on the lamina is shown in three different positions along the trajectory: $A_1B_1C_1$, $A_2B_2C_2$, and $A_3B_3C_3$.

Geometrically, a crank has the effect of constraining the center of its moving pivot to move on a circle. The fixed pivot is at the center of that circle. Consequently, the problem of synthesizing a four-bar linkage to move its coupler through the design positions is basically the problem of locating two points in the moving lamina. Successive positions of each point all lie on the same circle. These points are taken as the locations of the moving pivots of the two cranks. The centers of the two circles on which their successive positions lie become the fixed pivots of the cranks.

4.2.2 Two Positions

Because an infinite number of circles can be drawn through any two points, any point in the moving lamina can be chosen as a moving pivot when two positions are of interest. In the example shown in Fig. 4.6, the two positions of the lamina are defined relative to the fixed frame by the line segments A_1B_1 and A_2B_2, which are two positions of the line segment AB drawn on the moving lamina. Since any point in the lamina can be a moving pivot, we might as well choose A and B. In each case, we then have an infinite number of points that can be the fixed pivots, namely all points on the perpendicular bisector of A_1A_2 for the fixed pivot corresponding to A and all points on the perpendicular bisector of B_1B_2 for the fixed pivot corresponding to B.

The perpendicular bisector of A_1A_2 can be constructed by setting any convenient radius on a pair of compasses and drawing two arcs. The first arc is centered on A_1, and the second is centered on A_2. The perpendicular bisector is the line drawn through the two intersections of these two arcs. In practice, only small portions of the two arcs are drawn in the neighborhoods in which the intersections are expected, as shown in Fig. 4.6. This operation of constructing a perpendicular bisector will be used extensively in the following.

The four-bar linkage that results from this construction is, in its first position, $A^*A_1B_1B^*$. The base link is A^*B^*. The coupler is A_1B_1. That is, in this case, the coupler is simply the line segment used to define the positions of the moving lamina. This need not be so.

Any point in the moving lamina, not just A or B, can be chosen as a moving pivot. This is shown in Fig. 4.7, in which point C is chosen as the second moving pivot, rather than point B. The first step, in this case, is to locate the two positions C_1 and C_2 of this point. The convention for showing points on the moving plane which is used almost universally in the literature, and which is followed here, is that the moving lamina is drawn in its *first position.* Therefore, point C drawn on the moving lamina is identical to point C_1. In order

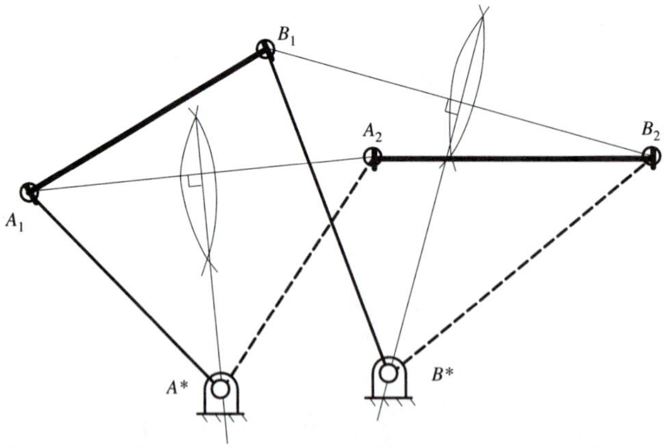

Figure 4.6 Construction of a four-bar linkage that moves its coupler plane through the positions A_1B_1 and A_2B_2.

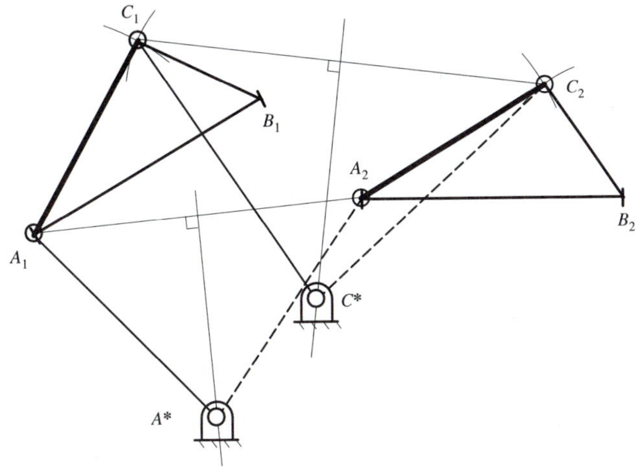

Figure 4.7 Solution of the same two-position problem shown in Fig. 4.6 with a different point: C_1 chosen as the second moving pivot.

to locate point C_2, we note that ABC is a triangle drawn on the rigid, moving lamina. Therefore it does not change shape, regardless of the motion. Therefore triangle $A_2B_2C_2$ is congruent to triangle $A_1B_1C_1$. Consequently, C_2 can be located by completing triangle $A_2B_2C_2$.

In practice, this is accomplished by setting radius A_1C_1 on a pair of compasses and drawing an arc with center A_2. The compasses are then set to radius B_1C_1, and an arc is drawn with center B_2. The intersection of the two arcs is point C_2. It is important to note that there are actually two possible intersections of these two arcs. One gives triangle $A_2B_2C_2$ congruent to triangle $A_1B_1C_1$, but the other gives the mirror image of that triangle. This second possibility will give incorrect results. Care is, therefore, necessary to ensure that the correct intersection is used. In some cases, the correct solution is not obvious. A simple check is to count off the vertices A_1, B_1, C_1 when proceeding in a counterclockwise direction around the triangle. Counting off A_2, B_2, C_2 when proceeding around $A_2B_2C_2$ in the same direction should give the same order. If the order is $A_2C_2B_2$, the triangle is the mirror image, and the solution is incorrect.

The problem can now be solved in exactly the same manner as it was before, except that C_1 and C_2 are used instead of B_1 and B_2. That is, the perpendicular bisector of A_1A_2 is constructed, and any point A^* is selected on that perpendicular bisector to be the fixed pivot corresponding to the moving pivot A_1. The perpendicular bisector of C_1C_2 is then constructed, and any point C^* on that bisector is chosen to be the fixed pivot corresponding to the moving pivot C_1. The resulting four-bar linkage is, in its first position, $A^*A_1C_1C^*$. In its second position it is $A^*A_2C_2C^*$.

Actually, for two positions, it is possible to locate a unique point such that the moving lamina can be attached to a single, fixed pivot at that point and will rotate through the two design positions. This is shown in Fig. 4.8. This point, P_{12}, is called the displacement pole for the two positions. One position can be reached from the other by means of a pure rotation about the pole.

The construction for locating the pole is as shown in Fig. 4.8. Since P_{12} lies on the perpendicular bisector of A_1 A_2, it is equidistant from A_1 and A_2. Similarly, it is equidistant from B_1 and B_2. Thus position 2 can be reached from position 1 by pure rotation about P_{12}.

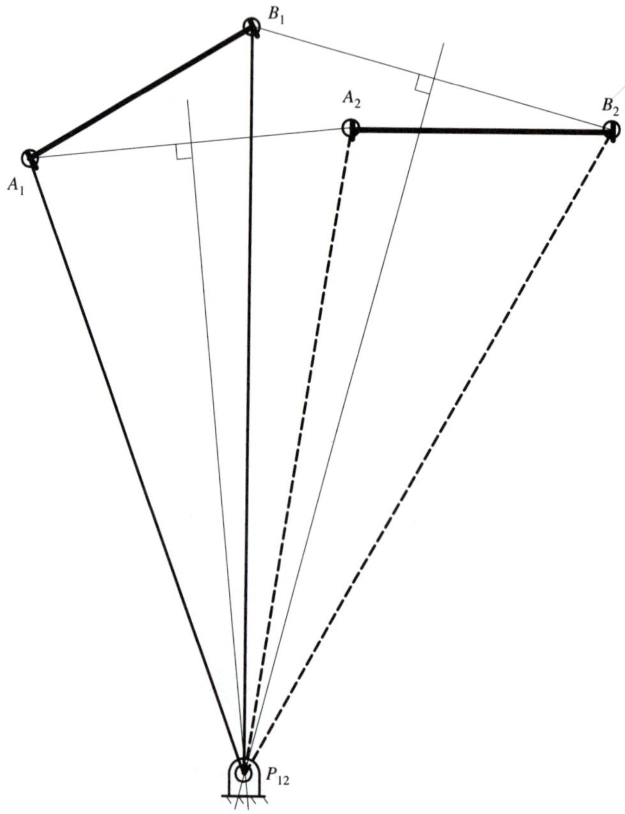

Figure 4.8 Location of the pole, P_{12}, of displacement of the moving lamina from position 1 to position 2. P_{12} is located at the intersection of the perpendicular bisectors of A_1A_2 and B_1B_2. The moving lamina can be displaced from position 1 to position 2 by a pure rotation about P_{12}.

4.2.3 Three Positions with Selected Moving Pivots

Because a circle can be drawn through any three points, any point on the moving lamina can be a moving pivot. The corresponding fixed pivot is at the center of the circle on which the three positions of the point lie. Taking A as one moving pivot, the corresponding fixed pivot A^* is located at the center of the circle upon which A_1, A_2, and A_3, the three positions of point A, lie. Notice that A_1, A_2, and A_3 represent the three positions of *a single point, A,* in the moving plane. They are the positions of that point *as seen from the fixed plane.* The positions of points and lines in the moving plane are, by convention, drawn on the first position of the moving plane. Thus, points A and A_1 can be regarded as being identical, as can B and B_1.

The center of the circle, A^*, can be found at the intersection of the perpendicular bisectors of A_1A_2 and A_2A_3. Similarly, B^* is located at the center of the circle on which B_1, B_2, and B_3 lie. That is, B^* is at the intersection of the perpendicular bisectors of B_1B_2 and B_2B_3. The solution linkage is then the four-bar $A^*A_1B_1B^*$ as shown in position 1. This construction is shown in Fig. 4.9.

As pointed out in the two-position case, it is not necessary for A and B to be chosen as the moving pivots. If a third point, C ($\equiv C_1$), is chosen as a moving pivot, its second and third positions may be found by constructing triangles $A_2B_2C_2$ and $A_3B_3C_3$ congruent

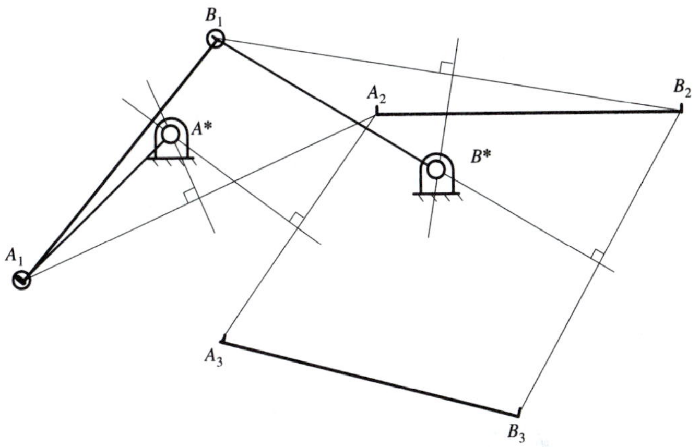

Figure 4.9 Synthesis of a four-bar linkage that moves its coupler plane through three nominated positions. The line segment AB defines the three positions of the moving plane. The points A and B are also chosen as the moving pivots of the two cranks. A^* and B^* are the fixed pivots of those cranks.

to triangle $A_1B_1C_1$. Figure 4.10 shows the synthesis of a four-bar linkage that moves its coupler through the three positions in Fig. 4.9. The points C and D that do not lie on the line AB are chosen as the moving pivots. Points C_2 and C_3 are located by constructing congruent triangles. Likewise, points D_2 and D_3 are located by constructing triangles $A_2B_2D_2$ and $A_3B_3D_3$ congruent to triangle $A_1B_1D_1$. Notice that, although we represent the moving lamina by means of the line segment AB, it is a *plane,* not a line, and we are at liberty to draw points and lines on it that do not lie on AB.

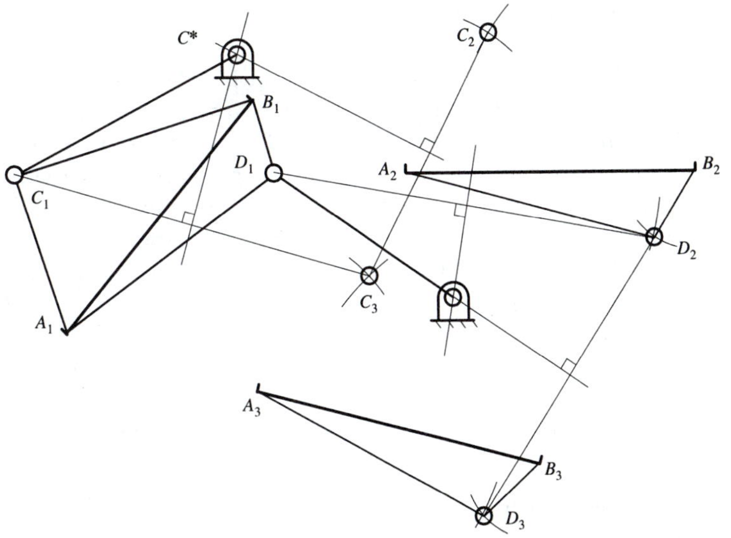

Figure 4.10 The same problem as that of Fig. 4.9 solved with points C and D selected as moving pivots, rather than A and B. Triangles $A_2B_2D_2$ and $A_3B_3D_3$ are congruent to $A_1B_1D_1$. The solution linkage, shown in its first position, is $C^*C_1D_1D^*$.

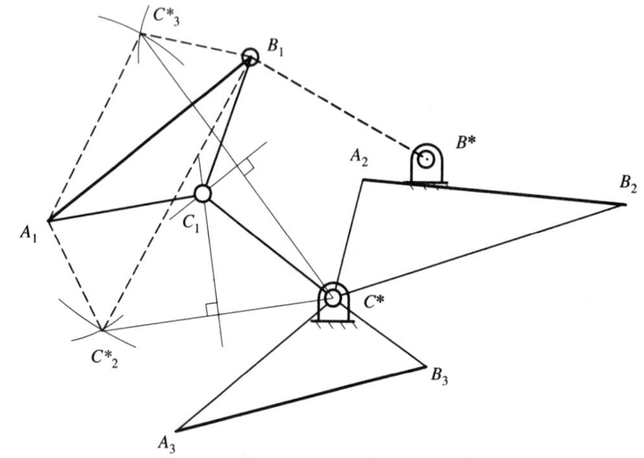

Figure 4.11 Synthesis of a crank with a selected fixed pivot C^*. C_2^* and C_3^* are, respectively, the second and third positions of point C^* as seen from the moving lamina. C_1 is the center of the circle passing through C^*, C_2^*, and C_3^*. After the crank C^*C_1 has been synthesized, the linkage may be completed by designing a second crank by any method. The dashed crank is the result of choosing B_1 as the moving pivot of the second crank.

4.2.4 Synthesis of a Crank with Chosen Fixed Pivots

The procedure given above allows us to synthesize a crank with any chosen moving pivot. If we wish to choose the fixed pivot rather than the moving pivot, the linkage must be inverted with the coupler becoming the reference frame. When this is done, the chosen fixed pivot is observed to move through three apparent positions of the fixed pivot as seen by the observer on the coupler. The resulting construction is shown in Fig. 4.11.

The three positions assumed by the chosen fixed pivot C^* relative to the moving lamina are plotted on the first position of that lamina. The apparent position of C^* when the lamina is in the first position is then its true position. Its apparent positions C_2^* and C_3^* when the lamina is in its second and third positions are obtained by constructing triangle $A_1B_1C_2^*$ congruent to $A_2B_2C^*$ and triangle $A_1B_1C_3^*$ congruent to triangle $A_3B_3C^*$. The location, C_1 of the moving pivot in the first position is obtained as the center of the circle on which C^*, C_2^*, and C_3^* lie. This defines the crank C^*C_1 in its first position. If needed, the second and third positions (C_2, C_3) of the moving pivot can be located by constructing triangle $A_2B_2C_2$ congruent to triangle $A_1B_1C_1$ and triangle $A_3B_3C_3$ congruent to triangle $A_1B_1C_1$.

This technique gives, of course, only one crank. If both cranks are to have nominated fixed pivots, the construction must be repeated for the second crank. If the moving pivot of the second crank is to be chosen, then the earlier construction is used.

4.2.5 Design of Slider Cranks and Elliptic Trammels

As was the case with a revolute crank, it is first necessary to consider the geometric effect of a prismatic-revolute link. The effect of the sliding joint is to constrain the center of the revolute joint to move along a straight line as shown in Fig. 4.12. Therefore, to design a linkage with a fixed slider, it is necessary to locate points whose three positions are collinear.

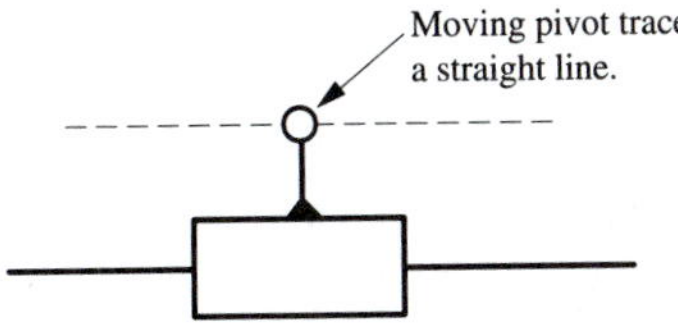

Figure 4.12 Geometric effect of replacing the fixed revolute of a crank by a prismatic joint.

The procedure for doing this is shown in Fig. 4.13 and described in the following:

1. Locate the poles P_{12}, P_{13}, and P_{23}.
2. Locate the image pole P'_{23} by making triangle $P_{12}P_{13}P'_{23}$ the mirror image of triangle $P_{12}P_{13}P_{23}$.
3. Locate the center of the circle $P_{12}P_{13}P'_{23}$ by drawing the perpendicular bisectors of $P_{12}P_{13}$ and $P_{13}P'_{23}$ or $P_{12}P'_{23}$.
4. Draw the circle through P_{12}, P_{13}, and P'_{23}. This circle is called the circle of sliders, and any point on this circle has all three of its positions collinear. Hence, any point on this circle can be used as the moving pivot of a slider-hinge link.
5. Construct the three positions of the moving pivot. The slide direction is parallel to the line on which all three positions lie.

That this construction will give slider points can be proved as follows (also see Hall[4] in the reference list for this chapter):

1. The angle subtended at the pole P_{1i} by any crank is $\theta_{1i}/2$ where θ_{1i} is the rotation of the moving plane between positions 1 and *i*. This is shown in Fig. 4.14. This follows because the circle point, *X*, being a point in the moving plane, rotates through angle θ_{1i} about P_{1i}

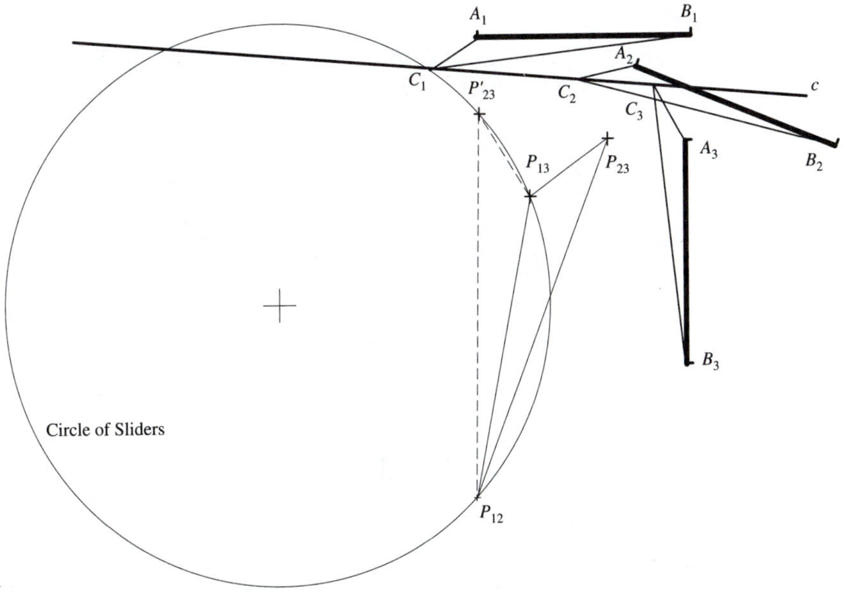

Figure 4.13 Construction of slider-hinge link. The three design positions are A_1B_1, A_2B_2, and A_3B_3. The slider point, C_1, is chosen from the circle which passes through the image poles P_{12}, P_{13}, and P'_{23}. C_2 and C_3 are its second and third positions. *c* is the line in the direction of sliding.

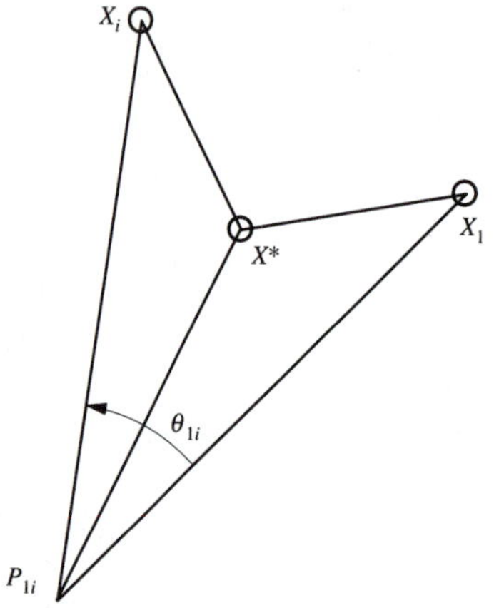

Figure 4.14 Relationship of angle subtended by a crank at a pole to the angular displacement about that pole.

in moving from position 1 to position i. Since $P_{1i}X^*$ is common for both positions, and $X^*X_i = X^*X_1$, and $P_{1i}X_1 = P_{1i}X_i$, it follows that triangle $P_{1i}X^*X_i$ is the mirror image of $P_{1i}X^*X$. Therefore, $\angle X_1P_{1i}X^* = \angle X^*P_{1i}X_i = \theta_{1i}/2$.

2. A slider-hinge link can be thought of as a crank with its center point at infinity. In this form, the preceding result 1 requires that the angle at the pole P_{1i} between the line joining P_{1i} to the circle point and a line normal to the slide be $\theta_{1i}/2$. Thus, given the direction c of the slide, the circle point C_1 whose three positions lie on a line parallel to c is located by drawing normals from c to P_{12} and P_{13} as shown in Fig. 4.15 and constructing lines at angles $\theta_{12}/2$ and $\theta_{13}/2$, respectively, to those normals. These lines intersect at the required point C_1 as shown in Fig. 4.15.

3. The pole triangle is shown in Fig. 4.16. The angle $P_{12}C_1P_{13}$ is $(\theta_{13}/2 - \theta_{12}/2) = \theta_{23}/2$ (the exterior angle of a triangle is equal to the sum of the interior opposite angles). Hence, $\angle P_{12}C_1P_{13}$ is independent of the direction of c and the locus of all points having three points collinear is the locus of points C_1 with $\angle(P_{12}C_1P_{13})/2$. This is a circle passing

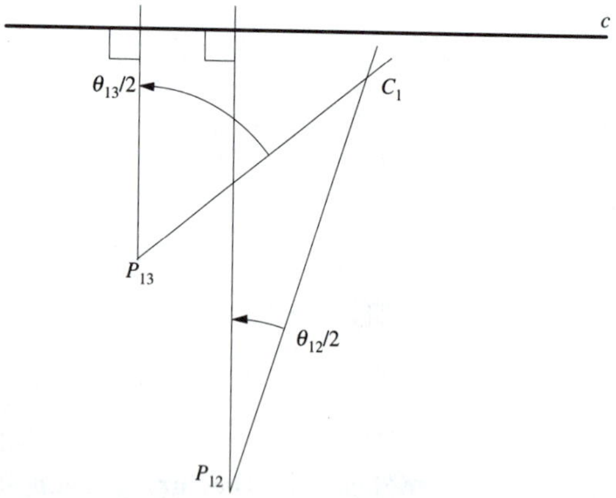

Figure 4.15 Relationship of sliding direction, c, and slider point C_1, to the poles: P_{12} and P_{13}.

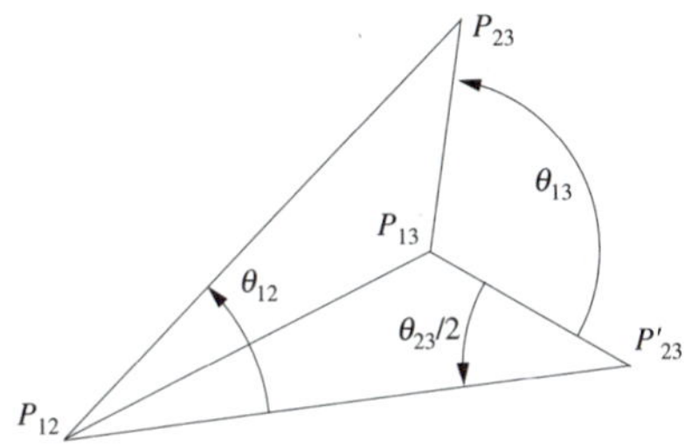

Figure 4.16 Angular relationships of pole triangle and image pole triangle. The triangles are mirror images of each other.

through poles P_{12} and P_{13} with the central angle subtended by $P_{12}P_{13}$ equal to θ_{23}. The image pole, P'_{23}, also lies on this circle because it rotates with the body to P_{23} through the angle θ_{13} about P_{13}. Thus, the angle $P_{12}P'_{23}P_{13}$ is $\theta_{23}/2$. Hence, the required circle is the circle which circumscribes $P_{12}P_{13}P'_{23}$.

4.2.6 Order Problem and Change of Branch

Note that the preceding techniques really only guarantee that the mechanism can be assembled in the design positions; they do not guarantee that the mechanism will function correctly between different design positions. It is confusing, but it is quite possible for the simple graphical procedures developed above to produce spurious solutions. These are solutions that do not physically pass through the design positions or pass through the design positions in the wrong order. Thus, two problems can occur that may make the design unacceptable.

The first problem is due to the fact that there are two possible positions of a four-bar linkage of given link lengths corresponding to a given value of the driving-crank angle. These are termed "assembly configurations" or "solution branches." If the solution linkage for a motion generation problem is such that some of the design positions lie on one assembly configuration and others on the other assembly configuration, it may not be possible to move the linkage through all design positions without physically disconnecting it and reassembling it in the other assembly configuration. Fortunately, there is a simple graphical test to identify this problem.

To detect whether a mechanism must change branches to pass through all of the positions, it is necessary only to assemble the mechanism in one position and determine whether it can be moved through the other two positions. This can be done conveniently if the linkage can be animated on a computer screen. If one position is missed, then a change of branch is indicated.

Another way to determine whether a change of branch is indicated is to examine the angle ψ between the coupler and the output link. Because an extreme position of the driving joint corresponds to the angle ψ passing through an angle of either 0 or π, the key to determining the branch change is the sign of that angle. A convenient method is to construct the cranks and coupler in all design positions and inspect the angle ψ between the driven crank (the longer of the two cranks) and the coupler. A change in direction of this angle indicates a change of branch in a crank-rocker or drag-link type of mechanism and a drive failure in a double-rocker type of linkage. In either case, the solution linkage is not usable. An example of this condition is shown in Fig. 4.17. There the direction of the angle $D^*D_1C_1$ is opposite to that of angles $D^*D_2C_2$ and $D^*D_3C_3$. Hence the linkage passes through a position in which the driving joint C^* is at a motion limit.

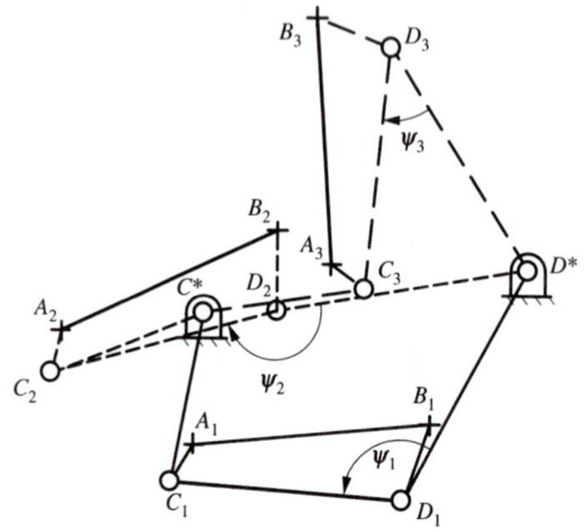

Figure 4.17 An example in which the solution linkage is not capable of moving through the design positions without being disconnected and reassembled. The solution linkage is shown in all three design positions as $C^*C_1D_1D^*$, $C^*C_2D_2D^*$, and $C^*C_3D_3D^*$, respectively. The angles between the driven (longer) crank and the coupler are examined in all three positions. These are the angles $D^*D_1C_1 = \psi_1$, $D^*D_2C_2 = \psi_2$, and $D^*D_3C_3 = \psi_3$ respectively. ψ_1 is counterclockwise, and ψ_2 and ψ_3 are clockwise. Thus the angle ψ changes sign, indicating a change of branch in the solution.

EXAMPLE 4.1 (*Position Synthesis of Four-Bar Linkage*)

PROBLEM

Design a four-bar linkage whose coupler moves through the three positions indicated by the line segment AB in Fig. 4.18. Point B is to be one moving pivot and point X^* is to be one fixed pivot.

SOLUTION

1. The procedure for locating the fixed pivot B^* is shown in Fig. 4.19. The construction used is that of Fig. 4.9.
2. The procedure for the location of moving pivot X is shown in Fig. 4.20. The construction used is that of Fig. 4.11.

Triangle $A_1B_1X_2^*$ is congruent to triangle $A_2B_2X^*$, and triangle $A_1B_1X_3^*$ is congruent to triangle $A_3B_3X^*$.

X_1 is located at the center of the circle $X^*X_2^*X_3^*$.

The other two positions of point X_1, X_2 and X_3, can then be located by constructing triangles $A_2B_2X_2$ and $A_3B_3X_3$ congruent to triangle $A_1B_1X_1$.

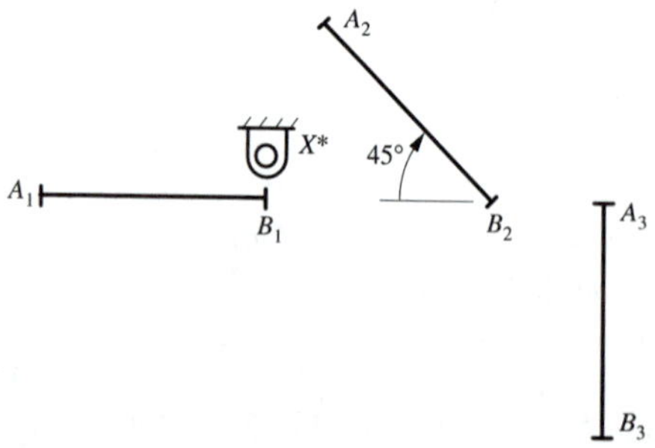

Figure 4.18 The problem of Example 4.1.

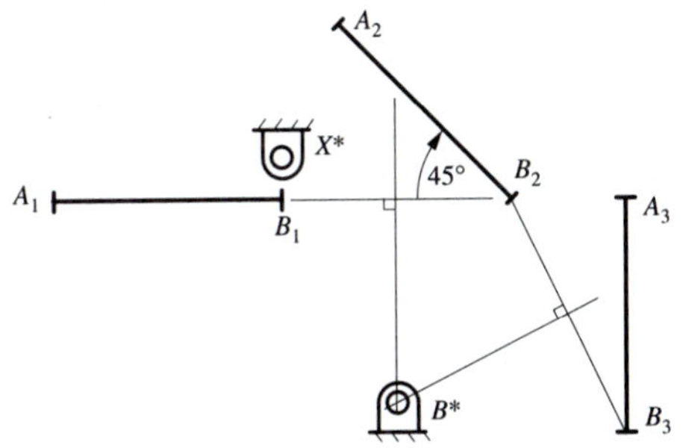

Figure 4.19 Location of the fixed pivot, B^*, given the moving pivot, B_1.

3. Check solution

We first check the Grashof type of the linkage:

$$X^*B^* = 2.41 = p$$
$$B_1X_1 = 4.28 = q$$
$$B^*B_1 = 2.06 = s$$
$$X^*X_1 = 4.62 = l$$
$$l + s = 6.68, \qquad p + q = 6.69$$

so

$l + s < p + q$ (barely)

The shortest link, s, is a crank, so the linkage is a crank rocker. ■

To check for change of branch, we draw the linkage in all three of the design positions, as shown in Fig. 4.21. All of the information necessary to do this has already been generated in previous stages of the construction procedure.

Because X^*X_1 is the longer of the two cranks, assume that it is the driven link, and B^*B is the driver link. We can then check the signs of the angles $X^*X_1B_1$, $X^*X_2B_2$,

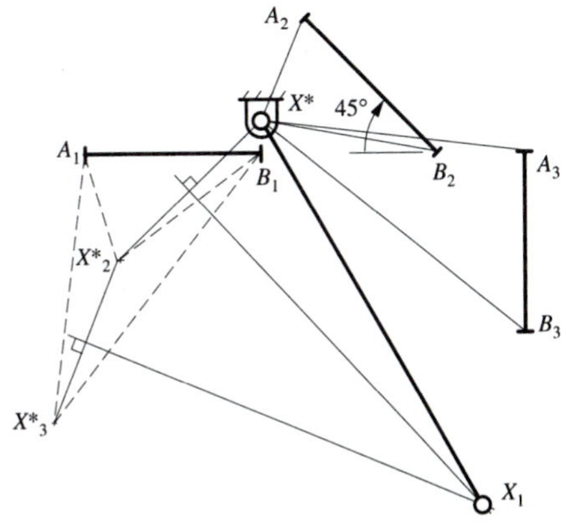

Figure 4.20 Location of moving pivot, X_1, given the location of the fixed pivot, X^*. Triangles $A_1B_1X_2^*$ and $A_2B_2X^*$ are congruent, as are triangles $A_1B_1X_3^*$ and $A_3B_3X^*$.

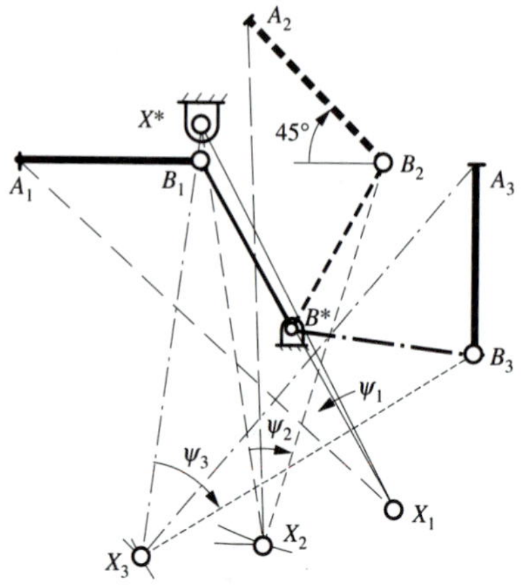

Figure 4.21 Construction of solution linkage and verification that it satisfies the design positions without disconnection. In this case, the linkage fails the test because $\psi_1 = \angle X^*X_1B_1$ is counterclockwise while ψ_2 and ψ_3 are clockwise. Hence, the linkage cannot be moved through the design positions by rotation of the crank B^*B.

$X^*X_3B_3$ to check for branching. $\angle X^*X_1B_1$ is counterclockwise. $\angle X^*X_2B_2$ and $\angle X^*X_3B_3$ are clockwise. Hence it is concluded that a change of branch occurs.

EXAMPLE 4.2 ***(Position Synthesis of Slider-Crank Mechanism)***

PROBLEM

Design a slider-crank mechanism to move a coupler containing the line AB through the three positions shown in Fig. 4.22. Use point A as a circle point.

SOLUTION

To design a slider-crank mechanism, it is necessary to identify a circle point and the corresponding center point (or vice versa) and a slider point. We must also locate the direction for the slider line. In this problem, point A has been identified as the circle point for the crank. Therefore, to locate the center point, we need only find the center of the circle on which the three positions of A lie. The construction for finding the center point (A^*) and the crank in position 1 is shown in Fig. 4.23.

To locate the slider point, we must locate the poles, the image pole P'_{23}, and the circle of sliders in position 1. This circle is attached to the coupler. The construction of the poles is

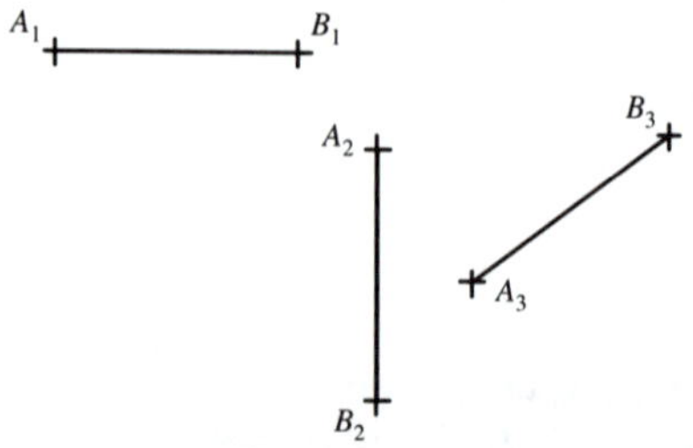

Figure 4.22 Design positions for Example 4.2.

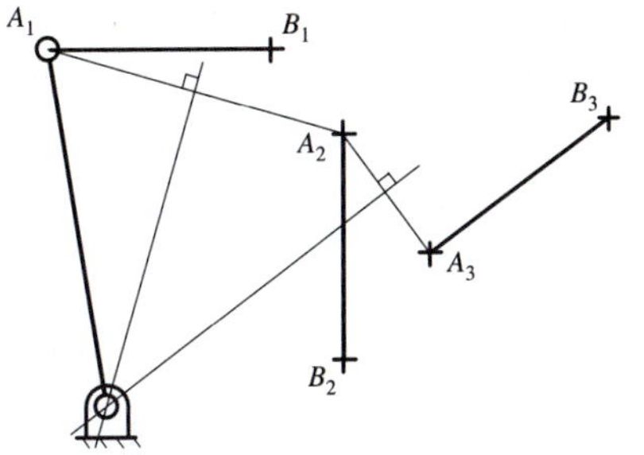

Figure 4.23 Construction of crank of slider-crank mechanism for Example 4.2.

shown in Fig. 4.24, and the locations of the image pole and circle of sliders are shown in Fig. 4.25. We can select any point on the slider circle as a slider point. The point chosen is C. To complete the design, we need to locate the slider point in positions 2 and 3. The three positions, C_1, C_2, and C_3, will be collinear on the slider line. The construction of the slider line is also shown in Fig. 4.25.

From the three positions of C shown in Fig. 4.25, it is clear that the linkage does not go through the three positions in the correct order. Therefore, the solution shown is not a good linkage and other circle points or slider points should be chosen. ■

4.2.7 Analytical Approach to Rigid Body Guidance

Rigid body guidance can be approached analytically in two ways. The first procedure requires coordinate transformations and is more general. It can be extended easily to four positions, and the different elements (such as the circle of sliders) developed in the graphical procedure arise naturally from the mathematics. However, for three positions, the procedure is more involved than the second procedure, which is a mathematical representation of the graphical procedure. Therefore, only the second procedure will be presented here.

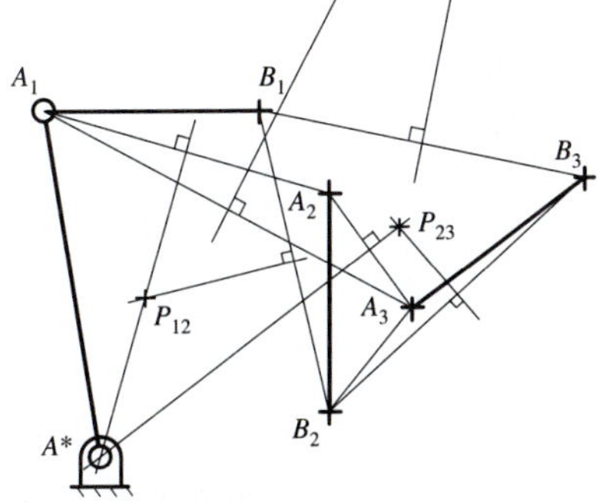

Figure 4.24 Construction of the poles for Example 4.2.

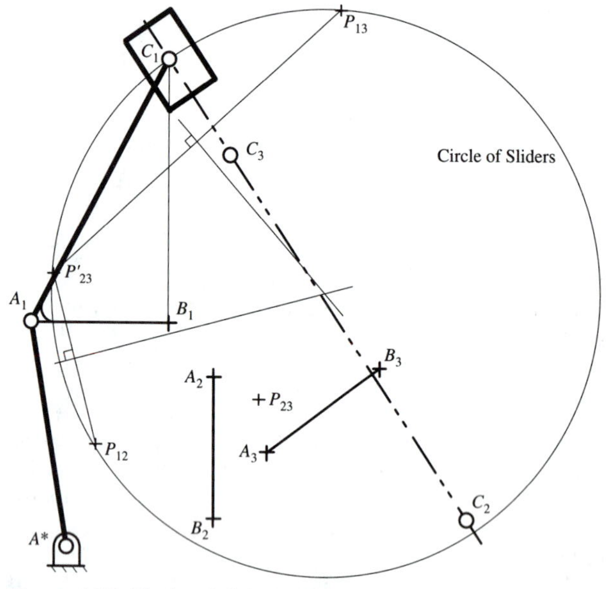

Figure 4.25 Circle of sliders and final linkage for Example 4.2.

Readers are referred to works by Sandor and Erdman[13], Waldron[15], or Suh and Radcliffe[14] for a more general analytical treatment.

The analytical approach to rigid body guidance involves coordinate transformations when center points are selected. Therefore, before addressing the topic directly, let us develop the equations for the needed coordinate transformations between the coupler and frame coordinate systems.

4.2.7.1 *Coordinate Transformations*

The general relationship between the coupler and frame systems is indicated in Fig. 4.26. From Fig. 4.26, we can write the vector equations as

$$\boldsymbol{r}_{P/O} = \boldsymbol{r}_{P/A} + \boldsymbol{r}_{A/O}$$

In this equation, $\boldsymbol{r}_{P/O}$ and $\boldsymbol{r}_{A/O}$ are defined in the frame coordinate system, and $\boldsymbol{r}_{P/A}$ is defined in the coupler coordinate system. Therefore, only $\boldsymbol{r}_{P/A}$ needs to be transformed to the frame coordinate system. This is shown in Fig. 4.27.

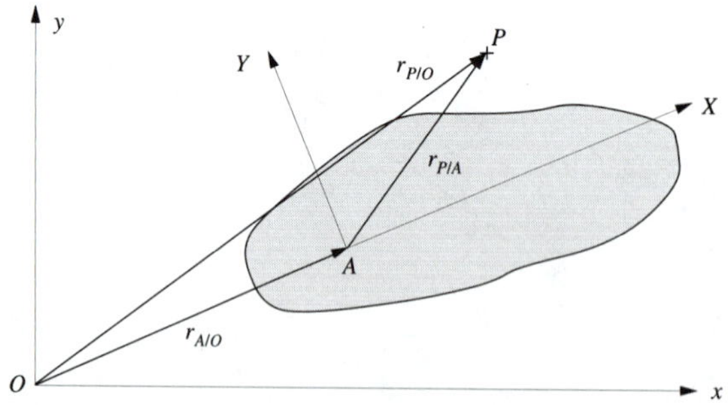

Figure 4.26 Relationship between coupler and frame coordinate systems.

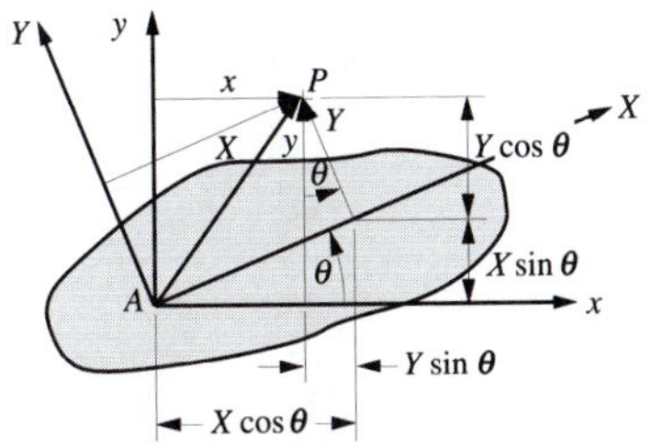

Figure 4.27 Transformation from coupler system XY to frame system xy.

In matrix form, the coordinate transformation from the coupler to the frame system is

$$\begin{Bmatrix} x \\ y \end{Bmatrix} = [R] \begin{Bmatrix} X \\ Y \end{Bmatrix} + \begin{Bmatrix} a_x \\ a_y \end{Bmatrix} \tag{4.1}$$

In Eq. (4.1), the matrix R indicates the orientation (rotation) of the coupler coordinate system relative to the frame coordinate system. The vector $\{a_x \; a_y\}^T$ gives the origin of the coupler coordinate system relative to the frame system. We need to determine the rotation matrix $[R]$ first. From Fig. 4.27,

$$\begin{aligned} x_{P/A} &= X \cos\theta - Y \sin\theta \\ y_{P/A} &= Y \cos\theta + X \sin\theta \end{aligned} \tag{4.2}$$

or

$$\begin{Bmatrix} x_{P/A} \\ y_{P/A} \end{Bmatrix} = \begin{bmatrix} \cos\theta & -\sin\theta \\ \sin\theta & \cos\theta \end{bmatrix} \begin{Bmatrix} X \\ Y \end{Bmatrix} \tag{4.3}$$

For any general point with coordinates (X, Y) relative to the coupler, the coordinates (x, y) relative to the frame are given by Eq. (4.1) or

$$\begin{Bmatrix} x \\ y \end{Bmatrix} = \begin{bmatrix} \cos\theta & -\sin\theta \\ \sin\theta & \cos\theta \end{bmatrix} \begin{Bmatrix} X \\ Y \end{Bmatrix} + \begin{Bmatrix} a_x \\ a_y \end{Bmatrix} = [R] \begin{Bmatrix} X \\ Y \end{Bmatrix} + \begin{Bmatrix} a_x \\ a_y \end{Bmatrix} \tag{4.4}$$

Therefore,

$$[R] = \begin{bmatrix} \cos\theta & -\sin\theta \\ \sin\theta & \cos\theta \end{bmatrix} \tag{4.5}$$

We can also transform from the frame coordinate system to the coupler coordinate system. This is shown in Fig. 4.28. From that figure it is clear that

$$\begin{aligned} X_{P/A} &= x_{P/A} \cos\theta + y_{P/A} \sin\theta \\ Y_{P/A} &= y_{P/A} \cos\theta - x_{P/A} \sin\theta \end{aligned} \tag{4.6}$$

or

$$\begin{Bmatrix} X_{P/A} \\ Y_{P/A} \end{Bmatrix} = \begin{bmatrix} \cos\theta & \sin\theta \\ -\sin\theta & \cos\theta \end{bmatrix} \begin{Bmatrix} x_{P/A} \\ x_{P/A} \end{Bmatrix} = [A] \begin{Bmatrix} x_{P/A} \\ y_{P/A} \end{Bmatrix} \tag{4.7}$$

where

$$[A] = \begin{bmatrix} \cos\theta & \sin\theta \\ -\sin\theta & \cos\theta \end{bmatrix} \tag{4.8}$$

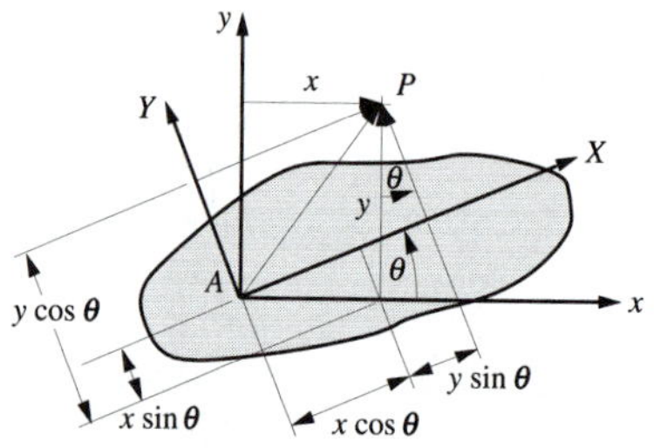

Figure 4.28 Transformation from frame system xy and coupler system XY.

When we compare Eqs. (4.5) and (4.8), it is clear that

$$\begin{bmatrix} \cos\theta & \sin\theta \\ -\sin\theta & \cos\theta \end{bmatrix} = \begin{bmatrix} \cos\theta & -\sin\theta \\ \sin\theta & \cos\theta \end{bmatrix}^T = \begin{bmatrix} \cos\theta & -\sin\theta \\ \sin\theta & \cos\theta \end{bmatrix}^{-1} \tag{4.9}$$

Therefore,

$$[A] = [R]^{-1} = [R]^T \tag{4.10}$$

Now assume that we have specified the position of the coupler coordinate system by the point (a_x, a_y) and the rotation angle θ. To transform from the coupler system (X, Y) to the frame coordinate system (x, y), use

$$\begin{Bmatrix} x \\ y \end{Bmatrix} = \begin{bmatrix} \cos\theta & -\sin\theta \\ \sin\theta & \cos\theta \end{bmatrix} \begin{Bmatrix} X \\ Y \end{Bmatrix} + \begin{Bmatrix} a_x \\ a_y \end{Bmatrix} \tag{4.11}$$

To transform from the frame system to the coupler system, use

$$\begin{Bmatrix} X \\ Y \end{Bmatrix} = \begin{bmatrix} \cos\theta & \sin\theta \\ -\sin\theta & \cos\theta \end{bmatrix} \begin{Bmatrix} x - a_x \\ y - a_y \end{Bmatrix} = \begin{bmatrix} \cos\theta & \sin\theta \\ -\sin\theta & \cos\theta \end{bmatrix} \begin{Bmatrix} x \\ y \end{Bmatrix} - \begin{bmatrix} \cos\theta & \sin\theta \\ -\sin\theta & \cos\theta \end{bmatrix} \begin{Bmatrix} a_x \\ a_y \end{Bmatrix} \tag{4.12}$$

In the following, we will assume that each position of the coupler is given by the coordinates of a point (a_x, a_y) and by the angle θ. If instead of the angle θ, the coordinates (b_x, b_y) of a second point B in the coupler are given, we must first compute the angle θ from the equation

$$\theta = \tan^{-1}\left[\frac{b_y - a_y}{b_x - a_x}\right] \tag{4.13}$$

When a circle point is selected, the coordinates of that point are given relative to the coupler coordinate system (X, Y). The coordinates of the three positions of that point relative to the frame coordinate system can be computed using Eq. (4.11). The corresponding center point is located by finding the center of the circle (analytically) on which the three positions of the circle point lie. The coordinates of the center of the circle will be defined in the frame coordinate system.

If a center point is given, the coordinates will be in the frame coordinate system (x, y). To find the corresponding circle point, the three apparent positions of the center point relative to the coupler coordinate system must be found. This is done using Eq. (4.12). The corresponding circle point is located by finding the center of the circle (analytically) on which the three positions of the center point lie. The coordinates of the center of the circle will be defined in the coupler coordinate system. The coordinates of this point in any of the positions can be found relative to the frame using Eq. (4.11). The crank length can be determined by computing the distance between the circle and center points once both are defined relative to the same coordinate system.

Locating center points given circle points and vice versa requires that the center of the circle corresponding to three positions of a point be found. To find the circle of sliders, the locations of the poles and the image pole must be found. We will discuss an analytical procedure for finding poles first. After this is done, it will be apparent that the same procedure can be used to find the center of a circle given three positions of a point.

4.2.7.2 Finding Poles

Let $\boldsymbol{A}_i$ and $\boldsymbol{A}_j$ and $\boldsymbol{B}_i$ and $\boldsymbol{B}_j$ be vectors defining the locations of two points in two positions. The x and y coordinates of each point are assumed to be known. As indicated in Fig. 4.8, the pole is the point that lets us move the rigid body from position i to position j by a simple rotation. To determine the location of the pole analytically, let r_A be the distance from the pole to point A and r_B be the distance from B to the pole P_{ij} as shown in Fig. 4.29a. The following geometric relationships then hold.

$$\begin{aligned}(A_{x_i} - p_{ij_x})^2 + (A_{y_i} - p_{ij_y})^2 = (r_A)^2 = (A_{x_j} - p_{ij_x})^2 + (A_{y_j} - p_{ij_y})^2 \\ (B_{x_i} - p_{ij_x})^2 + (B_{y_i} - p_{ij_y})^2 = (r_B)^2 = (B_{x_j} - p_{ij_x})^2 + (B_{y_j} - p_{ij_y})^2\end{aligned} \tag{4.14}$$

Expanding Eqs. (4.14) gives

$$A_{x_i}^2 - 2A_{x_i}p_{ij_x} + p_{ij_x}^2 + A_{y_i}^2 - 2A_{y_i}p_{ij_y} + p_{ij_y}^2 = A_{x_j}^2 - 2A_{x_j}p_{ij_x} + p_{ij_x}^2 + A_{y_j}^2 - 2A_{y_j}p_{ij_y} + p_{ij_y}^2$$

$$B_{x_i}^2 - 2B_{x_i}p_{ij_x} + p_{ij_x}^2 + B_{y_i}^2 - 2B_{y_i}p_{ij_y} + p_{ij_y}^2 = B_{x_j}^2 - 2B_{x_j}p_{ij_x} + p_{ij_x}^2 + B_{y_j}^2 - 2B_{y_j}p_{ij_y} + p_{ij_y}^2$$

These equations can be simplified to give

$$\begin{aligned}(A_{x_i}^2 + A_{y_i}^2) - (A_{x_j}^2 + A_{y_j}^2) = 2(A_{x_i} - A_{x_j})p_{ij_x} + 2(A_{y_i} - A_{y_j})p_{ij_y} \\ (B_{x_i}^2 + B_{y_i}^2) - (B_{x_j}^2 + B_{y_j}^2) = 2(B_{x_i} - B_{x_j})p_{ij_x} + 2(B_{y_i} - B_{y_j})p_{ij_y}\end{aligned} \tag{4.15}$$

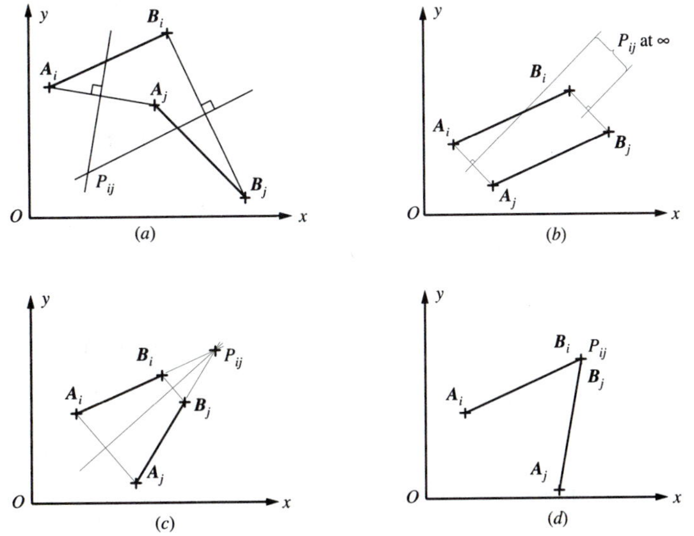

Figure 4.29 Location of pole P_{ij}: (a) general case, (b) parallel positions, (c) symmetric positions, (d) intersecting positions at B (or A)

These equations are linear in the unknown pole coordinates and can easily be solved. In matrix form, the equations become

$$\begin{bmatrix} 2(A_{x_i} - A_{x_j}) & 2(A_{y_i} - A_{y_j}) \\ 2(B_{x_i} - B_{x_j}) & 2(B_{y_i} - B_{y_j}) \end{bmatrix} \begin{Bmatrix} p_{ij_x} \\ p_{ij_y} \end{Bmatrix} = \begin{Bmatrix} (A_{x_i}^2 + A_{y_i}^2) - (A_{x_j}^2 + A_{y_j}^2) \\ (B_{x_i}^2 + B_{y_i}^2) - (B_{x_j}^2 + B_{y_j}^2) \end{Bmatrix} \tag{4.16}$$

Equations (4.16) can be solved using a calculator or a matrix equation solver such as MATLAB.

Equations (4.16) apply to most types of positions; however, there are three special cases that will make the matrix singular. These are shown in Fig. 4.29 and identified in the following:

1. Two positions of the coupler are parallel (Fig. 4.29b).
2. Lines linking the successive positions of two points are parallel (Fig. 4.29c).
3. Two successive positions of a point are coincident (Fig. 4.29d).

Each of these conditions can be handled separately.

4.2.7.2.1 *Two Parallel Positions*

When two positions are parallel, the resulting pole is at infinity in the direction given by the angle γ where

$$\gamma = \frac{\pi}{2} + \tan^{-1}\left(\frac{A_{y_j} - A_{y_i}}{A_{x_j} - A_{x_i}}\right) = \frac{\pi}{2} + \tan^{-1}\left(\frac{B_{y_j} - B_{y_i}}{B_{x_j} - B_{x_i}}\right)$$

4.2.7.2.2 *Lines A_iA_j and B_iB_j Are Parallel*

When this situation occurs, the pole is located at the intersection of the lines defined by A_iB_i and A_jB_j. The location of the pole is then given by solving the following simultaneous equations:

$$\left(\frac{p_{ij_x} - A_{x_i}}{B_{x_i} - A_{x_i}}\right) = \left(\frac{p_{ij_y} - A_{y_i}}{B_{y_i} - A_{y_i}}\right)$$

and

$$\left(\frac{p_{ij_x} - A_{x_j}}{B_{x_j} - A_{x_j}}\right) = \left(\frac{p_{ij_y} - A_{y_j}}{B_{y_j} - A_{y_j}}\right)$$

The equations can be simplified and rewritten as

$$p_{ij_x}(B_{y_i} - A_{y_i}) - p_{ij_y}(B_{x_i} - A_{x_i}) = A_{x_i}(B_{y_i} - A_{y_i}) - A_{y_i}(B_{x_i} - A_{x_i})$$
$$p_{ij_x}(B_{y_j} - A_{y_j}) - p_{ij_y}(B_{x_j} - A_{x_j}) = A_{x_j}(B_{y_j} - A_{y_j}) - A_{y_j}(B_{x_j} - A_{x_j})$$

or in matrix form,

$$\begin{bmatrix} (B_{y_i} - A_{y_i}) & -(B_{x_i} - A_{x_i}) \\ (B_{y_j} - A_{y_j}) & -(B_{x_j} - A_{x_j}) \end{bmatrix} \begin{Bmatrix} p_{ij_x} \\ p_{ij_y} \end{Bmatrix} = \begin{Bmatrix} A_{x_i}(B_{y_i} - A_{y_i}) - A_{y_i}(B_{x_i} - A_{x_i}) \\ A_{x_j}(B_{y_j} - A_{y_j}) - A_{y_j}(B_{x_j} - A_{x_j}) \end{Bmatrix} \tag{4.17}$$

Equations (4.17) can be solved using a calculator or a matrix equation solver such as MATLAB.

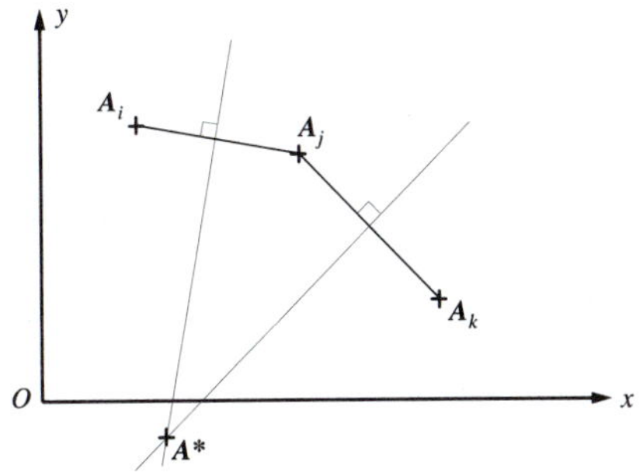

Figure 4.30 Locating center of circle using pole procedure. Compare this figure with Fig. 4.29a.

4.2.7.2.3 Two Successive Positions Coincident

When a point on the coupler does not move in successive coupler positions, that point becomes the pole directly. Therefore, if $\boldsymbol{A}_i = \boldsymbol{A}_j$ then both equal $\boldsymbol{p}_{ij}$, or if $\boldsymbol{B}_i = \boldsymbol{B}_j$ then both equal $\boldsymbol{p}_{ij}$.

4.2.7.3 Finding the Center of a Circle on Which Three Points Lie

The procedure given for finding poles can be used to find the center of the circle that passes through three points. To do this, simply treat $\boldsymbol{B}_i$ and $\boldsymbol{A}_j$ as the same point. This is shown schematically in Fig. 4.30 to find $\boldsymbol{A}^*$ given three positions of $\boldsymbol{A}$.

4.2.7.4 Image Pole

The image pole is found by reflecting the pole about a line through the two other poles. To find the image pole P'_{23}, we reflect the pole P_{23} about the line through poles P_{12} and P_{13}. This is shown in Fig. 4.31. Given the coordinates of poles P_{12}, P_{23}, and P_{13}, the coordinates of the image pole, P'_{23}, can be found as follows. First define

$$\boldsymbol{g} = (\boldsymbol{p}_{13} - \boldsymbol{p}_{12}) = (g_x, g_y)$$

and

$$\boldsymbol{h} = (\boldsymbol{p}_{23} - \boldsymbol{p}_{12}) = (h_x, h_y)$$

Then,

$$\beta = \tan^{-1}\left(\frac{h_y}{h_x}\right)$$

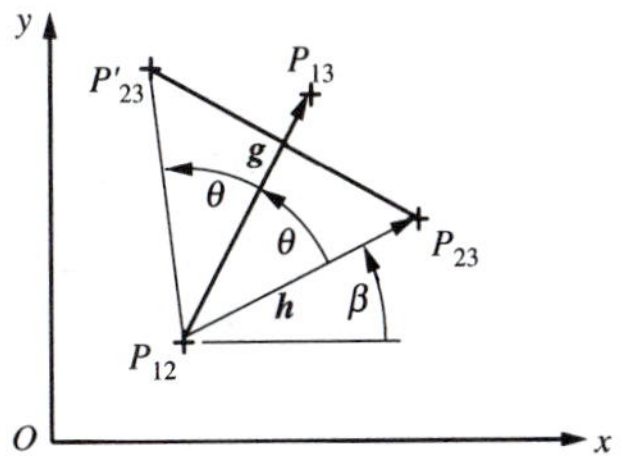

Figure 4.31 Image pole location.

and

$$\theta = \tan^{-1}\left(\frac{g_y}{g_x}\right) - \beta$$

Let $\boldsymbol{r} = \sqrt{h_x^2 + h_y^2}$. Then the coordinates of the image pole are given by

$$\boldsymbol{p}'_{23} = \boldsymbol{p}_{12} + (r\cos(\beta + 2\theta)\boldsymbol{i} + r\sin(\beta + 2\theta)\boldsymbol{j})$$

MATLAB functions for the pole and image pole routines are given on the disk included with this book.

4.2.7.5 *Crank Design Given Circle Point*

If the circle point is specified, the circle point coordinates (X, Y) will be given relative to the coupler coordinate system. The procedure for finding the corresponding center point and the resulting crank is given in the following.

1. Transform the coordinates of the circle point to the frame coordinate system using Eq. (4.11) for each position of the coupler. This will give three pairs of points (x_1, y_1), (x_2, y_2), and (x_3, y_3) relative to the frame coordinate system.
2. Set $\mathbf{A}_1 = (x_1, y_1)$, $\mathbf{A}_2 = (x_2, y_2)$, $\mathbf{A}_3 = (x_3, y_3)$. Then use the procedure in Section 4.2.7.2 to find the location of the center point in the frame coordinate system. Call this point (x^*, y^*).
3. The crank in position 1 is located by the line from (x^*, y^*) to (x_1, y_1).
4. Locate the second crank using the same procedure and complete the linkage.

4.2.7.6 *Crank Design Given Center Point*

If the center point is specified, the center point coordinates (x^*, y^*) will be given relative to the frame coordinate system. The procedure for finding the corresponding circle point relative to the coupler coordinate system and the subsequent crank is given in the following.

1. Transform the coordinates of the center point to the coupler coordinate system using Eq. (4.12) for each position of the coupler. This will give three pairs of points (X_1^*, Y_1^*), (X_2^*, Y_2^*), and (X_3^*, Y_3^*), relative to the coupler coordinates system.
2. Set $\mathbf{A}_1 = (X_1^*, Y_1^*)$, $\mathbf{A}_2 = (X_2^*, Y_2^*)$, and $\mathbf{A}_3 = (X_3^*, Y_3^*)$. Then use the procedure in Section 4.2.7.2 to find the location of the circle point relative to the coupler coordinate system. Call this point (X, Y).
3. Identify the position i in which the linkage is to be displayed. Transform the coordinates of point (X, Y) to the frame coordinate system using Eq. (4.11) for position i. Call the transformed point (x_i, y_i).
4. The crank in position i is located by the line from (x^*, y^*) to (x_i, y_i).
5. Locate the second crank and complete the linkage.

4.2.7.7 *Design of a Slider*

If a slider is to be used, we must find the circle of sliders and two of the three positions of the slider point. The procedure is given in the following:

1. Let two points in the coupler be given as $\boldsymbol{A} = (0, 0)$ and $\boldsymbol{B} = (1, 0)$ relative to the coupler coordinate system. Transform the coordinates of both points to the frame coordinate system using Eq. (4.11) for each position of the coupler. This will give three pairs of coordinates for $\mathbf{A}$ and three for $\boldsymbol{B}$ relative to the frame coordinate system. Call the point locations $\boldsymbol{A}_1$, $\boldsymbol{A}_2$, $\boldsymbol{A}_3$, $\boldsymbol{B}_1$, $\boldsymbol{B}_2$, and $\boldsymbol{B}_3$.

2. Compute the coordinates of the poles p_{12}, p_{13}, p_{23} using the coordinates of $\boldsymbol{A}$ and $\boldsymbol{B}$ and the procedure given in Section 4.2.7.2. The resulting coordinates will be in the frame coordinate system.
3. Compute the coordinates of the image pole p'_{23} using the procedure given in Section 4.2.7.4. The coordinates of p'_{23} will also be in the frame coordinate system.
4. Set $\boldsymbol{C}_i = p_{12}$, $\boldsymbol{C}_j = p_{13}$, and $\boldsymbol{C}_k = p'_{23}$. Then use the procedure in Section 4.2.7.2 to find the location of the center (x_c, y_c) of the slider circle in the frame coordinate system. Compute the radius of the circle using

$$r_c = \sqrt{(A_{1x} - x_c)^2 + (A_{1y} - y_c)^2}$$

The center (x_c, y_c) and radius r_c will correspond to the circle of sliders in position 1.
5. Select the x coordinate of the slider point. Call this coordinate x_s. Solve for the coordinate y_s using

$$y_s = y_c \pm \sqrt{r_c^2 - (x_s - x_c)^2}$$

Notice that there will be two possible values for y_s for each value of x_s. It is necessary to select the specific point desired by identifying which sign (+ or −) gives the proper configuration for the linkage. The resulting point $(x_s, y_s)_1$ will be the coordinates of the slider point in position 1 in the frame coordinate system.
6. Transform the point $(x_s, y_s)_1$ to the coupler cooordinate system using Eq. (4.12) and $(a_x, a_y)_1$ and the rotation angle θ_1. Call this point (X_s, Y_s). It identifies the slider point relative to the coupler coordinate system.
7. Determine the coordinates of point (X_s, Y_s) in the frame coordinate system for positions 2 and 3 of the coupler. Call these positions $(x_s, y_s)_2$ and $(x_s, y_s)_3$.
8. Define the slider line parametrically by

$$(x, y) = (x_s, y_s)_1 + \beta[(x_s, y_s)_3 - (x_s, y_s)_1] \tag{4.18}$$

where points along the slider line are a function of the single variable β. Note that $\beta = 0$ at the first position $(x_s, y_s)_1$ and $\beta = 1$ at the third position $(x_s, y_s)_3$
9. Compute β corresponding to the distance to the slider point in the second position using

$$\beta_2 = \frac{[(x_s, y_s)_2 - (x_s, y_s)_1]}{[(x_s, y_s)_3 - (x_s, y_s)_1]} \tag{4.19}$$

10. Check to ensure that the linkage goes through the positions in the correct order. For this to occur, position 2 must lie between positions 1 and 3 or $0 < \beta_2 < 1$.

4.2.7.8 Implementing the Analytical Approach to Rigid Body Guidance

The procedures given in Sections 4.2.7.1 through 4.2.7.7 can be used with a calculator to design four-bar linkages with revolute joints and sliders sliding on the frame. The procedures can also be programmed easily on a computer using any of the various languages. It is especially easy to program the procedure in MATLAB because of the ease with which matrix and vector manipulations may be carried out. A program that implements the procedure with limited graphical output is given on the disk included with this book.

4.3 FUNCTION GENERATION

The procedure developed here will use a four-bar linkage to generate the desired function; however, the general ideas presented can be used for any system for which a functional relationship can be derived between two variables. For example, assume that a "black box" is given such that the functional relationship between the two variables α and ρ is

$$f(\alpha, \rho, a_1, a_2, a_3, a_4) = 0 \tag{4.20}$$

where a_1, a_2, a_3, a_4 are design variables defining the system. To design the system to approximate the function

$$g(\alpha, \rho) = 0 \tag{4.21}$$

we simply need to solve Eq. (4.21) four times to obtain four pairs of values for α and ρ. We can designate these as (α_1, ρ_1), (α_2, ρ_2), (α_3, ρ_3), and (α_4, ρ_4). Next rewrite Eq. (4.20) four times (corresponding to the number of design variables), one for each pair of (α_i, ρ_i), and solve the resulting set of equations for a_1, a_2, a_3, a_4. Note that the equations may be nonlinear, requiring the use of numerical techniques.

In the example considered here, the function f in Eq. (4.20) will be the governing position equation for a four-bar linkage. The function g will be an arbitrary function which is specified at the beginning of the analysis.

4.3.1 Function Generation Using a Four-Bar Linkage

Function generation using a four-bar linkage was developed by Freudenstein[19], and the basic equation relating the input and output variables for the four-bar linkage is called Freudenstein's equation.

Given three pairs of values for θ and ϕ in Fig. 4.32, the objective for the case considered here is to find r_2, r_3, and r_4 for the four-bar linkage. This is the linkage that will approximate the function implied by the three pairs of values for θ and ϕ. In the three-position function generation problem, the size of the linkage does not affect the functional relationship between θ and ϕ. Therefore, the frame link (r_1) can be taken as 1 initially, and the entire linkage can be scaled after the basic design is established.

To develop the governing equation relating the input and output variables for the linkage, first determine expressions for the x and y components of the vectors corresponding to each link length. For the x direction,

$$r_2 \cos\theta + r_3 \cos\psi = 1 + r_4 \cos\phi \tag{4.22}$$

and for the y direction,

$$r_2 \sin\theta + r_3 \sin\psi = r_4 \sin\phi \tag{4.23}$$

We do not want ψ in the final equation. Therefore, isolate the terms involving ψ so that this angle can be eliminated. Then

$$\begin{aligned} r_3 \cos\psi &= 1 + r_4 \cos\phi - r_2 \cos\theta \\ r_3 \sin\psi &= r_4 \sin\phi - r_2 \sin\theta \end{aligned} \tag{4.24}$$

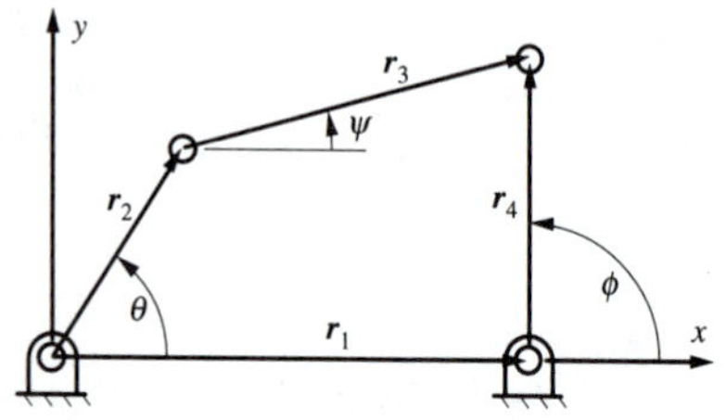

Figure 4.32 Four-bar linkage used for function generation.

Square both equations and add to get

$$r_3^2(\cos^2\psi + \sin^2\psi) = (1 + r_4\cos\phi - r_2\cos\theta)^2 + (r_4\sin\phi - r_2\sin\theta)^2$$

Expand the equation and simplify,

$$r_3^2 = 1 + r_4^2 + r_2^2 + 2r_4\cos\phi - 2r_2\cos\theta - 2r_4r_2\cos(\phi - \theta)$$

Since we have three pairs of values for θ and ϕ, this equation can be written three times as

$$\begin{aligned} r_3^2 &= 1 + r_4^2 + r_2^2 + 2r_4\cos\phi_1 - 2r_2\cos\theta_1 - 2r_4r_2\cos(\phi_1 - \theta_1) \\ r_3^2 &= 1 + r_4^2 + r_2^2 + 2r_4\cos\phi_2 - 2r_2\cos\theta_2 - 2r_4r_2\cos(\phi_2 - \theta_2) \\ r_3^2 &= 1 + r_4^2 + r_2^2 + 2r_4\cos\phi_3 - 2r_2\cos\theta_3 - 2r_4r_2\cos(\phi_3 - \theta_3) \end{aligned} \quad \textbf{(4.25)}$$

Equations (4.25) can be solved for r_2, r_3, and r_4. The procedure used to solve the equaitons depends on the tools that are available.

4.3.1.1 *Solution by Matrices*

The equations can be written simply in matrix form. To simplify the result, first divide each equation in Eq. (4.25) by $2r_2r_4$ and define the new unknowns as

$$z_1 = \frac{1 + r_2^2 + r_4^2 - r_3^2}{2r_2r_4} \quad \textbf{(4.26)}$$

$$z_2 = \frac{1}{r_2} \quad \textbf{(4.27)}$$

and

$$z_3 = \frac{1}{r_4} \quad \textbf{(4.28)}$$

We can then write Eqs. (4.25) as

$$\begin{aligned} z_1 + z_2\cos\phi_1 - z_3\cos\theta_1 &= \cos(\theta_1 - \phi_1) \\ z_1 + z_2\cos\phi_2 - z_3\cos\theta_2 &= \cos(\theta_2 - \phi_2) \\ z_1 + z_2\cos\phi_3 - z_3\cos\theta_3 &= \cos(\theta_3 - \phi_3) \end{aligned} \quad \textbf{(4.29)}$$

or in matrix form,

$$\begin{bmatrix} 1 & \cos\phi_1 & -\cos\theta_1 \\ 1 & \cos\phi_2 & -\cos\theta_2 \\ 1 & \cos\phi_3 & -\cos\theta_3 \end{bmatrix} \begin{Bmatrix} z_1 \\ z_2 \\ z_3 \end{Bmatrix} = \begin{Bmatrix} \cos(\theta_1 - \phi_1) \\ \cos(\theta_2 - \phi_2) \\ \cos(\theta_3 - \phi_3) \end{Bmatrix} \quad \textbf{(4.30)}$$

We can solve for z_1, z_2, z_3 using MATLAB or some other matrix solver. Symbolically,

$$\begin{Bmatrix} z_1 \\ z_2 \\ z_3 \end{Bmatrix} = \begin{bmatrix} 1 & \cos\phi_1 & -\cos\theta_1 \\ 1 & \cos\phi_2 & -\cos\theta_2 \\ 1 & \cos\phi_3 & -\cos\theta_3 \end{bmatrix}^{-1} \begin{Bmatrix} \cos(\theta_1 - \phi_1) \\ \cos(\theta_2 - \phi_2) \\ \cos(\theta_3 - \phi_3) \end{Bmatrix} \quad \textbf{(4.31)}$$

Knowing z_1, z_2, z_3, we can solve for the unknown link lengths in terms of r_1 using Eqs. (4.26), (4.27), and (4.28), since r_1 is arbitrary. Then,

$$r_2 = \frac{1}{z_2} \quad \textbf{(4.32)}$$

$$r_4 = \frac{1}{z_3} \tag{4.33}$$

and

$$r_3 = \sqrt{1 + r_2^2 + r_4^2 - 2r_2 r_4 z_1} \tag{4.34}$$

Note that the square root used to compute r_3 can be plus or minus. Only the plus sign has a physical meaning, however, since r_3 is physically the distance from the end of link 2 to the end of link 4.

4.3.1.2 *Unscaling the Solution*

In the preceding derivation, it is assumed that the base link (r_1) is 1. This is not generally the case. However, to determine the true size of the links, it is necessary to know the size of just one of the links initially. Through a scaling factor, we can determine the size of the other links.

Assume that the actual link lengths are R_1, R_2, R_3, R_4, where the R's are related to the computed r's through the following:

$$\left.\begin{aligned} R_1 &= Kr_1 \\ R_2 &= Kr_2 \\ R_3 &= Kr_3 \\ R_4 &= Kr_4 \end{aligned}\right\} \tag{4.35}$$

where K is the scale factor for the linkage. From Eq. (4.35),

$$K = R_1/r_1 = R_2/r_2 = R_3/r_3 = R_4/r_4 \tag{4.36}$$

After the design procedure is completed, we will know r_1, r_2, r_3, and r_4. Therefore, we need to specify only *one* of R_1, R_2, R_3, or R_4 to find K using Eq. (4.36). Knowing K, we can compute the actual link lengths using Eq. (4.35).

4.3.2 Design Procedure When $y = y(x)$ Is to Be Generated

Generally, in function generation, θ and ϕ will not be given directly. Instead, the linkage will be designed to approximate a function $y = y(x)$ where y corresponds to ϕ (the output) and x corresponds to θ (the input). The angles θ and ϕ will be related to x and y such that given θ and ϕ, x and y can be computed. The functional relationships between ϕ and y and θ and x are somewhat arbitrary; however, the problem is most easily solved if linear relationships are used. The most common relationships are

$$\frac{x - x_0}{x_f - x_0} = \frac{\theta - \theta_0}{\theta_f - \theta_0}$$

and

$$\frac{y - y_0}{y_f - y_0} = \frac{\phi - \phi_0}{\phi_f - \phi_0}$$

or

$$x = \frac{x_f - x_0}{\theta_f - \theta_0}(\theta - \theta_0) + x_0 \tag{4.37}$$

$$y = \frac{y_f - y_0}{\phi_f - \phi_0}(\phi - \phi_0) + y_0 \tag{4.38}$$

and

$$\theta = \frac{\theta_f - \theta_0}{x_f - x_0}(x - x_0) + \theta_0 \tag{4.39}$$

$$\phi = \frac{\phi_f - \phi_0}{y_f - y_0}(y - y_0) + \phi_0 \tag{4.40}$$

When the design is formulated, we will know $y = y(x)$, and the range for x ($x_0 \le x \le x_f$). Given x_0 and x_f, we can compute y_0 and y_f. We must then pick θ_0 and θ_f and ϕ_0 and ϕ_f. Then, given three design positions for x, three values for y, θ, and ϕ can be computed, and given the three values for θ and ϕ, the link lengths can be computed using the procedure given above.

Often, instead of selecting θ_0, θ_f, ϕ_0, and ϕ_f directly, θ_0, ϕ_0, and $\Delta\theta = \phi_f - \theta_0$ and $\Delta\phi = \phi_f - \phi_0$ are selected. Typically, choosing $\Delta\theta$ and $\Delta\phi$ to be between 60 and 120 degrees usually works best. It is also usually better to avoid having either the driver or the output link pass below the line defined by the two fixed pivots (line of centers) in the range where the function is to be matched; that is, make $0° \le \theta_0 \le \theta \le \theta_f \le 180°$ and $0° \le \phi_0 \le \phi \le \phi_f \le 180°$.

4.3.3 Selection of Design Positions

In general, the function actually generated by the linkage will match the actual function only at the precision points, and the error between the precision points will vary depending on where the precision points are placed in the range $x_0 \le x \le x_f$. Therefore, when trying to match the function $y = y(x)$ over the range $x_0 \le x \le x_f$, the objective is to select the precision points so that the deviation of the function actually generated from that desired between the design positions is minimized. The difference between the actual function generated and the desired function is called the *structural error e*. If this error is plotted as a function of x, it can be shown that the maximum structural error (e^*) is minimized when it takes the form shown in Fig. 4.33. Ideally, the maximum errors between the precision points are both equal in magnitude to the errors at the ends of the range.

It is usually difficult to locate the precision points so that this criterion for the error is met for an arbitrary function; however, a useful approximate solution is obtained by approximating the error function by a Chebyshev polynomial of order N, where N is equal to the number of precision points. If the approximation were exact, the optimum locations

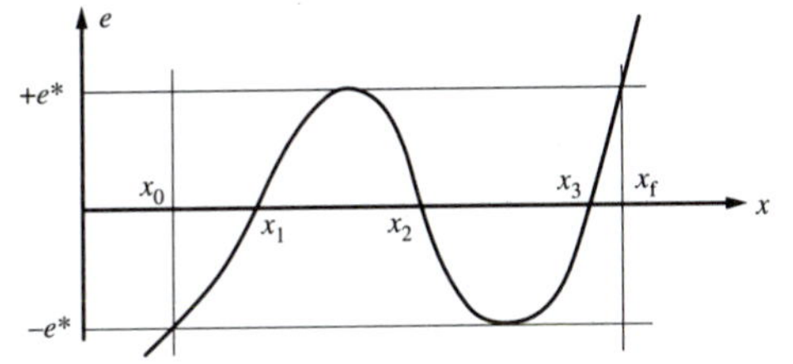

Figure 4.33 Optimum error distribution.

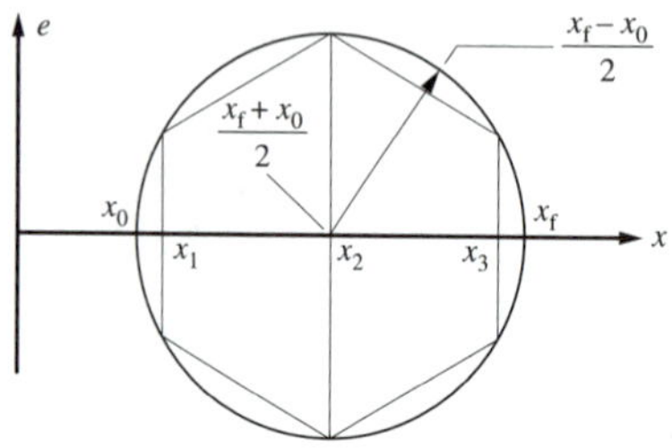

Figure 4.34 Chebyshev spacing for precision points.

of the precision points are given by

$$x_i = \frac{x_f + x_0}{2} - \frac{x_f - x_0}{2}\cos\left(\frac{\pi i}{N} - \frac{\pi}{2N}\right) \tag{4.41}$$

where $i = 1, 2, \ldots, N$. These values for x_i are the roots of the Chebyshev polynomial of order N. As noted above, Eq. (4.41) only approximates the optimum locations of the precision points, but it is still a useful starting solution to use, especially when there is no other basis upon which to choose design positions.

The roots of the Chebyshev polynomial can be given a geometric interpretation that makes it simple to derive Eq. (4.41) if the form of the equation is forgotten. For this, draw a circle of radius $(x_f - x_0)/2$ with its center at $(x_f + x_0)/2$. Then divide the circle into a regular polygon with $2N$ sides. The projection of the vertices of the polygon onto the x axis will give the locations of the precision points. This is shown in Fig. 4.34. When Chebyshev spacing is used, the center point, $(x_f + x_0)/2$, will be a precision point *only* when N is odd, and the extremes (x_f and x_0) of the range will *never* be chosen as precision points.

4.3.4 Summary of Solution Procedure

Let $y = y(x)$ be the function to be generated over the range $x_0 \le x \le x_f$. The design positions should be placed inside the range $x_0 \le x \le x_f$, and as a rule, use

$$x_1 = \frac{x_f + x_0}{2} - \frac{x_f - x_0}{2}\cos 30°$$

$$x_2 = \frac{x_f + x_0}{2}$$

$$x_3 = \frac{x_f + x_0}{2} - \frac{x_f - x_0}{2}\cos 150°$$

for these three positions.

Choose the angular range $\Delta\theta$ of the input crank which is to correspond to the range $x_0 \le x \le x_f$ for x. Also, choose the angle θ_0 corresponding to x_0 from which this range is to start.

Choose the angular range $\Delta\phi$ of the output crank which is to correspond to the range $y_0 \le y \le y_f$ for y where $y_0 = y(x_0)$ and $y_f = y(x_f)$. Also choose the angle ϕ_0 corresponding to y_0 from which the output range will start.

Compute the values of θ and ϕ which represent the precision points from the equations

$$\theta_i = \frac{\theta_f - \theta_0}{x_f - x_0}(x_i - x_0) + \theta_0 = \frac{x_i - x_0}{x_f - x_0}\Delta\theta + \theta_0$$

and

$$\phi_i = \frac{\phi_f - \phi_0}{y_f - y_0}(y_i - y_0) + \phi_0 = \frac{y_i - y_0}{y_f - y_0}\Delta\phi + \phi_0$$

where $y_i = y(x_i)$.

Next let the base length be unity and calculate the lengths for the driver, coupler, and output links using Eqs. (4.31)–(4.34). To do this, solve

$$\begin{Bmatrix} z_1 \\ z_2 \\ z_3 \end{Bmatrix} = \begin{bmatrix} 1 & \cos\phi_1 & -\cos\theta_1 \\ 1 & \cos\phi_2 & -\cos\theta_2 \\ 1 & \cos\phi_3 & -\cos\theta_3 \end{bmatrix}^{-1} \begin{Bmatrix} \cos(\theta_1 - \phi_1) \\ \cos(\theta_2 - \phi_2) \\ \cos(\theta_3 - \phi_3) \end{Bmatrix}$$

for z_1, z_2, and z_3. Then solve for r_2, r_3, and r_4 from

$$r_2 = \frac{1}{z_2}$$

$$r_4 = \frac{1}{z_3}$$

and

$$r_3 = \sqrt{1 + r_2^2 + r_4^2 - 2r_2r_4z_1}$$

Note that r_2 and r_4 can be negative. When r_2 or r_4 is negative, the direction for the vector representing the link length is reversed (see Fig. 4.35).

After the scaled link lengths are determined, determine the scale factor using

$$K = R_1/r_1 = R_2/r_2 = R_3/r_3 = R_4/r_4$$

and the true value of one of the link lengths. Then determine the true value for all of the link lengths using Eqs. (4.35).

Draw the linkage to scale, and check that a linkage with the calculated dimensions will pass through the design positions (θ_1, ϕ_1), (θ_2, ϕ_2), (θ_3, ϕ_3). The procedure guarantees only that the linkage can be assembled in the design positions. It may not be able to move from one position to another without changing branch. It is also important to check the force and torque transmission characteristics of the linkage at each design position. As will be discussed when crank-rocker mechanisms are considered, the force transmission characteristics of a four-bar linkage can change greatly from position to position.

If for some reason the linkage is unacceptable, change either the range or starting point for either θ or ϕ and determine another design.

Note that it is possible to choose different combinations of variables other than x_1, x_2, and x_3 when selecting the precision points. If we let

$$x_f - x_0 = \Delta x$$

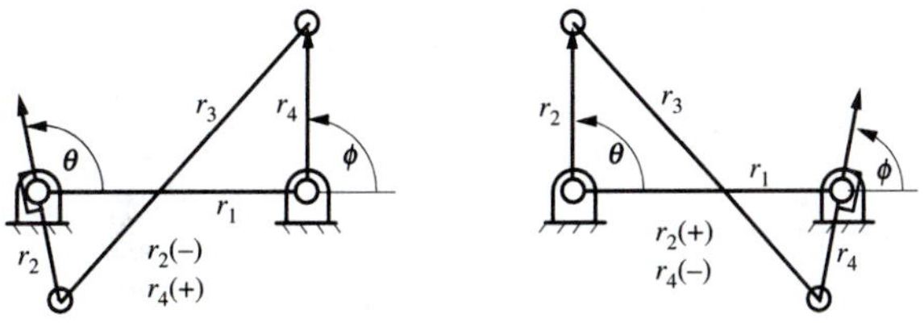

Figure 4.35 Interpretation of negative values for r_4 and r_2.

then the basic equations for selecting the precision points can be written as

$$x_1 = \frac{x_f + x_0}{2} - \frac{\Delta x}{2}\cos 30°$$

$$x_2 = \frac{x_f + x_0}{2}$$

$$x_3 = \frac{x_f + x_0}{2} + \frac{\Delta x}{2}\cos 30°$$

From this it is clear that we can select any three from the list of variables x_1, x_2, x_3, x_f, x_0, or Δx and solve for the other three.

The function generation equations have been programmed using MATLAB in the routine *fungen.m* provided on the disk with this book. This routine can be easily modified to handle a relatively wide range of function generation problems involving a four-bar linkage.

EXAMPLE 4.3 **(*Function Generation*)**

PROBLEM Design a linkage to generate the function $y = \log_{10} x$ over the range $1 \le x \le 2$.

SOLUTION The solution is given in the following. From the given information,

$$x_0 = 1, \qquad x_f = 2.$$

Using Chebyshev spacing for the precision points,

$$x_1 = \frac{x_f + x_0}{2} - \frac{x_f - x_0}{2}\cos 30° = \frac{1+2}{2} - \frac{2-1}{2}\cos 30° = 1.06699$$

Similarly, $x_2 = 1.5$ and $x_3 = 1.93301$. Then, the corresponding values for y are

$$y_f = \log_{10} 2 = 0.30103$$
$$y_0 = \log_{10} 1 = 0$$
$$y_1 = \log_{10} x_1 = 0.028160$$
$$y_2 = \log_{10} x_2 = 0.176091$$
$$y_3 = \log_{10} x_3 = 0.28623$$

Note that a minimum of five places of decimals is needed to ensure adequate solution accuracy. To identify the linkage angles, choose

$$\theta_0 = 45°, \qquad \Delta\theta = 60°$$

and

$$\phi_0 = 0°, \qquad \Delta\phi = 60°$$

Note that these values are somewhat arbitrary. If the resulting linkage is unacceptable, we can try other values. The precision points in terms of θ are

$$\theta_1 = \frac{x_1 - x_0}{x_f - x_0}\Delta\theta + \theta_0 = \frac{1.06699 - 1}{1}60° + 45° = 49.019°$$

$$\theta_2 = \frac{x_2 - x_0}{x_f - x_0}\Delta\theta + \theta_0 = \frac{1.5 - 1}{1}60° + 45° = 75.000°$$

$$\theta_3 = \frac{x_3 - x_0}{x_f - x_0}\Delta\theta + \theta_0 = \frac{1.93301 - 1}{1}60° + 45° = 100.981°$$

Similarly,

$$\phi_1 = \frac{y_1 - y_0}{y_f - y_0}\Delta\phi + \phi_0 = \frac{0.028160 - 1}{0.30103}60° + 0 = 5.612°$$

$$\phi_2 = \frac{y_2 - y_0}{y_f - y_0}\Delta\phi + \phi_0 = \frac{0.176091 - 1}{0.30103}60° + 0 = 35.098°$$

$$\phi_3 = \frac{y_3 - y_0}{y_f - y_0}\Delta\phi + \phi_0 = \frac{0.28623 - 1}{0.30103}60° + 0 = 57.050°$$

Using the matrix solution procedure,

$$\begin{Bmatrix} z_1 \\ z_2 \\ z_3 \end{Bmatrix} = \begin{bmatrix} 1 & \cos\phi_1 & -\cos\theta_1 \\ 1 & \cos\phi_2 & -\cos\theta_2 \\ 1 & \cos\phi_3 & -\cos\theta_3 \end{bmatrix}^{-1} \begin{Bmatrix} \cos(\theta_1 - \phi_1) \\ \cos(\theta_2 - \phi_2) \\ \cos(\theta_3 - \phi_3) \end{Bmatrix}$$

$$= \begin{bmatrix} 1 & 0.9952 & -0.6558 \\ 1 & 0.8182 & -0.2588 \\ 1 & 0.5439 & -0.1905 \end{bmatrix}^{-1} \begin{Bmatrix} 0.7265 \\ 0.7671 \\ 0.7202 \end{Bmatrix} = \begin{Bmatrix} -0.0900 \\ 1.2574 \\ 0.6631 \end{Bmatrix}$$

and

$$r_2 = \frac{1}{z_2} = \frac{1}{1.2574} = 0.7953$$

$$r_4 = \frac{1}{z_3} = \frac{1}{0.6631} = 1.5080$$

and

$$r_3 = \sqrt{1 + r_2^2 + r_4^2 - 2r_2r_4z_1} = \sqrt{1 + 0.7953^2 + 1.5080^2 - 2(0.7953)(1.5080)(-0.0900)} = 2.0304$$

For the overall size of the linkage, use a base link length of 2 in. Then the lengths of the other links become

$$R_1 = 1(2) = 2''$$

$$R_2 = 0.7953(2) = 1.5905 \text{ in}$$

$$R_4 = 1.5080(2) = 3.0160 \text{ in}$$

$$R_3 = 2.0304(2) = 4.0608 \text{ in}$$

The linkage is drawn to scale in Fig. 4.36. ■

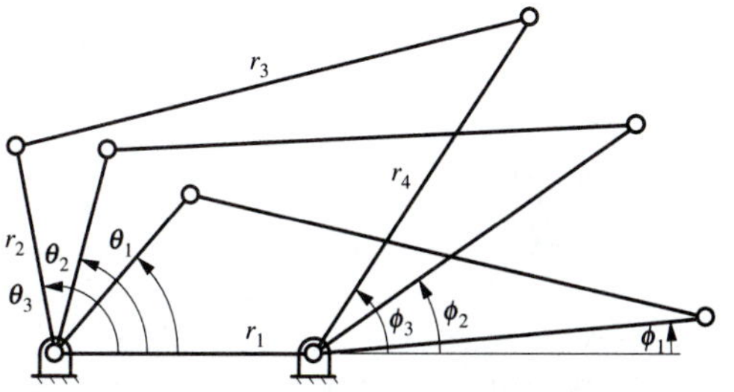

Figure 4.36 Final linkage for Example 4.3.

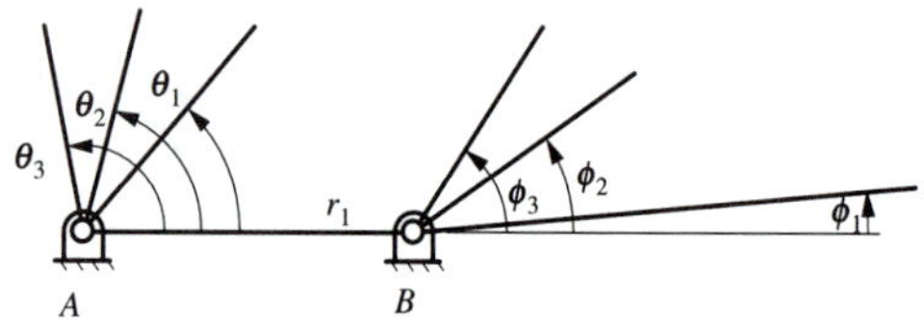

Figure 4.37 Three positions for input and output links for graphical synthesis.

4.3.5 Graphical Approach to Function Generation

The function generation problem can be solved graphically if the linkage is inverted so that the crank becomes the temporary frame. We can choose any position of the crank to start the construction, but position 1 is the most common position to choose. To illustrate the procedure, the problem in Example 4.3 will be used again. That is, it will be assumed that three pairs of points (θ_1, ϕ_1), (θ_2, ϕ_2), (θ_3, ϕ_3) are known. The input and output cranks in the three positions are shown in Fig. 4.37. Note that the base link (r_1) has been chosen, but the lengths of the input and output links (r_2 and r_4) have not been determined.

The next step is to invert the linkage so that the driver (r_2) becomes the frame. To do this, treat the group of links, r_1, r_2, and r_4, as a rigid body in each position and rotate the group of links such that r_2 is in the same location for each position. This is shown in Fig. 4.38. In the inverted linkage, r_2 is the frame, r_3, the original coupler, is the output link, and r_4, the original output link, becomes the coupler. For the inverted linkage, one crank is known, so we need only establish the other crank (r_3 for the original linkage), and the linkage geometry is established. The problem has therefore been converted to a rigid-body guidance problem where r_4 is the coupler to be guided. One crank (r_1) has already been established. To find the other crank, choose a point C on link 4. The center of the circle on which C_1, C_2, and C_3 lie is C^*. The synthesized linkage in position 1 is A, B_1, C_1, C^* as shown in Fig. 4.39. When inverted back to the original base, we have the solution of the function generation problem. The final solution linkage is shown in Fig. 4.40. Note that θ_0 and θ_f are different in this solution compared with the analytical solution, but $\Delta\theta$ is the same for both solutions. Here, we chose the position of C (that is, the length of r_4) rather than θ_0.

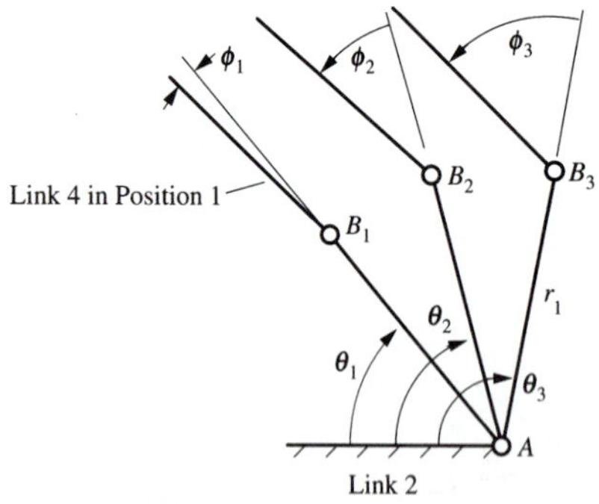

Figure 4.38 Inversion of linkage making r_2 the frame. The original linkage is inverted onto its driving crank in order to convert the function generation problem into a motion generation problem.

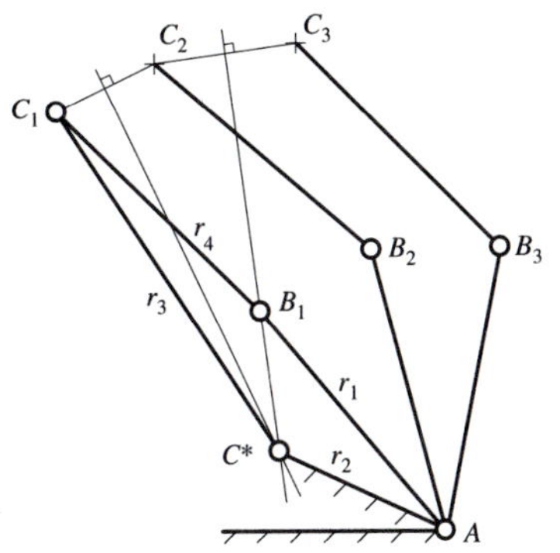

Figure 4.39 Construction of center point C* given three positions of C.

4.4 SYNTHESIS OF CRANK-ROCKER LINKAGES FOR SPECIFIED-ROCKER AMPLITUDE

In a crank-rocker mechanism, the crank rotates for 360°, and the rocker oscillates through an angle θ. This mechanism is often used interchangeably with cam mechanisms for the same function; however, there are many cases in which a crank-rocker mechanism is superior to a cam-follower mechanism. Among the advantages over cam systems are small forces involved, the elimination of the retaining spring, and the smaller clearances because of the use of revolute joints.

4.4.1 Extreme Rocker Positions and Simple Analytical Solution

The maximum and minimum rocker angles occur in the positions shown in Fig. 4.41. Using the cosine rule:

$$(r_2 + r_3)^2 = r_1^2 + r_4^2 - 2r_1 r_4 \cos\rho \tag{4.42}$$

and

$$(r_3 - r_2)^2 = r_1^2 + r_4^2 - 2r_1 r_4 \cos\beta \tag{4.43}$$

Since we have two equations in six variables, r_1, r_2, r_3, r_4, ρ, and β, the values of four variables can be specified. As in the case of function generation, the angles in the triangles are independent of the size of the triangles. Therefore, r_1 is usually taken as 1. Usually, the design problem requires only a specified value for the difference $\beta - \rho$. Nevertheless, the solution is much simpler if both ρ and β are specified. Since a crank rocker is sought, it is also helpful to specify r_2. Values in the range $0.1 \le r_2 \le 0.4$ will usually give good results when $r_1 = 1$. Assuming ρ, β, r_1, and r_2 are specified, it is possible to solve Eqs. (4.42) and (4.43) for r_3 and r_4. Adding the equations and simplifying, we get

$$r_2^2 + r_3^2 = r_1^2 + r_4^2 - r_1 r_4(\cos\rho + \cos\beta) \tag{4.44}$$

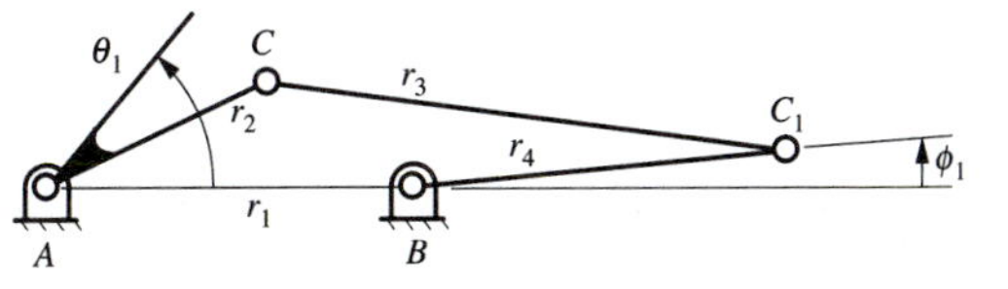

Figure 4.40 Final solution linkage.

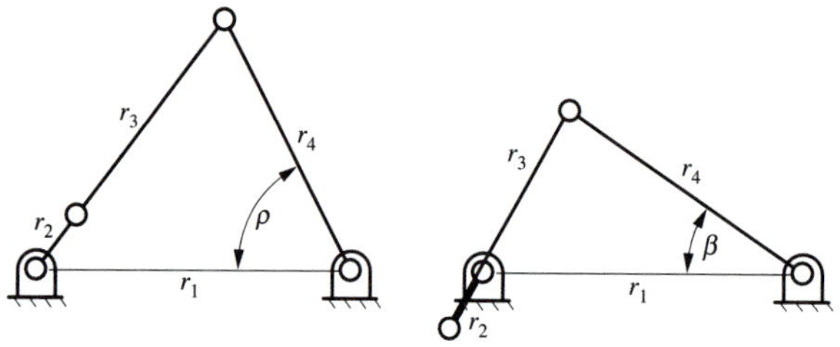

Figure 4.41 The positions of a crank-rocker linkage in which the rocker is at the extremes of its motion range.

Subtracting Eq. (4.43) from (4.42), we get

$$2r_2r_3 = r_1r_4(\cos\beta - \cos\rho) \tag{4.45}$$

Hence

$$r_3 = \frac{r_1r_4(\cos\beta - \cos\rho)}{2r_2} \tag{4.46}$$

Substitution into Eq. (4.44) gives, after some manipulation,

$$Pr_4^2 + Qr_4 + R = 0 \tag{4.47}$$

where

$$\begin{aligned} P &= 1 - \frac{r_1^2(\cos\beta - \cos\rho)^2}{4r_2^2} \\ Q &= -r_1(\cos\beta + \cos\rho) \\ R &= r_1^2 - r_2^2 \end{aligned} \tag{4.48}$$

The solution to the quadratic equation (Eq. 4.47) gives a positive root and a negative root for r_4. The negative root can be discarded. If no real roots exist, it is necessary to choose a new value of β and try again. Once r_4 is known, r_3 can be found from Eq. (4.46). The formulation of the problem guarantees that the joint between r_2 and r_3 will rotate completely, but this still does not guarantee a crank-rocker solution (it could be a type 1 double rocker). Thus, although it is a type 1 linkage, it is necessary to check that $r_3 > r_2$. The Grashof inequality can be used as a simple check on arithmetic.

The simplified procedure given above can be used if only the oscillation angle $(\rho - \beta)$ is of interest. The procedure gives no control over the time ratio of the forward oscillation to the reverse oscillation. This time ratio is often of interest, however, and the following procedure gives one a means of incorporating it in the basic design procedure.

4.4.2 The Rocker Amplitude Problem: Graphical Approach

As the crank in a crank-rocker mechanism rotates through 360°, the output link or rocker will oscillate through an angle θ. The limiting positions of the rocker occur when the crank and coupler are collinear as shown in the drawing (Fig. 4.42). In general, the time required for the rocker oscillation in one direction will be different from the time required for the other direction. As indicated above, the ratio of the times required for the forward and return motions is called the time ratio. An expression for the time ratio can be developed by using the nomenclature defined in Fig. 4.42.

In the crank rocker, the crank moves through the angle ψ while the rocker moves from

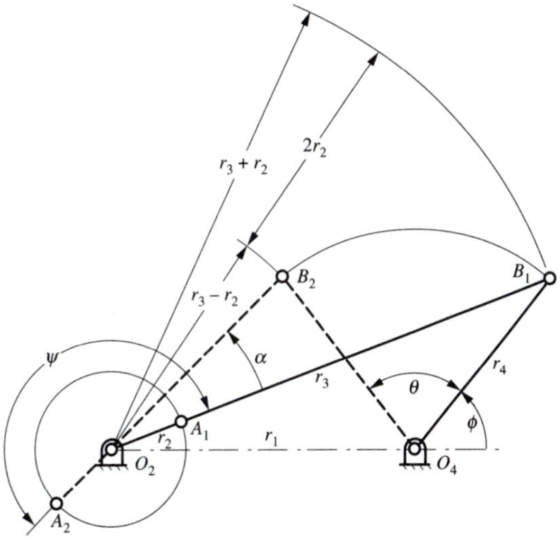

Figure 4.42 Crank-rocker mechanism in extreme positions.

B_1 to B_2 through the angle θ. On the return stroke, the crank moves through the angle $360° - \psi$ and the rocker moves from B_2 to B_1 through the same angle θ.

Assuming that the crank moves with constant angular velocity, the ratio of the times for the forward and reverse strokes of the follower can be related directly to the angles in Fig. 4.42. The crank angle for the forward stroke is ψ or $180° + \alpha$. The crank angle for the return stroke is $360° - \psi$ or $180° - \alpha$. Therefore, the time ratio, Q, can be written as

$$Q = \frac{180 + \alpha}{180 - \alpha} \tag{4.49}$$

where α is given in degrees.

The most common problem associated with the synthesis of crank-rocker mechanisms is that of designing the linkage for a given oscillation angle and a given time ratio. For the discussion here, assume that the time ratio, Q, has been given. The first step in the synthesis is to compute the angle α. This can be done by rewriting the basic equation for Q. Then,

$$\alpha = 180\frac{(Q - 1)}{(Q + 1)} \tag{4.50}$$

Note that α is positive when Q is greater than 1 and negative when Q is less than 1. Examples of positive and negative α are shown in Fig. 4.43.

Once α is known, there are a number of ways to proceed with the design. The simplest way is to choose a location for O_4, select ϕ, and draw the two positions of the rocker (r_4) separated by the angle θ. Draw any line x through the pivot at B_1, and construct a second line at an angle of α to the line x and through the pivot at B_2. Call the second line y. The intersection of the lines x and y defines the location of the second fixed pivot (O_2).

Next compute the values of r_2 and r_3. This is done by using the geometry relationships in Fig. 4.42. That is,

$$r_2 + r_3 = \overline{O_2B_1}$$

and

$$r_3 - r_2 = \overline{O_2B_2}$$

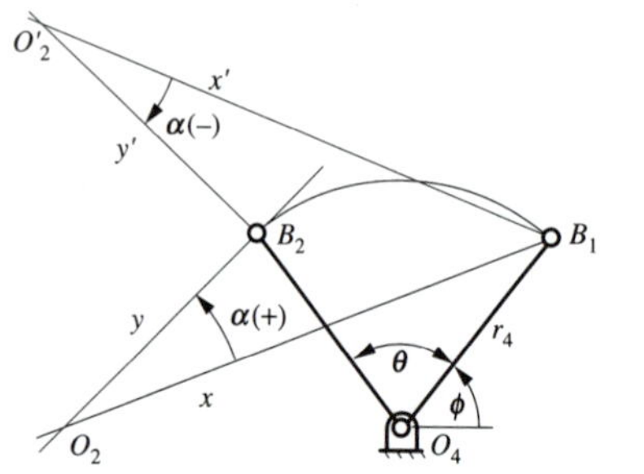

Figure 4.43 Locations of O_2 given θ and + or $-\alpha$.

Therefore,

$$r_2 = \frac{\overline{O_2B_1} - \overline{O_2B_2}}{2} \tag{4.51}$$

and

$$r_3 = \frac{\overline{O_2B_1} + \overline{O_2B_2}}{2} \tag{4.52}$$

Note that during the design procedure, several choices were made. Among these were the starting angle ϕ for the line O_4B_1 and the slope of the line x. There is an infinity of choices, and each choice will give a different linkage.

Note also that not all solutions are valid. In particular, the pivots B_1 and B_2 may not extend below the line of centers defined by a line through the fixed pivots O_2 and O_4. If this happens, the linkage must change branch to reach the two positions, and the desired oscillation angle will not be achieved. Once α and θ are known, the locus of acceptable positions for O_2 must lie on circle arcs represented by the heavy sections of the circles shown in Fig. 4.44. The locus of O_2 must be on a circle arc because the triangle $B_2B_1O_2$

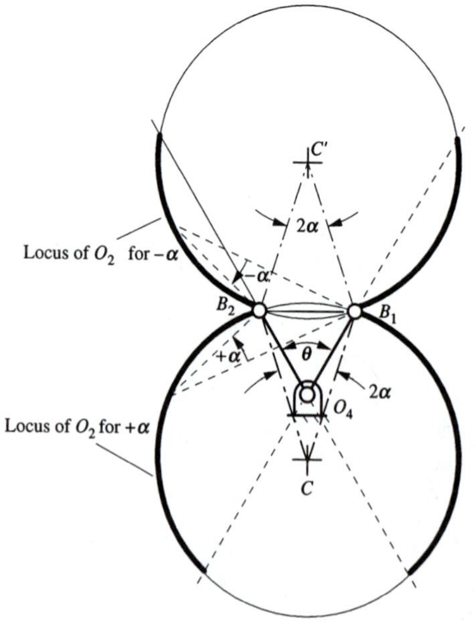

Figure 4.44 Possible locations for O_2 given θ and α.

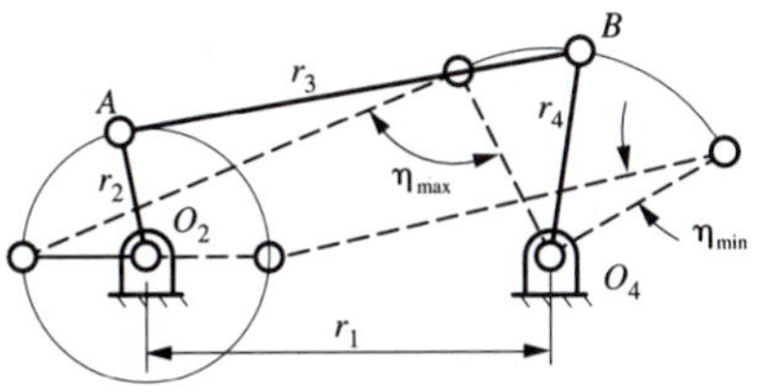

Figure 4.45 Maximum and minimum transmission angles.

has a fixed base and a constant apex angle (α). If O_2 is chosen in the light part of the circles, the two positions of B will be on opposite sides of the line of centers.

In some instances, the length of one of the links must be a specific length. However, the procedure outlined here will permit only the length of the rocker to be specified directly. If the length of one of the other links is known, the lengths of the links in the linkage can be scaled using the procedure given in Section 4.3.1.4.

4.4.3 Optimal Synthesis Approach Based on the Specification of O_2–O_4

Ultimately, the design of the crank rockers reduces to an optimization problem because a single design based on the construction will usually have poor transmission angle characteristics. The maximum and minimum transmission angles are shown in Fig. 4.45. Typically, a poor transmission angle corresponds to a large value of $|\pi/2 - \eta_{max/min}|$. The procedure given in Section 4.4.1 does not lend itself to exploration of the design in a procedural manner. For this, an approach presented by Hall[4] is easier to use. This approach reduces the design problem to a one-dimensional problem where well-defined limits are known for the design variable.

4.4.3.1 Alternative Graphical Design Procedure

Given the geometry in Fig. 4.46 and assuming θ and α are known, the following procedure provides a means for determining all of the linkages satisfying the design requirements.

1. Pick the base link and locate the ground pivots O_2 and O_4. The distance O_2O_4 determines the scale for the linkage.
2. Draw the line O_2G at an angle $\theta/2 - \alpha$ (positive clockwise) relative to O_2O_4.
3. Draw the line O_4G at an angle of $\theta/2$ (positive counterclockwise) relative to O_2O_4.
4. Draw the circle of radius GO_2 centered at G.
5. Draw a line ($O_4B_{2_m}$) through O_4 at an angle of θ (positive counterclockwise) relative to O_2O_4.

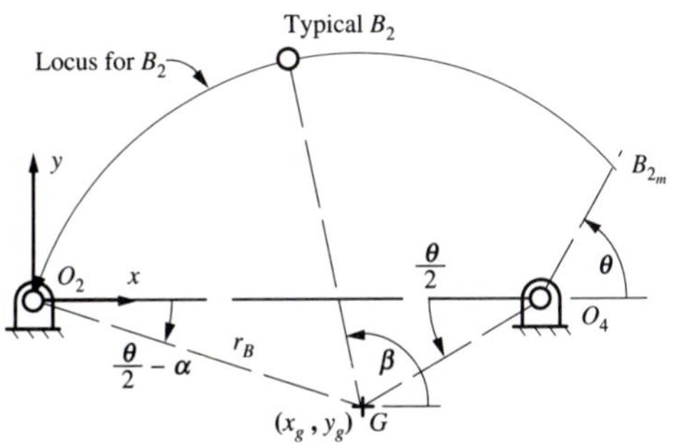

Figure 4.46 Construction of the circle arc giving the locus B_2.

6. The circle arc starting at O_2 and ending at the intersection with either $O_4B_{2_m}$ or O_2O_4 (whichever occurs first) gives the locus of the point B in the second extreme position of the rocker. The point (B_1) located at an angle of θ relative to the line O_4B is the other extreme position. The length r_4 is equal to O_4B.

After B_2 is chosen and B_1 is located, the remaining link lengths can then be computed by solving Eqs. (4.51) and (4.52). The transmission angle limits can be measured after drawing the linkage in the extreme positions shown in Fig. 4.45.

Reviewing the graphical procedure, it is apparent that once θ and α are known, the arc defining the loci for B_2 is defined. Locating a point on the arc requires the specification of only one additional parameter (β). Therefore, different designs can be developed by adjusting this single variable. Furthermore, because the locus for B_2 is the circle arc between O_2 and B_{2_m}, the limits for β can be established at the beginning of the design procedure.

4.4.3.2 Analytical Design Procedure

To determine the analytical equations for the arc giving the locus of B_2, it is necessary only to find the coordinates of the circle center G and the limits β_1 and β_2. Locating the x and y axes as shown in Fig. 4.46, the center (x_g, y_g) is defined by the triangle O_2O_4G, which has one known side (O_2O_4) and two known included angles ($\theta/2 - \alpha$ and $\theta/2$). By the law of sines, the B_2 circle radius is

$$r_B = O_2O_4\left[\frac{\sin(\theta/2)}{\sin(\pi - \theta + \alpha)}\right] \tag{4.53}$$

and the center of the circle is given by

$$x_g = r_B \cos\left(\frac{\theta}{2} - \alpha\right) \tag{4.54}$$

$$y_g = -r_B \sin\left(\frac{\theta}{2} - \alpha\right) \tag{4.55}$$

A given point on the B_2 circle can be found using r_B and the angle β; however, the allowable range for β must be determined first. The point B on the rocker cannot lie below the line defined by O_2O_4. One extreme position of the locus for B_2 is O_2. The other extreme position can occur in one of two ways as indicated in step 6 of the graphical procedure. The two conditions depend on the sign of α as indicated in Fig. 4.47.

In Fig. 4.47a, the extreme position is found by noting that B_2 cannot lie below the extension of O_2O_4. Therefore, the extreme position of the B_2 locus is at the intersection of the B_2 circle and the line through O_4 at an angle of θ with the horizontal. The coordinates of this intersection point are found most easily if we first scale the linkage by setting $O_2O_4 = 1$. Then

$$r_B \cos(\beta_{min}) = x_m - x_g \tag{4.56}$$

$$r_B \sin(\beta_{min}) = y_m - y_g \tag{4.57}$$

and

$$\frac{y_m}{x_m - 1} = \tan\theta \tag{4.58}$$

If Eqs. (4.56) and (4.57) are squared and added and Eq. (4.58) is used to eliminate x_m from the result, the following quadratic equation results for y_m:

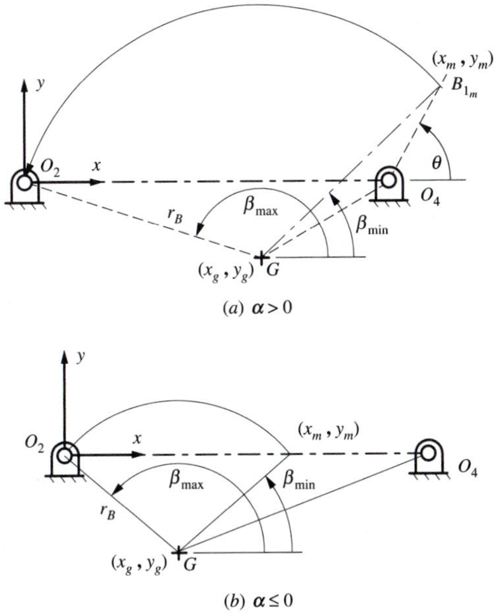

Figure 4.47 Extreme values for β.

$$\left[\frac{1}{\tan^2\theta}+1\right]y_m^2 - 2\left[\frac{x_g-1}{\tan\theta}+y_g\right]y_m + [1 + y_g^2 + x_g^2 - 2x_g - r_B^2] = 0 \quad \textbf{(4.59)}$$

or

$$y_m = [-B + \sqrt{B^2 - 4AC}]/2A \quad \textbf{(4.60)}$$

where

$$A = \frac{1}{\tan^2\theta} + 1$$

$$B = -2\left[\frac{x_g - 1}{\tan\theta} + y_g\right]$$

$$C = [1 + y_g^2 + x_g^2 - 2x_g - r_B^2]$$

Two values for y are mathematically possible, but there will be a maximum of one positive root. Only positive values for y_m are of interest. If both values of y are negative, the condition in Fig. 4.47b is indicated. If one y_m is positive, then x_m can be computed using Eq. (4.58) or

$$x_m = 1 + \frac{y_m}{\tan\theta} \quad \textbf{(4.61)}$$

For the case indicated in Fig. 4.47b,

$$y_m = 0 \quad \textbf{(4.62)}$$

and

$$x_m = 2r_B\cos\left(\frac{\theta}{2} - \alpha\right) \quad \textbf{(4.63)}$$

Three other special cases can cause numerical difficulties: when $\theta = 90°$, when $B^2 - 4AC < 0$, and when $\theta = \alpha$. The first condition gives an infinite value for tan θ. Then Eq. (4.59) can be solved directly. That is by solving

$$y_m^2 - [2y_g]y_m + [1 + y_g^2 + x_g^2 - 2x_g - r_B^2] = 0 \tag{4.64}$$

Again, only the positive root for y_m is of interersst. The second case implies that y_m is complex, so no real root is possible (i.e., no solution exists).

From Eq. (4.53), the third case gives an infinite value for r_B as $\sin(\pi - \theta + \alpha) = 0$. Therefore, the locus for B_2 is a straight line through O_2 at an angle of $\pi/2 + \theta/2 - \alpha$ to the horizontal. This is shown in Fig. 4.48. B_2 can be located anywhere along the straight line, including at infinity. When B_2 is located at infinity, the coupler becomes a slider and the mechanism becomes an inverted slider-crank mechanism (see Hall[4]). Computationally, this case can be treated separately, or a small positive value can be used for the expression $\sin(\pi - \theta + \alpha)$ when it would normally be zero. From a programming standpoint, using this limit is reasonable to avoid numerical instabilities. Once the coordinates for the extreme values for the B_2 circle are known, the limits for β can be found using Fig. 4.47a and b. The minimum β is

$$\beta_{min} = \tan^{-1}\left[\frac{y_m - y_g}{x_m - x_g}\right] \tag{4.65}$$

and the maximum value for β is

$$\beta_{max} = \pi + \alpha - \frac{\theta}{2} \tag{4.66}$$

For a valid design, $\beta_{min} \le \beta \le \beta_{max}$.

For any valid value of β, the link lengths are computed using Eqs. (4.51) and (4.52) together with the following:

$$x = r_B \cos\beta + x_g \tag{4.67}$$

$$y = r_B \sin\beta + y_g \tag{4.68}$$

$$r_4 = (x - 1)^2 + y \tag{4.69}$$

$$O_2B_1 = x^2 + y^2 \tag{4.70}$$

$$\phi = \tan^{-1}\left(\frac{y}{x - 1}\right) \tag{4.71}$$

$$O_2B_2 = [1 + r_4\cos(\phi - \theta)]^2 + [r_4\sin(\phi - \theta)]^2 \tag{4.72}$$

In Eq. (4.71), it is important to preserve the signs of both the numerator and denominator since ϕ may be greater than $\pi/2$. The crank-rocker design equations are coded in MATLAB and included on the disk with this book.

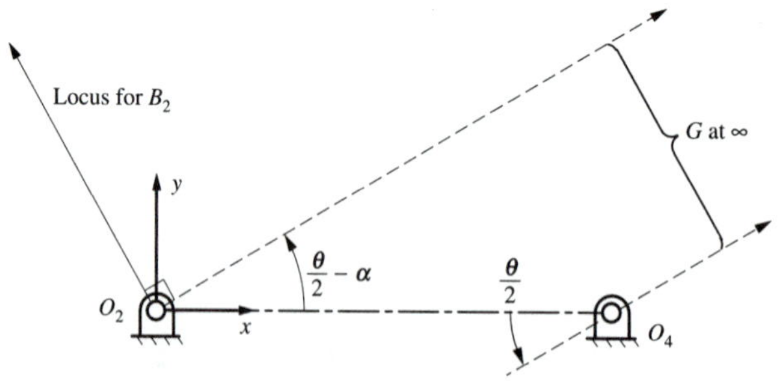

Figure 4.48 Locus for B_2 when $\theta = \alpha$.

4.4.3.3 Use of Analytical Design Procedure for Optimization

Using the procedure developed, any value of β satisfying $\beta_{min} \leq \beta \leq \beta_{max}$ will produce a linkage that satisfies the basic design requirements for θ and α (or Q). To choose the best linkage, it is reasonable to optimize the transmission angle, although other criteria can also be chosen. Referring to Fig. 4.45, the maximum and minimum values for the transmission angle can be computed from

$$\eta'_{max} = \cos^{-1}\left[\frac{r_4^2 - (r_1 + r_2)^2 + r_3^2}{2r_4 r_3}\right]$$

$$\eta'_{min} = \cos^{-1}\left[\frac{r_4^2 - (r_1 - r_2)^2 + r_3^2}{2r_4 r_3}\right]$$

If η'_{max} is negative, then $\eta_{max} = \pi + \eta'_{max}$. Otherwise, $\eta_{max} = \eta'_{max}$.

The basic objective function to be minimized is

$$U' = \text{max of } \left|\frac{\pi}{2} - \eta_{max}\right| \text{ and } \left|\frac{\pi}{2} - \eta_{min}\right|$$

However, if U' is used directly, the optimum linkage will occasionally be one where r_2 is very small compared with one of the other link lengths. Therefore, it is convenient to include the link length ratio as part of the objective function. Because r_2 will be the shortest link, this function can be written as

$$F = \text{max of } \left[\frac{r_1}{r_2}, \frac{r_3}{r_2}, \frac{r_4}{r_2}\right]$$

and

$$U'' = e^{(n-F)}$$

where n is an integer which represents the largest acceptable value for the link length ratio. Here, it is assumed that the length ratios less than n are acceptable. A typical value for n is 5.

The combined objective function is

$$U = U' + WU''$$

where W is a weighting factor that can be chosen to adjust the relative importance of the length ratio. In many problems, the value chosen for W is not important because linkages that have good transmission angles often have good link length ratios. When n is 5, values between 1 and 5 for W will generally give good results.

As written, once β is selected, the crank-rocker linkage is completely defined, and a value for U can be computed. Therefore, U is a function of β only, and well-established limits for β are known. The optimization can then be easily accomplished by varying β and computing U until U is minimized. This can be done manually, interactively, or by using any one-dimensional (line search) optimization routine. Several such one-dimensional routines are described by Arora[12].

4.5 PATH SYNTHESIS

The path synthesis problem is that of specifying the path taken by a single point fixed in a member of a mechanism. There may also be a requirement for faster or slower speeds along different portions of the path. Although many research papers have been written on the subject of path synthesis, designers usually use a trial-and-error approach in practice.

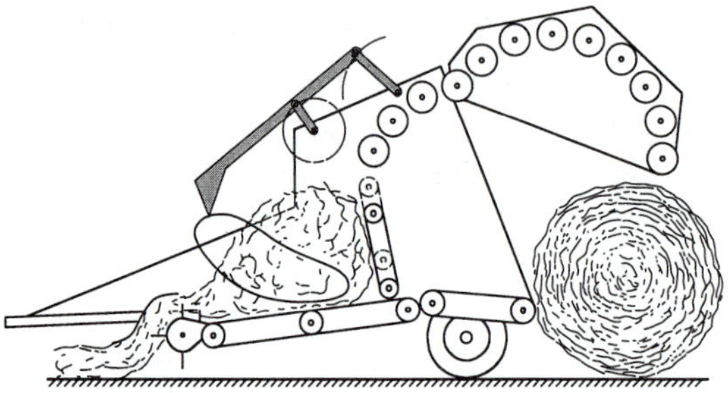

Figure 4.49 Coupler point used in packing mechanism in a round baler.[17]

The traditional tool for this purpose is the Hrones and Nelson coupler-curve atlas.[16] This is a large book containing plots of four-bar linkage coupler curves for a large variety of points located in the coupler plane and a large range of link length variations. The approach is to leaf through the coupler-curve atlas and pick out a curve that has more or less the right shape and then refine it by trial and error, testing the effect of small variations in the position of the coupler point or small variations in the link lengths. This gets quite laborious if done manually.

Use of either a simple program based on the theory presented in Chapter 3 or professionally written linkage analysis software can make this task much easier. Coupler curve programs written in MATLAB for both four-bar linkages and slider-crank mechanisms are included on the disk provided with this book. These programs use the same nomenclature as the Hrones and Nelson atlas. Using these programs, it is possible to quickly review the coupler curves available and determine the link lengths and coupler point that will generate the curve.

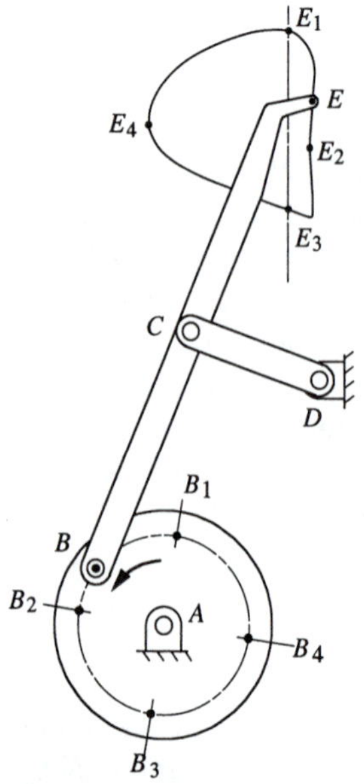

Figure 4.50 Coupler point used in film feed mechanism.[18]

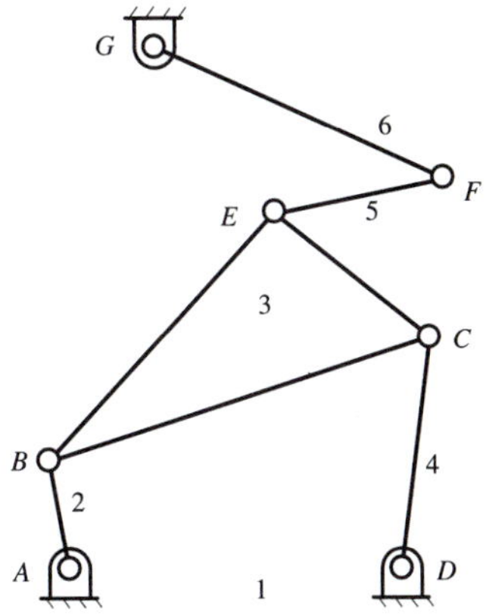

Figure 4.51 Six-link mechanism that can be designed using coupler curves.

4.5.1 Design of Six-Bar Linkages Using Coupler Curves

Coupler curves from four-bar linkages and slider-crank mechanisms are used in two main ways. The first is to use the motion of the coupler in the area of the curve to perform some function. A common use for such points is in packaging and conveying equipment. Figure 4.49 shows a coupler mechanism design for packing hay in a round bale.[17] Figure 4.50 shows a mechanism which has been used to feed film in a motion picture projector.[18]

A second use for coupler curves is to facilitate the design of six-link mechanisms where the output link is to have a prescribed motion relative to the input link. Such a mechanism is represented schematically in Fig. 4.51, where link 2 is the input and link 6 is the output. The output dyad (links 5 and 6) is driven by the coupler point (E) of the four-bar linkage. By properly selecting the coupler curve, different functional relationships between ϕ and θ can be achieved. This is illustrated in the following two examples.

EXAMPLE 4.4 ***(Design of a Six-Link Dwell Mechanism Using Coupler Curve)***

PROBLEM

A mechanism of the type shown in Fig. 4.51 is to be designed such that link 6 is an oscillating lever and link 2 rotates a full 360°. The output link is to oscillate through a range of 30° during the first 120° of crank rotation. Link 6 is then to dwell for 90° of crank rotation and return during the remaining 150° of crank rotation.

SOLUTION

To solve this problem, it is necessary to have access to an atlas of coupler curves or to use a program that can generate the coupler curves. Regardless of the procedure used, we must be able to determine the geometry of the curve and the travel distance along the curve as a function of input rotation. In the Hrones and Nelson atlas and in the programs *hr_crankrocker.m* and *hr_slidercrank.m* provided on the disk with this book, a dashed line is used for each 5° of crank rotation. We will use the four-bar program *hr_crankrocker.m* to generate candidate coupler curves. The first step is to visualize the shape of coupler curve that can be used to drive links 5 and 6. Several different geometries might be used, but the simplest is a curve of roughly elliptical shape. The coupler curves used are displayed in Figs. 4.52 and 4.53, and the design procedure is shown in Fig. 4.54. The procedure is described in the following:

1. Test different coupler curves to determine whether a portion of the curve in the vicinity of the minor axis is roughly circular in shape for the desired dwell period (90°). Figure 4.52

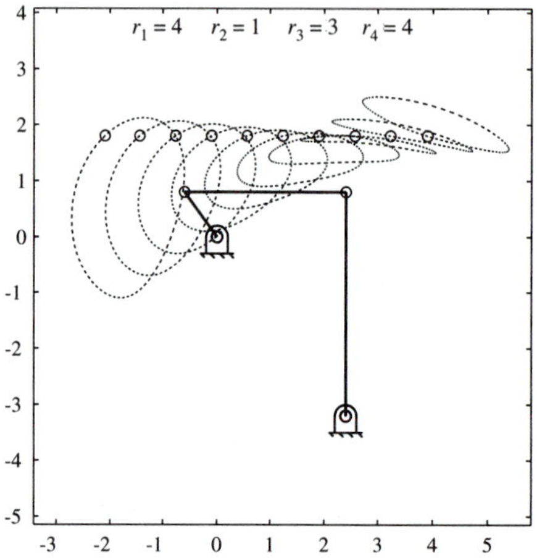

Figure 4.52 Coupler curves for Example 4.4.

gives a set of curves generated with the program *hr_crankrocker.m* when $r_1 = 4$, $r_2 = 1$, $r_3 = 3$, and $r_4 = 4$. From the curves displayed, we will select the curve shown in Fig. 4.53.

2. After a coupler curve is identified, find the center of the circle that best fits the circular region identified in step 1. The radius of the circle will be the length of link 5. Identify explicitly the beginning and end of the circular portion of the curve. The center of the circle arc will be one extreme position for point F. This is shown in Fig. 4.54, where the mechanism has been redrawn so that the frame link (r_1) is horizontal.
3. Point F' corresponds to the second extreme position of F. To locate F', identify the point on the coupler curve corresponding to 120° of crank rotation beyond the dwell. Locate a perpendicular line to the coupler point at this point, and locate F' on this line.
4. The pivot G must be located on the perpendicular bisector of the line FF'. Locate G such

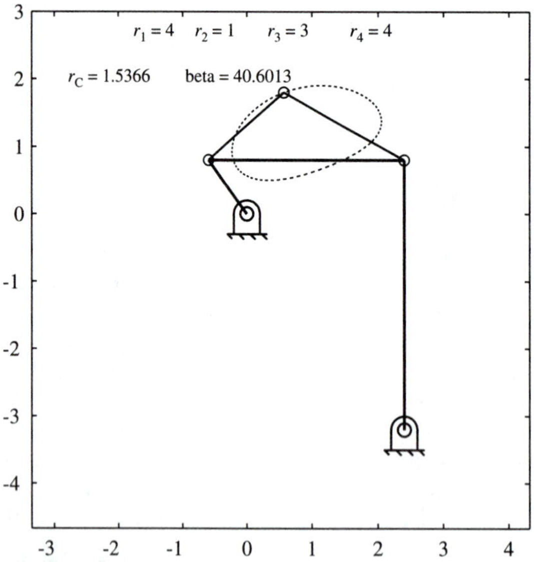

Figure 4.53 Coupler curve chosen for design.

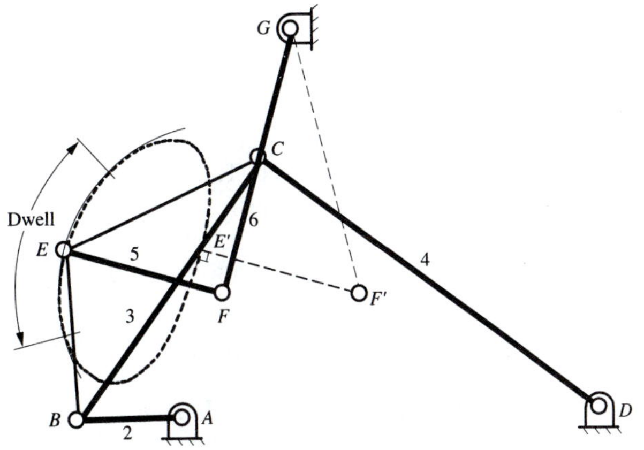

Figure 4.54 Design procedure for mechanism in Example 4.4.

that the angle FGF' is 30°. The parameters corresponding to the solution are

$r_1 = 4$	$r_5 = 1.682$	$BE = 1.537$
$r_2 = 1$	$r_6 = 2.694$	$\beta = 40.6°$
$r_3 = 3$	$G_x = 0.697$	$\theta_1 = 0$
$r_4 = 4$	$G_y = 3.740$	

5. Compute and plot the motion of link 6 relative to link 2 to evaluate the design. The resulting mechanism is simply a four-bar linkage with the addition of two more links (Watt's six-bar

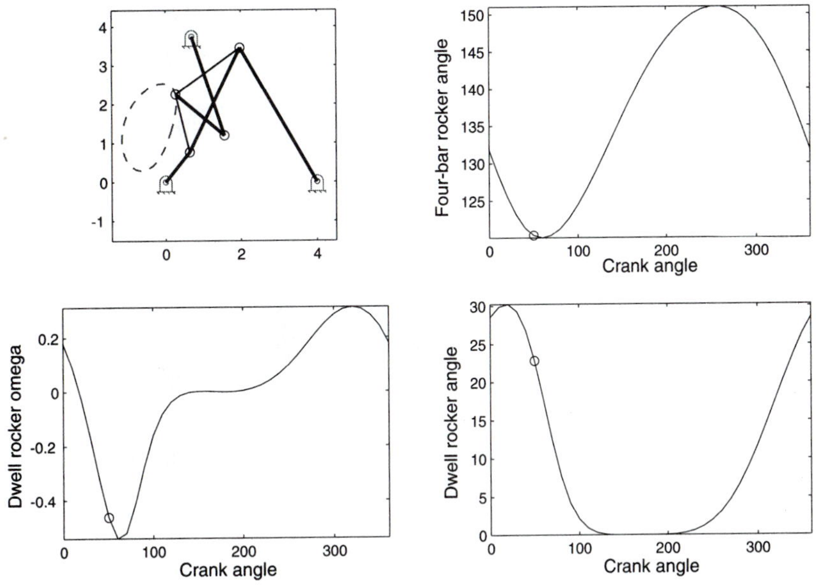

Figure 4.55 Analysis of linkage designed in Example 4.4. The velocity plot is based on a crank velocity of 1 rad/s.

mechanism). The two additional links are called a dyad and can be easily analyzed using the procedures given previously. A MATLAB routine (*sixbar.m*) for a six-bar linkage analysis based on the four-bar and dyad routines is provided on the disk with this book. Part of the analysis from this program is given in Fig. 4.55. The results are close to the design specifications, and the basic design is acceptable. Note in Fig. 4.55 that the angular velocity of link 6 is approximately zero during the dwell. The input (crank) velocity was 1 rad/s. ■

This design procedure can yield a large number of candidate designs. The best designs can be chosen by using some kind of evaluation criterion. Typical criteria are linkage size, force transmission characteristics, and acceleration characteristics.

EXAMPLE 4.5 **(*Design of a Six-Link Mechanism for Double Oscillation*)**

PROBLEM

A mechanism of the type shown in Fig. 4.51 is to be designed such that link 6 will make two complete 30° oscillations for each revolution of the driving link.

SOLUTION

For this problem, no timing information is required. Therefore, we need only ensure that the output link makes two complete oscillations for one oscillation of the input crank. Again, we will use the four-bar program (*hr_crankrocker.m*) to generate candidate coupler curves. The first step is to visualize the shape of coupler curve that can be used to drive links 5 and 6. One curve that will work for this type of problem is a figure-8 curve. The design procedure is shown in Fig. 4.56 and described in the following:

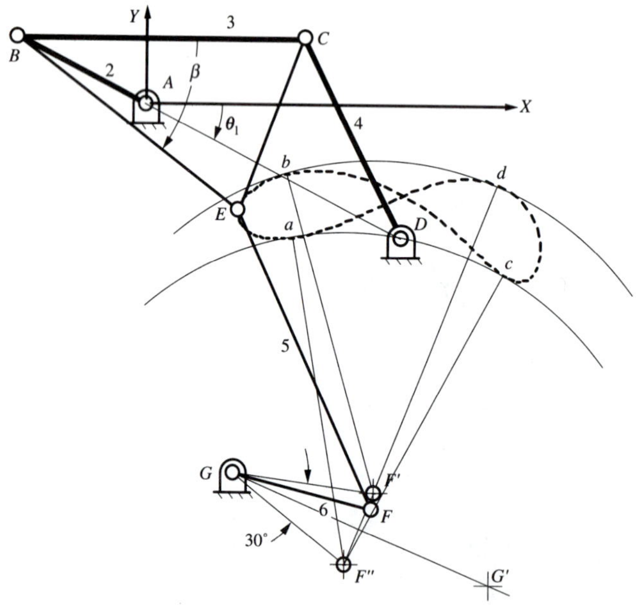

Figure 4.56 Procedure for designing linkage for Example 4.5.

1. Select a coupler curve that is a figure-8 curve with roughly equal lobes. The coupler curve selected is displayed in Fig. 4.57.

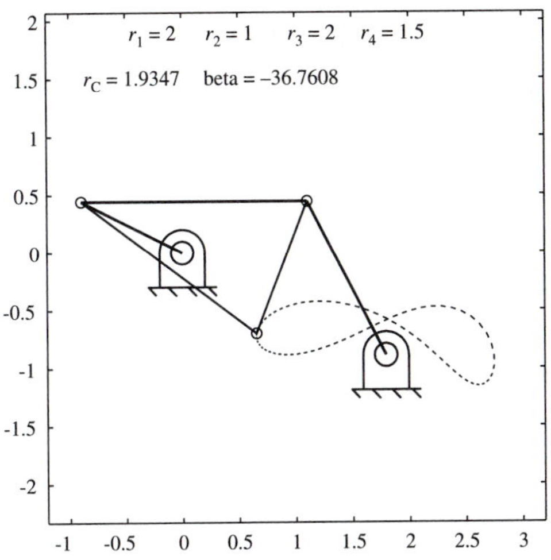

Figure 4.57 Curve that can be used for Example 4.5.

2. After the coupler curve is identified, select the length of link 5 and draw a circle or circle arc with a radius equal to the length of link 5 and tangent to the coupler curve at the two points b and d. The center, F', of this circle is one extreme position of point F. Draw another circle or circle arc of the same radius tangent to the coupler curve at points a and c. The center, F'', of this circle is the second extreme position of F.
3. The pivot point, G, must be located on the perpendicular bisector of the line $F'F''$. Locate G such that the angle $F'GF''$ is 30°. Link 6 is the link from F' to G (or from F'' to G). Note that there are two possible locations for point G. The location G is chosen in this example over G' because it will result in better transmission characteristics. If point G' is chosen, the linkage will lock up before it traverses its entire range of motion because the distance $G'b$ is slightly larger than $(EF + FG)$. The parameters corresponding to the solution are

$r_1 = 2$	$r_5 = 2.231$	$BE = 1.929$
$r_2 = 1$	$r_6 = 1.010$	$\beta = -36.44°$
$r_3 = 2$	$G_x = 0.633$	$\theta_1 = -26.57°$
$r_4 = 1.5$	$G_y = -2.292$	

4. Compute and plot the motion of link 6 relative to link 2 to evaluate the design. This is done in Fig. 4.58 based on the program (*sixbar.m*) provided on the disk with this book. Again, the input (crank) velocity was chosen as 1 rad/s.

The results given in Fig. 4.58 are very close to the design specifications. If more accurate results are desired, the location of G or the lengths of r_5 or r_6 could be adjusted slightly. This can be done manually or by using an optimization program that minimizes the error created by the linkage. However, even if an optimization program is used, the graphical procedure, which is very simple and quick to apply, is a good means of generating an initial estimate of the optimum solution. ■

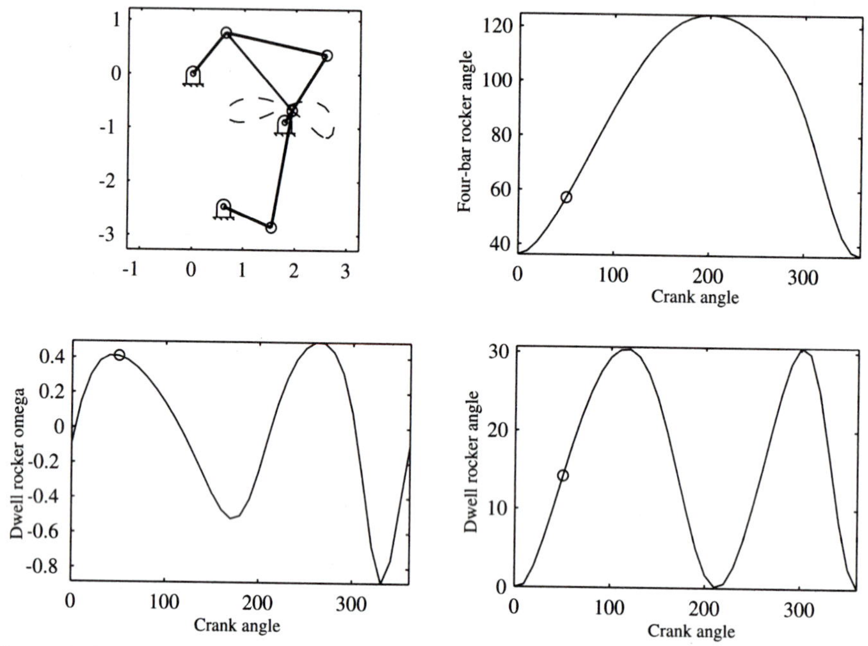

Figure 4.58 Analysis of mechanism designed for Example 4.5.

4.5.2 Four-Bar Cognate Linkages

A given four-bar linkage with coupler point C will generate a unique coupler curve. It is interesting to note that there are two other four-bar linkages that will generate exactly the same coupler curve. The three four-bar linkages that will generate the same coupler curve are called cognates. From a design standpoint, one of the linkages may have more desirable motion characteristics than the others. Therefore, it is useful to identify all three linkages once the coupler curve is defined in order to select the best one. The existence of the three cognate linkages was originally discovered by Roberts.[4,13] A general discussion of cognate linkages and a proof for Roberts' theorem which identifies the geometric relationships among cognates are given in Chapter 5. In this chapter, we will limit our discussion to a procedure for finding cognates.

The geometry of the cognate linkages can be determined by considering extreme versions

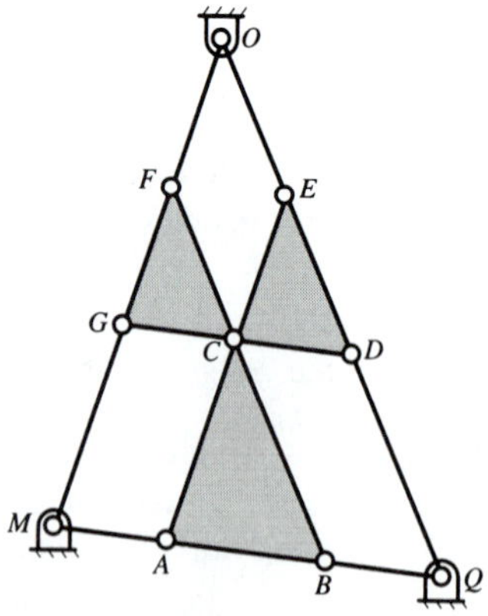

Figure 4.59 Roberts' linkage.

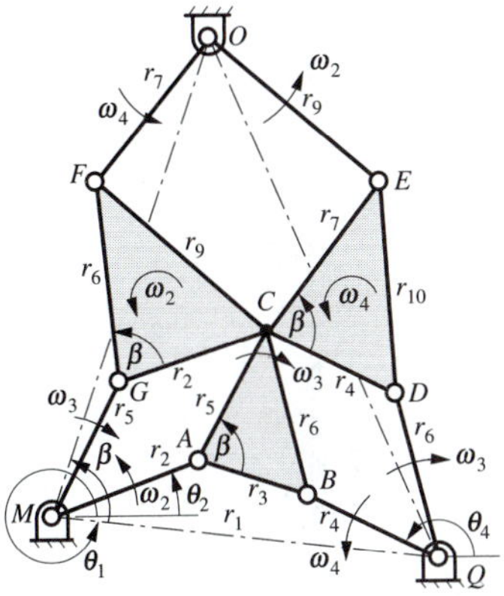

Figure 4.60 Three cognate linkages will generate the same coupler curve.

of the three linkages. The resulting diagram shown in Fig. 4.59 is called Roberts' linkage.[4] The mechanisms in this diagram will not move; however, it shows the relationships that must exist among the three cognate linkages. These relationships are maintained when triangle *MQO* is shrunk while maintaining similarity, thereby permitting the linkages to move. In particular, triangles *MQO, ABC, GCF,* and CDE are similar. Also, figures *MACG, BQDC,* and *FCEO* are parallelograms. The coupler point is *C,* and two cognate linkages share each of the pivots. Also, the couplers of each of the cognates are geometrically similar to each other and to the triangle formed by the pivots. If we identify the pivots as *M, Q,* and *O,* we can identify the three four-bar linkages by their pivots. That is, one four-bar is the *MQ* four-bar, one is the *MO* four-bar, and the third is the *QO* four-bar.

The geometric relationships among three general cognate linkages are shown in Fig. 4.60. When determining the cognate linkages, it is assumed that the *MQ* linkage is known along with the coupler point *C*. The cognate linkages can be identified with the aid of Roberts' linkage, which reveals the geometric relationships among the three linkages. Given the positions of *M* and *Q* and the lengths r_2, r_3, r_4, r_5, r_6, and r_7, the location of pivot *O* and the corresponding angles and lengths of the other cognate linkages are shown in Table 4.1. Note that the cognates will all be of the same Grashof type. When the

Table 4.1 Angle and link relationships permitting the cognate linkages to be determined. The variables refer to the diagram in Fig. 4.60. The coordinates (x_A, y_A), (x_B, y_B), and (x_C, y_C) are assumed to be known from the analysis of the MQ linkage. Alternatively, β, ϕ, and λ can be given separately.

$r_6 = r_2 \dfrac{r_5}{r_3}$	$r_7 = r_4 \dfrac{r_5}{r_3}$	$r_8 = r_1 \dfrac{r_5}{r_3}$
$\beta = \tan^{-1}\left[\dfrac{y_C - y_A}{x_C - x_A}\right] - \tan^{-1}\left[\dfrac{y_B - y_A}{x_B - x_A}\right]$		
$x_O = x_M + r_8 \cos(\theta_1 + \beta)$	$y_O = y_M + r_8 \sin(\theta_1 + \beta)$	
$x_G = x_M + (x_C - x_A)$	$y_G = y_M + (y_C - y_A)$	
$x_D = x_Q + (x_C - x_B)$	$y_D = y_Q + (y_C - y_B)$	
$\phi = \tan^{-1}\left[\dfrac{y_D - y_C}{x_D - x_C}\right]$	$\lambda = \tan^{-1}\left[\dfrac{y_C - y_G}{x_C - x_G}\right]$	
$x_E = x_D + r_7 \cos(\beta + \phi)$	$y_E = y_D + r_7 \sin(\beta + \phi)$	
$x_F = x_G + r_6 \cos(\lambda + \beta)$	$y_F = y_G + r_6 \sin(\lambda + \beta)$	
$r_9 = \sqrt{(x_E - x_O)^2 + (x_E - x_O)^2}$	$r_{10} = \sqrt{(x_E - x_D)^2 + (x_E - x_D)^2}$	

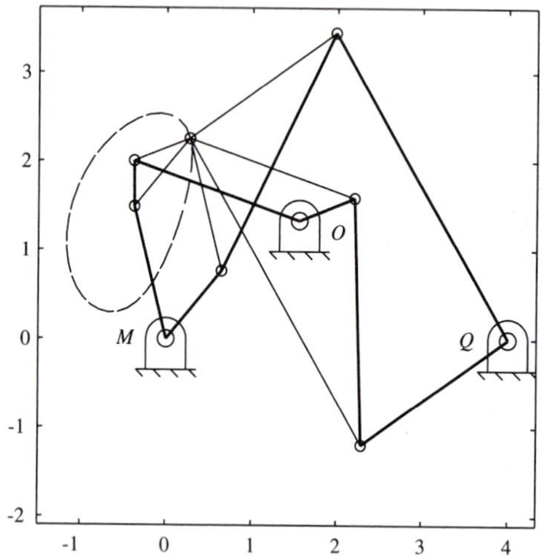

Figure 4.61 Three cognate linkages for the coupler curve in Fig. 4.53.

location of the coupler point is specified, the coordinates of A, B, and C must be given, or alternatively, the coordinates of A and B can be given along with the angle β and length r_5.

The equations in Table 4.1 can be easily programmed to determine the geometry of the cognate linkages, and this is done in a program (*cognates.m*) given on the disk with this book. Figure 4.61 shows the cognate linkages for the mechanism shown in Fig. 4.53.

4.6 REFERENCES

1. Kaufman, R. E., "Mechanism Design by Computer," Machine Design, pp. 94–100 (1978).
2. Waldron, K. J., "Improved Solutions of the Branch and Order Problems of Burmester Linkage Synthesis," Journal of Mechanism and Machine Theory, Vol. 13, pp. 199–207 (1978).
3. Erdman, A. G., and Riley, D., "Computer-Aided Linkage Design Using the LINCAGES Package," ASME Paper No. 81-DET-121 (1981).
4. Hall, A. S., *Kinematics and Linkage Design.* Balt Publishers, West Lafayette, IN (1961).
5. Soni, A. H., *Mechanism Synthesis and Analysis.* McGraw-Hill Book Co., New York, NY (1974).
6. Tao, D. C., *Applied Linkage Synthesis.* Addison-Wesley Publishing Co., Reading, MA (1964).
7. Hirschhorn, J., *Kinematics and Dynamics of Plane Mechanisms.* McGraw-Hill Book Co., New York, NY (1962).
8. Shigley, J. E., Uicker, and J. J., *Theory of Machines and Mechanisms.* McGraw-Hill Book Co., New York, NY (1980).
9. Chen, F. Y., "An Analytical Method for Synthesizing the Four-Bar Crank-Rocker Mechanism," *ASME Transactions, Journal of Engineering for Industry,* pp. 45–54 (1969).
10. Chen, F. Y., "On Closed Form Synthesis Equations of the Spherical Crank-Rocker Mechanism," *Proceedings of the Fourth World Congress on the Theory of Machines and Mechanisms,* Newcastle upon Tyne, England, pp. 707–710 (1975).
11. Brodell, R. J., and Soni, A. H., "Design of the Crank-Rocker Mechanism with Unit Time Ratio." *Jnl. Mechanism,* Vol. 5, pp. 1–4 (1970).
12. Arora, J. S., *Introduction to Optimal Design.* McGraw Hill Book Co., New York, NY (1989).
13. Sandor, G. N., and A. G. Erdman, *Advanced Mechanism Design: Analysis and Synthesis.* Prentice-Hall, Inc., Englewood Cliffs, NJ (1984).
14. Suh, C. H., and C. W. Radcliffe, *Kinematics and Mechanisms Design.* John Wiley & Sons, Inc., New York, NY (1978).
15. Waldron, K. J., "The Order Problem of Burmester Linkage Synthesis," *ASME Journal of Engineering for Industry,* pp. 1405, 1406 (1975).
16. Hrones, J. A., and G. L. Nelson, *Analysis of the Four-Bar Linkage,* The Technology Press of M.I.T. and John Wiley & Sons (1951).
17. Tooten, K., Entwicklung und Konstruktion einer neuen Rundballenpresse, *Konstruktion,* Vol. 39, pp. 285–290 (1987).
18. Hartenberg, R. S., and J. Denavit, *Kinematic Synthesis of Linkages.* McGraw-Hill Book Co., New York, NY (1964).
19. Freudenstein, F., "Approximate Synthesis of Four-Bar Linkages," *Transactions of ASME,* Vol. 77, pp. 853–861, August (1959).

4.7 CHAPTER 4 *Exercise Problems*

PROBLEM 4.1

Design a four-bar linkage to move a coupler containing the line AB through the three positions shown. The moving pivot (circle point) of one crank is at A and the fixed pivot (center point) of the other crank is at C^*. Draw the linkage in position 1 and use Grashof's equation to identify the type of four-bar linkage designed. Position A_1B_1 is horizontal, and positions A_2B_2 and A_3B_3 are vertical. $AB = 4$ in.

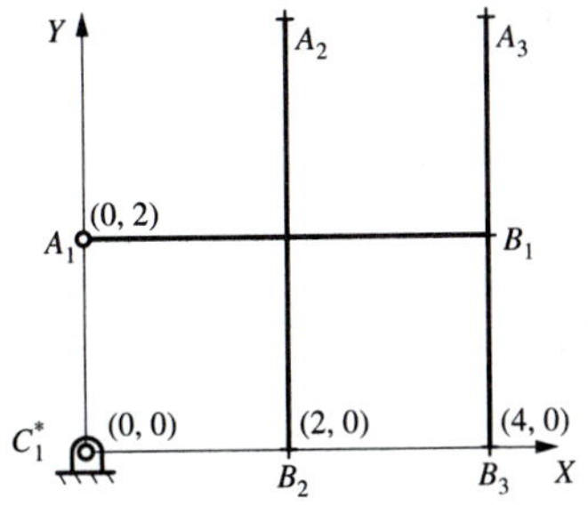

PROBLEM 4.2

In the drawing, $AB = 1.25$ cm. Use A and B as circle points, and design a four-bar linkage to move its coupler through the three positions shown. Use Grashof's equation to identify the type of four-bar linkage designed.

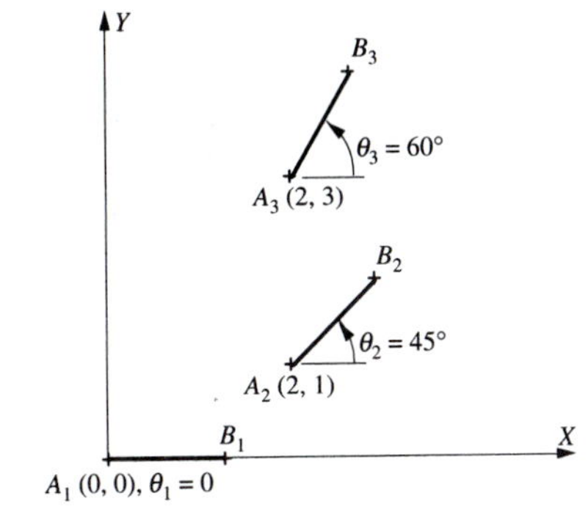

PROBLEM 4.3

Using points A and B as circle points, design a four-bar linkage that will position the body defined by AB in the three positions shown. Draw the linkage in position 1, and use Grashof's equation to identify the type of four-bar linkage designed. Position A_1B_1 is horizontal, and position A_2B_2 is vertical. $AB = 1.25$ in.

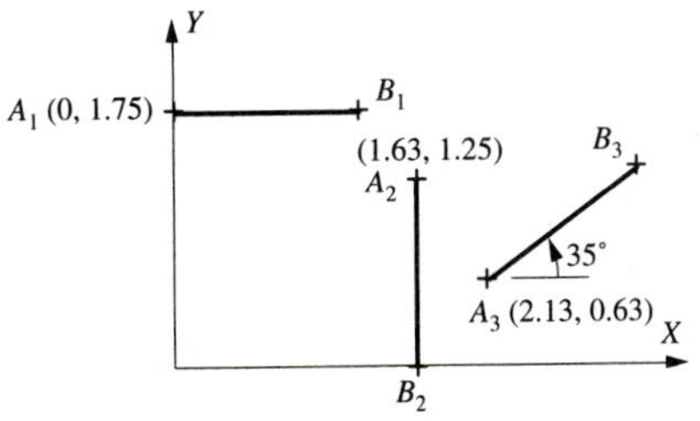

PROBLEM 4.4

A four-bar linkage is to be designed to move its coupler plane through the three positions shown. The moving pivot (circle point) of one crank is at A, and the fixed pivot (center point) of the other crank is at C^*. Draw the linkage in position 1 and use Grashof's equation to identify the type of four-bar linkage designed. Also determined whether the linkage changes branch in traversing the design positions. Positions A_1B_1 and A_2B_2 are horizontal, and position A_3B_3 is vertical. $AB = 3$ in.

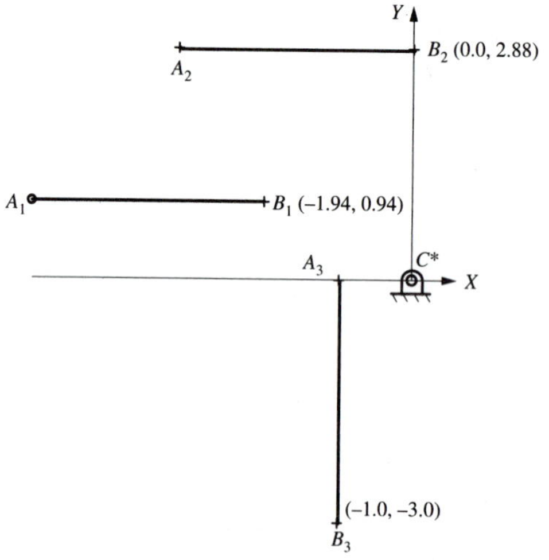

PROBLEM 4.5

Design a four-bar linkage to move the coupler containing line segment AB through the three positions shown. The moving pivot for one crank is to be at A, and the fixed pivot for the other crank is to be at C^*. Draw the linkage in position 1 and determine the classification of the resulting linkage (e.g., crank rocker, double crank). Positions A_2B_2 and A_3B_3 are horizontal, and position A_1B_1 is vertical. $AB = 3.5$ in.

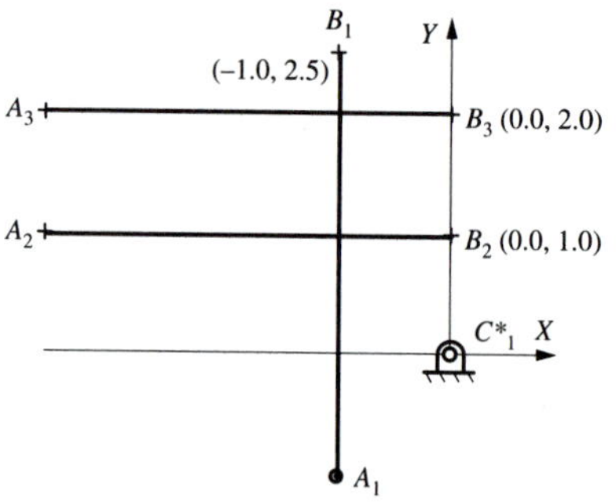

PROBLEM 4.6

A mechanism must be designed to move a computer terminal from under the desk to top level. The system will be guided by a linkage, and the use of a four-bar linkage will be tried first. As a first attempt at the design, do the following:

(a) Use C^* as a center point and find the corresponding circle point C in position 1.
(b) Use A as a circle point and find the corresponding center point A^*.
(c) Draw the linkage in position 1.
(d) Determine the type of linkage (crank rocker, double rocker, etc.) resulting.
(e) Evaluate the linkage to determine whether you would recommend that it be manufactured.

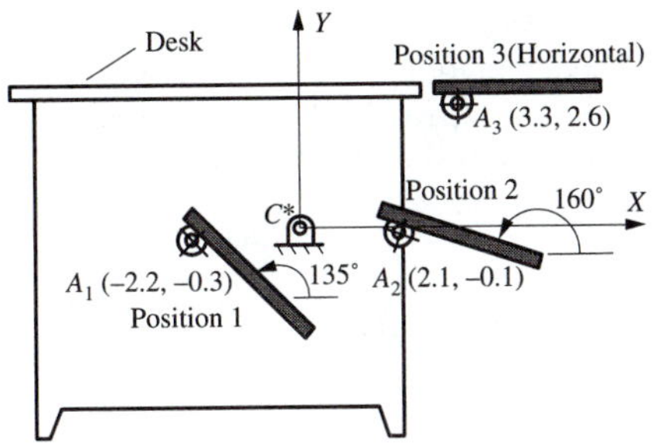

PROBLEM 4.7

Design a four-bar linkage to move the coupler containing line segment AB through the three positions shown. The moving pivot for one crank is to be at A, and the fixed pivot for the other crank is to be at C^*. Draw the linkage in position 1 and determine the classification of the resulting linkage (e.g., crank rocker, double crank). Also check to determine whether the linkage will change branch as it moves from one position to another. Position A_1B_1 is horizontal, and position A_3B_3 is vertical. $AB = 5.1$ cm.

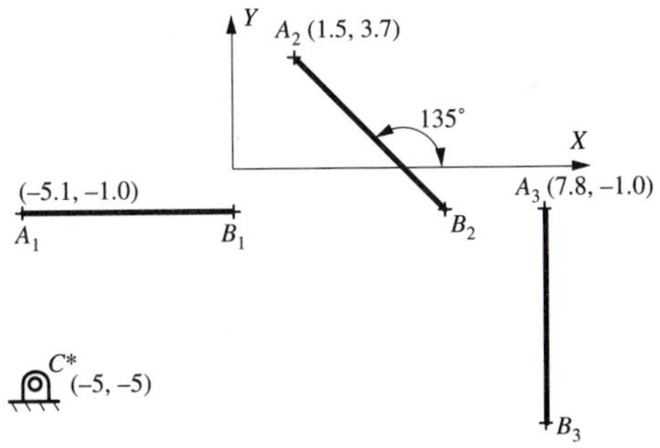

PROBLEM 4.8

A hardware designer wants to use a four-bar linkage to guide a door through the three positions shown. Position 1 is horizontal, and position 3 is vertical. As a tentative design, she selects point B^* as a center point and A as a circle point. For the three positions shown, determine the location of the circle point B corresponding to the center point B^* and the center point A^* corresponding to the circle point A. Draw the linkage in position 1 and determine the Grashof type for the linkage. Indicate whether you think that this linkage should be put into production.

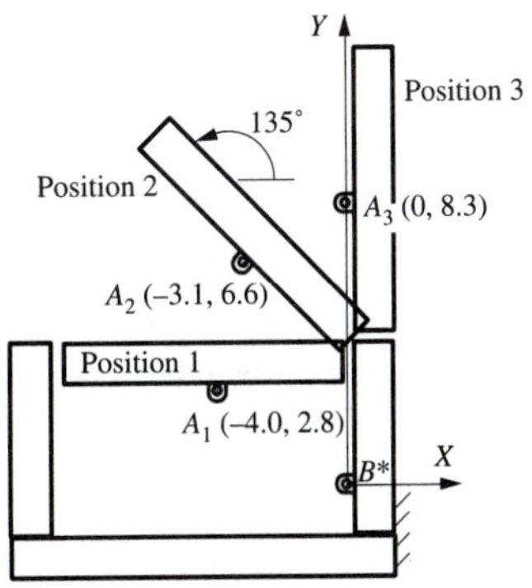

PROBLEM 4.9

Design a *slider-crank mechanism* to move the coupler containing line segment AB through the three positions shown. The moving pivot for the crank is to be at A. Determine the slider point, and draw the linkage in position 1. Also check to determine whether the linkage will move from one position to another without being disassembled. Position A_1B_1 is horizontal, and position A_3B_3 is vertical. $AB = 2.0$ in.

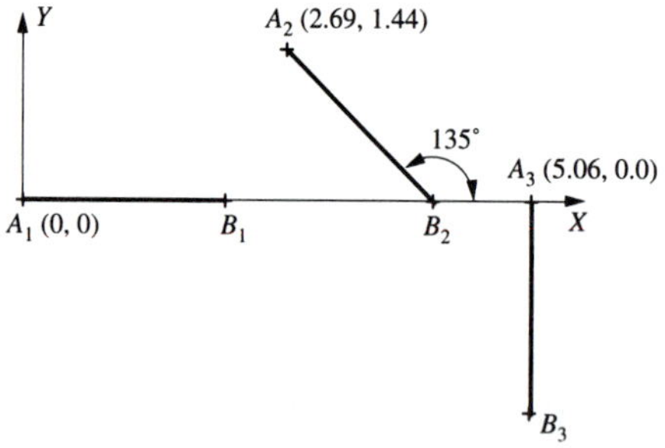

PROBLEM 4.10

Design a slider-crank mechanism to move a coupler containing the line AB through the three positions shown. The line AB is 1.25″ long. The moving pivot (circle point) of the crank is at A. The approximate locations of the three poles (P_{12}, P_{13}, P_{23}) are shown, but these should be determined accurately after the positions are redrawn. Find A^*, the slider point that lies above B_1 on a vertical line through B_1, and draw the linkage in position 1.

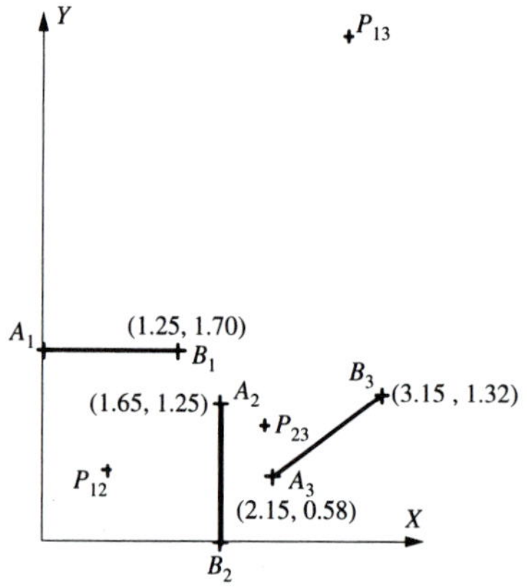

PROBLEM 4.11

Design a four-bar linkage that generates the function $y = \sqrt{x} - x + 3$ for values of x between 1 and 4. Use the Chebyshev spacing for three position points. The base length of the linkage must be 2 in. Use the following angle information:

$$\theta_0 = 25^\circ \qquad \Delta\theta = 60^\circ$$
$$\phi_0 = 30^\circ \qquad \Delta\phi = 60^\circ$$

PROBLEM 4.12

Design a four-bar linkage to generate the function $y = x^2 - 1$ for values of x between 1 and 5. Use Chebyshev spacing with three position points. The base length of the linkage must be 2 cm. Use the following angle information:

$$\theta_0 = 30^\circ \qquad \Delta\theta = 60^\circ$$
$$\phi_0 = 30^\circ \qquad \Delta\phi = 90^\circ$$

PROBLEM 4.13 Determine the link lengths and draw a four-bar linkage that will generate the function $\phi = \theta^2$ (θ and ϕ both in radians) for values of θ between 0.5 and 1.0 radians. Use Chebyshev spacing with three position points. The base length of the linkage must be 2 cm.

PROBLEM 4.14 A device characterized by the input-output relationship $\phi = a_1 + a_2 \cos \theta$ is to be used to generate (approximately) the function $\phi = \theta^2$ (θ and ϕ both in radians) over the range $0 \le \theta \le \pi/4$.

(a) Determine the number of precision points required to compute r_1 and r_2.
(b) Choose precision point values for θ from among 0, 0.17, 0.35, and 0.52, and determine the values of a_1 and a_2 that will allow the device to approximate the function.

PROBLEM 4.15 A mechanical device characterized by the input-output relationship $\phi = 2a_1 + a_2 \tan \theta + a_3^2$ is to be used to generate (approximately) the function $\phi = 2\theta^3$ (θ and ϕ both in radians) over the range $0 \le \theta \le \pi/3$. Exterior constraints on the design require that the parameter $a_3 = 1$.

(a) Determine the number of precision points required to complete the design of the system.
(b) Use Chebyshev spacing, and determine the values for the unknown design variables that will allow the device to approximate the function.

PROBLEM 4.16 The output arm of a lawn sprinkler is to rotate through an angle of 90°, and the ratio of the times for the forward and reverse rotations is to be 1 to 1. Design the crank-rocker mechanism for the sprinkler. If the crank is to be 1 inch long, give the lengths of the other links.

PROBLEM 4.17 Design a crank-rocker mechanism such that with the crank turning at constant speed, the oscillating lever will have a time ratio of advance to return of 3:2. The level is to oscillate through an angle of 80°, and the length of the base link is to be 2 in.

PROBLEM 4.18 A packing mechanism requires that the crank (r_2) rotate at a constant velocity. The advance part of the cycle is to take twice as long as the return to give a quick-return mechanism. The distance between fixed pivots must be 0.5 meter. Determine the lengths for r_2, r_3, and r_4.

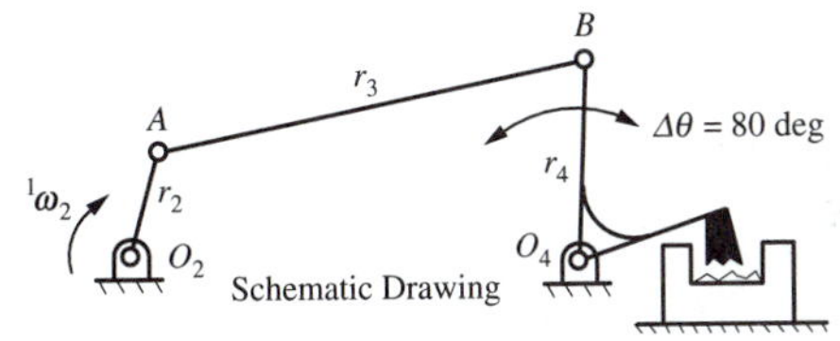

PROBLEM 4.19 The rocker O_4B of a crank-rocker linkage swings symmetrically about the vertical through a total angle of 70°. The return motion should take 0.75 the time that the forward motion takes. Assuming that the two pivots are 2.5 in apart, find the length of each of the links.

PROBLEM 4.20 A crank rocker is to be designed such that with the crank turning at a constant speed CCW, the rocker will have a time ratio of advance to return of 1.25. The rocking angle is to be 40°, and it rocks symmetrically about a vertical line through O_4. Assume that the two pivots are on the same horizontal line, 3 in apart. (Check the efficiency of the system.)

PROBLEM 4.21 Design a crank-rocker mechanism that has a base length of 2.0, a time ratio of 1.3, and a rocker oscillation angle of 100°. The oscillation is to be symmetric about a vertical line through O_4. Specify the length of each of the links.

PROBLEM 4.22 A crank-rocker mechanism with a time ratio of $2\frac{1}{3}$ and a rocker oscillation angle of 72° is to be designed. The oscillation is to be symmetric about a vertical line through O_4. Draw the mechanism in any position. If the length of the base link is 2 in, give the lengths of the other three links. Also show the transmission angle in the position in which the linkage is drawn.

PROBLEM 4.23

The mechanism shown is used to drive an oscillating sanding drum. The drum is rotated by a splined shaft that is cycled vertically. The vertical motion is driven by a four-bar linkage through a rack-and-pinion gear set (model as a rolling contact joint). The total vertical travel for the sander drum is 3 in, and the pinion has a 2 in radius. The sander mechanism requires that the crank (r_2) rotate at a constant velocity, and the advance part of the cycle is to take the same amount of time as the return part. The distance between fixed pivots must be 4 in. Determine the lengths for r_2, r_3, and r_4.

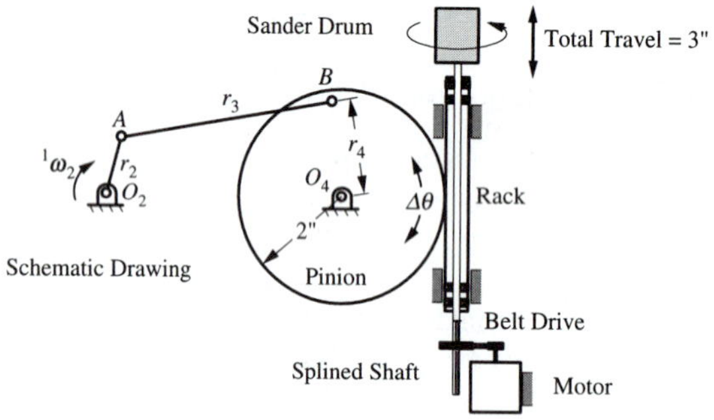

PROBLEM 4.24

The mechanism shown is proposed for a rock crusher. The crusher hammer rotates through an angle of 20°, and the gear ratio R_G/R_P is 4:1; that is, the radius r_G is four times the radius r_p. Contact between the two gears can be treated as rolling contact. The crusher mechanism requires that the crank (r_2) rotate at a constant velocity, and the advance part of the cycle is to take 1.5 times as much as the return part. The distance between fixed pivots O_2 and O_4 must be 4 ft. Determine the lengths for r_2, r_3, and r_4.

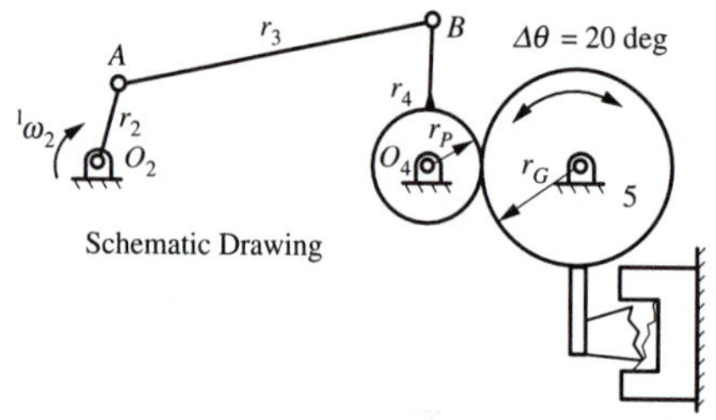

PROBLEM 4.25

The mechanism shown is proposed for a shaper mechanism. The shaper cutter moves back and forth such that the forward (cutting) stroke takes twice as much time as the return stroke. The crank (r_2) rotates at a constant velocity. The follower link (r_4) is to be 4 in and to oscillate through an angle of 80°. Determine the lengths for r_1, r_2, and r_3.

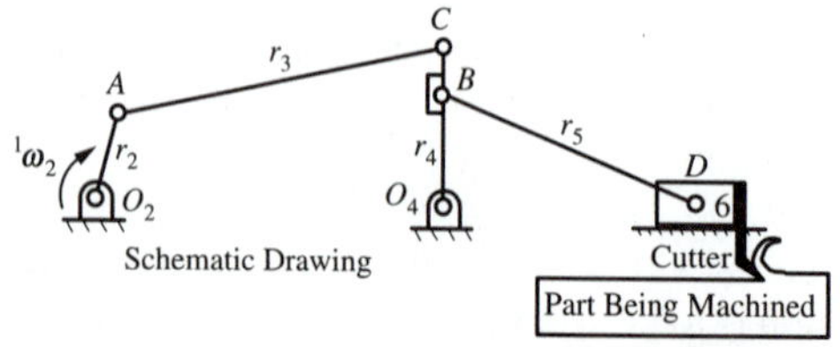

PROBLEM 4.26

A crank rocker is to be used in a door-closing mechanism. The door must open 100°. The crank motor is controlled by a timer mechanism such that it pauses when the door is fully open. Because of this, the mechanism can open and close the door in the same amount of time. If the crank (r_2) of the mechanism is to be 10 cm long, determine the lengths of the other links (r_1, r_3, and r_4). Sketch the mechanism to scale.

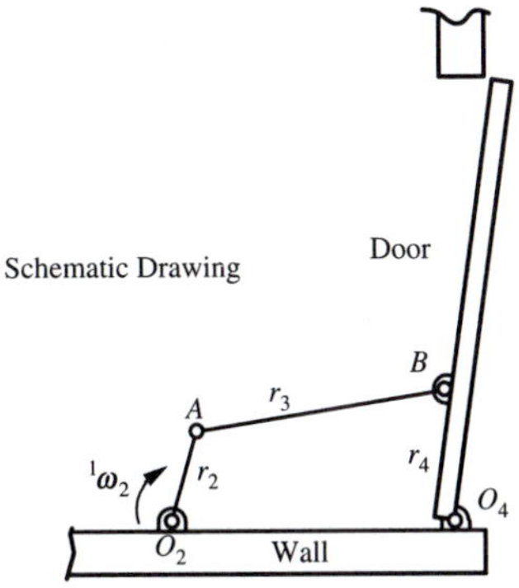

PROBLEM 4.27

A crank rocker is to be used for the rock crusher mechanism shown. The oscillation angle for the rocker is to be 80°, and the working (crushing) stroke for the rocker is to be 1.1 times the return stroke. If the frame link (r_1) of the mechanism is to be 10 ft long, determine the lengths of the other links (r_2, r_3, and r_4), Sketch the mechanism to scale.

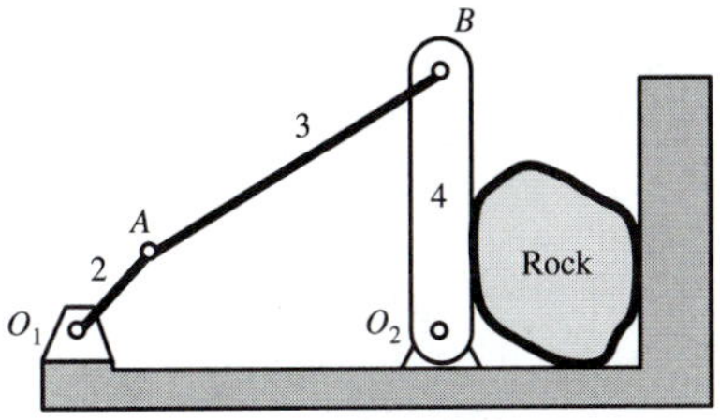

PROBLEM 4.28

A crank rocker is to be used in a windshield-wiping mechanism. The wiper must oscillate 80°. The time for the forward and return stroke for the wiper is the same. If the base link (r_1) of the mechanism is to be 10 cm long, determine the lengths of the other links (r_2, r_3, and r_4). Sketch the mechanism to scale.

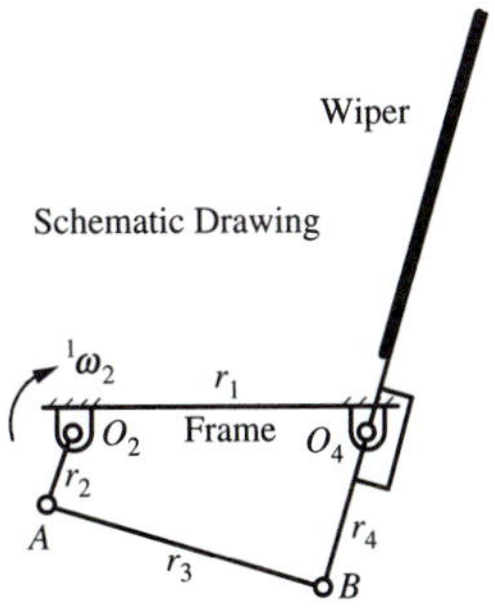

PROBLEM 4.29

A crank rocker is to be used in the transmission of an automatic washing machine to drive the agitator. The rocker link is attached to a gear sector which drives a pinion gear attached to the agitator shaft. The radius of the sector gear is 4 in, and the pinion radius is 1 in. The pivot for the output link is at O_4. The agitator is to oscillate 360°. The time for the forward and return stroke for the agitator is the same. If the base link (r_1) of the mechanism is to be 10 cm long, determine the lengths of the other links (r_2, r_3, and r_4). Sketch the mechanism to scale.

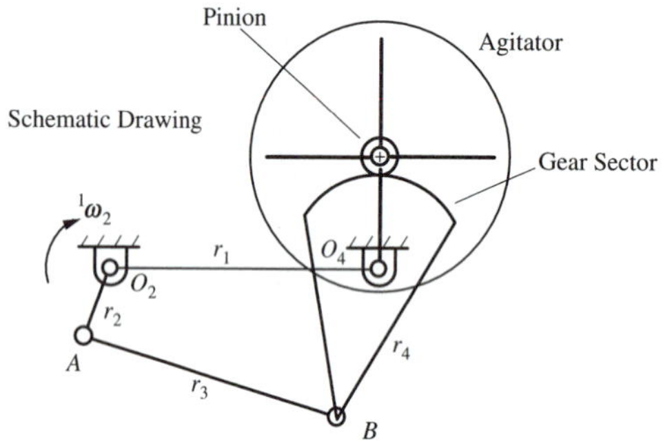

PROBLEM 4.30

Design a six-bar linkage like that shown in Fig. 4.51 such that the output link will make two complete 35° oscillations for each revolution of the driving link. (Hint: Select a coupler curve that is shaped like a figure 8.)

PROBLEM 4.31

Design a six-bar linkage like that shown in Fig. 4.51 such that the output link will do the following for one complete revolution of the input crank:

1. Rotate clockwise by 40°
2. Rotate counterclockwise by 35°
3. Rotate clockwise by 30°
4. Rotate counterclockwise by 35°

(Hint: Select a figure 8– or kidney bean–shaped coupler curve.)

PROBLEM 4.32

Determine the two 4-bar linkages cognate to the one shown below. The dimensions are $MA = 10$ cm, $AB = 16$ cm, $AC = 32$ cm, $QB = 21$ cm, and $MQ = 24$ cm. Draw the cognates in the position for $\theta = 90°$.

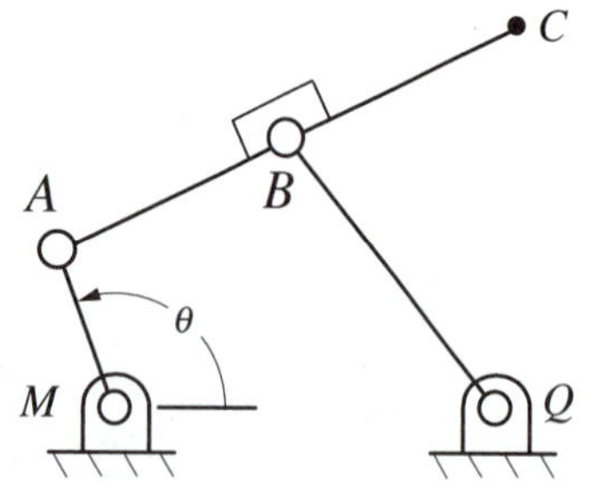

PROBLEM 4.33

Determine the two 4-bar linkages cognate to the one shown below. The dimensions are $MQ = 1.5$ in, $AB = BC = BQ = AC = 1$ in, and $AM = 0.5$ in. Draw the cognates in the position for $\theta = 90°$.

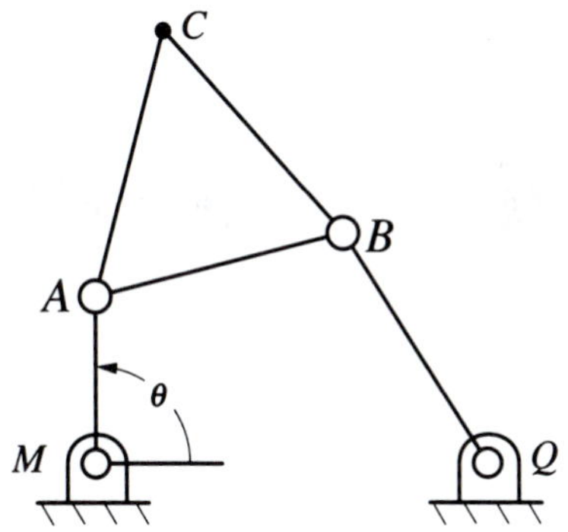

Chapter 5

Special Mechanisms

5.1 SPECIAL PLANAR MECHANISMS

5.1.1 Introduction

Although much of this book is concerned with the design of planar mechanisms, there are some that for one reason or another require special attention.

The classes of mechanisms discussed here meet a variety of common needs in mechanical engineering practice. For this reason they are important, but none require such extensive treatment as to justify a chapter to themselves, as was the case with cam or gear mechanisms.

Generation of a straight line by a simple linkage mechanism is a recurring theme. Slides or roller ways are not always acceptable for implementation in real mechanism designs, and there continues to be a place for simple, four-bar linkages that can approximate a straight line coupler point path with a high degree of accuracy. Likewise, linkages that can reproduce a path traced by one point at another tracing point with a change in scale find many uses ranging from machine tools for milling non–rotationally symmetric surfaces to remote actuation of robotic mechanisms.

Another recurring theme in mechanical engineering practice is transfer of torque and motion between shafts that are not coaxial, particularly when the relative alignment of the shafts must change. Very common examples occur in the drive shafts of automobiles that must accommodate movements resulting in changes of shaft alignment due to suspension movements and/or steering movements. There are also numerous examples of this situation in construction and manufacturing machinery.

Automotive steering and suspension mechanisms are among the most common linkage mechanisms in practical use. They are usually designed as decoupled, fundamentally planar linkages. However, misalignments are deliberately introduced to produce desirable effects such as a tendency for the steering to center itself. Thus, they become spatial linkages with complex interactions.

Yet another recurrring need in practical linkage design is for indexing: intermittent, timed advancement of a drive in a constant direction. This technology had very numerous and visible applications in the days of mechanical punched-card readers and similar business machines. The problem is of continuing practical importance with many applications in manufacturing and packaging machinery.

5.1.2 Approximate Straight Line Mechanisms

Approximate straight-line mechanisms occupy a very special place in the history of this subject. Toward the end of the eighteenth century, when James Watt and his contemporaries were developing the practical steam engines that powered the industrial revolution, there was no available means of smoothly guiding a point along a straight path. This was needed both to guide the crosshead of the piston rod and for the valve gear that opened the valves in coordination with the piston motion. The solution used by Watt and his contemporaries was to devise a four-bar mechanism with an acceptably long coupler-point trajectory that approximated a straight line to an acceptable degree of accuracy.

There are still many situations that arise in machine design in which a simple mechanism that generates an approximation to a straight line is a solution. A good example is the level luffing crane used on many docks to load and unload cargo. Here it is desirable that the path of the crane hook that carries the load be a horizontal straight line. This saves energy, because it is not necessary to raise and lower the load against gravity when it is being luffed or moved laterally. It also makes it easy for the human operators to extrapolate the path of the load and ensure that it will clear obstructions.

The jib of a typical level-luffing crane is arranged as a four-bar mechanism with a coupler point that approximately describes a horizontal straight line. The pulley at the end of the jib is placed at this point to produce the desired level luffing action.

Another common application of this class of mechanisms is in strip chart recorders and other instruments with which it is necessary to indicate on a linear scale.

5.1.2.1 Watt's Straight Line Mechanism

Watt's straight line mechanism has played an important role in the history of this subject. At the time of James Watt's invention of the high-pressure steam engine, there was not a capability of machining long ways to a high degree of straightness or of achieving low-friction linear motion. The solution was a four-bar mechanism that closely approximates a straight line over a portion of its coupler curve.

Watt's straight line mechanism continues to be of considerable practical importance. The linkage is simple, and the configuration is very flexible, allowing great freedom to the designer. For example, the ratio of the lengths a and b shown in Fig. 5.1 is not very critical. The linkage will produce reasonably straight motion over a wide range of dimensional ratios b/a. It is not even essential that the two cranks have the same length. The essential feature is that the dimensions be such that the linkage is capable of assuming a position like that shown in Fig. 5.1 with the two cranks being parallel and opposed, with the coupler normal to both. If the cranks are of equal length the tracing point is the midpoint of the coupler, and the line of the coupler in the position shown is the straight line which is approximated.

Because of its simplicity and ability to provide low friction, approximately linear guidance, Watt's straight line mechanism is useful anywhere exact conformance to linear motion is not essential. For example, it has been used in rear automotive suspensions of the live axle type to restrain lateral motion of the axle by constraining the center point of the axle to move along an approximate vertical straight line relative to the body.

The tracing point for the coupler curve shown in Fig. 5.1 is the midpoint of the coupler.

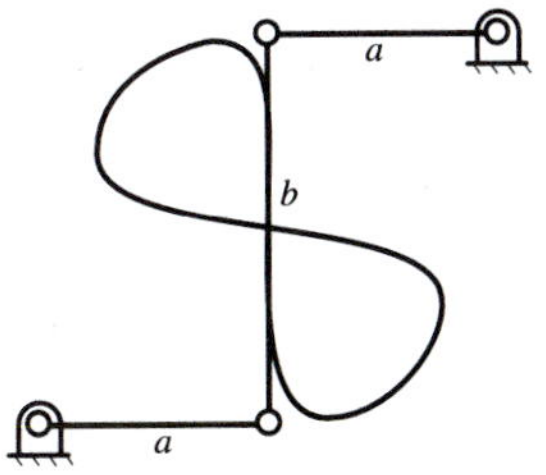

Figure 5.1 Watt's straight line mechanism.

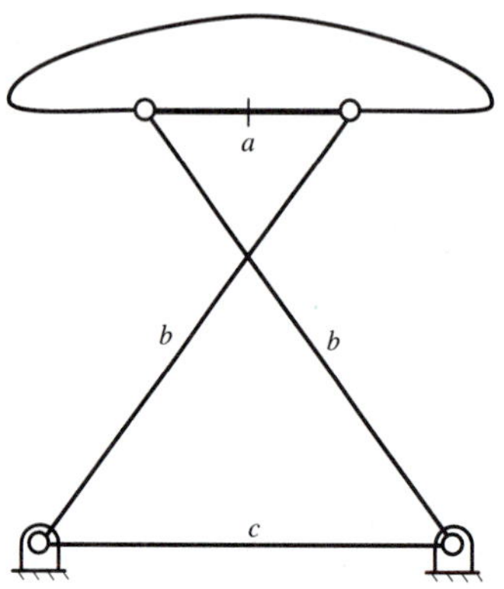

Figure 5.2 Chebyshev's approximate straight line mechanism.

The proportions of a and b are variable. In the case drawn, $a = 3$ and $b = 5$. The form of the coupler curve is known as a lemniscate. As can be seen, the central limbs of the lemniscate are good approximations to straight lines over a considerable length.

5.1.2.2 *Chebyshev's Straight Line Mechanism*

The Chebyshev approximate straight line mechanism is also a linkage that is both of historical importance and of continuing practical importance. Like the Watt mechanism, it is simple. Its advantages are that it provides a very long segment of the path of the coupler midpoint that is approximately linear and that both fixed pivots are on the same side of the linear path, as compared with the Watt mechanism, in which they are on opposite sides. However, the dimensions are more critical in this case. Referring to Fig. 5.2, the required proportions are $a = 1$, $b = 2.5$, and $c = 2$. As already noted, the tracing point is the midpoint of the coupler. As can be seen, it approximates a straight line for a considerable distance. It might be noted that these proportions require that the linkage is a type I double rocker. Since it is normally used for linear guidance of the tracing point, it is used in a coupler-driven mode.

5.1.2.3 *Roberts' Straight Line Mechanism*

Roberts' approximate straight line mechanism is also a symmetrical four-bar linkage as shown in Fig. 5.3. The coupler point indicated generates an approximate straight line for the motion between the fixed pivots. Referring to Fig. 5.3, the required proportions are $a = 1$, $b = 1.2$, $c = 2$, and $d = 1.09$. These dimensions make the mechanism a type II double rocker. It is normally used for linear guidance of the tracing point so that it is normally used in the coupler-driven mode.

5.1.2.4 *Other Approximate Straight Line Mechanisms*

There are many other four-bar linkage configurations that yield reasonable approximations to linear motion of a tracing point. Those used in level luffing cranes and similar devices need to have the tracing point outside the interval between the coupler pivots. The crane shown in Fig. 5.4 is a good example. The jib of the crane is configured as a four-bar mechanism that generates an approximate horizontal straight line at the axis of the sheave

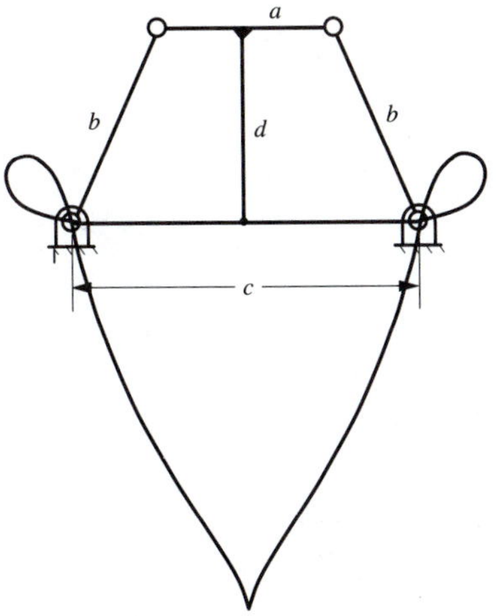

Figure 5.3 Roberts' approximate straight line mechanism.

over which the cable passes at the end of the jib. This means that the load moves approximately in a horizontal plane when only turret and jib movements are used. Vertical motion is accomplished by the crane's winch hauling or lowering the cable. Horizontal motion of the load has two very significant advantages. First, little energy is used for turret or jib motions if the cranks are counterweighted to eliminate work done against gravity in moving the mechanism itself. The drives for those motions do not need to have large capacity. Second, it is relatively easy for the crane operator to visualize a horizontal trajectory of the load and determine whether that trajectory will interfere with fixed obstacles such as the side of the ship.

5.1.3 Exact Straight Line Mechanisms

It is also possible, in principle, to generate a perfectly straight line with a linkage mechanism, but generally only at the cost of a relatively complex mechanism if large motions are desired. The first such mechanism to be invented was that of Peaucellier. Hart[1] devised a simpler mechanism that also generates an exact straight line, and several mechanisms based on the slider-crank mechanism will generate a straight line for limited motion.[2] There are several other known exact straight line–generating linkages with revolute joints, but all are much more complex than the four-bar approximate straight line generators discussed above.

A Peaucellier linkage is shown in Fig. 5.5. The linkage has a rhombic loop, *ABCD,* which forms a kite shape with the equal-length links *PB* and *PD*. The link *OA* is also equal in length to the base *OP*. Point *C* generates a true straight line normal to the base *OP*.

As may be seen, this is a much more complex linkage than the four-bar loops used above to generate approximate straight lines. It has eight members and six joints, four of which are ternary joints.

[1] Hartenberg, R. S., and J. Denavit, *Kinematic Synthesis of Linkages,* McGraw Hill Book Co., New York, NY (1964).

[2] Chironis, N. P., *Mechanisms, Linkages, and Mechanical Controls,* McGraw-Hill Book Co., New York, NY (1965).

Figure 5.4 A level-luffing crane. The jib of the crane is configured as a four-bar mechanism that generates an approximate horizontal straight line at the axis of the sheave over which the cable passes at the end of the jib. This means that the load moves approximately in a horizontal plane when only turret and jib movements are used.

If a slider is introduced, it is possible to generate an exact straight line using the slider-crank mechanism in Fig. 5.6. The range of motion is limited and a slider is required, but the basic mechanism is quite simple. Based on the geometry of the linkage, the output motion will be a simple sine function of the drive link (simple harmonic motion). As indicated in Fig. 5.6, the mechanism is made up of isosceles triangles.

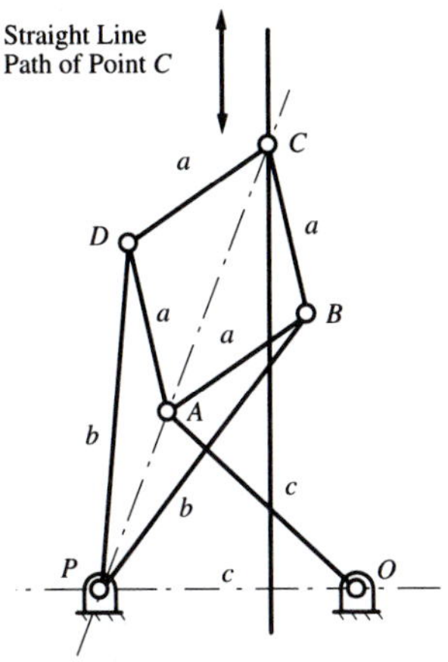

Figure 5.5 Peaucellier's exact straight line mechanism. This was the first and most famous exact straight line mechanism to be discovered. A number of others have since been discovered, including some that are a little simpler. *ABCD* is a rhombus, and links *PB* and *PD* have equal length. Link *OA* has the same length as the base *OP*. The path of point *C* is a true straight line normal to *OP*.

5.1.4 Pantographs

A plagiograph is a mechanism that exactly reproduces the path of a tracing point at a second tracer point, usually with a change of scale. The most common class of plagiographs is the family of pantograph mechanisms.

Pantographs have found many applications beyond that of plagiographs. These range from carrying contacts to overhead cables on electric trains and streetcars to legs of walking machines. As will be seen, pantographs are also of theoretical importance in that they lead to the theory of *cognate* linkages, which are different mechanisms capable of generating the same coupler point path. Cognate mechanisms, in turn, are of great usefulness in practical machine design.

5.1.4.1 The Planar Collinear Pantograph

The special properties of the pantograph linkage have been used in a variety of applications. They also have important theoretical implications leading to the theory of cognate linkages, which will be introduced briefly later in this chapter.

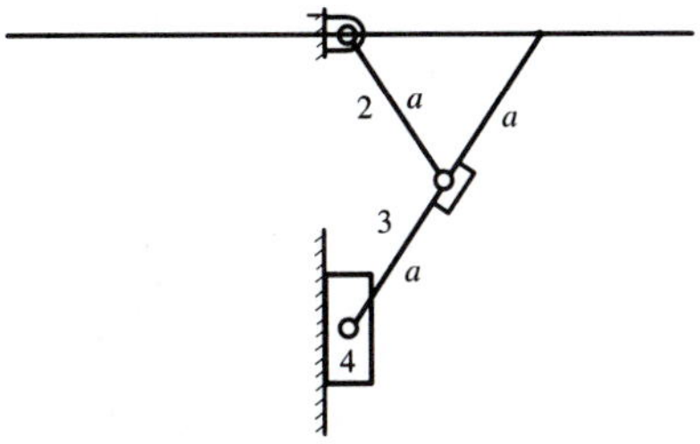

Figure 5.6 Straight line mechanism based on isosceles slider-crank mechanism. The entire range of the straight line can be reached if the mechanism is driven by the coupler.

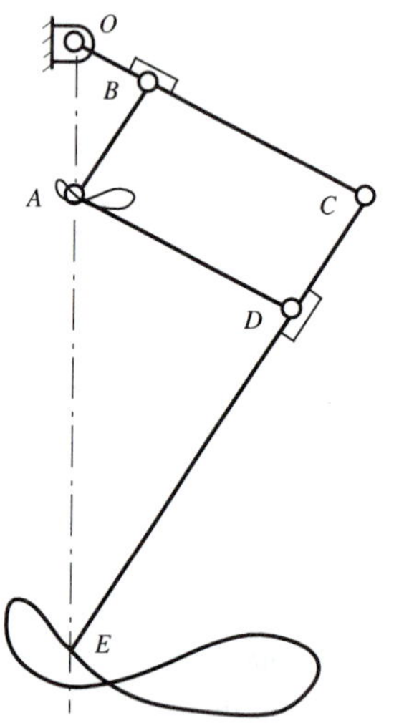

Figure 5.7 A simple form of planar pantograph linkage. Any path traced by point A is reproduced by point E at a magnification of 4:1. $ABCD$ is a parallelogram. The ratios of lengths CD to CE and OB to OC are both 1:4. Link OC is connected to the base by a fixed revolute joint at O.

A simple form of planar pantograph linkage is shown in Fig. 5.7. In Fig. 5.7, link AB has the length CD. Likewise, link AD has the length BC. Consequently, $ABCD$ is a parallelogram, regardless of the position of the linkage. Further, the lengths OB and OC are in the ratio 1:4, as are the lengths CD and CE. It follows that triangle OBA is similar to triangle OCE, because OB/OC is equal to BA/CE and angle OBA is equal to angle OCE. Consequently, the ratio of the lengths OA and OE is always 1:4. If point A traces any path in the plane of the linkage, point E will trace a geometrically similar path that is magnified by a factor of 4 compared with the path of point A. This is best understood by considering the path of point A to be a curve described in polar coordinates with origin at O. The position of the corresponding point on the path of point E is also described in polar coordinates centered on O. The angular coordinate of that point is the same as that of the corresponding point on the path of point A. Its radial coordinate is four times that of the corresponding point on the path of point A. Hence the curve is the same. It is simply scaled up by a factor of 4.

In consequence of the property described above, pantograph mechanisms have been used a great deal to copy and rescale text and other geometric figures. The magnification factor can be set to any desired value by varying the proportions of the links. In the form of the linkage that is shown in Fig. 5.7, it is always equal to the ratio of length OC to OB, which must also be equal to the ratio of CE to CD.

An example of the use of the pantograph mechanism to copy plane curves is a copying mill used to produce plate cams. The reader will find an in-depth discussion of cam geometry in Chapter 6. Most plate cams are bounded by mathematically complicated curves. In order to produce cams using a copy mill, a master cam is produced at an enlarged scale by hand. The profile of the master is traced by a roller with its central axis located at point E of Fig. 5.7. The axis of the milling cutter is at point A. The ratio of the roller diameter to the cutter diameter is the pantograph ratio. Consequently, the mill produces a cam that is geometrically similar to the master but which is reduced in size by the pantograph ratio. The use of point E, rather than point A, to trace the master provides improved accuracy,

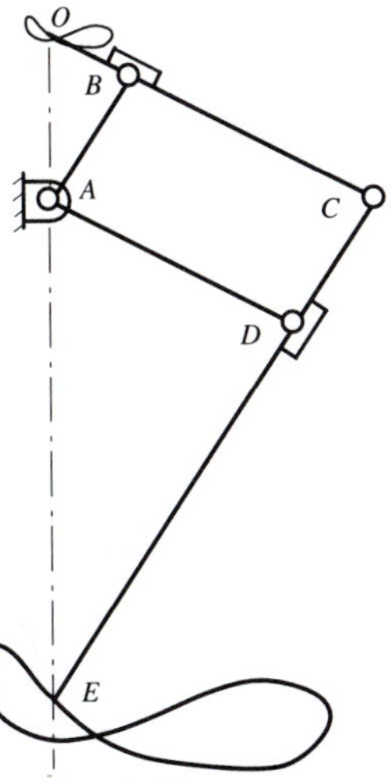

Figure 5.8 The pantograph of Fig. 5.7 inverted by mounting with a fixed revolute at point A. The paths of points O and E are geometrically similar. The magnification factor is now 3 : 1.

because errors in the master profile are reduced by the pantograph ratio. The large size of the master also facilitates its accurate manufacture.

The application described above is an example of inversion of the linkage by interchanging the tracing points A and E. The pantograph can also be inverted by hinging it to the base with a fixed revolute coincident with A, rather than at point O. This is shown in Fig. 5.8. The path of point O is now copied by a geometrically similar path of point E. However, the magnification ratio is now 3 : 1 rather than 4 : 1. This is because, with these dimensions, the ratio of length AE to AO is 3 : 1.

There are other variations on the same theme. Figure 5.9 shows the pantograph linkage used in the legs of the Adaptive Suspension Vehicle that was shown in Fig. 1.1. Here there is no fixed pivot. Rather, point O is on a vertical slide and point A is on a horizontal slide. Motion of point A alone, produced by a hydraulic cylinder, causes a horizontal rectilinear motion of the ankle point, E. Motion of point O alone, also produced by a hydraulic cylinder, causes a vertical rectilinear motion of point E. Simultaneous motion of points A and O results in motion of point E along a plane curve. This is what happens when the foot is picked up and the leg is swung back to its forward position. The magnification factor in this mechanism is 5 : 1 for the drive motion (point A) and 4 : 1 for the lift motion (point B).

5.1.4.2 Skew Pantographs

A more general form of pantograph is the skew pantograph shown in Fig. 5.10. $OLMN$ is a parallelogram, and triangle NMQ is similar to triangle LPM. As shown below, triangle OPQ is always similar to triangle LPM. Consequently, the path traced by point Q is similar to that traced by point P, is rotated through angle α from the path of P about O, and is magnified by the ratio LM/LP.

These properties are proved as follows. Note that since $OLMN$ is a parallelogram, $\angle MLO = \angle ONM = \phi$. Likewise, $\angle LON = \angle NML = \pi - \phi$.

Triangles PLO and ONQ are similar for the following reasons:

$$\angle PLO = \angle ONQ = \phi + \alpha.$$

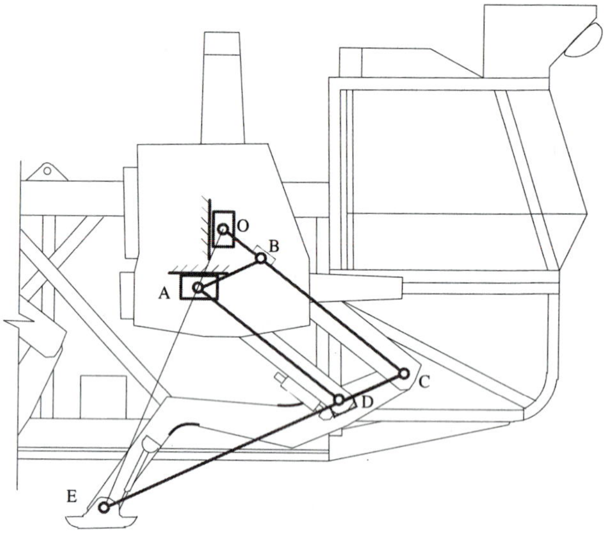

Figure 5.9 The leg mechanism of the Adaptive Suspension Vehicle. Point O moves on a slide that is vertical relative to the leg mounting structure to produce a corresponding vertical motion of the ankle point E. Point A moves on a slide that is horizontal relative to the vehicle body to drive point E along a horizontal path.

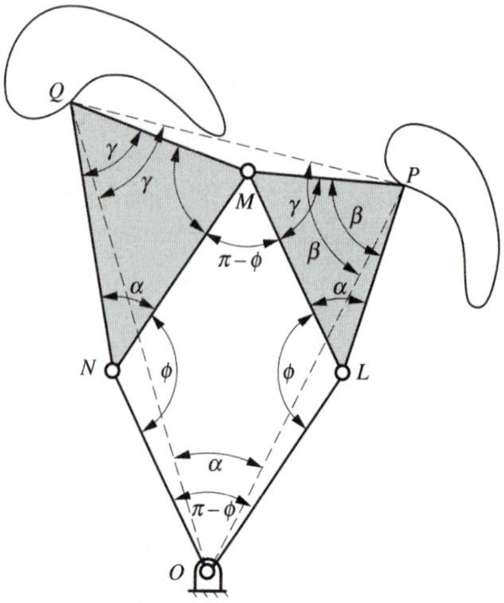

Figure 5.10 A skew pantograph. $OLMN$ is a parallelogram and triangles LPM and NMQ are similar. The triangle OPQ is always similar to triangles LPM and NMQ. Consequently, the path traced by point Q is similar to that traced by point P. The path traced by Q is rotated relative to that traced by P through angle α, and it is magnified by the ratio $OQ/OP = LM/LP$.

Also

$$\frac{NQ}{NM} = \frac{ML}{PL}$$

because triangles NMQ and LPM are similar and these are corresponding pairs of sides. Now $NM = OL$, and $ML = NO$ because $OLMN$ is a parallelogram. Making these substitutions, we get

$$\frac{NQ}{OL} = \frac{NO}{PL} \text{ or } \frac{NQ}{NO} = \frac{OL}{PL}$$

which establishes that triangles PLO and ONQ are similar since NQ and NO, and OL and PL, are corresponding side pairs and the equal angles $\angle PLO$ and $\angle ONQ$ are the included angles.

Also, triangle QMP is similar to PLO and ONQ for the following reasons:

$$\angle QMP = 2\pi - \beta - \gamma - (\pi - \phi) = \pi + \phi - \beta - \gamma$$

Also, because α, β, and γ are the vertex angles of triangle LPM,

$$\alpha + \beta + \gamma = \pi$$

so

$$\angle QMP = \pi + \phi - (\pi - \alpha) = \phi + \alpha = \angle ONQ$$

Because triangles NMQ and LPM are similar

$$\frac{MQ}{NQ} = \frac{PM}{LM},$$

or

$$\frac{PM}{MQ} = \frac{LM}{NQ} = \frac{ON}{NQ}$$

noting that $LM = ON$.

Therefore triangles QMP and ONQ are similar because the corresponding sides PM and MQ, and ON and NQ, are in the same ratio and the included angles QMP and ONQ are equal. Triangles QMP and PLO are similar because both are similar to ONQ.

It follows that

$$\angle NQO = \angle MQP$$

and so

$$\angle OQP = \angle NQM = \gamma$$

Likewise

$$\angle QPM = \angle OPL$$

and so

$$\angle QPO = \angle MPL = \beta$$

Consequently,

$$\angle POQ = \alpha$$

and triangle OPQ is similar to triangles LPM and NMQ.

The geometric similarity of the paths of points P and Q can be inferred from an argument similar to that employed in the case of the collinear pantograph. If the path of point P is considered to be a curve described in polar coordinates centered on O, the radial coordinate is OP. The path of Q is also described in polar coordinates centered on O. The radial coordinate is LM/LP times that of point P, and the angle reference is rotated through angle α from that used for the path of point P.

5.1.4.3 *Roberts' Theorem*

If the path of point P of the skew pantograph of the preceding section is a circle, then that of point Q will also be a circle as shown in Fig. 5.11. Thus, if P is constrained to move on a circle by a crank rotating about fixed pivot, O_P, then a crank can also be connected to point Q from a fixed pivot at the center of its path: O_Q. Because the path of Q is similar to that of P and triangle OPQ is always similar to triangle LPM, it follows that triangle OO_PO_Q is also similar to triangle LPM. This creates two planar four-bar linkages, $OLPO_P$ and O_QQNO, for each of which M is a coupler point. Thus the path generated by point M as a coupler point of $OLPO_P$ is identical to the path traced by M as a coupler point of O_QQNO. Thus we have generated two completely different four-bar mechanisms that generate identical coupler curves. Linkages that have this property are called *cognates.*

We can go further. If points R and S are located by constructing the parallelograms O_PPMR and O_QQMS, it can be shown that triangle MRS is similar to triangle LPM and hence that the four-bar linkage O_PRSO_Q is also cognate to $OLPO_P$ and O_QQNO, again with M as the tracing point. The assemblage shown in Fig. 5.12 is known as Roberts' mechanism.

Roberts' theorem states that if a planar four-bar mechanism is constructed, a coupler point is selected, and the corresponding coupler curve is traced, then there are two other

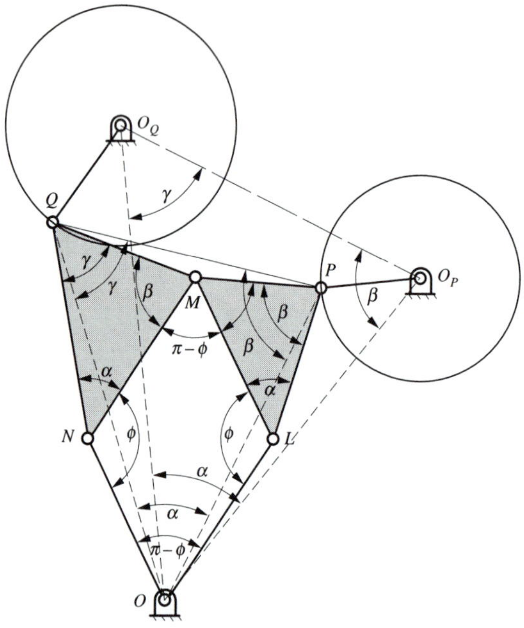

Figure 5.11 A pair of cognate linkages. The path of point P in the skew pantograph of Fig. 5.10 is a circle centered on O_P. Therefore the path of point Q is also a circle, with center O_Q where triangle OO_PO_Q is similar to triangle LPM. Therefore cranks O_PP and O_QQ can be added, and the assemblage will be mobile.

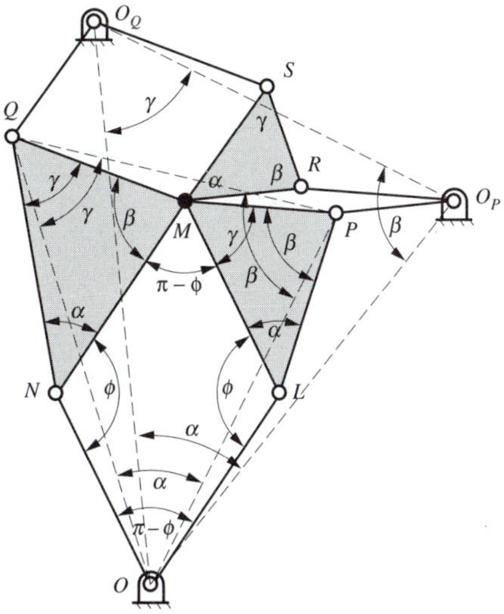

Figure 5.12 A general Roberts' mechanism. The three four-bar linkages $OLPO_P$, O_PRSO_Q, and O_QQNO are all cognate with M as the coupler point for each.

four-bar linkages that will generate the identical coupler curve. That is, there are two four-bar linkages that are cognate to the original four-bar. They may be constructed by constructing the Roberts' mechanism based on the original four-bar. In the case of Fig. 5.12, if we view $OLPO_P$ as the original four-bar linkage, with M being the selected coupler point, then the cognates are O_QQNO and O_PRSO_Q with M being the coupler point in both cases. Starting with points O, L, P, O_P, and M, the remainder of the figure may be constructed by first completing the parallelograms $OLMN$ and O_PRMP to locate points N and R. Triangles OO_PO_Q, NMQ, and MRS may then be constructed similar to triangle LPM to complete the figure.

If the original four-bar linkage is of Grashof type I, then the cognates will also be type I. Likewise, if the original four-bar is type II, then the cognates are also type II. Further, if the original four-bar is type I and is a crank-rocker linkage, then one of the cognates will also be a crank-rocker linkage. The other will be a type I double-rocker. The cognates of a drag-link linkage are both also drag links. The cognates of a type I double rocker are both crank rockers.

As indicated in Chapter 4, cognate linkages can be very useful when a linkage has been found that generates a desired path but that solution linkage has undesirable properties such as interference with other components. Often one of the cognates will produce the desired path without the problems of the original linkage.

EXAMPLE 5.1 (*Using Roberts' Theorem to Generate Cognates of Chebyshev Mechanism*)

PROBLEM

The Chebyshev linkage of Fig. 5.2 is a type I double-rocker. As was discussed in Section 1.18, it is difficult to transfer motion from the tracing point of this linkage due to interference with the cranks, since the coupler tumbles between the cranks. As discussed above, the cognates of a type I double-rocker are both crank rockers and should be free of this problem. Construct the cognates and, hence, produce a crank-rocker linkage with the same approximate straight-line coupler curve segment as the Chebyshev linkage.

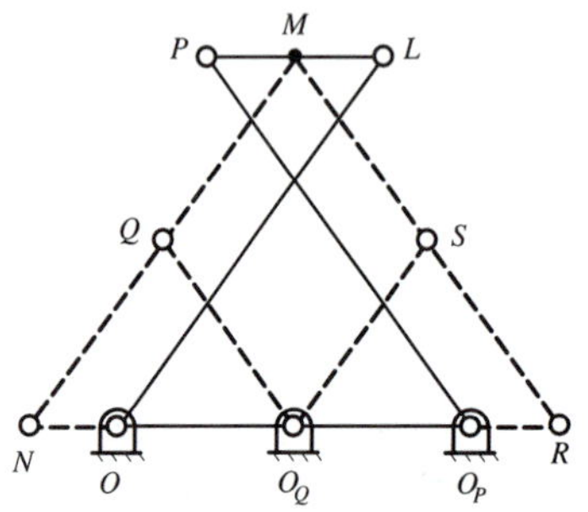

Figure 5.13 Construction of the cognates of the Chebyshev linkage shown in Fig. 5.2. $OLPO_P$ is the original Chebyshev four-bar, and M is the coupler point. The cognates are $ONQO_Q$ and O_PRSO_Q. Their symmetry with one another is a result of the bilateral symmetry of the original linkage.

SOLUTION

Examination of Fig. 5.2 indicates that the coupler point is the midpoint of the line between the coupler pivots. That is, the triangle LPM of Fig. 5.12 has collapsed into a line. Therefore triangle OO_PO_Q will also be collapsed to a line. Since O corresponds to L, O_P corresponds to P, and O_Q corresponds to M in these triangles (corresponding vertices have the same vertical angles), it follows that O_Q will be midway between O and O_P, as shown in Fig. 5.13. Similarly, triangles RMS and MNQ will collapse to line segments.

Parallelograms $OLMN$ and O_PPMR are constructed as shown in Fig. 5.13 to locate points N and R. The line MQN is drawn. Note that in Fig. 5.12, N corresponds to L, M corresponds to P, and Q corresponds to M in the two similar triangles LPM and NMQ. Therefore Q will be at the midpoint of NM in Fig. 5.13. Similarly, the line RSM is drawn to represent the coupler of the second cognate. The cranks O_QQ and O_QS are drawn in to complete the two cognates shown by the dashed lines in Fig. 5.13.

The cognate $ONQO_Q$ is drawn on its own in Fig. 5.14 with the path of point M plotted. Not only is it much easier to transfer motion from this linkage than from the original Chebyshev linkage, but the linkage can be driven by continuous rotation of the crank ON, if desired.

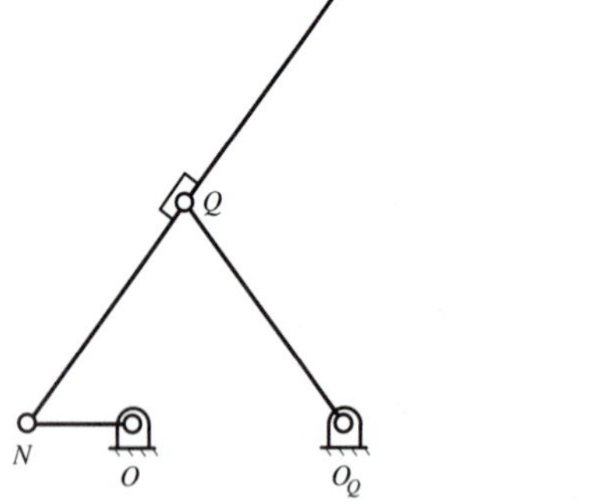

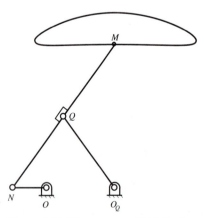

Figure 5.14 The cognate $ONQO_Q$ from Fig. 5.12 with its coupler curve plotted.

5.2 SPHERICAL LINKAGES

Although spatial linkages, in general, will be discussed in Chapter 7, there are other classes of mechanisms that are not general spatial linkages in the sense of satisfying the spatial Kutzbach criterion (Eq. 1.3) and are certainly not planar mechanisms. One of the most extensive and practically important such groups is the class of spherical mechanism, that includes not only linkages but also spherical cam mechanisms and gears, namely bevel gears, and rolling contact bearings, namely tapered roller bearings.

Although it is beyond the scope of this book, spherical mechanism theory is an important component of spatial mechanism theory. This is because the rotational equations defined for spatial mechanisms in Chapter 7 are identical for spherical mechanisms. However, the translation equations, also discussed in Chapter 7, are absent in the case of spherical mechanisms. This allows inferences to be made on the basis of a spherical analog and applied to spatial mechanisms. There is also a way of generating valid translation equations directly from the rotational equations.

5.2.1 Introduction

Spherical linkages form a family much like planar linkages. However, whereas in a planar linkage all the revolute joint axes are parallel, in a spherical linkage they all intersect at a common point, called the concurrency point. Actually, planar linkages can be thought of as spherical linkages for which the concurrency point is at infinity.

There are many similarities in the properties of spherical and planar linkages. For example, spherical linkages obey the same form of the Kutzbach criterion that planar linkages do (Section 1.7):

$$M = 3(m - j - 1) + \sum_{i=1}^{j} f_i \qquad \textbf{(1.1)}$$

Consequently, the simplest nontrivial spherical linkage is a four-bar linkage, just as in the planar case.

Also, there is a form of the Grashof inequality governing rotatability of joints that works for spherical linkages:

$$\alpha_l + \alpha_s < \alpha_p + \alpha_q$$

Here, instead of dealing with the lengths of the links, as in the case of a planar linkage, we work with the angles between successive joint axes. α_s is the smallest angle between two successive joints, α_l is the largest such angle, and α_p and α_q are the other two angles.

As in the planar case, the inequality governs the presence of joints in a four-bar linkage that can be completely rotated. If the inequality is satisfied, there are two completely rotatable joints. They are the joints whose axes bound the angle α_s. Depending on which link is chosen as the base, the linkage will have characteristics similar to those of a crank rocker planar four-bar, or a drag link, or a type 1 double rocker. If the inequality is not satisfied, there is no completely rotatable joint, and the linkage behaves like a type 2 planar linkage.

There is one variation from the planar analog. Whereas there is no limit on the length of a link in a planar linkage, beyond the fact that it must be less than the sum of the lengths of the other three links for it to be possible to assemble the loop, no side angle of a spherical linkage can be greater than 90°. This is because there are, in fact, always two angles between two lines which are supplements of one another. Either the angle or its supplement can be viewed as the angle between two axes in a spherical four-bar linkage. If the angle is greater than 90°, its supplement is less than 90°, so side angles in a spherical linkage can be said to have an upper limit of 90°.

The closure equations for a spherical four-bar linkage, such as that shown schematically in Fig. 5.15, may, in principle, be developed using a procedure analogous to that used to derive the closure equations for a planar four-bar linkage in Chapter 3. However, this becomes very complex because the entities being dealt with are angles rather than lengths. A more convenient procedure is to use the loop matrix transformations defined in Chapter 7. Using either method, the relationship between angle ϕ_1, considered to be the input angle, and ϕ_2, considered to be the output angle of the linkage of Fig. 5.15, can be expressed as follows:

$$\begin{aligned}\sin\phi_1 \sin\phi_2 \sin\alpha_2 \sin\alpha_4 &- \cos\phi_1 \cos\phi_2 \cos\alpha_1 \sin\alpha_2 \sin\alpha_4 \\ &+ \cos\phi_1 \sin\alpha_1 \cos\alpha_2 \sin\alpha_4 + \cos\phi_2 \sin\alpha_1 \sin\alpha_2 \cos\alpha_4 \\ &+ \cos\alpha_1 \cos\alpha_2 \cos\alpha_4 - \cos\alpha_3 = 0\end{aligned} \quad \textbf{(5.1)}$$

where ϕ_1, ϕ_2, ϕ_3, and ϕ_4 are the joint angles and α_1, α_2, α_3, and α_4 are the angles between the joint axes of the spherical four-bar loop as shown in Fig. 5.15.

If ϕ_1 is regarded as having a known value, this rather intimidating-looking equation has the form

$$P\cos\phi_2 + Q\sin\phi_2 + R = 0 \quad \textbf{(5.2)}$$

for which a solution was developed in Chapter 3. Here

$$\begin{aligned}P &= -\cos\phi_1 \cos\alpha_1 \sin\alpha_2 \sin\alpha_4 + \sin\alpha_1 \sin\alpha_2 \cos\alpha_4 \\ Q &= \sin\phi_1 \sin\alpha_2 \sin\alpha_4 \\ R &= \cos\phi_1 \sin\alpha_1 \cos\alpha_2 \sin\alpha_4 + \cos\alpha_1 \cos\alpha_2 \cos\alpha_4 - \cos\alpha_3\end{aligned} \quad \textbf{(5.3)}$$

Hence, referring to Table 3.1, we can obtain values for ϕ_2 given ϕ_1 from

$$t = \frac{-Q + \sigma\sqrt{P^2 + Q^2 - R^2}}{R - P} \quad \textbf{(5.4)}$$

where $\sigma = \pm 1$ is a sign variable and

$$\phi_2 = 2\tan^{-1}(t) \quad \textbf{(5.5)}$$

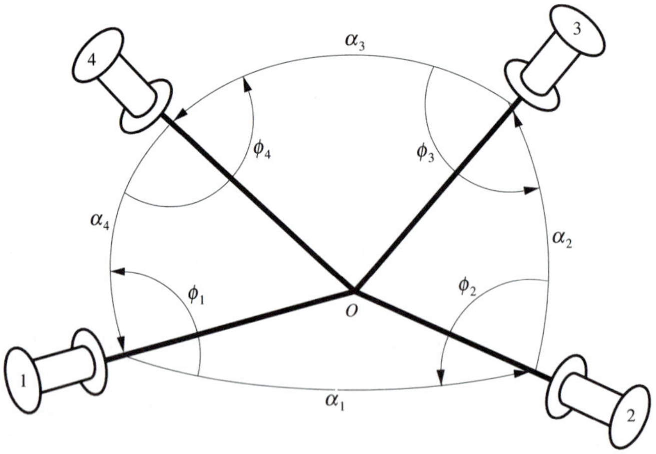

Figure 5.15 A schematic representation of a spherical four-bar mechanism. The heavy lines represent the joint axes with concurrency point O. α_1, α_2, α_3, and α_4 are the angles between the successive axes. ϕ_1, ϕ_2, ϕ_3, and ϕ_4 are the joint angles.

We can also develop relationships between the angular velocities and accelerations about joints 1 and 2 by differentiation of Eq. (5.1). Differentiation of Eq. (5.1) with respect to time gives, after rearrangement,

$$\dot{\phi}_2 = -\dot{\phi}_1 \frac{\sin\alpha_4(\cos\phi_1 \sin\phi_2 \sin\alpha_2 + \sin\phi_1 \cos\phi_2 \cos\alpha_1 \sin\alpha_2 - \sin\phi_1 \sin\alpha_1 \cos\alpha_2)}{\sin\alpha_2(\sin\phi_1 \cos\phi_2 \sin\alpha_4 + \cos\phi_1 \sin\phi_2 \cos\alpha_1 \sin\alpha_4 - \sin\phi_2 \sin\alpha_1 \cos\alpha_4)} \tag{5.6}$$

Further differentiation gives

$$\ddot{\phi}_2 = \frac{\ddot{\phi}_1 \dot{\phi}_2}{\dot{\phi}_1} + \frac{B}{A}\dot{\phi}_1\dot{\phi}_2 + \frac{C}{A}\dot{\phi}_1^2 + \frac{D}{A}\dot{\phi}_2^2 \tag{5.7}$$

where

$$\begin{aligned}
A &= \sin\alpha_2(\sin\phi_1 \cos\phi_2 \sin\alpha_4 + \cos\phi_1 \sin\phi_2 \cos\alpha_1 \sin\alpha_4 - \sin\phi_2 \sin\alpha_1 \cos\alpha_4)\\
B &= 2\sin\alpha_2 \sin\alpha_4(\sin\phi_1 \sin\phi_2 \cos\alpha_1 - \cos\phi_1 \cos\phi_2)\\
C &= \sin\alpha_4(\sin\phi_1 \sin\phi_2 \sin\alpha_2 - \cos\phi_1 \cos\phi_2 \cos\alpha_1 \sin\alpha_2 + \cos\phi_1 \sin\alpha_1 \cos\alpha_2)\\
D &= \sin\alpha_2(\sin\phi_1 \sin\phi_2 \sin\alpha_4 - \cos\phi_1 \cos\phi_2 \cos\alpha_1 \sin\alpha_4 + \cos\phi_2 \sin\alpha_1 \cos\alpha_4)
\end{aligned} \tag{5.8}$$

EXAMPLE 5.2 ***(Analysis of Spherical Four-Bar Mechanism)***

PROBLEM

A spherical four-bar linkage is constructed with the angle between the axes of joints 1 and 2 (α_1) being 120°, the angle between axes 2 and 3 (α_2) 90°, that between axes 3 and 4 (α_3) 75°, and that between axes 1 and 4 (α_4) 30°. Member 1 is the base and the mechanism is a spherical crank-rocker linkage. Find the output angle, ϕ_2, when the driving joint angle, ϕ_1 is 90°.

If the input crank is driven at a constant angular velocity of 10 rad/s, find the angular velocity, $\dot{\phi}_2$, of the driven crank, and its angular acceleration, $\ddot{\phi}_2$, in the same position.

SOLUTION

Substitution of the values $\alpha_1 = 120°$, $\alpha_2 = 90°$, $\alpha_3 = 75°$, $\alpha_4 = 30°$, and $\phi_1 = 90°$ into Eq. (5.3) gives

$$P = 0.75, \quad Q = 0.5, \quad R = -0.2588.$$

Substitution of these values into Eq. (5.4) gives

$$t = -0.3603 \quad \text{or} \quad t = 1.3515$$

Application of Eq. (5.5) gives

$$\phi_2 = -39.63° \quad \text{or} \quad \phi_2 = 107.00°$$

The two solutions correspond to the two solutions obtained in the position problem of a planar four-bar and have the same source in the reflection of the driven crank and coupler about the plane of the moving joint axis of the driving crank and the fixed joint axis of the driven crank.

Substitution of these values plus $\dot{\phi}_1 = 10$ rad/s into Eq. (5.6) gives the values $\dot{\phi}_2 = 2.2302$ rad/s and $\dot{\phi}_2 = 0.8467$ rad/s, respectively corresponding to the two solutions for ϕ_2 given above. Further substitution into Eq. (5.8) gives the following sets of values:

For $\phi_2 = -39.63°$: $A = 0.8634, B = 0.3189, C = -0.3189, D = 0.2588$

For $\phi_2 = 107.00°$: $A = -0.8634, B = -0.4781, C = 0.4781, D = 0.2588$

When substituted into Eq. (5.7) with $\ddot{\phi}_1 = 0$ and the preceding values for $\dot{\phi}_1$ and $\dot{\phi}_2$, we get the following two values for the acceleration of the driven crank: $\ddot{\phi} = -27.30$ rad/s^2 and $\ddot{\phi} = -50.90$ rad/s^2, respectively. Once again, these correspond to the two possible solutions of the position problem. ■

5.2.2 Gimbals

A set of gimbals is a spherical serial chain that allows an axis through the concurrency point to be placed in any possible direction. Gimbals are often used in the mounts of directional instruments such as theodolites or telescopes. They are also used in gyroscopes to allow the rotor axis freedom to assume any direction relative to the base of the instrument.

5.2.3 Universal Joints

The simplest means of transferring motion between noncoaxial shafts is by means of one or two universal joints, also known as Cardan joints in Europe and Hooke joints in Britain. For this reason, this very simple spherical mechanism appears in an enormous variety of applications. They may be found as components of the Stewart platform and 3-2-1 platform parallel mechanisms discussed in Chapter 7 and in many other situations.

5.2.3.1 Properties of the Universal Joint

A common need in machinery is to transfer rotation between two shafts that are not parallel to one another and that may be free to move relative to one another. A universal joint is a simple spherical four-bar mechanism that transfers rotary motion between two shafts whose axes pass through the concurrency point. The joint itself consists of two revolute joints whose axes are orthogonal to one another. They are often configured in a cross-shaped member as shown in Fig. 5.16. One of these joints is arranged with its axis at 90° to that of the driving shaft, and the other has its axis at 90° to that of the driven shaft. In practice, the ends of the shafts are often configured as clevises to mate with the cruciform shafts of the intermediate member. Together with the bearings in which the two shafts turn, the universal joint forms a spherical four-bar linkage with three sides being 90° angles. The fourth side is, in general, not 90°. This may be better seen in Fig. 5.17, in which only one side of each of the crossed intermediate shafts is shown.

In general, the angular motion is not uniformly transferred from the driving shaft to the driven shaft. The relationship between the angles of the driving shaft, θ_1, and the driven shaft, θ_2, is

$$\cos \gamma = \tan \theta_1 \tan \theta_2 \tag{5.9}$$

where γ is the angular misalignment of the shafts. This relationship can be quickly derived from Eq. (5.1). As is indicated in Fig. 5.17, $\alpha_2 = \alpha_3 = \alpha_4 = 90°$. Also, $\alpha_1 = \gamma$ and $\phi_1 = \theta_1$, $\phi_2 = \theta_2$ where α_i and ϕ_i, $i = 1, 2, 3, 4$, are consistent with Eq. (5.1). Substituting these values Eq. (5.1) reduces to

$$\sin \theta_1 \sin \theta_2 - \cos \theta_1 \cos \theta_2 \cos \gamma = 0$$

which can be rearranged into Eq. (5.9).

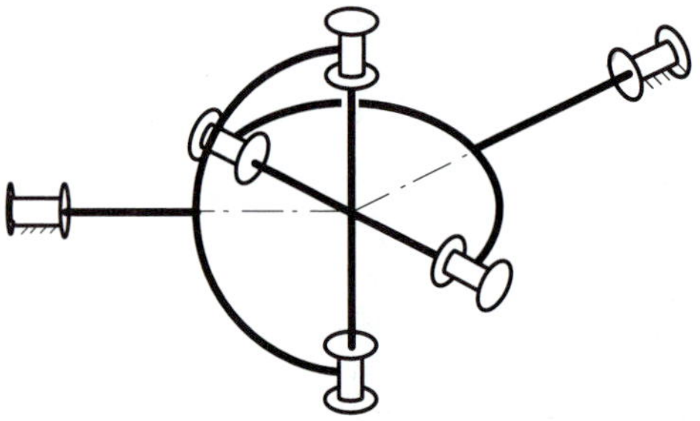

Figure 5.16 Universal, Cardan, or Hooke joint.

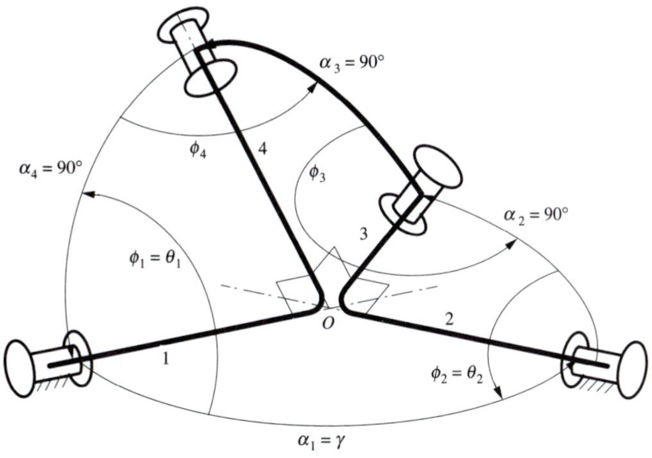

Figure 5.17 Universal joint geometry: γ is the angular misalignment of the shafts; θ_1 is the angle of the input shaft; θ_2 is the angle of the output shaft.

If the driving shaft turns with a uniform angular velocity, the rotation of the driven shaft is not uniform but fluctuates. That is, a single universal joint is not a constant-velocity coupling such as those that will be discussed in the next section. However, if the angle between the shaft axes is small, the fluctuation will also be small and is acceptable in many applications.

The angular velocity relationship can be obtained by differentiating Eq. (5.9) written in the form

$$\tan\,\theta_2 = \cos\,\gamma\,\cot\,\theta_1$$

Differentiation with respect to time gives

$$\dot{\theta}_2\,\sec^2\,\theta_2 = -\dot{\theta}_1\,\csc^2\,\theta_1\,\cos\,\gamma$$

Hence the ratio of the magnitudes of the shaft velocities is

$$\frac{\omega_2}{\omega_1} = \frac{\dot{\theta}_2}{\dot{\theta}_1} = \frac{\cos^2\theta_2\cos\gamma}{\sin^2\theta_1}$$

It is helpful to work in terms of the input angle, θ_1, alone. Hence the angle equation is used to eliminate $\tan\,\theta_2$:

$$\frac{\omega_2}{\omega_1} = \frac{\cos\gamma}{\sin^2\theta_1(1+\tan^2\theta_2)} = \frac{\cos\gamma}{\sin^2\theta_1(1+\cos^2\gamma\cot^2\theta_1)} = \frac{\cos\gamma}{\sin^2\theta_1+\cos^2\gamma\cos^2\theta_1}$$

This expression can be further simplified by replacing $\cos^2\theta$ by $1-\sin^2\theta$ as follows:

$$\frac{\omega_4}{\omega_2} = \frac{\cos\gamma}{1-\cos^2\theta_1+\cos^2\gamma\cos^2\theta_1} = \frac{\cos\gamma}{1-\sin^2\gamma\cos^2\theta_1} \qquad \textbf{(5.10)}$$

It may be seen that the velocity ratio is a function of θ_1, so that for constant input velocity the output velocity will fluctuate. The velocity ratio varies from

$$\frac{1}{\cos\gamma} \quad \text{to} \quad \cos\gamma$$

during the motion cycle. This relationship is plotted in Fig. 5.18.

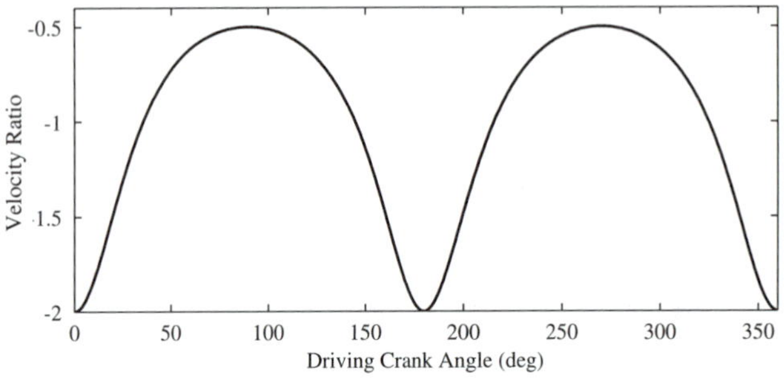

Figure 5.18 Velocity ratio fluctuation for a universal joint with $\gamma = 120°$. The negative values of the velocity ratio are an artifact of the way these angles are defined in Fig. 5.17. Examination of that figure indicates that θ_2 decreases when θ_1 increases. Looking from the driving shaft toward the driven shaft, this indicates that both shafts are rotating in the same direction.

EXAMPLE 5.3 ***(Analysis of Universal Joint for Front Wheel–Driven Car)***

PROBLEM

A simple automotive vehicle is driven via the front wheels. Universal joints are used in the shafts connecting the differential to the front wheels, as a low-cost alternative to the constant-velocity joints that are normally used to allow rotation of the front wheels about vertical axes for steering. At full steering lock, the inside front wheel is rotated 30° from the straight-ahead position. Calculate the percentage fluctuation in wheel velocity in this position.

SOLUTION

If shaft 1 in Fig. 5.17 is viewed as the shaft from the engine and shaft 2 is viewed as the half-shaft driving the wheel, in the full lock position the angle between the axis of shaft 1 and the axis of shaft 2 will be 30°. That is, $\gamma = 180° - 30° = 150°$. Applying Eq. (5.10),

$$\frac{\omega_4}{\omega_2} = \frac{-0.8660}{1 - 0.25\cos^2\theta_1}$$

Thus, the maximum magnitude of the velocity ratio is $0.8660/0.75 = 1.155$, and the minimum magnitude is $0.8660/1.25 = 0.693$. Thus the maximum is 115% of the mean value of 1.0, and the minimum is 69% of the mean. The maximum percentage fluctuation is 31%. ■

5.2.3.2 *Dual Universal Joints*

By using two universal joints in a symmetric combination it is possible to have the second joint cancel out the fluctuation generated by the first. This combination then produces a constant-velocity action. If the joints are aligned so that axis 3 of the first coupling is parallel to axis 2 of the second, as shown in Fig. 5.19, then $\theta_1' = \theta_2$. Hence, using Eq. (5.9),

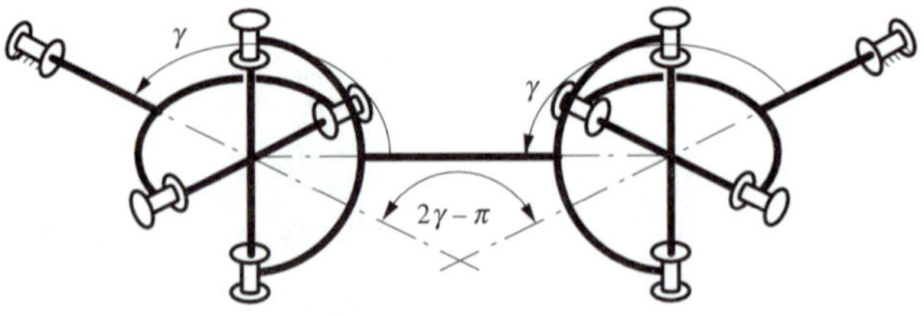

Figure 5.19 Dual universal joints arranged symmetrically. The combination provides a true constant-velocity coupling, as described in the text.

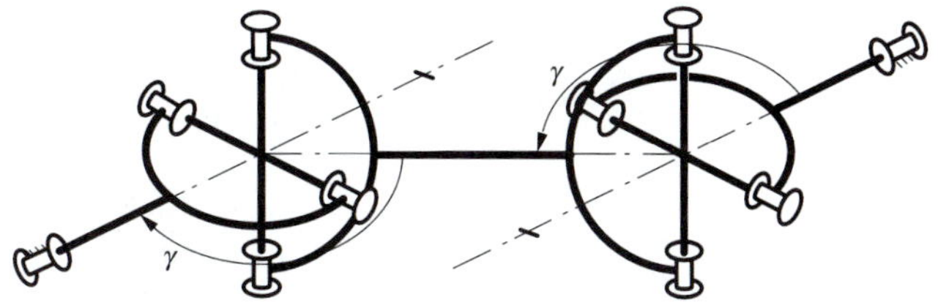

Figure 5.20 Dual universal joints on parallel, offset shafts. This arrangement also gives motion transfer between the input and output shafts at a constant velocity ratio.

$$\cos\gamma = \tan\theta_1 \tan\theta_2$$
$$\cos\gamma = \tan\theta_1' \tan\theta_2' \qquad \textbf{(5.11)}$$

and

$$\tan\theta_2' = \frac{\cos\gamma}{\tan\theta_1'} = \frac{\cos\gamma}{\tan\theta_2} = \frac{\cos\gamma\tan\theta_1}{\cos\gamma} = \tan\theta_1 \qquad \textbf{(5.12)}$$

Hence the output angle of the combined joint, θ_2', is always equal to the input angle θ_1. The same relationship is true if the shafts are not angulated, as in Fig. 5.19, but are parallel and offset, as in Fig. 5.20. This is also a configuration of considerable practical importance. In fact, the drive shafts of almost all front engine, rear wheel–driven automobiles feature this arrangement.

EXAMPLE 5.4 ***(Analysis of Universal Joint of Rear Wheel–Driven Car)***

PROBLEM

A front engine, rear wheel-driven automobile employs a drive shaft with two universal joints in the alignment of Fig. 5.20 to transmit torque from the output shaft of the gearbox to the differential. The differential is mounted on the rear axle, and the suspension is of the live axle type (solid rear axle). The universal joints accommodate movement of the rear axle permitted by the suspension. The differential shaft is nominally parallel to the gearbox shaft. However, the suspension setup maintains this relationship only to a good approximation. Also, some fore-aft rocking of the differential housing occurs because of elastic deflection and backlash in suspension components. The angle γ, as defined in Fig. 5.20, varies from 175° to 160° between the suspension stops. The error in γ at the rear universal joint is estimated to be ±0.5°. Estimate the maximum percentage fluctuation in the velocity ratio between the gearbox shaft and the differential shaft.

SOLUTION

Since the error in γ is small, we should be able to use a small-angle approximation with acceptable accuracy.

Equations (5.11) become

$$\cos\gamma = \tan\theta_1 \tan\theta_2$$

and

$$\cos(\gamma + \delta\gamma) = \tan\theta_1' \tan\theta_2'$$

or

$$\cos\gamma - \delta\gamma\sin\gamma = \tan\theta_1' \tan\theta_2'$$

Noting that

$$\tan\theta_1' = \tan\theta_2 = \frac{\cos\gamma}{\tan\theta_1}$$

$$\cos\gamma - \delta\gamma\sin\gamma = \frac{\cos\gamma\tan\theta_2'}{\tan\theta_1}$$

or

$$\tan\theta_2' = \tan\theta_1(1 - \delta\gamma\tan\gamma)$$

Differentiation of this expression with respect to time gives

$$\dot\theta_2'\sec^2\theta_2' = \dot\theta_1\sec^2\theta_1(1 - \delta\gamma\tan\gamma)$$

Using

$$\sec^2\theta_2' = 1 + \tan^2\theta_2' = 1 + \tan^2\theta_1(1 - \delta\gamma\tan\gamma)^2$$

the velocity ratio is given by

$$\frac{\dot\theta_2'}{\dot\theta_1'} = \frac{1 - \delta\gamma\tan\gamma}{\cos^2\theta_1\{1 + \tan^2\theta_1(1 - \delta\gamma\tan\gamma)^2\}} = \frac{1 - \delta\gamma\tan\gamma}{1 - 2\delta\gamma\tan\gamma\sin^2\theta_1}$$

Here the small-angle approximation has been used by dropping the $\delta\gamma^2$ term in the expansion of the denominator. This expression may be further simplified by multiplying top and bottom by the factor $1 + 2\,\delta\gamma\,\tan\gamma\,\sin^2\theta_1$ and again applying the small-angle approximation. Then

$$\frac{\dot\theta_2'}{\dot\theta_1'} = (1 + 2\delta\gamma\tan\gamma\sin^2\theta_1)(1 - \delta\gamma\tan\gamma) = 1 - \delta\gamma\tan\gamma + 2\delta\gamma\tan\gamma\sin^2\theta_1$$

or

$$\frac{\dot\theta_2'}{\dot\theta_1'} = 1 - \delta\gamma\tan\gamma\cos 2\theta_1$$

It is clear from this expression that the maximum magnitude of the velocity ratio, R, is $1 + \delta\gamma\tan\gamma$, and the minimum value is $1 - \delta\gamma\tan\gamma$. Applying the values given in this particular problem, $\tan\gamma$ will be at a maximum when $\gamma = 160°$ and $\delta\gamma = 15°$. Then

$$\delta\gamma = 0.5 \times \pi/180 = 0.00873 \text{ rad}$$
$$\tan\gamma = -0.364$$

so

$$R_{max} = 1.0032 \quad \text{and} \quad R_{min} = 0.9968.$$

The maximum percentage fluctuation of the velocity ratio is 0.32%. ■

5.3 CONSTANT-VELOCITY COUPLINGS

As may be seen in the preceding subsections, universal joints are not constant-velocity joints. Although paired universal joints can function as constant-velocity joints, the arrangement must satisfy special geometric conditions. There is a need for single joints that can provide true constant-velocity action and that can accommodate other changes of alignment such as plunging (movement in the direction of the shaft axis) of one shaft relative to the other.

5.3.1 Geometric Requirements of Constant-Velocity Couplings

It has been shown that an essential requirement for constant-velocity transfer of rotation between nonaligned shafts is that the coupling mechanism be symmetric relative to the plane that bisects the spatial angle between the shaft axes. Examination of Fig. 5.19 indicates that this condition is satisfied by the double universal joint. However, in many situations, such as the drive trains of front wheel–driven automobiles, a more compact joint is needed.

5.3.2 Practical Constant-Velocity Couplings

A common commercial constant-velocity coupling uses bearing balls moving in shaped races between inner and outer journals to transmit torque. The races are shaped so that the centers of the balls are always in the plane of symmetry. The arrangement is shown in Fig. 5.21.

Figure 5.21 shows a ball-type constant-velocity coupling with six balls, and Fig. 5.22 shows a photograph of the coupling. The inner journal has a spherical outer surface with six equally spaced races with semicircular cross sections cut into it. The centerline of each race is a great circle of a neutral sphere that is slightly larger than the surface of the journal. The planes of the great circles are inclined at equal angles to the journal axis, and alternate races are cut at opposing angles. The outer journal has a spherical inner surface slightly larger than the neutral sphere. Races are also cut into it with their centerlines being great circles in the neutral sphere. They are cut at the same angle to the journal axis as the races in the inner journal, and successive races are again cut at opposing angles. The joint is assembled with each ball rolling in inner and outer races that are at opposing angles. Therefore, the ball center is always at the intersection of the race centerlines. This

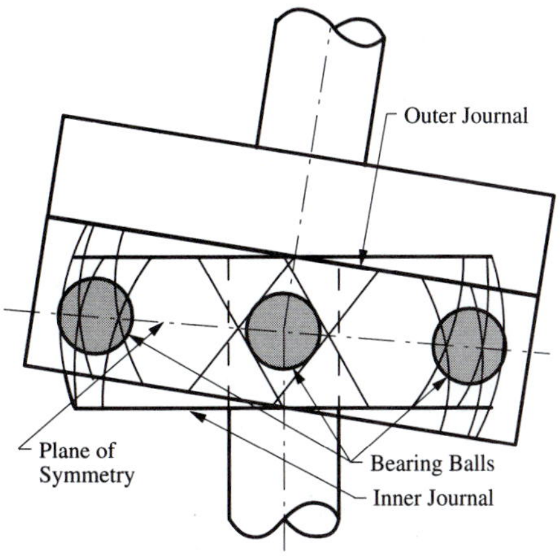

Figure 5.21 Ball-type constant-velocity coupling. The balls, six in the configuration shown, roll in races cut in the inner and outer journals. The centerlines of the races are great circles with their planes inclined to the journal axes at the same angle. Alternate races are cut with opposing angles. The angles of the races in the inner and outer races are also opposite. In this way the center of each ball is always at the intersection of the centerlines of the races in the inner and outer journals, which lies in the plane of symmetry of the angulated joint. Because the ball centers always lie in a common plane of symmetry, the condition for constant-velocity action is satisfied.

Figure 5.22 Ball-type constant-velocity coupling used in front wheel–driven automobile.

ensures that all ball centers lie in a common plane at all times. This plane bisects the angle between the two journal axes and is, therefore, a plane of symmetry. Since the ball centers all lie in a common plane of symmetry at all times, the symmetry condition is satisfied, and the joint transmits motion with constant velocity.

This type of joint can be made relatively compact and is commonly used in automotive drive shafts to allow smooth torque transmission despite the movements of the wheels permitted by the suspension.

5.4 AUTOMOTIVE STEERING AND SUSPENSION MECHANISMS

5.4.1 Introduction

Automotive steering and suspension mechanisms are primarily designed as separate, planar mechanisms acting in different planes. However, they are interconnected because they have common links. Also, both have modifications that which make them spatial linkages. For example, the axis about which the wheel turns in response to movements of the steering linkage is not vertical. The inclination of the axis, called camber, creates a tendency for the steering to center itself at low speed, since it results in the vehicle body being raised slightly whenever the wheels are turned away from the straight-ahead position. Camber is not effective in providing centering at high speed. However, another modification, called caster, provides this action. The wheel steering axis is moved forward relative to the wheel a little way. The distance the wheel rotation axis trails the steering axis is the caster.

The interconnection, together with modifications such as camber that create a truly spatial character, can lead to undesirable dynamic interactions. It is very undesirable for suspension movements to be felt through the steering, or for the position of the steering linkage to influence suspension performance.

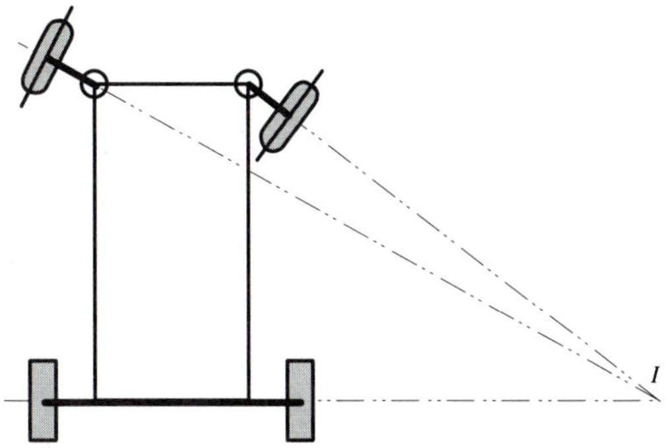

Figure 5.23 The Ackermann steering condition. Since the axes of all four wheels meet at a common instantaneous center, the wheels can roll without any lateral scuffing action. This is the ideal steering geometry at low speeds.

5.4.2 Steering Mechanisms

From a purely kinematic viewpoint, the essential geometry of an automotive steering linkage is that the axes of the front wheels should, at all times, be concurrent at the axis of the rear wheels. It is possible to synthesize a four-bar linkage that will constrain the front wheel axes to approximate this condition very closely. This is the basis of the Ackermann steering gear. As can be seen from Fig. 5.23, it is necessary that the front wheels be "toed out" to an increasing extent as the radius of curvature of the vehicle path is reduced.

However, a close approximation to Ackermann geometry is often not used on modern automobiles, particularly on high-performance vehicles and race cars. The reason is that steering at high speeds is a dynamic problem. To change direction, it is necessary to develop lateral forces at the tire contacts with the road. The production of lateral force requires some slip between the wheel and the road. By using less toe-out than would be required by the Ackermann geometry, more lateral slip is generated at the outside front wheel, which also carries a greater share of the vehicle weight due to dynamic load transfer and is, therefore, able to generate more lateral force. Some race car steering setups go so far as to reverse the kinematically ideal relationship by actually toeing the front wheels in by a small amount during turns. This very aggressive geometry produces very strong cornering action at the expense of tire wear, which is not, of course, such a concern in a race over a limited distance as it is in general automotive use.

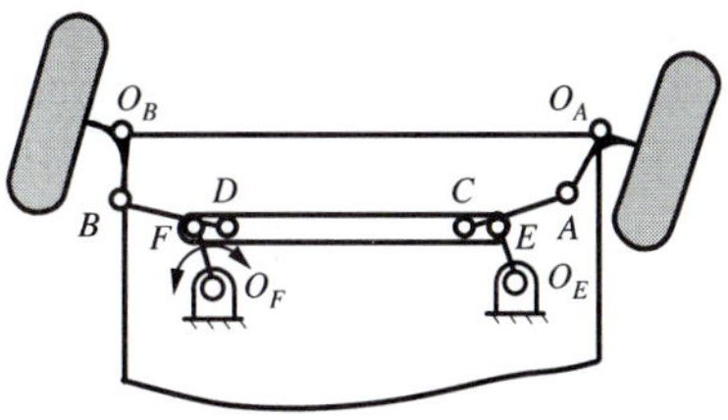

Figure 5.24 A typical steering gear arrangement. The Pitman arm, $O_F F$, is turned by the steering column. The four-bar loop $O_E E F O_F$ is a parallelogram. $O_E E$ is the idler arm, and EF is the relay rod. AC and BD are the tie rods, and $O_A A$ and $O_B B$, which are fixed to the structures that carry the stub axles, are called the steering arms. The steering arms turn about the steering knuckles O_A and O_B. Note that the linkage is bilaterally symmetric about the centerline of the vehicle.

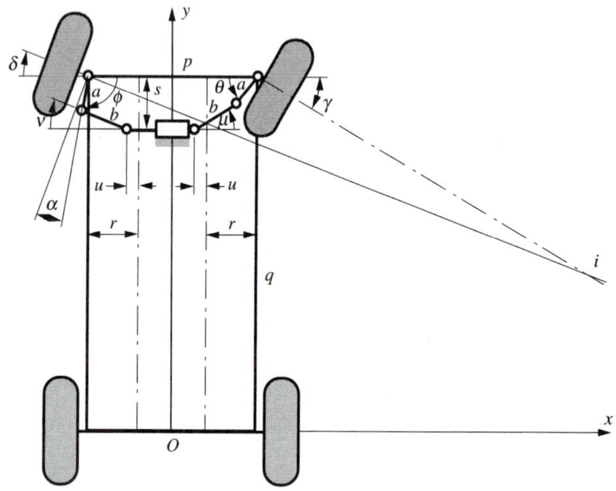

Figure 5.25 The rack-and-pinion steering linkage geometry analyzed in Example 5.5. The position of the intersection of the front wheel axes as a function of the rack displacement, u, and the values of the wheel angles γ and δ are tabulated in Table 5.1. The coordinates of i are plotted as a function of u in Fig. 5.26.

Ackermann action, or any other desired relationship between the steering angles of the wheels, can be adequately approximated by an eight-bar steering linkage such as that shown in Fig. 5.24. Since the wheels move vertically with suspension travel, it is necessary for the joints at the ends of the tie rods to be spherical joints. Thus, the linkage becomes spatial, although still approximating the designed planar behavior.

In modern cars, it is more common to use a linear input to the steering linkage. This is typically produced by a rack and pinion type of steering box. This linear input is applied directly to the relay rod. This arrangement may be thought of as the limiting case of the mechanism in Fig. 5.24 as the arms $O_E E$ and $O_F F$ become infinitely long, producing the configuration of Fig. 5.25. It has the advantages of being simpler: six members versus eight, more compact, and potentially lighter.

Table 5.1 Numerical Values Obtained by Solution of Example 5.5

u	γ	δ	x	y
0.2	4.13	3.99	1475.17	−4.65
0.3	6.25	5.94	973.19	−3.91
0.4	8.43	7.87	718.74	−2.86
0.5	10.67	9.79	563.51	−1.50
0.6	12.99	11.68	457.89	0.17
0.7	15.38	13.56	380.59	2.18
0.8	17.88	15.43	320.98	4.52
0.9	20.50	17.29	273.11	7.23
1.0	23.28	19.13	233.42	10.34
1.1	26.25	20.98	199.57	13.90
1.2	29.50	22.81	169.98	17.99
1.3	33.13	24.65	143.39	22.74
1.4	37.38	26.48	118.66	28.44
1.5	42.87	28.31	94.09	35.86

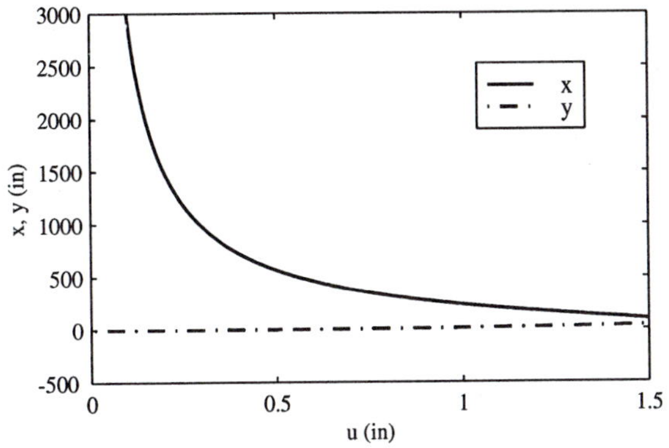

Figure 5.26 The coordinates of the intersection of the front wheel axes, *i*, plotted against the rack displacement, *u*. The *x* approximates the radius of curvature of the vehicle's path, and *y* is the error in location of the intersection relative to the rear axle axis. That is, *y* is the deviation from the Ackermann condition. When $u = 0$, $x = \infty$ and $y = 0$. The values used in the plot are included in Table 5.1.

EXAMPLE 5.5 (Analysis of Rack and Pinion Type of Steering Linkage)

PROBLEM

A steering linkage for an automobile is shown in Fig. 5.25. The wheel base of the automobile (distance between front and rear wheel axes) is $q = 100$ in. The distance between the steering knuckles is $p = 50$ in. The length of the steering arm is $a = 3$ in, and it is inclined at angle $\alpha = 9°$ to the plane of the wheel. The length of the tie rods is $b = 10$ in. When the wheels are in the straight-ahead position shown in Fig. 5.25, the inner ends of the tie rods are distant $r = 10.08$ in from the steering knuckles in the lateral direction, and $s = 5.72$ in in the longitudinal direction.

Plot the *x* and *y* coordinates of the intersection of the front wheel axes for increments of 0.1 in of the rack displacement, *u*, in the range $0 < u \leq 1.5$ in, where the reference frame has its origin at the middle of the rear axle, as shown. The *x* coordinate can be interpreted as the radius of curvature of the path followed by the vehicle, and the *y* coordinate is the error from perfect Ackermann geometry. As indicated in Fig. 5.23, if the Ackermann condition were exactly met, *y* would be zero at all times. Also calculate the angles of the inner and outer front wheels relative to the straight-ahead position throughout this range.

SOLUTION

The linkage can be analyzed as two slider-crank linkages acting in parallel with a common input, *u*, applied to the sliders. Resolving in the *x* and *y* directions, respectively, we have for the right side:

$$a \cos \theta + b \cos \mu = r + u \qquad \textbf{(5.13)}$$

$$a \sin \theta + b \sin \mu = s \qquad \textbf{(5.14)}$$

where μ is the tie rod angle as shown in Fig. 5.25.

Similarly, for the left side

$$a \cos \phi + b \cos \nu = r - u \qquad \textbf{(5.15)}$$

$$a \sin \phi + b \sin \nu = s \qquad \textbf{(5.16)}$$

μ may be eliminated from Eqs. (5.13) and (5.14) by segregating the μ terms on one side of each equation, squaring both sides of both equations, and adding to give

$$b^2 = (r + u - a \cos \theta)^2 + (s - a \sin \theta)^2$$

or

$$b^2 = r^2 + s^2 + a^2 + u^2 + 2ru - 2au\cos\theta - 2ar\cos\theta - 2as\sin\theta \tag{5.17}$$

This equation has the form

$$P\cos\theta + Q\sin\theta + R = 0 \tag{5.18}$$

where

$$\begin{aligned} P &= 2a(u + r) \\ Q &= 2as \\ R &= b^2 - a^2 - r^2 - s^2 - u^2 - 2ru \end{aligned} \tag{5.19}$$

Hence the standard solution of Table 3.1 may be applied to obtain values of θ corresponding to given values of u. Two values of θ are obtained for each value of u, one positive and one negative. Only the negative value is consistent with the configuration shown in Fig. 5.25, so the positive value is discarded.

Similarly, elimination of v from Eqs. (5.15) and (5.16) gives

$$b^2 = (r - u - a\cos\phi)^2 + (s - a\sin\phi)^2$$

or

$$b^2 = r^2 + s^2 + a^2 + u^2 - 2ru + 2au\cos\phi - 2ar\cos\phi - 2as\sin\phi \tag{5.20}$$

This equation has the form

$$P'\cos\phi + Q'\sin\phi + R' = 0 \tag{5.21}$$

where

$$\begin{aligned} P' &= 2a(r - u) \\ Q &= 2as \\ R &= b^2 - a^2 - r^2 - s^2 - u^2 + 2ru \end{aligned} \tag{5.22}$$

for which the solution is also given by Table 3.1. Values of ϕ for incremental values of u throughout the specified range can be calculated. As was the case for θ, two values of ϕ are obtained for each value of u, one positive and one negative. Only the negative solution is consistent with the configuration drawn in Fig. 5.25, so the positive solution is discarded.

Now $\gamma = \pi/2 - \theta - \alpha$, and $\delta = \phi + \alpha - \pi/2$, where γ and δ are the steering angles of the inner and outer front wheels, as shown in Fig. 5.25, and values of γ and δ may now be calculated. The resulting values of γ and δ throughout the range of values of u are listed in Table 5.1. Also, γ and δ determine the location of the intersection, i, of the axes of the wheels:

$$\tan\gamma = \frac{q - y}{x - p/2}, \qquad \tan\delta = \frac{q - y}{x + p/2} \tag{5.23}$$

Hence,

$$(x - p/2)\tan\gamma = (x + p/2)\tan\delta$$

which, when solved for x, gives

$$x = \frac{p}{2}\left(\frac{\tan\gamma + \tan\delta}{\tan\gamma - \tan\delta}\right) \tag{5.24}$$

Substitution for x into either of Eqs. (5.23) allows solution for y to give

$$y = q - p\left(\frac{\tan\gamma\tan\delta}{\tan\gamma - \tan\delta}\right)$$

The results are tabulated in Table 5.1 and are plotted in Fig. 5.26. It may be seen that the linkage gives a reasonable approximation to the Ackermann condition, except at very large wheel angles. ■

5.4.3 Suspension Mechanisms

An automotive suspension performs the function of a vibration filter, reducing the amplitudes of vibrations excited by geometric variations in the road surface. This is the function of the spring damper arrangements that are integral components of the suspension. Analysis of this vibration filtering action is normally covered in texts on mechanical vibrations and is beyond the scope of this book. Here we confine ourselves to the kinematic requirements of suspension mechanisms.

Automotive suspension mechanisms must allow controlled, single-degree-of-freedom motion of the wheel axis relative to the body of the vehicle. The travel allowed needs to be as close as possible to normal to the plane of the ground at the wheel contact. Also, it is necessary for the suspension mechanism to maintain the plane of the wheel as perpendicular as possible to the ground at all times. This is because automobile tires are designed to develop maximum lateral force when they are in the upright position, as opposed to motorcycle tires, which must function in inclined positions during hard cornering. Since the center of mass of an automotive vehicle is almost always higher than the wheel axes, there is a tendency for the body to roll toward the outside of a turn. Another objective of suspension design is to attempt to control this tendency to roll.

Automotive steering and suspension mechanisms are truly spatial mechanisms. However, their initial design generally rests on planar principles.

When viewed from the front, the instantaneous center of motion of the body of the vehicle relative to the ground is called the roll center. The location of the roll center for a typical independent suspension geometry is shown in Fig. 5.27. The center is located by using the Kennedy-Aronholdt theorem as described in Chapter 2.

Of course, the roll center moves as the position of the vehicle body moves. Whereas the roll center will be on the vehicle centerline for a road vehicle at rest on a level surface, it will shift off that line in the asymmetric positions that result from cornering. There is also a roll center for the rear suspension, so one can think of a roll axis, which is the line that passes through both roll centers.

The location of the roll center relative to the center of mass of the vehicle governs the effect of inertial forces due to cornering on the system. Obviously, if the vertical distance between the roll center and the center of mass is large, the moment produced by lateral acceleration will be large. A suspension geometry that brings the roll center progressively

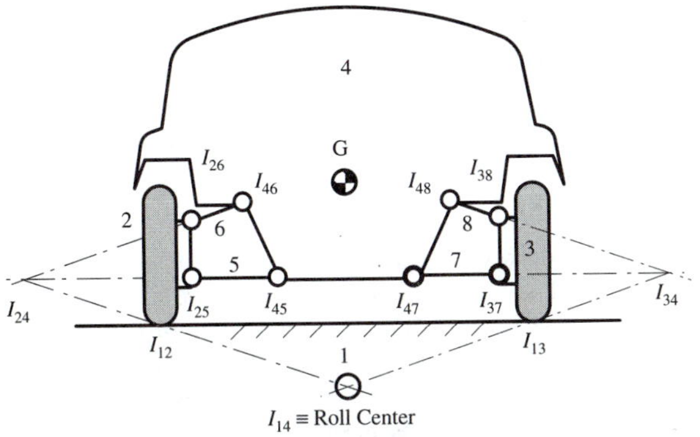

Figure 5.27 Roll center geometry for an automotive independent suspension geometry. The roll center is the instantaneous center of relative motion of the vehicle body and the ground.

closer to the center of mass with increasing body roll might be attractive, because if the action of the suspension springs was linear, it would lead to increasing roll stiffness with increasing roll angle.

Suspension designers think of the roll center as the point of transfer of the inertial force between the sprung and unsprung masses of the vehicle. The unsprung mass is the wheels and suspension members directly attached to them whose position is directly determined by the road surface. The sprung mass is everything that moves when the springs are deflected.

5.5 INDEXING MECHANISMS

Indexing mechanisms are intermittent motion mechanisms that hold position alternately with a timed, unidirectional motion of the output member. This is distinct from other types of intermittent motion mechanisms such as dwell cams, which alternate forward and return motion with holding position. The output member of an indexing mechanism always advances in the same direction. Indexing mechanisms are practically important in such applications as weaving looms, in advancing workpieces in repetitive manufacturing operations, and in many instrument mechanisms.

5.5.1 Geneva Mechanisms

The most common type of indexing mechanism is a Geneva mechanism. Geneva mechanisms come in many varieties, both planar and spherical. When advancing, it is kinematically similar to an inverted slider crank. When holding position, it functions as a simple journal bearing.

The name Geneva mechanism originated because these mechanism were used in mechanical watch and clock movements in the days when mechanical movements were dominant, and Switzerland was the world center of the industry.

A simple example of a Geneva mechanism is shown in Fig. 5.28. The pin, *P*, on the

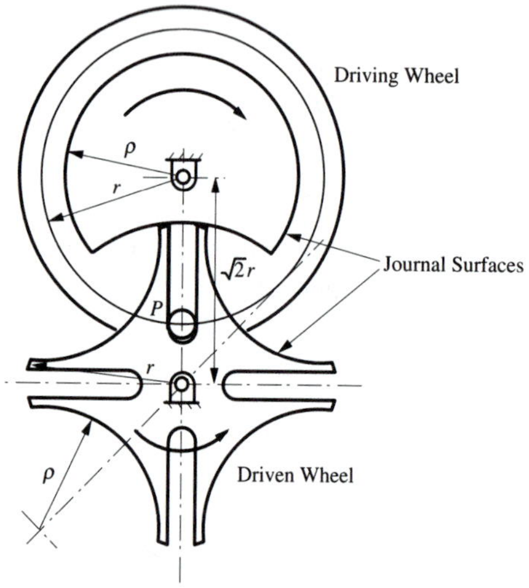

Figure 5.28 A four-station Geneva mechanism. The output member is the star wheel. The star wheel is advanced by the pin in the input wheel. The star wheel is advanced one-quarter of a revolution counterclockwise for every revolution of the input wheel. The advance movement occurs during one-quarter of a cycle with the star wheel being locked by the journal surface on the input wheel for the other three-quarters of the cycle.

driving wheel engages the slots in the star-shaped driven wheel to advance the driven wheel one-quarter turn for every rotation of the driving wheel. In between the advance movements, the eccentric cylindrical journal surfaces cut into the star wheel engage with the journal surface on the driving wheel to lock the star wheel in position, although the driving wheel continues to rotate. The centerline of the slot must be tangent to the circle, with radius r, described by the center of the pin at the position in which the pin enters or leaves the slot. If this condition is not satisfied, there will be infinite acceleration at the beginning of advancement and infinite deceleration at the end. This condition dictates that the center distance of the two wheels should be $\sqrt{2}r$. It also requires that the outer radius of the star wheel be r. The radius of the journal surfaces is flexible. The centers of the cylindrical cutouts on the star wheel lie on a circle with radius $\sqrt{2}r$.

During the advancing phase of the cycle, the mechanism is kinematically equivalent to an inverted slider crank. One of its attractions is that it smoothly accelerates and then decelerates the star wheel.

The motion of the star wheel may be analyzed by reference to Fig. 5.29. Resolving the sides of the triangle whose vertices are the two shaft axes and the pin axis in the vertical and horizontal directions:

$$r \sin \theta = x \sin \phi$$
$$r \cos \theta + x \cos \phi = \sqrt{2}r \tag{5.25}$$

Elimination of x by substitution from the first of these equations into the second gives

$$\cos \theta + \frac{\sin \theta}{\tan \phi} = \sqrt{2}$$

after canceling the common factor, r. Rearrangement of this expression gives

$$\tan \phi = \frac{\sin \theta}{\sqrt{2} - \cos \theta} \tag{5.26}$$

or

$$\phi = \tan^{-1}\left(\frac{\sin \theta}{\sqrt{2} - \cos \theta}\right) \tag{5.27}$$

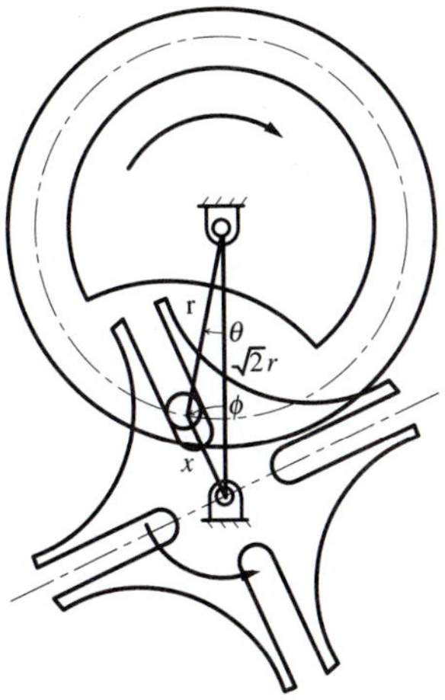

Figure 5.29 Kinematic modeling of the Geneva mechanism of Fig. 5.28. θ is the angle of rotation of the driving wheel, measured from the line of centers; ϕ is the angle of rotation of the star wheel.

Differentiation of Eq. (5.26) with respect to time followed by simplification gives

$$\dot{\phi}(1 + \tan^2 \phi) = \dot{\theta}\frac{(\sqrt{2}\cos\theta - 1)}{(\sqrt{2} - \cos\theta)^2}$$

Substitution for $\tan \phi$ from Eq. (5.26) gives, after rearrangement and simplification,

$$\dot{\phi} = \dot{\theta}\left(\frac{\sqrt{2}\cos\theta - 1}{3 - 2\sqrt{2}\cos\theta}\right) \quad \textbf{(5.28)}$$

Differentiation again with respect to time gives, after simplification,

$$\ddot{\phi} = \ddot{\theta}\left(\frac{\sqrt{2}\cos\theta - 1}{3 - 2\sqrt{2}\cos\theta}\right) - \dot{\theta}^2\frac{\sqrt{2}\sin\theta}{(3 - 2\sqrt{2}\cos\theta)^2}$$

In the usual case in which the driving wheel is driven at constant angular velocity, the first term disappears and

$$\ddot{\phi} = \dot{\theta}^2\frac{\sqrt{2}\sin\theta}{(3 - 2\sqrt{2}\cos\theta)^2} \quad \textbf{(5.29)}$$

Equations (5.27), (5.28), and (5.29) are plotted versus θ (in degrees) in Fig. 5.30; ϕ is plotted in radians. Of course, ϕ varies from $-45°$ to $45°$ during the advancement. The angular velocity curve is actually $\dot{\phi}/\dot{\theta}$, and the angular acceleration curve is $\ddot{\phi}/\dot{\theta}^2$.

As can be seen from Fig. 5.30, the velocity and acceleration curves are smooth and well behaved, but the derivative of the acceleration (jerk) is infinite at the beginning and end of the advancement. So far, we have considered only the simplest version of the Geneva mechanism: the four-station planar variety. The number of stations is the number of slots in the star wheel and may, in principle, be any number, although the geometric lower limit is three. There is also a practical upper limit at which the journal surfaces on the star wheel become too short to effectively lock the output between advancements. The number of pins on the driving wheel is usually one, but drivers with two or more are possible.

The essential geometry for relating the number of stations to the duration of the advancement is shown in Fig. 5.31. Here α is the angle between the slot centerline and the line of centers of the two wheels at the moment of engagement or disengagement of the pin.

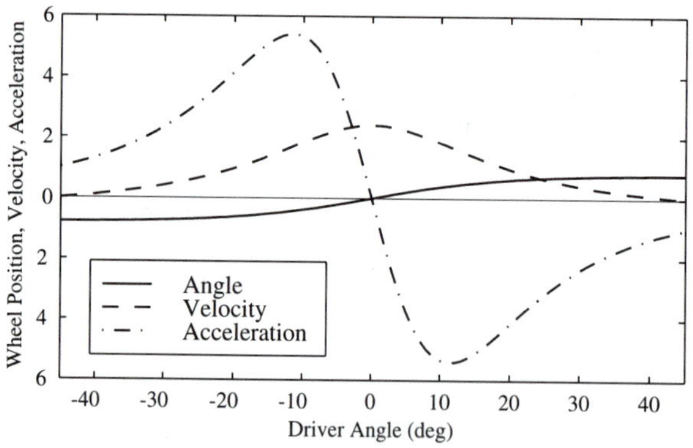

Figure 5.30 Position, velocity, and acceleration of the driven wheel of the Geneva mechanism shown in Figs. 5.28 and 5.29 during the advancement phase of the motion cycle. The angular position of the star wheel is in radians. The angular velocity and acceleration curves are respectively normalized to the driver angular velocity and driver angular velocity squared.

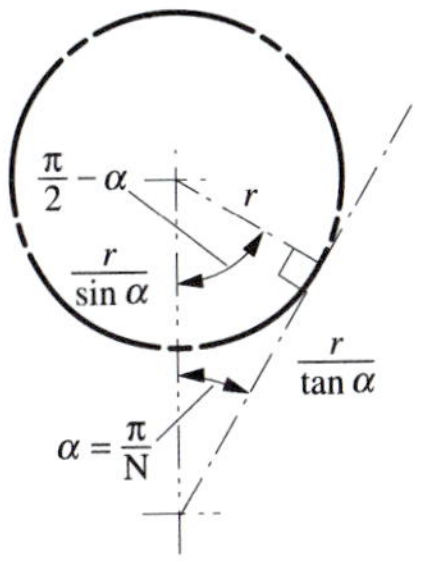

Figure 5.31 Critical geometry for a Geneva mechanism with N stations. α is the angle between the slot centerline and the line of centers at the moment of engagement of the pin; α is half the angle between successive slots on the star wheel.

That is, α is half the angle between successive slots, or $360°/(2N)$, where N is the number of stations. As already noted, the slot axis must be tangent to the circle traversed by the pin center at these positions in order to avoid infinite accelerations. This determines the relationship between N and the duration of the advancement, which is $\pi - 2\alpha$ by inspection of the figure. Consequently, the duration of the advancement increases with the number of stations, approaching a limit of 180° as the number of stations becomes very large. This has the advantage of making the advancement motion more gentle but the possible disadvantage of decreasing the duration of the period for which the output is stationary. The trade-off between these effects and with the desirability of avoiding gearing downstream of the indexing mechanism determine the choice of the number of stations. Gearing downstream of an indexing mechanism should be avoided due to the inaccuracy and uncertainty in position introduced by necessary backlash in the gear train. Gear backlash is not usually a problem if the gears are in uniform motion. However, the discontinuous motion output from an indexing mechanism and consequent reversals of acceleration result in slapping across the backlash interval. Hence, any speed reduction should be done upstream of the indexing mechanism.

The number of stations also determines the ratio of the center distance of the wheel axes to the pin radius and the outside diameter of the star wheel. By inspection of Fig. 5.31, the former ratio is $1/\sin \alpha$ and the latter is $1/\tan \alpha$.

Noting that $\alpha = \pi/N$, Eqs. (5.27)–(5.29), respectively, become for this more general case:

$$\phi = \tan^{-1}\left(\frac{\sin \alpha \sin \theta}{1 - \sin \alpha \cos \theta}\right) \quad \textbf{(5.30)}$$

$$\dot{\phi} = \dot{\theta} \sin \alpha \left(\frac{\cos \theta - \sin \alpha}{1 + \sin^2\alpha - 2 \sin \alpha \cos \theta}\right) \quad \textbf{(5.31)}$$

$$\ddot{\phi} = -\dot{\theta}^2 \frac{\sin \alpha \cos^2 \alpha \sin \theta}{(1 + \sin^2\alpha - 2 \sin \alpha \cos \theta)^2} \quad \textbf{(5.32)}$$

Spherical Geneva mechanisms allow indexed motion transfer between angulated shafts. More important, a large number of stations can be accommodated without losing positive locking action between advances.

EXAMPLE 5.6 ***(Analysis of Geneva Wheel)***

PROBLEM

An indexing drive is to be driven by a synchronous electric motor turning at 360 rpm (the speed of a synchronous motor is locked to the alternating-current cycle frequency and so is essentially constant). The single pin driver is to turn a six-station Geneva wheel. Compute the following:

(a) The number of advances per second
(b) The angle through which the Geneva wheel advances during every revolution of the driving wheel
(c) The duration in seconds of the dwell in the output motion
(d) The peak angular velocity of the output shaft
(e) The peak angular acceleration of the output shaft

SOLUTION

(a) The number of advances per second is the number of revolutions of the driver per second, which is 360/60 = 6.
(b) The angle advanced is $2\alpha = 360°/N = 60°$, with N, the number of stations, being 6 in this case. Hence $\alpha = 30°$.
(c) The fraction of the cycle during which the output is locked (dwelling) is

$$\lambda = \frac{180 - 2\alpha}{360}$$

with α in degrees giving $\lambda = 1/3$. The duration of the complete cycle is $T = 1/6$ s from part (a). Hence the duration of the dwell is

$$\tau = \lambda T = 1/18 = 0.0555 \text{ s}$$

(d) Referring to Eq. (5.32), $\dot{\phi}$ is at its maximum value when $\theta = 0$. Also, for $N = 6$,

$$\sin \alpha = 0.5$$

so, substituting this value and $\theta = 0$ in Eq. (5.31),

$$\dot{\phi}_{\max} = \dot{\theta}$$

$\dot{\theta}$ is the angular velocity of the drive wheel, so

$$\dot{\theta} = 2\pi \times 6 = 37.70 \text{ rad/s}$$

Therefore,

$$\dot{\phi}_{\max} = 37.70 \text{ rad/s}$$

Note that ϕ is positive in the CCW direction while θ is positive in the CW direction (see Fig. 5.29). Therefore the positive values for both $\dot{\phi}$ and $\dot{\theta}$ indicate that the star wheel rotates in the opposite direction to the driver.
(e) It is necessary to determine the value of θ that maximizes $\ddot{\phi}$. A straightforward way to do this would be to plot Eq. (5.32) in the same way as in Fig. 5.30, but with $\alpha = 30°$. $\ddot{\phi}$ and the angle θ at which it occurs could then be read directly from the plot.

Alternatively, we can differentiate Eq. (5.32) to identify the extrema of $\ddot{\phi}$. Noting that $\dot{\theta}$ is constant,

$$\frac{d\ddot{\phi}}{dt} = \frac{-\dot{\theta}^3 \sin \alpha \cos^2\alpha}{(1 + \sin^2\alpha - 2 \sin \alpha \cos \theta)^3}\{(1 + \sin^2\alpha - 2 \sin \alpha \cos \theta) \cos \theta - 4 \sin \alpha \sin^2\theta\}$$

and so

$$\frac{d\ddot{\phi}}{dt} = 0$$

when

$$(1 + \sin^2\alpha) \cos \theta - 2 \sin \alpha \cos^2 \theta - 4 \sin \alpha \sin^2\theta = 0$$

Replacement of $\sin^2\theta$ by $1 - \cos^2\theta$ and rearrangement of the equation give

$$\cos^2\theta + \gamma \cos\theta - 2 = 0$$

where

$$\gamma = \frac{1 + \sin^2\alpha}{2 \sin\alpha} \tag{5.33}$$

The preceding equation can be treated as a quadratic equation in the variable $\cos\theta$. Solving for $\cos\theta$,

$$\cos\theta = \frac{-\gamma \pm \sqrt{\gamma^2 + 8}}{2}$$

It is possible to show that only the positive value of the square root gives a value of $\cos\theta$ with magnitude between 0 and 1 in the allowable range of α, $0 < \alpha < 60°$, so only that solution is valid. Hence, $\ddot{\phi}$ is at a maximum when

$$\theta = \pm \cos^{-1}\left(\frac{-\gamma + \sqrt{\gamma^2 + 8}}{2}\right) \tag{5.34}$$

where the $\pm$ sign now comes from inversion of the cosine, not from the quadratic solution. Equations (5.33) and (5.34) are of general validity for locating the maximal values of $\ddot{\phi}$. In the present case, substituting $\sin\alpha = 0.5$ in Eq. (5.33) gives

$$\gamma = 1.25$$

Hence Eq. (5.34) gives

$$\theta = \pm 22.90°$$

Substitution of these values into Eq. (5.32) gives

$$\frac{\ddot{\phi}}{\dot{\theta}^2} = \pm 1.372$$

Hence, since $\dot{\theta} = 6 \times 2\pi = 37.70$ rad/s, the peak angular acceleration is 1,950 rad/s^2.

■

5.6 CHAPTER 5 *Exercise Problems*

PROBLEM 5.1

A coupler curve has the approximate straight-line section shown in the figure below. Design a four-bar linkage that will generate the portion of the curve shown. Describe the linkage in sufficient detail that it can be manufactured.

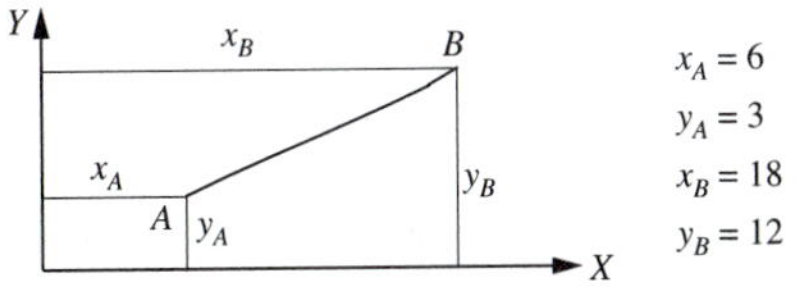

PROBLEM 5.2

Resolve Problem 5.1 if $x_A = 3$, $y_A = 3$, $x_B = 20$, and $y_B = 15$.

PROBLEM 5.3

Determine the cognate linkages that will trace the same coupler curve as that traced by point C in the figure below.

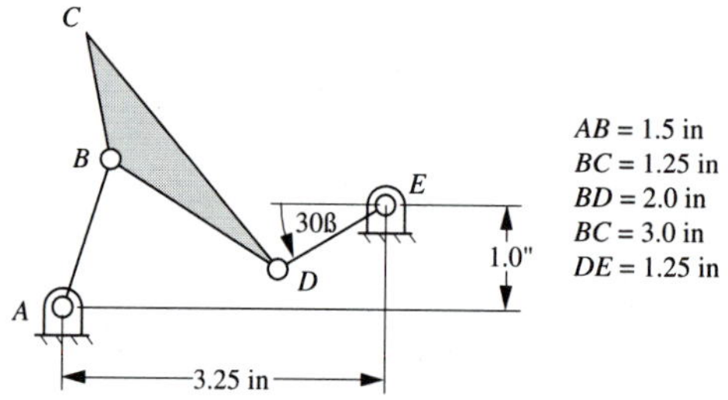

PROBLEM 5.4

Determine the cognate linkages that will trace the same coupler curve as that traced by point C in the figure below.

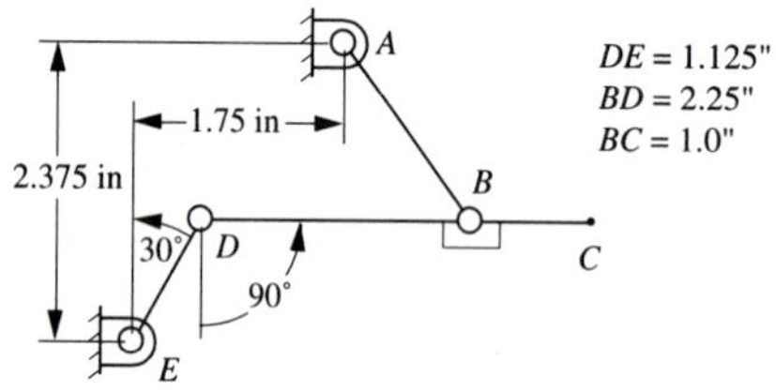

PROBLEM 5.5

A spherical four-bar linkage is shown in the figure below. If the angular velocity of link 2 is 100 rad/s (constant), find the angular velocity and angular acceleration of link 4 as a function of the rotation of link 2. Plot the angular velocity and angular acceleration of link 4 for a full rotation of link 2. Make the calculations for the assembly mode shown in the figure.

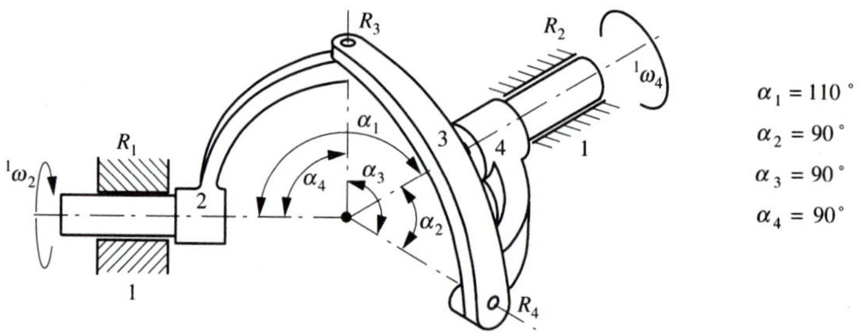

PROBLEM 5.6

Resolve Problem 5.5 if $\alpha_1 = 150°$ but all other data remain the same.

PROBLEM 5.7

The mechanism shown is used for a steering linkage for an automobile. The wheel base is 110 inches, and link $O_F F$ is driven by the steering column. The toe-in angle (α) is 9°. If the link dimensions are given as shown, determine the y error in the Ackermann steering condition (see Figs. 5.23 and 5.26) for a 10° CCW rotation of $O_F F$. Recall that the linkage $O_E EFO_F$ is a parallelogram.

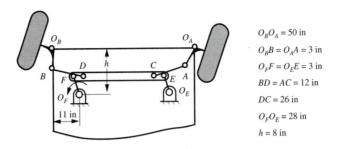

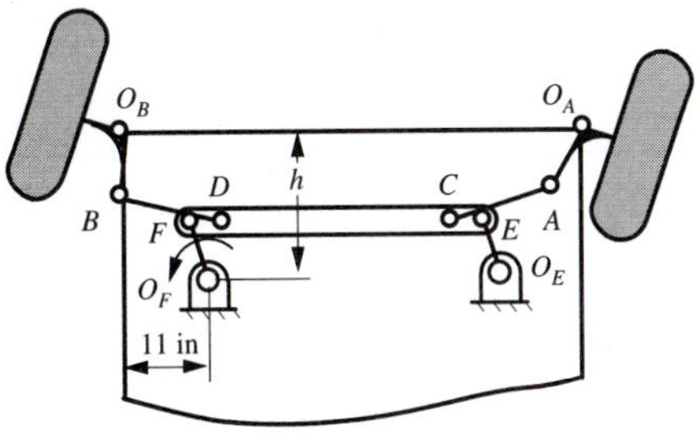

$O_BO_A = 50$ in
$O_BB = O_AA = 3$ in
$O_FF = O_EE = 3$ in
$BD = AC = 12$ in
$DC = 26$ in
$O_FO_E = 28$ in
$h = 8$ in

PROBLEM 5.8 Write a computer program to analyze the steering linkage shown in Problem 5.7. If only h can change, determine the optimum value for h that will give the least error in y for the Ackermann steering condition for a $\pm 15°$ rotation of O_FF.

PROBLEM 5.9 In the rack-and-pinion mechanism shown in Fig. 5.25, the wheel base is 125 in. If the link dimensions are as given below, determine the y error in the Ackermann steering condition as a function of the rotation of the displacement u (see Fig. 5.23) for a ± 1.5 in displacement of u.

$p = 55$ in $\quad b = 12$ in
$a = 3.5$ in $\quad r = 11$ in
$\alpha = 10°$ $\quad s = 6.0$ in

PROBLEM 5.10 A new subcompact automobile is being designed for rack-and-pinion steering. Assume that the wheel base is 90 in. Determine the other dimensions such that the error in the Ackermann steering condition is as small as possible for a ± 1.5-in displacement of the rack.

PROBLEM 5.11 The center distance between the driver and follower of a Geneva mechanism is to be 3 in. The driver is to rotate five revolutions for each rotation of the follower. The driving pin is to enter the slot tangentially so that there will be no impact load. Do the following:

1. Design the Geneva mechanism and draw it.
2. Determine the angular velocity and acceleration of the Geneva wheel for one-fifth of a revolution if the angular velocity of the driver is 100 rpm CCW. Plot the results.

PROBLEM 5.12 Resolve Problem 5.11 if the input link rotates three revolutions for each rotation of the follower. Conduct the velocity and acceleration analysis for one-third of a rotation.

Chapter 6

Profile Cam Design

6.1 INTRODUCTION

Cams are used for essentially the same purpose as linkages, that is, generation of irregular motion. Cams have an advantage over linkages because cams can be designed for much tighter motion specifications. In fact, in principle, any desired motion program can be exactly reproduced by a cam. Cam design is also, at least in principle, simpler than linkage design, although, in practice, it can be very laborious. Automation of cam design using interactive computing has not, at present reached the same level of sophistication as linkage design.

The disadvantages of cams are manufacturing expense, poor wear resistance, and relatively poor high-speed capability. Although numerical control (NC) machining does cut the cost of cam manufacture in small lots, they are still quite expensive in comparison with linkages. In large lots, molding or casting techniques cut cam costs, but not to the extent that stamping and so forth cut linkages costs for similar lot sizes.

Unless roller followers are used, cams wear quickly. However, roller followers are bulky and require larger cams, creating size and dynamic problems. In addition, the bearings in roller followers create their own reliability problems.

The worst problems with cams are, however, noise and follower bounce at high speeds. The result of this is a preoccupation with dynamic optimization in cam design.

Cam design usually requires two steps (from a geometric point of view):

1. Synthesis of the motion program for the follower
2. Generation of the cam profile

If the motion program is fully specified throughout the motion cycle, as is the case, for example, with the stitch pattern cams in sewing machines, the first step is not needed. More usually, the motion program is specified only for portions of the cycle, allowing the synthesis of the remaining portions for optimal dynamic performance. An example is the cam controlling the valve opening in an automotive engine. Here the specification is that the valve should be fully closed for a specified interval and more or less fully open for another specified interval. For the portions of the cycle between those specified, a suitable program must be synthesized. This can be done, with varying levels of sophistication, to make the operation of the cam as smooth as possible. In general, the higher the level of dynamic performance required, the more difficult the synthesis process.

The second stage of the process, profile generation, is achieved by kinematic inversion. The cam is taken as the fixed link, and a number of positions of the follower relative to the cam are constructed. A curve tangent to the various follower positions is drawn and becomes the cam profile. If the process is performed analytically, any level of accuracy can be achieved.

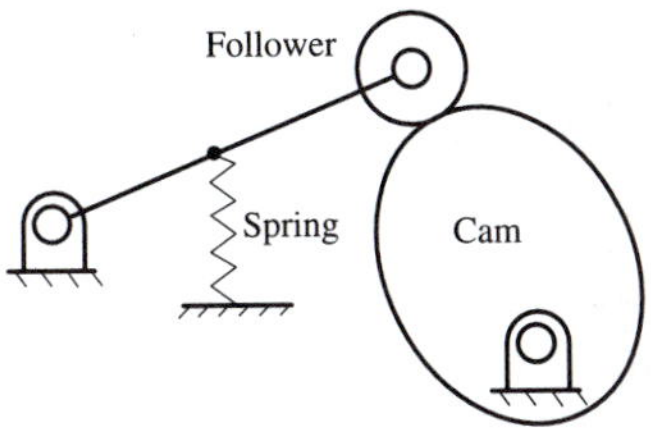

Figure 6.1 Elements of a cam-follower system.

6.2 CAM-FOLLOWER SYSTEMS

A general cam-follower system consists of three elements as shown in Fig. 6.1. The first two are the cam and follower, and the third is a spring or other means of ensuring that the follower remains in contact with the cam. The function of the spring can be replaced by gravity or by constraining the follower between the two surfaces on the cam or constraining the cam between two surfaces on the follower. Both of these approaches are usually more expensive than using a spring and therefore are not commonly used.

A follower is characterized by its motion relative to the ground link and by the geometry of their face that makes contact with the cam. The cam-follower motion may be either rotational or translational, and translating followers may be either radial or offset. Examples of these are shown in Fig. 6.2. The follower surfaces may be either knife edged, flat, spherical (or cylindrical), or roller as shown in Fig. 6.3.

Actually, these geometries are all of the same class depending on the radius of curvature of the follower face. That is, the knife edge has a radius of curvature that is zero, the flat face has a radius of curvature that is infinite, and the general roller and cylindrical followers have a finite (but not zero) radius of curvature. In this discussion, only planar cams will be considered, so no distinction between spherical and cylindrical follower faces will be made. Also, if only geometric information is of interest, no distinction needs to be made between roller and rigid cylindrical-faced followers. Obviously, there is a significant difference from an overall design standpoint, however.

Although here we will consider only planar, rotating cams, in practice a large number of different cam geometries are found. Some of the different types of cams and follower systems are shown in Fig. 6.4.

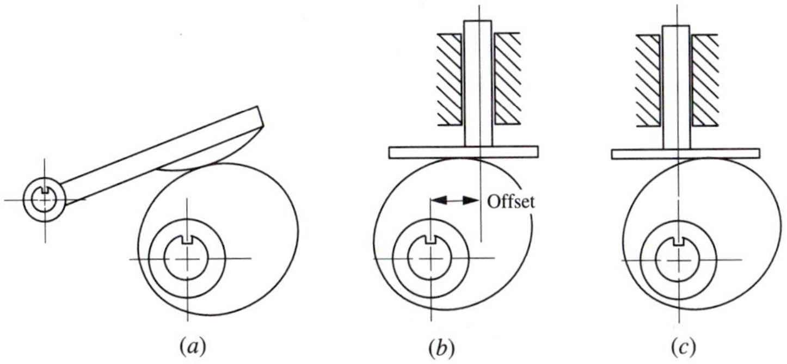

Figure 6.2 (a) Cylindrical-faced oscillating follower. (b) Offset flat-faced translating follower. (c) Radial flat-faced translating follower.

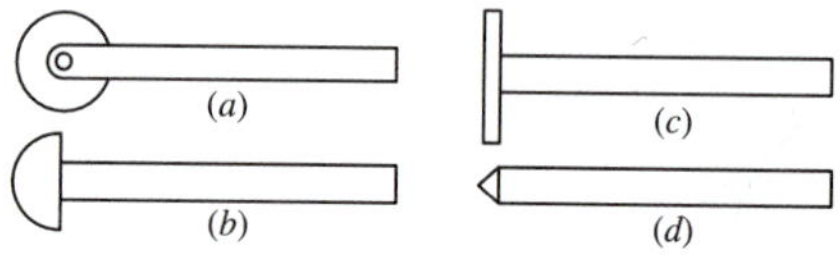

Figure 6.3 (a) Roller follower. (b) Cylindrical-faced follower. (c) Flat-faced follower. (d) Knife-edged follower.

6.3 SYNTHESIS OF MOTION PROGRAMS

This is the problem of filling in, in an optimal way, the portions of the motion cycle that are not completely specified. The characteristics of the problem may be demonstrated by consideration of a cam that is required to drive a follower that dwells at 0 for a cam rotation of 60°, dwells at 1.0 in for a cam rotation of 110° to 150°, and is required to move with constant velocity from a displacement of 0.8 in to 0.2 in for 200° to 300° of cam rotation. The specified portions of the motion program are displayed in Fig. 6.5.

A simple solution to the problem of filling the gaps is simply to move the cam at constant velocity between the specified segments, giving a time-motion diagram as shown in Fig. 6.6.

Notice, however, that if this is done, the acceleration becomes infinite at cam angles of 60°, 110°, 150°, 200°, 300°, and 360°. Since the follower cannot follow an infinite acceleration, this leads to loss of contact and/or excessive local stresses and resultant noise and wear problems.

The preceding motion program matches only the *displacements* at the ends of the segments. The infinite acceleration problem can be removed by matching both displacement

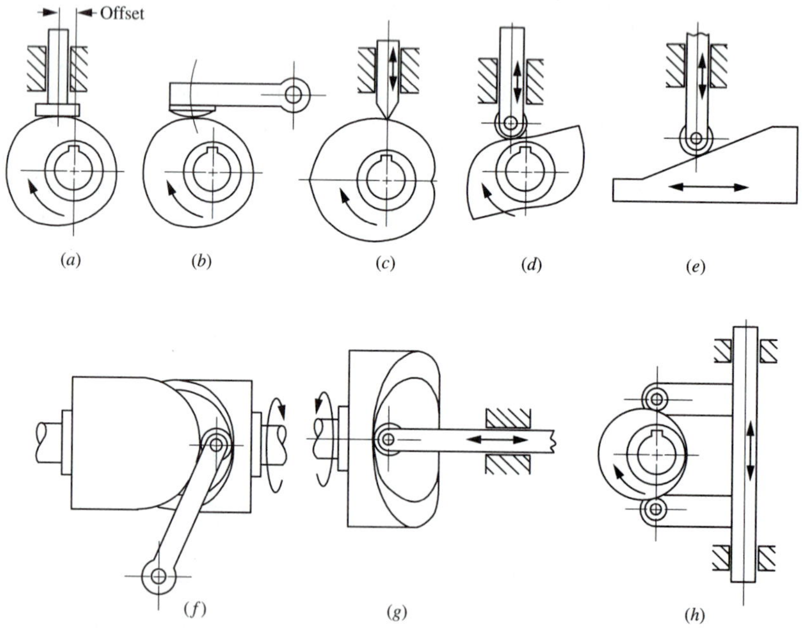

Figure 6.4 Some types of cams. (a) Radial cam and flat-faced offset translating follower. (b) Radial cam and spherical-faced oscillating follower. (c) Radial (heart) cam and translating knife-edged follower. (d) Radial two-lobe frog cam and translating offset roller follower. (e) Wedge cam and translating roller follower. (f) Cylindrical cam and oscillating roller follower. (g) End or face cam and translating roller follower. (h) Yoke cam and translating roller follower.

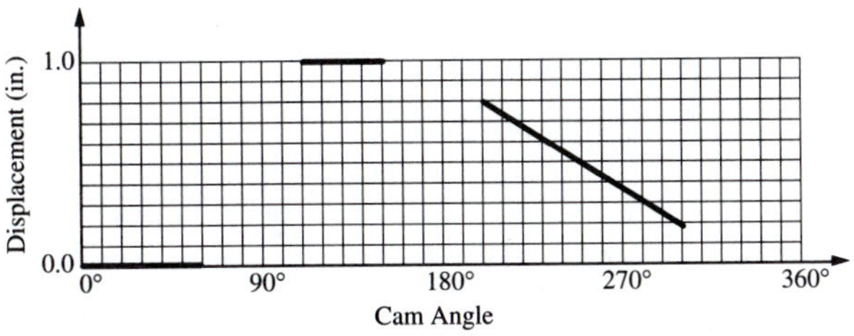

Figure 6.5 The statement of the required displacements of a cam design problem in graphical form.

and velocity at the ends of segments of the program. One way to do this is to subdivide the synthesized segments into two parts with a constant acceleration on the first and constant deceleration on the second. On such a subsegment, if the acceleration is a, the velocity is given by

$$v = v_0 + at$$

where v_0 is the velocity at the beginning of the segment. The displacement is given by

$$y = s_0 + v_0 t + \frac{a}{2} t^2$$

where s_0 is the displacement at the beginning of the segment. Now, if the cam is driven at constant velocity,

$$\theta = \theta_0 + \omega t$$

where θ is the cam angle, θ_0 is the cam angle at the beginning of the segment, and ω is the angular velocity. Hence:

$$t = \frac{(\theta - \theta_0)}{\omega}$$

$$v = v_0 + a \frac{(\theta - \theta_0)}{\omega}$$

$$y = s_0 + v_0 \frac{(\theta - \theta_0)}{\omega} + a \frac{(\theta - \theta_0)^2}{2\omega^2}$$

Therefore, the relationship between s and θ, as plotted on the time-motion diagram, is parabolic (see Fig. 6.7). Cam time-motion programs that use this type of transition are called parabolic. The cam profiles developed from them are also called "parabolic." It is important to understand that a so-called parabolic cam *does not* have a parabolic curve

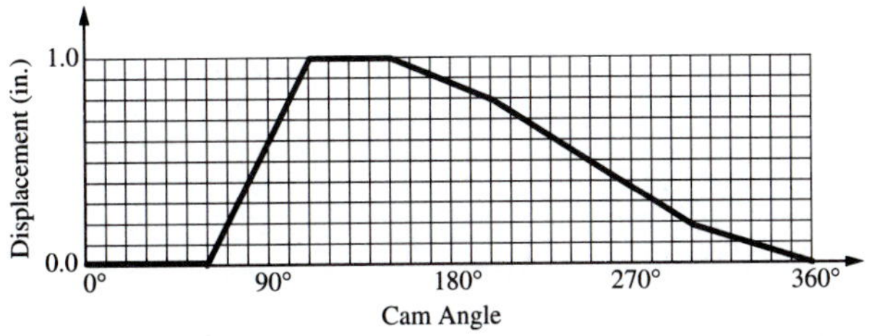

Figure 6.6 A time-motion program that satisfies the displacement requirements specified in Fig. 6.5.

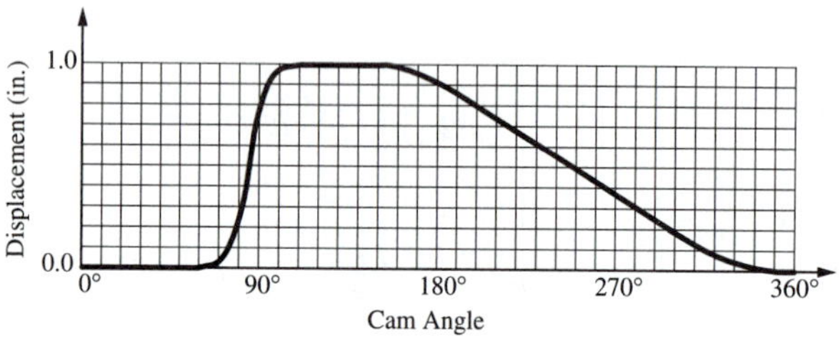

Figure 6.7 A time-motion program that satisfies the displacement requirements of Fig. 6.5 using parabolic transitions. This is called a *parabolic* time-motion program.

in its profile. Rather, the parabolas are in the transition curves used in the time-motion program.

6.4 ANALYSIS OF DIFFERENT TYPES OF FOLLOWER DISPLACEMENT FUNCTIONS

Several different standard functions can be used to connect the parts of the displacement diagram where a specific type of motion is required. These displacement profiles ultimately determine the shape of the cam. Many different types of motions have been used in practice, and some have been extensively studied. These include:

1. Uniform motion
2. Parabolic motion
3. Simple harmonic motion
4. Cycloidal motion
5. General polynomial motion

The first two types of program have already been introduced. The first four types of program can be generated graphically as well as analytically, but the fifth type is generated only analytically. Both graphical and analytical development will be considered here, where possible. Both methods assume that the angular velocity, ω, of the cam is constant. If this is the case, then

$$y = y(\theta)$$

and

$$\theta = \theta_0 + \omega t$$

Here, y is used as generic output variable. It may correspond to either a linear or angular displacement. Note that if the cam motion is given as a function of time, the motion can easily be represented as a function of the cam rotation in degrees using the preceding expressions.

The higher derivatives are given by

$$\dot{y} = \frac{dy(\theta)}{dt} = \frac{dy}{d\theta}\frac{d\theta}{dt} = y'\,\omega$$

and

$$\ddot{y} = \frac{d^2y(\theta)}{dt^2} = \frac{d}{dt}\left(\frac{dy}{d\theta}\frac{d\theta}{dt}\right) = \frac{d^2y}{d\theta^2}\left(\frac{d\theta}{dt}\right)^2 + \frac{dy}{d\theta}\frac{d^2\theta}{dt^2} = \frac{d^2y}{d\theta^2}\omega^2 + \frac{dy}{d\theta}\alpha$$

But because ω is constant, $\alpha = 0$ and

$$\ddot{y} = y''\,\omega^2$$

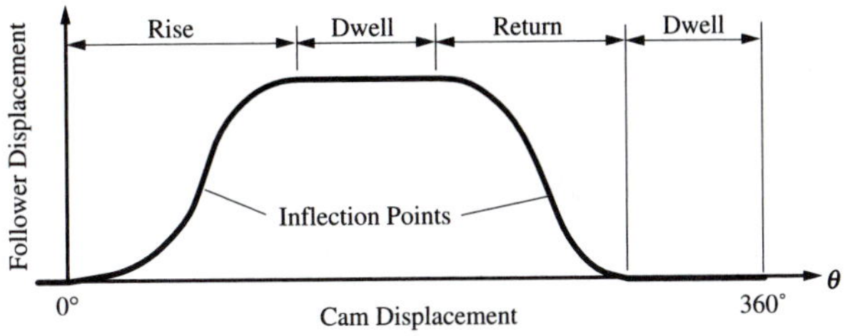

Figure 6.8 Terminology used when discussing time-motion programs.

Therefore, $\dot{y}$ is a simple constant times y', and $\ddot{y}$ is also a constant times y''. Consequently, even though we ultimately want to know the response to the time derivatives ($\dot{y}$ and $\ddot{y}$), we may work directly with the derivatives (y' and y'') with respect to the cam displacement. If the cam velocity is not a constant, then the cam profile can be designed for only one operating situation if higher derivatives are important. In the following, a constant-velocity cam is assumed, and y is again used to represent either an angular or linear displacement of the follower. Similarly, θ is used for the displacement of the cam and it may be either an angular or linear displacement.

The follower curves can be studied in terms of the simple diagram shown in Fig. 6.8. A general displacement diagram will be made up of three or more parts:

1. Rises (1 or more)
2. Returns (1 or more)
3. Dwells (0 or more)

Both the rise and return parts will contain one or more inflection points. These are points where a maximum slope is reached, and they correspond to points on the cam surface with maximum steepness. These points are identified by the locations where the curvature of the diagram changes sign. At the inflection points, the radius of curvature of the curve is infinite.

In each of the standard curve cases, we will look at only the rise part of the follower profile. The return part can be determined using the mirror images of the curves considered.

6.5 UNIFORM MOTION

Uniform motion is represented in Fig. 6.9. To derive the equations for the follower displacement, a general form for the mathematical expression corresponding to the type of motion is assumed. The general equation will have undetermined constants in it, and these constants can be determined by matching boundary conditions at the two ends of the curve. For

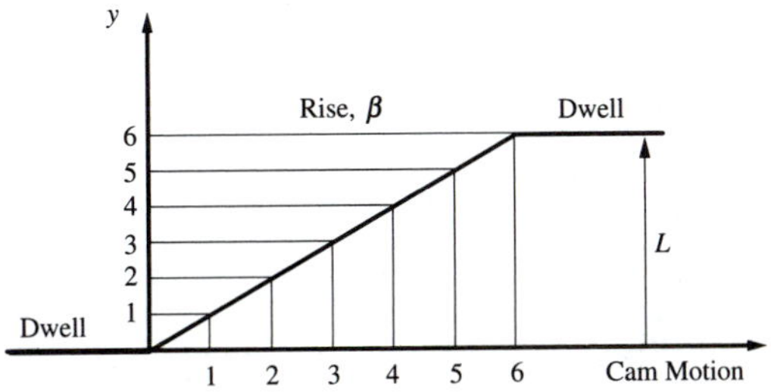

Figure 6.9 Uniform motion.

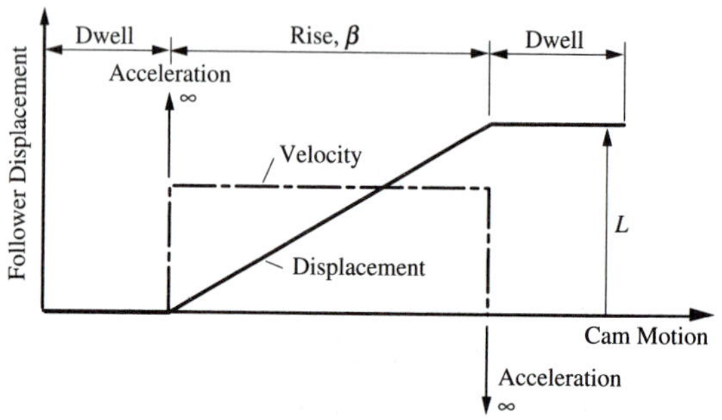

Figure 6.10 Displacement, velocity, and acceleration relations for uniform motion.

uniform motion, the general form of the curve used is

$$y = C\theta$$

If L is the amount of the rise and β is the cam rotation required for the rise, then the constant C must be L/β, and y becomes

$$y = \frac{L}{\beta}\theta$$

During the rise, the velocity and acceleration are

$$\dot{y} = \frac{L}{\beta}\omega$$

and

$$\ddot{y} = 0$$

This displacement, velocity, and acceleration are plotted in Fig. 6.10. As noted earlier, the acceleration is infinite at the points where the uniform motion meets the dwells. Therefore, even for low speeds and elastic members, the forces transmitted will be very large. For *very* low speeds, however, this type of displacement diagram might be acceptable.

Graphically, the uniform motion-displacement diagram is characterized by a uniform change in y for a uniform change in the cam motion. This condition is shown in Fig. 6.9.

6.6 PARABOLIC MOTION

The equations for parabolic motion can be derived using the same procedure as described in Section 6.3. However, two parabolas must be used for each transition. The two parabolas meet at the point midway between the ends of the two dwell regions. The general form for both parabolas is

$$y = C_0 + C_1\theta + C_2\theta^2 \tag{6.1}$$

and

$$y' = C_1 + 2C_2\theta \tag{6.2}$$
$$y'' = 2C_2$$

If the cam displacement is taken as 0 at the beginning of the rise, then at $\theta = 0$, $y = y' = 0$. Then, $C_0 = C_1 = 0$. Also, at $\theta = \beta/2$, $y = L/2$. Therefore, the displacement and first and second derivatives with respect to θ are

$$y = 2L\left(\frac{\theta}{\beta}\right)^2$$

$$y' = 4\frac{L}{\beta^2}\theta \tag{6.3}$$

$$y'' = 4\frac{L}{\beta^2}$$

and the velocity and acceleration are

$$\dot{y} = 4\frac{L\omega}{\beta^2}\theta$$

$$\ddot{y} = 4\frac{L\omega^2}{\beta^2}$$

At the point at which the curve meets the first dwell, the velocity and acceleration are continuous, but the third derivative, or jerk, is infinite. This derivative is proportional to the change in force and for high-speed cams is an important aspect of the motion. Although not so serious as having an infinite acceleration pulse, an infinite jerk can excite vibratory behavior in the system.

For the second half of the rise, the conditions to match are at $\theta = \beta/2$, $y = L/2$, and at $\theta = \beta$, $y = L$ and $y' = 0$. Then,

$$\frac{L}{2} = C_0 + C_1\frac{\beta}{2} + C_2\left(\frac{\beta}{2}\right)^2$$

$$L = C_0 + C_1\beta + C_2\beta^2$$

$$0 = C_1 + 2C_2\beta$$

The solution to this linear set of equations yields

$$C_0 = -L$$

$$C_1 = \frac{4L}{\beta}$$

$$C_2 = -\frac{2L}{\beta^2}$$

so that

$$y = L\left[1 - 2\left(1 - \frac{\theta}{\beta}\right)^2\right] \tag{6.4}$$

and

$$y' = \frac{4L}{\beta}\left(1 - \frac{\theta}{\beta}\right)$$

$$y'' = -\frac{4L}{\beta^2}$$

Finally, the velocity and acceleration are given by

$$\dot{y} = \frac{4L\omega}{\beta}\left(1 - \frac{\theta}{\beta}\right)$$

$$\ddot{y} = -\frac{4L\omega^2}{\beta^2}$$

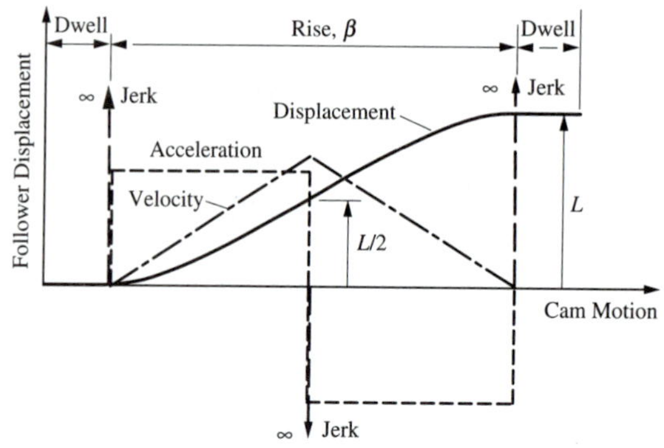

Figure 6.11 Displacement, velocity, acceleration, and jerk relations for parabolic motion during rise.

These equations apply to the segment of the program to the right of the inflection point shown in Fig. 6.11.

Graphically, the part of the curve up to the inflection point can be generated using the construction shown in Fig. 6.12. For the construction, the horizontal axis is divided into uniform increments, and the maximum rise is evenly divided into the same number of equal increments. The point at the origin is then connected to each of the points on the line of the maximum rise. Points on the displacement curve are given by the intersection of the diagonal lines with the corresponding vertical lines.

A cam return using parabolic motion is shown in Fig. 6.13. To determine the equations for the return from $y = L$ to 0 during the angular displacement β, we can use Eq. (6.1) again but with different boundary conditions. To simplify the equations, we will shift the origin of the coordinate system to the end of the dwell at the beginning of the return. For the first part of the return, $y = L$ and $y' = 0$ at $\theta = 0$ and $y = L/2$ at $\theta = \beta/2$. For these conditions, $C_0 = L$, $C_1 = 0$, and $C_2 = -2L/\beta^2$.

The displacement equation is

$$y = L\left[1 - 2\left(\frac{\theta}{\beta}\right)^2\right] \tag{6.5}$$

For the second half of the fall, the conditions to match are at $\theta = \beta/2$, $y = L/2$, and at $\theta = \beta$, $y = 0$ and $y' = 0$. For these conditions,

$$y = 2L\left(1 - \frac{\theta}{\beta}\right)^2 \tag{6.6}$$

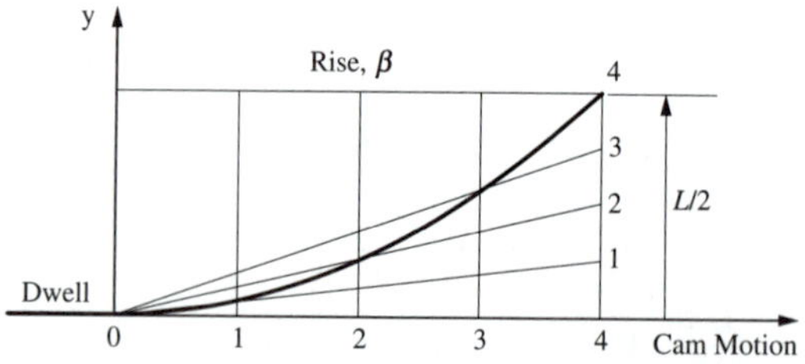

Figure 6.12 Construction of parabolic segment of time-motion program.

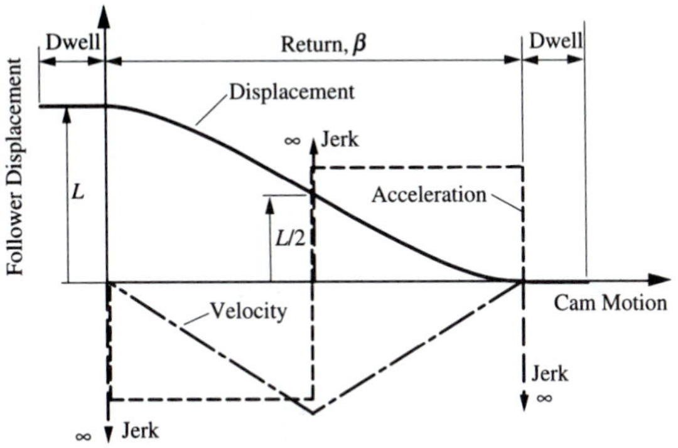

Figure 6.13 Displacement, velocity, acceleration, and jerk relations for parabolic motion during return.

In general, the rise and return will not always start at $\theta = 0$. However, in these cases, a simple coordinate transformation can be used. If the rise or fall actually starts at $\theta = \gamma$, substitute $(\theta - \gamma)$ wherever θ appears in Eqs. (6.3)–(6.6).

EXAMPLE 6.1 *(Design for Parabolic Motion)*

PROBLEM

Design a parabolic cam time-motion program to provide a dwell at zero lift for the first 120° of the motion cycle and to dwell at 0.8 in lift for cam angles from 180° to 210°. The cam profile will be laid out using 10° plotting intervals. Assume that the cam rotates with constant angular velocity.

SOLUTION

The motion specification is as shown in Fig. 6.14. For the first part of the rise ending at θ = 150° in the interval 120° to 180°, Eq. (6.3) applies if we use $(\phi - 120) = \theta$ and $0.8 = L$. The resulting expression for the first part of the rise is

$$y = 2L\left(\frac{\theta}{\beta}\right)^2 = 1.6\left(\frac{\phi - 120}{60}\right)^2 \tag{6.7}$$

For the second part of the rise starting at θ = 150°, Eq. (6.4) applies if we again use $(\phi - 120) = \theta$ and $0.8 = L$. The resulting expression is

$$y = L\left[1 - 2\left(1 - \frac{\theta}{\beta}\right)^2\right] = 0.8\left[1 - 2\left(1 - \frac{(\phi - 120)}{60}\right)^2\right] \tag{6.8}$$

Using Eqs. (6.7) and (6.8), the successive lifts are given in Table 6.1. For the first part of the return ending at θ = 285° in the interval 210° to 360°, Eq. (6.5) applies if we use $(\phi - 210) = \theta$

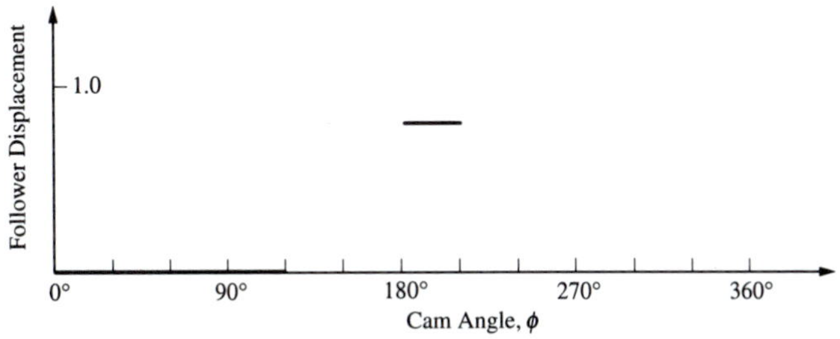

Figure 6.14 The motion specification for Example 6.1.

Table 6.1 Cam Follower Data for Rise in Example 6.1

θ	120°	130°	140°	150°	160°	170°	180°
y	0.0000	0.0444	0.1778	0.4000	0.6222	0.7556	0.8000

Table 6.2 Cam Follower Data for Return in Example 6.1

θ	210°	220°	230°	240°	250°	260°	270°	280°
y	0.8000	0.7929	0.7716	0.7360	0.6862	0.6222	0.5440	0.4516
θ	290°	300°	310°	320°	330°	340°	350°	360°
y	0.3484	0.2560	0.1778	0.1138	0.0640	0.0284	0.0071	0.0000

and $0.8 = L$. The resulting expression for the first part of the return is

$$y = L\left[1 - 2\left(\frac{\theta}{\beta}\right)^2\right] = 0.8\left[1 - 2\left(\frac{(\phi - 210)}{150}\right)^2\right] \tag{6.9}$$

For the second part of the return starting at $\theta = 285°$, Eq. (6.6) applies if we use $(\phi - 210) = \theta$ and $0.8 = L$. The resulting expression is

$$y = 2L\left(1 - \frac{\theta}{\beta}\right)^2 = 1.6\left(1 - \frac{\phi - 210}{150}\right)^2 \tag{6.10}$$

Using Eqs. (6.9) and (6.10), points on the return curve are given in Table 6.2. The resulting transition curves are plotted in Fig. 6.15. Notice that the lift values are tabulated to four decimal places. Cam and follower systems normally use very rigid components and even small profile variations are important. For this reason, we normally work with at least four decimal places when doing cam calculations. Gears are another type of profile mechanism in which the components are very rigid and, consequently, even tiny profile variations can be important. ■

6.7 HARMONIC TIME-MOTION PROGRAMS

Simple harmonic motion can be generated by an offset (eccentric) circular cam with a radial follower and is therefore a common form to use for a displacement diagram. Cams with this type of transition curve are commonly referred to as "harmonic cams." The

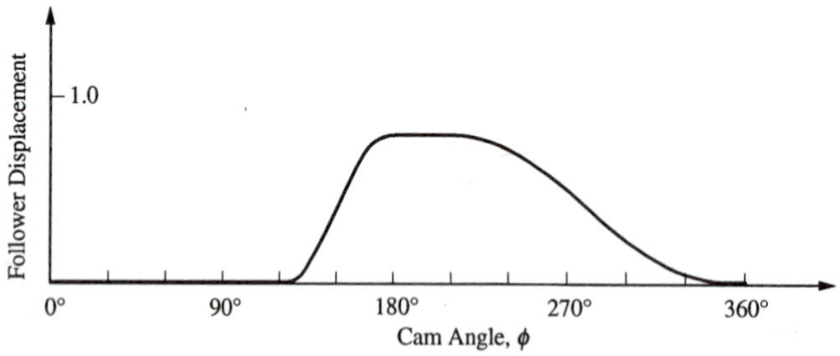

Figure 6.15 The parabolic time-motion program generated in Example 6.1.

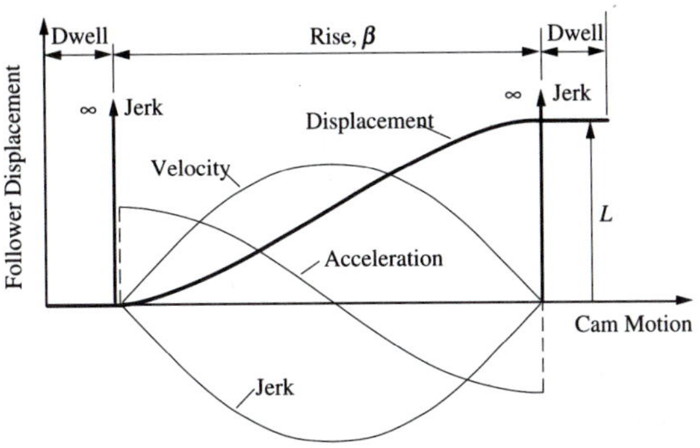

Figure 6.16 Displacement, velocity, acceleration, and jerk relations for simple harmonic motion.

equations for simple harmonic motion are formed from the basic equation:

$$y = C_0 + C_1 \cos C_2\theta = C_0 \left(1 + \frac{C_1}{C_0} \cos C_2\theta\right)$$

The displacement, velocity, acceleration, and jerk diagrams are shown in Fig. 6.16. Simple harmonic motion produces a sine velocity curve and a cosine acceleration curve. There is no discontinuity at the transition point, so θ is defined for all angles between zero and β. The equations for a rise starting from $\theta = 0$ and ending at $\theta = \beta$ and $y = L$ are

$$\begin{aligned} y &= \frac{L}{2}\left(1 - \cos\frac{\pi\theta}{\beta}\right) \\ y' &= \frac{\pi L}{2\beta}\sin\frac{\pi\theta}{\beta}, \qquad \dot{y} = \frac{\pi L\omega}{2\beta}\sin\frac{\pi\theta}{\beta} \\ y'' &= \frac{L}{2}\left(\frac{\pi}{\beta}\right)^2\cos\frac{\pi\theta}{\beta}, \qquad \ddot{y} = \frac{L}{2}\left(\frac{\pi\omega}{\beta}\right)^2\cos\frac{\pi\theta}{\beta} \end{aligned} \tag{6.11}$$

The equations for the return from $\theta = 0$, $y = L$ to $\theta = \beta$, $y = 0$ are

$$\begin{aligned} y &= \frac{L}{2}\left(1 + \cos\frac{\pi\theta}{\beta}\right) \\ y' &= -\frac{\pi L}{2\beta}\sin\frac{\pi\theta}{\beta}, \qquad \dot{y} = -\frac{\pi L\omega}{2\beta}\sin\frac{\pi\theta}{\beta} \\ y'' &= -\frac{L}{2}\left(\frac{\pi}{\beta}\right)^2\cos\frac{\pi\theta}{\beta}, \qquad \ddot{y} = -\frac{L}{2}\left(\frac{\pi\omega}{\beta}\right)^2\cos\frac{\pi\theta}{\beta} \end{aligned} \tag{6.12}$$

A simple harmonic displacement diagram can be generated graphically by drawing a semicircle on the vertical axis and dividing it into an even number of segments. The cam motion axis is then divided into the same number of even increments, and horizontal lines are drawn from the points on the semicircle axis. The intersections of the horizontal lines with the corresponding vertical lines give the location of points on the simple harmonic curve. This construction is shown in Fig. 6.17. For the construction, note that

$$\frac{\Delta\alpha}{\Delta\theta} = \frac{180}{\beta}$$

where β is the cam rotation for the follower to move from lift 0 to L. With the advent of computers, the graphical procedure is typically used only for schematic representations of simple harmonic motion.

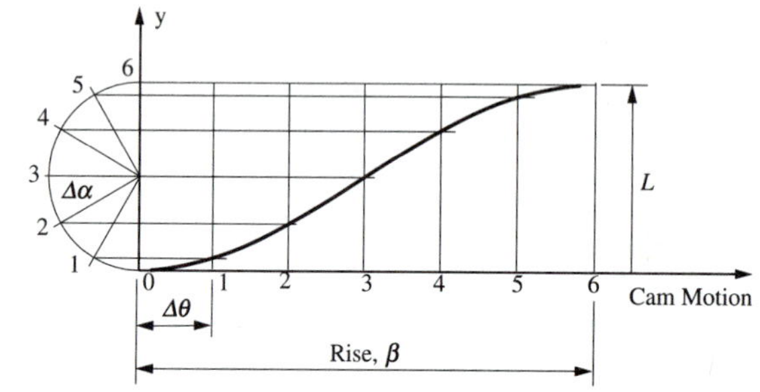

Figure 6.17 Graphical construction of displacement diagram for simple harmonic motion.

EXAMPLE 6.2 ***(Design for Harmonic Motion)***

PROBLEM

Design a harmonic cam to satisfy the same motion specifications as for Example 6.1. That is, the motion program is to provide a dwell at zero lift for the first 120° of the motion cycle and to dwell at 0.8 in lift for cam angles from 180° to 210°. The cam profile will be laid out using 10° plotting intervals.

SOLUTION

The motion specification is as shown in Fig. 6.14.

The rise in the interval 120° to 180° can be computed using Eq. (6.11) if we use $\theta = (\phi - 120)$ and $0.8 = L$. The resulting expression for the rise is

$$y = \frac{L}{2}\left(1 - \cos\frac{\pi\theta}{\beta}\right) = 0.4\left(1 - \cos\frac{\pi(\phi - 120)}{60}\right) \tag{6.13}$$

The results are given in Table 6.3.

For the return in the interval 210° to 360°, Eq. (6.12) applies if we use $\theta = (\phi - 210)$ and $0.8 = L$. The resulting expression for the return is

$$y = \frac{L}{2}\left(1 + \cos\frac{\pi\theta}{\beta}\right) = 0.4\left(1 + \cos\frac{\pi(\phi - 210)}{150}\right) \tag{6.14}$$

Using this equation, the successive values for y are given in Table 6.4.

The tabulated lift values may be compared with those of Example 6.1 to observe the differences between comparable parabolic and harmonic transition curves. If plotted, the time-motion program would be difficult to distinguish from Fig. 6.15. However, there are important differences in the values. ■

Table 6.3 Cam Follower Data for Rise Using Simple Harmonic Motion in Example 6.2

θ	120°	130°	140°	150°	160°	170°	180°
y	0.0000	0.0536	0.2000	0.4000	0.6000	0.7464	0.8000

Table 6.4 Cam Follower Data for Return in Example 6.2

θ	210°	220°	230°	240°	250°	260°	270°	280°
y	0.8000	0.7913	0.7654	0.7236	0.6677	0.6000	0.5236	0.4418
θ	290°	300°	310°	320°	330°	340°	350°	360°
y	0.3582	0.2764	0.2000	0.1323	0.0764	0.0346	0.0087	0.0000

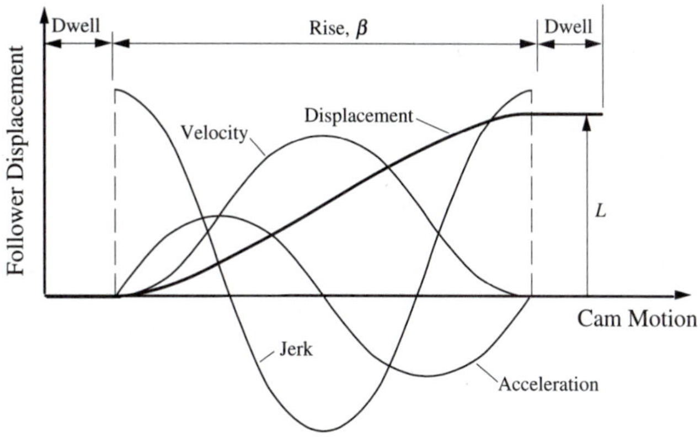

Figure 6.18 Displacement, velocity, acceleration, and jerk relations for cycloidal motion.

6.8 CYCLOIDAL TIME-MOTION PROGRAMS

All of the motions given above have nonzero values of acceleration (and therefore infinite jerk) at the beginnings and ends of the motion and therefore are limited to relatively low speeds. Cycloidal motion has zero acceleration at the beginnings and ends of the motion and so is useful for relatively high speeds.

A cycloidal transition produces a sinusoidal acceleration curve. The equations for the rise are

$$y = L\left(\frac{\theta}{\beta} - \frac{1}{2\pi}\sin\frac{2\pi\theta}{\beta}\right)$$

$$y' = \frac{L}{\beta}\left(1 - \cos\frac{2\pi\theta}{\beta}\right); \qquad \dot{y} = \frac{L\omega}{\beta}\left(1 - \cos\frac{2\pi\theta}{\beta}\right)$$

$$y'' = \frac{2L\pi}{\beta^2}\sin\frac{2\pi\theta}{\beta}; \qquad \ddot{y} = 2L\pi\left(\frac{\omega}{\beta}\right)^2\sin\frac{2\pi\theta}{\beta}$$

These curves are plotted in Fig. 6.18. There is no discontinuity at the inflection point, and therefore the equations are valid for values of θ from zero to β. The curve is symmetric, and the return is given by $\bar{y} = L - y$. Therefore, $\bar{y}' = -y'$ and $\bar{y}'' = -y''$.

Cycloidal motion may be obtained by rolling a circle of radius $L/2\pi$, where L is the total rise, on the displacement axis as shown in Fig. 6.19. However, to construct the curve graphically, a more convenient, alternative way is shown in Fig. 6.19. First, a circle with diameter L/π and center at (β, L) is divided into an even number of increments, and the resulting points are projected onto the displacement axis. The cam motion axis is divided

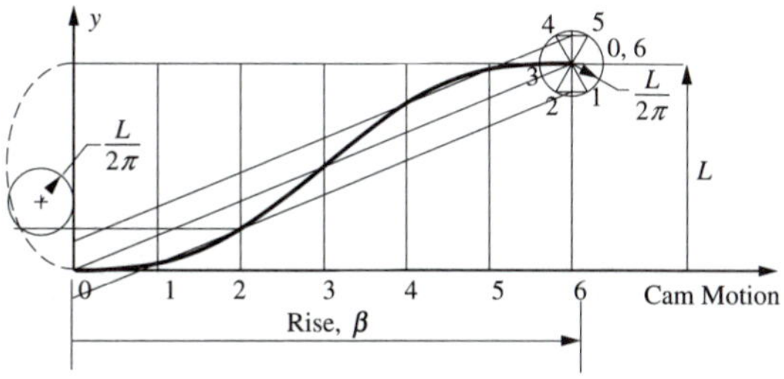

Figure 6.19 Graphical construction of displacement diagram for cycloidal motion.

into the same number of increments. A series of lines parallel to the line from the origin to the point (β, L) are then drawn from the projected points on the circle diameter. The intersections of these lines with the corresponding vertical lines from the cam-motion axis give points on the cycloidal curve.

6.9 GENERAL POLYNOMIAL TIME-MOTION PROGRAMS

For high-speed machines, it is common to specify a general polynomial profile for a cam. Depending on the order of the polynomial chosen, it is theoretically possible to match almost any conditions posed by the designer. A polynomial curve is fitted to the rise or return. Only odd-order polynomials are appropriate for rises or returns between dwells if the same conditions are to be matched at both ends of the polynomial. A first-order polynomial gives constant velocity and infinite acceleration at the beginning and end of the transition. This is the uniform motion profile discussed above. A third-order polynomial gives a parabolic velocity variation, linear acceleration, and infinite jerk at the beginning and end of the transition. A fifth-order polynomial gives finite acceleration and jerk. The derivative of jerk is infinite at the ends of the transition. A fifth-order fit is the practical maximum unless great care is taken during manufacturing. Dynamic effects due to manufacturing errors tend to become more important than those due to curve fitting at this stage.

For a general polynomial follower displacement, the displacement function is given by

$$y = f(\theta) = \sum_{i=0}^{n} A_i \theta^i$$

where θ is the cam angle and the A's must be determined from the conditions to be matched. The equation permits us to match the same number of conditions as there are A's, that is, $n + 1$ conditions. When n is large and the angles are measured in degrees, the terms in the summation can vary greatly in size. For example, if θ is 100° and n is 10, the coefficients of θ^i in the equation can vary from 1 to 10^{20}, and round-off error will make it difficult to obtain an accurate solution. Therefore, it is convenient to rewrite the displacement equation in terms of the cam rotation angle β, which gives the range over which the equation is to be used. The resulting equation is

$$y = f(\theta) = \sum_{i=0}^{n} C_i \left(\frac{\theta}{\beta}\right)^i$$

Now the coefficients of the constants are always numbers between zero and 1, and round-off error problems have been greatly reduced. The constants in the two equations are related by the simple expression

$$A_i = \frac{C_i}{\beta^i}$$

The conditions to be matched will typically involve at least the velocity and acceleration of the follower, and the required equations for these conditions can be written as

$$\dot{y} = \dot{f}(\theta) = \frac{1}{\beta}\frac{d\theta}{dt}\sum_{i=1}^{n} iC_i \left(\frac{\theta}{\beta}\right)^{(i-1)}$$

and

$$\ddot{y} = \ddot{f}(\theta) = \frac{1}{\beta}\frac{d^2\theta}{dt^2}\sum_{i=1}^{n} iC_i \left(\frac{\theta}{\beta}\right)^{(i-1)} + \frac{1}{\beta^2}\left(\frac{d\theta}{dt}\right)^2 \sum_{i=2}^{n} i(i-1)C_i \left(\frac{\theta}{\beta}\right)^{(i-2)}$$

Notice that the summation on the velocity term starts at 1 because C_0 does not appear in the equation, and the summation on the acceleration term starts at 2 because neither C_0 nor C_1 appears in the equation.

Now if a constant-velocity cam is used,

$$\frac{d\theta}{dt} = \omega$$

and

$$\frac{d^2\theta}{dt^2} = 0$$

where ω is the angular veocity of the cam. The equations may then be reduced to the following:

$$y = f(\theta) = \sum_{i=0}^{n} C_i \left(\frac{\theta}{\beta}\right)^i$$

$$\dot{y} = \dot{f}(\theta) = \frac{\omega}{\beta} \sum_{i=1}^{n} iC_i \left(\frac{\theta}{\beta}\right)^{(i-1)}$$

$$\ddot{y} = \ddot{f}(\theta) = \left(\frac{\omega}{\beta}\right)^2 \sum_{i=2}^{n} i(i-1)C_i \left(\frac{\theta}{\beta}\right)^{(i-2)}$$

As an example of the use of the polynomial profile, assume that we begin and end the follower displacement with a dwell as shown in Fig. 6.20 and assume that we want to match the position, velocity, and acceleration at both the beginning and end of the period being considered.

For points A and B in Fig. 6.20, the following conditions apply:

$$\theta = 0 \Rightarrow y = \dot{y} = \ddot{y} = 0$$

$$\theta = \beta \Rightarrow y = L$$

$$\dot{y} = \ddot{y} = 0$$

There are six conditions, so the position equation must have six constants. The resulting equations for position, velocity, and acceleration are

$$y = C_0 + C_1\left(\frac{\theta}{\beta}\right) + C_2\left(\frac{\theta}{\beta}\right)^2 + C_3\left(\frac{\theta}{\beta}\right)^3 + C_4\left(\frac{\theta}{\beta}\right)^4 + C_5\left(\frac{\theta}{\beta}\right)^5$$

$$\dot{y} = \frac{\omega}{\beta}\left[C_1 + 2C_2\left(\frac{\theta}{\beta}\right) + 3C_3\left(\frac{\theta}{\beta}\right)^2 + 4C_4\left(\frac{\theta}{\beta}\right)^3 + 5C_5\left(\frac{\theta}{\beta}\right)^4\right]$$

$$\ddot{y} = \left(\frac{\omega}{\beta}\right)^2\left[2C_2 + 6C_3\left(\frac{\theta}{\beta}\right) + 12C_4\left(\frac{\theta}{\beta}\right)^2 + 20C_5\left(\frac{\theta}{\beta}\right)^3\right]$$

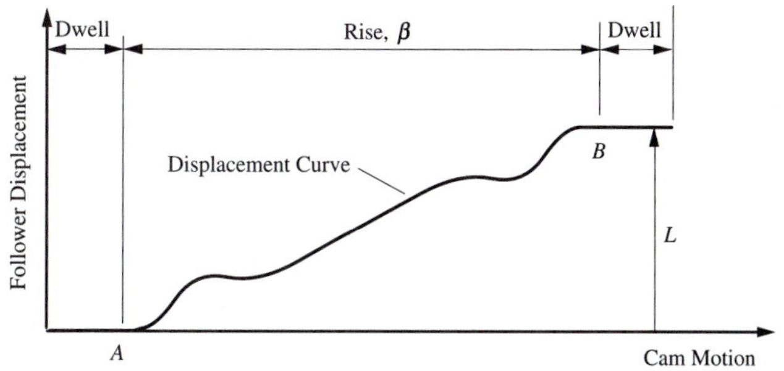

Figure 6.20 Initial information for polynomial profile example.

Evaluation of these equations at the beginning and end of the rise period gives the following six equations which must be solved.

$$0 = C_0$$

$$0 = \frac{\omega}{\beta} C_1$$

$$0 = \left(\frac{\omega}{\beta}\right)^2 2C_2$$

$$L = C_0 + C_1 + C_2 + C_3 + C_4 + C_5$$

$$0 = \frac{\omega}{\beta}[C_1 + 2C_2 + 3C_3 + 4C_4 + 5C_5]$$

$$0 = \left(\frac{\omega}{\beta}\right)^2 [2C_2 + 6C_3 + 12C_4 + 20C_5]$$

Solution for the unknown constants C_0 through C_6 gives

$$C_0 = C_1 = C_2 = 0$$

$$C_3 = 10L, \qquad C_4 = -15L; \qquad C_5 = 6L$$

The displacement equation can then be written in the form

$$y = 10L\left(\frac{\theta}{\beta}\right)^3 - 15L\left(\frac{\theta}{\beta}\right)^4 + 6L\left(\frac{\theta}{\beta}\right)^5$$

This is the so-called 3-4-5 polynomial transition because of the powers of the terms that remain in the expression. The first three derivatives and the velocity, acceleration, and jerk are given by

$$y' = \frac{30L}{\beta}\left[\left(\frac{\theta}{\beta}\right)^2 - 2\left(\frac{\theta}{\beta}\right)^3 + \left(\frac{\theta}{\beta}\right)^4\right], \qquad \dot{y} = \frac{30\omega L}{\beta}\left[\left(\frac{\theta}{\beta}\right)^2 - 2\left(\frac{\theta}{\beta}\right)^3 + \left(\frac{\theta}{\beta}\right)^4\right]$$

$$y'' = \frac{60L}{\beta^2}\left[\left(\frac{\theta}{\beta}\right) - 3\left(\frac{\theta}{\beta}\right)^2 + 2\left(\frac{\theta}{\beta}\right)^3\right], \qquad \ddot{y} = 60L\left(\frac{\omega}{\beta}\right)^2\left[\left(\frac{\theta}{\beta}\right) - 3\left(\frac{\theta}{\beta}\right)^2 + 2\left(\frac{\theta}{\beta}\right)^3\right]$$

$$y''' = \frac{60L}{\beta^3}\left[1 - 6\left(\frac{\theta}{\beta}\right) + 6\left(\frac{\theta}{\beta}\right)^2\right]; \qquad \dddot{y} = 60L\left(\frac{\omega}{\beta}\right)^3\left[1 - 6\left(\frac{\theta}{\beta}\right) + 6\left(\frac{\theta}{\beta}\right)^2\right]$$

These general relationships are plotted in Fig. 6.21. The displacement results are visually similar to the cycloidal curve, but the velocity, acceleration, and jerk are significantly

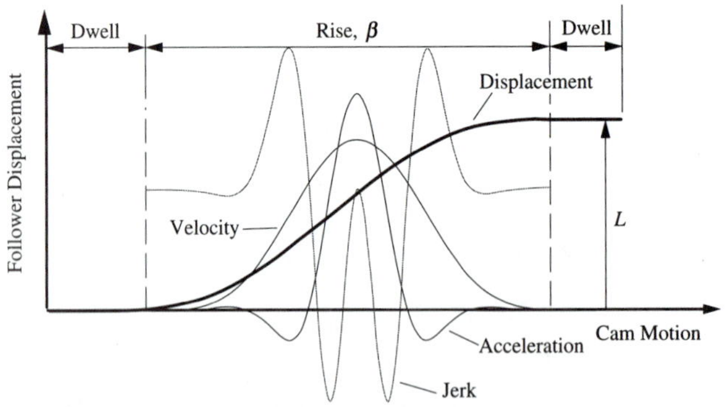

Figure 6.21 Displacement, velocity, acceleration, and jerk relations for the 3-4-5 polynomial motion.

Table 6.5 Comparison of the Different Types of cam Follower Motion for $\beta = L = 1$

θ	y (Linear)	y (Parabolic)	y (Harmonic)	y (Cycloidal)	y (Polynomial)
0.00	0.0000	0.0000	0.0000	0.0000	0.0000
0.05	0.0500	0.0050	0.0062	0.0008	0.0012
0.10	0.1000	0.0200	0.0245	0.0065	0.0086
0.15	0.1500	0.0450	0.0545	0.0212	0.0266
0.20	0.2000	0.0800	0.0955	0.0486	0.0579
0.25	0.2500	0.1250	0.1464	0.0908	0.1035
0.30	0.3000	0.1800	0.2061	0.1486	0.1631
0.35	0.3500	0.2450	0.2730	0.2212	0.2352
0.40	0.4000	0.3200	0.3455	0.3065	0.3174
0.45	0.4500	0.4050	0.4218	0.4008	0.4069
0.50	0.5000	0.5000	0.5000	0.5000	0.5000
0.55	0.5500	0.5950	0.5782	0.5992	0.5931
0.60	0.6000	0.6800	0.6545	0.6935	0.6826
0.65	0.6500	0.7550	0.7270	0.7788	0.7648
0.70	0.7000	0.8200	0.7939	0.8514	0.8369
0.75	0.7500	0.8750	0.8536	0.9092	0.8965
0.80	0.8000	0.9200	0.9045	0.9514	0.9421
0.85	0.8500	0.9550	0.9455	0.9788	0.9734
0.90	0.9000	0.9800	0.9755	0.9935	0.9914
0.95	0.9500	0.9950	0.9938	0.9992	0.9988
1.00	1.0000	1.0000	1.0000	1.0000	1.0000

different. In general, this type of cam will begin and end its motion more slowly than the other types, and in order to produce such a cam, extreme machining accuracy is required, especially at the beginning and end of the motion. The machining is commonly done on a computer numerically controlled milling machine.

To compare the profiles generated by the different time-motion programs, let $\beta = L = 1$ and vary θ from 0 to β. We can then compute y as a function of θ in increments of 0.05. The results are shown in Table 6.5. Notice that the variation among the different profiles is very small in most cases. This emphasizes that extreme accuracy must be achieved if the benefits of using the different time-motion programs are to be realized.

6.10 DETERMINING THE CAM PROFILE

Once the follower motion is determined as a function of the cam displacement, the cam surface can be found either graphically or analytically. For extremely accurate cams, the geometry must be determined analytically and the machining done using numerically controlled milling machines. For low-speed cams, however, a graphical layout and manual machining are adequate.

In both the graphical and analytical approaches to determining the cam geometry, the cam mechanism must be inverted. That is, the cam is taken as the reference system, and the frame and follower are considered to move relative to the cam. To maintain the correct relative motion, the follower will move relative to the cam in a direction opposite to the motion of the cam relative to the follower.

If we restrict our discussions to planar, rotating cams, four general types of followers are possible. They are a translating, cylindrical-faced follower; a translating, flat-faced follower; a rotating, cylindrical-faced follower; and a rotating, flat-faced follower (Fig. 6.22).

Notice that the cam geometry is independent of the type of joint between the cylindrical-faced follower and the cam. The kinematic design procedure is exactly the same when a

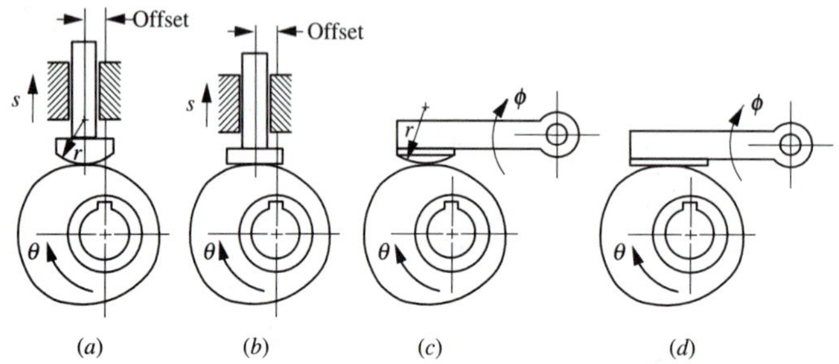

Figure 6.22 Common follower configurations for planar, rotating cams. (a) Translating, cylindrical-faced follower; (b) translating, flat-faced follower; (c) rotating, cylindrical-faced follower; (d) rotating, flat-faced follower.

roller follower or a solid cylindrical-faced follower is involved. We will consider both graphical and analytical approaches to the design of the cam for each type of follower shown in Fig. 6.22.

6.10.1 Graphical Cam Profile Layout

As indicated above, cam profiles are laid out graphically using inversion. That is, the cam is viewed as stationary, and the successive positions of the follower are located relative to it. This results in a polar plot of successive follower positions. The cam profile is then filled in as the envelope curve of the follower positions.

The first step in laying out the cam profile is to select a base circle radius. The base circle represents the position of the follower at zero lift. Successive lift values are plotted radially outward from the base circle.

Choosing a large base circle radius results in a large cam. However, if the base circle is too small the cam profile may have hollows of smaller radius than the follower. Since the follower will bridge across such a hollow, it will not follow the desired lift program. Obviously, this situation must be avoided.

The pressure angle of a cam is the angle between the contact normal and the velocity of the point on the follower at the contact location. Reducing the pressure angle reduces the contact loads and promotes smoother operation with less wear. Increasing the base circle radius decreases the maximum value of pressure angle. Thus, it is good practice to use the largest base circle that the design constraints will allow. As a general rule of thumb, the base circle radius should be two to three times the maximum lift value.

EXAMPLE 6.3 ***(Layout of Cam Profile for Radial Roller Follower)***

PROBLEM

Lay out a cam profile using the harmonic follower displacement profile of Example 6.2. That is, the follower is to dwell at zero lift for the first 120° of the motion cycle and to dwell at 0.8 in lift for cam angles from 180° to 210°. The cam is to have a translating, roller follower with a 1-in roller diameter. The cam will rotate clockwise. Lay out the cam profile using 10° plotting intervals.

SOLUTION

The basic motion specification is as shown in Fig. 6.14. Using the results of Example 6.2, the lift values to be plotted are as given in Table 6.6. Notice that the dwells correspond to locations on the cam where the radius is constant.

Table 6.6 Follower Displacements for Example 6.3

θ	0, 360°	10°	20°	30°	40°	50°	60°	70°	80°
y	0.0000	0.0000	0.0000	0.0000	0.0000	0.0000	0.0000	0.0000	0.0000
θ	90°	100°	110°	120°	130°	140°	150°	160°	170°
y	0.0000	0.0000	0.0000	0.0000	0.0536	0.2000	0.4000	0.6000	0.7464
θ	180°	190°	200°	210°	220°	230°	240°	250°	260°
y	0.8000	0.8000	0.8000	0.8000	0.7913	0.7654	0.7236	0.6677	0.6000
θ	270°	280°	290°	300°	310°	320°	330°	340°	350°
y	0.5236	0.4418	0.3582	0.2764	0.2000	0.1323	0.0764	0.0346	0.0087

The layout of the cam is accomplished by drawing radial lines at 10-degree increments. Because the cam rotates clockwise, the radial lines are laid off and labeled in the counterclockwise direction as shown in Fig. 6.23. Next, the base circle and the prime circle are drawn. The *base circle* is chosen to have a 1.5-inch radius, and it is the largest circle that can be drawn inside the cam profile and be tangent to the cam profile. The radius of the *prime circle* is equal to $r_b + r_0$, where r_b is the base circle radius and r_0 is the radius of the roller follower. In this

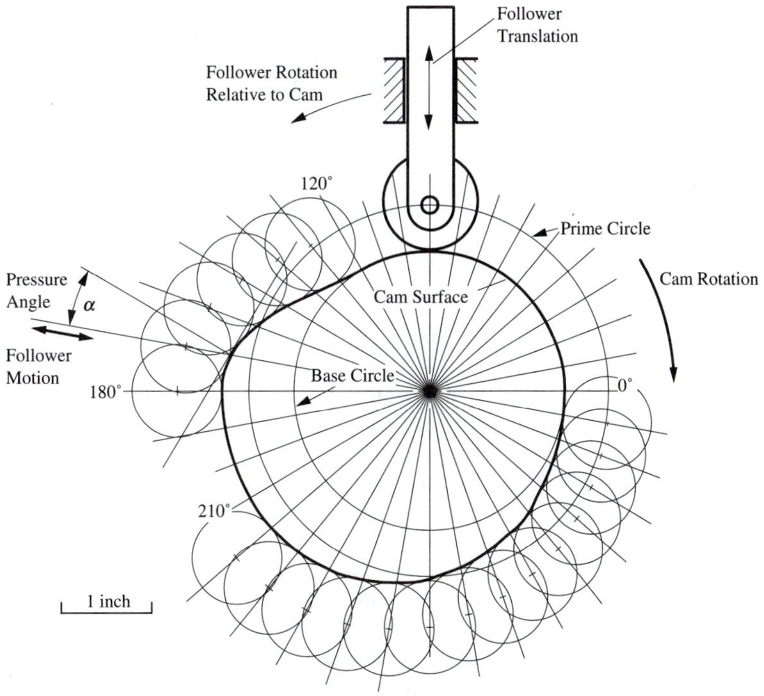

Figure 6.23 Layout of the cam profile for Example 6.3. The process of laying out a cam profile is one of inversion. That is, the cam is viewed as being stationary and successive positions of the follower are plotted relative to it. In this case, a prime circle of 2.0-in radius was chosen. This represents the location of the follower center at zero lift. Positive lift values are plotted outward from the base circle. The successive positions of the follower are then drawn using the plotted points as centers. Finally, the profile is plotted as an envelope curve of the successive follower positions. Because of the inversion, if the cam is to rotate clockwise, the positions of the follower must be plotted in the opposite direction, that is, counterclockwise.

problem, the prime-circle radius is 2.0 in. The cam profile is initially laid off from the prime circle to give the pitch curve. The *pitch curve* is the curve traced by the center of the roller follower. Notice that the pitch curve will be the cam profile if r_0 is zero. This corresponds to the case of a knife-edged follower.

Once the radial lines and prime circle are established, the displacements can be laid off from the prime circle as shown in Fig. 6.23. The radius of the follower is drawn with its center located on the pitch curve at a series of locations. The cam can be defined by drawing a curve tangent to roller locations as shown in Fig. 6.23.

As indicated before, an important parameter for cam motion is the pressure angle. In the case of the translating, roller follower, this is the angle between the follower travel and the normal to the curve. For a given force on the follower roller, the force in the direction of travel of the cam will be proportional to the cosine of the pressure angle. The force normal to the direction of travel of the follower is proportional to the sine of the pressure angle. Wear on the follower stem will increase with the normal force; therefore, from design considerations, we want the pressure angle (α) to be as small as possible.

The maximum pressure angle will occur at the *pitch points.* These correspond to the inflection points on the follower displacement curves (see Fig. 6.8). If the torque on the cam is more or less constant, the pressure angles at the pitch points will correspond to the parts of the cycle where the maximum normal force occurs and hence the times when the follower stem wear will be greatest. It will also correspond to the parts of the cycle where the follower will tend to bind in the stem bearing. Because of problems with wear and binding, the pressure angle is usually limited to angles on the order of $\pm 30°$. If the pressure angle becomes excessive, the base circle should be increased or the follower displacement profile changed.

The problem statement indicated that a roller follower was to be designed. However, the construction would be *exactly* the same if a solid, cylindrical-faced follower were involved. From the standpoint of the cam geometry, the important issues are the radius of the cylindrical face and the direction of translation relative to the cam. ■

EXAMPLE 6.4 *(Layout of Cam Profile for Radial Flat-Faced Follower)*

PROBLEM

Again, lay out a cam profile using the harmonic displacement profile of Examples 6.2 and 6.3. The cam is to have a translating, flat-faced follower that is offset by 0.2 in. The cam will rotate clockwise. Lay out the cam profile using 10° plotting intervals.

SOLUTION

The basic motion specification is the same as in Example 6.3 (Table 6.6). The layout of the cam is again accomplished by drawing radial lines at 10-degree increments. Because the cam rotates clockwise, the radial lines are laid off and labeled in the counterclockwise direction as was done in Fig. 6.23. Next the base circle is drawn. Because a flat-faced follower is being designed, there is no prime circle. However, selection of the base circle requires careful consideration.

A major restriction on the cam profile driving a flat-faced follower is that the profile must form a convex surface. This means that the vectors from every point on the cam to the corresponding center of curvature must point toward the interior of the cam. An alternative way to approach the convexity problem is to imagine an arbitrary line drawn across the face of the cam. If it is possible to select an arbitrary line that intersects the cam at more than two points, the cam profile is not convex. If the cam is not convex, the flat-faced follower cannot contact the cam at all points, and the desired motion will not be generated. This condition will be illustrated mathematically when an analytical approach to cam synthesis is discussed. Clearly, the cam generated in Fig. 6.23 does not satisfy the convexity condition. When the resulting cam is not convex, we must increase the size of the base circle.

To begin the construction, we can select a base circle somewhat arbitrarily. Next we construct lines parallel to the original radial lines but offset by 0.2 in. We then lay off the displacements in Table 6.6 from the base circle but along the lines that are offset from the radial lines. Next, draw a line perpendicular at the locations on the offset lines corresponding to the displacement locations. These perpendicular lines correspond to the face of the follower. This is illustrated

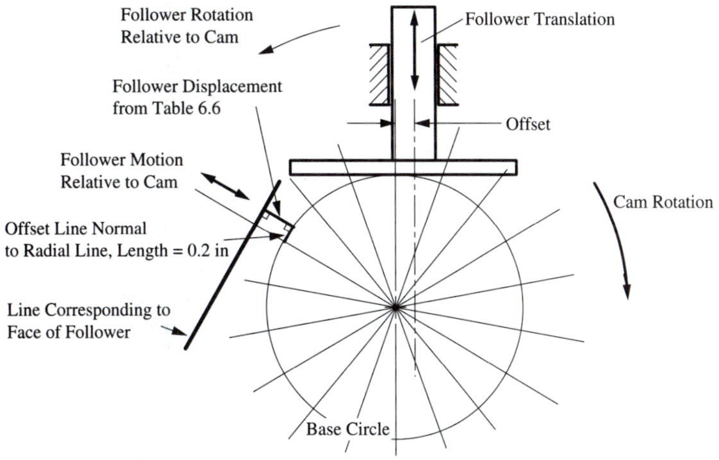

Figure 6.24 Basic construction lines for determining cam profile for flat-faced follower.

in Fig. 6.24. The lines for different positions of the follower will form an envelope that defines the geometry of the cam surface. We construct the outline of the cam by drawing a curve that contacts the lines corresponding to the different positions of the follower face at tangent points.

As the lines corresponding to the different positions of the follower face are drawn, successive lines will intersect. For the geometry to be valid, the angle increment for successive intersections must be positive. If an intersection requires a negative angle increment, it will not be possible to generate the cam, and a larger base circle must be used. This situation is illustrated in the current problem in Fig. 6.25 for the positions corresponding to rotations angles of 150°, 160°, 170°, and 180°. In Fig. 6.25, a base circle radius of 1.5 in was chosen. The angle increment for the intersection corresponding to 160° and 170° is positive, but the increment corresponding to 170° and 180° is negative. This situation indicates that the base circle is too small and must be increased.

The smallest base circle is the one for which the angle increment corresponding to 170° and 180° is no longer negative, that is, when it is zero. This occurs when the follower-face lines corresponding to 160°, 170°, and 180° intersect. This occurs for a base circle radius of approximately 5.5 in. For this base circle, the cam will have a point or cusp corresponding to the

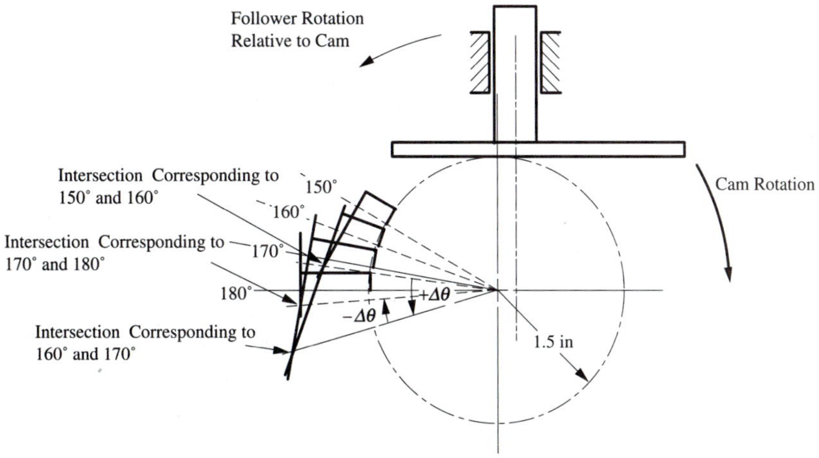

Figure 6.25 Condition when the base circle is too small to generate an acceptable cam for a flat-faced follower.

location where the face lines intersect. The envelope of the face lines for the nondwell part of the cam is shown in Fig. 6.26, and the resulting cam is shown in Fig. 6.27. The cam designed for the roller follower and for the same displacement profile is shown in Fig. 6.27 for comparison. Based on the size of the cam required, a flat-faced follower would not be a good choice for this type of displacement profile.

As indicated before, an important parameter for cam motion is the pressure angle. When a flat-faced follower is used, the normal to the follower profile is always in the direction of the follower travel. This makes the pressure angle always zero; however, there can be significant lateral loads on the follower bearings caused by the frictional force at the cam-follower interface and by the moment generated by the normal force at the cam-follower interface and the offset line of action. The friction force can be reduced by lubrication but never completely eliminated, and the bearing couple that opposes the moment from the normal force must be addressed in the design of the cam and follower system. Depending on the lubrication and design, the lateral forces can be as high as or higher than the corresponding lateral force with a roller follower. Also, the cam may be so large to avoid the convexity condition that the roller follower would be preferred from size considerations.

Another important parameter that must be determined for the design of a cam for a flat-faced follower is the size (length) of the follower face. It is essential that the face be long

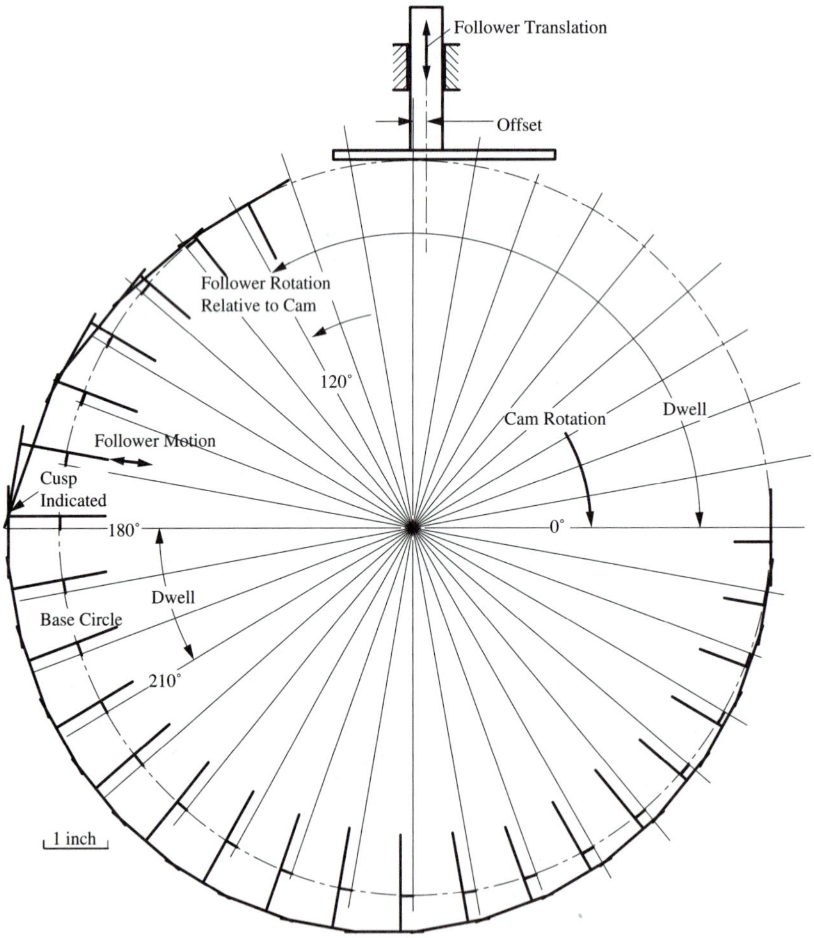

Figure 6.26 Envelope generated by drawing lines corresponding to the follower face for Example 6.4. The base circle is the minimum possible for the flat-faced follower to generate the profile indicated in Table 6.6.

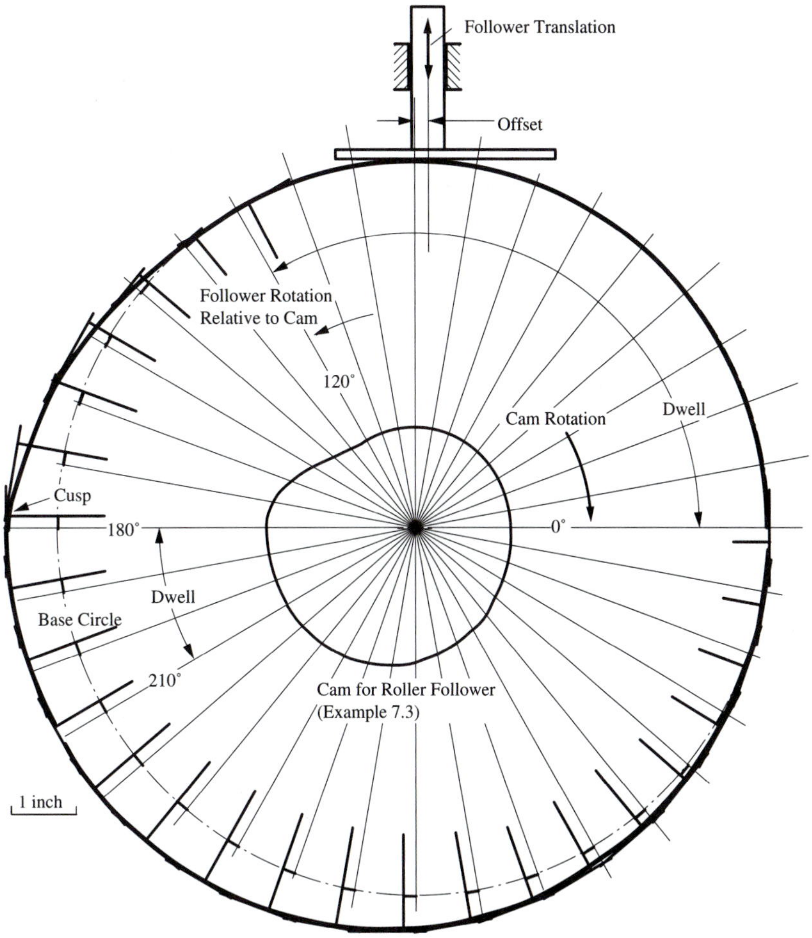

Figure 6.27 Cam generated for Example 6.4. The cam generated for the roller follower in Example 6.3 is included for comparison.

enough on both sides to maintain contact with the cam on a tangent line. The minimum length of the follower face can be established by direct measurement. The actual length would be equal to the minimum length plus a small margin. ■

EXAMPLE 6.5 *(Layout of Cam Profile for Oscillating, Cylindrical Follower)*

PROBLEM

Lay out a cam profile assuming that the follower starts from a dwell for 0° to 120° of cam rotation and the cam rotates clockwise. The rise occurs during the cam rotation from 120° to 200°. The follower then dwells for 40° of cam rotation, and the return occurs for the cam rotation from 240° to 360°. The amplitude of the follower rotation is 30°, and the follower radius is 0.75 inch. Lay out the cam profile using 20° plotting intervals. (This plotting interval is too coarse for the development of an accurate cam; however, it will be used to simplify the resulting drawing).

SOLUTION

The basic motion specification is visually similar to that shown in Fig. 6.15; however, the follower motion is a rotation instead of a translation. To begin the design, we must determine the follower rotation, ϕ, as a function of the cam rotation, θ. Assume parabolic motion for the follower.

Table 6.7 Follower Displacements for Example 6.5

θ	0, 360°	20°	40°	60°	80°	100°	120°	140°	160°
y	0.0000	0.0000	0.0000	0.0000	0.0000	0.0000	0.0000	3.7500°	15.0000°
θ	180°	200°	220°	240°	260°	280°	300°	320°	340°
y	26.2500°	30.0000°	30.0000°	30.0000°	28.3333°	23.3333°	15.0000°	6.6666°	1.6666°

From Section 6.6, the equations for each part of the motion are

$$0 < \theta \leq 120° \qquad \phi = 0°$$

$$120° \leq \theta \leq 160° \qquad \phi = 2L\left(\frac{\Delta\theta}{\beta}\right)^2 = 60\left(\frac{\theta - 120}{80}\right)^2$$

$$160° \leq \theta \leq 200° \qquad \phi = L\left[1 - 2\left(1 - \frac{\Delta\theta}{\beta}\right)^2\right] = 30\left[1 - 2\left(1 - \frac{\theta - 120}{80}\right)^2\right]$$

$$200° \leq \theta \leq 240° \qquad \phi = 30°$$

$$240° \leq \theta \leq 300° \qquad \phi = L\left[1 - 2\left(\frac{\Delta\theta}{\beta}\right)^2\right] = 30\left[1 - 2\left(\frac{\theta - 240}{120}\right)^2\right]$$

$$300° \leq \theta \leq 360° \qquad \phi = 2L\left(1 - \frac{\Delta\theta}{\beta}\right)^2 = 60\left(1 - \frac{\theta - 240}{120}\right)^2$$

Here, L is treated as the generic amplitude of the follower motion. In this case, L is given as 30°.

Using these equations, the lift values to be plotted are as given in Table 6.7. Notice that the dwells correspond to locations on the cam where the radius is constant.

The cam-follower system is similar to that shown in Fig. 6.28, and the points shown in Fig. 6.28 will be used to describe the layout of the cam. In particular, A is the location of the axis of rotation of the cam, B is the center of curvature of the follower, and C is the rotation axis of the follower.

The first step in the cam layout is to draw the base circle and prime circle. The *base circle* is chosen to have a 1.25-inch radius, and the radius of the *prime circle* is equal to $r_b + r_0 = 1.25 + 0.75 = 2.0$ in. The base circle radius is the radius of the cam during the dwell for the first 120° of cam rotation.

The second step is to select the distance from the cam rotation axis to the pivot of the follower (AC). The larger the value chosen, the smaller the pressure angle; however, this distance also directly affects the size of the cam-follower system. We will choose the distance between pivots to be 4 inches. When we invert the motion, the follower pivot will appear to rotate around the cam. Therefore, we must draw a circle with a radius of 4 inches about the cam for the follower pivot circle. The pivot circle is shown in Fig. 6.29.

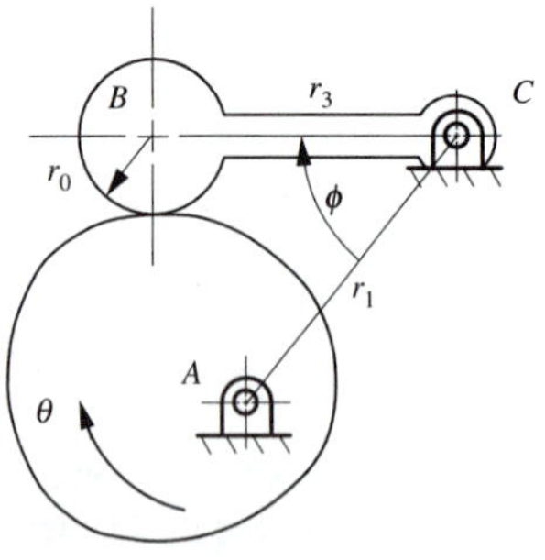

Figure 6.28 Cam-follower system to be designed in Example 6.5.

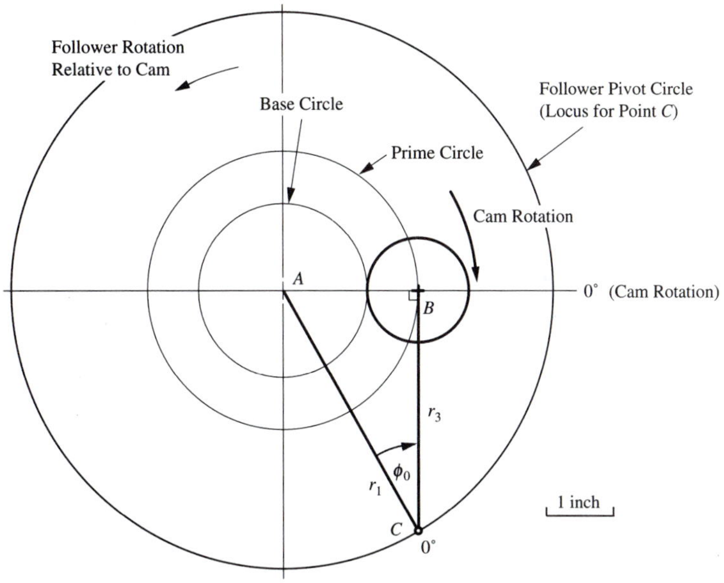

Figure 6.29 Location of the follower in the initial position. This determines the reference angle ϕ_0.

The third step is to determine the length of the follower ($BC = r_3$) and the reference angle ϕ_0 for ϕ. We will locate the follower in the initial dwell position in such a way that the pressure angle is zero. This is done by drawing a line tangent to the prime circle through the ray corresponding to $\theta = 0$. The intersection of this tangent line with the follower pivot circle will give the location of the follower pivot for the initial position of the cam. Two intersections will be given. One will correspond to a clockwise rotation of the follower, and the second will correspond to a counterclockwise rotation. Based on the problem statement, we will choose the intersection corresponding to the clockwise rotation as shown in Fig. 6.29. Because a right triangle is involved, the follower length is given by

$$r_3 = \sqrt{r_1^2 - (r_b + r_0)^2} = \sqrt{4^2 - 2^2} = 3.464$$

We also need to determine the base angle, ϕ_0, because all subsequent displacements of the follower will be measured relative to this angle. If we let the distance between the cam and follower pivots be r_1, the angle ϕ_0 is given directly by

$$\phi_0 = \tan^{-1}\left(\frac{r_b + r_0}{r_3}\right) = \tan^{-1}\left(\frac{2}{3.464}\right) = 30.000°$$

The location of pivot C for the follower when the cam angle is 0° gives the first position of the follower pivot. Subsequent positions of the pivot will be at angle increments of 20°. Therefore, the fourth step in the cam layout is to draw radial lines at 20-degree increments from the cam rotation axis to the follower pivot circle starting from the initial position of AC corresponding to the cam rotation angle of 0°. Label the radial lines corresponding to the beginning and end of dwells. These radial lines are shown in Fig. 6.30.

The fifth step is to draw a line tangent to the prime circle from the intersection of the radial lines and the follower pivot circle as shown in Fig. 6.30. These lines give the position of the follower relative to the cam, if the cam is a simple cylinder. These lines will give the base lines from which to measure the ϕ angles given in Table 6.7. Next lay off lines 3.464 inches in length at the angles indicated in Table 6.7 from the corresponding base lines. The ends of these lines will be the centers of the cylindrical cam follower in the different positions. Draw circles corresponding to the follower, and construct the cam surface tangent to the follower positions. The final cam is shown in Fig. 6.30.

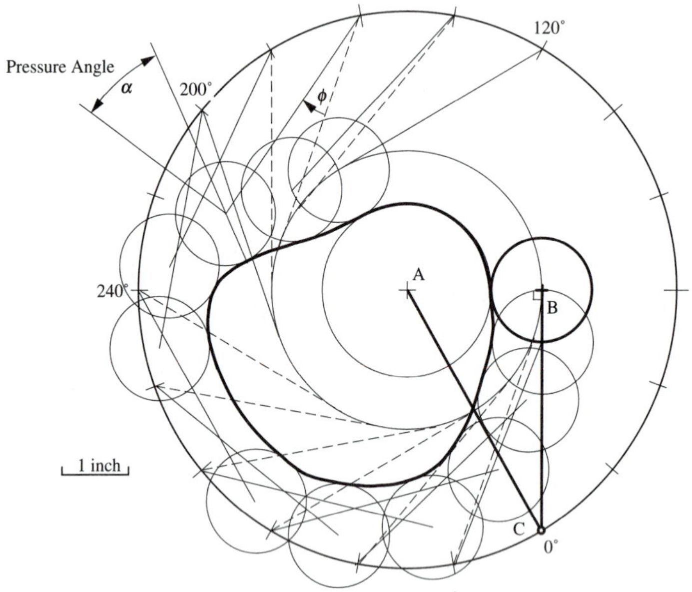

Figure 6.30 Final layout of the cam profile for Example 6.5.

As in the case of an axial roller follower, an important parameter for cam motion is the pressure angle. For the oscillating cylindrical follower, this is the angle between the velocity of the contact point on the follower (vector normal to line from contact point and point C) and the normal to the cam at the contact point. To reduce wear on the follower pivot, we want the pressure angle, α, to be as small as possible. In the design shown in Fig. 6.30, the pressure angle will become fairly high (greater than 30°) in the rise region. To improve the design, the diameter of the base circle should be increased.

The problem statement indicated that a cylindrical follower was to be designed. However, the construction would be *exactly* the same if a roller follower were involved. From the standpoint of the cam geometry, the important issues are the radius of the cylindrical face and the direction of motion relative to the cam. ■

EXAMPLE 6.6 *(Layout of Cam Profile for Oscillating Flat-Faced Follower)*

PROBLEM

Lay out the rise portion of the cam profile for the follower motion indicated in Table 6.7. The cam will rotate *counterclockwise.* Assume that the follower face angle is 170°, and lay out the cam profile using 20° plotting intervals. Again, this plotting interval is too coarse for an accurate cam; however, it will be used to illustrate the procedure.

SOLUTION

The cam-follower system is similar to that shown in Fig. 6.31, and the points shown in that figure will be used to describe the layout of the cam. Point A is the location of the axis of rotation of the cam, C is the rotation axis of the follower, and β is the follower face angle. Point B is the location of the intersection of the centerline of the follower stem with the face of the follower.

The first step in the cam layout is to select and draw the base circle. The base circle is chosen to be 2.0 in.

The second step is to select the distance from the cam rotation axis to the pivot of the follower (AC). The value chosen will affect the size of the cam, and typically the smallest value possible is chosen. The pivot distance must be large enough that the cam does not contact the follower pivot. Also, the force between the cam and follower will increase as the distance decreases. Therefore, it may be necessary to increase the pivot distance from machine design considerations.

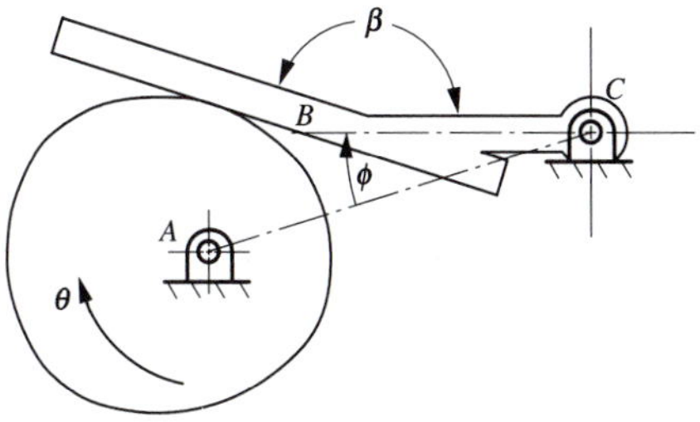

Figure 6.31 Cam follower system to be designed in Example 6.6.

We will choose the distance between pivots to be 4 inches. When we invert the motion, the follower pivot will appear to rotate around the cam. Therefore, we must draw a circle with a radius of 4 in about the cam for the follower pivot circle as shown in Fig. 6.32.

The third step is to determine the reference angle ϕ_0 for ϕ. The location of the follower to determine ϕ_0 will affect the length of the follower and therefore the cost of the system. For simplicity, we will locate the follower in the initial dwell position (cam angle $\theta = 0$) in such a way that point B contacts the cam in that position. This is done by drawing a line tangent to the base circle at B. Then construct a line at an angle of 170° to the tangent line at B. The intersection of this line with the follower pivot circle will locate C in the initial position. This is shown in Fig. 6.32. The length BC can then be computed using the law of cosines. That is,

$$BC^2 + AB^2 = AC^2 + 2(BC)(AB)\cos 100°$$

or

$$BC^2 - 2(BC)(AB)\cos 100° + (AB^2 - AC^2) = 0$$

and

$$BC = \frac{2(AB)\cos 100° + \sqrt{[2(AB)\cos 100°]^2 - 4(AB^2 - AC^2)}}{2} = \frac{-0.694 + \sqrt{48.482}}{2} = 3.134$$

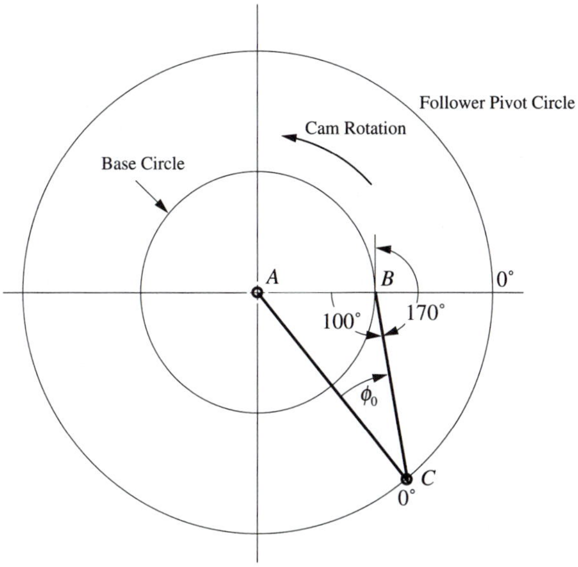

Figure 6.32 Determining the initial position of the follower.

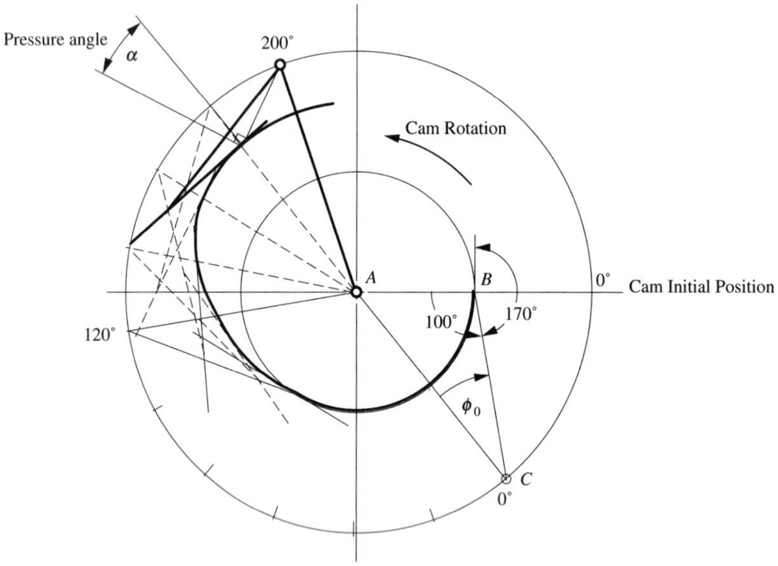

Figure 6.33 Final layout of the rise portion of the cam profile for Example 6.6.

The base angle, ϕ_0, can be computed using the law of sines. That is,

$$\phi_0 = \sin^{-1}\left[\frac{AB}{AC}\sin 100°\right] = \sin^{-1}\left[\frac{2}{4}\sin 100°\right] = 29.499°$$

This angle is important because all subsequent displacements of the follower will be measured relative to this angle.

The location of pivot C for the follower when the cam angle is 0 gives the first position of the follower pivot. Subsequent positions of the pivot will be at angle increments of 20° in the clockwise direction (opposite the cam rotation). Therefore, the fourth step in the cam layout is to draw radial lines at 20° increments from the cam rotation axis to the follower pivot circle starting from the initial position of AC corresponding to the cam rotation angle of 0°. These will be the positions of AC for each 20° rotation of the cam. Label the radial lines corresponding to the beginning and end of the first dwell and the second dwell. In the dwell regions, the cam will have a circular contour. These radial lines are shown in Fig. 6.33.

The fifth step is to draw the line CB relative to each position of AC where the angle between CB and AC is $(\phi_0 + \phi)$ measured in the clockwise direction about C. This will locate the successive positions of B. Next, draw a line through each position of B at an angle of 170° to BC. This will locate the successive positions of the face of the cam follower. Construct the cam surface tangent to the follower positions. The rise portion of the final cam is shown in Fig. 6.33.

Because of the follower face angle β, the pressure angle will not be constant as in the case of the axial flat-faced follower. The pressure angle will depend on the contact point as shown in Fig. 6.33 and will be equal to $10° \pm \zeta$, where ζ is the angle between a line from the contact point to C and the line BC. If the follower face angle is 180°, the pressure angle will be a constant, that is, 0°. Because of this, the angle β would be 180° unless specific design conditions dictated otherwise. ■

6.10.2 Analytical Determination of Cam Profile

Although the graphical approach works well for low-speed cams, for high speeds greater accuracy is required, making analytical techniques necessary. The analytical approach to determining the cam profiles also uses inversion. The general procedure is to establish a coordinate system on both the cam and the cam follower. The position of the follower is stationary relative to the follower coordinate system, and it should be possible to define the geometry of the follower relative to this coordinate system. Therefore, to determine the location of the follower relative to the cam, it is necessary only to determine the location of the moving (follower) coordinate system relative to the fixed (cam) coordinate system. Successive positions of the follower will generate an envelope that will define the geometry of the cam profile.

The approach to the design of each of the four different follower systems considered is slightly different, and in each case the analytical approach will follow much the same procedure as was illustrated in the graphical approach.

6.10.2.1 Analytical Determination of Cam Profile for an Offset, Radial Roller Follower

Prior to determining the geometry of the cam, it is assumed that the basic motion specification for the follower is known in the form $y = f(\theta)$, where y is the displacement of the follower and θ is the cam rotation angle. The base circle radius (r_b), the follower radius (r_0), and the offset δ are also assumed to be known. The procedure is the same for both a roller follower and a cylindrical-faced follower. We will first locate successive positions of the roller center and then determine the envelope formed by the rollers.

Two positions of the follower relative to the cam are shown in Fig. 6.34. In the figure, it is assumed that the cam rotates clockwise, which means that the follower moves relative to the cam in the counterclockwise direction. The radial position of the center of the follower from the origin of the coordinate system located at the center of the cam is given by

$$R = r_0 + r_b + f(\theta) \tag{6.15}$$

where $f(\theta)$ is the function defining the follower displacement as a function of the cam rotation angle, θ. Referring to Fig. 6.34,

$$x = R\cos\theta - \delta\sin\theta = [r_0 + r_b + f(\theta)]\cos\theta - \delta\sin\theta \tag{6.16}$$

and

$$y = R\sin\theta + \delta\cos\theta = [r_0 + r_b + f(\theta)]\sin\theta + \delta\cos\theta \tag{6.17}$$

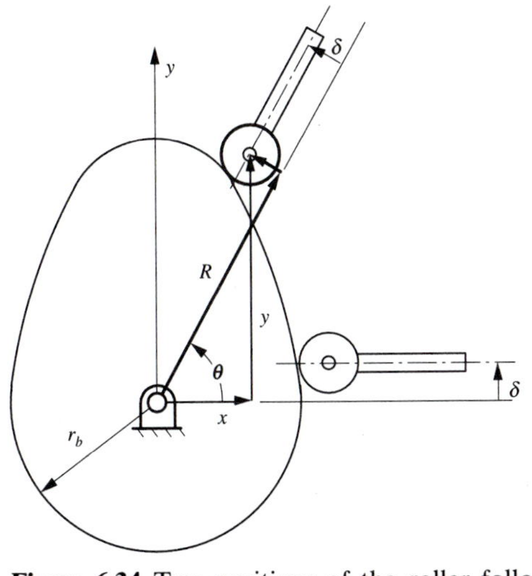

Figure 6.34 Two positions of the roller follower relative to cam.

Given the follower displacement, the center of the roller can be easily computed as a function of the cam rotation angle θ. These equations will also give the cam profile if a knife-edged follower is involved. Also, if the cutter radius is the same as the roller radius, the coordinates can be used to generate the cam profile directly. However, for other cases, we must determine the cam profile indirectly.

As the cam angle θ is incremented, the roller can be represented in a series of positions as shown in Fig. 6.35. In any given position, the coordinates of all points on the roller can be defined relative to the cam. The cam surface is tangent to the successive positions of the rollers. Therefore, to determine the cam profile, we must locate points on a curve that is tangent to the series of circles.

The series of circles will form an envelope that will define the cam profile. Using envelope theory, it is possible to define the profile exactly, and envelope theory is discussed briefly at the end of this chapter. However, envelope theory is somewhat complex except in the case of translating followers. Fortunately, if fine enough increments (less than 1°) are used, the cam profile can be determined very accurately using numerical techniques, and this is the approach that we will use here. The numerical techniques are much simpler than envelope theory, and once the path of the center of the follower is known, the cam profile can be determined without knowledge of whether an oscillating or translating follower was used.

Referring to Fig. 6.35, we can approximate the cam profile by a series of points defined in a variety of ways. Some of these are:

1. Intersections of roller circles (A)
2. Average tangent location formed by successive tangent lines (B)
3. Intersections of successive tangent lines (C)
4. Tangent points formed by a circle tangent to three successive positions of the roller (D)

Other schemes based on fitting other curves to the circles could also be developed. However, it is apparent that each of the procedures will converge to the true tangent points defining the cam profile if the successive positions of the follower are close enough together.

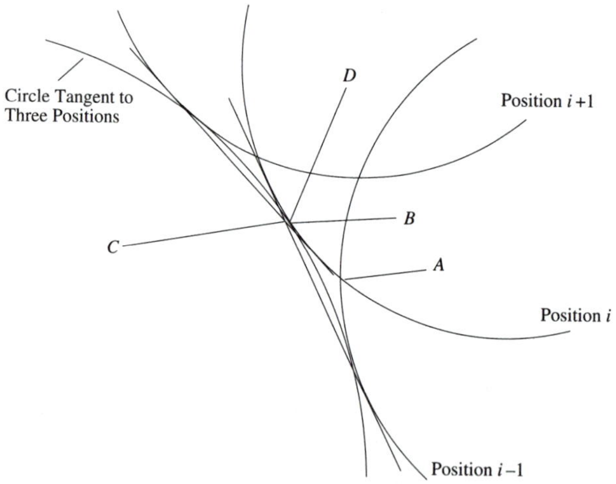

Figure 6.35 Approximate procedures for locating a point on the cam profile given three successive positions of the roller follower. Point A is found by the simple intersection of two successive positions. Point B is the average tangent location in position i, point C is the intersection of two successive tangents, and point D is the tangent location in position i of a circle tangent to positions $i - 1$, i, and $i + 1$.

Of the approximate methods given, procedures 2 and 4 above will be the most accurate, and the simpler of these is procedure 2. This is the procedure which will be developed here. This procedure is especially simple to program, and it can be made as accurate as desired by using increasingly smaller increments of the cam rotation θ. The basic procedure is illustrated in Fig. 6.36. For the procedure, assume that we know the coordinates of the center of each position of the roller circle relative to the cam. Number the center points from 1 to n, where n is the total number of positions.

To identify the point associated with position i, we need the (x, y) coordinates of the center of the roller in positions $i - 1$, i, and $i + 1$ as shown in Fig. 6.36. Given these coordinates, the point on the cam corresponding to position i can be approximated using the following procedure.

1. Determine the angles σ_i associated with the lines connecting successive positions of the roller center from the following:

$$\sigma_i = \tan^{-1}\left[\frac{y_{i+1} - y_i}{x_{i+1} - x_i}\right] \qquad (i = 1, 2, 3, \ldots, n - 1)$$

and

$$\sigma_n = \tan^{-1}\left[\frac{y_1 - y_n}{x_1 - x_n}\right]$$

2. Compute the angle β_i for the line from (x_i, y_i) to the tangent point on the roller for each position i. These are

$$\beta_i = \sigma_i + \frac{\pi}{2}$$

3. Compute the angle γ_i that bisects β_i and β_{i-1} at each point. Do this by first locating a vector in the direction of γ_i. Assuming the difference between β_i and β_{i-1} is small,

$$\mathbf{v}_i = \left(\frac{r_0}{2}(\cos\beta_i + \cos\beta_{i-1}), \frac{r_0}{2}(\sin\beta_i + \sin\beta_{i-1})\right)$$

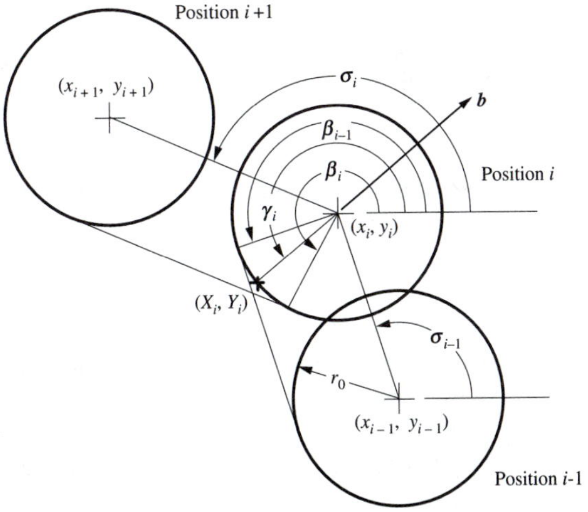

Figure 6.36 Approximating a point on the cam profile by using the average tangent location.

Then,

$$\gamma_1 = \tan^{-1}\left[\frac{\sin\beta_1 + \sin\beta_n}{\cos\beta_1 + \cos\beta_n}\right]$$

and

$$\gamma_i = \tan^{-1}\left[\frac{\sin\beta_i + \sin\beta_{i-1}}{\cos\beta_i + \cos\beta_{i-1}}\right] \qquad (i = 2, 3, \ldots, n)$$

Notice that we could compute the angle γ_i using $\gamma_i = \frac{1}{2}(\beta_i + \beta_{i-1})$. Although this is technically correct, it requires extra logic when the procedure is programmed. Problems arise because a given angle can be defined by a plus or minus number (e.g., the angle π is the same as $-\pi$) or by the angle plus 2π. If β_i uses one convention and β_{i-1} uses another, the expression $\gamma_i = \frac{1}{2}(\beta_i + \beta_{i-1})$ will not give the desired result. For example if $\beta_i = 0.001$, $\beta_{i-1} = 6.28$, the average will be 3.24, which will give a vector in the opposite direction to the one desired.

4. Compute the coordinates (X_i, Y_i) of the point on the cam using

$$X_i = x_i + r_0 \cos\gamma_i$$

and

$$Y_i = y_i + r_0 \sin\gamma_i$$

To compute the pressure angle at any given position, we must find the angle between a vector in the direction of the follower travel and a vector that is normal to the follower roller at each location. The direction of travel of the follower relative to the cam is simply θ, the cam rotation angle (see Fig. 6.34); therefore, a vector in that direction is

$$\boldsymbol{a} = 1\angle\theta$$

A vector in the direction normal to the roller is the line from (X_i, Y_i) to (x_i, y_i), which is simply $\gamma_i - \pi$ (see Fig. 6.36); therefore, the resulting vector is

$$\boldsymbol{b} = 1\angle(\gamma_i - \pi)$$

The resulting pressure angle is simply the angle between the vectors $\boldsymbol{a}$ and $\boldsymbol{b}$. That is,

$$\alpha_i = \cos^{-1}\left[\frac{\boldsymbol{a}\cdot\boldsymbol{b}}{|\boldsymbol{a}||\boldsymbol{b}|}\right] = \cos^{-1}[\cos\theta\cos(\gamma_i - \pi) + \sin\theta\sin(\gamma_i - \pi)] \qquad \textbf{(6.18)}$$

Notice that we could also compute the pressure angle directly as the angle $\gamma_i - \pi - \theta$. However, because both γ_i and θ can change sign as well as be larger than 2π, it is easier to compute the pressure angle using the slightly more complicated expression in Eq. (6.18). Using Eq. (6.18), all of the special cases are directly accommodated without special programming logic. If the calculation is to be done manually, however, the straightforward expression $(\gamma_i - \pi - \theta)$ is obviously preferable.

The equations necessary to determine points on the cam profile are summarized in Table 6.8. These can be easily programmed to determine the (X_i, Y_i) coordinates of points on the cam profile, and a MATLAB program for doing this is included on the disk with this book. Given the (X_i, Y_i) coordinates, the cam can be machined on a CNC milling machine. The accuracy of the profile will be determined in part by the angle increment chosen for the cam rotation angle θ.

Table 6.8 Summary of Equations for Determining the Cam Profile Coordinates and Pressure Angle for a Radial Cylindrical-Faced or Roller Follower. The Follower Displacement Is Assumed To Be Given by $f(\theta)$, the Radius of the Follower Is r_0, and the Base Circle Radius Is r_b. There Are Assumed To Be n Points on the Cam Profile

Cam Coordinates

$$x_i = [r_0 + r_b + f(\theta_i)]\cos\theta_i - \delta\sin\theta_i$$

$$y_i = [r_0 + r_b + f(\theta_i)]\sin\theta_i + \delta\cos\theta_i$$

$$\sigma_i = \tan^{-1}\left[\frac{y_{i+1} - y_i}{x_{i+1} - x_i}\right] \qquad (i = 1, 2, 3, \ldots, n-1),\ \sigma_n = \tan^{-1}\left[\frac{y_1 - y_n}{x_1 - x_n}\right]$$

$$\beta_i = \sigma_i + \frac{\pi}{2}$$

$$\gamma_1 = \tan^{-1}\left[\frac{\sin\beta_1 + \sin\beta_n}{\cos\beta_1 + \cos\beta_n}\right], \qquad \gamma_i = \tan^{-1}\left[\frac{\sin\beta_i + \sin\beta_{i-1}}{\cos\beta_i + \cos\beta_{i-1}}\right] \qquad (i = 2, 3, \ldots, n)$$

$$X_i = x_i + r_0\cos\gamma_i$$

$$Y_i = y_i + r_0\sin\gamma_i$$

Pressure Angle

$$\alpha_i = \cos^{-1}[\cos\theta\cos(\gamma_i - \pi) + \sin\theta\sin(\gamma_i - \pi)]$$

Radius of Curvature

$$\begin{Bmatrix} x_c \\ y_c \end{Bmatrix} = \begin{bmatrix} 2(X_{i+1} - X_i) & 2(Y_{i+1} - Y_i) \\ 2(X_{i-1} - X_i) & 2(Y_{i-1} - Y_i) \end{bmatrix}^{-1} \begin{Bmatrix} (X_{i+1}^2 - X_i^2) + (Y_{i+1}^2 - Y_i^2) \\ (X_{i-1}^2 - X_i^2) + (Y_{i-1}^2 - Y_i^2) \end{Bmatrix}$$

$$\rho = \frac{[(X_i - X_{i-1})(Y_{i+1} - Y_i) - (X_{i+1} - X_i)(Y_i - Y_{i-1})]}{|(X_i - X_{i-1})(Y_{i+1} - Y_i) - (X_{i+1} - X_i)(Y_i - Y_{i-1})|}\sqrt{(X_i - x_c)^2 + (Y_i - y_c)^2}$$

6.10.2.2 *Cam Radius of Curvature*

The radius of curvature for a cam is important for several reasons. These include:

- If the cam is concave in a given area, the radius of curvature determines the maximum diameter of the cutter which can be used to machine the cam. The radius of curvature of the cam cannot be smaller than the cutter radius if the cam is concave in the area being machined.
- If the cam is concave in a given area, the radius of curvature defines the maximum diameter of the follower which can be used with the cam.
- The contact stresses between the cam and the follower are a function of the cam radius of curvature.

If we have a parametric expression for the cam geometry in terms of $x(\theta)$ and $y(\theta)$, an expression for the radius of curvature, ρ, of any curve in the xy plane is given from calculus as

$$\rho = \frac{\sqrt{[(dx/d\theta)^2 + (dy/d\theta)^2]^3}}{(dx/d\theta)(d^2y/d\theta^2) - (dy/d\theta)(d^2x/d\theta^2)} \qquad \textbf{(6.19)}$$

To determine ρ, we need only differentiate the given expressions for $x(\theta)$ and $y(\theta)$ and substitute the resulting expressions into Eq. (6.19).

The radius of curvature given in Eq. (6.19) will have a magnitude and direction, and both are important. When the cam profile is convex, the radius of curvature is positive, and when the cam is concave, the radius of curvature is negative. The radius can also be zero, which corresponds to a cusp. Technically, the cusp can be concave or convex, but

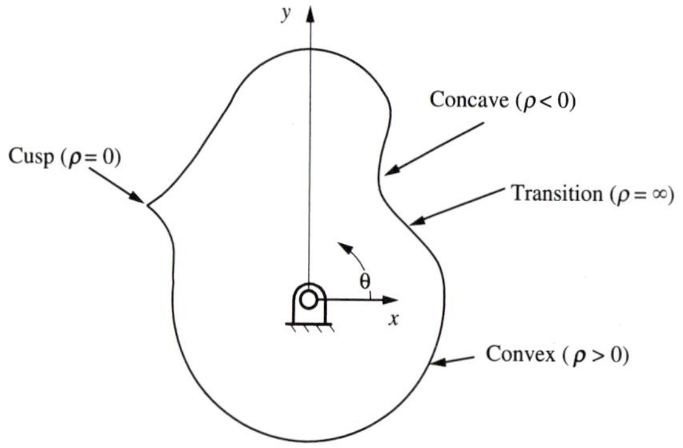

Figure 6.37 Interpretation of the sign of ρ.

from a practical standpoint, cusps are of interest only for convex surfaces. The transition between concave and convex areas of the cam results in the radius of curvature becoming infinite. The physical interpretations of the various signs of ρ are shown in Fig. 6.37.

Equation (6.19) applies to any cam profile regardless of the type of follower. However, to apply it directly, an analytical expression for $x(\theta)$ and $y(\theta)$ must be available. Except in the case of a translating flat-faced follower, simple expressions for $x(\theta)$ and $y(\theta)$ generally will not be available. Because of this, the purely analytical approach to determining the radius of curvature will not be addressed here except for this special case. Alternatively, a simpler numerical approach will be used.

If the points defining the cam profile are relatively close together, the radius of curvature can be determined numerically. This can be done by evaluating the derivatives in Eq. (6.19) numerically or by fitting circles to the points of the cam profile. Often an approximate value for the radius of curvature is sufficient because in most cases we do not wish to design cams that are close to the operating limits defined by the radii of curvature.

To determine an approximate location for the center of curvature $p_c = (x_c, y_c)$ corresponding to any point (X_i, Y_i) on the cam profile, we will fit a circle to three successive points. This can be done using the procedures given in Section 4.2.7. Given three successive points $p_{i-1} = (X_{i-1}, Y_{i-1})$, $p_i = (X_i, Y_i)$, and $p_{i+1} = (X_{i+1}, Y_{i+1})$ on the cam profile, Eq. (4.16) in Chapter 4 can be rewritten as

$$2\begin{bmatrix}(X_{i+1} - X_i) & (Y_{i+1} - Y_i)\\(X_{i-1} - X_i) & (Y_{i-1} - Y_i)\end{bmatrix}\begin{Bmatrix}x_c\\y_c\end{Bmatrix} = \begin{Bmatrix}(X_{i+1}^2 - X_i^2) + (Y_{i+1}^2 - Y_i^2)\\(X_{i-1}^2 - X_i^2) + (Y_{i-1}^2 - Y_i^2)\end{Bmatrix} \tag{6.20}$$

This linear matrix equation can be solved easily for (x_c, y_c). The magnitude of the radius of curvature is given by

$$\rho = |\boldsymbol{r}_{p_i/p_c}| = \sqrt{(X_i - x_c)^2 + (Y_i - y_c)^2}$$

Once the center of curvature is known, the sign of the radius of curvature can be found by taking the cross-product of the vector $\boldsymbol{r}_{p_i/p_{i-1}}$ with the vector $\boldsymbol{r}_{p_{i+1}/p_i}$. The vectors are shown in Fig. 6.38, and the cross-product is given by

$$\boldsymbol{r}_{p_i/p_{i-1}} \times \boldsymbol{r}_{p_{i+1}/p_i} = [(X_i - X_{i-1})(Y_{i+1} - Y_i) - (X_{i+1} - X_i)(Y_i - Y_{i-1})]\boldsymbol{k}$$

If the cross-product is in the positive $\boldsymbol{k}$ direction, the radius of curvature is positive, and the cam is convex at (X_i, Y_i). If the cross-product is in the negative $\boldsymbol{k}$ direction, the cam is concave. If the determinant of the coefficient matrix in Eq. (6.20) is zero, then (x_c, y_c) is at infinity, and $\boldsymbol{r}_{p_i/p_c}$ is infinite, and an inflection point is indicated. If $(x_c, y_c) = (X_i, Y_i)$, the radius of curvature is zero.

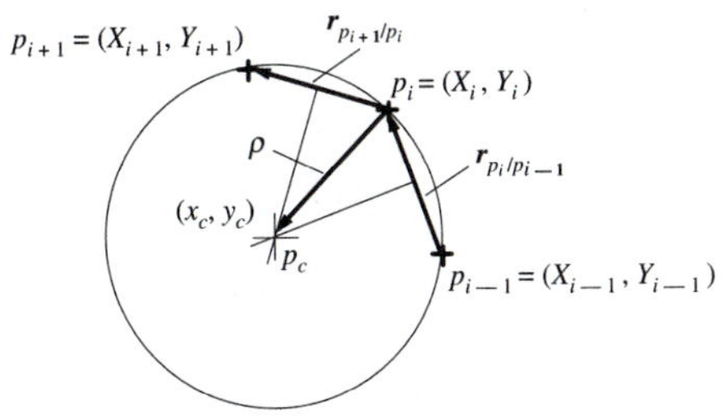

Figure 6.38 Use of vectors to determine the sign of ρ.

The equations for the magnitude and direction of the radius of curvature are summarized in Table 6.8.

EXAMPLE 6.7 ***(Cam Profile Coordinates for Radial, Roller Follower)***

PROBLEM

Determine the cam profile assuming that the follower starts from a dwell from 0° to 90° and rotates clockwise. The rise occurs with cycloidal motion during the cam rotation from 90° to 180°. The follower then dwells for 60° of cam rotation, and the return occurs with simple harmonic motion for the cam rotation from 240° to 360°. The amplitude of the follower translation is 2 cm, and the follower radius is 1 cm. The base circle radius is 4 cm, and the offset is 0.5 cm.

SOLUTION

To solve the problem, we must identify the equations for the follower motion as a function of the cam rotation angle θ and then select an increment for θ. From Sections 6.7 and 6.8, the

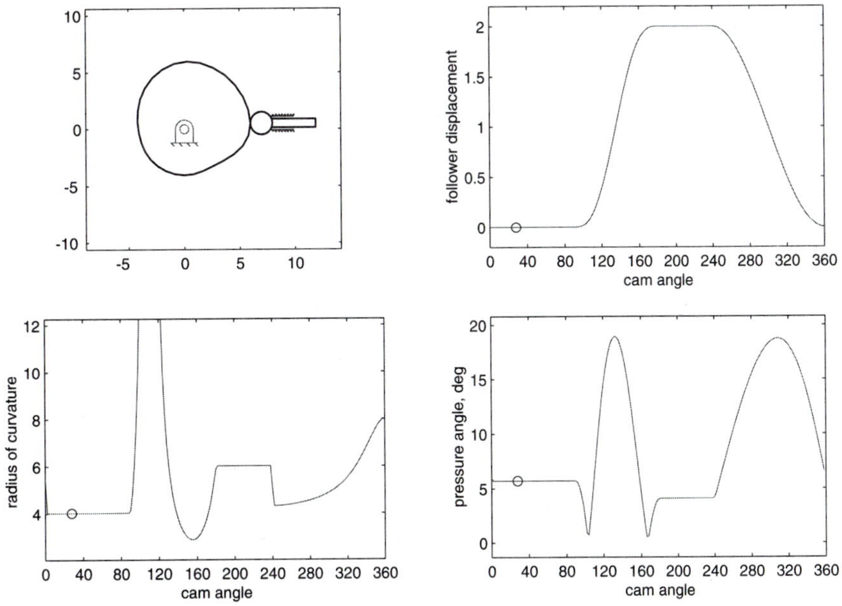

Figure 6.39 Cam profile, follower displacement, radius of curvature, and pressure angle for Example 6.7.

equations (expressed in terms of radians) for each part of the motion are

$$0 \le \theta \le \pi/2 \qquad y = 0$$

$$\pi/2 \le \theta \le \pi \qquad f(\theta) = \frac{2L}{\pi}\left(\left(\theta - \frac{\pi}{2}\right) - \frac{1}{4}\sin 4\left(\theta - \frac{\pi}{2}\right)\right)$$

$$\pi \le \theta \le 4\pi/3 \qquad f(\theta) = 2, \qquad f'(\theta) = f''(\theta) = 0$$

$$4\pi/3 \le \theta \le 2\pi \qquad f(\theta) = \frac{L}{2}\left(1 + \cos\frac{3}{2}\left(\theta - \frac{4\pi}{3}\right)\right)$$

These equations correspond to $f(\theta)$ in Table 6.8, and using the equations in Table 6.8 we can compute the coordinates of the cam as accurately as we wish. For this problem, an angle increment of 0.1° was used. The problem was solved using the program included on the disk with this book. With this program, it is possible to evaluate quickly the cam profile for areas where the follower roller is too large. We can also quickly evaluate the pressure angle using the equation in Table 6.8. The cam, displacement diagram, radius of curvature, and pressure angle plots are shown in Fig. 6.39. ■

6.10.2.3 Analytical Determination of the Cam Profile for a Radial Flat-Faced Follower

The development of the equations for a radial flat-faced follower can be accomplished using a procedure similar to that for a roller follower. Again, we must invert the mechanism so that the follower appears to rotate about the cam.

The radial displacement of the follower from the origin of the coordinate system located at the center of the cam is given by

$$R = r_b + f(\theta) \tag{6.21}$$

where $f(\theta)$ is the function defining the follower displacement as a function of the cam rotation angle, θ. Referring to Fig. 6.40, Eq. (6.21) can also be written as

$$R = y \sin\theta + x \cos\theta$$

If t is the distance from the follower axis to the point of contact,

$$t = y\cos\theta - x\sin\theta = \frac{dR}{d\theta} = f'(\theta)$$

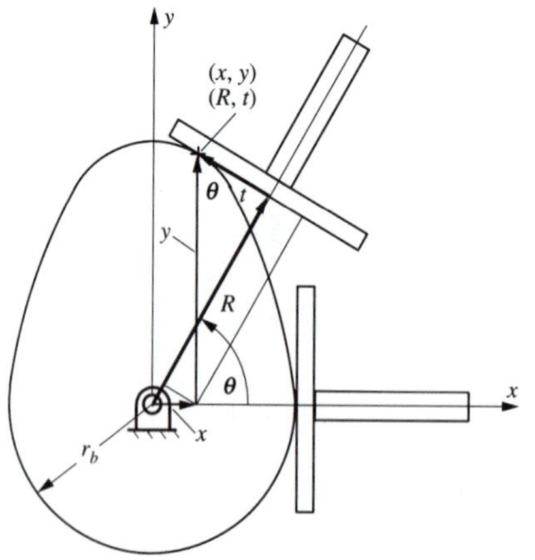

Figure 6.40 Radial flat-faced follower.

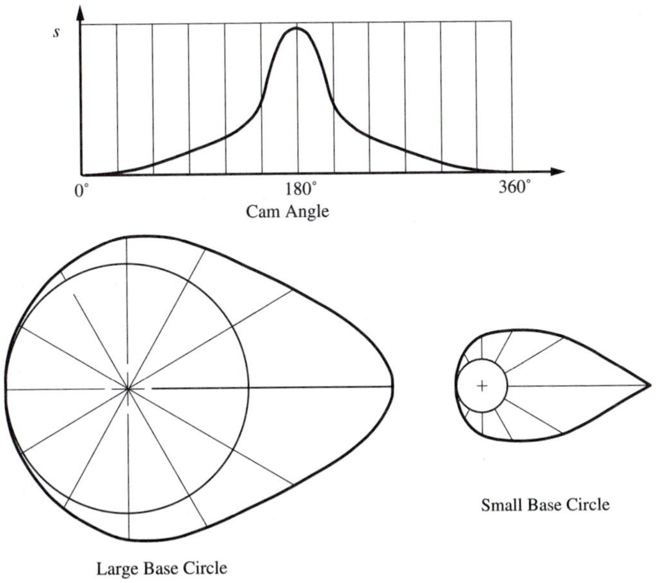

Figure 6.41 Formation of a cusp on the surface of the cam.

The maximum and minimum values for t will give the minimum limits for the length of the follower face. Solving for x and y gives

$$x = R\cos\theta - t\sin\theta$$
$$y = R\sin\theta + t\cos\theta$$

and in terms of $f(\theta)$,

$$\begin{aligned} x &= [r_b + f(\theta)]\cos\theta - f'(\theta)\sin\theta \\ y &= [r_b + f(\theta)]\sin\theta + f'(\theta)\cos\theta \end{aligned} \quad \textbf{(6.22)}$$

In general, we want to use the smallest base circle that will satisfy the geometric constraints. Normally, the base circle is determined in part by the pressure angle; however, for a radial flat-faced follower, the pressure angle *is always zero.* Therefore, we must select another criterion for determining the base circle radius. The minimum base circle radius will be the one that avoids cusps in the cam profile.

For a given displacement profile, the cam becomes sharper and sharper as the base circle radius decreases. This is shown in Fig. 6.41. Eventually, the cam surface will generate a sharp point or cusp. This condition gives the limiting radius for the base circle.

To establish an equation for the base circle, we can use Eq. (6.19) directly because we have analytical expressions for $x(\theta)$ and $y(\theta)$. For Eq. (6.19), we need the first and second derivatives of $x(\theta)$ and $y(\theta)$. These are given in the following:

$$\frac{\partial x}{\partial\theta} = f'(\theta)\cos\theta - [r_b + f(\theta)]\sin\theta - f''(\theta)\sin\theta - f'(\theta)\cos\theta = -[r_b + f(\theta) + f''(\theta)]\sin\theta$$

and

$$\frac{\partial^2 x}{\partial\theta^2} = -[f'(\theta) + f'''(\theta)]\sin\theta - [r_b + f(\theta) + f''(\theta)]\cos\theta$$

Also,

$$\frac{\partial y}{\partial\theta} = f'(\theta)\sin\theta + [r_b + f(\theta)]\cos\theta + f''(\theta)\cos\theta - f'(\theta)\sin\theta = [r_b + f(\theta) + f''(\theta)]\cos\theta$$

and

$$\frac{\partial^2 y}{\partial \theta^2} = [f'(\theta) + f'''(\theta)] \cos \theta - [r_b + f(\theta) + f''(\theta)] \sin \theta$$

Substitution into Eq. (6.19) gives

$$\rho = \frac{\sqrt{[(dx/d\theta)^2 + (dy/d\theta)^2]^3}}{(dx/d\theta)(d^2y/d\theta^2) - (dy/d\theta)(d^2x/d\theta^2)} = [r_b + f(\theta) + f''(\theta)]$$

For there to be no cusp (and for the cam to be convex) at all locations

$$\rho \geq 0$$

or

$$[r_b + f(\theta) + f''(\theta)] \geq 0 \tag{6.23}$$

for all θ. Both $f(\theta)$ and $f''(\theta)$ will be determined by the follower displacement schedule. Therefore, only r_b can be externally controlled in Eq. (6.23); that is,

$$r_b \geq -f(\theta) - f''(\theta)$$

If $-f(\theta) - f''(\theta)$ is negative, then any positive or zero value of the base circle radius will be acceptable from a cusp standpoint.

If the cam rotates counterclockwise, the equations developed must be modified slightly. In general, we will treat the follower displacement as a function of the angular displacement rather than the absolute angle. That is, the follower is assumed to translate a given distance regardless of the direction of rotation of the cam. The equations of Sections 6.3–6.9 were derived assuming that the cam angles were positive. When the cam rotates in the counterclockwise direction relative to the frame, the follower will rotate clockwise relative to the cam. This results in negative values for θ. However, the expressions for $f(\theta)$ must use the absolute values of θ. The effect of this is that the correct sign of the derivative $f'(\theta)$ is not preserved for counterclockwise rotation of the cam. The problem is resolved by changing the sign of the derivative term in Eq. (6.22). That is, when a counterclockwise

Table 6.9 Summary of Equations for Determining the Cam Profile Coordinates, Minimum Face Length, and Minimum Base Circle Radius for a Radial Flat-Faced Follower. The Follower Displacement is Assumed to be Given by $f(\theta)$, and the Base Circle Radius is r_b.

Cam coordinates—clockwise rotation of cam
$x_i = [r_b + f(\theta_i)] \cos \theta_i - f'(\theta_i) \sin \theta_i$
$y_i = [r_b + f(\theta_i)] \sin \theta_i + f'(\theta_i) \cos \theta_i$
Cam coordinates—counterclockwise rotation of cam
$x = [r_b + f(\theta)] \cos \theta + f'(\theta) \sin \theta$
$y = [r_b + f(\theta)] \sin \theta - f'(\theta) \cos \theta$
Minimum face length
$l_{\max,\min} = f'(\theta_i)\vert_{\max,\min}$
Radius of curvature
$\rho = [r_b + f(\theta) + f''(\theta)]$

rotation of the cam is involved, the cam coordinates are given by

$$x = [r_b + f(\theta)] \cos \theta + f'(\theta) \sin \theta$$
$$y = [r_b + f(\theta)] \sin \theta - f'(\theta) \cos \theta$$

The equations necessary to determine points on the cam profile and to determine the length of the follower face are summarized in Table 6.9. These can easily be programmed to determine the (x_i, y_i) coordinates of points on the cam profile, and a MATLAB program for doing this is included on the disk with this book. Given the (x_i, y_i) coordinates, the cam can be machined on a CNC milling machine. The accuracy of the profile will be determined in part by the angle increment chosen for the cam rotation angle θ.

EXAMPLE 6.8 *(Cam Profile Coordinates for Radial, Flat-Faced Follower)*

PROBLEM

Determine the cam profile for the follower motion given in Example 6.7. First find the minimum base circle radius based on avoiding cusps, and use that base circle to design the cam.

SOLUTION

To solve the problem, we must determine the derivatives of the functions for the follower displacement. Because derivatives are involved, the angles in the displacement functions will

Table 6.10 Values for $f(\theta)$ and Its Derivatives for Follower Displacement Specified for Example 6.8

θ	$f(\theta)$ (cm)	$f'(\theta)$ (cm)	$f''(\theta)$ (cm)	r_{bmin} (cm)	t (cm)
0°	0.000	0.000	0.000	0.000	0.000
			–dwell–		
90°	0.000	0.000	0.000	0.000	0.000
100°	0.018	0.298	3.274	−3.291	0.298
110°	0.131	1.052	5.016	−5.147	1.052
120°	0.391	1.910	4.411	−4.802	1.910
130°	0.780	2.470	1.742	−2.522	2.470
140°	1.220	2.470	−1.742	0.522	2.470
150°	1.609	1.910	−4.411	2.802	1.910
160°	1.869	1.052	−5.016	3.147	1.052
170°	1.982	0.298	−3.274	1.291	0.298
180°	2.000	0.000	0.000	−2.000	0.000
			–dwell–		
240°	2.000	−0.000	−2.250	0.250	−0.000
250°	1.966	−0.388	−2.173	0.207	−0.388
260°	1.866	−0.750	−1.949	0.083	−0.750
270°	1.707	−1.061	−1.591	−0.116	−1.061
280°	1.500	−1.299	−1.125	−0.375	−1.299
290°	1.259	−1.449	−0.582	−0.676	−1.449
300°	1.000	−1.500	0.000	−1.000	−1.500
310°	0.741	−1.449	0.582	−1.324	−1.449
320°	0.500	−1.299	1.125	−1.625	−1.299
330°	0.293	−1.061	1.591	−1.884	−1.061
340°	0.134	−0.750	1.949	−2.083	−0.750
350°	0.034	−0.388	2.173	−2.207	−0.388

be converted to radians. For the different intervals, the functions and derivatives are summarized in the following:

$0 \le \theta \le \pi/2$

$$f(\theta) = f'(\theta) = f''(\theta) = 0$$

$\pi/2 \le \theta \le \pi$

$$f(\theta) = \frac{2L}{\pi}\left(\left(\theta - \frac{\pi}{2}\right) - \frac{1}{4}\sin 4\left(\theta - \frac{\pi}{2}\right)\right), \qquad f'(\theta) = \frac{2L}{\pi}\left(1 - \cos 4\left(\theta - \frac{\pi}{2}\right)\right),$$

$$f''(\theta) = \frac{8L}{\pi}\left(\sin 4\left(\theta - \frac{\pi}{2}\right)\right)$$

$\pi \le \theta \le 4\pi/3$

$$f(\theta) = 2, \qquad f'(\theta) = f''(\theta) = 0$$

$4\pi/3 \le \theta \le 2\pi$

$$f(\theta) = \frac{L}{2}\left(1 + \cos\frac{3}{2}\left(\theta - \frac{4\pi}{3}\right)\right), \qquad f'(\theta) = -\frac{3L}{4}\sin\frac{3}{2}\left(\theta - \frac{4\pi}{3}\right),$$

$$f''(\theta) = -\frac{9L}{8}\cos\frac{3}{2}\left(\theta - \frac{4\pi}{3}\right)$$

Given the equations for the follower displacement, we now need only to increment θ and to evaluate the expressions in Table 6.9 to determine the minimum base circle radius and limiting values for the face length. We can then determine the cam coordinates. The values are computed for 10° increments of θ in Table 6.10. From that table, it is clear that the base circle radius must be at least 3.147 cm and the follower face needs to be at least 2.47 cm above the centerline

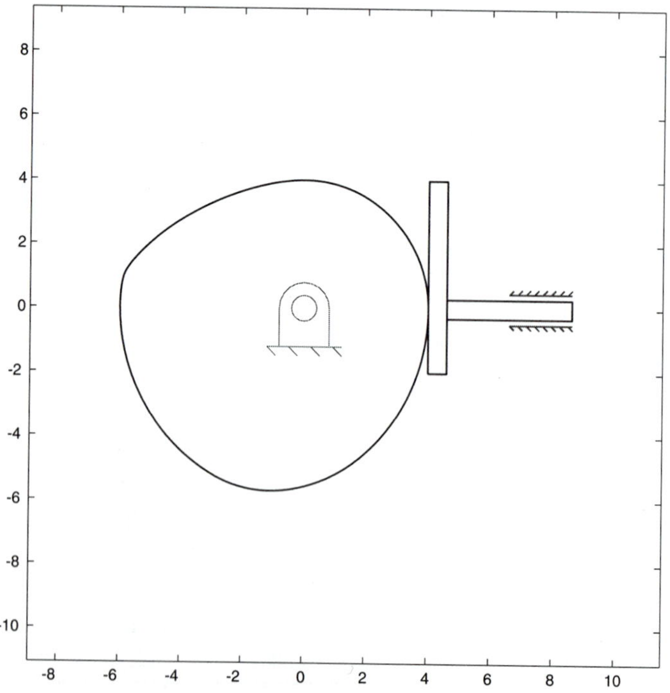

Figure 6.42 Cam designed for Example 6.8.

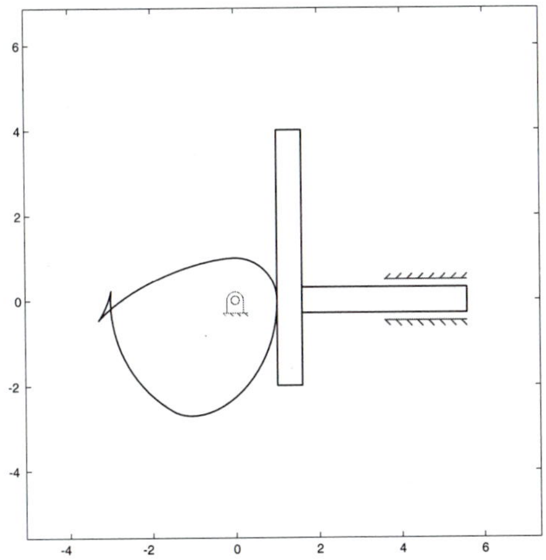

Figure 6.43 Results for Example 6.8 when a base circle radius of 1 cm is used. This cam cannot be manufactured. The smallest value for the base circle radius to avoid cusps is 3.147.

and 1.5 cm below it. Note that a negative value for t implies a distance below the stem centerline, and a positive value implies a distance above the centerline.

The cam will be designed with a base circle radius of 3.2 cm, and the follower face will be 2.6 cm above the stem centerline and 1.6 cm below it. The results obtained from the MATLAB program included on the disk with this book are shown in Fig. 6.42. To illustrate the effect of choosing a base circle radius less than the critical value, the program was rerun with the base circle radius of 1 cm. The results are shown in Fig. 6.43. The result is clearly a cam that cannot be manufactured. ■

6.10.2.4 Analytical Determination of a Cam Profile for an Oscillating Cylindrical Follower

The equations for an oscillating cylindrical follower can be derived on the basis of the graphical procedure. Again, we must invert the mechanism so that the follower appears to rotate about the cam. To begin the procedure, we will assume that the cam base radius r_b, the radius r_0 of the cylindrical follower, the pivot distance r_1, and the distance r_3 from the follower pivot to the center of the cylindrical contour are known.

In the procedure, the prime curve will be located first. This is the path traced by the center of curvature of the cylindrical follower. The cam profile can then be determined using the procedure developed in Section 6.10.2.1. The follower is assumed to begin in its lowest position when the follower contacts the base circle on the cam as shown in Fig. 6.44. For this position, we need to determine the initial angles, θ_0 and ϕ_0, where θ_0 gives the initial angle for the cam and ϕ_0 gives the initial angle for the follower. The motion of the cam and follower will be measured relative to these initial angles, respectively. From Fig. 6.44a, these angles can be computed using the law of cosines. That is,

$$\phi_0 = \cos^{-1}\left[\frac{r_1^2 + r_3^2 - (r_b + r_0)^2}{2r_1r_3}\right] \tag{6.24}$$

and

$$\theta_0 = \cos^{-1}\left[\frac{r_1^2 + (r_b + r_0)^2 - r_3^2}{2r_1(r_b + r_0)}\right] \tag{6.25}$$

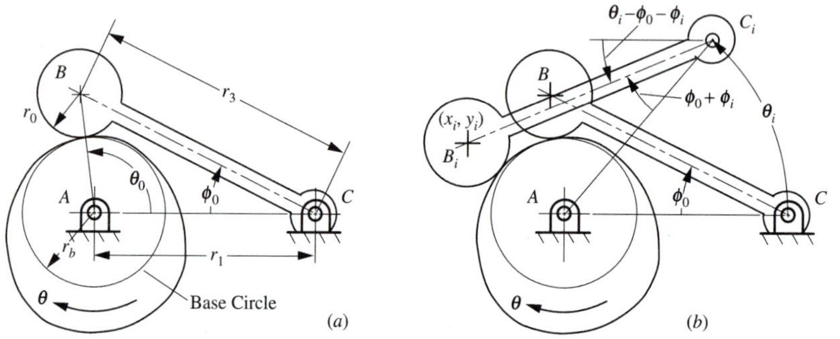

Figure 6.44 Oscillating roller-raced follower. (a) Initial geometry; (b) displaced geometry.

To compute the coordinates of the center of the follower (B) relative to a coordinate system at point A on the cam, we will assume that the cam rotates clockwise so that the follower rotates *counterclockwise* relative to the cam.

Referring to Fig. 6.44b, for a given cam rotation angle θ_i, the coordinates of point B_i are given by

$$x_i = r_1 \cos\theta_i + r_3 \cos[\pi + \theta_i - \phi_i - \phi_0] \tag{6.26}$$

and

$$y_i = r_1 \sin\theta_i + r_3 \sin[\pi + \theta_i - \phi_i - \phi_0] \tag{6.27}$$

Given the coordinates of B_i for a series of cam rotation angles (θ_i), we can compute the coordinates of the corresponding contact points (X_i, Y_i) on the cam face using the procedure in Section 6.10.2.1.

To compute the pressure angle at any given position, we must find the angle between a vector in the direction of the follower travel and a vector that is normal to the follower roller at each location as shown in Fig. 6.45. The direction of travel of the follower pivot (pointc) relative to the cam is simply θ, the cam rotation angle (see Fig. 6.45); therefore, a vector in the direction of travel of point B relative to the cam is

$$\boldsymbol{a} = 1\angle\psi_i$$

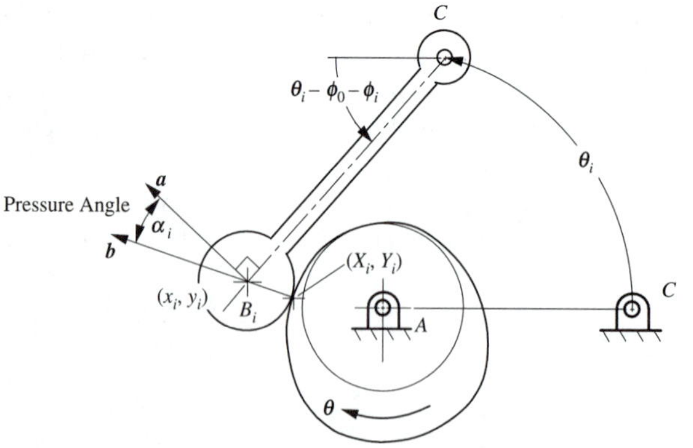

Figure 6.45 Pressure angle for oscillating roller-raced follower.

where

$$\psi_i = \theta_i - \phi_i - \phi_0 + \pi/2$$

A vector in the direction normal to the roller is the line from (X_i, Y_i) to (x_i, y_i); therefore, the resulting vector is

$$\boldsymbol{b} = 1\angle\lambda_i$$

where

$$\lambda_i = \tan^{-1}\left[\frac{y_i - Y_i}{x_i - X_i}\right]$$

The resulting pressure angle is simply the angle between the vectors $\boldsymbol{a}$ and $\boldsymbol{b}$. That is,

$$\alpha_i = \cos^{-1}\left[\frac{\boldsymbol{a}\cdot\boldsymbol{b}}{|\boldsymbol{a}||\boldsymbol{b}|}\right] = \cos^{-1}[\cos\psi_i \cos\lambda_i + \sin\psi_i \sin\lambda_i] \tag{6.28}$$

The equations to determine points on the cam profile are summarized in Table 6.11. In Table 6.11, the equations for computing (X_i, Y_i) and for the radius of curvature of the cam are taken from Table 6.8. These can be programmed easily to determine the (X_i, Y_i)

Table 6.11 Summary of Equations for Determining the Cam Profile Coordinates and Pressure Angle for an Oscillating Cylindrical-Faced or Roller Follower. The Follower Oscillation ϕ Is Assumed To Be Given by $f(\theta)$, the Radius of the Follower Is r_0, the Base Circle Radius Is r_b, the Distance Between the Cam and Follower Pivots Is r_1, and the Length of the Follower Is r_3. There Are Assumed To Be n Points on the Cam Profile.

Cam coordinates

$$\phi_0 = \cos^{-1}\left[\frac{r_1^2 + r_3^2 - (r_b + r_0)^2}{2r_1 r_3}\right]$$

$$x_i = r_1\cos\theta_i + r_3\cos[\pi + \theta_i - \phi_i - \phi_0], \qquad y_i = r_1\sin\theta_i + r_3\sin[\pi + \theta_i - \phi_i - \phi_0]$$

$$\sigma_i = \tan^{-1}\left[\frac{y_{i+1} - y_i}{x_{i+1} - x_i}\right] \quad (i = 1, 2, 3, \ldots, n-1), \qquad \sigma_n = \tan^{-1}\left[\frac{y_1 - y_n}{x_1 - x_n}\right]$$

$$\beta_i = \sigma_i + \frac{\pi}{2}$$

$$\gamma_1 = \tan^{-1}\left[\frac{\sin\beta_1 + \sin\beta_n}{\cos\beta_1 + \cos\beta_n}\right], \qquad \gamma_i = \tan^{-1}\left[\frac{\sin\beta_i + \sin\beta_{i-1}}{\cos\beta_i + \cos\beta_{i-1}}\right] \quad (i = 2, 3, \ldots, n)$$

$$X_i = x_i + r_0\cos\gamma_i, \qquad Y_i = y_i + r_0\sin\gamma_i$$

Pressure angle

$$\psi_i = \theta_i - \phi_i - \phi_0 + \pi/2$$

$$\lambda_i = \tan^{-1}\left[\frac{y_i - Y_i}{x_i - X_i}\right]$$

$$\alpha_i = \cos^{-1}[\cos\psi_i \cos\lambda_i + \sin\psi_i \sin\lambda_i]$$

Radius of curvature

$$\begin{Bmatrix} x_c \\ y_c \end{Bmatrix} = \begin{bmatrix} 2(X_{i+1} - X_i) & 2(Y_{i+1} - Y_i) \\ 2(X_{i-1} - X_i) & 2(Y_{i-1} - Y_i) \end{bmatrix}^{-1} \begin{Bmatrix} (X_{i+1}^2 - X_i^2) + (Y_{i+1}^2 - Y_i^2) \\ (X_{i-1}^2 - X_i^2) + (Y_{i-1}^2 - Y_i^2) \end{Bmatrix}$$

$$\rho = \frac{[(X_i - X_{i-1})(Y_{i+1} - Y_i) - (X_{i+1} - X_i)(Y_i - Y_{i-1})]}{|(X_i - X_{i-1})(Y_{i+1} - Y_i) - (X_{i+1} - X_i)(Y_i - Y_{i-1})|}\sqrt{(X_i - x_c)^2 + (Y_i - y_c)^2}$$

coordinates of points on the cam profile, and a MATLAB program for doing this is included on the disk with this book. Given the (X_i, Y_i) coordinates, the cam can be machined on a CNC milling machine. The accuracy of the profile will be determined in part by the increment chosen for the cam rotation angle θ.

EXAMPLE 6.9 (*Cam Profile Coordinates for a Roller Follower That Oscillates*)

PROBLEM

Determine the cam profile assuming that the follower dwells while the cam rotates *counter*clockwise from 0° to 90°. The rise occurs with 3-4-5 polynomial motion during the cam rotation from 90° to 180°. The follower then dwells for 90° of cam rotation, and the return occurs with simple harmonic motion for the cam rotation from 270° to 360°. The amplitude of the follower oscillation is 30°, and the follower radius is 1 in. The base circle radius is 2 in, and the distance between pivots is 6 in. The length of the follower is to be determined such that the pressure angle starts out at zero.

SOLUTION

To solve the problem, we must identify the equations for the follower motion as a function of the cam rotation angle θ and then select an increment for θ. Because the cam is rotating counterclockwise, the follower rotates clockwise relative to the cam. Mathematically, this is accomplished by using negative angles for the displacements starting from 0. However, the equations developed for the follower displacements give the follower position as a function of the cam position and do not depend on the direction of cam rotation. For a given angular displacement, the rise will be the same if the cam is to be designed for clockwise or counterclockwise rotation. Therefore, we must use the magnitude of the cam angle from the start of the interval over which the function is defined. In the equations, we will let $\bar{\theta}$ be the magnitude of θ. Then from Sections 6.7 and 6.9, the equations (expressed in terms of radians) for each part of the motion become:

$0 < \bar{\theta} \le \pi/2$

$$\phi = 0$$

$\pi/2 \le \bar{\theta} \le \pi$, $(\beta = \pi/2;\ L = \pi/6)$

$$\phi = 10L\left(\frac{\bar{\theta} - \pi/2}{\beta}\right)^3 - 15L\left(\frac{\bar{\theta} - \pi/2}{\beta}\right)^4 + 6L\left(\frac{\bar{\theta} - \pi/2}{\beta}\right)^5$$

$$= \left(\frac{5\pi}{3}\right)\left(\frac{\bar{\theta} - \pi/2}{\pi/2}\right)^3 - \left(\frac{5\pi}{2}\right)\left(\frac{\bar{\theta} - \pi/2}{\pi/2}\right)^4 + \pi\left(\frac{\bar{\theta} - \pi/2}{\pi/2}\right)^5$$

$\pi \le \bar{\theta} \le 3\pi/2$

$$\phi = \frac{\pi}{6}$$

$3\pi/2 \le \bar{\theta} \le 2\pi$, $(\beta = \pi/2;\ L = \pi/6)$

$$\phi = \frac{L}{2}\left(1 + \cos\frac{\pi}{\beta}\left(\bar{\theta} - \frac{3\pi}{2}\right)\right) = \frac{\pi}{12}\left(1 + \cos 2\left(\bar{\theta} - \frac{3\pi}{2}\right)\right)$$

To determine the length, r_3, of the follower that will give zero pressure angle in the initial position, refer to Fig. 6.45. From that figure, the angle between the $(r_0 + r_b)$ and r_3 must be 90° for the pressure angle to be zero during a dwell. Therefore,

$$r_3 = \sqrt{r_1^2 - (r_b + r_0)^2} = \sqrt{6^2 - (2 + 1)^2} = 5.196 \text{ in}$$

The equations for ϕ and r_3 can be used to determine ϕ_0 in Table 6.11, and using the equations in Table 6.11 we can compute the coordinates of the cam as accurately as we wish by using a small increment for the cam rotation angle θ.

For the calculations, the angle θ is incremented in the negative direction (the follower rotates

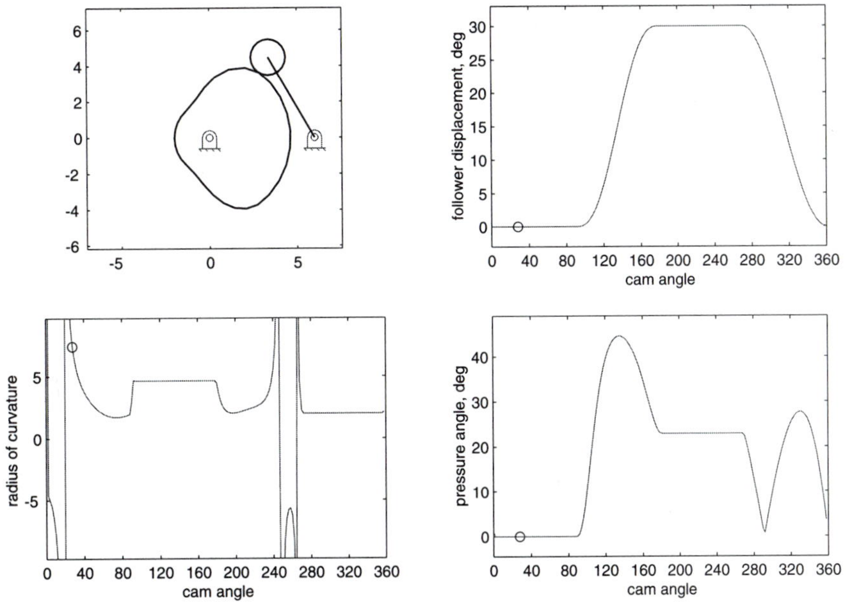

Figure 6.46 Cam profile, follower displacement, radius of curvature, and pressure angle for Example 6.9.

clockwise relative to the cam). Therefore, the points corresponding to the center of the roller (prime curve points) will be generated and labeled initially in the clockwise direction. However, the equations in Table 6.11 were derived assuming that the points are ordered in the counterclockwise direction. Therefore, after all of the prime points are generated, they must be reordered in the counterclockwise direction. After the prime points are reordered in this way, the corresponding points on the cam surface can be computed directly as was done in Example 6.7.

For plotting purposes, we have used the program on the disk included with this book with an angle increment of $-1°$, which is adequate for visual purposes. If we want to machine the cam with high accuracy, a finer increment for θ would be used. It should be noted, however, that when very fine increments are used, the computations using MATLAB can be significant. The cam, displacement diagram, radius of curvature, and pressure angle plots are shown in Fig. 6.46. ■

6.10.2.5 Analytical Determination of a Cam Profile for a Flat-Faced Follower That Oscillates

The cam-follower system is shown in Fig. 6.47, and the points shown in that figure will be used to describe the design of the cam. Point A is the location of the axis of rotation of the cam, C is the axis of rotation of the follower, and d is the follower offset. Point B is the location of the tangent point between the face of the follower and the cam surface, and D is located at the intersection of the follower face and a line from C perpendicular to the follower face. The distance R is the perpendicular distance from the follower face to the center of rotation of the cam.

The development of the equations for an oscillating flat-faced follower is similar to the graphical procedure in Example 6.6. As in the other cases, we must invert the mechanism so that the follower appears to rotate about the cam. To begin the procedure, we will assume that the cam base radius r_b, the pivot distance r_1, and the follower angle β are known.

In the procedure, we will locate the line representing the face of the cam first. We will then locate the points on the cam by locating successive circles that are tangent to three lines.

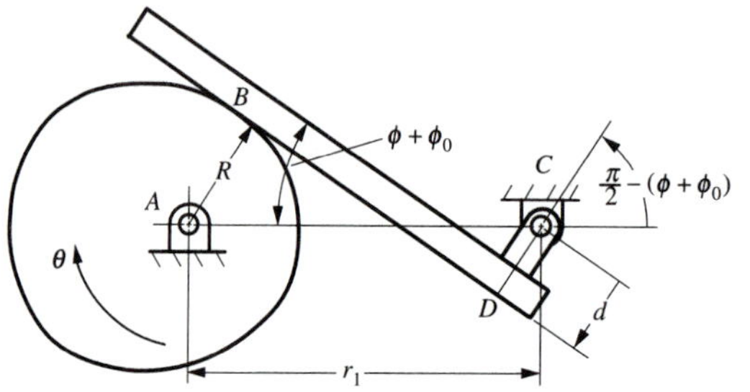

Figure 6.47 Geometry for oscillating flat-faced follower.

The first step is to determine the reference angle ϕ_0 that gives the orientation of the follower in the initial position. As shown in Fig. 6.47, in this position,

$$R = r_b = AB$$

From the geometry in Fig. 6.48,

$$AE = \frac{r_1}{1 + d/R} = \frac{r_1}{1 + d/r_b}$$

Then

$$BE = \sqrt{AE^2 - r_b^2}$$

and

$$\phi_0 = \tan^{-1} \frac{r_b}{BE}$$

For each position of the cam given by θ, we will know $\phi(\theta)$ from the follower displacement equations. The orientation of the follower face relative to the cam is then given by the angle $\phi + \phi_0$.

To locate the follower relative to the cam as the cam rotates, we can invert the motion relative to the cam. For simplicity, we will again assume that the cam rotates clockwise relative to the frame, which means that the frame rotates counterclockwise relative to the cam. A typical position is shown in Fig. 6.49.

For a given rotation of the cam, the coordinates of point C relative to the cam are

$$x_C = r_1 \cos \theta$$

and

$$y_C = r_1 \sin \theta$$

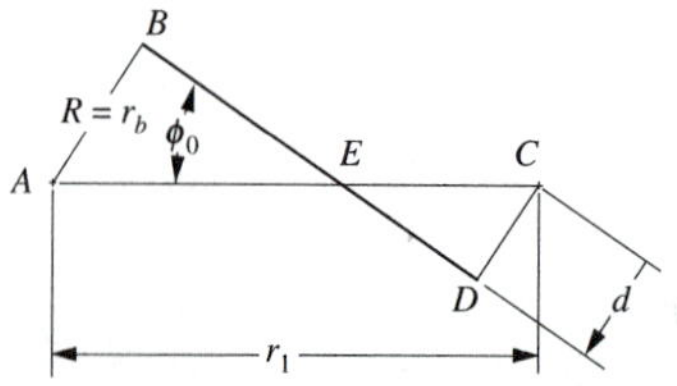

Figure 6.48 Geometry for finding ϕ_0.

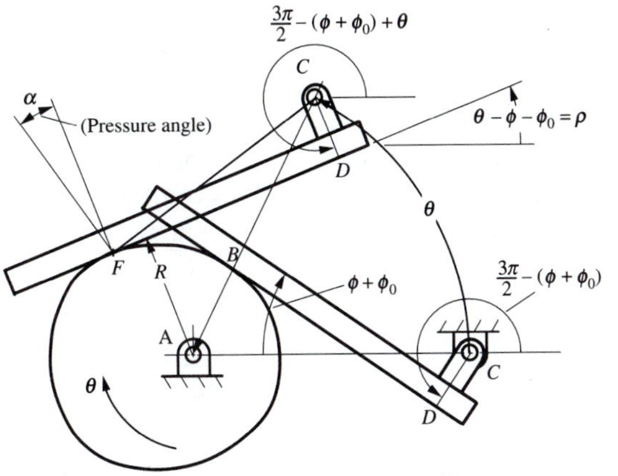

Figure 6.49 Motion of oscillating flat-faced follower relative to cam.

Similarly, the coordinates of point D are

$$x_D = x_C + d\cos\left(\frac{3\pi}{2} - \phi - \phi_0 + \theta\right)$$

and

$$y_D = y_C + d\sin\left(\frac{3\pi}{2} - \phi - \phi_0 + \theta\right)$$

To locate the line defining the follower face, we need to locate a line oriented at an angle of $\theta_i - \phi_i - \phi_0$ and passing through D_i. The equation of the line defining the follower face can be defined parametrically in terms of the distance τ from D_i as shown in Fig. 6.50. The parametric equations are

$$x_f = x_D + \tau\cos\rho \tag{6.29}$$

and

$$y_f = y_D + \tau\sin\rho \tag{6.30}$$

where

$$\rho = \theta - \phi - \phi_0$$

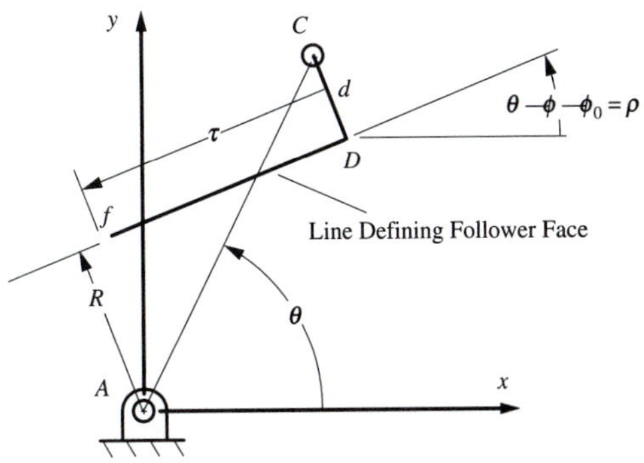

Figure 6.50 Parametric line defining follower face. Here τ would be negative.

Using Eqs. (6.29) and (6.30), we can define a line corresponding to the follower face for each position chosen for the cam. To define the cam profile, we must locate a series of points that are tangent to the follower face in each position. The coordinates of the tangent points can be approximated using the geometry shown in Fig. 6.51. Assume that we have three positions ($i - 1$, i, and $i + 1$) of the follower relative to the cam. Each follower face line can be defined in terms of the coordinates of D and τ and ρ. To determine the location of the cam contact point on the centerline, approximate the cam by a circle that is tangent to each of the lines simultaneously. Referring to Fig. 6.51, this can be done as follows:

1. Find the successive intersections (G_i and G_{i+1}) of the face lines. To find G_i, use Eqs. (6.29) and (6.30), set $x_{f_i} = x_{f_{i-1}}$ and $y_{f_i} = y_{f_{i-1}}$, and solve the resulting set of linear equations for τ_{i-1} and τ_i.

$$\begin{Bmatrix} x_{D_i} - x_{D_{i-1}} \\ y_{D_i} - y_{D_{i-1}} \end{Bmatrix} = \begin{bmatrix} -\cos\rho_i & \cos\rho_{i-1} \\ -\sin\rho_i & \sin\rho_{i-1} \end{bmatrix} \begin{Bmatrix} \tau_i \\ \tau_{i-1} \end{Bmatrix}$$

The location of G_i is then given by

$$x_{G_i} = x_{D_i} + \tau_i \cos\rho_i$$

and

$$y_{G_i} = y_{D_i} + \tau_i \sin\rho_i$$

2. Find the equation of the line that bisects the angle between positions i and $i - 1$ of the face line. This line passes through point G_i and is oriented at an angle of:

$$\gamma_i = \frac{\rho_i + \rho_{i-1}}{2} + \frac{3\pi}{2}$$

The parametric equation of the line is then given by

$$x_{m_i} = x_{G_i} + \nu_i \cos\gamma_i$$

and

$$y_{m_i} = y_{G_i} + \nu_i \sin\gamma_i$$

were ν_i is the distance from G_i in the positive γ_i direction.

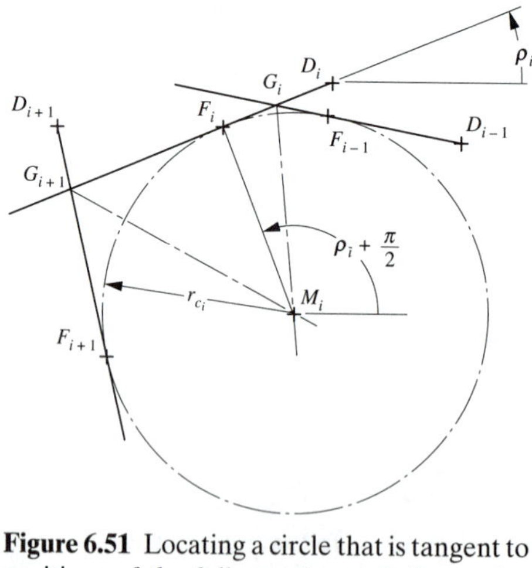

Figure 6.51 Locating a circle that is tangent to three successive positions of the follower face relative to the cam.

3. Find the intersection of the lines bisecting the angles at G_i and G_{i+1}. For this, set $x_{m_i} = x_{m_{i+1}}$ and $y_{m_i} = y_{m_{i+1}}$ and solve the resulting set of linear equations for ν_i and ν_{i+1}.

$$\begin{Bmatrix} x_{G_i} - x_{G_{i+1}} \\ y_{G_i} - y_{G_{i+1}} \end{Bmatrix} = \begin{bmatrix} -\cos\gamma_i & \cos\gamma_{i+1} \\ -\sin\gamma_i & \sin\gamma_{i+1} \end{bmatrix} \begin{Bmatrix} \nu_i \\ \nu_{i+1} \end{Bmatrix}$$

The coordinates of the center of the circle tangent to the three positions of the face line are then given by

$$x_{M_i} = x_{G_i} + \nu_i \cos\gamma_i$$

and

$$y_{M_i} = y_{G_i} + \nu_i \sin\gamma_i$$

4. Find the radius of the tangent circle and coordinates of the tangent points. These are obtained by locating a line through M_i and perpendicular to each face line. The coordinates of the point F_i are given by

$$x_{F_i} = x_{M_i} + r_{c_i} \cos\left(\rho_i + \frac{\pi}{2}\right) = x_{G_i} + \eta_i \cos\rho_i$$

and

$$y_{F_i} = y_{M_i} + r_{c_i} \sin\left(\rho_i + \frac{\pi}{2}\right) = y_{G_i} + \eta_i \sin\rho_i$$

where η_i is the distance from G_i to the tangency point. Both r_{c_i} and η_i can be found by solving the resulting set of simultaneous linear equations.

$$\begin{Bmatrix} x_{M_i} - x_{G_i} \\ y_{M_i} - y_{G_i} \end{Bmatrix} = \begin{bmatrix} -\cos\left(\rho_i + \frac{\pi}{2}\right) & \cos\rho_i \\ -\sin\left(\rho_i + \frac{\pi}{2}\right) & \sin\rho_i \end{bmatrix} \begin{Bmatrix} r_{c_i} \\ \eta_i \end{Bmatrix} \tag{6.31}$$

The location of the tangent point is then given by

$$x_{F_i} = x_{G_i} + \eta_i \cos\rho_i \tag{6.32}$$

and

$$y_{F_i} = y_{G_i} + \eta_i \sin\rho_i \tag{6.33}$$

To find the tangency points on the other faces, substitute $i - 1$ for i and then $i + 1$ for i in Eqs. (6.32) and (6.33).

5. The cam coordinates can be approximated by the points F_i computed in step 4. To determine the minimum face length, we can compute the distance from F_i to D_i for each position of the cam.

To compute the pressure angle at any given position, we must find the angle between a vector in the direction of the follower travel and a vector that is normal to the follower face at each location as shown in Fig. 6.49. The direction of travel of the follower is given by a unit vector $\boldsymbol{a}$ perpendicular to line F_iC_i. This vector is given by

$$\boldsymbol{a} = 1\angle\psi_i$$

where

$$\psi_i = \tan^{-1}\left[\frac{y_{F_i} - y_{C_i}}{x_{F_i} - x_{C_i}}\right] - \frac{\pi}{2}$$

We can approximate the direction that is perpendicular to the follower face by a unit vector $\boldsymbol{b}$ from M_i to F_i or

$$\boldsymbol{b} = 1\angle\lambda_i$$

where

$$\lambda_i = \tan^{-1}\left[\frac{y_{F_i} - y_{M_i}}{x_{F_i} - x_{M_i}}\right] = \rho_i + \frac{\pi}{2}$$

Table 6.12 Summary of Equations for Determining the Cam Profile Coordinates for an Oscillating Flat-Faced Follower. The Follower Oscillation ϕ Is Assumed To Be Given by $f(\theta)$, the Follower Offset Is d, the Base Circle Radius Is r_b, and the Distance Between the Cam and Follower Pivots Is r_1. There Are Assumed To Be n Points on the Cam Profile.

Cam coordinates

$$AE = \frac{r_1}{1 + d/r_b}, \qquad BE = \sqrt{AE^2 - r_b^2}$$

$$\phi_0 = \tan^{-1}\frac{r_b}{BE}$$

$$x_{C_i} = r_1 \cos\theta_i, \qquad y_{C_i} = r_1 \sin\theta_i$$

$$x_{D_i} = x_{C_i} + d\cos\left(\frac{3\pi}{2} - \phi_i - \phi_0 + \theta_i\right), \qquad y_{D_i} = y_{C_i} + d\sin\left(\frac{3\pi}{2} - \phi_i - \phi_0 + \theta_i\right)$$

$$\rho_i = \theta_i - \phi_i - \phi_0$$

$$\begin{Bmatrix} x_{D_i} - x_{D_{i-1}} \\ y_{D_i} - y_{D_{i-1}} \end{Bmatrix} = \begin{bmatrix} -\cos\rho_i & \cos\rho_{i-1} \\ -\sin\rho_i & \sin\rho_{i-1} \end{bmatrix} \begin{Bmatrix} \tau_i \\ \tau_{i-1} \end{Bmatrix} \qquad (i = 2, 3, \ldots, n)$$

$$\begin{Bmatrix} x_{D_1} - x_{D_n} \\ y_{D_1} - y_{D_n} \end{Bmatrix} = \begin{bmatrix} -\cos\rho_1 & \cos\rho_n \\ -\sin\rho_1 & \sin\rho_n \end{bmatrix} \begin{Bmatrix} \tau_1 \\ \tau_n \end{Bmatrix}$$

$$x_{G_i} = x_{D_i} + \tau_i \cos\rho_i, \qquad y_{G_i} = y_{D_i} + \tau_i \sin\rho_i$$

$$\gamma_i = \frac{\rho_i + \rho_{i-1}}{2} + \frac{3\pi}{2}$$

$$\begin{Bmatrix} x_{G_i} - x_{G_{i+1}} \\ y_{G_i} - y_{G_{i+1}} \end{Bmatrix} = \begin{bmatrix} -\cos\gamma_i & \cos\gamma_{i+1} \\ -\sin\gamma_i & \sin\gamma_{i+1} \end{bmatrix} \begin{Bmatrix} \nu_i \\ \nu_{i+1} \end{Bmatrix} \qquad (i = 1, 2, \ldots, n-1)$$

$$\begin{Bmatrix} x_{G_n} - x_{G_1} \\ y_{G_n} - y_{G_1} \end{Bmatrix} = \begin{bmatrix} -\cos\gamma_n & \cos\gamma_1 \\ -\sin\gamma_n & \sin\gamma_1 \end{bmatrix} \begin{Bmatrix} \nu_n \\ \nu_1 \end{Bmatrix}$$

$$x_{M_i} = x_{G_i} + \nu_i \cos\gamma_i, \qquad y_{M_i} = y_{G_i} + \nu_i \sin\gamma_i$$

$$\begin{Bmatrix} x_{M_i} - x_{G_i} \\ y_{M_i} - y_{G_i} \end{Bmatrix} = \begin{bmatrix} -\cos\left(\rho_i + \frac{\pi}{2}\right) & \cos\rho_i \\ -\sin\left(\rho_i + \frac{\pi}{2}\right) & \sin\rho_i \end{bmatrix} \begin{Bmatrix} r_{c_i} \\ \eta_i \end{Bmatrix}$$

$$x_{F_i} = x_{G_i} + \eta_i \cos\rho_i, \qquad y_{F_i} = y_{G_i} + \eta_i \sin\rho_i$$

Table 6.13 Summary of Equations for Determining the Pressure Angle and Radius of Curvature for an Oscillating Flat-Faced Follower

Pressure angle

$$\psi_i = \tan^{-1}\left[\frac{y_{F_i} - y_{C_i}}{x_{F_i} - x_{C_i}}\right] - \frac{\pi}{2}$$

$$\lambda_i = \tan^{-1}\left[\frac{y_{F_i} - y_{M_i}}{x_{F_i} - x_{M_i}}\right]$$

$$\alpha_i = \cos^{-1}[\cos\psi_i \cos\lambda_i + \sin\psi_i \sin\lambda_i]$$

Radius of curvature

$$\begin{Bmatrix} x_c \\ y_c \end{Bmatrix} = \begin{bmatrix} 2(X_{i+1} - X_i) & 2(Y_{i+1} - Y_i) \\ 2(X_{i-1} - X_i) & 2(Y_{i-1} - Y_i) \end{bmatrix}^{-1} \begin{Bmatrix} (X_{i+1}^2 - X_i^2) + (Y_{i+1}^2 - Y_i^2) \\ (X_{i-1}^2 - X_i^2) + (Y_{i-1}^2 - Y_i^2) \end{Bmatrix}$$

$$\rho = \frac{[(X_i - X_{i-1})(Y_{i+1} - Y_i) - (X_{i+1} - X_i)(Y_i - Y_{i-1})]}{|(X_i - X_{i-1})(Y_{i+1} - Y_i) - (X_{i+1} - X_i)(Y_i - Y_{i-1})|}\sqrt{(X_i - x_c)^2 + (Y_i - y_c)^2}$$

The resulting pressure angle is simply the angle between the vectors ***a*** and ***b***. That is,

$$\alpha_i = \cos^{-1}\left[\frac{\boldsymbol{a}\cdot\boldsymbol{b}}{|\boldsymbol{a}||\boldsymbol{b}|}\right] = \cos^{-1}[\cos\psi_i \cos\lambda_i + \sin\psi_i \sin\lambda_i] \qquad \textbf{(6.34)}$$

The equations necessary to determine points on the cam profile are summarized in Table 6.12, and the equations for the pressure angle and radius of curvature of the cam are summarized in Table 6.13. The equations for the radius of curvature are taken from Table 6.8. The equations in Table 6.12 can be easily programmed to determine the coordinates of points on the cam profile, and a MATLAB program for doing so is included on the disk with this book. Given the coordinates of a set of sequential points on the profile, the cam can be machined on a CNC milling machine. The accuracy of the profile will be determined in part by the increment chosen for the cam rotation angle θ.

EXAMPLE 6.10 (*Cam Profile Coordinates for a Flat-Faced Follower That Oscillates*)

PROBLEM

Assume that the follower starts from a dwell from 0° to 45° and rotates clockwise. The rise occurs with simple harmonic motion during the cam rotation from 45° to 180°. The follower then dwells for 90° of cam rotation, and the return occurs with simple harmonic motion for the cam rotation from 270° to 360°. The amplitude of the follower oscillation is 20°, and the follower offset is 0.5 in. The base circle radius is 2 in, and the distance between pivots is 6 in.

SOLUTION

To solve the problem, we must specify the length of the follower face, identify the equations for the follower motion as a function of the cam rotation angle θ, and then select an increment for θ. The length of the follower face is somewhat arbitrary as long as it is large enough to maintain contact with the cam. The minimum value can be calculated by computing the distance F_i to D_i for each position of the follower relative to the cam. However, in this example, we will select the length to be 9 in, which is large enough to ensure that the follower will maintain contact with the cam.

The second part of the displacement schedule is similar to that used in Example 6.9, and the equation for the rise portion can be obtained from Section 6.7. The resulting equations (expressed in terms of radians) are

$0 \leq \theta \leq \pi/2$

$$\phi = 0$$

$\pi/4 \leq \theta \leq \pi$ ($\beta = 3\pi/4$; $L = \pi/9$)

$$\phi = \frac{L}{2}\left(1 - \cos\frac{\pi}{\beta}\left(\theta - \frac{\pi}{4}\right)\right) = \frac{\pi}{18}\left(1 - \cos\frac{4}{3}\left(\theta - \frac{\pi}{4}\right)\right)$$

$\pi \leq \theta \leq 3\pi/2$

$$\phi = \frac{\pi}{6}$$

$3\pi/2 \leq \theta \leq 2\pi$ ($\beta = \pi/2$; $L = \pi/9$)

$$\phi = \frac{L}{2}\left(1 + \cos\frac{\pi}{\beta}\left(\theta - \frac{3\pi}{2}\right)\right) = \frac{\pi}{18}\left(1 + \cos 2\left(\theta - \frac{3\pi}{2}\right)\right)$$

Given the equations for the follower displacement, we now need only increment θ and evaluate the expressions in Table 6.12 and 6.13 to determine the cam coordinates and pressure angle. The values are computed for 20° increments of θ in Table 6.12. The results obtained from the MATLAB program included on the disk with this book are shown in Fig. 6.52.

It is much more difficult to design a cam for an oscillating flat-faced follower than for the other types of followers. For example, if the amplitude of oscillation in Example 6.10 is changed to 30°, a cam will be developed that has a discontinuity and therefore cannot be manufactured. Changing the base circle radius may not always improve the situation, and the extent to which the base circle can be changed is limited by the distance between pivots. The problem can be improved by reducing the amplitude of oscillation, by changing the function chosen for the rise/return, or by increasing the range chosen for the rise/return. However, in some cases, it may be necessary to choose another type of follower. ■

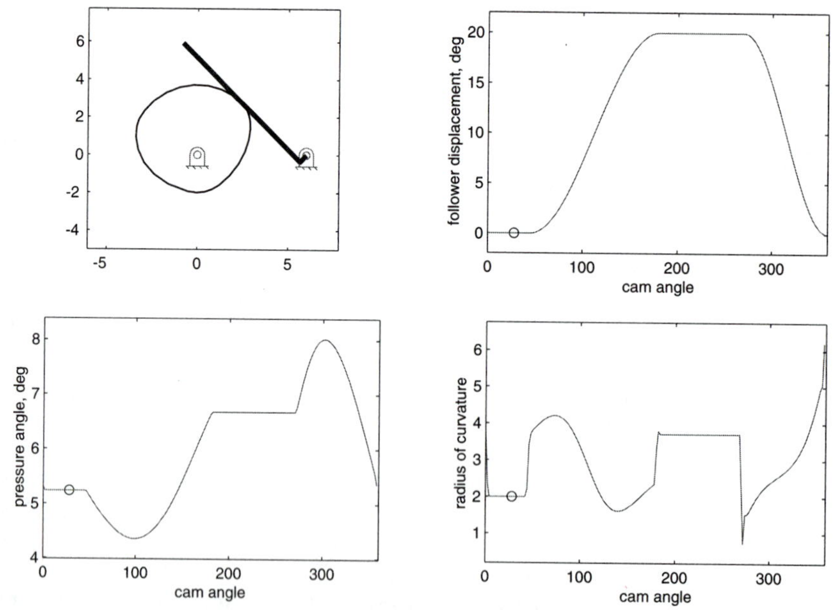

Figure 6.52 Cam profile, follower displacement, radius of curvature, and pressure angle for Example 6.10.

6.11 CHAPTER 6 *Exercise Problems*

PROBLEM 6.1

Assume that s is the cam follower displacement and θ is the cam rotation. The rise is h after β degrees of rotation, and the rise begins at a dwell and ends with a constant-velocity segment. The displacement equation for the follower during the rise period is

$$s = h \sum_{i=0}^{n} C_i \left(\frac{\theta}{\beta}\right)^i$$

(a) If the position, velocity, and acceleration are continuous at $\theta = 0$ and the position and velocity are continuous at $\theta = \beta$, determine the n required in the equation, and find the coefficients C_i that will satisfy the requirements if $s = h = 1.0$.

(b) If the follower is a radial flat-faced follower and $\beta = \pi/2$, what is the minimum radius for the base circle for the cam if cusps are to be avoided for the region involved?

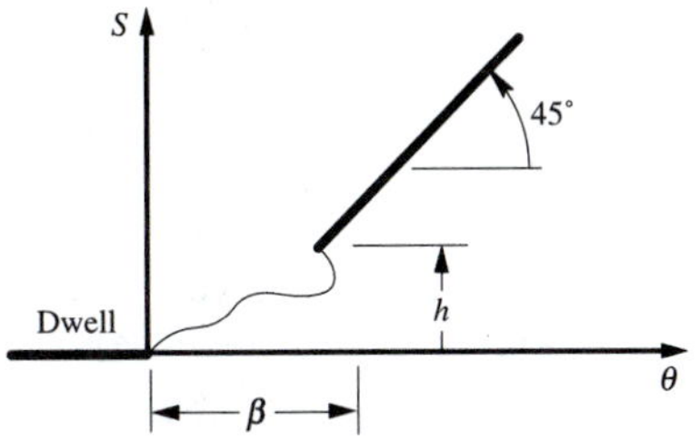

PROBLEM 6.2

Construct the part of the profile of a disk cam that follows the displacement diagram shown below. The cam has a 5-cm-diameter prime circle and is rotating counterclockwise. The follower is a knife-edged, radial, translating follower. Use 10-degree increments for the construction.

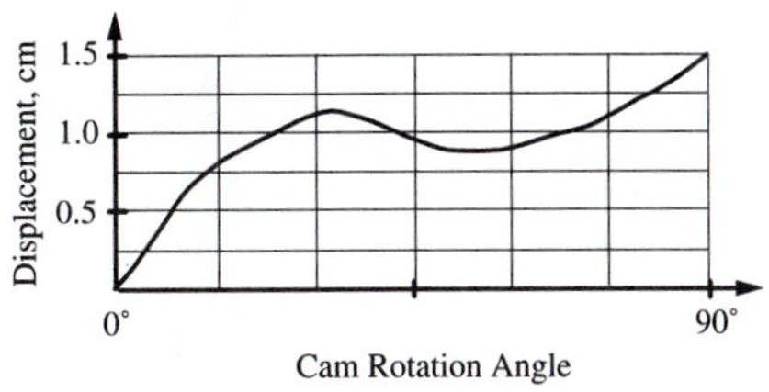

PROBLEM 6.3

Construct the profile of a disk cam that follows the displacement diagram shown below. The follower is a radial roller and has a diameter of 10 mm. The base circle diameter of the cam is to be 40 mm and the cam rotates clockwise.

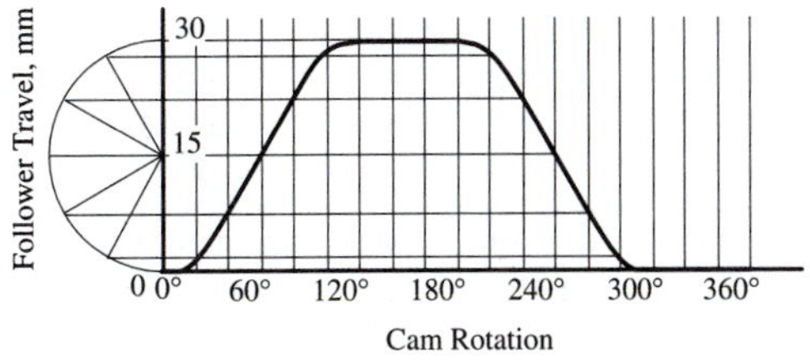

PROBLEM 6.4

A cam returns from a full lift of 1.2 in during its initial 60° rotation. The first 0.4 in of the return is half-cycloidal. This is followed by a half-harmonic return. Determine β_1 and β_2 so that the motion has continuous first and second derivatives. Draw a freehand sketch of y', y'', and y''' indicating any possible mismatch in the third derivative.

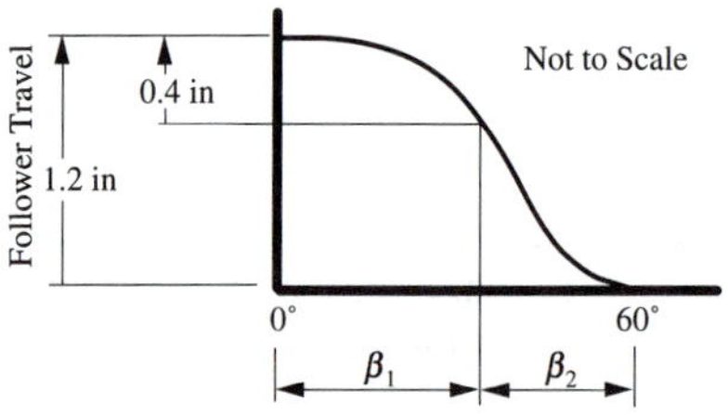

PROBLEM 6.5

A cam that is designed for cycloidal motion drives a flat-faced follower. During the rise, the follower displaces 1 in for 180° of cam rotation. If the cam angular velocity is constant at 100 rpm, determine the displacement, velocity, and acceleration of the follower at a cam angle of 60°.

PROBLEM 6.6

A constant-velocity cam is designed for simple harmonic motion. If the flat-faced follower displaces 2 in for 180° of cam rotation and the cam angular velocity is 100 rpm, determine the displacement, velocity, and acceleration when the cam angle is 45°.

PROBLEM 6.7

A cam drives a radial, knife-edged follower through a 1.5-in rise in 180° of cycloidal motion. Give the displacement at 60° and 100°. If this cam is rotating at 200 rpm, what are the velocity (ds/dt) and the acceleration (d^2s/dt^2) at $\theta = 60°$?

PROBLEM 6.8

Assume that s is the cam follower displacement and θ is the cam rotation. The rise is 1.0 cm after 1.0 radian of rotation, and the rise begins and ends at a dwell. The displacement equation for the follower during the rise period is

$$s = h \sum_{i=0}^{n} C_i \left(\frac{\theta}{\beta}\right)^i$$

(a) If the position, velocity, and acceleration are continuous at $\theta = 0$, and the position and velocity are continuous at $\theta = 1.0$ rad, determine the value of n required in the equation, and find the coefficients C_i if $\dot{\theta} = 2$ rad/s. Note: Use the minimum possible number of terms.

(b) If the follower is a radial flat-faced follower, what is the minimum radius for the base circle for the cam if cusps are to be avoided when $\theta = 45°$?

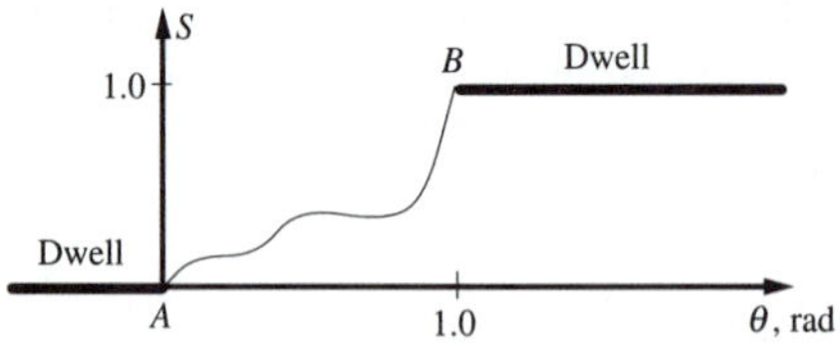

PROBLEM 6.9

Accurately sketch one-half of the cam profile (stations 0–6) for the cam follower, base circle, and displacement diagram given below. The base circle diameter is 1.2 in.

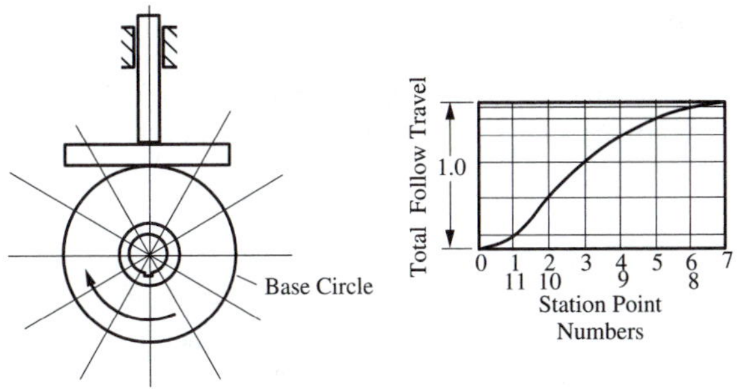

PROBLEM 6.10

In the sketch shown, the disk cam is used to position the radial flat-faced follower in a computing mechanism. The cam profile is to be designed to give a follower displacement S for a counterclockwise cam rotation θ according to the function $S = k\theta^2$ starting from dwell. For 60° of cam rotation from the starting position, the lift of the follower is 1.0 cm. By analytical methods, determine the distances R and L when the cam has been turned 45° from the starting position. Also calculate whether cusps in the cam profile would occur in the total rotation of 60°.

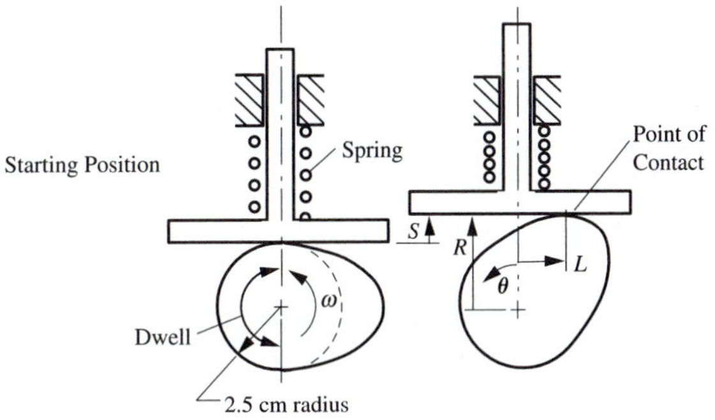

PROBLEM 6.11

Lay out a cam profile using a harmonic follower displacement (both rise and return). Assume that the cam is to dwell at zero lift for the first 100° of the motion cycle and to dwell at a 1 in lift for cam angles from 160° to 210°. The cam is to have a translating, radial, roller follower with a 1-in roller diameter, and the base circle radius is to be 1.5 in. The cam will rotate clockwise. Lay out the cam profile using 20° plotting intervals.

PROBLEM 6.12

Lay out a cam profile using a cycloidal follower displacement (both rise and return) if the cam is to dwell at zero lift for the first 80° of the motion cycle and to dwell at 2 in lift for cam angles from 120° to 190°. The cam is to have a translating, radial, roller follower with a roller diameter of 0.8 in. The cam will rotate counterclockwise, and the base circle diameter is 2 in. Lay out the cam profile using 20° plotting intervals.

PROBLEM 6.13

Lay out a cam profile assuming that an oscillating, roller follower starts from a dwell for 0° to 140° of cam rotation, and the cam rotates clockwise. The rise occurs with parabolic motion during the cam rotation from 140° to 220°. The follower then dwells for 40° of cam rotation, and the return occurs with parabolic motion for the cam rotation from 260° to 360°. The amplitude of the follower rotation is 35°, and the follower radius is 1 in. The base circle radius is 2 in, and the distance between the cam axis and follower rotation axis is 4 in. Lay out the cam profile using 20° plotting intervals such that the pressure angle is 0° when the follower is in the bottom dwell position.

PROBLEM 6.14 Lay out the rise portion of the cam profile if a flat-faced, translating, radial follower's motion is uniform. The total rise is 1.5 in, and the rise occurs over 100° of can rotation. The follower dwells for 90° of cam rotation prior to the beginning of the rise and dwells for 80° of cam rotation at the end of the rise. The cam will rotate counterclockwise, and the base circle radius is 3 in.

PROBLEM 6.15 Determine the cam profile assuming that the translating cylindrical-faced follower starts from a dwell from 0° to 80° and rotates clockwise. The rise occurs with cycloidal motion during the cam rotation from 80° to 180°. The follower than dwells for 60° of cam rotation, and the return occurs with simple harmonic motion for the cam rotation from 240° to 360°. The amplitude of the follower translation is 3 cm, and the follower radius is 0.75 cm. The base circle radius is 5 cm, and the offset is 0.5 cm.

PROBLEM 6.16 Resolve Problem 6.15 if the amplitude of the follower translation is 4 cm and the follower radius is 1 cm. The base circle radius is 5 cm, and the offset is 1 cm.

PROBLEM 6.17 Solve Problem 6.15 if the cam rotates counterclockwise.

PROBLEM 6.18 Determine the cam profile assuming that the translating flat-faced follower starts from a dwell from 0° to 80° and rotates clockwise. The rise occurs with parabolic motion during the cam rotation from 80° to 180°. The follower then dwells for 60° of cam rotation, and the return occurs with simple harmonic motion for the cam rotation from 240° to 360°. The amplitude of the follower translation is 3 cm. First find the minimum base circle radius based on avoiding cusps, and use that base circle to design the cam.

PROBLEM 6.19 Solve Problem 6.18 if the cam rotates counterclockwise.

PROBLEM 6.20 Determine the cam profile assuming that an oscillating, cylindrical-faced follower dwells while the cam rotates counterclockwise from 0° to 100°. The rise occurs with 3-4-5 polynomial motion during the cam rotation from 100° to 190°. The follower then dwells for 80° of cam rotation, and the return occurs with simple harmonic motion for the cam rotation from 270° to 360°. The amplitude of the follower oscillation is 25°, and the follower radius is 0.75 in. The base circle radius is 2 in, and the distance between pivots is 6 in. The length of the follower is to be determined such that the pressure angle starts out at zero.

PROBLEM 6.21 Solve Problem 6.20 if the cam rotates clockwise.

PROBLEM 6.22 Design a cam and oscillating roller follower assuming that the follower starts from a dwell for 0° to 80° of cam rotation and the cam rotates clockwise. The rise occurs with cycloidal motion during the cam rotation from 80° to 200°. The follower then dwells for 40° of cam rotation, and the return occurs with cycloidal motion for the cam rotation from 240° to 360°. The amplitude of the follower rotation is 45°. Determine the cam base circle radius, distance between cam and follower pivots, the length of the follower, and the radius of the follower for acceptable performance.

PROBLEM 6.23 Determine the cam profile assuming that an oscillating, flat-faced follower starts from a dwell from 0° to 45° and rotates counterclockwise. The rise occurs with simple harmonic motion during the cam rotation from 45° to 180°. The follower then dwells for 90° of cam rotation, and the return occurs with simple harmonic motion for the cam rotation from 270° to 360°. The amplitude of the follower oscillation is 20°, and the follower offset is 0.5 in. The base circle radius is 5 in, and the distance between pivots is 8 in.

PROBLEM 6.24 Solve Problem 6.23 if the cam rotates clockwise.

PROBLEM 6.25

Design the cam system assuming that an oscillating, flat-faced follower starts from a dwell for 0° to 100° of cam rotation and the cam rotates counterclockwise. The rise occurs with uniform motion during the cam rotation from 100° to 200°. The follower then dwells for 40° of cam rotation, and the return occurs with parabolic motion for the cam rotation from 240° to 360°. The oscillation angle is 20°.

PROBLEM 6.26

Design the cam system assuming that an oscillating, flat-faced follower starts from a dwell for 0° to 50° of cam rotation and the cam rotates clockwise. The rise occurs with cycloidal motion during the cam rotation from 50° to 200°. The follower then dwells for 90° of cam rotation, and the return occurs with harmonic motion for the cam rotation from 290° to 360°. The oscillation angle is 25°.

PROBLEM 6.27

Determine the cam profile assuming that the translating knife-edged follower starts from a dwell from 0° to 80° and rotates clockwise. The rise occurs with cycloidal motion during a cam rotation from 80° to 180°. The follower then dwells for 60° of cam rotation, and the return occurs with simple harmonic motion for the cam rotation from 240° to 360°. The amplitude of the follower translation is 4 cm. The base circle radius is 5 cm, and the offset is 0.5 cm.

Chapter 7

Spatial Linkage Analysis

7.1 SPATIAL MECHANISMS

7.1.1 Introduction

There are many mechanisms that do not conform to planar, spherical, or other relatively simple motion domains. The landing gear of many aircraft involves a complex sequence of unfolding and rotation movements. The example of automotive suspension and steering linkages has already been discussed in Chapter 5. Control linkages used on farm and construction machinery are additional examples. Many problems involving spatial mechanisms can be attacked using the vector analysis techniques of courses in statics and dynamics. This is true if the information to be elicited is in the velocity or acceleration domain. Unfortunately, if the position of the linkage must be determined, such methods are likely to fall short in any but the simplest mechanism configurations. In order to develop systematic methods for handling position problems in spatial mechanism, it is necessary to develop new tools. This chapter provides a brief introduction to the most commonly used methods. More advanced texts specifically directed as kinematics and dynamics of spatial mechanisms or robotics may be consulted for more extensive treatment of these topics.

In this chapter we will develop methods for modeling simple spatial mechanisms. Among other applications, the material presented in this chapter is fundamental to the construction of coordination software for robotic mechanisms.

It is appropriate to consider, first of all, the behavior of serial chain mechanisms. A serial chain is simply a set of members connected in series. Each member has two joints connecting it to its neighbors, except for the end members, which have only one joint. This is actually a very important type of structure from a practical viewpoint, since it is the configuration of most industrial robots. We will start by developing the methods necessary for systematic position, velocity, and acceleration analysis of serial chains in which all joints are actively controlled and then introduce cases involving closed-loop structures and passive joints later in the chapter.

The study of spatial mechanisms is a large and complex subject. Even a very basic treatment would be much more extensive than we can supply in this one chapter. Consequently, this presentation is limited to an introduction to some of the more important fundamental concepts and the provision of some tools that can be employed to solve relatively simple spatial linkage problems.

7.1.2 Velocity and Acceleration Relationships

The relationship between the velocities of two points embedded in the same moving body that was derived for planar motion in Chapter 2 is actually perfectly general for three-dimensional motion:

$$\boldsymbol{v}_B = \boldsymbol{v}_A + \boldsymbol{\omega} \times \mathbf{r}_{B/A} \tag{2.1}$$

Similarly, the relationship between the accelerations of those two points

$$\boldsymbol{a}_B = \boldsymbol{a}_A + \boldsymbol{\alpha} \times \mathbf{r}_{B/A} + \boldsymbol{\omega} \times (\boldsymbol{\omega} \times \mathbf{r}_{B/A}) \tag{2.7}$$

was also shown to be valid for three-dimensional motion. These expressions, for which three-dimensional proofs were developed in Chapter 2, form the basis for velocity and acceleration analysis of simple spatial mechanisms.

In many cases these relationships can be applied directly. Examples are developed in the next two subsections. The limitations of this approach become evident only when the mechanism position is not known a priori and the configuration is sufficiently complex that the position cannot be determined by inspection. As pointed out for planar mechanisms in Chapters 2 and 3, analysis must proceed in the order position analysis, velocity analysis, and acceleration analysis. Spatial system geometry does not lend itself to graphical analysis because the essential geometry cannot be captured in a single, two-dimensional representation. Consequently, any systematic approach to spatial linkage analysis must be analytically based. As was the case with planar mechanisms, in the analytical domain the position problem tends to be more complex than the velocity and acceleration problems. It is, therefore, advantageous to develop effective means of dealing with the position solution of spatial mechanisms and then to develop velocity and acceleration formulations that are fully compatible with the technique used for position modeling. This will be done in Section 7.2 and subsequent sections of this chapter.

However, as noted above, simple vector analysis based on Eqs. (2.1) and (2.7) does work if the position of the mechanism is not an issue. It is, therefore, worthwhile studying some examples using this approach before moving to a more systematic and powerful approach.

Keeping track of the reference frame to which points in a given body are referred is even more important in spatial mechanism analysis than in planar mechanisms. Consequently, the notation used in Chapter 2, in which the reference frame to which a vector is referred is indicated by a superscript placed before the symbol, will be used again here.

EXAMPLE 7.1 *(Analysis of Simple Spatial Linkage)*

PROBLEM

Given that the disk shown in Fig. 7.1 is rotating about a horizontal axis at A with an angular velocity with magnitude Ω, and the entire assembly is rotating about the y axis with an angular velocity with magnitude ω, determine the velocity of point q located on the perimeter of the disk at an angle of β from the y axis. The radius r of the disk is given.

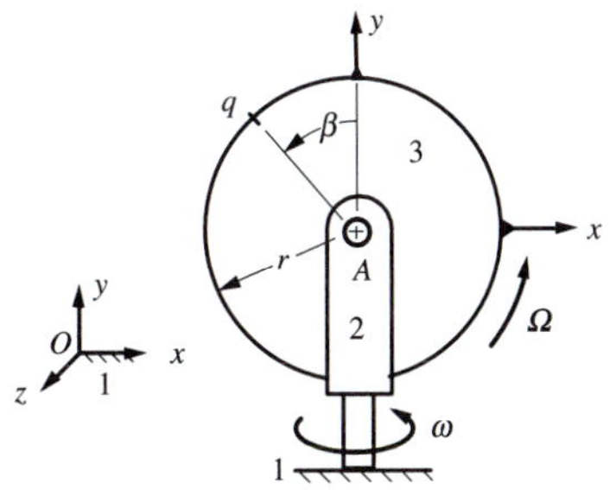

Figure 7.1 The mechanism analyzed in Example 7.1.

SOLUTION

The problem may be restated in the following form: Given ${}^1\boldsymbol{\omega}_2 = \omega \boldsymbol{j}$, and ${}^2\boldsymbol{\omega}_3 = \Omega \boldsymbol{k}$, together with the quantities r and β, find $\boldsymbol{v}_q = {}^1\boldsymbol{v}_{q_3/O_1}$.

We can use the chain rule developed in Section 3.3.3, but it is first necessary to recognize that location A gives the points that are easiest to use as a basis for the analysis.

$${}^1\boldsymbol{v}_{q_3/O_1} = {}^1\boldsymbol{v}_{q_3/A_3} + {}^1\boldsymbol{v}_{A_3/A_1} + {}^1\boldsymbol{v}_{A_1/O_1}$$

Because A_3 and A_1 are permanently coincident and ${}^1\boldsymbol{v}_{A_1} = 0$, and because A_1 and O_1 are both fixed to link 1, then

$${}^1\boldsymbol{v}_{A_3/A_1} = {}^1\boldsymbol{v}_{A_1/O_1} = 0$$

and

$${}^1\boldsymbol{v}_{q_3/O_1} = {}^3\boldsymbol{v}_{q_3/A_3} + {}^1\boldsymbol{\omega}_3 \times {}^3\boldsymbol{r}_{q_3/A_3}$$

Now,

$${}^3\boldsymbol{v}_{q_3/A_3} = 0$$

and

$${}^1\boldsymbol{\omega}_3 = \omega \boldsymbol{j} + \Omega \boldsymbol{k} = {}^1\boldsymbol{\omega}_2 + {}^2\boldsymbol{\omega}_3$$

Also

$${}^3\boldsymbol{r}_{q_3/A_3} = -r \sin\beta\, \boldsymbol{i} + r\cos\beta\, \boldsymbol{j}$$

At this instant, the coordinate frames are all parallel. Therefore the original cross-product can be written as

$${}^1\boldsymbol{\omega}_3 \times {}^3\mathbf{r}_{q_3/A_3} = \begin{vmatrix} \boldsymbol{i} & \boldsymbol{j} & \boldsymbol{k} \\ 0 & \omega & \Omega \\ -r\sin\beta & r\cos\beta & 0 \end{vmatrix} = -(\Omega r \cos\beta)\boldsymbol{i} - (\Omega r \sin\beta)\boldsymbol{j} + (\omega r \sin\beta)\boldsymbol{k}$$

Therefore, the solution is

$$\boldsymbol{v}_q = {}^1\boldsymbol{v}_{q_3/O_1} = -(\Omega r \cos\beta)\boldsymbol{i} - (\Omega r \sin\beta)\boldsymbol{j} + (\omega r \sin\beta)\boldsymbol{k}$$

■

EXAMPLE 7.2 *(Velocity Analysis of Spatial Manipulator)*

PROBLEM

A simple, three-degree-of-freedom spatial manipulator arm is shown in Fig. 7.2. In this arm, $AB = BC = CD = 2$ meters, and AB is perpendicular to the y axis, CB is perpendicular to AB, and CD is perpendicular to BC. The pair variables associated with the three revolute joints are the angles γ, θ, and ϕ defined as shown. At a given instant in time, rotary potentiometers and tachometers integral with each revolute joint servomotor (A, B, and C) indicate that $\gamma = 30°$, $\theta = 30°$, $\phi = 75°$, $\dot{\gamma} = 20$ rpm, $\dot{\theta} = -10$ rpm, and $\dot{\phi} = 15$ rpm. (Positive angles and angular velocities are in the same directions as the angles shown in Fig. 7.2.) At this instant find the velocity of point D located on link 4 relative to the fixed reference frame.

SOLUTION

In our nomenclature, we want to find

$$\boldsymbol{v}_D = {}^1\boldsymbol{v}_{D_4} = {}^1\boldsymbol{v}_{D_4/A_1}$$

The velocity equation for point D is

$${}^1\boldsymbol{v}_{D_4/A_1} = {}^1\boldsymbol{v}_{D_4/C_4} + {}^1\boldsymbol{v}_{C_3/B_3} + {}^1\boldsymbol{v}_{B_2/A_2}$$

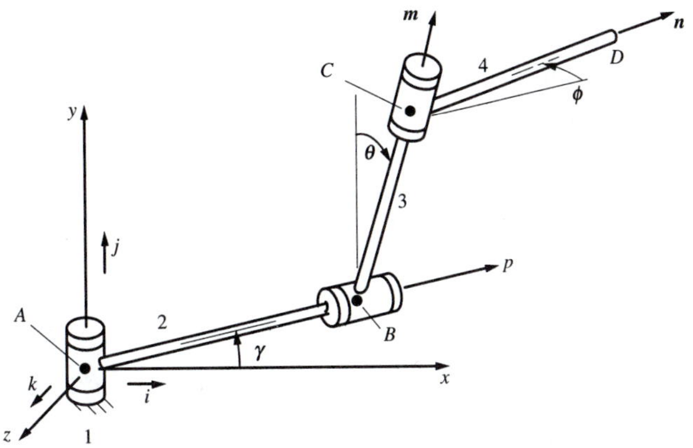

Figure 7.2 The three-degree-of-freedom manipulator analyzed in Example 7.2.

which can be expanded to give

$$^1\boldsymbol{v}_{D_4/A_1} = {}^1\boldsymbol{\omega}_4 \times {}^4\boldsymbol{r}_{D_4/C_4} + {}^1\boldsymbol{\omega}_3 \times {}^3\boldsymbol{r}_{C_3/B_3} + {}^1\boldsymbol{\omega}_2 \times {}^2\boldsymbol{r}_{B_2/A_2}$$

Let $\boldsymbol{p}$ and $\boldsymbol{m}$ be unit vectors along lines AB and BC, respectively. The components of $\boldsymbol{p}$ and $\boldsymbol{m}$ resolved in the fixed frame are

$$\boldsymbol{p} = \cos\gamma\,\boldsymbol{i} - \sin\gamma\,\boldsymbol{k}$$

and referring to Fig. 7.3,

$$\boldsymbol{m} = \cos\theta\,\boldsymbol{j} + \sin\theta(\cos\gamma\,\boldsymbol{k} + \sin\gamma\,\boldsymbol{i}) = \sin\theta\sin\gamma\,\boldsymbol{i} + \cos\theta\,\boldsymbol{j} + \sin\theta\cos\gamma\,\boldsymbol{k}$$

In order to decompose $\boldsymbol{n}$, think of it being initially parallel to $\boldsymbol{p}$. Then imagine that $\boldsymbol{n}$ moves in a plane inclined at an angle of θ to the x-z plane, as shown in Fig. 7.4.

$$\begin{aligned}\boldsymbol{n} &= \cos\phi\,\boldsymbol{p} + \sin\phi(\sin\theta\,\boldsymbol{j} - \cos\theta[\cos\gamma\,\boldsymbol{k} + \sin\gamma\,\boldsymbol{i}])\\ &= (\cos\phi\cos\gamma - \sin\phi\cos\theta\sin\gamma)\boldsymbol{i} + (\sin\phi\sin\theta)\boldsymbol{j} - (\cos\phi\sin\gamma + \sin\phi\cos\theta\cos\gamma)\boldsymbol{k}\end{aligned}$$

The relative angular velocity expressions that are given in the problem statement may now be expressed in the following forms:

$$^1\boldsymbol{\omega}_2 = \dot{\gamma}\boldsymbol{j} = 20(2\pi)/60\boldsymbol{j} = 2.094\boldsymbol{j}\ \text{rad/s}$$

$$\begin{aligned}^2\boldsymbol{\omega}_3 &= \dot{\theta}\boldsymbol{p} = -10(2\pi)/60\boldsymbol{p} = -1.047\boldsymbol{p}\ \text{rad/s}\\ &= -1.047(\cos\gamma\,\boldsymbol{i} - \sin\gamma\,\boldsymbol{k}) = (-0.9069\boldsymbol{i} + 0.5236\boldsymbol{k})\ \text{rad/s}\end{aligned}$$

$$\begin{aligned}^3\boldsymbol{\omega}_4 &= \dot{\phi}\boldsymbol{m} = 15(2\pi)/60\boldsymbol{m} = 1.571\boldsymbol{m}\ \text{rad/s} = 1.571(\sin\theta\sin\gamma\,\boldsymbol{i} + \cos\theta\,\boldsymbol{j} + \sin\theta\cos\gamma\,\boldsymbol{k})\\ &= (0.3927\boldsymbol{i} + 1.360\boldsymbol{j} + 0.6802\boldsymbol{k})\ \text{rad/s}\end{aligned}$$

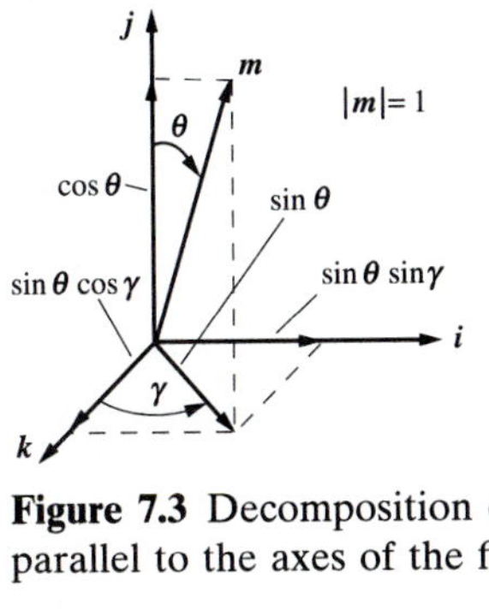

Figure 7.3 Decomposition of the vector $\boldsymbol{m}$ into components parallel to the axes of the fixed reference frame.

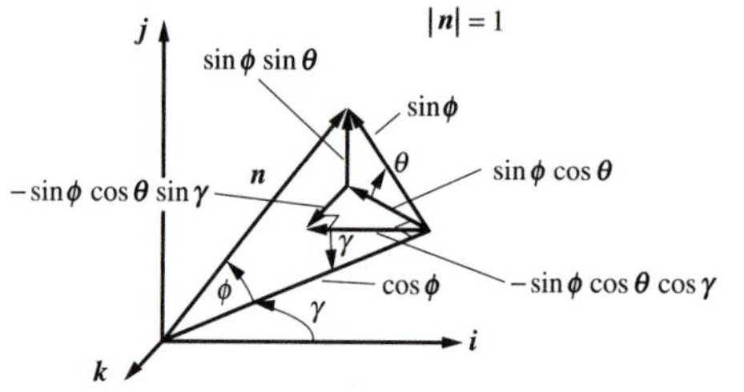

Figure 7.4 Decomposition of the vector ***n*** into components parallel to the axes of the fixed reference frame.

The angular velocities relative to the fixed reference frame necessary for evaluating the cross-products are then found using the chain rule (Section 3.3.3). That is,

$$
\begin{aligned}
{}^1\boldsymbol{\omega}_3 &= {}^1\boldsymbol{\omega}_2 + {}^2\boldsymbol{\omega}_3 \\
&= 2.094\boldsymbol{j} - 0.9069\boldsymbol{i} + 0.5236\boldsymbol{k} = (-0.9069\boldsymbol{i} + 2.094\boldsymbol{j} + 0.5236\boldsymbol{k}) \text{ rad/s}
\end{aligned}
$$

and

$$
\begin{aligned}
{}^1\boldsymbol{\omega}_4 &= {}^1\boldsymbol{\omega}_3 + {}^3\boldsymbol{\omega}_4 \\
&= -\ 0.9069\boldsymbol{i}\, 2.094\boldsymbol{j} + 0.5236\boldsymbol{k} + 0.3927\boldsymbol{i} + 1.360\boldsymbol{j} + 0.6802\boldsymbol{k} \\
&= (-0.514\boldsymbol{i} + 3.454\boldsymbol{j} + 1.204\boldsymbol{k}) \text{ rad/s}
\end{aligned}
$$

The required displacement vectors are

$$
{}^2\boldsymbol{r}_{B_2/A_2} = 2\boldsymbol{p} = 2(\cos\gamma\, \boldsymbol{i} - \sin\gamma\, \boldsymbol{k}) = (1.732\boldsymbol{i} - \boldsymbol{k}) \text{ m}
$$

$$
{}^3\boldsymbol{r}_{C_3/B_3} = 2\boldsymbol{m} = 2(\sin\theta \sin\gamma\, \boldsymbol{i} + \cos\theta\, \boldsymbol{j} + \sin\theta\cos\gamma\, \boldsymbol{k}) = (0.5\boldsymbol{i} + 1.732\boldsymbol{j} + 0.866\boldsymbol{k}) \text{ m}
$$

$$
\begin{aligned}
{}^4\boldsymbol{r}_{D_4/C_4} &= 2\boldsymbol{n} = 2[(\cos\phi\cos\gamma - \sin\phi\cos\theta\sin\gamma)\boldsymbol{i} + (\sin\phi\sin\theta)\boldsymbol{j} \\
&\quad -(\cos\phi\sin\gamma + \sin\phi\cos\theta\cos\gamma)\boldsymbol{k}] = (-0.388\boldsymbol{i} + 0.966\boldsymbol{j} - 1.708\boldsymbol{k}) \text{ m}
\end{aligned}
$$

The terms in the velocity equation can now be computed as follows:

$$
{}^1\boldsymbol{v}_{B_2/A_2} = {}^1\boldsymbol{\omega}_2 \times {}^2\boldsymbol{r}_{B_2/A_2} = 2.094\boldsymbol{j} \times (1.732\boldsymbol{i} - \boldsymbol{k}) = (-2.094\boldsymbol{i} - 3.627\boldsymbol{k}) \text{ m/s}
$$

$$
\begin{aligned}
{}^1\boldsymbol{v}_{C_3/B_3} &= {}^1\boldsymbol{\omega}_3 \times {}^3\boldsymbol{r}_{C_3/B_3} = (-0.9069\boldsymbol{i} + 2.094\boldsymbol{j} + 0.5236\boldsymbol{k}) \times (0.5\boldsymbol{i} + 1.732\boldsymbol{j} + 0.866\boldsymbol{k}) \\
&= (0.906\boldsymbol{i} + 1.047\boldsymbol{j} - 2.618\boldsymbol{k}) \text{ m/s}
\end{aligned}
$$

$$
\begin{aligned}
{}^1\boldsymbol{v}_{D_4/C_4} &= {}^1\boldsymbol{\omega}_4 \times {}^4\boldsymbol{r}_{D_4/C_4} = (-0.514\boldsymbol{i} + 3.454\boldsymbol{j} + 1.204\boldsymbol{k}) \times (-0.388\boldsymbol{i} + 0.966\boldsymbol{j} - 1.708\boldsymbol{k}) \\
&= (-7.062\boldsymbol{i} - 1.345\boldsymbol{j} + 0.844\boldsymbol{k}) \text{ m/s}
\end{aligned}
$$

Therefore, the velocity of point D relative to the fixed reference frame is

$$
\begin{aligned}
{}^1\boldsymbol{v}_{D_4/A_1} &= {}^1\boldsymbol{v}_{B_2/C_2} + {}^1\boldsymbol{v}_{C_3/B_3} + {}^1\boldsymbol{v}_{D_4/C_4} = (-2.094\boldsymbol{i} - 3.627\boldsymbol{k}) + (0.906\boldsymbol{i} + 1.047\boldsymbol{j} - 2.618\boldsymbol{k}) \\
&\quad + (-7.062\boldsymbol{i} - 1.345\boldsymbol{j} + 0.844\boldsymbol{k}) \\
&= (-8.250\boldsymbol{i} - 0.298\boldsymbol{j} - 5.401\boldsymbol{k}) \text{ m/s} \\
&= 9.865(-0.836\boldsymbol{i} - 0.030\boldsymbol{j} - 0.547\boldsymbol{k}) \text{ m/s}
\end{aligned}
$$

The final result has been expressed in the form of the magnitude of velocity multiplied by a unit vector having the direction of the velocity.

Note that the complexities of this example originated in the necessity of resolving position vectors into components in the axis directions of the fixed reference frame. This resolution can be difficult to visualize and is a potential source of error. Removing those difficulties is a primary motivation of the matrix transformation approach to solving the position problem that is presented in Section 7.3. ■

EXAMPLE 7.3 ***(Acceleration Analysis of Spatial Manipulator)***

PROBLEM

In the spatial manipulator arm shown in Fig. 7.2 and considered in Example 7.2, the acceleration readings at the instant examined are $\ddot{\gamma} = 3$ rad/s^2, $\ddot{\theta} = 1$ rad/s^2, and $\ddot{\phi} = -2$ rad/s^2. As in the previous example, $AB = BC = CD = 2$ meters and AB is perpendicular to the y axis, CB is perpendicular to AB, and CD is perpendicular to BC. Also $\gamma = 30°$, $\theta = 30°$, $\phi = 75°$, $\dot{\gamma} = 20$ rpm, $\dot{\theta} = -10$ rpm, and $\dot{\phi} = 15$ rpm.

Find the acceleration of point D relative to the fixed reference frame.

SOLUTION

We shall make use of the following results from Example 7.2:

$$\boldsymbol{p} = \cos\gamma\,\boldsymbol{i} - \sin\gamma\,\boldsymbol{k}$$

$$\boldsymbol{m} = \sin\theta\sin\gamma\,\boldsymbol{i} + \cos\theta\boldsymbol{j} + \sin\theta\cos\gamma\,\boldsymbol{k}$$

$$\boldsymbol{n} = (\cos\phi\cos\gamma - \sin\phi\cos\theta\sin\gamma)\boldsymbol{i} + (\sin\phi\sin\theta)\boldsymbol{j} - (\cos\phi\sin\gamma + \sin\phi\cos\theta\cos\gamma)\boldsymbol{k}$$

$${}^1\boldsymbol{\omega}_2 = 2.094\boldsymbol{j}\text{ rad/s}$$

$${}^1\boldsymbol{\omega}_3 = (-0.9069\boldsymbol{i} + 2.094\boldsymbol{j} + 0.5236\boldsymbol{k})\text{ rad/s}$$

$${}^1\boldsymbol{\omega}_4 = (-0.514\boldsymbol{i} + 3.454\boldsymbol{j} + 1.204\boldsymbol{k})\text{ rad/s}$$

$${}^2\boldsymbol{r}_{B_2/A_2} = 2\boldsymbol{p} = (1.732\boldsymbol{i} - \boldsymbol{k})\text{ m}$$

$${}^3\boldsymbol{r}_{C_3/B_3} = 2\boldsymbol{m} = (0.5\boldsymbol{i} + 1.732\boldsymbol{j} + 0.866\boldsymbol{k})\text{ m}$$

$${}^4\boldsymbol{r}_{D_4/C_4} = 2\boldsymbol{n} = (-0.388\boldsymbol{i} + 0.966\boldsymbol{j} - 1.708\boldsymbol{k})\text{ m}$$

$${}^1\boldsymbol{v}_{B_2/A_2} = {}^1\boldsymbol{\omega}_2 \times {}^2\boldsymbol{r}_{B_2/A_2} = (-2.094\boldsymbol{i} - 3.627\boldsymbol{k})\text{ m/s}$$

$${}^1\boldsymbol{v}_{C_3/B_3} = {}^1\boldsymbol{\omega}_3 \times {}^3\boldsymbol{r}_{C_3/B_3} = (0.906\boldsymbol{i} + 1.047\boldsymbol{j} - 2.618\boldsymbol{k})\text{ m/s}$$

$${}^1\boldsymbol{v}_{D_4/C_4} = {}^1\boldsymbol{\omega}_4 \times {}^4\boldsymbol{r}_{D_4/C_4} = (-7.062\boldsymbol{i} - 1.345\boldsymbol{j} + 0.844\boldsymbol{k})\text{ m/s}$$

The fundamental equation that must be solved to obtain ${}^1\boldsymbol{a}_{D_4/A_1}$ is

$${}^1\mathbf{a}_{D_4/A_1} = {}^1\mathbf{a}_{D_4/C_4} + {}^1\mathbf{a}_{C_3/B_3} + {}^1\mathbf{a}_{B_2/A_2}$$

The individual terms in this equation will be evaluated separately. However, it is first necessary to evaluate the angular accelerations relative to the fixed reference frame:

The relative angular acceleration expressions are

$${}^1\boldsymbol{\alpha}_2 = \ddot{\gamma}\boldsymbol{j} = (3\boldsymbol{j})\text{ rad/s}^2$$

$${}^2\boldsymbol{\alpha}_3 = \ddot{\theta}\boldsymbol{p} = 1\boldsymbol{p} = 1(\cos\gamma\,\boldsymbol{i} - \sin\gamma\,\boldsymbol{k}) = (0.866\boldsymbol{i} - 0.500\boldsymbol{k})\text{ rad/s}^2$$

$$\begin{aligned}{}^3\boldsymbol{\alpha}_4 &= \ddot{\phi}\boldsymbol{m} = -2\boldsymbol{m} = -2(\sin\theta\sin\gamma\,\boldsymbol{i} + \cos\theta\boldsymbol{j} + \sin\theta\cos\gamma\,\boldsymbol{k})\\ &= (-0.500\boldsymbol{i} - 1.732\boldsymbol{j} - 0.866\boldsymbol{k})\text{ rad/s}\end{aligned}$$

so the angular accelerations relative to the fixed reference frame are

$${}^1\boldsymbol{\alpha}_2 = \ddot{\gamma}\boldsymbol{j} = (3\boldsymbol{j})\text{ rad/s}^2$$

$$\begin{aligned}{}^1\boldsymbol{\alpha}_3 &= {}^1\boldsymbol{\alpha}_2 + {}^2\boldsymbol{\alpha}_3 + {}^1\boldsymbol{\omega}_2 \times {}^2\boldsymbol{\omega}_3 = 3\boldsymbol{j} + 0.866\boldsymbol{i} - 0.500\boldsymbol{k} + 2.094\boldsymbol{j} \times (-0.9069\boldsymbol{i} + 0.5236\,\boldsymbol{k})\\ &= (1.962\boldsymbol{i} + 3\boldsymbol{j} + 1.399\boldsymbol{k})\text{ rad/s}^2\end{aligned}$$

$$\begin{aligned}{}^1\boldsymbol{\alpha}_4 &= {}^1\boldsymbol{\alpha}_3 + {}^3\boldsymbol{\alpha}_4 + {}^1\boldsymbol{\omega}_3 \times {}^3\boldsymbol{\omega}_4 = 1.962\boldsymbol{i} + 3\boldsymbol{j} + 1.399\boldsymbol{k} - 0.500\boldsymbol{i} - 1.732\boldsymbol{j} - 0.866\boldsymbol{k}\\ &\quad + (-0.907\boldsymbol{i} + 2.094\boldsymbol{j} + 0.524\boldsymbol{k}) \times (0.3927\boldsymbol{i} + 1.360\boldsymbol{j} + 0.6802\boldsymbol{k})\\ &= (2.174\boldsymbol{i} + 2.091\boldsymbol{j} - 1.523\boldsymbol{k})\text{ rad/s}^2\end{aligned}$$

Now the individual acceleration terms are

$${}^1\boldsymbol{a}^r_{B_2/A_2} = {}^1\boldsymbol{\omega}_2 \times {}^1\boldsymbol{v}_{B_2/A_2} = (2.094\boldsymbol{j}) \times (-2.094\boldsymbol{i} - 3.627\boldsymbol{k}) = (-7.595\boldsymbol{i} + 4.385\boldsymbol{k})\text{ m/s}^2$$

$${}^1\boldsymbol{a}^t_{B_2/A_2} = {}^1\boldsymbol{\alpha}_2 \times {}^2\boldsymbol{r}_{B_2/A_2} = (3\boldsymbol{j}) \times (1.732\boldsymbol{i} - \boldsymbol{k}) = (-3\boldsymbol{i} - 5.196\boldsymbol{k})\text{ m/s}^2$$

$$
\begin{aligned}
{}^1\boldsymbol{a}^r_{C_3/B_3} &= {}^1\boldsymbol{\omega}_3 \times {}^1\boldsymbol{v}_{C_3/B_3} = (-0.907\boldsymbol{i} + 2.094\boldsymbol{j} + 0.524\boldsymbol{k}) \times (0.906\boldsymbol{i} + 1.047\boldsymbol{j} - 2.618\boldsymbol{k}) \\
&= (-6.031\boldsymbol{i} - 1.900\boldsymbol{j} - 2.847\boldsymbol{k})\ \mathrm{m/s^2} \\
{}^1\boldsymbol{a}^t_{C_3/B_3} &= {}^1\boldsymbol{\alpha}_3 \times {}^3\boldsymbol{r}_{C_3/B_3} = (1.962\boldsymbol{i} + 3\boldsymbol{j} + 1.399\boldsymbol{k}) \times (0.5\boldsymbol{i} + 1.732\boldsymbol{j} + 0.866\boldsymbol{k}) \\
&= (0.175\boldsymbol{i} - 1.00\boldsymbol{j} + 1.898\boldsymbol{k})\ \mathrm{m/s^2} \\
{}^1\boldsymbol{a}^r_{D_4/C_4} &= {}^1\boldsymbol{\omega}_4 \times {}^1\boldsymbol{v}_{D_4/C_4} = (-0.514\boldsymbol{i} + 3.454\boldsymbol{j} + 1.204\boldsymbol{k}) \times (-7.062\boldsymbol{i} - 1.345\boldsymbol{j} + 0.844\boldsymbol{k}) \\
&= (4.535\boldsymbol{i} - 8.069\boldsymbol{j} + 25.083\boldsymbol{k})\ \mathrm{m/s^2} \\
{}^1\boldsymbol{a}^t_{D_4/C_4} &= {}^1\boldsymbol{\alpha}_4 \times {}^4\boldsymbol{r}_{D_4/C_4} = (2.174\boldsymbol{i} + 2.091\boldsymbol{j} - 1.523\boldsymbol{k}) \times (-0.388\boldsymbol{i} + 0.966\boldsymbol{j} - 1.708\boldsymbol{k}) \\
&= (-2.100\boldsymbol{i} + 4.304\boldsymbol{j} + 2.911\boldsymbol{k})\ \mathrm{m/s^2}
\end{aligned}
$$

The total acceleration of point D is now given by

$$
\begin{aligned}
{}^1\boldsymbol{a}_{D_4/A_1} &= (-7.595 - 3.000 - 6.031 + 0.175 + 4.535 - 2.100)\boldsymbol{i} \\
&\quad + (-1.900 - 1.000 - 8.069 + 4.304)\boldsymbol{j} \\
&\quad + (4.385 - 5.196 - 2.847 + 1.898 + 25.083 + 2.911)\boldsymbol{k} \\
&= (-14.02\boldsymbol{i} - 6.67\boldsymbol{j} + 26.23\boldsymbol{k})\ \mathrm{m/s^2} = 30.48(-0.460\boldsymbol{i} - 0.219\boldsymbol{j} + 0.861\boldsymbol{k})\ \mathrm{m/s^2}
\end{aligned}
$$

The final form of the result is expressed as the magnitude of the acceleration times a unit vector in the direction of the acceleration. Problems of this type can be solved relatively quickly with the aid of a programmable hand calculator set up to solve three-dimensional cross-products. Less powerful calculators will require some repetitious mathematical manipulations by the user but still enable such problems to be solved in a reasonable time without resorting to the use of a computer. ■

7.2 ROBOTIC MECHANISMS

Robotic mechanisms form a very important class of spatial mechanisms. They are distinct from most of the mechanisms discussed elsewhere in this book in that they have a large number of actuated degrees of freedom. Whereas many mechanisms have only one degree of freedom, actuated by a single drive, robotic mechanisms have four, five, six, or more with independent actuation of each degree of freedom. This leads to the problem of *coordination,* which may be stated: given the desired motion of the robot "hand," what should the commanded values of the joint actuator variables be to achieve that motion? In a complex mobility one mechanism there may be a considerable number of output motions all driven by a single input. These motions are coordinated kinematically via the mechanism. In a robotic mechanism, on the other hand, the coordination is done electronically by a computer integrated into the machine. The need for a theory for robotic coordination has led to the formulation of the following important subproblems.

Two important types of problems in robotic coordination have come to be known as the direct and inverse kinematics problems. They may be stated as follows.

The direct kinematics problem for a serial chain such as that shown in Fig. 7.5 is

Given the positions of all of the joints, find the position of the "hand" relative to the "base."

We will commonly refer to the free end member of a serial chain as the "hand," following from the common application of this type of analysis to robotic systems. In robotics, the free end member is also often called the "end effector." This term results from its function as being the tool with which a manipulator interacts with the workpiece. We use "hand" for convenience, although the end member may not resemble a human hand and need not be a tool. Likewise, it is convenient to refer to the other end member as the "base," even though in some situations it may not be fixed relative to the Earth.

A more precise and general statement of the direct kinematics problem is

Given the values of a number of joint parameters equal to the mobility (number of degrees of freedom) of a mechanism, find the relative position of any two designated members.

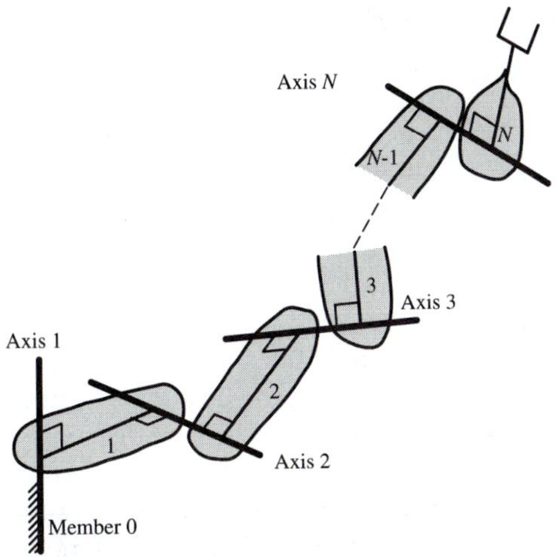

Figure 7.5 A general serial chain with axial joints. The chain has N joint axes, labeled as 1 to N, and $N + 1$ members labeled as 0 to N. Member 0 is the fixed base member.

The inverse kinematics problem for the serial chain shown in Fig. 7.5 is

Given the position of the hand relative to the base, find the positions of all of the joints.

Once again, this is a simplified statement applying only to the serial chain. A more general statement is

Given the relative positions of two members of a mechanism, find all of the joint parameters.

As will be seen later, it is necessary to qualify further the statements of the inverse kinematics problem. Those given above apply without qualification only to the case in which the mechanism has six degrees of freedom.

Analogs of the direct and inverse kinematics problems can be stated in terms of joint rates and member velocities. These are also very important problems in the development of robotic software. They can be stated for the serial chain as follows.

The direct rate kinematics problem for the serial chain is

Given the positions of all members of the chain and the rates of motion about all joints, find the total velocity of the hand.

Here the rate of motion about the joint is the angular velocity of rotation about a revolute joint or the translational velocity of sliding along a prismatic joint. The total velocity of a member is the velocity of the origin of the reference frame fixed to it combined with its angular velocity. That is, the total velocity has six independent components and, therefore, completely represents the velocity field of the member.

The inverse rate kinematics problem for a serial chain is

Given the positions of all members of the chain and the total velocity of the hand, find the rates of motion of all joints.

It is important to notice that these definitions include an assumption that the position of the mechanism is completely known. In many situations, this means that either the direct or inverse position kinematics problem must be solved before the direct or inverse rate kinematics problem can be addressed.

7.3 DIRECT POSITION KINEMATICS OF SERIAL CHAINS

7.3.1 Introduction

The problem to be addressed here is that of finding the hand position of a serial chain when the joint positions are known. This is a very important problem from the point of view of constructing manipulator coordination algorithms. The reason is that the joint positions are directly measured by sensors mounted on the joints. It is necessary to compute the positions of the joint axes relative to the fixed reference frame. This is done by means of the solution presented below.

The first problem in developing a mathematical model of a spatial mechanism, such as a robot, is that of algebraically describing the position of a body in space. There are many ways to do this. The approach we will use here is the most common one in which the position of a body is represented by the elements of a matrix coordinate transformation. This will be done by writing a position transformation between two Cartesian coordinate frames, one in the body and the other fixed. The parameters that describe this transformation can then be said to describe the position of the body relative to the fixed frame.

Let $[x, y, z]$ be a fixed reference frame and $[x', y', z']$ a reference frame fixed in the moving body (Fig. 7.6). The coordinates of the point P in the fixed reference frame may be obtained from its coordinates in the moving reference frame by a transformation of the form

$$\boldsymbol{p} = \boldsymbol{Q}\boldsymbol{p}' + \boldsymbol{q}$$
$$\boldsymbol{p} = \begin{bmatrix} x \\ y \\ z \end{bmatrix}, \qquad \boldsymbol{p}' = \begin{bmatrix} x' \\ y' \\ z' \end{bmatrix} \tag{7.1}$$

Here $\boldsymbol{Q}$ is a 3×3 matrix, and $\boldsymbol{q}$ is the position of the origin of the moving reference frame relative to the fixed reference frame.

Now, if the geometry of the moving body is known in terms of the reference frame (x', y', z') fixed in it, the position of any point in that body with respect to the fixed frame is also known provided the matrix $\boldsymbol{Q}$ and the vector $\boldsymbol{q}$ are known. The transformation can be regarded as describing the position of the moving body in space.

The matrix $\boldsymbol{Q}$ has a very special property: It is an orthogonal matrix. For an orthogonal matrix:

$$\boldsymbol{Q}^T\boldsymbol{Q} = \boldsymbol{I} \quad \text{or} \quad \boldsymbol{Q}^T = \boldsymbol{Q}^{-1} \tag{7.2}$$

The property of orthogonality follows from the assumption that the body is rigid.

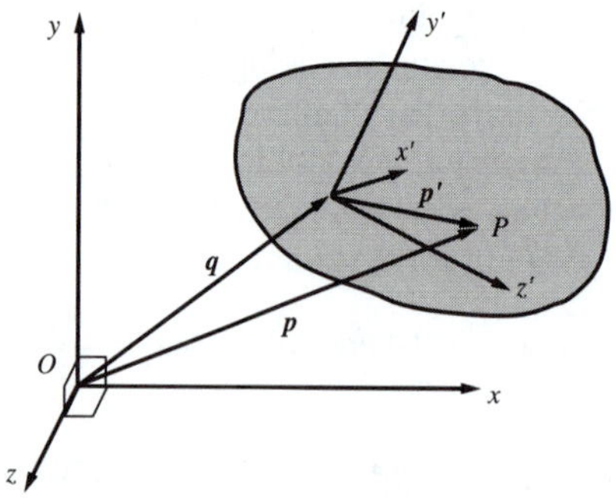

Figure 7.6 Relationship between body reference frame $[x', y', z']$ and fixed reference frame $[x, y, z]$. P is an arbitrary point, $\boldsymbol{p}'$ is the position of P in the body-fixed reference frame, and $\boldsymbol{p}$ is the position of that point in the fixed reference frame.

An important feature of the description of the position of a body in space by coordinate transformation elements, that is, the elements of $\boldsymbol{Q}$ and $\boldsymbol{q}$, is that there is a total of 12 elements in these two entities, 9 in $\boldsymbol{Q}$:

$$\boldsymbol{Q} = \begin{bmatrix} q_{11} & q_{12} & q_{13} \\ q_{21} & q_{22} & q_{23} \\ q_{31} & q_{32} & q_{33} \end{bmatrix}$$

and 3 in $\boldsymbol{q}$:

$$\boldsymbol{q} = \begin{bmatrix} q_1 \\ q_2 \\ q_3 \end{bmatrix}$$

Since only 6 coordinates are needed to fix the position of a body in space, these 12 quantities cannot be independent. There must be at least six independent relationships between them. The orthogonality condition, $\boldsymbol{Q}^T\boldsymbol{Q} = \boldsymbol{I}$, provides these relationships.

$$\boldsymbol{Q}^T\boldsymbol{Q} = \begin{bmatrix} q_{11} & q_{21} & q_{31} \\ q_{12} & q_{22} & q_{32} \\ q_{13} & q_{23} & q_{33} \end{bmatrix} \begin{bmatrix} q_{11} & q_{12} & q_{13} \\ q_{21} & q_{22} & q_{23} \\ q_{31} & q_{32} & q_{33} \end{bmatrix} = \begin{bmatrix} 1 & 0 & 0 \\ 0 & 1 & 0 \\ 0 & 0 & 1 \end{bmatrix}$$

giving

$$\begin{aligned}
q_{11}^2 + q_{21}^2 + q_{31}^2 &= 1 && \textbf{(i)} \\
q_{11}q_{12} + q_{21}q_{22} + q_{31}q_{32} &= 0 && \textbf{(ii)} \\
q_{11}q_{13} + q_{21}q_{23} + q_{31}q_{33} &= 0 && \textbf{(iii)} \\
q_{12}q_{11} + q_{22}q_{21} + q_{32}q_{31} &= 0 && \\
q_{12}^2 + q_{22}^2 + q_{32}^2 &= 1 && \textbf{(iv)} \quad \textbf{(7.3)} \\
q_{12}q_{13} + q_{22}q_{23} + q_{32}q_{33} &= 0 && \textbf{(v)} \\
q_{13}q_{11} + q_{23}q_{21} + q_{33}q_{31} &= 0 && \\
q_{13}q_{12} + q_{23}q_{22} + q_{33}q_{32} &= 0 && \\
q_{13}^2 + q_{23}^2 + q_{33}^2 &= 1 && \textbf{(vi)}
\end{aligned}$$

As can be seen, three of these equations are repeated, so only the six marked are distinct. Therefore, only three of the elements of $\boldsymbol{Q}$ are independent. The rest can be generated from these three by means of Eqs. (i) to (vi). It follows that six variables are needed to specify the transformation (7.1): three elements of $\boldsymbol{Q}$ and the three elements of $\boldsymbol{q}$. That is, as expected, six coordinates are needed to describe the position of the body.

7.3.2 Concatenation of Transformations

In Section 7.3.1 the concept of representing the position of a body in space by means of a coordinate transformation was introduced. A reference frame is fixed to the body, and the transformation is written that converts the coordinates of points relative to that frame into their coordinates relative to the fixed frame. It has the form

$$\boldsymbol{p} = \boldsymbol{Q}\boldsymbol{p}' + \boldsymbol{q} \qquad \textbf{(7.1)}$$

Here $\boldsymbol{p}'$ is the position of a point relative to the reference frame fixed in the body, $\boldsymbol{p}$ is the position of the same point referred to the fixed reference frame, $\boldsymbol{Q}$ is a 3×3 matrix that rotates the body reference frame until it is parallel to the fixed frame, and $\boldsymbol{q}$ is the position of the origin of the body reference frame relative to the fixed frame.

We will now consider the application of this model to a serial chain in which $N + 1$

members are connected by N joints with each member having two joints formed in it except for the two end members, which have only one joint each. Figure 7.5 shows a general serial chain.

The important feature here is to establish a convention that defines a consistent location for the reference frame in each member of the chain. The transformations representing the positions of the joints can then be applied successively to produce the transformation relating the end members of the chain.

Figure 7.7 shows the important geometric features of a binary link on which are mounted two axial joints. The term "axial joint" is used to define any joint that has a fixed axis of motion. Among lower pairs, revolute, screw, and cylindric joints are axial. A prismatic joint can also be regarded as axial with an axis in the direction of translation. The axis is, in this case, not unique and may be specified as lying along any convenient line parallel to the direction of translation. As will be seen, the restriction to axial joints allows simple expression of the transformation matrices. Because the ovewhelming majority of kinematic joints in practical use are axial joints, this simplification is not unduly restrictive.

The two joints mounted on member N are named joints N and $N + 1$. The common normal to the two joint axes, $S_N T_{N+1}$, has length a_N and the distance $T_N S_N$ along axis N between the foot of the common normal to joints $N - 1$ and N, and that to joints N and $N + 1$, is the offset, r_N. The angle between those two normals is the joint angle, θ_N. The positive direction of each normal for the purpose of defining this angle is directed from the lower numbered axis to the higher. The angle between axes N and $N + 1$ is the twist angle, α_N. The reference frame fixed to the Nth member has its origin located at T_{N+1}: the foot of the common normal on axis $N + 1$. The x_N axis is aligned with the common

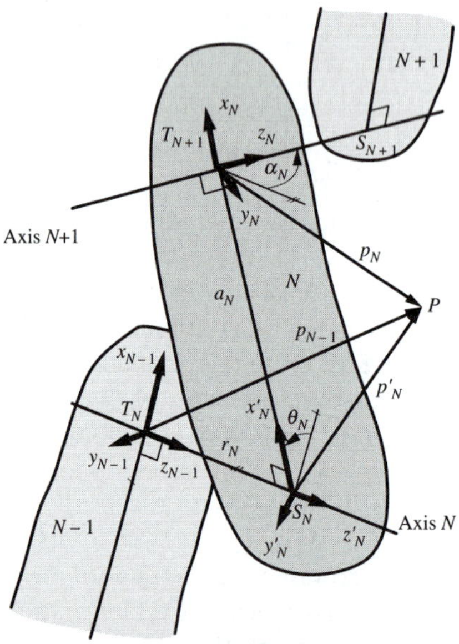

Figure 7.7 The "skeleton" of a binary link. A binary link is one with two axial joints. The geometry of the link and the position of the joint on axis N are completely defined by the parameters a_N, the length of the common normal of the joint axes; r_N, the distance along axis N from the foot of the common normal of link $N - 1$ to that of link N; α_N, the twist angle between axis N and axis $N + 1$; and θ_N, the angle between the normal of link $N - 1$ and that of link N.

normal, as shown, and the z_N axis lies along axis $N + 1$. Reference frame $N - 1$ is located in the corresponding position on member $N - 1$. An intermediate reference frame N', $[x', y', z']$, is also fixed to body N. Its origin is at S_N, the foot of the common normal on axis N. The x'_N axis lies along that normal, and the z'_N axis lies along axis N.

The joint axes together with the common normals on the links can be thought of as defining the "skeleton" of the linkage. The parameters a_N, r_N, α_N, and θ_N together completely define link N and the position of joint N.

Let us consider the transformation from reference frame N to frame N'. For the position of an arbitrary point P we have

$$\boldsymbol{p} = [x_N, y_N, z_N]^T \quad \text{and} \quad \mathbf{p}' = [x'_N, y'_N, z'_N]^T$$

where

$$x'_N = x_N + a_N$$
$$y'_N = y_N \cos\alpha_N - z_N \sin\alpha_N$$
$$z'_N = y_N \sin\alpha_N + z_N \cos\alpha_N$$

The last two equations are readily derived with the aid of Fig. 7.8, which shows a view of the system directed along the $x_N x'_N$ axis. The preceding equations can be written in the form

$$\boldsymbol{p}'_N = \boldsymbol{V}_N \boldsymbol{p}_N + \boldsymbol{v}_N$$

where

$$\boldsymbol{V}_N = \begin{bmatrix} 1 & 0 & 0 \\ 0 & \cos\alpha_N & -\sin\alpha_N \\ 0 & \sin\alpha_N & \cos\alpha_N \end{bmatrix} \quad \text{and} \quad \boldsymbol{v}_N = \begin{bmatrix} a_N \\ 0 \\ 0 \end{bmatrix}$$

The transformation from the $N - 1$ frame to the N' frame is obtained from

$$\boldsymbol{p}_{N\text{-}1} = [x_{N\text{-}1}, y_{N\text{-}1}, z_{N\text{-}1}]^T$$

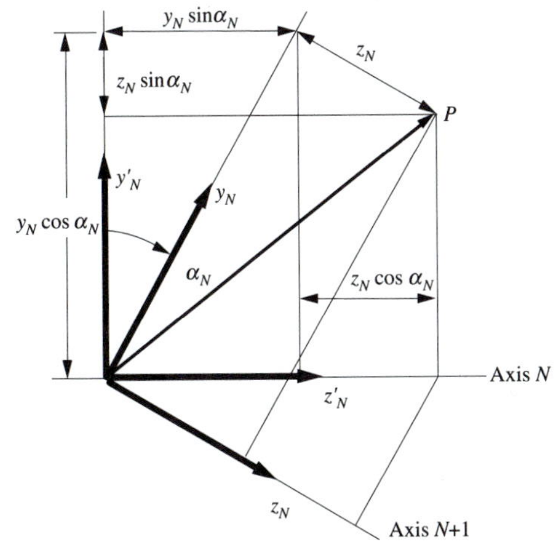

Figure 7.8 View along the common normal of axes N and $N + 1$. The coordinates of point P in the N and N' frames can be related by taking components as shown.

where

$$x_{N-1} = x'_N \cos\theta_N - y'_N \sin\theta_N$$
$$y_{N-1} = x'_N \sin\theta_N + y'_N \cos\theta_N$$
$$z_{N-1} = z'_N + r_N$$

Figure 7.9 might be of some assistance in understanding this. It shows frames $N-1$ and N' viewed along axis N. The equations can be written in the form

$$\boldsymbol{p}_{N-1} = \boldsymbol{U}_N \boldsymbol{p}'_N + \boldsymbol{u}_N \tag{7.5}$$

where

$$\boldsymbol{U}_N = \begin{bmatrix} \cos\theta_N & -\sin\theta_N & 0 \\ \sin\theta_N & \cos\theta_N & 0 \\ 0 & 0 & 1 \end{bmatrix}, \qquad \boldsymbol{u}_N = \begin{bmatrix} 0 \\ 0 \\ r_N \end{bmatrix}$$

Expanding Eq. (7.5),

$$\boldsymbol{p}_{N-1} = \boldsymbol{U}_N(\boldsymbol{V}_N \boldsymbol{p}_N + \boldsymbol{v}_N) + \boldsymbol{u}_N$$

which can be further rearranged to

$$\boldsymbol{p}_{N-1} = \boldsymbol{U}_N(\boldsymbol{V}_N \boldsymbol{p}_N + \boldsymbol{s}_N) \tag{7.6}$$

where

$$\boldsymbol{s}_N = \boldsymbol{v}_N + \boldsymbol{U}_N^T \boldsymbol{u}_N = [a_N, 0, r_N]^T$$

Equation 7.6 has the same form as Eq. (7.1) but is specialized for an axial joint.

Depending on which of the parameters of the transformation in Eq. (7.6) are allowed to vary, it can represent different types of axial joint. A revolute is obtained by allowing θ_N to vary while keeping all other dimensions constant. A prismatic joint is obtained by allowing r_N to vary while θ_N and all other dimensions remain constant.

It is now possible to write a transformation expressing the position of member N of a chain such as that shown in Fig. 7.5. The transformation is expressed in terms of the joint variables. For a chain containing only axial joints

$$\boldsymbol{p}_0 = \boldsymbol{U}_1(\boldsymbol{V}_1\boldsymbol{U}_2(\boldsymbol{V}_2\boldsymbol{U}_3(\ldots \boldsymbol{U}_N(\boldsymbol{V}_N\boldsymbol{p}_N + \boldsymbol{s}_N) + \cdots) + \boldsymbol{s}_2) + \boldsymbol{s}_1) \tag{7.7}$$

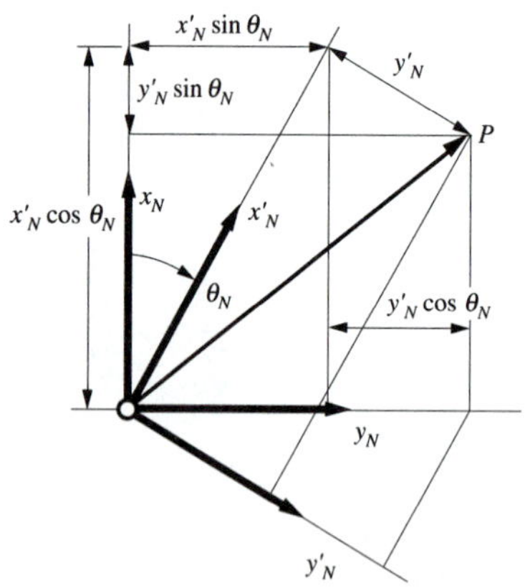

Figure 7.9 View along axis N. The coordinates of point P in the $N-1$ and N' frames can be related by taking components as shown.

This equation can be used to find the position in a fixed frame, $\boldsymbol{p}_0$, of any point given its position $\boldsymbol{p}_N$, relative to the end member, and the position coordinates of the joints in the chain. Alternatively, it may be used to derive equations between the joint position coordinates given the position of one end member relative to the other. In this latter case, the position of member N relative to member 0 is given by a known transformation

$$\boldsymbol{p}_0 = \boldsymbol{Q}\boldsymbol{p}_N + \boldsymbol{q} \qquad \textbf{(7.1)}$$

Since this transformation and Eq. (7.7) must be identical, we have

$$\boldsymbol{Q} = \boldsymbol{U}_1\boldsymbol{V}_1\boldsymbol{U}_2\boldsymbol{V}_2\boldsymbol{U}_3 \cdots \boldsymbol{U}_N\boldsymbol{V}_N \qquad \textbf{(7.8a)}$$

The scalar component equations from Eq.(7.8a) are called the rotation equations. Also

$$\boldsymbol{q} = \boldsymbol{U}_1\boldsymbol{V}_1\boldsymbol{U}_2\boldsymbol{V}_2\boldsymbol{U}_3 \ldots \boldsymbol{V}_{N-1}\boldsymbol{U}_N\boldsymbol{s}_N + \cdots + \boldsymbol{U}_1\boldsymbol{V}_1\boldsymbol{U}_2\boldsymbol{s}_2 + \boldsymbol{U}_1\boldsymbol{s}_1 \qquad \textbf{(7.8b)}$$

The scalar component equations from Eq. (7.8b) are called the translation equations.

7.3.3 Homogeneous Transformations

The transformation

$$\boldsymbol{p}_{N-1} = \boldsymbol{U}_N(\boldsymbol{V}_N\boldsymbol{p}_N + \boldsymbol{s}_N) \qquad \textbf{(7.6)}$$

that converts from the coordinate frame on link N to that on link $N - 1$ can be written as a single, four-dimensional matrix-vector multiplication:

$$\boldsymbol{p}'_{N-1} = \boldsymbol{A}_N\boldsymbol{p}'_N \qquad \textbf{(7.9)}$$

where

$$\boldsymbol{A}_N = \begin{bmatrix} \cos\theta_N & -\sin\theta_N\cos\alpha_N & \sin\theta_N\sin\alpha_N & a_N\cos\theta_N \\ \sin\theta_N & \cos\theta_N\cos\alpha_N & -\cos\theta_N\sin\alpha_N & a_N\sin\theta_N \\ 0 & \sin\alpha_N & \cos\alpha_N & r_N \\ 0 & 0 & 0 & 1 \end{bmatrix} \qquad \textbf{(7.10)}$$

and

$$\boldsymbol{p}'_N = \begin{bmatrix} x_N \\ y_N \\ z_N \\ 1 \end{bmatrix} \qquad \textbf{(7.11)}$$

The three-by-three submatrix in the top left-hand corner of $\boldsymbol{A}_N$ can be recognized as $\boldsymbol{U}_N\boldsymbol{V}_N$, while the fourth column is $\boldsymbol{U}_N\boldsymbol{s}_N$ with 1 added as the fourth element. The reader may verify that, when expanded, Eqs. (7.6) and (7.9) give the same three component equations. The fourth equation given by the latter is $1 = 1$.

Using this formulation, Eqs. (7.8a) and (7.8b), which relate hand position to the joint variables, assume the form (For $N = 6$)

$$\boldsymbol{Q}' = \boldsymbol{A}_1\boldsymbol{A}_2\boldsymbol{A}_3\boldsymbol{A}_4\boldsymbol{A}_5\boldsymbol{A}_6 \qquad \textbf{(7.12)}$$

where

$$\boldsymbol{Q}' = \begin{bmatrix} q_{11} & q_{12} & q_{13} & q_1 \\ q_{21} & q_{22} & q_{23} & q_2 \\ q_{31} & q_{32} & q_{33} & q_3 \\ 0 & 0 & 0 & 1 \end{bmatrix} \qquad \textbf{(7.13)}$$

This very compact formulation is attractive and variants of it have been extensively used in spatial mechanism theory, computer graphics, and robotics. It is particularly attractive when ease of programming is the most important consideration.

Equation (7.9) is called a homogeneous transformation because it has the form of a transformation of the homogeneous coordinates used in projective geometry.

EXAMPLE 7.4 *(Computation of Homogeneous Transformation)*

PROBLEM

The fixed geometric parameters of the manipulator shown in Fig. 7.10 are shown in Table 7.1. If the joint variables have the values $\theta_1 = 60°$, $\theta_2 = 120°$, $\theta_3 = 135°$, $r_4 = 4.0$, $\theta_5 = -60°$, and $\theta_6 = 45°$, find the homogeneous transformation $\boldsymbol{Q}'$ that specifies the position of the hand.

Notice that, because joint 4 is a prismatic joint, the joint variable is r_4, and θ_4 is a fixed geometric parameter. All of the other joints are revolutes, so that θ_i is the joint variable, and r_i is a fixed geometric parameter for all i except 4.

SOLUTION

Substitution of a_1, α_1, and r_1 from Table 7.1, together with the value of θ_1 in Eq. (7.10), gives

$$\mathbf{A}_1 = \begin{bmatrix} 0.5 & 0 & 0.8660 & 0 \\ 0.8660 & 0 & -0.5 & 0 \\ 0 & 1 & 0 & 0 \\ 0 & 0 & 0 & 1 \end{bmatrix}$$

Similarly,

$$\mathbf{A}_2 = \begin{bmatrix} -0.5 & 0 & 0.8660 & 0 \\ 0.8660 & 0 & 0.5 & 0 \\ 0 & 1 & 0 & 0 \\ 0 & 0 & 0 & 1 \end{bmatrix}, \quad \mathbf{A}_3 = \begin{bmatrix} -0.7070 & -0.7070 & 0 & 0 \\ 0.7070 & -0.7070 & 0 & 0 \\ 0 & 0 & 1 & 0 \\ 0 & 0 & 0 & 1 \end{bmatrix},$$

$$\mathbf{A}_4 = \begin{bmatrix} 1 & 0 & 0 & 0 \\ 0 & 0 & -1 & 0 \\ 0 & 1 & 0 & 4 \\ 0 & 0 & 0 & 1 \end{bmatrix}$$

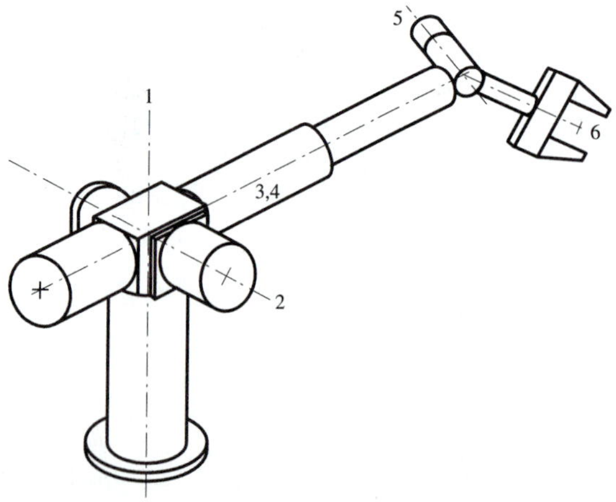

Figure 7.10 The manipulator of Example 7.4. The geometric parameters are summarized in Table 7.1.

Table 7.1 Geometry Parameters for Example 7.4

i	a_i	r_i	α_i	θ_i
1	0	0	90°	60°
2	0	0	90°	120°
3	0	0	0	135°
4	0	4.0	90°	0
5	0	0	90°	−60°
6	0	1.0	0	45°

$$\boldsymbol{A}_5 = \begin{bmatrix} 0.5 & 0 & -0.8660 & 0 \\ -0.8660 & 0 & -0.5 & 0 \\ 0 & 1 & 0 & 0 \\ 0 & 0 & 0 & 1 \end{bmatrix}, \quad \boldsymbol{A}_6 = \begin{bmatrix} 0.7070 & -0.7070 & 0 & 0 \\ 0.7070 & 0.7070 & 0 & 0 \\ 0 & 0 & 1 & 1 \\ 0 & 0 & 0 & 1 \end{bmatrix}$$

Applying Eq. (7.12), we multiply these matrices together to get

$$\boldsymbol{Q}' = \boldsymbol{A}_1\boldsymbol{A}_2\boldsymbol{A}_3\boldsymbol{A}_4\boldsymbol{A}_5\boldsymbol{A}_6 = \begin{bmatrix} 0.3220 & 0.2945 & -0.8998 & 0.8323 \\ -0.9423 & 0.0096 & -0.3340 & 2.6660 \\ -0.0897 & 0.9555 & 0.2803 & 2.2803 \\ 0 & 0 & 0 & 1 \end{bmatrix}$$

Then $\boldsymbol{Q}'$ is the transformation matrix that describes the position of the hand. ■

An inconvenience of working with the $\boldsymbol{A}_N$ matrices, rather than the $\boldsymbol{U}_N$ and $\boldsymbol{V}_N$ matrices from which they are derived, is that, unlike $\boldsymbol{U}_N$ and $\boldsymbol{V}_N$, $\boldsymbol{A}_N$ is not an orthogonal matrix. That is,

$$\boldsymbol{A}_N^{-1} \neq \boldsymbol{A}_N^T$$

In order to derive an expression for $\boldsymbol{A}_N^{-1}$ it is most efficient to return to Eq. (7.6)

$$\boldsymbol{p}_{N-1} = \boldsymbol{U}_N(\boldsymbol{V}_N\boldsymbol{p}_N + \boldsymbol{s}_N) \tag{7.6}$$

Premultiplication of both sides by $\boldsymbol{U}_N^T$ gives

$$\boldsymbol{U}_N^T\boldsymbol{p}_{N-1} = \boldsymbol{V}_N\boldsymbol{p}_N + \boldsymbol{s}_N$$

Moving $\boldsymbol{s}_N$ to the other side of the equation and then premultiplying by $\boldsymbol{V}_N^T$ gives

$$\boldsymbol{p}_N = \boldsymbol{V}_N^T(\boldsymbol{U}_N^T\boldsymbol{p}_{N-1} - \boldsymbol{s}_N) \tag{7.14}$$

In the same manner as Eq. (7.10), this expression can be written as a 4×4 matrix expression:

$$\boldsymbol{p}'_N = \boldsymbol{A}_N^{-1}\boldsymbol{p}'_{N-1} \tag{7.15}$$

where

$$\boldsymbol{A}_N^{-1} = \begin{bmatrix} \cos\theta_N & \sin\theta_N & 0 & -a_N \\ -\sin\theta_N\cos\alpha_N & \cos\theta_N\cos\alpha_N & \sin\alpha_N & -r_N\sin\alpha_N \\ \sin\theta_N\sin\alpha_N & -\cos\theta_N\sin\alpha_N & \cos\alpha_N & -r_N\cos\alpha_N \\ 0 & 0 & 0 & 1 \end{bmatrix} \tag{7.16}$$

and

$$p'_N = \begin{bmatrix} x_N \\ y_N \\ z_N \\ 1 \end{bmatrix} \tag{7.11}$$

as before.

The reader may verify that the expressions for A_N and A_N^{-1} given by Eqs. (7.10) and (7.16) satisfy the relationship

$$A_N^{-1}A_N = I$$

7.4 INVERSE POSITION KINEMATICS

As stated earlier, the inverse kinematics problem for the serial chain shown in Fig. 7.5 is

Given the position of the hand relative to the base, find the positions of all of the joints.

This amounts to finding the joint parameters if the homogeneous matrix, Q', of Eq. (7.12) is given. Solution requires expansion of Eq. (7.12) using the expression given by Eq. (7.10) for each of the matrices A_N. N of the resulting set of scalar equations are selected to be solved for the variables θ_i, $i = 1, \ldots, N$, describing the positions of revolute joints, or r_i describing the positions of prismatic joints. The resulting equations are nonlinear and are sometimes very difficult to solve, although the simple geometries commonly used are usually tractable. Nevertheless, the solution of the inverse position equations is beyond the scope of this book. Fortunately, this problem is of less practical importance than the direct position kinematics and inverse rate kinematics problems, which must be solved on line many times per second in most manipulator coordination schemes.

7.5 RATE KINEMATICS

7.5.1 Introduction

As stated earlier, the direct rate kinematics problem for the serial chain is

Given the positions of all members of the chain and the rates of motion about all joints, find the total velocity of the hand.

The rate of motion about the joint is the angular velocity of rotation about a revolute joint or the translational velocity of sliding along a prismatic joint. The total velocity of a member is the velocity of the origin of the reference frame fixed to it combined with its angular velocity. That is, the total velocity has six independent components and therefore completely represents the velocity field of the member.

The inverse rate kinematics problem for a serial chain is

Given the positions of all members of the chain and the total velocity of the hand, find the rates of motion of all joints.

Once again, it is important to notice that these definitions include an assumption that the position of the mechanism is completely known. In many situations, this means that either the direct or inverse position kinematics problem must be solved before the direct or inverse rate kinematics problem can be addressed.

When controlling a movement of an industrial robot that operates in the point-to-point mode, it is not only necessary to compute the final joint positions needed to assume the desired final hand position. It is also necessary to generate a smooth trajectory for motion between the initial and final positions. There are, of course, infinitely many possible trajectories for this purpose. However, the most straightforward and successful approach employs algorithms based on the solution of the inverse rate kinematics problem. This technique originated in the work of Whitney[1] and of Pieper and Roth.[2]

7.5.2 Direct Rate Kinematics

According to the chain rule for angular velocities developed in Section 2.11.3.2, the angular velocity of the hand of a serial chain, such as that shown in Fig. 7.5, with $N = 6$ joints (seven members including the base) is

$$^0\boldsymbol{\omega}_6 = {}^0\boldsymbol{\omega}_1 + {}^1\boldsymbol{\omega}_2 + {}^2\boldsymbol{\omega}_3 + {}^3\boldsymbol{\omega}_4 + {}^4\boldsymbol{\omega}_5 + {}^5\boldsymbol{\omega}_6 \tag{7.15}$$

Here the base member to which the fixed reference frame is attached is referred to as member 0. Each member has a reference frame attached to it, and so $^i\boldsymbol{\omega}_j$ is the angular velocity of member j relative to a reference frame fixed to member i. If i and j are consecutive members, $^i\boldsymbol{\omega}_j$ becomes the angular velocity of one member relative to the other about the joint that connects them. That is,

$$^{i-1}\boldsymbol{\omega}_i = \dot{\theta}_i\,{}^0\mathbf{w}_i \tag{7.16}$$

where $^0\mathbf{w}_i$ is a unit vector having the direction of axis i.

Combining Eq. (7.16) with Eq. (7.15), the angular velocity $^0\boldsymbol{\omega}_N$, of the hand is related to the joint rates by

$$^0\boldsymbol{\omega}_N = \sum_{k=1}^{N} \dot{\theta}_k\,{}^0\mathbf{w}_k \tag{7.17}$$

where $^0\mathbf{w}_k$ is a unit vector having the direction of joint axis k.

The velocity of the origin, O_N, of the reference frame fixed to the hand (frame 6) relative to the fixed frame (frame 0) can be obtained by summing the components of that velocity due to motion about the respective joints. The velocity of point O_N relative to frame 0 produced by motion about joint i is

$$^0\mathbf{v}_{O_N,i} = {}^{i-1}\boldsymbol{\omega}_i \times {}^0\boldsymbol{r}_i \tag{7.18}$$

where $^0\boldsymbol{r}_i$ is the vector from any point on axis i to point O_N. Following the convention of Fig. 7.7, the origin, O_{i-1} of frame $i - 1$ is on axis i. Hence $^0\boldsymbol{r}_i$ can be interpreted as the vector from O_{i-1} to O_N referred to the fixed reference frame. That is, when $N = 6$,

$$^0\mathbf{v}_{O_6} = {}^0\mathbf{v}_{O_6,1} + {}^0\mathbf{v}_{O_6,2} + {}^0\mathbf{v}_{O_6,3} + {}^0\mathbf{v}_{O_6,4} + {}^0\mathbf{v}_{O_6,5} + {}^0\mathbf{v}_{O_6,6}$$

More generally,

$$^0\mathbf{v}_{O_N} = \sum_{i=1}^{N} {}^{i-1}\boldsymbol{\omega}_i \times {}^0\boldsymbol{r}_i = \sum_{i=1}^{N} \dot{\theta}_i\,{}^0\mathbf{w}_i \times {}^0\boldsymbol{r}_i$$

Now, if $^0\boldsymbol{\rho}_i$ is the position, relative to the fixed reference frame, of the origin of reference frame $i - 1$ on joint axis i, then

$$^0\boldsymbol{r}_i = {}^0\boldsymbol{\rho}_{N+1} - {}^0\boldsymbol{\rho}_i$$

Here the position of O_N is indicated by $^0\boldsymbol{\rho}_{N+1}$. In general, reference frame 6 is not placed with its origin on a joint axis, since there is no axis $N + 1$. However, for consistency, it is convenient to use the same notation for the location of the origin of frame N as for the other reference frames.

Therefore,

$$\begin{aligned} ^0\mathbf{v}_{O_N} &= \sum_{i=1}^{N} \dot{\theta}_i\,{}^0\mathbf{w}_i \times ({}^0\boldsymbol{\rho}_{N+1} - {}^0\boldsymbol{\rho}_i) \\ ^0\mathbf{v}_{O_N} &= {}^0\boldsymbol{\omega}_N \times {}^0\boldsymbol{\rho}_{N+1} - \sum_{i=1}^{N} \dot{\theta}_i\,{}^0\mathbf{w}_i \times {}^0\boldsymbol{\rho}_i \end{aligned} \tag{7.19}$$

In order to make use of Eqs. (7.17) and (7.19) it is necessary to be able to calculate $\boldsymbol{w}_i$ and $\boldsymbol{\rho}_i$ for each joint axis. This is actually a problem of direct position kinematics.

The notation used is that of Fig. 7.7. Joint axes k and $k+1$ are fixed in member k. The reference frame on member k is placed with its z axis along joint axis $k+1$ and with its x axis in the direction of the common normal from joint axis k to $k+1$. The members and joints are numbered serially outward, the fixed frame being 0 and the hand 6.

In reference frame k, $\boldsymbol{w}_{k+1} = \boldsymbol{k} = [0, 0, 1, 0]^T$. A convenient point on axis $k+1$ to use is the origin of frame k. Hence, in frame k, $\boldsymbol{\rho}_{k+1} = \boldsymbol{0} = [0, 0, 0, 1]^T$.

Notice that in the 4×4 representation, the fourth element of a unit vector, such as $\boldsymbol{k}$, is zero. This is because a unit vector expresses only a direction, and hence the position information in the fourth column of the $\boldsymbol{A}_k$ matrix has no relevance. If the fourth element of the unit vector w were not zero, $|w|$ would no longer be one, and would change with the transformation by the $\boldsymbol{A}_k$ matrices. With zero as the fourth element, the magnitude of a unit vector is one and remains one when the vector is multiplied by $\boldsymbol{A}_k$.

Using Eq. (7.9), position vectors of a point ${}^k\boldsymbol{\rho}_k$ in frame k and the same point ${}^{k-1}\boldsymbol{\rho}_k$ in frame $k-1$ are related by

$$^{k-1}\boldsymbol{\rho}_k = \boldsymbol{A}_k {}^k\boldsymbol{\rho}_k \tag{7.20}$$

where

$$\boldsymbol{A}_k = \begin{bmatrix} \cos\theta_k & -\sin\theta_k\cos\alpha_k & \sin\theta_k\sin\alpha_k & a_k\cos\theta_k \\ \sin\theta_k & \cos\theta_k\cos\alpha_k & -\cos\theta_k\sin\alpha_k & a_k\sin\theta_k \\ 0 & \sin\alpha_k & \cos\alpha_k & r_k \\ 0 & 0 & 0 & 1 \end{bmatrix} \tag{7.10}$$

As indicated in Fig. 7.7, α_k is the angle between axes k and $k+1$, a_k is the length of the common normal between those axes, and r_k is the distance along axis k from the origin of frame $k-1$ to the foot of that common normal. A unit vector ${}^k\boldsymbol{u}_k$ in frame k is related to the same vector, ${}^{k-1}\boldsymbol{u}_k$ in frame $k-1$ by

$$^{k-1}\boldsymbol{u}_k = \boldsymbol{A}_k {}^k\boldsymbol{u}_k \tag{7.21}$$

Equations (7.18) and (7.19) can now be used recursively to obtain the vectors $\boldsymbol{\rho}_k$ and $\boldsymbol{w}_k$ referred to frame 0.

EXAMPLE 7.5 (*Direct Rate Kinematics of a Three-Axis Manipulator*)

PROBLEM

The inner three joint axes of a manipulator are shown in Fig. 7.11. The geometric parameters of the links, the joint positions, and the joint rates are tabulated in Table 7.2. Calculate the angular velocity of member 3 and the velocity of the wrist concurrency point, P, relative to the fixed reference frame.

SOLUTION

Since there are only three joint axes, we substitute $N = 3$ in Eqs. (7.17) and (7.19).

$$^0\boldsymbol{\omega}_3 = \sum_{k=1}^{3} \dot{\theta}_k {}^0\boldsymbol{w}_k \tag{a}$$

$$^0\boldsymbol{v}_{O_3} = {}^0\boldsymbol{\omega}_3 \times {}^0\boldsymbol{\rho}_4 - \sum_{i=1}^{3} \dot{\theta}_i {}^0\boldsymbol{w}_i \times {}^0\boldsymbol{\rho}_i \tag{b}$$

$$\boldsymbol{\rho}_1 = \boldsymbol{0}$$

$$^0\boldsymbol{w}_1 = \boldsymbol{k}$$

so

$$^0\boldsymbol{\omega}_1 = 1\boldsymbol{k} = \boldsymbol{k} \text{ rad/s}$$

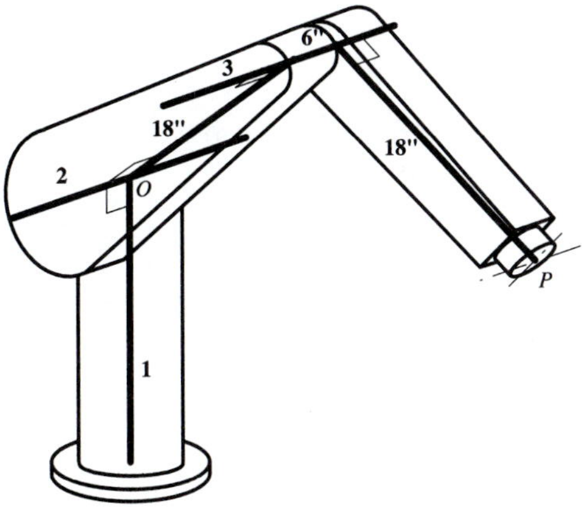

Figure 7.11 The three-axis manipulator analyzed in Example 7.5.

Applying Eq. (7.20)

$$\boldsymbol{\rho}_2 = \mathbf{A}_1\mathbf{0} = \begin{bmatrix} \cos(45°) & 0 & \sin(45°) & 0 \\ \sin(45°) & 0 & -\cos(45°) & 0 \\ 0 & 1 & 0 & 0 \\ 0 & 0 & 0 & 1 \end{bmatrix} \begin{bmatrix} 0 \\ 0 \\ 0 \\ 1 \end{bmatrix} = \mathbf{0}$$

and from Eq. (7.21)

$$^0\boldsymbol{w}_2 = \boldsymbol{A}_1\boldsymbol{k} = \begin{bmatrix} \cos(45°) & 0 & \sin(45°) & 0 \\ \sin(45°) & 0 & -\cos(45°) & 0 \\ 0 & 1 & 0 & 0 \\ 0 & 0 & 0 & 1 \end{bmatrix} \begin{bmatrix} 0 \\ 0 \\ 1 \\ 0 \end{bmatrix} = \begin{bmatrix} \sin(45°) \\ -\cos(45°) \\ 0 \\ 0 \end{bmatrix} = \begin{bmatrix} 0.7071 \\ -0.7071 \\ 0 \\ 0 \end{bmatrix} = 0.7071(\boldsymbol{i} - \boldsymbol{j})$$

Therefore

$$^1\boldsymbol{\omega}_2 = -0.75 \times 0.7071\,(\boldsymbol{i} - \boldsymbol{j}) = 0.5303(\boldsymbol{j} - \boldsymbol{i}) \text{ rad/s}$$

Similarly, applying Eq. (7.20) twice to transform from frame 2 to frame 0, we get

$$\boldsymbol{\rho}_3 = \boldsymbol{A}_1\boldsymbol{A}_2\mathbf{0} = \begin{bmatrix} \cos(45°) & 0 & \sin(45°) & 0 \\ \sin(45°) & 0 & -\cos(45°) & 0 \\ 0 & 1 & 0 & 0 \\ 0 & 0 & 0 & 1 \end{bmatrix} \begin{bmatrix} \cos(30°) & -\sin(30°) & 0 & 18\cos(30°) \\ \sin(30°) & \cos(30°) & 0 & 18\sin(30°) \\ 0 & 0 & 1 & 0 \\ 0 & 0 & 0 & 1 \end{bmatrix} \begin{bmatrix} 0 \\ 0 \\ 0 \\ 1 \end{bmatrix}$$

$$= \begin{bmatrix} 11.023 \\ 11.023 \\ 9.000 \\ 1 \end{bmatrix}$$

$$= 11.023\boldsymbol{i} + 11.023\boldsymbol{j} + 9.000\boldsymbol{k}$$

Table 7.2 Parameter for Robot in Fig. 7.11

i	a_i (in)	r_i (in)	α_i	θ_i	$\dot{\theta}_i$ (rad/s)
1	0	0	90°	45°	1
2	18	0	0°	30°	−0.75
3	18	6	0°	−60°	0.5

and applying Eq. (7.21) twice

$$
\begin{aligned}
{}^0\boldsymbol{w}_3 &= \boldsymbol{A}_1\boldsymbol{A}_2\boldsymbol{k} \\
&= \begin{bmatrix} \cos(45^\circ) & 0 & \sin(45^\circ) & 0 \\ \sin(45^\circ) & 0 & -\cos(45^\circ) & 0 \\ 0 & 1 & 0 & 0 \\ 0 & 0 & 0 & 1 \end{bmatrix} \begin{bmatrix} \cos(30^\circ) & -\sin(30^\circ) & 0 & 18\cos(30^\circ) \\ \sin(30^\circ) & \cos(30^\circ) & 0 & 18\sin(30^\circ) \\ 0 & 0 & 1 & 0 \\ 0 & 0 & 0 & 1 \end{bmatrix} \begin{bmatrix} 0 \\ 0 \\ 1 \\ 0 \end{bmatrix} \\
&= \begin{bmatrix} \sin(45^\circ) \\ -\cos(45^\circ) \\ 0 \\ 0 \end{bmatrix} = \begin{bmatrix} 0.7071 \\ -0.7071 \\ 0 \\ 0 \end{bmatrix} \\
&= 0.7071(\boldsymbol{i} - \boldsymbol{j}) = \boldsymbol{w}
\end{aligned}
$$

Since joint axes 2 and 3 are parallel, ${}^0\boldsymbol{w}_3$ should be equal to ${}^0\boldsymbol{w}_2$. Hence

$$
{}^2\boldsymbol{\omega}_3 = 0.5 \times 0.7071(\boldsymbol{i} - \boldsymbol{j}) = 0.3536(\boldsymbol{i} - \boldsymbol{j}) \text{ rad/s}
$$

Finally, we need the location of point P. Point P is the origin of reference frame 3, that, is point $\boldsymbol{\rho}_4$:

$$
\begin{aligned}
\boldsymbol{\rho}_4 &= \boldsymbol{A}_1\boldsymbol{A}_2\boldsymbol{A}_3\boldsymbol{0} \\
&= \begin{bmatrix} \cos(45^\circ) & 0 & \sin(45^\circ) & 0 \\ \sin(45^\circ) & 0 & -\cos(45^\circ) & 0 \\ 0 & 1 & 0 & 0 \\ 0 & 0 & 0 & 1 \end{bmatrix} \begin{bmatrix} \cos(30^\circ) & -\sin(30^\circ) & 0 & 18\cos(30^\circ) \\ \sin(30^\circ) & \cos(30^\circ) & 0 & 18\sin(30^\circ) \\ 0 & 0 & 1 & 0 \\ 0 & 0 & 0 & 1 \end{bmatrix} \\
&\quad \times \begin{bmatrix} \cos(60^\circ) & \sin(60^\circ) & 0 & 18\cos(60^\circ) \\ -\sin(60^\circ) & \cos(60^\circ) & 0 & -18\sin(60^\circ) \\ 0 & 0 & 1 & 6 \\ 0 & 0 & 0 & 1 \end{bmatrix} \begin{bmatrix} 0 \\ 0 \\ 0 \\ 1 \end{bmatrix} \\
&= \begin{bmatrix} 0.6123 & -0.3536 & 0.7071 & 11.022 \\ 0.6123 & -0.3536 & -0.7071 & 11.022 \\ 0.5000 & 0.8660 & 0 & 9.000 \\ 0 & 0 & 0 & 1 \end{bmatrix} \begin{bmatrix} 9.000 \\ -15.588 \\ 6 \\ 1 \end{bmatrix} \\
&= \begin{bmatrix} 26.288 \\ 17.803 \\ 0.001 \\ 1 \end{bmatrix} = 26.288\boldsymbol{i} + 17.803\boldsymbol{j} + 0.001\boldsymbol{k}
\end{aligned}
$$

Now, using Eq. (a)

$$
{}^0\boldsymbol{\omega}_3 = \boldsymbol{k} + 0.5303(\boldsymbol{j} - \boldsymbol{i}) + 0.3536(\boldsymbol{i} - \boldsymbol{j}) = -0.1767\boldsymbol{i} + 0.1767\boldsymbol{j} + \boldsymbol{k} \text{ rad/s}
$$

and using Eq. (b)

$$
\begin{aligned}
{}^0\boldsymbol{v}_{O_3} &= {}^0\boldsymbol{\omega}^3 \times {}^0\boldsymbol{\rho}_4 - \sum_{i=1}^{3} \dot{\theta}_i\, {}^0\boldsymbol{w}_i \times {}^0\boldsymbol{\rho}_i \\
&= {}^0\boldsymbol{\omega}_3 \times {}^0\boldsymbol{\rho}_4 - {}^0\boldsymbol{\omega}_1 \times {}^0\boldsymbol{\rho}_1 - {}^1\boldsymbol{\omega}_2 \times {}^0\boldsymbol{p}_2 - {}^2\boldsymbol{\omega}_3 \times {}^0\boldsymbol{\rho}_3
\end{aligned}
$$

or

$$
\begin{aligned}
{}^0\boldsymbol{v}_{O_3} &= (-0.1767\boldsymbol{i} + 0.1767\boldsymbol{j} + \boldsymbol{k}) \times (26.288\boldsymbol{i} + 17.803\boldsymbol{j} + 0.001\boldsymbol{k}) \\
&\quad -0.3536(\boldsymbol{i} - \boldsymbol{j}) \times (11.023\boldsymbol{i} + 11.023\boldsymbol{j} + 9.000\boldsymbol{k}) \\
&= -14.62\boldsymbol{i} + 29.47\boldsymbol{j} - 15.59\boldsymbol{k}
\end{aligned}
$$

■

7.5.3 Inverse Rate Kinematics

As is well known, the position of the hand relative to the fixed frame can be described by means of a set of six independent algebraic equations containing the joint position variables $\theta_1, \ldots, \theta_N$. There are numerous ways to formulate such a set of equations. Differentiation with respect to time of the direct position kinematics equations yields a set of equations of the form

$$\boldsymbol{v} = \boldsymbol{\Gamma}\,\dot{\boldsymbol{\theta}} \tag{7.22}$$

where $\boldsymbol{v}$ is a six-component veolcity vector, $\dot{\boldsymbol{\theta}} = [\dot{\theta}_1, \dot{\theta}_2, \ldots, \dot{\theta}_N]^{\mathrm{T}}$ is an N-dimensional vector composed of the joint rates, and $\boldsymbol{\Gamma}$ is a $6 \times N$ matrix whose elements are, in general, nonlinear functions of $\theta_1, \ldots, \theta_N$. $\boldsymbol{\Gamma}$ is called the Jacobian matrix of this algebraic system. If the joint positions $\theta_1, \ldots, \theta_N$ are known, Eq. (7.22) yields six linear algebraic equations in the joint rates $\dot{\theta}_1, \ldots, \dot{\theta}_N$. If the joint rates are given, solution of Eq. (7.22) for $\boldsymbol{v}$ is a solution of the direct rate kinematics problem. Notice that $\boldsymbol{\Gamma}$ can be regarded as a known matrix for this purpose provided all the joint positions are known.

Solution of Eq. (7.22) for the joint rates is possible if $N = 6$. In the remainder of this chapter discussion will be limited to this case. A mode of control for $N > 6$ compatible with the work presented here makes use of pseudoinversion of the Jacobian to eliminate the ambiguity due to the additional degrees of freedom. This is beyond the scope of this book.

The ease of computation of the elements of the Jacobian matrix from the joint positions, and of inverse solution of Eq. (7.22), is very strongly dependent on the way in which this equation is formulated. Since these computations must be repeated at each time step in a real-time rate coordination algorithm, the formulation becomes critical to the efficiency of the software. A very elegant and efficient geometric relationship between the rates of motion about the joints of a linkage and the velocity states of the members is available in the form of screw system theory. Therefore, in order to make use of geometric information to give an efficient formulation of Eq. (7.22) an algebraic formulation based on Eqs. (7.17) and (7.19) will be used.

The angular velocity, $\boldsymbol{\omega}$, of the hand of a six-axis serial manipulator is related to the joint rates by

$$^0\boldsymbol{\omega}_6 = \sum_{k=1}^{6} \dot{\theta}_k\,{}^0\boldsymbol{w}_k \tag{7.17}$$

where $^0\boldsymbol{w}_k$ is a unit vector having the direction of joint axis k.

Using Eq. (7.19),

$$^0\boldsymbol{v}_{\mathrm{O}_6} = {}^0\boldsymbol{\omega}_6 \times {}^0\boldsymbol{\rho}_7 - \sum_{i=1}^{6} \dot{\theta}_i\,{}^0\boldsymbol{w}_i \times {}^0\boldsymbol{\rho}_i \tag{7.19}$$

it is convenient in this case to combine $^0\boldsymbol{\omega}_6 \times {}^0\boldsymbol{\rho}_7$ with $^0\boldsymbol{v}_{\mathrm{O}_6}$ because for inverse rate kinematics we may assume that $^0\boldsymbol{\omega}_6$ will be known, and $^0\boldsymbol{\rho}_7$ may be calculated from the readings of the joint position sensors. Let

$$^0\boldsymbol{\mu} = {}^0\boldsymbol{v}_{\mathrm{O}_6} - {}^0\boldsymbol{\omega}_6 \times {}^0\boldsymbol{\rho}_7 \tag{7.23}$$

Then

$$^0\boldsymbol{\mu} = \sum_{i=1}^{6} \dot{\theta}_i\,{}^0\boldsymbol{\rho}_i \times {}^0\boldsymbol{w}_i \tag{7.24}$$

${}^0\boldsymbol{\mu}$ can be thought of as the velocity of the point in the hand, possibly extended, which is instantaneously coincident with the origin of frame 0.

Equation (7.24) can be written in the form

$$ {}^0\boldsymbol{\mu} = \sum_{k=1}^{6} \dot{\theta}_k {}^0\boldsymbol{\lambda}_k \tag{7.25}$$

where ${}^0\boldsymbol{\lambda}_k = {}^0\boldsymbol{\rho}_k \times {}^0\boldsymbol{w}_k$. If joint k is a slider, the kth term is replaced by $\dot{r}_k {}^0\boldsymbol{w}_k$. Note that the joint rates for sliders do not appear in Eq. (7.17).

Equations (7.17) and (7.25) can now be combined to form a system of the form of Eq. (7.22):

$$ \boldsymbol{v} = \begin{bmatrix} \boldsymbol{\omega} \\ \boldsymbol{\mu} \end{bmatrix}, \qquad \boldsymbol{\Gamma} = \begin{bmatrix} \boldsymbol{w}_1, \ldots, \boldsymbol{w}_6 \\ \boldsymbol{\lambda}_1, \ldots, \boldsymbol{\lambda}_6 \end{bmatrix} \tag{7.26}$$

Again, if the kth joint is a slider, the kth column of $\boldsymbol{\Gamma}$ is replaced by $[\boldsymbol{0}, \boldsymbol{w}_k]^{\mathrm{T}}$.

To compute the elements of $\boldsymbol{\Gamma}$ it is necessary to have expressions for $\boldsymbol{w}_k$ and $\boldsymbol{\rho}_k$ in terms of the joint angles. These are easily obtained in a recursive form convenient for computation by means of a direct position kinematics solution in the same way as was done in Example 7.5.

As already noted, the direct rate kinematics problem is that of finding $\boldsymbol{v} = [\boldsymbol{\omega}, \boldsymbol{\mu}]^{\mathrm{T}}$ given the joint rates $\dot{\theta}_1, \ldots, \dot{\theta}_6$. It is solved by direct substitution of the joint rates in Eq. (7.22) to give $\boldsymbol{v}$ directly. The direct rate kinematics problem is important when doing acceleration analysis for the purpose of studying dynamics. The total velocities of the members are needed for the computation of Coriolis and centripetal acceleration components.

The important problem from the point of view of robotic coordination is the inverse rate kinematics problem. As will be outlined below, it is the basis of all robot software that generates prescribed trajectories relative to the world or fixed coordinate frame.

To solve the linear system of equations in the joint rates obtained by decomposing Eq. (7.22) into its component equations when $\boldsymbol{v}$ is known, it is necessary to invert the Jacobian matrix $\boldsymbol{\Gamma}$. The equation becomes

$$ \dot{\boldsymbol{\theta}} = \boldsymbol{\Gamma}^{-1}\boldsymbol{v} $$

Since $\boldsymbol{\Gamma}$ is a 6×6 matrix, numerical inversion is not very attractive in real-time software that must run at computation cycle rates of the order of 100 Hz. Worse, it is quite possible for $\boldsymbol{\Gamma}$ to become singular ($|\boldsymbol{\Gamma}| = 0$). The inverse does not then exist. Even when the Jacobian matrix does not become singular, and, in fact, singularity is a rare occurrence in practice, it may become ill conditioned, leading to degraded performance in significant portions of the manipulator's working envelope.

Most industrial robot geometries are simple enough that the Jacobian matrix can be inverted analytically, leading to a set of explicit equations for the joint rates. This greatly reduces the number of computation operations needed as compared with numerical inversion.

7.6 CLOSED-LOOP LINKAGES

As was the case for planar linkages, closing a loop in the linkage results in the generation of a set of closure equations. In order to find the positions of all joints, given the position of one joint chosen as the input joint, it is necessary to solve the closure equations. The mobility criterion

$$ M = 6(m - j - 1) + \sum_{i=1}^{j} f_i \tag{1.3}$$

can be used to determine the connectivity sum of the joints in a single closed-loop spatial linkage that has mobility one. In a single closed loop, $m = j$, so Eq. (1.3) gives

$$1 = -6 + \sum f_i$$

or

$$\sum f_i = 7$$

If all joints have connectivity one, this implies that the linkage has seven joints and hence seven members.

The closure equations can be generated using the transformation Eq. (7.9):

$$\mathbf{p}_{N-1} = \mathbf{A}_N \mathbf{p}_N \qquad \textbf{(7.9)}$$

If this transformation is applied between successive pairs of members around the loop, one eventually returns to the original reference frame. Hence:

$$\mathbf{I} = \mathbf{A}_1\mathbf{A}_2\mathbf{A}_3\mathbf{A}_4\mathbf{A}_5\mathbf{A}_6\mathbf{A}_7 \qquad \textbf{(7.27)}$$

Expansion of Eq. (7.27) produces a set of equations in the seven joint variables. By selecting six of these equations and solving them, it is possible to find all joint variables given the value of one input variable. The problem is actually very similar to the inverse position problem of the open chain linkage and presents similar difficulties of solution.

EXAMPLE 7.6 ***(Derivation of Input-Output Relationship of Universal Joint)***

PROBLEM

Develop an equation (Eq. 5.10) relating the input and output shaft angles of the universal joint discussed in Section 5.2.3.1.

The mechanism is shown in Fig. 7.12, which is similar to Fig. 5.17. Its geometric parameters and kinematic variables are listed in Table 7.3. Here θ_1 and θ_2 are considered to be the input and output angles. The angle between shafts 1 and 2 is $\alpha_1 = \gamma$. All the other angles between successive joint angles: $\alpha_2 = \alpha_3 = \alpha_4 = 90°$.

Applying Eq. (7.27) to the four-bar loop:

$$\mathbf{I} = \mathbf{A}_1\mathbf{A}_2\mathbf{A}_3\mathbf{A}_4 \qquad \textbf{(7.28)}$$

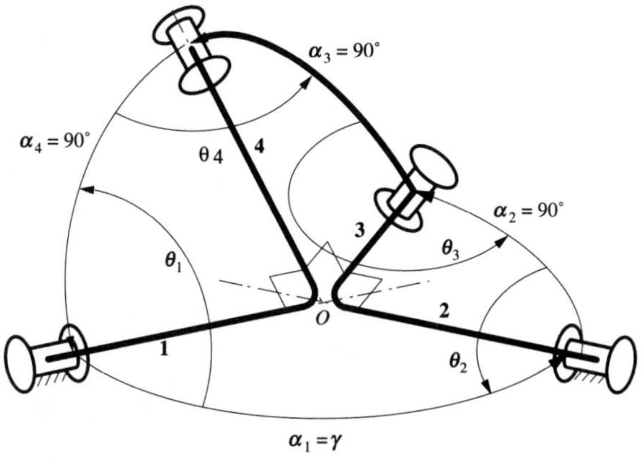

Figure 7.12 The geometric parameters and kinematic variables of the universal joint analyzed in Example 7.7. Note that this figure is similar to Fig. 5.17.

Table 7.3 Geometric Properties of Universal Joint in Example 7.6

i	a_i	r_i	α_i	θ_i
1	0	0	γ	θ_1
2	0	0	90°	θ_2
3	0	0	90°	θ_3
4	0	0	90°	θ_4

There is great advantage to trying to simplify the equations as much as possible before expansion. In this case, we can separate the variables θ_1 and θ_2 from θ_3 and θ_4 by postmultiplying both sides of the equation first by $\mathbf{A}_4^{-1}$ and then by $\mathbf{A}_3^{-1}$ to get

$$\mathbf{A}_1\mathbf{A}_2 = \mathbf{A}_4^{-1}\mathbf{A}_3^{-1} \tag{7.29}$$

It is now time to expand into component equations by substituting from Eqs. (7.10) and (7.16):

$$\begin{bmatrix} \cos\theta_1 & -\sin\theta_1\cos\gamma & \sin\theta_1\sin\gamma & 0 \\ \sin\theta_1 & \cos\theta_1\cos\gamma & -\cos\theta_1\sin\gamma & 0 \\ 0 & s\gamma & c\gamma & 0 \\ 0 & 0 & 0 & 1 \end{bmatrix} \begin{bmatrix} \cos\theta_2 & 0 & \sin\theta_2 & 0 \\ \sin\theta_2 & 0 & -\cos\theta_2 & 0 \\ 0 & 1 & 0 & 0 \\ 0 & 0 & 0 & 1 \end{bmatrix}$$

$$= \begin{bmatrix} \cos\theta_4 & \sin\theta_4 & 0 & 0 \\ 0 & 0 & 1 & 0 \\ \sin\theta_4 & -\cos\theta_4 & 0 & 0 \\ 0 & 0 & 0 & 1 \end{bmatrix} \begin{bmatrix} \cos\theta_3 & \sin\theta_3 & 0 & 0 \\ 0 & 0 & 1 & 0 \\ \sin\theta_3 & -\cos\theta_3 & 0 & 0 \\ 0 & 0 & 0 & 1 \end{bmatrix}$$

Performing the matrix multiplications and equating corresponding elements, we get the following system of scalar equations. Note that only the three-by-three submatrix in the top left-hand corner gives nontrivial equations. The remaining equations, which are $0 = 0$ or $1 = 1$, are not written out below.

$$\cos\theta_1\cos\theta_2 - \sin\theta_1\sin\theta_2\cos\gamma = \cos\theta_3\cos\theta_4 \tag{1,1}$$

$$\sin\theta_1\sin\gamma = \sin\theta_3\cos\theta_4 \tag{1,2}$$

$$\cos\theta_1\sin\theta_2 + \sin\theta_1\cos\theta_2\cos\gamma = \sin\theta_4 \tag{1,3}$$

$$\sin\theta_1\cos\theta_2 + \cos\theta_1\sin\theta_2\cos\gamma = \sin\theta_3 \tag{2,1}$$

$$-\cos\theta_1\sin\gamma = -\cos\theta_3 \tag{2,2}$$

$$\sin\theta_1\sin\theta_2 - \cos\theta_1\cos\theta_2\cos\gamma = 0 \tag{2,3}$$

$$\sin\theta_2\sin\gamma = \cos\theta_3\sin\theta_4 \tag{3,1}$$

$$\cos\gamma = \sin\theta_3\sin\theta_4 \tag{3,2}$$

$$-\cos\theta_2\sin\gamma = -\cos\theta_4 \tag{3,3}$$

These are all valid relationships between the angle variables θ_1, θ_2, θ_3, and θ_4. They are not independent, since the corresponding elements on each side of the equation are related by Eqs. (7.3). Inspection of these equations reveals that Eq. (*2,3*) contains only the two variables of interest, θ_1 and θ_2. This equation can be readily manipulated into the form

$$\tan\theta_1\tan\theta_2 = \cos\gamma \tag{5.10}$$

which was the result to be derived. As was shown in Section 5.2.3.1, this equation gives the input-output relationship of the shaft angles of the universal joint. ■

7.7 LOWER PAIR JOINTS

Linkages containing any combination of lower pair joints (see Table 1.1) can be analyzed by the methods presented above. For the *axial* joints, namely revolute, prismatic, screw, and clindrical joints, a single transform is needed for each joint. Referring to Eq. (7.10),

$$\mathbf{A}_N = \begin{bmatrix} \cos\theta_N & -\sin\theta_N\cos\alpha_N & \sin\theta_N\sin\alpha_N & a_N\cos\theta_N \\ \sin\theta_N & \cos\theta_N\cos\alpha_N & -\cos\theta_N\sin\alpha_N & a_N\sin\theta_N \\ 0 & \sin\alpha_N & \cos\alpha_N & r_N \\ 0 & 0 & 0 & 1 \end{bmatrix} \tag{7.10}$$

the same transform works in all four cases. The only difference is in which of the parameters are regarded as being variable. In the case of a revolute, θ_N is the variable and r_N is a constant. For a prismatic joint, r_N is variable and θ_N is a constant. In the case of a screw joint, θ_N and r_N both vary, but they are not independent of one another, being related by the equation

$$r_N = R_N + h_N\theta_N \tag{7.30}$$

where R_N and h_N are constant. Here h_N is the pitch of the screw, and R_N is the constant value of the joint offset when θ_N is zero. Finally, in the case of a cylindric joint, both r_N and θ_N are variable, and they may vary completely independently of one another.

The two remaining lower pair joint types, spherical and planar joints, may each be modeled by means of kinematically equivalent chains of three revolute joints. In the case of the spherical joint, the three revolute axes are concurrent at the center of the spherical joint being modeled, and the joint axes are successively orthogonal. That is, the middle joint axis is orthogonal to each of the other two. Those other two joint axes are not, in general, orthogonal to each other. In fact, the angle between them varies with motion about the middle joint. In the case of a planar joint, the three revolute axes are parallel and orthogonal to the plane of motion of the plane joint being modeled. These two equivalent chains are shown in Figs. 7.13 and 7.14, respectively.

The primary application of these equivalent mechanisms is in linkages that include closed loops. This is because spherical and plane joints are always passive joints that cannot be used in open chain structures.

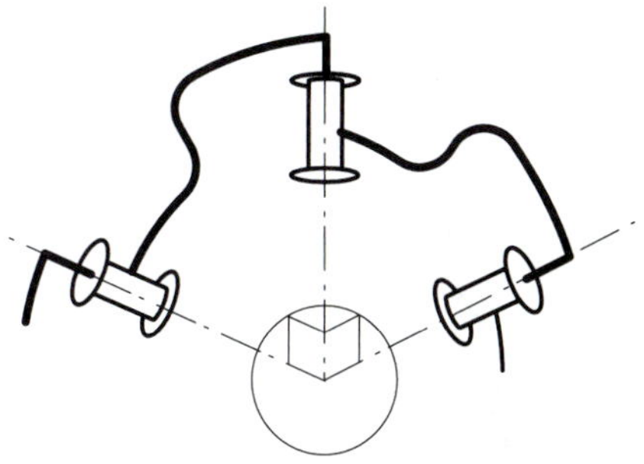

Figure 7.13 Chain of three revolute joints that is kinematically equivalent to a spherical joint.

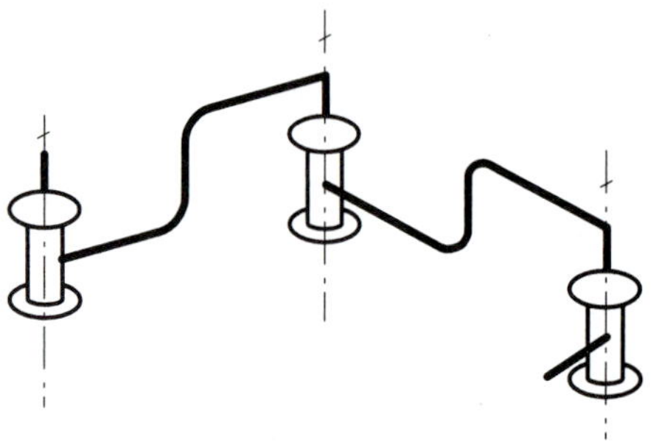

Figure 7.14 Chain of three revolute joints that is kinematically equivalent to a planar joint.

EXAMPLE 7.7 ***(Input-Output Relationship of a Simple Coupling)***

PROBLEM

Two nonintersecting shafts turn in cylindric joints, as shown in Fig. 7.15. The shaft axes are at angle μ to one another, and the length of their common normal is m. The two shafts are coupled by means of an offset spherical joint, as shown in the figure. The center of the spherical joint is distant p and q from the two shafts as shown. Develop equations relating the angle of shaft 2 to that of shaft 1 and the axial displacements of both shafts to the angular displacement of shaft 1.

SOLUTION

As indicated above, the spherical joint can be replaced by the kinematically equivalent combination of three successively orthogonal revolutes, with concurrency point at the center of the sphere. This substitution is shown in Fig. 7.16. We are at liberty to arrange the directions of the axes of joints 3 and 5 relative to 2 and 1, respectively, for convenience. Hence we choose to make axis 3 parallel to 2 and axis 5 parallel to 1. The linkage geometric parameters and joint variables are then as shown in Table 7.4.

In this case we have five joints, so the closure equation is

$$\mathbf{A}_1\mathbf{A}_2\mathbf{A}_3\mathbf{A}_4\mathbf{A}_5 = \boldsymbol{I}$$

As in the case of the previous example, it is advantageous to rearrange the equation to split up the variables before expanding. Postmultiplying in succession by $\mathbf{A}_5^{-1}, \mathbf{A}_4^{-1}, \mathbf{A}_3^{-1}$, gives the form

$$\mathbf{A}_1\mathbf{A}_2 = \mathbf{A}_5^{-1}\mathbf{A}_4^{-1}\mathbf{A}_3^{-1}$$

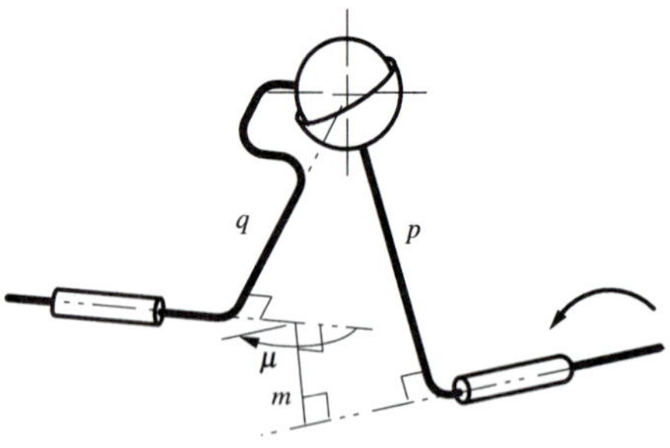

Figure 7.15 The simple coupling between two skew shafts that is analyzed in Example 7.7. The linkage has only three members connected in a single loop via two cylindric joints and a spherical joint.

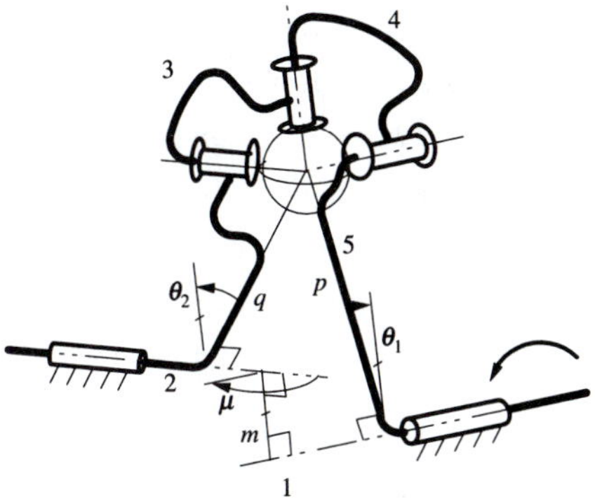

Figure 7.16 The linkage of Fig. 7.15 with a chain of three successively orthogonal revolutes substituted for the spherical joint. The concurrency point of the revolute axes is at the center of the spherical joint that is replaced.

We now embark on expansion of this equation, again referring to Eqs. (7.10) and (7.16).

$$\mathbf{A}_1\mathbf{A}_2 = \begin{bmatrix} \cos\theta_1 & -\sin\theta_1\cos\mu & \sin\theta_1\sin\mu & m\cos\theta_1 \\ \sin\theta_1 & \cos\theta_1\cos\mu & -\cos\theta_1\sin\mu & m\sin\theta_1 \\ 0 & \sin\mu & \cos\mu & r_1 \\ 0 & 0 & 0 & 1 \end{bmatrix} \begin{bmatrix} \cos\theta_2 & -\sin\theta_2 & 0 & q\cos\theta_2 \\ \sin\theta_2 & \cos\theta_2 & 0 & q\sin\theta_2 \\ 0 & 0 & 1 & r_2 \\ 0 & 0 & 0 & 1 \end{bmatrix}$$

and

$$\mathbf{A}_5^{-1}\mathbf{A}_4^{-1}\mathbf{A}_3^{-1} = \begin{bmatrix} \cos\theta_5 & \sin\theta_5 & 0 & -p \\ -\sin\theta_5 & \cos\theta_5 & 0 & 0 \\ 0 & 0 & 1 & 0 \\ 0 & 0 & 0 & 1 \end{bmatrix} \begin{bmatrix} \cos\theta_4 & \sin\theta_4 & 0 & 0 \\ 0 & 0 & 1 & 0 \\ \sin\theta_4 & -\cos\theta_4 & 0 & 0 \\ 0 & 0 & 0 & 1 \end{bmatrix} \begin{bmatrix} \cos\theta_3 & \sin\theta_3 & 0 & 0 \\ 0 & 0 & 1 & 0 \\ \sin\theta_3 & -\cos\theta_3 & 0 & 0 \\ 0 & 0 & 0 & 1 \end{bmatrix}$$

Combining these expressions, performing the matrix multiplication, and equating corresponding elements of the matrices on either side give the following set of equations:

$$\cos\theta_1\cos\theta_2 - \sin\theta_1\sin\theta_2\cos\mu = \cos\theta_3\cos\theta_4\cos\theta_5 + \sin\theta_3\sin\theta_5 \quad (1,1)$$

$$-\cos\theta_1\sin\theta_2 - \sin\theta_1\cos\theta_2\cos\mu = \sin\theta_3\cos\theta_4\cos\theta_5 - \cos\theta_3\sin\theta_5 \quad (1,2)$$

$$\sin\theta_1\sin\mu = \sin\theta_4\cos\theta_5 \quad (1,3)$$

Table 7.4 Parameters for the Linkage in Example 7.7

i	a_i	r_i	α_i	θ_i
1	m	r_1	μ	θ_1
2	q	r_2	0	θ_2
3	0	0	90°	θ_3
4	0	0	90°	θ_4
5	p	0	0	θ_5

$$q(\cos\theta_1\cos\theta_2 - \sin\theta_1\sin\theta_2\cos\mu) + r_2\sin\theta_1\sin\mu + m\cos_1 = -p \quad (1,4)$$

$$\sin\theta_1\cos\theta_2 + \cos\theta_1\sin\theta_2\cos\mu = -\cos\theta_3\cos\theta_4\sin\theta_5 + \sin\theta_3\cos\theta_5 \quad (2,1)$$

$$-\sin\theta_1\sin\theta_2 + \cos\theta_1\cos\theta_2\cos\mu = -\sin\theta_3\cos\theta_4\sin\theta_5 - \cos\theta_3\cos\theta_5 \quad (2,2)$$

$$\cos\theta_1\sin\mu = \sin\theta_4\sin\theta_5 \quad (2,3)$$

$$q(\sin\theta_1\cos\theta_2 + \cos\theta_1\sin\theta_2\cos\mu) - r_2\cos\theta_1\sin\mu + m\sin\theta_1 = 0 \quad (2,4)$$

$$\sin\theta_2\sin\mu = \cos\theta_3\sin\theta_4 \quad (3,1)$$

$$\cos\theta_2\sin\mu = \sin\theta_3\sin\theta_4 \quad (3,2)$$

$$\cos\mu = -\cos\theta_4 \quad (3,3)$$

$$q\sin\theta_2\sin\mu + r_2\cos\mu + r_1 = 0 \quad (3,4)$$

The equations corresponding to the fourth row of the matrices are trivial.

Now we have no interest in the angles of the "virtual" joints: θ_3, θ_4, and θ_5. The variables of interest are the independent variable θ_1 and the dependent variables θ_2, r_1, and r_2. Inspection of these equations reveals that the *(1,4)*, *(2,4)*, and *(3,4)* equations contain only these latter four variables. Therefore, it should be possible to solve these three equations to obtain expressions for θ_2, r_1, and r_2 in terms of θ_1.

If the *(1,4)* equation is multiplied by $\cos\theta_1$ and added to the *(2,4)* equation multiplied by $\sin\theta_1$, the r_2 terms cancel, and using $\cos\theta_1^2 + \sin\theta_1^2 = 1$, the resulting equation reduces to

$$q\cos\theta_2 + m = -p\cos\theta_1$$

or

$$\theta_2 = \pm\cos^{-1}\left\{\frac{m + p\cos\theta_1}{q}\right\}$$

Here the second solution corresponds to physically sliding the shaft out of cylindric joint 2, turning member 2 through 180° and sliding the shaft back into the bushing in the opposite direction, and then sliding along the joint axis until the spherical joint returns to its original position. This would not be physically possible without encountering interference unless the bushing and shaft are duplicated, in which case member 2 would be symmetric and the two solutions would be indistinguishable from one another. Therefore, nothing is lost by ignoring the second solution. Hence

$$\theta_2 = \cos^{-1}\left\{\frac{m + p\cos\theta_1}{q}\right\} \quad \textbf{(a)}$$

Multiplying the *(1,4)* equation by $\sin\theta_1$ and subtracting from it the *(2,4)* equation multiplied by $\cos\theta_1$ gives, after similar simplification,

$$-q\sin\theta_2\cos\mu + r_2\sin\mu = -p\sin\theta_1$$

or

$$r_2 = \frac{q\sin\theta_2\cos\mu - p\sin\theta_1}{\sin\mu} \quad \textbf{(b)}$$

Consequently, once θ_2 has been obtained from Eq. (a), this expression can be used to compute r_2. Finally, after θ_2 and r_2 have been computed, the *(3,4)* equation can be used to solve for r_1.

$$r_1 = \frac{p\sin\theta_1\cos\mu - q\sin\theta_2}{\sin\mu} \quad \textbf{(c)}$$

Solution of Eqs. (a), (b), and (c) in sequence solves the problem of obtaining the values of θ_2, r_1, and r_2 for any given value of θ_1. ■

7.8 MOTION PLATFORMS

7.8.1 Mechanisms Actuated in Parallel

The mechanisms discussed in this section are spatial mechanisms and detailed analysis is complex and requires methods developed from those presented in earlier sections of this chapter. However, we will give a simple presentation here, since these are mechanisms that have considerable practical importance.

The mechanisms in this class provide multiple actuated degrees of freedom, typically three or six, in the same way that the serial chain mechanisms discussed earlier do. As compared with those serial mechanisms, parallel mechanisms are stronger and stiffer but are usually more restricted in the range of motion they provide.

7.8.2 The Stewart Platform

Although it is far from the simplest parallel mechanism, we will discuss the Stewart platform first because of its practical and historic importance. The Stewart platform has six limbs acting in parallel to connect the "platform" member to the base. Each limb has a linear actuator, such as a hydraulic cylinder or ball screw, that provides the kinematic equivalent of an actuated prismatic joint. Each limb is connected to the base and the platform by a spherical joint at one end and a universal joint at the other. In Fig. 7.17 the spherical joints are indicated as being at the upper ends of the limbs and the universal joints at the lower ends, but they may be reversed.

Stewart platforms have been very widely used as motion bases for aircraft simulators and similar devices. The great strength and stiffness of the mechanism, together with its ability to produce universally controlled spatial motion, make it ideal for this application. Its relatively restricted motion range is not usually a serious limitation in this instance.

Stewart platforms have mathematical properties that, in some respects, bear an inverse relationship to the corresponding properties of serial chains. The *inverse* position problem for a Stewart platform is very straightforward. Restating the problem, it is: given the transformation describing the position of the floating platform reference frame relative to the base reference frame, find the limb lengths.

Referring to Fig. 7.18, the problem may be restated as follows:

Given the 4×4 transformation:

$$p = Qp' \tag{7.31}$$

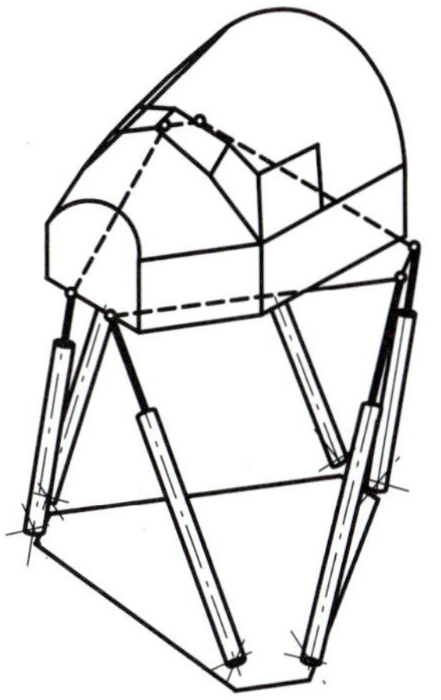

Figure 7.17 A Stewart platform used as the motion base of a flight simulator.

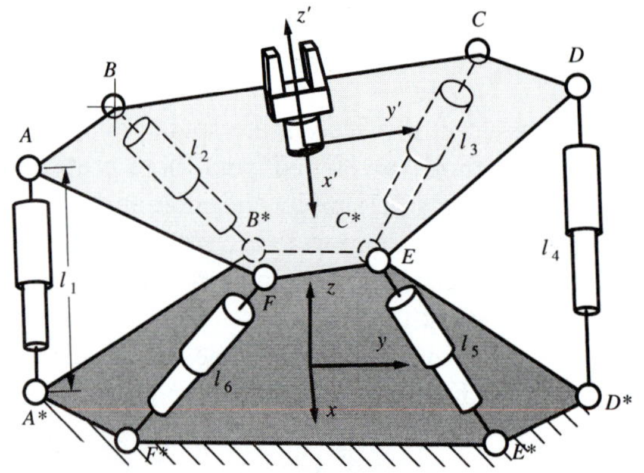

Figure 7.18 The Stewart platform model used in describing its kinematic properties in the text.

relating coordinates referred to the fixed and floating frames, respectively, find the lengths l_i, $i = 1, \ldots 6$.

Since the positions of the points A, B, C, D, E, F can be assumed to be known relative to the primed reference frame, Eq. (7.31) can be used to obtain their positions relative to the fixed frame. Similarly, the positions of the points A^*, B^*, C^*, D^*, E^* can be assumed to be known relative to the fixed frame. Then Pythagoras' theorem can be used to find

$$l_1^2 = (\boldsymbol{p}_A - \boldsymbol{p}_{A^*})^2 \quad \textbf{(7.32)}$$

where $\boldsymbol{p}_A$ is the position of A relative to the fixed frame, and $\boldsymbol{p}_{A^*}$ is the position of A^* relative to that frame. Similarly, the lengths of the other limbs can be calculated.

In contrast, the *direct* position problem—namely, given l_i, $i = 1, \ldots 6$, find the transformation matrix $\boldsymbol{Q}$ in Eq. (7.31)—is quite difficult and has multiple solutions: 16 for the configuration shown in Fig. 7.18. This is in contrast to the serial chain as described earlier. There the *direct* problem was straightforward and single valued, and the *inverse* problem was demanding and multivalued.

The inverse symmetry is even more marked in the velocity domain. The *inverse* rate kinematics may be solved directly. The *direct* rate problem requires inversion of a 6×6 matrix with structure closely related to the Jacobian matrix described in Section 7.5.3.

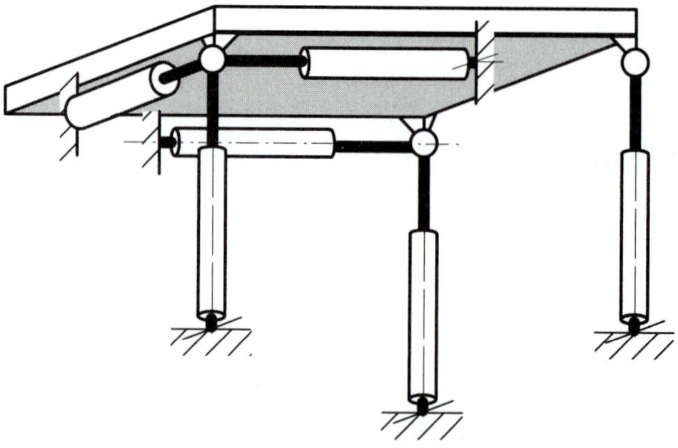

Figure 7.19 A 3-2-1 motion platform. This is actually a variant of the Stewart platform. For small displacements the motion components are decoupled.

7.8.3 The 3-2-1 Platform

The 3-2-1 platform (Fig. 7.19) is a special configuration that has some practical importance. It is arranged so that the motion and force equations decouple for small displacements. For this reason it has been used in three-dimensional vibration testing tables and for similar functions. The same configuration can be used as a six-component force sensor, with the actuators replaced by uniaxial load cells.

7.9 CHAPTER 7 *Exercise Problems*

PROBLEM 7.1 For the information given, find the velocity of point C. Show all equations used with terms properly labeled.

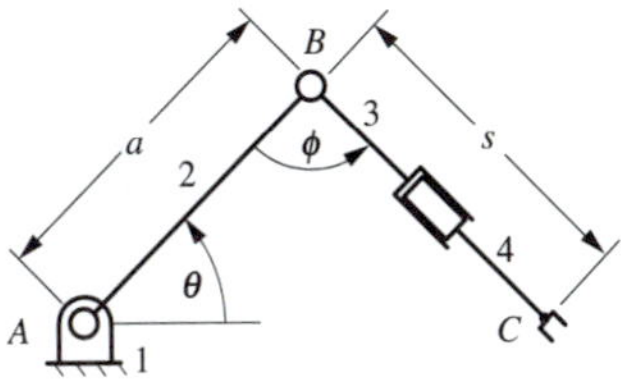

$\theta = 45°,\ \dot{\theta} = 0.1\ \frac{\text{rad}}{\text{s}}$

$\phi = 90°,\ \dot{\phi} = -0.1\ \frac{\text{rad}}{\text{s}}$

$s = 2\ m,\ \dot{s} = 0.5\ \frac{m}{\text{s}},\ a = 2\ m$

PROBLEM 7.2 In the manipulator shown, the joint axes at A and B are oriented along the z and x axes, respectively, and link 3 is perpendicular to link 2. For the position shown, find the velocity of point C. Links 2 and 3 lie in the y-z plane.

$$\dot{\theta} = 100\frac{\text{rad}}{\text{s}} \qquad \dot{\phi} = 10\frac{\text{rad}}{\text{s}} \qquad \overline{AB} = \overline{BC} = 10\,\text{m}$$

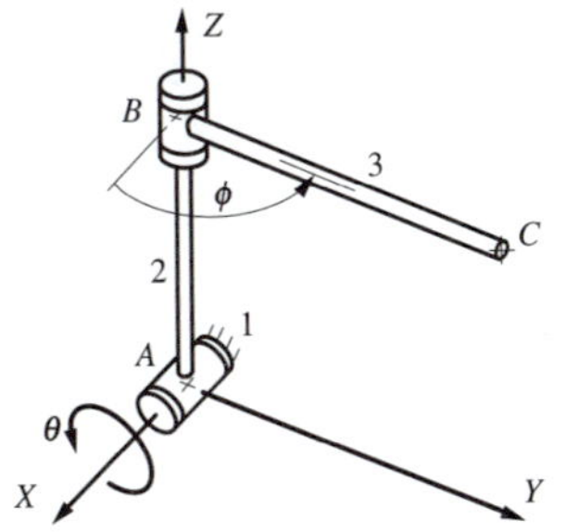

PROBLEM 7.3 For the information given, find the velocity and acceleration of point C.

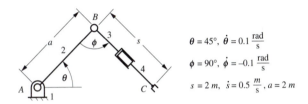

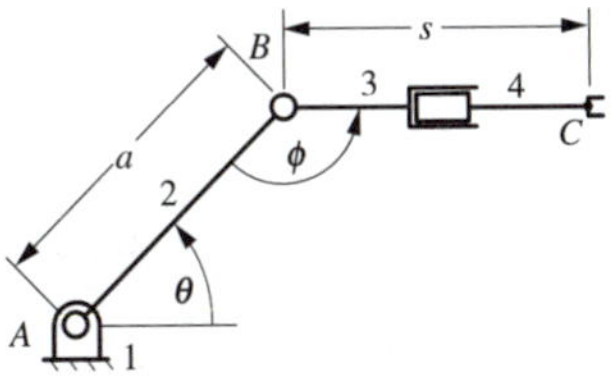

$\theta = 45°,\ \dot{\theta} = -0.1\ \frac{\text{rad}}{\text{s}},\ \ddot{\theta} = 0.1\ \frac{\text{rad}}{\text{s}^2}$

$\phi = 135°,\ \dot{\phi} = -0.1\ \frac{\text{rad}}{\text{s}},\ \ddot{\phi} = 2.0\ \frac{\text{rad}}{\text{s}^2}$

$s = 2\ m,\ \dot{s} = 0.5\ \frac{m}{\text{s}},\ \ddot{s} = 1.0\ \frac{m}{\text{s}^2},\ a = 2\ m$

PROBLEM 7.4

For the information given, find the velocity and acceleration of point C.

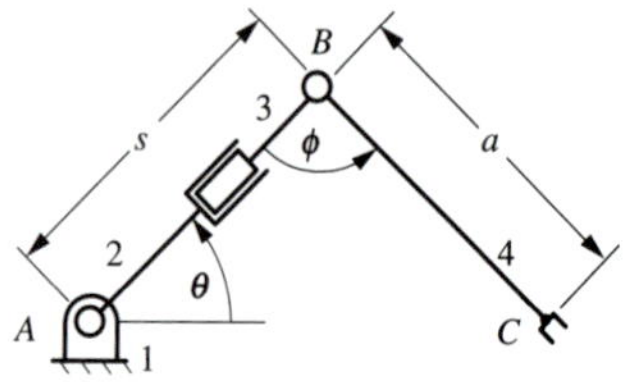

$\theta = 45°,\ \dot{\theta} = 0.1\,\frac{\text{rad}}{\text{s}},\ \ddot{\theta} = 0$

$\phi = 90°,\ \dot{\phi} = -0.1\,\frac{\text{rad}}{\text{s}},\ \ddot{\phi} = 0$

$s = 2\ m,\ \dot{s} = 0.5\,\frac{m}{\text{s}},\ \ddot{s} = 0,\ a = 2\ m$

PROBLEM 7.5

For the planar manipulator given, compute the following:

1. $\boldsymbol{r}_{B2/A2},\ \boldsymbol{r}_{C3/B3},\ \boldsymbol{r}_{D4/C4},\ \boldsymbol{r}_{D4/A1}$
2. ${}^1\boldsymbol{\omega}_2,\ {}^1\boldsymbol{\omega}_3,\ {}^1\boldsymbol{\omega}_4;\ \boldsymbol{v}_{B2/A1},\ \boldsymbol{v}_{C3/A1},\ \boldsymbol{v}_{D4/A1}$
3. ${}^1\boldsymbol{\alpha}_2,\ {}^1\boldsymbol{\alpha}_3,\ {}^1\boldsymbol{\alpha}_4;\ \boldsymbol{a}_{B2/A1},\ \boldsymbol{a}_{C3/A1},\ \boldsymbol{a}_{D4/A1}$

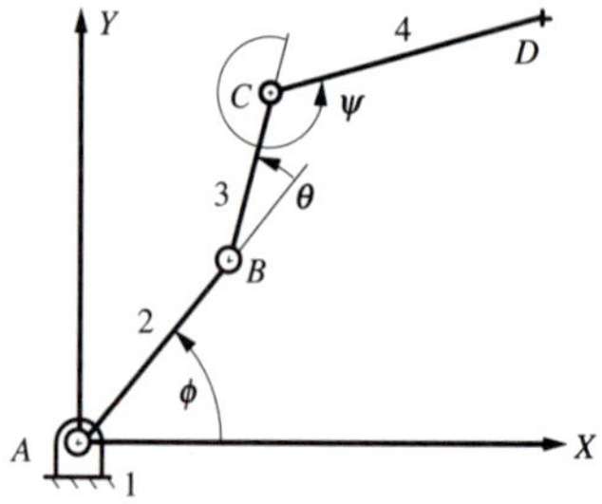

$AB = 4$ in, $BC = 3$ in, $CD = 5$ in

$\phi = 50°,\ \theta = 25°\ \psi = 300°$

$\dot{\phi} = 1\,\frac{\text{rad}}{\text{s}},\ \dot{\theta} = 2\,\frac{\text{rad}}{\text{s}},\ \dot{\psi} = -1.5\,\frac{\text{rad}}{\text{s}}$

$\ddot{\phi} = 0.2\,\frac{\text{rad}}{\text{s}^2},\ \ddot{\theta} = -0.5\,\frac{\text{rad}}{\text{s}^2},\ \ddot{\psi} = 0.4\,\frac{\text{rad}}{\text{s}^2}$

PROBLEM 7.6

In the spatial manipulator, link 2, the unit vector $\boldsymbol{l}$ and the unit vector $\boldsymbol{m}$ are all in the X-Z plane. The vector $\boldsymbol{m}$ is perpendicular to $\boldsymbol{l}$, and the unit vector $\boldsymbol{n}$ is perpendicular to $\boldsymbol{m}$. The angle ϕ is measured positive CCW about the $\boldsymbol{m}$ axis. Compute the following:

1. $\boldsymbol{r}_{B2/A2},\ \boldsymbol{r}_{C3/B3},\ \boldsymbol{r}_{C3/A1}$
2. ${}^1\boldsymbol{\omega}_2,\ {}^1\boldsymbol{\omega}_3;\ \boldsymbol{v}_{B2/A1},\ \boldsymbol{v}_{C3/A1}$
3. ${}^1\boldsymbol{\alpha}_2,\ {}^1\boldsymbol{\alpha}_3;\ \boldsymbol{a}_{B2/A1},\ \boldsymbol{a}_{C3/A1}$

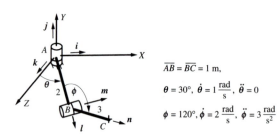

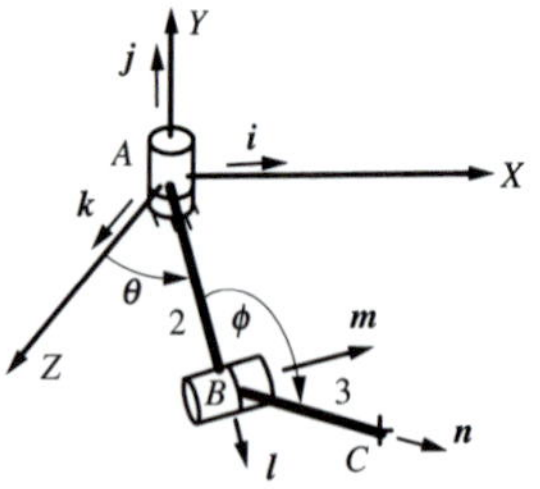

$\overline{AB} = \overline{BC} = 1$ m,

$\theta = 30°,\ \dot{\theta} = 1\,\frac{\text{rad}}{\text{s}},\ \ddot{\theta} = 0$

$\phi = 120°,\ \dot{\phi} = 2\,\frac{\text{rad}}{\text{s}},\ \ddot{\phi} = 3\,\frac{\text{rad}}{\text{s}^2}$

PROBLEM 7.7

In the manipulator shown, the joint axes at A and B are oriented along the z and $\boldsymbol{m}$ axes, respectively. Link 2 lies in the XY plane, and link 3 is perpendicular to link 2. For the position shown, link 3 is vertical (parallel to z). Determine ${}^1\boldsymbol{v}_{C3/A1}$, ${}^1\boldsymbol{v}_{B3/A1}$, and ${}^1\boldsymbol{\alpha}_3$.

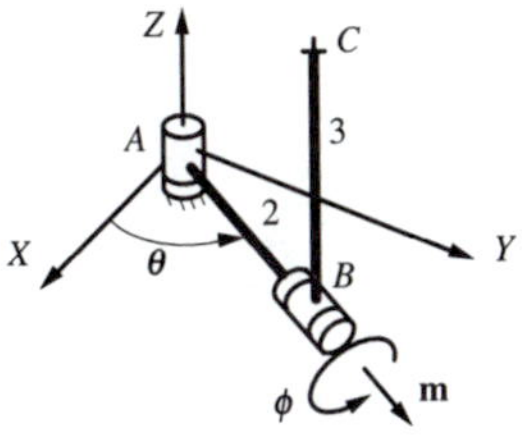

$\overline{AB} = 3$ in, $\overline{BC} = 4$ in

$\theta = 60°,\ \dot{\theta} = 5\,\dfrac{\text{rad}}{\text{s}},\ \ddot{\theta} = 10\,\dfrac{\text{rad}}{\text{s}^2}$

$\dot{\phi} = 5\,\dfrac{\text{rad}}{\text{s}},\ \ddot{\phi} = 0\,\dfrac{\text{rad}}{\text{s}^2}$

PROBLEM 7.8

In the manipulator shown, the joint axes at A and C are oriented along the $\mathbf{Z}$ axis, and the axis at b is oriented along the $\boldsymbol{m}$ axis. In the position to be analyzed, link 2 lies in a plane parallel to the XY plane and points along a line parallel to the Y axis. Link 3 is perpendicular to link 2. Link 4 is perpendicular to link 3 and lies along a line parallel to the Y axis. For the position to be analyzed, link 3 is vertical (parallel to Z). The joints between link 2 and the frame at A and between links 3 and 4 are revolute joints, and that at B is a cylindrical joint. From a previous analysis, we know that ${}^1\boldsymbol{v}_{C3/A1} = 13\boldsymbol{i} + \boldsymbol{j}$, ${}^1\boldsymbol{a}_{B3/A1} = -4\boldsymbol{i}$, and ${}^1\boldsymbol{\alpha}_3 = \boldsymbol{k} - 5\boldsymbol{i}$. Using this information, determine ${}^1\boldsymbol{v}_{D4/A1}$, ${}^1\boldsymbol{a}_{C3/A1}$, and ${}^1\boldsymbol{\alpha}_4$.

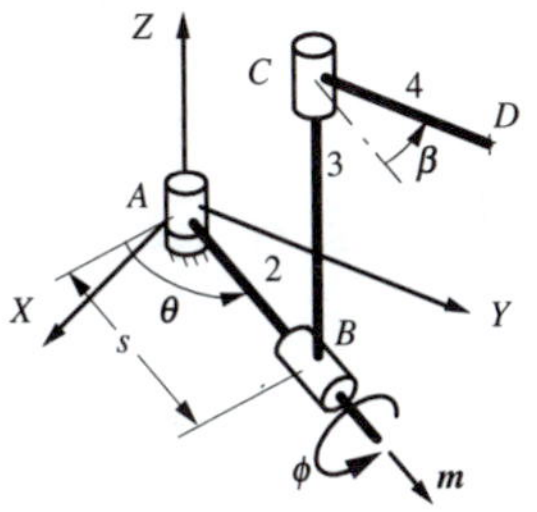

$\theta = 90°,\ \dot{\theta} = 1\,\dfrac{\text{rad}}{\text{s}},\ \ddot{\theta} = 1\,\dfrac{\text{rad}}{\text{s}^2}$

$\dot{\phi} = 5\,\dfrac{\text{rad}}{\text{s}},\ \ddot{\phi} = 0\,\dfrac{\text{rad}}{\text{s}^2},\ \dot{s} = 1\,\dfrac{\text{in}}{\text{s}},\ \ddot{s} = 2\,\dfrac{\text{in}}{\text{s}^2}$

$\dot{\beta} = 1\,\dfrac{\text{rad}}{\text{s}},\ \ddot{\beta} = 1\,\dfrac{\text{rad}}{\text{s}^2}$

$\overline{AB} = 2$ in, $\overline{BC} = 3$ in, $\overline{CD} = 2$ in

PROBLEM 7.9

In the manipulator shown, the joint axes at A and B are oriented along the $\mathbf{Z}$ and $\boldsymbol{m}$ axes, respectively. In the position to be analyzed, link 2 lies in a plane parallel to the XY plane and points along a line parallel to the Y axis, and link 3 is perpendicular to link 2. For the position to be analyzed, link 3 is vertical (parallel to Z). The joint between link 2 and the frame at A is a revolute joint, and that at B is a cylindrical joint. Determine ${}^1\boldsymbol{v}_{C3/A1}$, ${}^1\boldsymbol{a}_{B3/A1}$, and ${}^1\boldsymbol{\alpha}_3$.

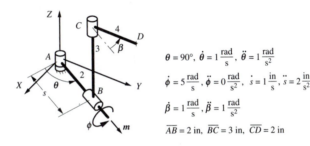

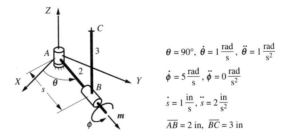

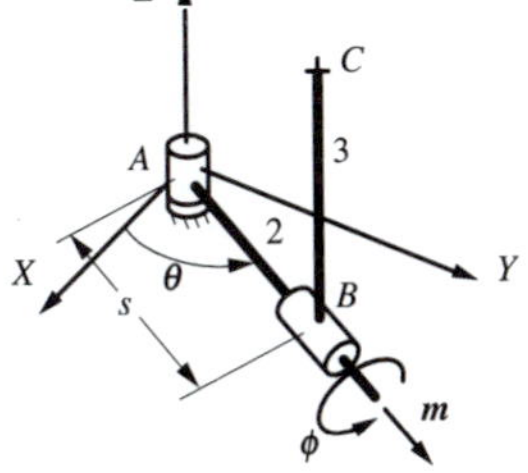

$\theta = 90°,\ \dot{\theta} = 1\,\dfrac{\text{rad}}{\text{s}},\ \ddot{\theta} = 1\,\dfrac{\text{rad}}{\text{s}^2}$

$\dot{\phi} = 5\,\dfrac{\text{rad}}{\text{s}},\ \ddot{\phi} = 0\,\dfrac{\text{rad}}{\text{s}^2}$

$\dot{s} = 1\,\dfrac{\text{in}}{\text{s}},\ \ddot{s} = 2\,\dfrac{\text{in}}{\text{s}^2}$

$\overline{AB} = 2$ in, $\overline{BC} = 3$ in

PROBLEM 7.10

In the manipulator shown, the joint axis at A is oriented along the $\mathbf{Z}$ axis. *In the position to be analyzed,* link 2 lies in a plane parallel to the YZ plane and points along a line parallel to the Y axis. Link 3 is perpendicular to link 2 and lies parallel to the XY plane. The joint between link 2 and the frame is a cylindrical joint, and the joint between links 2 and 3 is a revolute joint. Using the information given, determine ${}^1\boldsymbol{v}_{C3/A1}$, ${}^1\boldsymbol{a}_{C3/A1}$, and ${}^1\boldsymbol{\alpha}_3$.

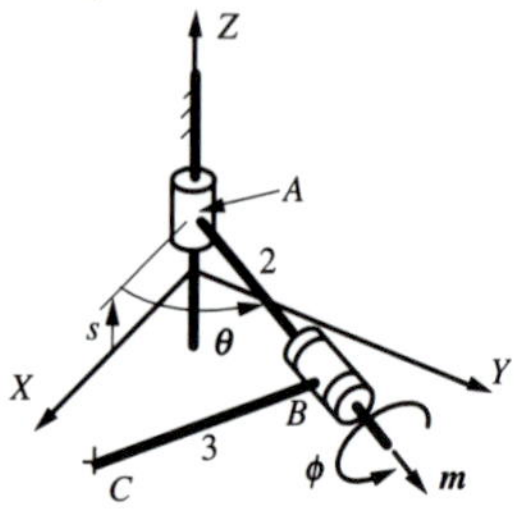

$\theta = 90°,\ \dot{\theta} = 1\frac{\text{rad}}{\text{s}},\ \ddot{\theta} = 0\frac{\text{rad}}{\text{s}^2}$

$\dot{\phi} = 1\frac{\text{rad}}{\text{s}},\ \ddot{\phi} = 0\frac{\text{rad}}{\text{s}^2},$

$s = 1\text{ in},\ \dot{s} = 0\frac{\text{in}}{\text{s}},\ \ddot{s} = 1\frac{\text{in}}{\text{s}^2}$

$\overline{AB} = 1\text{ in},\ \overline{BC} = 1\text{ in}$

PROBLEM 7.11

In the manipulator shown, all joints are revolutes. The link parameters and joint angles are as tabulated. Find the 4 × 4 matrix, $\boldsymbol{Q}'$, that expresses the position of the mounting flange for the end effector in the position specified by the values of θ_k given in the table.

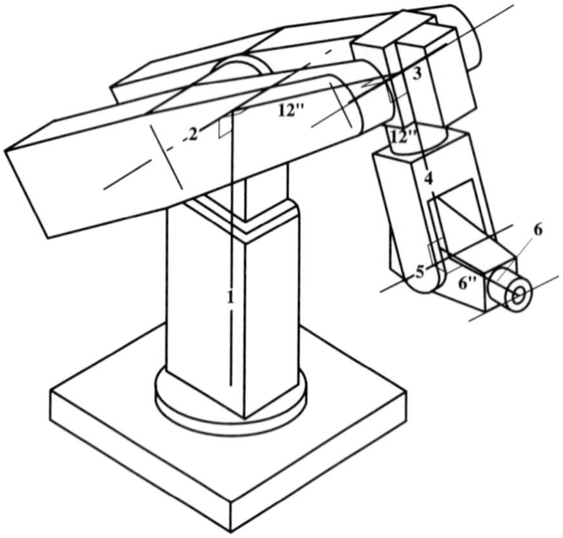

i	a_i	r_i	α_i	θ_i
1	0	0	90°	30°
2	12 in	0	0°	45°
3	0	0	90°	−50°
4	0	12 in	90°	−20°
5	0	0	90°	40°
6	0	6 in	0°	30°

PROBLEM 7.12

The SCARA robot geometry shown in the figure is often used for assembly robots because it is suited to moving and placing parts vertically over a relatively large horizontal area. All of its joint axes, except axis 4, are revolutes. Axis 4 is a prismatic joint. Axes 1, 2, 3, and 4 are parallel, and axes 3 and 4 are coincident (note that this does not produce degeneracy, since 3 is a revolute joint and 4 is a prismatic joint). For a given position of the manipulator, the linkage parameters are as tabulated. Find the matrix $\boldsymbol{Q}'$ that expresses the position of the end effector for the tabulated values of the joint variables.

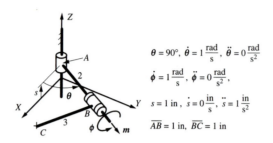

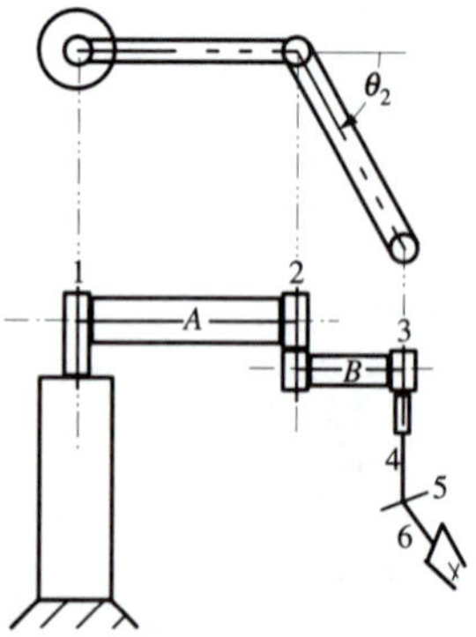

i	a_i	r_i	α_i	θ_i
1	15 in	0	0°	60°
2	15 in	4 in	0°	45°
3	0	0	0°	−50°
4	0	9 in	90°	0°
5	0	0	90°	0°
6	0	6 in	0°	30°

PROBLEM 7.13

For the manipulator of Example 7.4, Fig. 7.10, and Table 7.1, find the 4 × 4 matrix, $\boldsymbol{Q}'$, that expresses the position of the gripper when the joint positions are as follows: $\theta_1 = -30°$, $\theta_2 = 150°$, $\theta_3 = 30°$, $r_4 = 2.5$, $\theta_5 = -45°$, $\theta_6 = 0$.

PROBLEM 7.14

For the three-axis manipulator of Example 7.5, Fig. 7.11, and Table 7.2, find the angular velocity of member 3 and the velocity of the wrist concurrency point, P, relative to the fixed reference frame when the joint positions and rates are as follows: $\theta_1 = -60°$, $\theta_2 = 45°$, $\theta_3 = -30°$, $\dot{\theta}_1 = -0.5$ rad/s, $\dot{\theta}_2 = 0.5$ rad/s, $\dot{\theta}_3 = 1.0$ rad/s.

PROBLEM 7.15

For a SCARA robot of basic geometry similar to that of Problem 7.2, the geometric parameters and joint variables in a general position are as tabulated below. The Jacobian matrix that relates the velocity of the end effector to the joint velocities is also given.

a. Find the angular velocity of the end effector when the joint positions and rates are:

$$\theta_1 = 30°, \quad \theta_2 = 15°, \quad \theta_3 = -15°, \quad r_4 = 0.1, \quad \theta_5 = 0°, \quad \theta_6 = 45°$$

$$\dot{\theta}_1 = -0.2 \text{ rad/s}, \quad \dot{\theta}_2 = 0.5 \text{ rad/s}, \quad \dot{\theta}_3 = -0.2 \text{ rads/}, \quad \dot{r}_4 = 0.05 \text{ unit/s}, \quad \dot{\theta}_5 = 0, \quad \dot{\theta}_6 = 0.1 \text{ rad/s}$$

b. Verify that the system is singular whenever either $\theta_2 = 0$ or $\theta_5 = 0$.

i	a_i	r_i	α_i	θ_i
1	0.5	0	0°	θ_1
2	0.4	0	0°	θ_2
3	0	0	0°	θ_3
4	0	r_4	90°	θ_4
5	0	0	90°	θ_5
6	0.1	0	0°	θ_6

$$\Gamma = \begin{bmatrix} 0 & 0 & 0 & 0 & 0 & \sin\theta_5 \\ 1 & 1 & 1 & 0 & 0 & -\cos\theta_5 \\ 0 & 0 & 0 & 0 & 1 & 0 \\ 0.4\sin\theta_3 + 0.5\sin(\theta_2+\theta_3) & 0.4\sin\theta_3 & 0 & 0 & 0 & 0 \\ 0 & 0 & 0 & 1 & 0 & 0 \\ -0.4\cos\theta_3 - 0.5\cos(\theta_2+\theta_3) & -0.4\cos\theta_3 & 0 & 0 & 0 & 0 \end{bmatrix}$$

Chapter 8

Spur Gears

8.1 INTRODUCTION

The mechanisms discussed in previous chapters are used primarily as nonuniform-motion or force transformers. In the case in which uniform motion (constant-velocity ratio) or force transmission is required, circular gears, friction drives, belt drives, and chain drives are preferred. In such devices, if the input shaft turns at a constant speed, then the output shaft will turn at a constant angular speed. In practice, however, there may be a small but undesirable oscillatory motion superposed on the output motion because of imperfections in the system.

When low power is to be transmitted, constant-velocity transfer can be achieved by friction drives, and a simplified version of such a drive is shown in Fig. 8.1. In the device shown, disk 2 is the driver and disk 3 is driven, and both disks are assumed to be perfect circular cylinders in contact at point P. Link 1 is fixed, and there is a spring that forces disk 3 against disk 2. The disks rotate about points A and B. Assuming that there is no slip at P, the two disks will roll on each other, and the condition for rolling can be written as

$$\mathbf{v}_{P_2} = \mathbf{v}_{P_3}$$

or

$$\boldsymbol{\omega}_2 \times \mathbf{r}_{P/A} = \boldsymbol{\omega}_3 \times \mathbf{r}_{P/B}$$

The vector form is needlessly cumbersome for simple friction drives, and only the magnitudes need to be considered. The directions can be determined easily by inspection; that

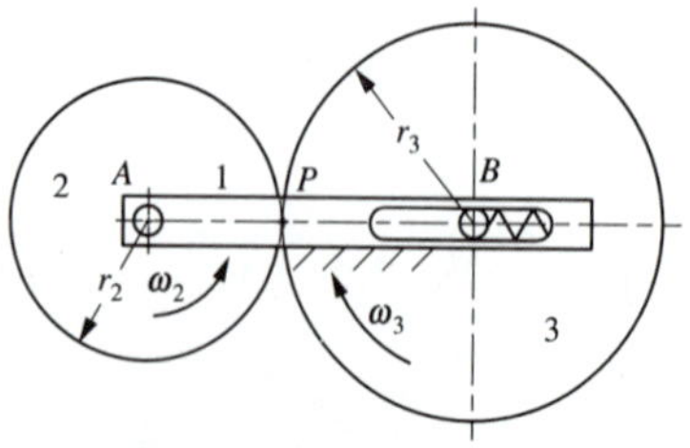

Figure 8.1 Friction disks.

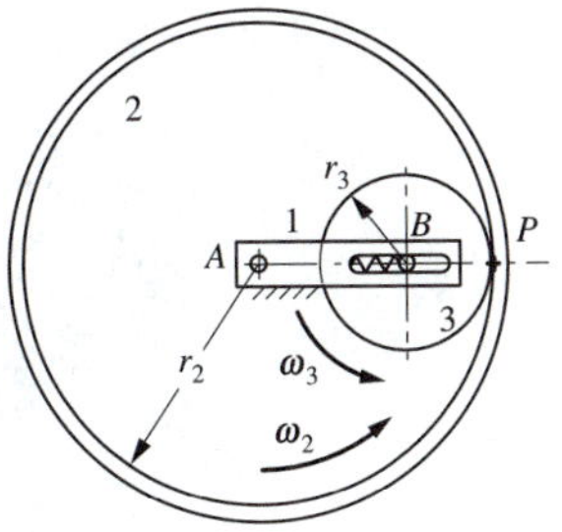

Figure 8.2 Internal friction drive.

is, the two disks in Fig. 8.1 rotate in opposite directions. Then,

$$\omega_2 r_2 = \omega_3 r_3$$

or

$$\frac{\omega_2}{\omega_3} = \frac{r_3}{r_2} \qquad \textbf{(8.1)}$$

where $\omega_2 = |\boldsymbol{\omega}_2|$, $\omega_3 = |\boldsymbol{\omega}_3|$, r_2 is the radius of disk 2, and r_3 is the radius of disk 3. Thus, the friction drive velocity ratio (the input angular velocity divided by the output angular velocity) is constant and inversely proportional to the ratio of the disk radii. Note that if a plus sign were assigned to one of the ω's, the other ω would have a negative sign.

If an external disk drives an internal disk as shown in Fig. 8.2, the velocity ratio given by Eq. (8.1) would still apply; however, in that case, the two angular velocities would be in the same direction.

A friction drive can transmit torque only if the normal force at the contact point is sufficient to prevent slippage. If large torques are involved, large Hertzian contact stresses will be created at the contact point and subsequent pitting and galling of the contact surfaces are inevitable. The wear can be lessened by using very hard contact surfaces. In addition, special lubricants have been developed that prevent actual molecular contact of the frictional surfaces but have a sufficiently high shear strength in very thin sections to provide a usable coefficient of friction.

Friction drives are used for their quietness and smoothness of operation relative to gear and chain drives. In addition, they are easily adaptable to situations in which a variable-speed transmission is required. One major disadvantage of friction drives is their relatively low power capacities. Another is that for maximum efficiency and life, they must be operated in extremely clean environments because particles can cause rapid wear rates and accelerated failure.

In the rugged environments in which most industrial machinery must operate, the use of friction drives is not practical. Instead, gear systems are often used. In gears, slip between disks is prevented by the gear teeth of mating gears. The gears contact each other like cammed surfaces; however, for smooth, low-maintenance operation, the gear teeth have a special geometry that permits constant-velocity transfer. If designed properly, two meshing gears will behave very much the same as two friction disks, which is why gears are usually modeled grossly as two disks rolling on each other.

8.2 SPUR GEARS

Gears are used to transmit power from one shaft to another, and the power is usually transferred in such a way that the velocity ratio is constant. If the velocity ratio is not

Figure 8.3 Meshing spur gears.

Figure 8.4 Pinion and rack.

constant, the driven gear will be continuously accelerated and decelerated when the driving gear rotates at constant velocity. This results in cyclic stresses, vibration, noise, and other problems. Profiles of meshing gear teeth that give constant velocity ratio are termed conjugate. As discussed later in this chapter, given the geometry of any gear tooth profile, it is possible to construct a conjugate profile using graphical and analytical techniques similar to those used for finding cam profiles. However, there are relatively few profile types that are useful for most applications.

Spur gears are the simplest type of gear commonly used in industry. The characteristic of spur gears is that the gear rotation axes are all parallel and the gear teeth are parallel to the rotation axes. The gear may be equivalent to a rolling cylinder of any radius called the pitch cylinder. When the pitch cylinder radius becomes infinitely large, the teeth are located on a plane, and such a gear is called a rack. A simple pair of meshing gears is shown in Fig. 8.3. When two gears of unequal size are meshed, the smaller gear is referred to as the pinion and the larger gear as the "gear" or "wheel." A pinion and rack are shown in Fig. 8.4.

8.3 CONDITION FOR CONSTANT VELOCITY RATIO

Figure 8.5 represents the teeth of two gears that rotate about O_1 and O_2, respectively. The tooth profiles contact each other at point Q. The instantaneous center for the motion of gear 3 relative to gear 2 lies on the normal to the profiles at the point of contact. By Kennedy's theorem, it also lies on the line joining the centers of rotation of gears 2 and

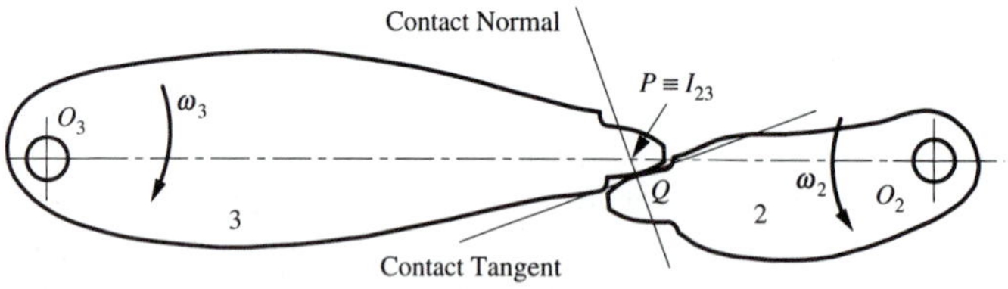

Figure 8.5 The location of the instantaneous center of relative rotation of members 2 and 3 which respectively rotate relative to the base about O_2 and O_3. The instantaneous center is P (I_{23}). The point P is also called the pitch point. The angular velocities of members 2 and 3 relative to the base are, respectively, $\boldsymbol{\omega}_2$ and $\boldsymbol{\omega}_3$. For the external gears shown, the senses of $\boldsymbol{\omega}_2$ and $\boldsymbol{\omega}_3$ are opposite.

3 relative to the base. Thus, the instantaneous center of motion of gear 3 relative to gear 2 must be at point I_{23}, which is also known as the pitch point, P.

Hence

$$\omega_2 O_2 P = \omega_2 O_2 I_{23} = -\omega_3 O_3 I_{23} = -\omega_3 O_3 P$$

or considering magnitudes only,

$$R = \frac{\omega_2}{\omega_3} = \frac{O_3 P}{O_2 P} \tag{8.2}$$

where R is the velocity ratio. Thus, a constant velocity ratio implies that O_3P/O_2P is constant. That is, P is a fixed point on the line of centers between the two gears.

Summarizing, for conjugate profiles, the normal to the profiles at the point of contact always intersects the line of centers at the same point. This is the fundamental law of gearing. This point (P) is called the pitch point. It is the instantaneous center of the relative motion of the gears.

The velocity, v_s, with which the gear teeth slide over one another is important because the rate of wear of the gear teeth depends on it. The sliding velocity of point Q is the relative velocity between the coincident points at the contact location Q; that is, it is the velocity v_{Q_2/Q_3}. From Chapter 2, the relative velocity between two coincident points can be written as

$$v_s = {}^1v_{Q_2/Q_3} = {}^3v_{Q_2/Q_3} = {}^3v_{Q_2/P_2} + {}^3v_{P_2/P_3} + {}^3v_{P_3/Q_3} = {}^3v_{Q_2/P_2} = {}^3\omega_2(r_{Q/P})$$

where "1" is used to designate the ground link. The relative velocity can be written as

$${}^3\omega_2 = {}^1\omega_2 - {}^1\omega_3$$

Dropping the superscripts that are associated with the ground link,

$$v_s = (r_{Q/P})(\omega_2 - \omega_3)$$

or using the simpler nomenclature

$$v_s = PQ|\omega_2 - \omega_3| \tag{8.3}$$

The sliding velocity is proportional to the distance between the contact point Q and the pitch point P. When P and Q coincide, the sliding velocity becomes zero and the teeth are instantaneously rolling on one another.

8.4 INVOLUTES

Of the many shapes that are possible for tooth geometries, the involute and cycloid are the most common. Of these two forms, the involute is used in almost all cases except in watches and clocks, in which the cycloidal form is still found. The involute form has several advantages, but two are most important. The first is that it is very easy to manufacture involute gears with simple tooling. The second is that the constant velocity ratio is maintained even when the center distance between the two gears is changed. This is important in manufacturing, because it is never possible to mount the gears precisely at the designed center distance.

The involute profile is developed from a circle called a base circle. The involute of a given base circle is the curve traced by any fixed point on a taut string that is being unwound from the base circle as shown in Fig. 8.6. The involute curve has the following important properties:

1. The normal to the involute at any point of the curve is tangent to the base circle.
2. The length of the normal is equal to the corresponding arc of the base circle; that

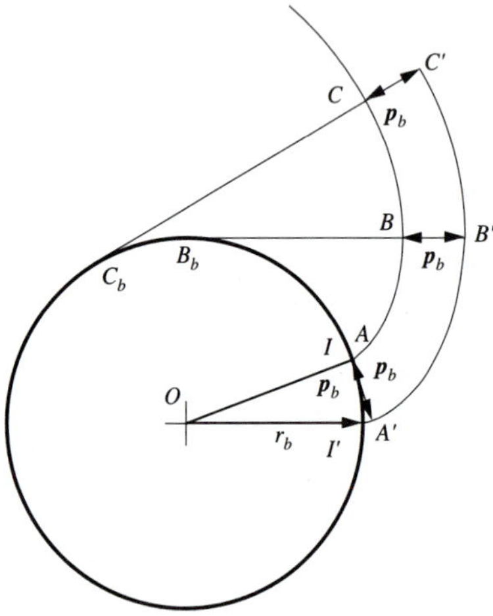

Figure 8.6 Generation of an involute by a point on a stretched string unwound from a cylinder. ABC and $I'A'B'C'$ are two involutes generated off the same base circle that has radius r_b. The string is originally wrapped around the base circle with its end at I'. When it has been unwrapped until it is tangent to the base circle at A, its end is at A'. Therefore line AA' has the same length as the curvilinear distance II'. ($I \equiv A$.) When it has been unwrapped so that its end is at point B', it is tangent to the base circle at B_b. Similarly, when the end is at C', the string is tangent to the base circle at C_b. Similarly, the point on the string that is initially at A is at B when the end of the string is at B', and it is at C when the end is at C'. Hence the normal distance between the two involutes is constant.

is, B_bB = arc B_bI, C_bC = arc C_bI. Consequently, the normal distance between two involutes of the same base circle is equal to their base pitch; that is, $AA' = BB' = CC'$ = arc $II' = p_b$.

3. The length of the normal is equal to the local radius of curvature of the involute; that is, $B_bB = \rho_b$, $C_bC = \rho_C$.
4. Any two involute profiles are mutually conjugate regardless of the base-circle diameters.
5. The path of contact between two involute profiles is rectilinear. Consequently, the pressure angle is constant.
6. A gear with involute tooth profiles can be generated by a straight-sided rack cutter.
7. Two involute profiles remain conjugate if their center distance is changed.

Proofs

1,2,3. Obtained by inspection of Fig. 8.6.

4. See Fig. 8.7. This figure shows two involute profiles in contact in an arbitrary position. The normal at the point of contact Q is tangent to both base circles. Clearly, this normal remains the same regardless of the motion of the profiles. Hence, its intersection, P, with the line of centers is fixed and the velocity ratio is constant. The velocity ratio is

$$R = \frac{O_3P}{O_2P} = \frac{r_{P_3}}{r_{P_2}} \tag{8.4}$$

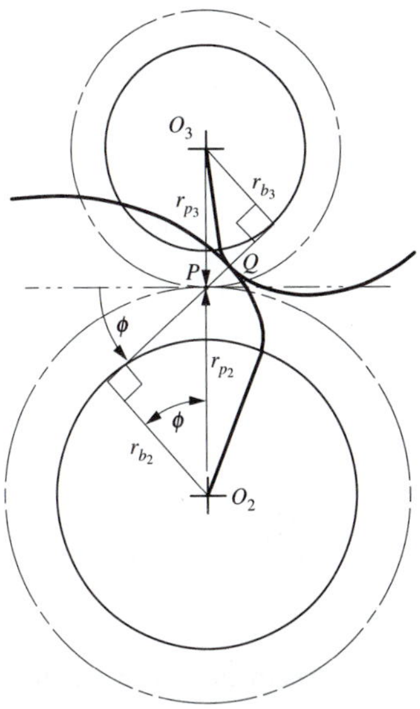

Figure 8.7 Two involutes in contact. The normal to both curves at the point of contact, Q, is tangent to both base circles. Hence it always passes through the fixed point P. Therefore two involutes are always conjugate.

where $r_{p_2} = O_2P$ is the radius of the *pitch circle* of gear 2, and $r_{p_3} = O_3P$ is the radius of the *pitch circle* of gear 3. The pitch circles are the dashed circles shown in Fig. 8.7. The radii of the pitch circles are related to those of the corresponding base circles by the equations

$$r_{p_2} = \frac{r_{b_2}}{\cos\phi} \tag{8.5}$$

and

$$r_{p_3} = \frac{r_{b_3}}{\cos\phi}$$

These equations follow from the fact illustrated in Fig. 8.7 that r_{p_2} is the hypotenuse of a right-angled triangle. r_{b_2} is another side of that triangle, and ϕ is the angle included between them. The same condition holds for gear 3. Therefore,

$$R = \frac{r_{b_3}}{r_{b_2}} \tag{8.6}$$

where r_{b_2}, r_{b_3} are the pitch circle radii. The constant angle ϕ made by the contact normal with the tangent to the pitch circles at point P is called the pressure angle.

5. Since the normal at the point of contact is always the same line, it follows that the point of contact simply moves backward and forward along this line. This profile is conjugate to a straight-sided rack. That is, if we have a linear profile inclined at angle ϕ to the line OP that simply slides back and forth without rotating, the involute profile is conjugate to it. The normal at the point of contact is always normal to the linear profile as indicated in Fig. 8.8. Because it is also always tangent to the base circle, it is always the line shown. The point P is fixed. This implies that a gear with involute teeth can be cut by a cutter in

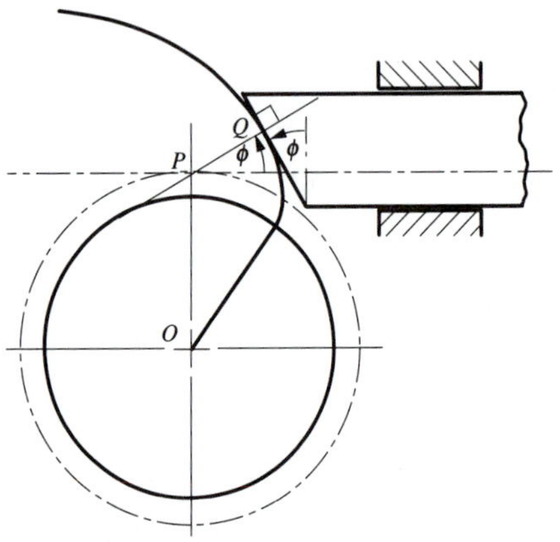

Figure 8.8 Conjugacy of involute to a straight-sided rack.

the shape of a straight-sided rack that reciprocates in a direction parallel to the axis of the gear blank.

7. The proof of item 7 follows from the proof of item 4, because the involute profiles are arbitrary and their center distance is arbitrary.

8.5 GEAR TERMINOLOGY AND STANDARDS

8.5.1 Terminology

Several terms commonly used for describing gears can be defined with the aid of Figs. 8.9 and 8.10. The *pitch circle* is the circle centered on the gear axis passing through the pitch point. The pitch circles of meshing gears roll on one another, and the point of contact is the *pitch point.* The *pitch circle diameter* or *pitch diameter* will be designated as d_p.

Although a gear will be designed to have a particular pitch circle, the actual pitch circle will depend on the gear with which it meshes and the center distance.

The *circular pitch,* p_c, is the curvilinear distance measured on the pitch circle from a point on one tooth to the corresponding point on the next tooth.

The *base pitch,* p_b, is the curvilinear distance measured on the base circle from a point on one tooth to the corresponding point on the next tooth.

The *diametral pitch,* P_d, is the number of teeth on the gear per unit of pitch diameter.

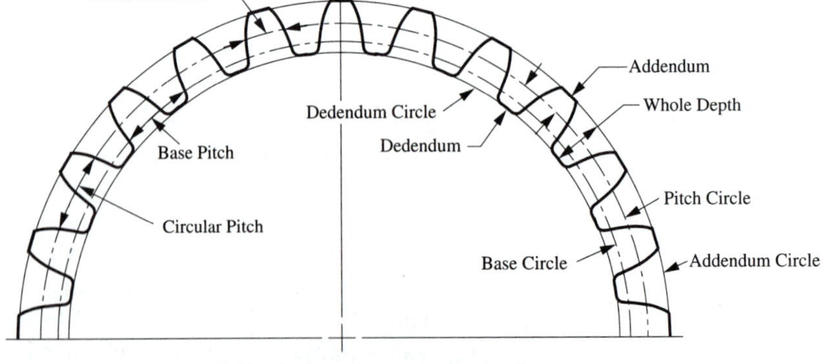

Figure 8.9 Gear tooth in-plane geometry terminology.

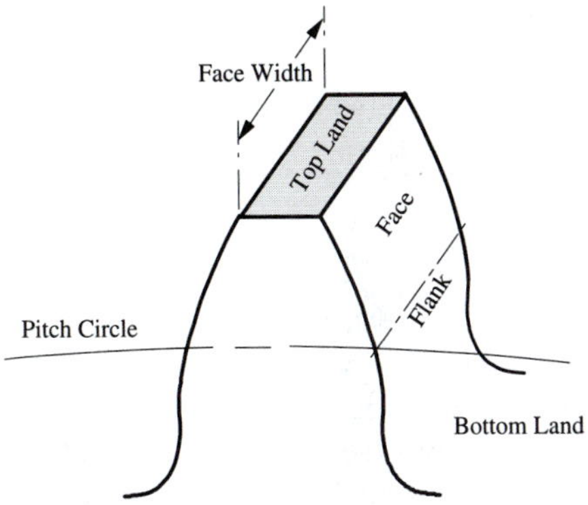

Figure 8.10 Gear tooth axial geometry terminology.

That is,

$$P_d = N/d_p \tag{8.7}$$

where N is the number of teeth on the gear. Also, from the definition of circular pitch

$$p_c = \pi d_p/N \tag{8.8}$$

so

$$P_d = \pi/p_c \tag{8.9}$$

or

$$P_d p_c = \pi \tag{8.10}$$

The base pitch is given by

$$p_b = \pi d_b/N \tag{8.11}$$

From Eq. (8.5) the base circle diameter and the pitch circle diameter are related by

$$d_p = \frac{d_b}{\cos \phi}$$

Therefore, Eq. (8.8) can be rewritten as

$$p_c = \pi d_p/N = \frac{\pi d_b}{N \cos \phi} = \frac{p_b}{\cos \phi} \tag{8.12}$$

or

$$p_b = p_c \cos \phi \tag{8.13}$$

In the metric system, gears are specified by the ratio of the pitch diameter in mm to the number of teeth. This ratio is called the *module, m*, that is expressed as follows:

$$m = \frac{d_p}{N} \tag{8.14}$$

From Eq. (8.8),

$$p_c = m\pi \tag{8.15}$$

The *addendum circle* passes through the tips of the gear teeth, and the *dedendum circle* passes through the base of the gear teeth. The *addendum* is the radial distance from the pitch circle to the top land (see Fig. 8.9) of the gear, and the *dedendum* is equal to the radial distance from the pitch circle to the bottom land of the tooth. The *whole depth* is the sum of addendum and dedendum.

With involute gears, the contact between the two gears occurs on the pressure line. This line then is called the *line of action.*

8.5.2 Standards

If two gears are to mesh, it is necessary that their pitches be the same. Since the base pitch, circular pitch, diametral pitch, and module are uniquely related, if any one of these pitch measures is the same for both gears, all will be.

Gears cut to standard dimensions (see Table 8.1) are interchangeable in the sense that any two standard gears can be meshed together. For this to be possible, the following conditions are required:

1. The pressure angles must be the same.
2. The diametral pitches must be the same.
3. The gears must have the same addendum and same dedendum.
4. The tooth thicknesses must be equal to one-half the circular pitch.

This is what permits vendors to offer off-the-shelf gears. It is also what makes possible applications such as the change gears used in a lathe to set up different cutting and feed rates. However, nonstandard gears are also used in many applications, for a variety of reasons, some of which will be discussed later in this chapter. In general, nonstandard gears have to be designed as a meshing pair, and neither member of the pair will mesh satisfactorily with a standard gear.

Table 8.1 shows the standard equations for the 20° and 25° pressure angles that are the most commonly used standard gears. A gear system that is now little used is the $14\frac{1}{2}°$ pressure angle system. This system was used extensively when gears were cast because the sine of $14\frac{1}{2}°$ is approximately 1/4, which facilitated pattern layout. As will be shown later, the $14\frac{1}{2}°$ system is inferior to the systems with higher pressure angles because the gears must have a larger number of teeth to avoid interference than are required in the 20° and 25° systems. Therefore, for a relatively small numbers of teeth, gears in the $14\frac{1}{2}°$

Table 8.1 Standard AGMA[a] and USASI[b] Tooth Systems for Involute Spur Gears[1]

System[c]	Coarse Pitch (1*P* to 19.99*P*) Full Depth		Fine Pitch (20*P* to 200*P*) Full Depth	Stub Teeth
Pressure angle, ϕ	20°	25°	25°	20°
Addendum, a	$1/P_d$	$1/P_d$	$1/P_d$	$0.8/P_d$
Dedendum, b	$1.25/P_d$	$1.25/P_d$	$1.20/P_d + 0.002$ in	$1/P_d$
Working depth, h_k	$2/P_d$	$2/P_d$	$2/P_d$	$1.6/P_d$
Whole depth, h_t (min)	$2.25/P_d$	$2.25/P_d$	$2.25/P_d + 0.002$ in	$1.8/P_d$
Circular tooth thickness, t	$\pi/2P_d$	$\pi/2P_d$	$\pi/2P_d$	$\pi/2P_d$
Fillet radius of basic rack, r_f	$0.3/P_d$	$0.3/P_d$	Not standardized	
Basic clearance, c (min)	$0.25/P_d$	$0.25/P_d$	$0.2/P_d + 0.002$ in	$0.2/P_d$
Clearance, c (shaved or ground teeth)	$0.35/P_d$	$0.35/P_d$	$0.35/P_d + 0.002$ in	
Minimum width of top land, t_0	$0.25/P_d$	$0.25/P_d$	Not standardized	

[a] American Gear Manufacturers' Association

[b] United States of America Standards Institute

[c] The standard pitches in common use are: 1 to 2 varying by 1/4 pitch, 2 to 4 varying by 1/2 pitch, 4 to 10 varying by 1 pitch, 10 to 20 varying by 2 pitch, and 20 to 200 varying by 4 pitch.

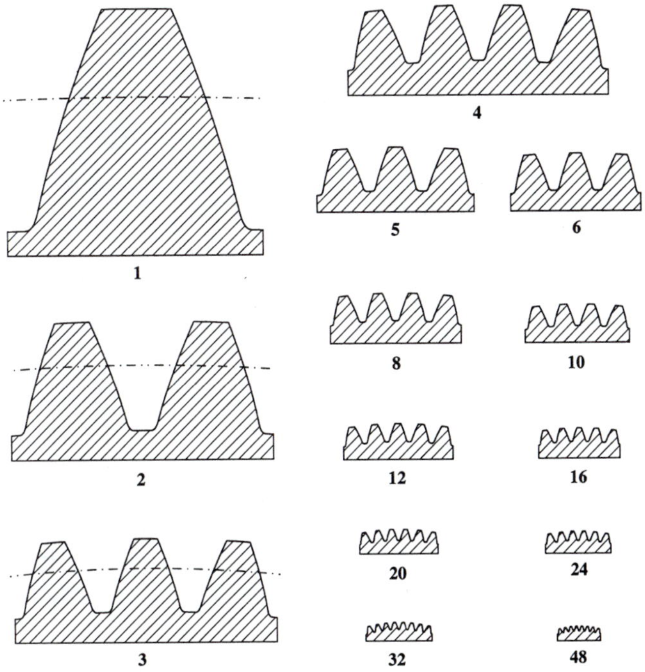

Figure 8.11 Actual relative sizes of gear teeth with diametral pitches shown.

system will have a lower beam strength and a lower load rating than corresponding gears in the 20° and 25° systems.

While any diametral pitch is technically possible, the standard pitches in common use are 1 to 2 varying by 1/4 pitch, 2 to 4 varying by 1/2 pitch, 4 to 10 varying by 1 pitch, 10 to 20 varying by 2 pitch and 20 to 200 varying by 4 pitch.

Based on the equations in Table 8.1, the physical size of the gear teeth decreases as the diametral pitch, P, increases. This is also shown in Fig. 8.11. In general, for a given set of pitch diameters, the load transfer will be smoother and undesirable affects such as noise will be lower if a large number of small teeth is used rather than a small number of larger teeth. This is because more teeth share the load when smaller teeth are used.

8.6 CONTACT RATIO

An important feature of gear action is the *contact ratio,* which can be thought of as the *average* number of tooth pairs that are in contact through the gear cycle. Thus, a contact ratio of 1.2 indicates that one tooth pair is contacting 80% of the time and two pairs 20% of the time, and this is generally considered to be the minimum contact ratio that should be used for typical designs. If the contact ratio is less than one, contact will be lost part of the time, and the gear pair cannot function properly. A larger contact ratio normally means smoother running gears and enhanced load-carrying capacity and stiffness, since the load is transferred in parallel between several tooth pairs. However, to take full advantage of a high contact ratio, very accurately cut tooth profiles are necessary.

Figure 8.12 shows a pair of mating teeth in the two positions in which they first come in contact and are at the point of losing contact. Between these two positions, the point of contact moves along the line of contact through the pitch point as described in Section 8.4. As can be seen from the figure, the locations of first and last contact are determined by the addenda of the teeth. Contact is initiated at point I, which is the intersection of

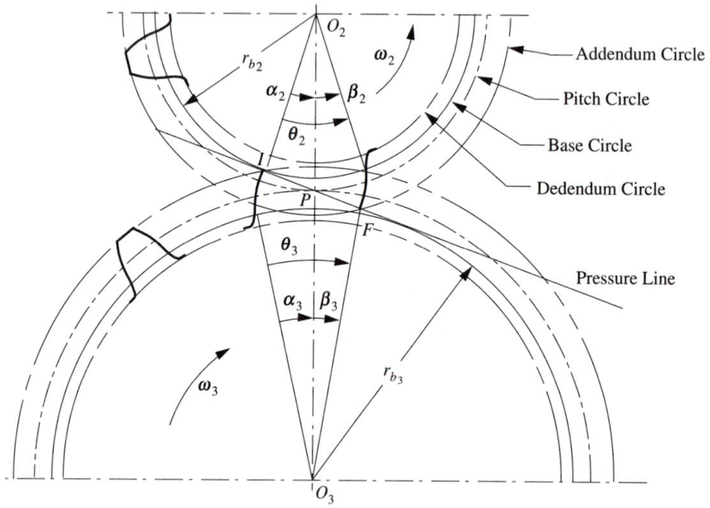

Figure 8.12 Contact angles: the contact angle of gear 2 is θ_2 and that of gear 3 is θ_3.

the pressure line with the addendum circle of gear 3. Contact is terminated at point F, which is the intersection of the pressure line with the addendum circle of gear 2.

For gear 2, the angle of approach, α_2, is the angle through which the gear rotates from the time of first contact between a pair of teeth and the time at which the contact point reaches the pitch point. Correspondingly, the angle of recess, β_2, is the angle the gear rotates through from the time at which contact is at the pitch point to the time at which contact is lost. The angle $\theta_2 = \alpha_2 + \beta_2$ is the angle of contact of gear 2 when in mesh with gear 3. Meshing gear 2 with another gear will change both α_2 and β_2 if the size of the second gear changes.

Because the distance the contact point moves along the path of contact is equal to the curvilinear distance around the base circle, according to the relationship illustrated in Fig. 8.12,

$$\alpha_2 = IP/r_{b_2}$$

and

$$\beta_2 = PF/r_{b_2}$$

Hence

$$\theta_2 = IF/r_{b_2}$$

The distance IP from the point of initial contact to the pitch point can be computed from the geometry shown in Fig. 8.13, where r_{p_3} and r_{a_3} are, respectively, the pitch circle radius and addendum (tip) circle radius of the driven gear (gear 3), ϕ is the pressure angle, and u is the length IP. Application of the cosine rule to the triangle IPO_3 gives:

$$r_{a_3}^2 = u^2 + r_{p_3}^2 - 2ur_{p_3}\cos(\phi + \pi/2)$$

or

$$u^2 + r_{p_3}^2 + 2ur_{p_3}\sin\phi - r_{a_3}^2 = 0$$

Solution of this quadratic equation for u and simplifying using $\sin^2\phi + \cos^2\phi = 1$ gives

$$u = -r_{p_3}\sin\phi \pm \sqrt{r_{a_3}^2 - r_{p_3}^2\cos^2\phi}$$

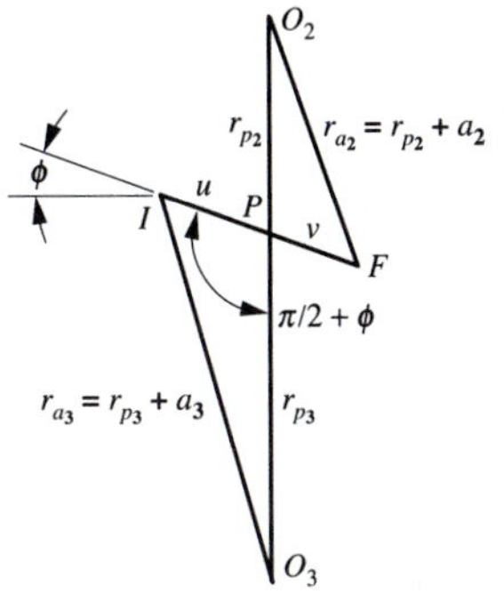

Figure 8.13 The geometry used in computing the length of the path of contact, *IF*. The points O_2, O_3, P, I, and F are the same as those shown in Fig. 8.12.

Now $r_{a_3} = r_{P_3} + a_3$, where a_3 is the addendum of gear 3. Therefore this equation can be written in the form:

$$u = -r_{P_3} \sin \phi \pm \sqrt{a_3^2 + 2a_3 r_{P_3} + r_{P_3}^2 \sin^2 \phi}$$

It can be seen that the square root term is always larger than $r_{P_3} \sin \phi$, so there will always be a positive root and a negative root. Only the positive root is consistent with the geometry presented. Therefore:

$$u = -r_{P_3} \sin \phi + \sqrt{a_3^2 + 2a_3 r_{P_3} + r_{P_3}^2 \sin^2 \phi} \tag{8.16}$$

Because of the symmetry between the triangles in Fig. 8.13, Eq. (8.16) may also be used to compute $v = PF$ by replacing r_{P_3} with r_{P_2} and a_3 with a_2. That is,

$$v = -r_{P_2} \sin \phi + \sqrt{a_2^2 + 2a_2 r_{P_2} + r_{P_2}^2 \sin^2 \phi}$$

As shown in Fig. 8.14, the distance along the respective base circles between the involutes in the initial and final positions is

$$\lambda = IF = IP + PF = u + v$$

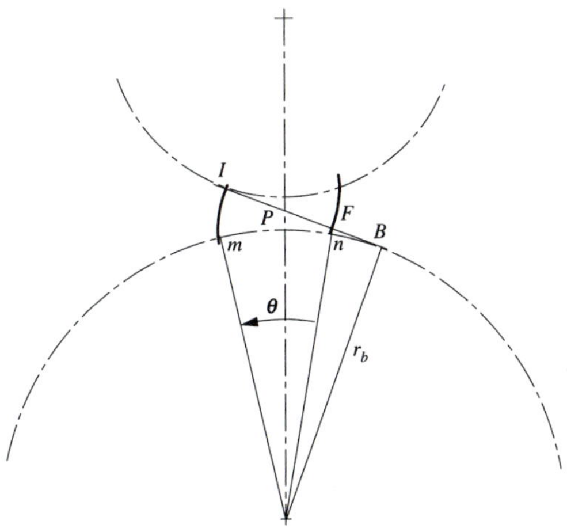

Figure 8.14 Relationship of the length of the path of contact to the angle of contact. The length *IF* is equal to the curvilinear distance *mn* around the base circle. Hence the angle of contact is $\theta = IF/r_b$.

or

$$\lambda = -r_{p_2} \sin\phi + \sqrt{a_2^2 + 2a_2 r_{p_2} + r_{p_2}^2 \sin^2\phi} - r_{p_3} \sin\phi + \sqrt{a_3^2 + 2a_3 r_{p_3} + r_{p_3}^2 \sin^2\phi} \quad \textbf{(8.17)}$$

The angle of recess of the driving gear is, therefore,

$$\eta_2 = v/r_{b_2} = v/(r_{p_2} \cos\phi)$$

Similarly, the angle of approach of the driven gear is

$$\eta_3 = u/r_{b_3} = u/(r_{p_3} \cos\phi)$$

Using Eqs. (8.7)–(8.13) and (8.17), the contact ratio can now be computed as the ratio of the length of the path of contact to the base pitch. That is,

$$m_c = \frac{\lambda}{p_b} = \frac{\lambda}{p_c \cos\phi} = \frac{\lambda P_d}{\pi \cos\phi}$$
$$= \frac{P_d\{-r_{p_2} \sin\phi + \sqrt{a_2^2 + 2a_2 r_{p_2} + r_{p_2}^2 \sin^2\phi} - r_{p_3} \sin\phi + \sqrt{a_3^2 + 2a_3 r_{p_3} + r_{p_3}^2 \sin^2\phi}\}}{\pi \cos\phi}$$

(8.18)

As stated above, the significance of the contact ratio (m_c) is that it determines the load sharing among the teeth. If the contact ratio is less than 1, there will be periods in which contact is completely lost, and the gears cannot function. If the contact ratio is close to 1, variations caused by errors in mounting or wear may cause loss of contact. In practice, contact ratios less than 1.2 should be avoided.

EXAMPLE 8.1 ***(Contact Ratio for Two 20° Pressure-Angle Gears)***

PROBLEM

Two gears are in mesh such that one gear (gear 2) has 20 teeth and the other (gear 3) has 30. The diametral pitch for each gear is 4, and the working pressure angle is 20°. Standard gears are involved in each case, and the addendum constant is 1. Determine the length of the contact line and the contact ratio.

SOLUTION

To compute the contact ratio, we need to determine the base pitch and all of the terms in Eq. (8.17). From Table 8.1, the addendum for both gears is given by

$$a = \frac{1}{P_d} = \frac{1}{4} = 0.25 \text{ in} = a_2 = a_3$$

Similarly, the circular pitch for both gears is given by

$$p_c = \frac{\pi}{P_d} = \frac{\pi}{4} = 0.785 \text{ in}$$

and from Eq. (8.13), the base pitch is related to the circular pitch by

$$p_b = p_c \cos\phi = 0.785 \cos 20° = 0.738 \text{ in}$$

From Eq. (8.7), the two pitch radii are given by

$$r_{p_2} = \frac{N_2}{2P_d} = \frac{20}{2(4)} = 2.5 \text{ in}$$

and

$$r_{p_3} = \frac{N_3}{2P_d} = \frac{30}{2(4)} = 3.75 \text{ in}$$

The length of the line of contact is given by Eq. (8.17) as

$$\lambda = -r_{p_2}\sin\phi + \sqrt{a_2^2 + 2a_2 r_{p_2} + r_{p_2}^2\sin^2\phi} - r_{p_3}\sin\phi + \sqrt{a_3^2 + 2a_3 r_{p_3} + r_{p_3}^2\sin^2\phi}$$

$$= -2.5\sin 20^\circ + \sqrt{0.25^2 + 2(0.25)(2.5) + 2.5^2\sin^2 20^\circ}$$
$$-3.75\sin 20^\circ + \sqrt{0.25^2 + 2(0.25)(3.75) + 3.75^2\sin^2 20^\circ}$$
$$= 1.185 \text{ in}$$

From Eq. (8.18), the contact ratio is

$$m_c = \frac{\lambda}{p_b} = \frac{1.185}{0.7380} = 1.6052$$

Therefore, on average, approximately 1.6 teeth are in contact as the gears mesh. ■

8.7 INVOLUTOMETRY

It is important, for the purpose of analyzing stress and deflection of gear teeth, to be able to compute the thickness of a tooth at any radius. From Fig. 8.15, the arc distance AB is equal to the linear distance BT. Therefore,

$$AB = BT = r_b \tan\xi$$

and

$$\chi = AB/r_b - \xi = \tan\xi - \xi$$

χ is called the involute function of ξ, written inv(ξ)

$$\text{inv}(\xi) = \tan\xi - \xi \tag{8.19}$$

Involute functions can be obtained from tables. They are also easily calculated on a pocket calculator. The thickness of the tooth at any radius r can be computed using involute functions.

Referring to Fig. 8.16,

$$\frac{t_p}{2r_p} = \alpha - \text{inv}\,\phi$$

$$\frac{t}{2r} = \alpha - \text{inv}\,\beta$$

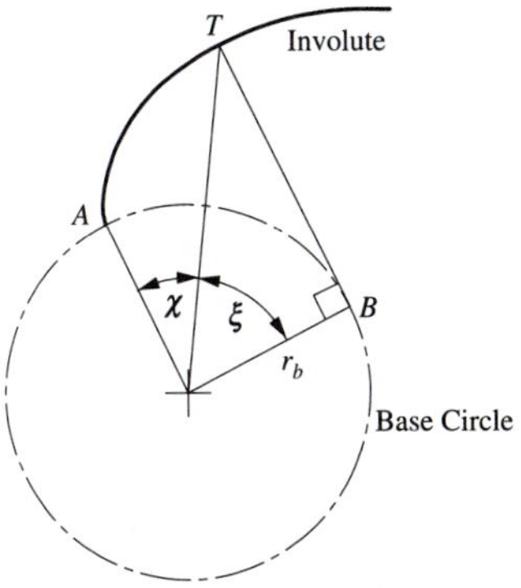

Figure 8.15 The basic geometry for the definition of the involute function.

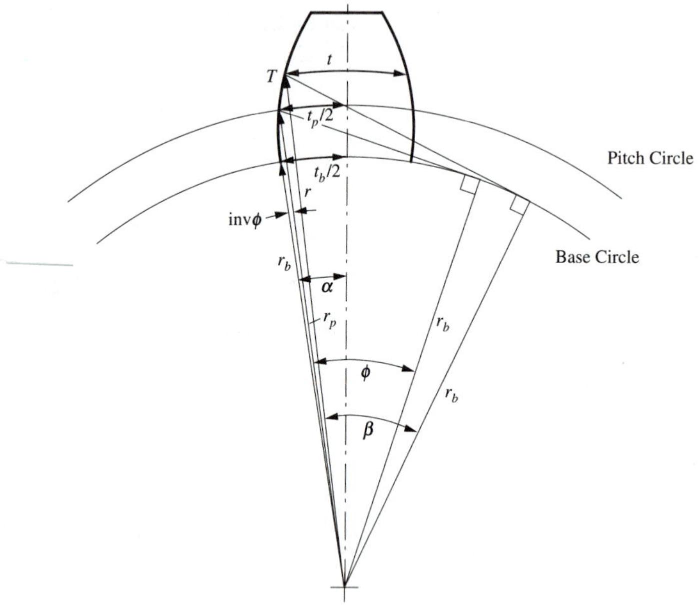

Figure 8.16 Computation of the tooth thickness, t, at any radius, r. Tooth thickness is computed as a *curvilinear* length along the circumference of the circle of radius r through point T.

Hence

$$\frac{t}{2r} = \frac{t_p}{2r_p} + \text{inv}\,\phi - \text{inv}\,\beta \tag{8.20}$$

where

$$\cos\beta = \frac{r_b}{r} = \frac{r_p\cos\phi}{r} \tag{8.21}$$

EXAMPLE 8.2 ***(Thickness of Gear Tooth)***

PROBLEM

Find the thickness at the addendum and base circles of a tooth of diametral pitch 4 on a gear with 30 teeth cut with a 20° pressure angle, standard dimensions, and with the tooth thickness at the pitch circle equal to half the circular pitch.

SOLUTION

First compute the radii at the three locations for which the tooth thickness is desired. The radius of the pitch circle is given by Eq. (8.7) as

$$r_p = \frac{N}{2P_d} = \frac{30}{2(4)} = 3.75 \text{ in}$$

Using Table 8.1, the addendum radius is given by

$$r_a = r_p + a = r_p + \frac{1}{P_d} = 3.75 + \frac{1}{4} = 4.0 \text{ in}$$

The base circle radius is given by Eq. (8.5) as

$$r_b = r_p\cos\phi = 3.75\cos 20^\circ = 3.524 \text{ in}$$

From Table 8.1, the circular tooth thickness is

$$t_p = \frac{\pi}{2P_d} = \frac{\pi}{2(4)} = 0.393 \text{ in}$$

Next compute the angle β at the addendum circle. From Eq. (8.21),

$$\beta_a = \cos^{-1}\left[\frac{r_p \cos\phi}{r_a}\right] = \cos^{-1}\left[\frac{3.75 \cos 20°}{4.0}\right] = 28.241°$$

The value of β at the base circle is zero.

From Eq. (8.20), the equation for the tooth thickness at the addendum is

$$t_a = 2r_a\left(\frac{t_p}{2r_p} + \text{inv}\,\phi - \text{inv}\,\beta_a\right) = 2(4.0)\left(\frac{0.393}{2(3.75)} + \text{inv}\,20° - \text{inv}\,28.241°\right) = 0.184 \text{ in}$$

The value of β at the base circle is zero. Therefore, the tooth thickness at the base circle is

$$t_b = 2r_b\left(\frac{t_p}{2r_p} + \text{inv}\,\phi - \text{inv}\,\beta_b\right) = 2(3.524)\left(\frac{0.393}{2(3.75)} + \text{inv}\,20° - 0\right) = 0.474 \text{ in}$$

■

EXAMPLE 8.3 (*Minimum Thickness of Gear Tooth*)

PROBLEM

From Example 8.2, it is clear that the gear tooth thickness becomes smaller as the radius increases from the base circle radius. In the limiting case, the tooth thickness is zero. For the gear in Example 8.2, determine the maximum radius that could be specified for the addendum circle if a nonstandard gear were used.

SOLUTION

From Eq. (8.20) and Fig. 8.17, the gear tooth thickness becomes zero when

$$2r\left(\frac{t_p}{2r_p} + \text{inv}\,\phi - \text{inv}\,\beta\right) = 0 \qquad \textbf{(8.22)}$$

Equation (8.22) is satisfied when

$$\text{inv}\,\beta = \frac{t_p}{2r_p} + \text{inv}\,\phi = \frac{0.393}{2(3.75)} + \text{inv}\,20° = 0.0673$$

This equation can be solved for β by using a table of values for the involute function or by simply computing a series of values for inv β on a calculator or with a program such as MATLAB.

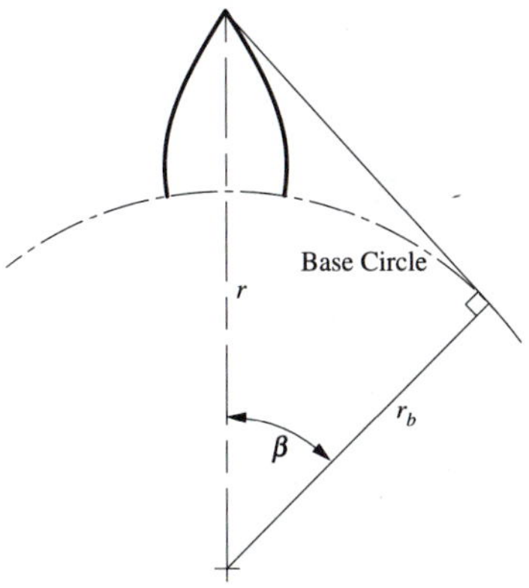

Figure 8.17 Limiting case for r (tooth thickness = 0).

The solution is $\beta = 32.13°$. From Fig. 8.17, it is clear that β and r are related by

$$r = r_b/\cos\beta \tag{8.23}$$

Therefore,

$$r = r_b/\cos\beta = 3.524/\cos 32.13° = 4.161 \text{ in} \tag{8.24}$$

■

8.8 INTERNAL GEARS

An internal or annular gear is a gear that has its center on the same side of the pitch circle as the pinion meshing with it as shown in Fig. 8.18. The addendum circle for the internal gear is inside the pitch circle, and the teeth are concave rather than convex. Internal gears are commonly used in planetary gear systems and compact gear boxes. The primary advantage of an internal gear set is the compactness of the drive. Also, both the pinion and gear rotate in the same direction. Other advantages are the lower contact stresses because the surfaces conform better than external gear sets. There are also lower relative sliding between teeth and a greater length of contact possible between mating teeth since there is no limit to the involute profile on the flank of the internal gear. Because of the tooth shape, the bending strength of the internal teeth is much greater than the strength of the teeth on the pinion. Therefore, the pinion is always the weaker member unless different materials are used for the two gears.

As in the case of external gears, the contact occurs along the line of action that is tangent to both base circles and passes through the pitch point or point of tangency between the two pitch circles. Referring to Fig. 8.18, the line of action is tangent to the base circle of the internal gear at C and to the base circle of the pinion at B. If contact occurs between B and C, interference will result, because the involute part of the pinion tooth does not cross the line of action until B is reached. Contact continues until point A is reached, where A is the intersection of the line of action with the addendum of the pinion. Therefore, the length of the line of action is AB, and the addendum of the gear should not extend beyond point B.

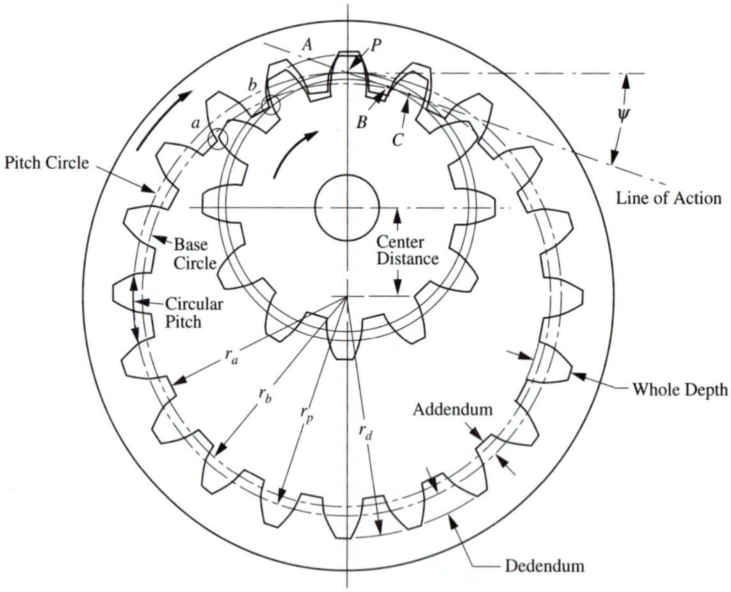

Figure 8.18 Internal gear and pinion.

A different type of interference between the gear and pinion can also occur when the number of teeth on both gears varies only slightly. The interference is called fouling and it occurs at inactive profiles when the teeth of the pinion withdraw from the space of the gear. Potential sites for fouling are locations *a* and *b* in Fig. 8.18. To remove the potential for fouling, internal gears are usually cut with a shaper cutter that has two teeth fewer than the internal gear being cut. This automatically relieves the tips of the internal gear and eliminates the potential for fouling for any pinion with fewer teeth than the cutter.[2]

Standard tooth proportions are not used for internal gears because the addendum of the gears must be shorter than those given in Table 8.1 in order to avoid interference. However, the basic equations for circular and diametral pitch apply, and the contact ratio and angles of action can be determined in the same manner as for external gears.

8.9 GEAR MANUFACTURING

Gear teeth can be formed in a large number of ways including various forms of casting: sand casting, investment casting, and die casting. They can be cut from flat stock using electron-discharge machining (EDM), CNC milling, and even precision sawing (and secondary machining or grinding). Gears made from polymers, aluminum, magnesium, and so on can be extruded and cut to width. Thin gears can be blanked from sheet stock.

Gears that must carry large loads relative to their overall size are usually made of high-strength materials such as steel. Some gears, such as bevel gears, can be forged; however, the vast majority of production gears are machined from blanks. Very small quantities can be machined using EDM, CNC milling, or horizontal milling with a formed cutter; however, when large volumes are involved, the machining is usually done with a generating cutter. In formed cutters, the tooth takes the exact shape of the cutter. Therefore, a separate cutter is technically required for each gear pitch and each number of teeth because the shape of the space between the teeth varies with both pitch and tooth number. In reality, the change in space is not significant in many cases. For a given pitch, only eight cutters are required to cut any gear in the range of 12 teeth to a rack with reasonable accuracy. However, such gears are usually not accurate enough for high speeds. A separate set of cutters is required for each pitch.[3] Figure 8.19 shows an example of milling using a formed cutter.

In a generating cutter, the tool has a shape different from the tooth profile, and the tool is moved relative to the gear blank to produce the desired gear shape.

When large volumes are involved, the fabrication of gears normally involves the following steps:

1. Blank fabrication
2. Forming of teeth
3. Refining of teeth
4. Heat treatment
5. Grinding, deburring, and cleaning
6. Finish coating

Blank fabrication involves all of the general and special features of the gear blank. This includes forming the hub and keyways. Tooth generation includes machining the gear teeth using one of the processes discussed below. The refining operations include shaving, grinding, burnishing, and lapping and are used to improve the accuracy of the gear teeth. These are necessary to remove machining marks and to improve the gear quality. The higher the gear quality, the greater the power rating and the lower the noise generated by meshing gear pairs. Heat treatment includes case hardening, which is necessary for high-performance gears to improve the resistance to surface pitting and tooth fracture. If heat treated, the gears must be reground if high accuracy is required. Deburring and

Figure 8.19 Milling gear teeth using a formed cutter. For a given gear pitch, eight cutters are required to form (approximately) gears with from 12 teeth to a rack.

cleaning are essential for all gears regardless of how they have been manufactured or the accuracy desired. Finish coatings include processes such as anodizing aluminum or depositing diamond films, but they may involve only grease or paint. The objective is to improve corrosion resistance, to reduce friction and wear, or simply to improve appearance.

Machining by shaping and hobbing are the most common methods of gear tooth generation. The objective is to slowly mesh a gear blank into a cutting tool that has teeth that will be conjugate with the teeth cut into the blank. The cutting action is always orthogonal to the side of the gear blank. In the shaping process, the cutter looks like a gear but is made of much harder material (see Figs. 8.20 and 8.21). When the gear is shaped, the reciprocating shaper cutter is moved radially into the blank until the pitch circle of the cutter and the gear blank are tangent. After each cutting stroke, the cutter is raised above

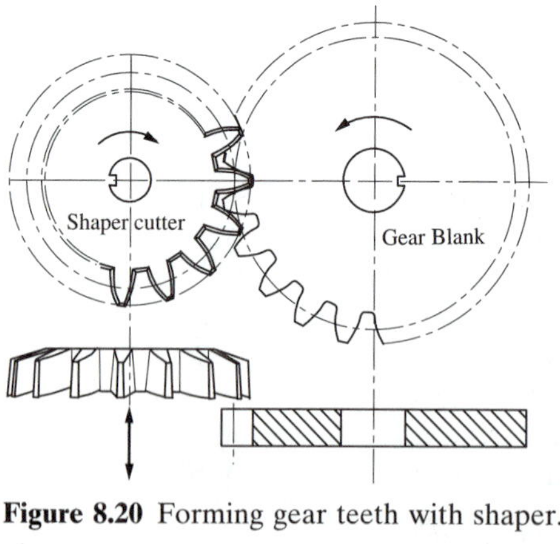

Figure 8.20 Forming gear teeth with shaper.

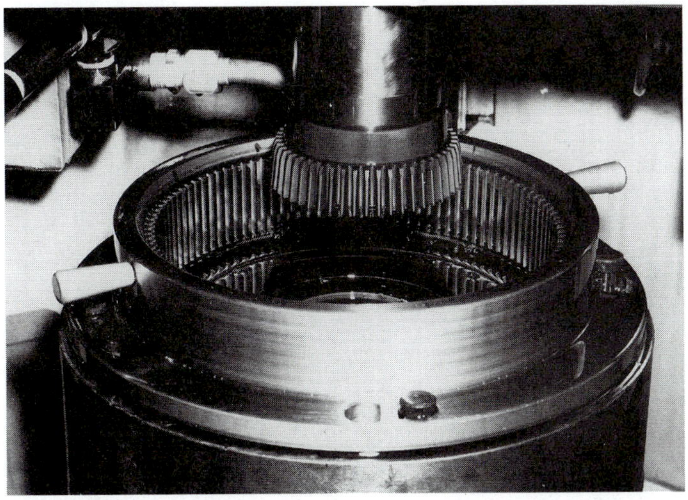

Figure 8.21 Shaping an internal gear.

the blank, and both the blank and the cutter roll a very small amount on their pitch circles. The process continues until all of the teeth are cut. Shaping is the main method used to produce internal gears and gears integral with a shaft that has a shoulder next to the gear.

The shaping cutter can also be in the form of a rack. In this case, the pitch circle of the cutter is a straight line, and the cutter moves on a straight line tangent to the pitch circle of the gear. Relative to the gear, the pitch line of the rack appears to roll around the pitch circle of the gear. The gear teeth are formed by the envelope of the rack teeth as shown in Fig. 8.22.

Hobbing is a method of generating gear teeth that is geometrically similar to generating the teeth with a rack cutter. The teeth of a hob are of the same shape as those on a rack cutter. However, the teeth are attached to a helical path on a cylindrical cutter. The hob looks like a worm gear with horizontal slices taken out of it. It is similar in appearance to a machine screw tap. The hob action is shown in Fig. 8.23, and a hobbing machine is shown in Fig. 8.24. In the hobbing machine, the hob teeth are aligned with the axis of the

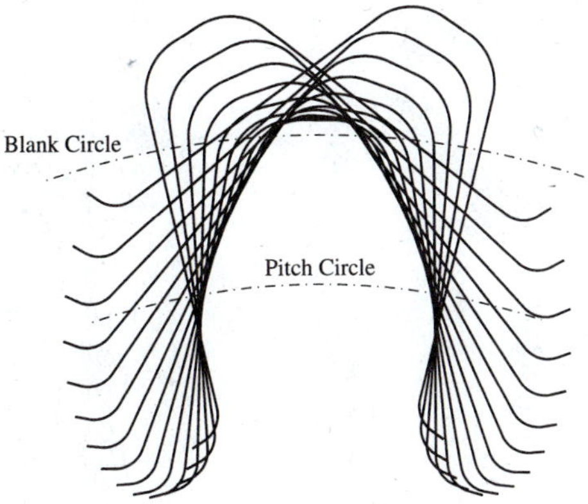

Figure 8.22 Envelope of rack teeth relative to gear blank.

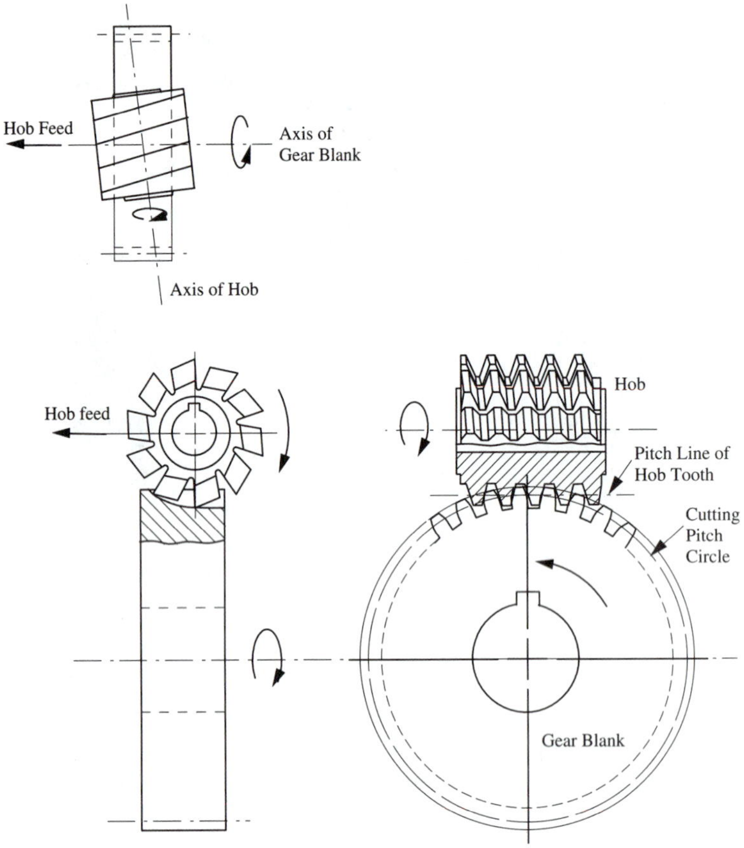

Figure 8.23 Tooth generation with hob.

Figure 8.24 Hobbing spur gear teeth.

gear teeth. The hob and blank are rotated continuously at the proper angular velocity ratio, and the hob is fed slowly across the face of the width of the blank from one side to the other to form the teeth. The hobbing process is the most popular among the machining processes because it produces the most teeth in a given time and because the same tool can be used to cut helical as well as spur gear teeth. However, for internal gears, a gear shaper must be used.

8.10 INTERFERENCE AND UNDERCUTTING

If the path of contact extends beyond the point of tangency with the base circle of the pinion, called the interference point, the tips of the teeth on the gear come into contact with portions of the pinion tooth profile inside the base circle. Because the involute is not defined inside the base circle, conjugate action is lost, and, in fact, the tips of the gear teeth will interfere with the lower portion of the pinion tooth flank if the tooth is machined with a formed cutter. If the pinion is generated with a rack-type cutter or hob, the cutter teeth will undercut the pinion teeth and weaken them. This situation is shown in Fig. 8.25. Thus, it is very undesirable to have the path of contact extend past the interference point.

In Fig. 8.26, two gears are shown in mesh. The interference point on gear 2 is point A and that on gear 3 is point D. The addendum circle of gear 3 intersects the line of action at B, and the addendum circle of gear 2 intersects the line of action at C. If the addendum circle for gear 2 extends beyond point D or the addendum circle of gear 3 extends beyond point A, interference will occur. Actually, we need only investigate one of these conditions because, as indicated in Fig. 8.26, interference will always occur first on the smaller gear. Therefore, we need only check point D. In Fig. 8.26, if we visualize the pitch diameter of gear 2 becoming larger, point C will gradually move toward point D. Eventually, there will be a diameter that causes point C to move beyond point D, and interference will occur. If gear 2 is actually a shaper cutter, the material in the interference region will simply be removed. If, however, the teeth on gear 3 were formed by a smaller shaper cutter or cut directly, for example, by using a conforming milling cutter, then material might exist in the problem region. In that case, when gear 2 is meshed with gear 3, there will be volumetric interference, and the gear loads could be high enough to cause premature failure.

The worst case for interference or undercutting is when the pinion is meshed with a rack or generated with a rack cutter. This condition is shown in Fig. 8.27. Therefore, if we design the gear to avoid interference with a rack, the gear will mesh with all other standard gears satisfying the criteria in Section 8.5.2. Undercutting becomes more severe as the number of teeth on the gear is reduced because for a given diametral pitch, the pitch diameter decreases when the number of teeth is reduced. Therefore, there will be a critical number of teeth on the gear to just avoid undercutting. The minimum number of teeth to avoid undercutting can be determined by identifying when the addendum circle

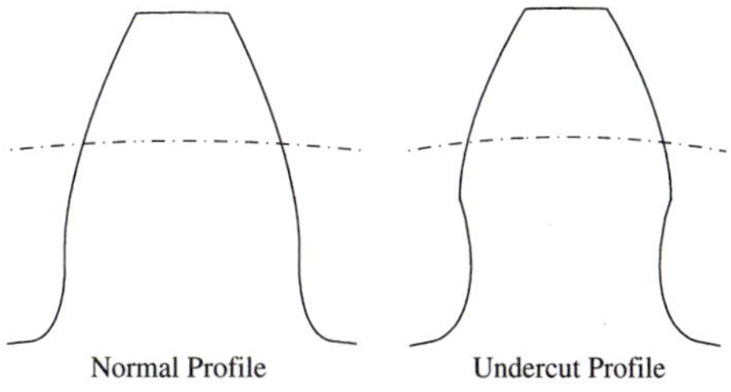

Figure 8.25 Normal and undercut gears.

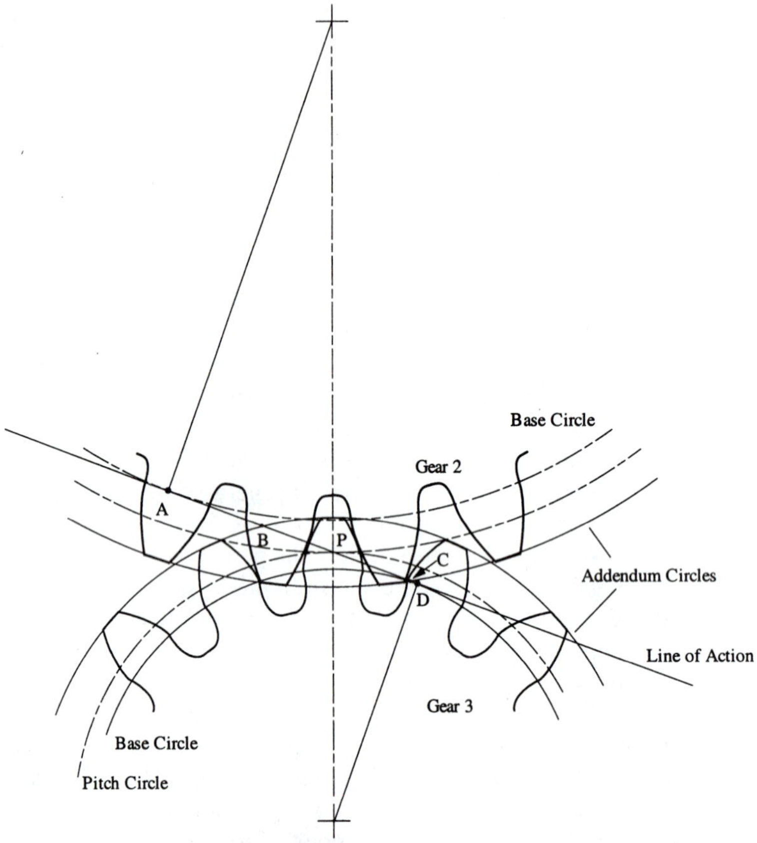

Figure 8.26 Meshing conditions for two gears.

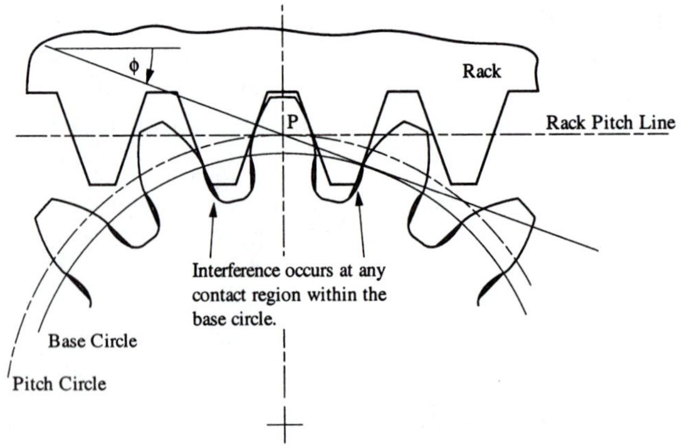

Figure 8.27 Interference will occur at all contact locations within the base circle.

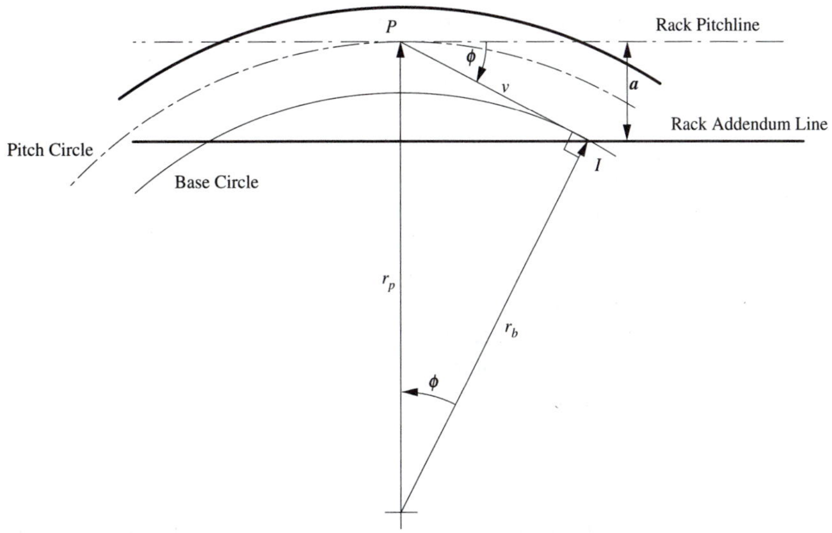

Figure 8.28 Geometry for determining undercutting.

of the rack cutter extends beyond the interference point on the gear. This can be done using the simplified geometry in Fig. 8.28. Using that figure,

$$v = \frac{a}{\sin \phi} = r_p \sin \phi$$

or

$$r_p = \frac{a}{\sin^2 \phi} \tag{8.25}$$

We may use k/P_d as the addendum for a standard gear, where k is a constant (1 for full depth gears and 0.8 for stub tooth gears), and from Eq. (8.7)

$$2r_p = \frac{N}{P_d}$$

Substituting into Eq. (8.25) gives

$$N = \frac{2k}{\sin^2 \phi} \tag{8.26}$$

Results for some of the common pressure angles and systems are summarized in Table 8.2. Note that the number of teeth must be a whole number for a continuously rotating gear. Also notice that the minimum number of teeth varies inversely with the pressure

Table 8.2 Minimum Number of Teeth to Avoid Undercutting for Standard Gears

System	Full Depth	Full Depth	Full Depth	Stub
ϕ	$14\frac{1}{2}°$	20°	25°	20°
k	1	1	1	0.8
N	31.9	17.10	11.20	13.68
N_{min}	32	18	12	14

angle. Therefore, the minimum number of teeth for a $14\frac{1}{2}°$ pressure angle is 32, while for a 20° pressure angle it is 18. This is one of the reasons why the $14\frac{1}{2}°$ system is rarely used in modern machinery.

8.11 NONSTANDARD GEARING

Sometimes, to save space, the preceding minima are violated. Interference can still be avoided by offsetting the cutting rack so that the addendum line of the rack passes through the interference point or outside it. This requires a larger blank for the pinion. The net effect is to increase the addendum of the pinion and decrease its dedendum. The removal of undercutting is accompanied by other beneficial effects; the tooth shape is stronger and the contact ratio is improved. However, a nonstandard gear results.

Referring to Fig. 8.29, the cutter can be offset a distance e to bring the addendum line through the pitch point. This offset distance can be computed from

$$e = a + r_b \cos\phi - r_p \tag{8.27}$$

where ϕ is the cutting pressure angle and r_p is the cutting pitch radius. We can eliminate r_b from Eq. (8.27) using Eq. (8.5) and the expression $\sin^2\phi + \cos^2\phi = 1$. Then,

$$e = a + r_p \cos^2\phi - r_p = a - r_p \sin^2\phi$$

The expression for the tooth thickness can be used to relate tooth thickness at any radius to the actual meshing pressure angle (ϕ_m)

$$t_{m_2} = 2r_{m_2}\left(\frac{t_2}{2r_2} + \text{inv}\,\phi - \text{inv}\,\phi_m\right)$$

$$t_{m_3} = 2r_{m_3}\left(\frac{t_3}{2r_3} + \text{inv}\,\phi - \text{inv}\,\phi_m\right)$$

where t_{m_2} and t_{m_3} are the thicknesses at the meshing pitch circle, and r_{m_2} and r_{m_3} are the meshing pitch radii.

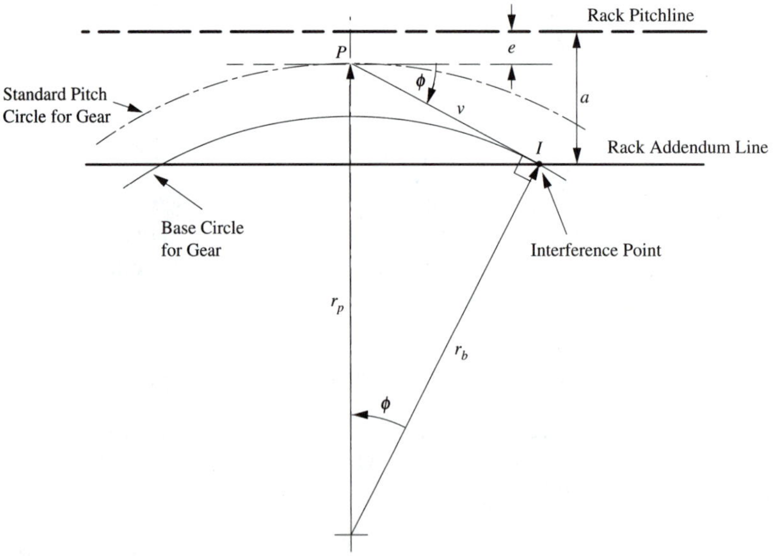

Figure 8.29 Geometry for determining undercutting when the hob is withdrawn by an amount e.

Table 8.3 Minimum Values of Addendum for Unequal Addendum and Dedendum Gears to Avoid Undercut[a]

Number of Pinion Teeth	AGMA 20-deg Course-Pitch System		Minimum Number of Gear Teeth	AGMA 20-deg Fine Pitch System		Minimum Number of Gear Teeth
	a_P	a_G		a_P	a_G	
7	—	—	—	1.4143[b]	0.4094	42
8	—	—	—	1.43669[b]	0.4679	39
9	—	—	—	1.4190[b]	0.5264	36
10	1.468	0.532	25	1.4151	0.5849	33
11	1.409	0.591	24	1.3566	0.6434	30
12	1.351	0.649	23	1.2982	0.7019	27
13	1.292	0.708	22	1.2397	0.7604	25
14	1.234	0.766	21	1.1812	0.8189	23
15	1.117	0.825	20	1.1227	0.8774	21
16	1.117	0.883	19	1.0642	0.9358	19
17	1.058	0.942	18	1.0057	0.9943	18
18	1.000	1.000	18	1.0000	1.0000	

[a] The values in this table are for one diametral pitch; divide them by the desired diametral pitch to obtain the addendum.

[b] Not proportional to the increase in tooth thickness; a reasonable top land must be provided.

Now

$$t_{m_2} + t_{m_3} = \frac{2\pi r_{m_2}}{N_2}$$

and

$$\frac{r_{m_3}}{r_{m_2}} = \frac{r_3}{r_2} = \frac{N_3}{N_2}$$

Therefore,

$$t_{m_2} + t_{m_3} = 2r_{m_2}\left(\frac{t_2}{2r_2} + \text{inv}\,\phi - \text{inv}\,\phi_m\right) + 2r_{m_3}\left(\frac{t_3}{2r_3} + \text{inv}\,\phi - \text{inv}\,\phi_m\right)$$

gives

$$\frac{2\pi r_{m_2}}{N_2} = 2r_{m_2}\left(\frac{t_2}{2r_2} + \text{inv}\,\phi - \text{inv}\,\phi_m\right) + \frac{2N_3}{N_2} r_{m_2}\left(\frac{t_3}{2r_3} + \text{inv}\,\phi - \text{inv}\,\phi_m\right)$$

or

$$\frac{2\pi}{N_2} = \frac{t_2 + t_3}{r_2} + \frac{2(N_2 + N_3)}{N_2}(\text{inv}\,\phi - \text{inv}\,\phi_m)$$

giving

$$\text{inv}\,\phi_m = \text{inv}\,\phi - \frac{2\pi r_2 - N_2(t_2 + t_3)}{2r_2(N_2 + N_3)} \quad \textbf{(8.28)}$$

This equation permits computation of ϕ_m given the cutting pitch radius, r_2, of the gear, the cutting pressure angle, ϕ, the tooth numbers, N_2 and N_3, the tooth thickness, t_2, at

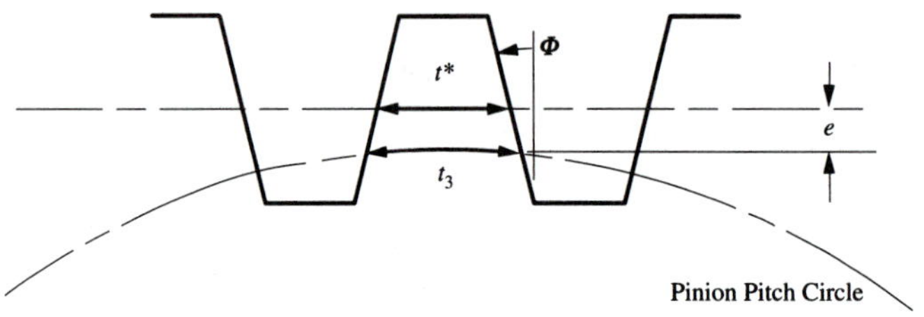

Figure 8.30 Cutting nonstandard pinion.

the cutting pitch circle of the gear, and the tooth thickness, t_3, of the pinion at the cutting pitch radius.

Referring to Fig. 8.30,

$$t_3 = t^* + 2e \tan \phi$$

Table 8.4 Summary of Spur Gear Formulas

Quantity	Formula
Diametral pitch	$P_d = \dfrac{N}{d_p} = \dfrac{N}{2r_p} = \dfrac{\pi}{p_c}$
Circular pitch	$p_c = \dfrac{\pi}{P_d} = \dfrac{\pi d_p}{N} = \pi m = \dfrac{p_b}{\cos \phi} = \dfrac{\pi d_b}{N \cos \phi}$
Base pitch	$p_b = \dfrac{\pi d_b}{N} = p_c \cos \phi = \dfrac{\pi d_p \cos \phi}{N}$
Module	$m = \dfrac{d_p}{N} = \dfrac{1}{P_d} = \dfrac{p_c}{\pi}$
Tooth thickness at pitch circle	$t = \dfrac{p_c}{2} = \dfrac{\pi}{2P_d} = \dfrac{\pi d_p}{2N} = \dfrac{\pi m}{2} = \dfrac{p_b}{2 \cos \phi} = \dfrac{\pi d_b}{2N \cos \phi}$
Pitch diameter	$d_p = \dfrac{N}{P_d} = \dfrac{N p_c}{\pi} = Nm$
Outside diameter	$d_o = d_p + 2a$ (typically $a = k/P_d$ where $k = 1$ or 0.8)
Root diameter	$d_r = d_p - 2b$ (typically $b = q/P_d$ where $q = 1.25$, 1.2, or 1)
Base circle diameter	$d_b = d_p \cos \phi$
Center distance	$C = r_{p_2} + r_{p_3} = \dfrac{d_{p_2} + d_{p_3}}{2} = \dfrac{N_2 + N_3}{2P_d} = \dfrac{p_c(N_2 + N_3)}{2\pi}$
Length of line of contact	$\lambda = -r_{p_2} \sin \phi + \sqrt{a_2^2 + 2a_2 r_{p_2} + r_{p_2}^2 \sin^2 \phi}$ $-r_{p_3} \sin \phi + \sqrt{a_3^2 + 2a_3 r_{p_3} + r_{p_3}^2 \sin^2 \phi}$
Contact ratio	$m_c = \dfrac{\lambda}{p_b} = \dfrac{\lambda}{p_c \cos \phi} = \dfrac{\lambda P_d}{\pi \cos \phi}$
Velocity ratio	$R = \dfrac{\omega_2}{\omega_3} = \dfrac{r_{p_3}}{r_{p_2}} = \dfrac{d_{p_3}}{d_{p_2}} = \dfrac{N_3}{N_2}$
Tooth thickness at radius r	$t = 2r\left[\dfrac{t_p}{2r_p} + \text{inv}\,\phi - \text{inv}\,\beta\right]; \cos \beta = \dfrac{r_b}{r} = \dfrac{r_p \cos \phi}{r}$
Involute function	$\text{inv}\,\phi = \tan \phi - \phi$
No. of teeth at undercutting	$N = \dfrac{2aP_d}{\sin^2 \phi}$
Hob withdrawal for no undercutting	$e = \dfrac{1}{P_d} - \dfrac{N_3}{2P_d} \sin^2 \phi$

and

$$t^* = \frac{p_c}{2} = \frac{\pi}{2P_d}$$

Hence

$$t_3 = \frac{\pi}{2P_d} + 2e \tan \phi \tag{8.29}$$

but

$$e = a - r_3 \sin^2 \phi$$

or

$$e = \frac{1}{P_d} - \frac{N_3}{2P_d} \sin^2 \phi \tag{8.30}$$

The equations for spur tooth gearing are summarized in Table 8.4.

EXAMPLE 8.4 **(*Computing Nonstandard Gear Geometry*)**

PROBLEM

A 13-tooth pinion with diametral pitch 6 and 20° cutting pressure angle is to mate with a 50-tooth gear. Find the center distance, meshing pressure angle, and contact ratio if the pinion is cut with a standard cutter offset so that the addendum line passes through the interference point. (Compare with the corresponding values for a standard pinion.)

SOLUTION

From Eq. (8.30)

$$e = \frac{1}{P_d} - \frac{N_3}{2P_d} \sin^2 \phi = \frac{1}{6} - \frac{13}{2 \times 6} \sin^2 20° = 0.03994$$

From (8.29),

$$t_3 = \frac{\pi}{2P_d} + 2e \tan \phi = \frac{\pi}{2 \times 6} + 2 \times 0.03994 \tan 20° = 0.29087$$

From Eq. (8.7)

$$r_2 = \frac{N_2}{2P_d} = \frac{50}{2 \times 6} = 4.1667$$

and from Table 8.1

$$t_2 = \frac{2\pi r_2}{2N_2} = 0.26180$$

From Eq. (8.28)

$$\text{inv}\,\phi_m = \text{inv}(20°) - \frac{2\pi \times 4.1667 - 50(0.29087 + 0.26180)}{2 \times 4.1667(50 + 13)} = 0.017673$$

To find the value for ϕ_m, it is possible to use tables,[3] or the value can be found using a simple program. A MATLAB function, *inverse_inv*(y), is given on the disk with this book for doing this. The result is

$$\phi_m = 21.127°$$

Next use Eq. (8.5) to compute

$$r_{m_2} = \frac{r_{b_2}}{\cos \phi_m} = \frac{r_2 \cos \phi}{\cos \phi_m} = 4.1975$$

and

$$r_{m_3} = \frac{r_{b_3}}{\cos \phi_m} = \frac{r_3 \cos \phi}{\cos \phi_m} = \frac{N_3 \cos \phi}{2P_d \cos \phi_m} = \frac{13 \cos 20^\circ}{2 \times 6 \times \cos 21.127^\circ} = 1.0914 \text{ in}$$

The meshing center distance is

$$C_m = 4.1975 + 1.0914 = 5.2889 \text{ in}$$

The standard center distance

$$C_s = r_2 + r_3 = r_2 + \frac{N_3}{2P_d} = 4.1667 + \frac{13}{2 \times 6} = 5.2500 \text{ in}$$

■

8.12 CARTESIAN COORDINATES OF AN INVOLUTE TOOTH GENERATED WITH A RACK

It is often desirable to know the coordinates of the tooth profile for a given generating rack. This is of interest for drawing the gear or for machining the gear on a standard CNC milling machine without form cutters. If only the involute portion is required, it is relatively easy to compute the coordinates of points on the gear contact surface. However, if the entire profile is desired, the procedure is considerably more difficult. The problem can be approached directly if the geometry of the rack is known analytically.[4,5] However, an indirect approach developed by Vijayakar[6] is more general and applicable to a wide range of gear generators. This approach, which is relatively easy to program, is given here. The data required to generate the gear are a description of the rack, the number of teeth on the gear, and the outer diameter of the gear.

8.12.1 Coordinate Systems

Figure 8.31 shows a coordinate system attached to the rack that is assumed to be fixed. The origin of the system is at the intersection of the centerline of a rack tooth and the pitch line of the rack. The coordinates of any point P on the surface of the rack can be defined by (x_r, y_r) relative to the coordinate system fixed to the rack. Also, the outer normal to the rack at this point can be represented by (n_x, n_y). For any specified rack

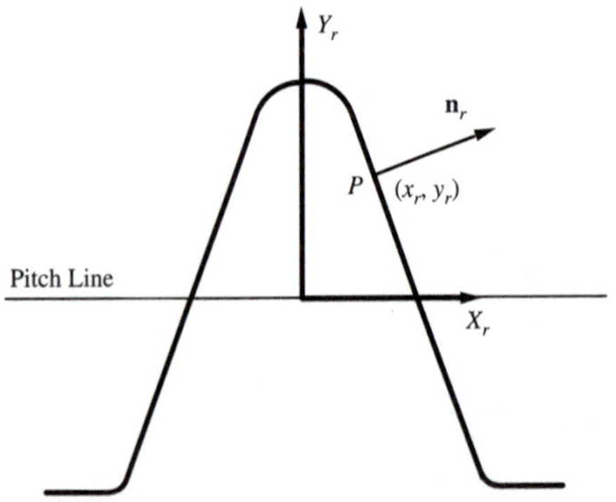

Figure 8.31 Rack and its attached coordinate system.

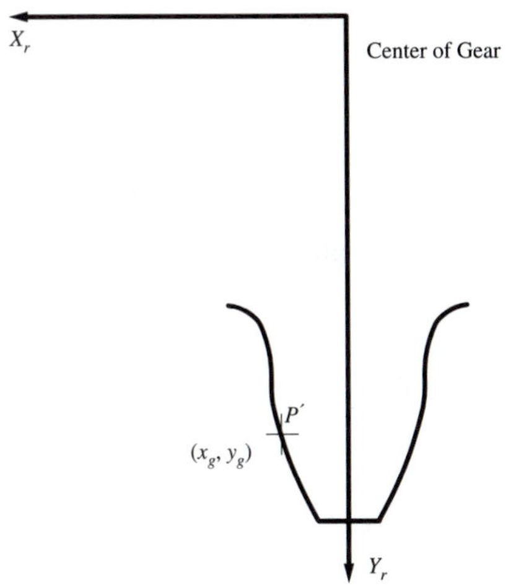

Figure 8.32 Gear and its attached coordinate system.

geometry, the coordinates and outer normal at any point on the rack profile are easily obtained.

Figure 8.32 shows a gear tooth with an attached coordinate system with its origin at the center of the gear. Let point P' be a point on the gear profile that corresponds to point P on the rack. As the gear pitch circle rolls on the rack pitch line, the point P on the rack makes sliding (cam) contact with the point P' on the gear.

Figure 8.33 shows the relative position of the gear and rack after the gear has rolled through the angle θ. The pitch circle radius of the gear is r_p. For this arbitrary orientation of the gear, the transformation from the rack coordinate system to the gear coordinate system is defined by the matrix equation

$$\begin{Bmatrix} x_g \\ y_g \end{Bmatrix} = \begin{bmatrix} -\cos\theta & \sin\theta \\ -\sin\theta & -\cos\theta \end{bmatrix} \begin{Bmatrix} x_r - r_p\theta \\ y_r - r_p \end{Bmatrix} \tag{8.31}$$

As the gear rolls, the relative velocity of the point P' on the gear with respect to the rack (fixed system) is given by

$${}^r\mathbf{v}_{P'} = {}^r\mathbf{v}_{B/C} + {}^r\mathbf{v}_{P'/B}$$

Recognizing that C is the instant center between the gear and the rack and points P', A, B, and C are all fixed to the gear,

$${}^r\mathbf{v}_{P'} = {}^r\boldsymbol{\omega}_g \times \mathbf{r}_{B/C} + {}^r\boldsymbol{\omega}_g \times \mathbf{r}_{P'/B} = {}^r\boldsymbol{\omega}_g \times (\mathbf{r}_{B/C} + \mathbf{r}_{P'/B}) = {}^r\boldsymbol{\omega}_g \times \mathbf{r}_{P'/C}$$

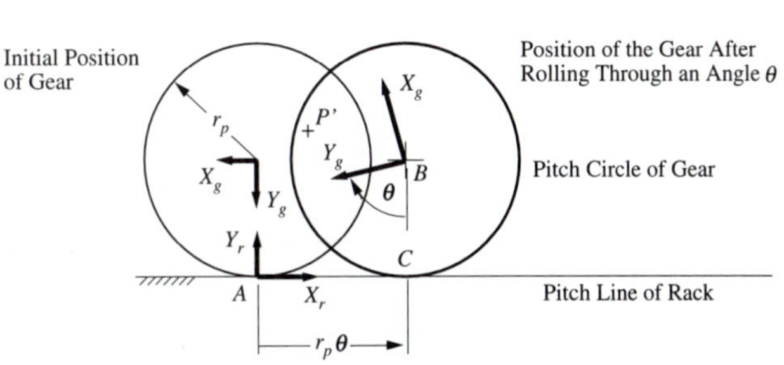

Figure 8.33 Relative orientation of gear and rack coordinate systems during generation.

where ${}^{r}\boldsymbol{\omega}_g$ is the angular velocity of the gear relative to the rack. In matrix form, this equation becomes

$$^{r}\mathbf{v}_{P'} = -{}^{r}\omega_g \begin{Bmatrix} y_r \\ r_p\theta - x_r \end{Bmatrix}$$

Because cam contact is involved at the contact point (assumed to be P'), the relative velocity of the point on the gear should have no component normal to the rack. Hence the dot product between the normal ${}^{r}\mathbf{v}_{P'}$ must be zero. Then,

$$^{r}\mathbf{v}_{P'} \cdot \begin{Bmatrix} n_x \\ n_y \end{Bmatrix} = 0$$

or

$$-n_x y_r + (x_r - r_p\theta)n_y = 0$$

The roll angle for the gear at which the point P makes contact with a point on the gear is given by

$$\theta = \frac{n_y x_r - n_x y_r}{r_p n_y} \tag{8.32}$$

Given any point P on the rack, its coordinates, and its normal vector, the roll angle at which it makes contact with the gear can be computed using Eq. (8.32). The coordinates of the corresponding point P' on the gear can be obtained by substituting for θ in Eq. (8.31). Therefore, a sequence of points on the gear profile can be found that corresponds to a sequence of points on the rack profile.

The procedure is to begin with points at the tip of the hob tooth and to sequence through the points until the corresponding points on the gear are located beyond the addendum circle. If r_o is the outer radius of the gear, the extreme points are reached when the coordinates given in Eq. (8.31) satisfy the following

$$\sqrt{x_g^2 + y_g^2} \geq r_o$$

Before drawing the gear tooth profile, it is necessary to check for undercutting. If undercutting occurs, the tooth profile generated using the foregoing procedure will look like that shown in Fig. 8.34. The part B-C-D-B must be detected, and the parts associated with this region eliminated from the sequence of points used to define the gear profile.

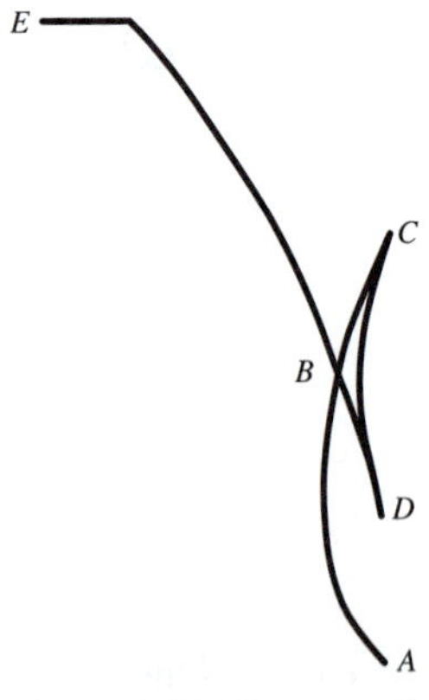

Figure 8.34 Gear profile with undercutting.

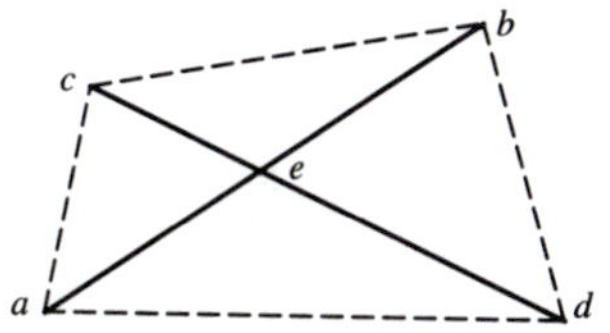

Figure 8.35 Determination of the point of intersection of two line segments.

Let $\mathbf{r}_i$, $i = 1, 2, \ldots, n$, be a sequence of vectors corresponding to the points on the gear tooth profile. To detect the crossover point B shown in Fig. 8.34, we need to check whether there are integers i and j such that the line segment $(i, i + 1)$ that joins $\mathbf{r}_i$ with $\mathbf{r}_{i+1}$ intersects the line segment $(j, j + 1)$. Figure 8.35 shows two line segments (a, b) and (c, d). These line segments will intersect if and only if[6]

$$\mathbf{A}(a, b, c) \cdot \mathbf{A}(a, b, d) < 0$$

and

$$\mathbf{A}(c, d, a) \cdot \mathbf{A}(c, d, b) < 0$$

where

$$\mathbf{A}(a, b, c) = (\boldsymbol{r}_b - \boldsymbol{r}_a) \times (\boldsymbol{r}_c - \boldsymbol{r}_a)$$

and

$$\boldsymbol{r}_a = \begin{Bmatrix} x_a \\ y_a \\ 0 \end{Bmatrix}, \quad \boldsymbol{r}_b = \begin{Bmatrix} x_b \\ y_b \\ 0 \end{Bmatrix}, \quad \boldsymbol{r}_c = \begin{Bmatrix} x_c \\ y_c \\ 0 \end{Bmatrix}, \quad \boldsymbol{r}_d = \begin{Bmatrix} x_d \\ y_d \\ 0 \end{Bmatrix}$$

The location of the point of intersection will be given by $\boldsymbol{r}_e$ where

$$\boldsymbol{r}_e = \alpha \boldsymbol{r}_a + (1 - \alpha)\boldsymbol{r}_b$$

where

$$\alpha = \frac{|\mathbf{A}(c, d, b)|}{|\mathbf{A}(c, d, a)| + |\mathbf{A}(c, d, b)|}$$

Using this method, the whole profile can be searched for segments $(i, i + 1)$, $i = 1, 2, \ldots, n - 2$, and $(j, j + 1)$, $j = i + 1, i + 2, \ldots, n - 1$, that intersect. If such an i and j are found, then all points on the profile between $i + 1$ and $j + 1$ are discarded and replaced by the single point $\boldsymbol{r}_e$ at the point of intersection.

8.12.2 Gear Equations

Once the set of points is defined for the single tooth, this set of coordinates can be rotated in increments of $2\pi/N$ to form the other tooth. If θ is the rotation angle, the coordinates (X_i, Y_i) of successive teeth are related to the original coordinates (x_i, y_i) by

$$X_i = x_i \cos\theta - y_i \sin\theta$$
$$Y_i = x_i \sin\theta + y_i \cos\theta$$

The rotation angle is incremented for each tooth to be drawn. For example, for the first tooth, $\theta = 0$; for the second tooth, $\theta = 2\pi/N$; for the third tooth, $\theta = 2(2\pi/N)$; and for the jth tooth, $\theta = (j - 1)(2\pi/N)$. A MATLAB routine for drawing the hob, gear tooth, and gear is given on the disk with this book.

EXAMPLE 8.5 ***(Geometry of Simple Rack)***

PROBLEM

Half of the tooth for a standard rack is represented in Fig. 8.36. Determine the equations for each region and the equations for the corresponding normal vectors.

SOLUTION

Before developing the equations for the different regions, define the following terms:

D_p = diametral pitch

A = addendum of rack

B = dedendum of rack

ϕ = pressure angle

r_t = radius of tip of rack tooth

r_f = radius of fillet of rack tooth

Miscellaneous terms can be computed based on geometry to be

$$\Gamma = \frac{\pi}{2} - \phi$$

$$\ell_t = \frac{\pi}{2D_p} - 2A \tan \phi - 2r_t \tan\left(\frac{\Gamma}{2}\right)$$

$$\ell_b = \frac{\pi}{2D_p} - 2B \tan \phi - 2r_f \tan\left(\frac{\Gamma}{2}\right)$$

It is assumed that the maximum radius, r_t, at the tip of the hob is a full radius that occurs when ℓ_t is zero. Then the maximum tip radius is

$$r_t = \frac{(\pi/2D_p) - 2A \tan \phi}{2 \tan (\Gamma/2)}$$

Similarly, it is assumed that the maximum radius, r_b, at the fillet of the hob is a full radius that occurs when ℓ_b is zero. Then, the maximum fillet radius is

$$r_f = \frac{(\pi/2D_p) - 2B \tan \phi}{2 \tan (\Gamma/2)}$$

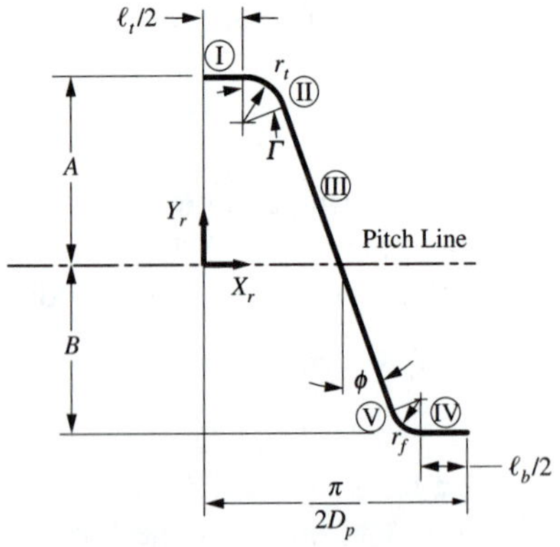

Figure 8.36 Geometry of a basic rack.

The coordinates of the points and normal vectors along the rack profile in each region are given by:

Region I (top land)

$$\begin{Bmatrix} x_r \\ y_r \end{Bmatrix} = \begin{Bmatrix} \beta\ell_t/2 \\ A \end{Bmatrix}, \quad \begin{Bmatrix} n_x \\ n_y \end{Bmatrix} = \begin{Bmatrix} 0 \\ 1 \end{Bmatrix} \quad 0 < \beta \le 1$$

Region II (tip radius)

$$\begin{Bmatrix} x_r \\ y_r \end{Bmatrix} = \begin{Bmatrix} \ell_t/2 + r_t\sin(\beta\Gamma) \\ A - r_t[1 - \cos(\beta\Gamma)] \end{Bmatrix}, \quad \begin{Bmatrix} n_x \\ n_y \end{Bmatrix} = \begin{Bmatrix} \sin(\beta\Gamma) \\ \cos(\beta\Gamma) \end{Bmatrix} \quad 0 < \beta \le 1$$

Region III (tooth flank)

$$\begin{Bmatrix} x_r \\ y_r \end{Bmatrix} = \begin{Bmatrix} (1-\beta)[\ell_t/2 + r_t\sin(\Gamma)] + \beta[\pi/(2D_p) - \ell_b/2 - r_f\sin\Gamma] \\ (1-\beta)[A - r_t\{1 - \cos(\Gamma)\}] + \beta[-B + r_f(1 - \cos\Gamma)] \end{Bmatrix}, \quad \begin{Bmatrix} n_x \\ n_y \end{Bmatrix} = \begin{Bmatrix} \cos\phi \\ \sin\phi \end{Bmatrix}$$

$$0 < \beta \le 1$$

Region IV (root fillet)

$$\begin{Bmatrix} x_r \\ y_r \end{Bmatrix} = \begin{Bmatrix} \pi/(2D_p) - \ell_b/2 - r_f\sin[(1-\beta)\Gamma] \\ -B + r_f\{1 - \cos[(1-\beta)\gamma]\} \end{Bmatrix}, \quad \begin{Bmatrix} n_x \\ n_y \end{Bmatrix} = \begin{Bmatrix} \sin[(1-\beta)\Gamma] \\ \cos[(1-\beta)\Gamma] \end{Bmatrix} \quad 0 < \beta \le 1$$

Region V (bottom land)

$$\begin{Bmatrix} x_r \\ y_r \end{Bmatrix} = \begin{Bmatrix} \pi/(2D_p) - \ell_b/2(1-\beta) \\ -B \end{Bmatrix}, \quad \begin{Bmatrix} n_x \\ n_y \end{Bmatrix} = \begin{Bmatrix} 0 \\ 1 \end{Bmatrix} \quad 0 < \beta \le 1$$

■

EXAMPLE 8.6 ***(Generation of Gear)***

PROBLEM

Assume that the rack in Example 8.5 is used to generate a gear. The pressure angle is 20° with a diametral pitch (D_p) of 10 teeth per inch. The tip and root radii are both 0.01 in. The addendum constant for the rack is 1.25, and the dedendum constant is 1.1. There are 15 teeth on the gear, and the addendum constant for the gear is 1.0. Determine the shape of the gear tooth.

SOLUTION

Because of the number of calculations required to solve the problem, it is efficient to program the equations given in Example 8.5. This was done using the program *geardr.m* given on the

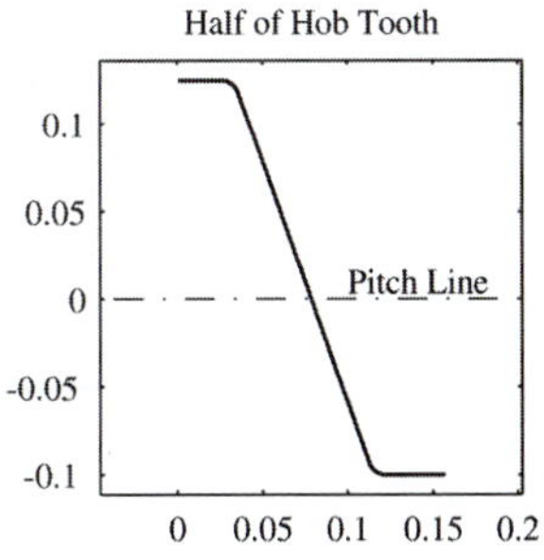

Figure 8.37 Profile of half of a hob tooth.

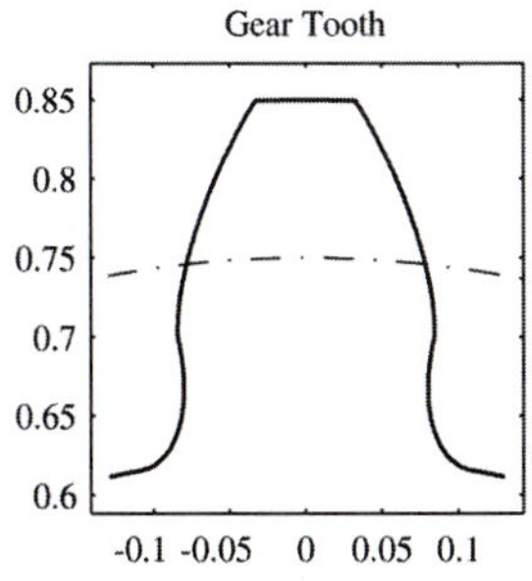

Figure 8.38 Generated gear tooth.

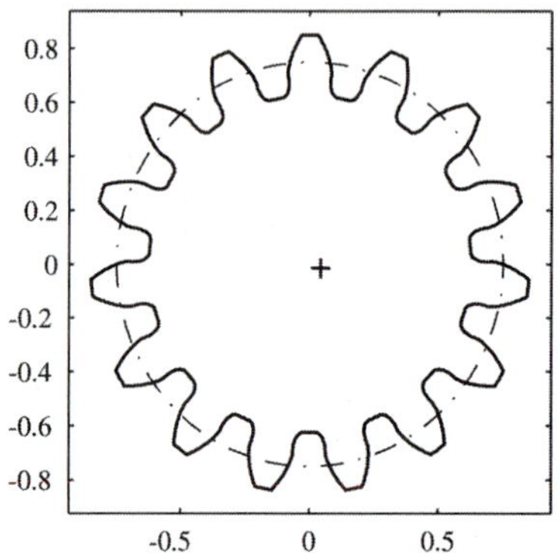

Figure 8.39 Profile of generated gear.

disk with this book. The hob profile (half of a tooth) is shown in Fig. 8.37, and the tooth and gear profiles are shown in Figs. 8.38 and 8.39. Notice that the gear is undercut. ■

8.13 REFERENCES

1. American Gear Manufacturers' Association (AGMA), "Tooth Proportions for Coarse Pitch Involute Spur Gears," AGMA Publication 201.02 and 201.02A (1968).
2. Mabie, H. H., and F. W. Ocvirk, *Mechanisms and Dynamics of Machinery,* 3rd ed., John Wiley & Sons, New York (1975).
3. Shigley, J. E., *Kinematic Analysis of Mechanisms,* McGraw-Hill Book Co., New York (1975).
4. Kinzel, G. L., and D. M. Farley, "A Computer Approach to the Generation of Conjugate Tooth Forms," *Proceedings of the Sixth IFToMM Congress on Theory of Machines and Mechanisms,* New Delhi, December 15–20, 1983, pp. 778–782.
5. Mitchiner, R. G, and H. H. Mabie, "The Determination of the Lewis Form Factor and the AGMA Geometry Factor *J* for External Spur Gear Teeth," *Transactions of ASME,* Vol. 104, January 1982, pp. 148–158.
6. Vijayakar, S., *Finite Element Methods for Quasi-Prismatic Bodies with Application to Gears,* Ph.D. dissertation, The Ohio State University (1987).

8.14 CHAPTER 8 *Exercise Problems*

PROBLEM 8.1

Two spur gears have a diametral pitch of 6. Gear 2 has 24 teeth, and gear 3 has 48. The working pressure angle is 20°, and both gears are standard involutes. Determine the length of the contact line and the contact ratio.

PROBLEM 8.2

For the gear pairs given below and meshing at their correct center-to-center distance, determine whether any interference is present and determine the contact ratio for each case. Assume that the addendum is $1/P_d$ in each case, and if any interference is present, assume the interference is removed by cutting off the ends of the gear teeth before determining the contact ratio.

(a) $14\frac{1}{2}°$ involute gears having 30 and 45 teeth
(b) 20° involute gears having 20 teeth and a rack
(c) 25° involute gears having 30 and 60 teeth

PROBLEM 8.3

A 20° involute pinion having 30 teeth is meshing with a 60-tooth internal gear. The addendum of the pinion is $1.25/P_d$, and the addendum of the gear is $1/P_d$. Is there any interference? Determine the contact ratio. If there is interference, assume that the interfering portion is removed by cutting off the ends of the teeth.

PROBLEM 8.4 What is the largest gear that will mesh with a 20° standard full-depth gear of 22 teeth with no interference?

PROBLEM 8.5 What is the smallest gear that will mesh with a 20° standard full-depth gear of 22 teeth with no interference?

PROBLEM 8.6 Assume a gear has a diametral pitch of 6. Determine the addendum, dedendum, and clearance if the pressure angle is 20° full depth, 25° full depth, 20° stub teeth.

PROBLEM 8.7 Assume two meshing gears have a diametral pitch of 6 and a 20° pressure angle. The gear has 38 teeth, and the pinion has 24. Determine the design center distance. Now assume that the center distance is increased by 0.01 in. Determine the pressure angle for the new center distance.

PROBLEM 8.8 Assume a standard full-depth rack has a diametral pitch of 2. Determine the smallest gear that will mesh with the rack without interference if the pressure angle is (a) 20°, (b) 25°.

PROBLEM 8.9 Assume two meshing gears have a diametral pitch of 8 and a 25° pressure angle. The gear has 60 teeth, and the pinion has 30. Determine the design center distance. Now assume that the center distance is increased by 0.012 in. Determine the pressure angle for the new center distance.

PROBLEM 8.10 Two standard gears have a diametral pitch of 2 and a pressure angle of $14\frac{1}{2}°$. The tooth numbers are 14 and 16. Determine whether interference occurs. If it does, compute the amount that the addendum(s) must be shortened to remove the interference, and the new contact ratio.

PROBLEM 8.11 Two standard gears have 18 and 32 teeth, respectively. The diametral pitch is 10, and the pinion rotates at 1000 rpm. Determine the following: (a) center distance, (b) pitch diameters, (c) circular pitch, (d) pitch line velocity, (e) angular velocity of the gear.

PROBLEM 8.12 Two standard gears have a diametral pitch of 10 and a velocity ratio of 2.5. The center distance is 3.5 in. Determine the number of teeth on each gear.

PROBLEM 8.13 Is it possible to specify arbitrary values for the velocity ratio, center distance, and diametral pitch in a problem such as that given in Problem 8.12? Explain.

PROBLEM 8.14 Two standard meshing gears are to have 9 and 36 teeth, respectively. They are to be cut with a 20° full-depth cutter with a diametral pitch of 3.

(a) Determine the amount that the addendum of the gear is to be shortened to eliminate interference.
(b) If the addendum of the pinion is increased the same amount, determine the new contact ratio.

PROBLEM 8.15 Two standard meshing gears are to have 13 and 20 teeth, respectively. They are to be cut with a 20° full-depth cutter with a diametral pitch of 2. To reduce interference, the center distance is increased by 0.1 in. Determine the following:

(a) Whether the interference is completely eliminated
(b) The pitch diameters of both gears
(c) The new pressure angle
(d) The new contact ratio

PROBLEM 8.16 Assume that you have a 13-tooth pinion and a 50-tooth gear. What is the smallest (nonstandard) pressure angle that can be used if interference is to be avoided? What is the smallest pressure angle that can be used if only standard pressure angles can be considered?

PROBLEM 8.17 A standard, full-depth spur gear has a pressure angle of 25° and an outside diameter of 225 mm. If the gear has 48 teeth, find the module and circular pitch.

PROBLEM 8.18 The pinion of a pair of spur gears has 16 teeth and a pressure angle of 20°. The velocity ratio is to be 3:2, and the module is 6.5 mm. Determine the initial center distance. If the center distance is increased by 3 mm, find the resulting pressure angle.

PROBLEM 8.19 Two standard, full-depth spur gears are to have 10 teeth and 35 teeth, respectively. The cutter has a 20° pressure angle with a module of 10 mm. Determine the amount by which the addendum of the gear must be reduced to avoid interference. Then determine the length of the new path of contact and the contact ratio.

PROBLEM 8.20 A standard, full-depth spur gear tooth has been cut with a 20° hob and has a diametral pitch of 6. The tooth thickness at a radius of 2.1 in. is 0.1860 in. Determine the thickness of the gear tooth at the base circle.

PROBLEM 8.21 A pair of standard, full-depth spur gears has been cut with a 25° hob. The pinion has 31 teeth and the gear 60 teeth. The diametral pitch is 4. Find the velocity ratio, the pitch circle radii, the outside diameters, the center distance, and the contact ratio.

PROBLEM 8.22 A pair of standard, full-depth spur gears has been cut with a 25° hob. The pinion has 14 teeth and the gear 51 teeth. The diametral pitch is 5. Find the velocity ratio, the pitch circle radii, the outside diameters, the center distance, and the contact ratio.

PROBLEM 8.23 A pair of standard, full-depth spur gears has been cut with a 20° hob. The pinion has 27 teeth and the gear 65 teeth. The diametral pitch is 2. Find the velocity ratio, the pitch circle radii, the outside diameters, the center distance, and the contact ratio.

Chapter 9

Helical, Bevel, and Worm Gears

9.1 HELICAL GEARS

Helical gears have the same involute tooth form as spur gears but are cut with the teeth inclined to the gear axes so that the intersections of tooth faces with cylindrical surfaces about the gear axis are helices. This produces a progressive contact action with contact beginning at a point at one end of the tooth, enlarging progressively to an inclined line across the tooth face, and then diminishing progressively to a point at the other end of the tooth. This progressive action carries important benefits in reductions of gear noise and vibration. Because of the gradual load transfer and longer teeth, helical gears will also have a higher load rating for a given size gear than spur gears. Another advantage of helical gears is that the gear axes do not need to be parallel. Because of these advantages, helical gears are almost always used in high-speed transmissions where noise and high power are issues. Examples of helical gears are shown in Figs. 9.1–9.4.

The helical tooth geometry results in a reaction thrust component along the shaft axis, which therefore requires a thrust bearing. For this reason, gears are sometimes cut with

Figure 9.1 Helical gears.

Figure 9.2 Herringbone gears.

two sets of helical teeth with opposite pitch. The resulting chevron-shaped teeth lead to the name "herringbone gears" for this type. Herringbone gears do not have an axial reaction thrust component.

9.1.1 Helical Gear Terminology

As indicated in Fig. 9.5, the terminology for helical gears is defined in two planes, the transverse plane and the normal plane. The transverse plane is viewed along the axis of the gear shaft, and the normal plane is viewed along the tooth of the gear. The rotation

Figure 9.3 Helical rack and pinion.

Figure 9.4 Crossed helical gears.

of the gear is in the transverse plane; however, the gear is manufactured by a cutting action along the normal plane.

Three pitches must be considered when evaluating the geometry of helical gears. These are the normal pitch, the transverse pitch, and the axial pitch.

When a helical gear is generated, the properties in the normal plane will correspond to those of the hob or shaper cutter. In particular, the normal pitch, p_n, is related to the normal diametral pitch, P_n, by

$$p_n = \frac{\pi}{P_n} \tag{9.1}$$

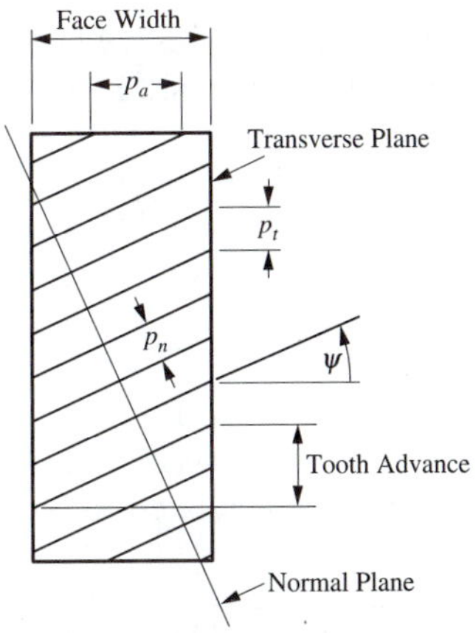

Figure 9.5 End view of helical gear showing normal and transverse planes. ψ is the helix angle, the angle at which the teeth are inclined to the axis of the gear; p_n is the circular pitch normal to the tooth face; p_t is the transverse circular pitch in a plane normal to the gear axis.

and both p_n and P_n are the same as those for the cutter if the gear is cut with a hob. The normal pitch is usually the one specified in gear catalogs.

Also, from Fig. 9.5,

$$p_n = p_t \cos \psi \tag{9.2}$$

where p_t is the transverse circular pitch and ψ is the helix angle. The transverse circular pitch is the distance from tooth to tooth on the pitch cylinder when observed from the transverse plane. The transverse circular pitch is related to the transverse diametral pitch through an equation similar to Eq. (9.1). That is,

$$p_t = \frac{\pi}{P_t} \tag{9.3}$$

and

$$P_t = \frac{N}{d_t} \tag{9.4}$$

where d_t is the diameter of the pitch cylinder. From Eqs. (9.1)–(9.4),

$$\frac{\pi}{P_n} = \frac{\pi}{P_t} \cos \psi$$

or

$$P_t = P_n \cos \psi \tag{9.5}$$

The transverse diametral pitch has the traditional meaning of being the number of teeth per inch of pitch diameter in the transverse plane. However, the normal diametral pitch is less obvious. We can still think of P_n as the number of teeth per pitch diameter if the gear were a spur gear with a pitch diameter equal to the transverse pitch diameter. Obviously, the helical gear appears as an ellipse when viewed from the normal plane; however, the circular approximation is useful when considering the local geometry in the vicinity of an individual tooth.

In the metric system, the module, m_t, is used, and

$$m_t = \frac{d_t}{N} = \frac{m_n}{\cos \psi}$$

where m_n is the module for the hob. Comparing Eqs. (9.1)–(9.5), the metric system and the conventional system used in the United States are related by

$$m_t = \frac{1}{P_t} \quad \text{and} \quad m_n = \frac{1}{P_n}$$

and

$$p_t = \pi m_t \quad \text{and} \quad p_n = \pi m_n$$

The third pitch is the axial pitch, p_a. This is the distance from a point on one tooth to the corresponding point on an adjacent tooth measured at the pitch cylinder and in the direction of the rotation axis. The axial pitch is related to the other pitches by

$$p_a = \frac{p_t}{\tan \psi} = \frac{p_n}{\sin \psi} \tag{9.6}$$

There are two pressure angles for helical gears, one for each of the two principal planes, and the relationship among the pressure angles and the helix angle is shown in Fig. 9.6. The transverse pressure angle, ϕ_t, is measured in the transverse plane, and the normal pressure angle, ϕ_n, is measured in the normal plane.

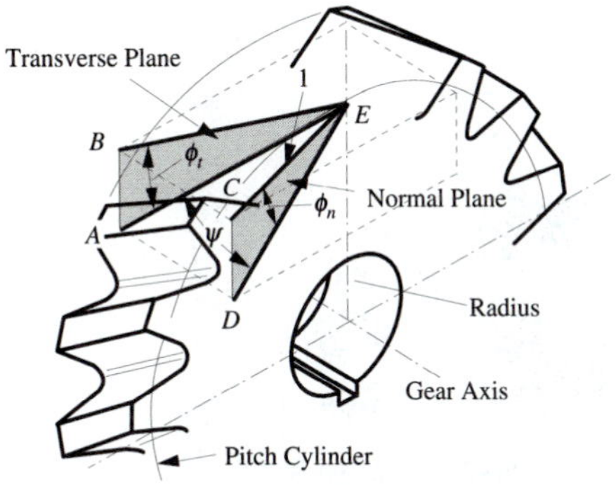

Figure 9.6 Relationship among normal and transverse pressure angles and helix angle.

Because helical gears are usually cut with the same hob cutters as spur gears, the normal diametral pitch and pressure angle usually have standard spur gear values, but the transverse diametral pitch and pressure angle do not. Since the transverse diametral pitch and pressure angle determine the kinematic characteristics of the gear pair, it is important to be able to obtain them from the corresponding normal values and the helix angle. Conversely, the tooth stiffness and strength are determined by the thickness in the normal plane. Thus, this is the plane used when analyzing tooth stresses.

The relationship among ϕ_t, ϕ_n, and ψ can be determined by looking at the lines AE, BE, CE, and DE in Fig. 9.6. Assume that line CE is 1 unit long and normal to the gear tooth at the pitch cylinder. Triangle CED is in the normal plane, BEA is in the transverse plane, AED is in the plane tangent to the pitch circle at point E, and $BACD$ is a rectangle.

Then DE is the projection of CE in the tangent plane, BE is the projection of CE in the transverse plane, and AE is the projection of BE in the tangent plane. From simple geometry,

$$AB = CD$$

Also,

$$\begin{aligned} DE &= \cos \phi_n \\ CD &= \sin \phi_n \\ AE &= DE \cos \psi = \cos \phi_n \cos \psi \\ AB &= AE \tan \phi_t = \cos \phi_n \cos \psi \tan \phi_t \end{aligned}$$

And because $AB = CD$,

$$\sin \phi_n = \cos \phi_n \cos \psi \tan \phi_t$$

or

$$\tan \phi_t = \frac{\tan \phi_n}{\cos \psi} \tag{9.7}$$

9.1.2 Helical Gear Manufacturing

As was the case for spur gears, the vast majority of helical gears are manufactured by hobbing (Fig. 9.7) or shaping (Fig. 9.8). When gears are hobbed, the hob axis is inclined at an angle equal to the gear helix angle relative to the gear rotation axis. The hob properties correspond to the gear properties in the normal plane. Therefore, the normal

Figure 9.7 Hobbing helical gear.

Figure 9.8 Shaping a helical gear.

diametral pitch (P_n) of the gear is equal to the diametral pitch (P_d) of the hob. The same hob can be used to cut both spur and helical gears.

If the gear is an internal gear, if it has a shoulder, or if it is a herringbone gear (Fig. 9.2), the gear cannot be hobbed. Such gears are normally fabricated by shaping. The shaper axis is parallel to the gear rotation axis. Therefore, the shaper properties correspond to the transverse plane. The transverse circular pitch of the gear is equal to the transverse circular pitch of the shaper cutter, so Eqs. (9.3) and (9.4) apply to both the cutter and gear. When the gear is shaped, the same cutter cannot be used for both helical and spur gears.

9.1.3 Minimum Tooth Number to Avoid Undercutting

One of the advantages of helical gears is that they can have fewer teeth on the pinion than can the corresponding spur gear and still avoid undercutting. The minimum number of teeth to avoid undercutting is given by using Fig. 8.28 and Eq. (8.25) in Chapter 8, where all quantities are for the transverse plane. Then,

$$r_p = \frac{a}{\sin^2 \phi_t} \tag{9.8}$$

where a is the addendum for the cutter used to form the gear. The cutter properties are associated with the *normal* plane. Therefore,

$$a = \frac{k}{P_n}$$

and Eq. (9.8) can be rewritten as

$$r_p = \frac{k}{P_n \sin^2 \phi_t}$$

The pitch radius (r_p) is related to the number of teeth (N) and transverse diametral pitch (P_t) by

$$r_p = \frac{N}{2P_t} \tag{9.9}$$

Therefore,

$$\frac{N}{2P_t} = \frac{k}{P_n \sin^2 \phi_t}$$

Using Eq. (9.5) and simplifying, the minimum number of teeth to avoid undercutting is given by

$$N = \frac{2kP_t}{P_n \sin^2 \phi_t} = \frac{2k \cos \psi}{\sin^2 \phi_t} \tag{9.10}$$

Values for N as a function of helix angle and *normal* pressure angle are given in Table 9.1 for standard full-depth ($k = 1$) gears cut with a hob. The values were computed by first solving Eq. (9.7) for ϕ_t for selected values of ψ and ϕ_n and then substituting the results into Eq. (9.10). Notice that for very large helix angles, the number of teeth on the gear can be reduced to 1. This is the case that exists with worm gears, which will be discussed later in this chapter. The helix angle for helical gears is usually limited to 45°.

Unlike spur gears, two helical gears can be meshed with either parallel shafts or crossed shafts. The properties in Eqs. (9.1)–(9.10) apply to individual helical gears regardless of the axis orientation of the meshing gear.

Table 9.1 Minimum Tooth Numbers as a Function of the Helix Angle for Standard Values of the Normal Pitch

$\psi \downarrow$	ϕ_n (deg.)→ $14\frac{1}{2}°$	20°	25°
0° (spur gear)	32	18	12
5°	32	17	12
10°	31	17	11
15°	29	16	11
20°	27	15	10
25°	25	14	9
30°	22	12	8
35°	19	10	7
40°	15	9	6
45°	12	7	5
50°	10	6	4
55°	7	4	3
60°	5	3	3
65°	4	2	2
70°	2	2	2
75°	2	1	1
80°	1	1	1

9.1.4 Helical Gears with Parallel Shafts

Two parallel-shaft gears will mesh if the following conditions are satisfied.

1. Both gears have equal pitches.
2. Both gears have equal helix angles.
3. The helix angles on the two gears are opposite hand. This means that one helix is left handed and the other is right handed (see Fig. 9.9).

9.1.4.1 Velocity Ratio and Center Distance

The expressions for the angular velocity ratio and the center distance for two parallel-shaft helical gears are the same as the corresponding expressions for spur gears. The velocity ratio is given by

$$\frac{\omega_2}{\omega_3} = \frac{r_{p_3}}{r_{p_2}} = \frac{D_{t_3}}{D_{t_2}} = \frac{N_3/P_t}{N_2/P_t} = \frac{N_3}{N_2} \tag{9.11}$$

where ω_2, r_{p_2}, D_{t_2}, and N_2 are the angular velocity magnitude, pitch radius, pitch diameter, and number of teeth, respectively, on gear 2. The corresponding values are for gear 3.

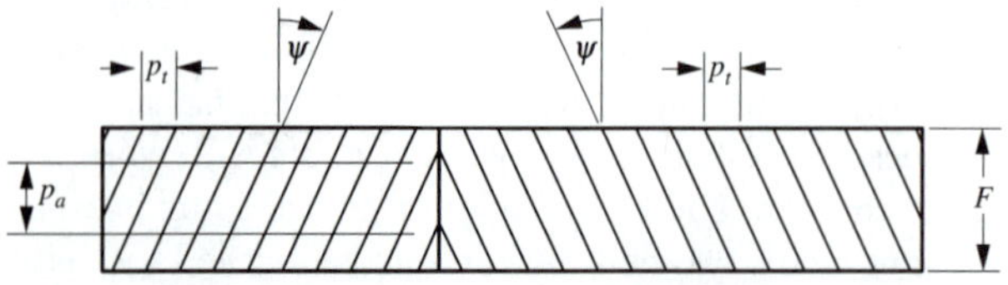

Figure 9.9 Conditions for two helical gears to mesh.

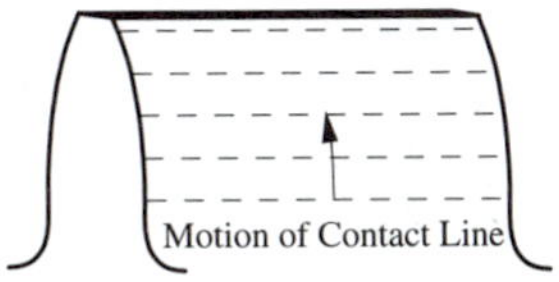

Figure 9.10 Contact lines for spur gear.

The center distance is given by

$$C = r_{p_2} + r_{p_3} = \frac{d_{t_2} + d_{t_3}}{2} \tag{9.12}$$

or in terms of the pitches,

$$C = \frac{N_2 + N_3}{2P_t} = \frac{N_2 + N_3}{2P_n \cos \psi} \tag{9.13}$$

9.1.4.2 Minimum Face Width

The maximum face width of a pair of spur gears is limited by the accuracy of alignment; however, the minimum face width can be much smaller than the circular pitch. Helical gears are much more expensive to produce than spur gears, and therefore, from practical considerations, the face width must be large to achieve the stated benefits as compared with spur gears. In general, the face width (F) should be larger than the axial pitch (p_a), and the AGMA[1] recommends that the face width be at least 15% larger than the axial pitch. Typically, the face width is at least two times the axial pitch.[2] The axial pitch is given by Eq. (9.6). Therefore the limiting condition for the face width is

$$F \geq \frac{1.15 p_t}{\tan \psi} \tag{9.14}$$

or

$$F \geq \frac{1.15 p_n}{\sin \psi} \tag{9.15}$$

9.1.4.3 Contact Ratio

Mating spur gears make contact on a line that is parallel to the axis of rotation as shown in Fig. 9.10. Helical gear teeth make contact on a line that is diagonal across the gear teeth as shown in Fig. 9.11. Contact between the two teeth begins at A, and as the gears mesh the contact moves diagonally across the gear tooth. When contact is being lost at B on one side of the tooth, it is just beginning at C on the other side. This gradual contact is one of the benefits of helical gears.

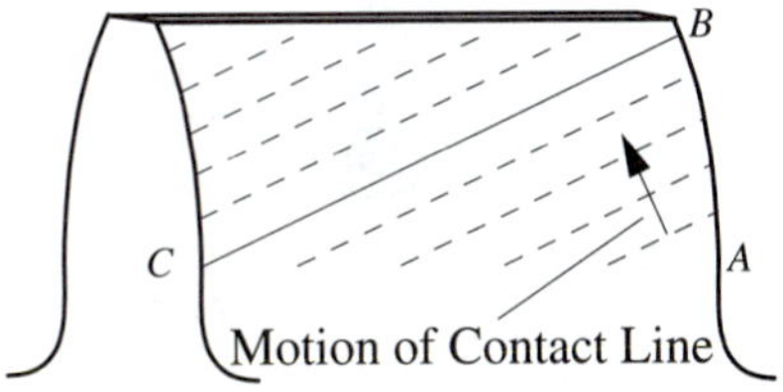

Figure 9.11 Contact lines for helical gear.

The contact ratio of helical gears is increased over that of equivalent spur gears by the axial overlap of the teeth. Therefore, the contact ratio (m_c) is the sum of the transverse contact (m_{c_t}) ratio and the axial or face contact ratio (m_{c_a}). The transverse contact ratio is computed in the same manner as for spur gears. Therefore, the transverse contact ratio is given as

$$m_{c_t} = \frac{p_t \lambda}{\pi \cos \phi_t} \tag{9.16}$$

where

$$\lambda = \sum_{i=2}^{3} \{-r_{p_i} \sin \phi_t + \sqrt{a_i^2 + 2a_i r_{p_i} + r_{p_i}^2 \sin^2 \phi_t}\}$$

The axial contact ratio is the ratio of the face width of the gear to the axial pitch. This is

$$m_{c_a} = \frac{F}{p_a} = \frac{F \tan \psi}{p_t} \tag{9.17}$$

Notice that the transverse contact ratio is not defined for a single gear; however, the axial contact ratio depends on the properties of a single gear. The normal and tangential pitches will be the same for both gears. If the face widths are different, the face width of the narrower gear is used in Eq. (9.17).

The total contact ratio is

$$m_c = m_{c_t} + m_{c_a} \tag{9.18}$$

As in the case of spur gears, m_c gives the average number of teeth in contact during the gear action. The contact ratio for helical gears is always higher than for comparable spur gears.

EXAMPLE 9.1 ***(Helical Gear Geometry)***

PROBLEM

A pair of helical gears with 19 and 34 teeth, respectively, and a 30° helix angle are cut with a standard spur gear hob that has a diametral pitch of 4 and a pressure angle of 20°. Find the transverse diametral pitch, the pitch cylinder radii, and the axial, transverse, and total contact ratios. The face width is 12 in.

SOLUTION

The properties of the gears in the normal direction will be the same as those of the hob. The transverse diametral pitch (P_t) for both gears is related to the normal diametral pitch by Eq. (9.5). Based on the diametral pitch of the hob,

$$P_t = P_n \cos \psi = 4 \cos 30° = 3.464$$

The pitch cylinder diameters are related to the diametral pitch through Eq. (9.4). Therefore,

$$d_t = \frac{N_t}{P_t}$$

and

$$r_{p_2} = \frac{N_2}{2P_t} = \frac{19}{2(3.464)} = 2.742 \text{ in}$$

and

$$r_{p_{3_t}} = \frac{N_3}{2P_t} = \frac{34}{2(3.464)} = 4.908 \text{ in}$$

The transverse contact ratio is given by Eq. (9.16) as

$$m_{c_t} = \frac{p_t \lambda}{\pi \cos \phi_t}$$

where

$$\lambda = \sum_{i=2}^{3} \{-r_{p_i} \sin \phi_t + \sqrt{a_i^2 + 2a_i r_{p_i} + r_{p_i}^2 \sin^2 \phi_t}\}$$

The addenda are determined by the hob. Because a standard hob is used, both addenda are given by

$$a = \frac{1}{P_n} = \frac{1}{4} = 0.25 \text{ in}$$

The transverse circular pitch is given by Eq. (9.3) as

$$p_t = \frac{\pi}{P_t} = \frac{\pi}{3.464} = 0.907$$

and the transverse pressure angle is given by Eq. (9.7) as

$$\phi_t = \tan^{-1}\left(\frac{\tan \phi_n}{\cos \psi}\right) = \tan^{-1}\left(\frac{\tan 20°}{\cos 30°}\right) = 22.796°$$

Now,

$$\begin{aligned}\lambda &= -r_{p_2} \sin \phi_t + \sqrt{a_2^2 + 2a_2 r_{p_2} + r_{p_2}^2 \sin^2 \phi_t} - r_{p_3} \sin \phi_t + \sqrt{a_3^2 + 2a_3 r_{p_3} + r_{p_3}^2 \sin^2 \phi_t} \\ &= -2.742 \sin(22.796°) + \sqrt{(0.25)^2 + 2(0.25)(2.742) + [2.742 \sin 22.796°]^2} \\ &\quad -4.908 \sin 22.796° + \sqrt{(0.25)^2 + 2(0.25)(4.908) + [4.908 \sin 22.796°]^2} = 1.1131\end{aligned}$$

Therefore, the transverse contact ratio is

$$m_{c_t} = \frac{p_t \lambda}{\pi \cos \phi_t} = \frac{0.907(1.1131)}{\pi \cos 22.796°} = 0.3486$$

The axial contact ratio is given by Eq. (9.17) as

$$m_{c_a} = \frac{F \tan \psi}{p_t} = \frac{12 \tan 30°}{0.907} = 7.6394$$

The total contact ratio is

$$m_c = m_{c_t} + m_{c_a} = 0.3486 + 7.6394 = 7.99$$

Therefore, an average of approximately 8 teeth will be in contact as the gears mesh. ■

EXAMPLE 9.2 *(Helical Gear Replacement of Spur Gears)*

PROBLEM

Sometimes it is desirable to replace an existing set of spur gears to eliminate a noise problem or to increase the capacity of a gear box. Assume the original spur gears are 14 pitch, 20° full-depth gears with 30 and 85 teeth. These are to be replaced by a set of helical gears without any major modifications to the gear box. The size and angular velocity ratio are to remain the same. Determine the helix angle, outside diameters of the blanks, and the face width of the replacement gears. The new helical gears can be cut with the same hob as was used for the spur gears.

SOLUTION If there are to be no major modifications to the gear box, the center distance must remain the same. The original pitch radii are given by

$$r_{p_2} = \frac{N_2}{2P_d} = \frac{30}{2(14)} = 1.071 \text{ in}$$

and

$$r_{p_3} = \frac{N_3}{2P_d} = \frac{85}{2(14)} = 3.036 \text{ in}$$

and the center distance is

$$C = r_{p_2} + r_{p_3} = \frac{N_2 + N_3}{2P_d} = \frac{30 + 85}{2(14)} = 4.109 \text{ in}$$

The velocity ratio is also to remain constant. The original velocity ratio was

$$\frac{\omega_2}{\omega_3} = \frac{N_3}{N_2} = \frac{85}{30} = \frac{17}{6} \tag{9.19}$$

For the helical gears, we need to determine the helix angle and the number of teeth on each gear. We will find the number of teeth and transverse diametral pitch first and then determine the helix angle. The velocity ratio is given by Eq. (9.11) as

$$\frac{\omega_2}{\omega_3} = \frac{r_{p_3}}{r_{p_2}} = \frac{d_{t_3}}{d_{t_2}} = \frac{N_3/P_t}{N_2/P_t} = \frac{N_3}{N_2}$$

and the center distance is given by Eq. (9.12)

$$C = r_{p_2} + r_{p_3} = \frac{d_{t_2} + d_{t_3}}{2}$$

From these two equations, it is clear that the pitch radii for the helical gears must be the same as the corresponding radii for the spur gears. However, the tooth numbers can and will be different. The transverse diametral pitch is related to the normal diametral pitch by

$$P_t = P_n \cos \psi \tag{9.5}$$

and to the teeth numbers by

$$P_t = \frac{N_2}{2r_{p_2}} = \frac{N_3}{2r_{p_3}} \tag{9.20}$$

From Eq. (9.5), it is clear that $P_t < P_n$. Therefore, based on Eq. (9.20), the tooth numbers on the helical gears must be less than those on the spur gears. As a result, when we investigate tooth numbers that satisfy Eq. (9.19), we need only consider values that are lower than the corresponding values for the spur gears. A set of values is

N_2	$N_3 = \frac{17}{6}N_2$	
30	85	(spur gear)
29	82.167	
28	79.333	
27	76.500	
26	73.667	
25	70.833	
24	68	

From the table, the first set of teeth numbers that are integers are $N_2 = 24$ and $N_3 = 68$. For these numbers, the transverse pitch is given by Eq. (9.20) as

$$P_t = \frac{N_2}{2r_{p_2}} = \frac{N_3}{2r_{p_3}} = \frac{24}{2(1.071)} = \frac{68}{2(3.036)} = 11.199$$

From Eq. (9.5),

$$\psi = \cos^{-1}\left(\frac{P_t}{P_n}\right) = \cos^{-1}\left(\frac{11.199}{14}\right) = 36.877°$$

Notice that this is the lowest helix angle possible (other than 0°) if the center distance and velocity ratio are to be maintained. The blank diameters of the two gears are given by

$$D_{o_2} = 2r_{p_2} + 2a_2 = 2r_{p_2} + 2\frac{k}{P_n} = 2(1.071) + 2\frac{1}{14} = 2.286 \text{ in}$$

and

$$D_{o_3} = 2r_{p_3} + 2a_3 = 2r_{p_3} + 2\frac{k}{P_n} = 2(3.036) + 2\frac{1}{14} = 6.214 \text{ in}$$

The minimum face width is given by Eq. (9.14):

$$F \geq \frac{1.15p_t}{\tan\psi} \quad \text{or} \quad F \geq \frac{1.15\pi}{P_t \tan\psi}$$

Therefore,

$$F \geq \frac{1.15\pi}{11.199 \tan 36.877°} \quad \text{or} \quad F > 0.430$$

Use $F = \frac{7}{16}$ in ■

9.1.4.4 Designing for Axial Force

When spur gears transmit power, the force between the gears lies entirely in the transverse plane. However, when helical gears mesh, there will be an axial force component because of the helix angle. This axial load must be considered when designing the shaft bearings. If the load is too large to be carried by the bearings, two helical gears of opposite hand can be used instead of a single gear. Alternatively, and more commonly, a herringbone gear (Fig. 9.2) can be used. A herringbone gear is in essence two helical gears of opposite hand cut on the same gear blank.

A summary of the equations specific to parallel-shaft helical gears is given in Table 9.2. Most of the equations in Table 8.4 apply to helical gears if the transverse plane is used. Therefore, most of the equations in Table 8.4 have not been repeated in Table 9.2.

9.1.5 Crossed Helical Gears

Helical gears need not be used on parallel shafts but can be used to transfer power between nonparallel, nonintersecting shafts (Fig. 9.4). When the shafts are not parallel, the gears are called crossed helical gears.

The only requirement for crossed helical gears to mesh properly is that they have the same normal pitch. The transverse pitches need not be the same, and the helix angles need not be the same. The helix angles can be of the same or opposite hand.

The velocity ratio for crossed helical gears can be developed from Eq. (9.11) as

$$R = \frac{\omega_2}{\omega_3} = \frac{N_3}{N_2} = \frac{d_{t_3}P_{t_3}}{d_{t_2}P_{t_2}} = \frac{d_{t_3}P_n \cos\psi_3}{d_{t_2}P_n \cos\psi_2} = \frac{d_{t_3}\cos\psi_3}{d_{t_2}\cos\psi_2} \qquad \textbf{(9.21)}$$

If the angle between the shafts of the meshing crossed helical gears is Σ and the helix angles are ψ_2 and ψ_3, the relationship among the three angles is

$$\Sigma = \psi_2 \pm \psi_3 \qquad \textbf{(9.22)}$$

where the plus sign applies if the gears are of the same hand, and the minus sign applies if the gears are of the opposite hand. This is shown in Fig. 9.12.

The center distance in the U.S. system between crossed helical gears is given by

$$C = r_{p_2} + r_{p_3} = \frac{N_2}{2P_{t_2}} + \frac{N_3}{2P_{t_3}} = \frac{N_2}{2P_n\cos\psi_2} + \frac{N_3}{2P_n\cos\psi_3} = \frac{1}{2P_n}\left[\frac{N_2}{\cos\psi_2} + \frac{N_3}{\cos\psi_3}\right] \qquad \textbf{(9.23)}$$

and recalling that

$$m_n = \frac{1}{P_n}$$

the center distance in the metric system is

$$C = r_{p_2} + r_{p_3} = \frac{N_2}{2P_{t_2}} + \frac{N_3}{2P_{t_3}} = \frac{N_2 m_n}{2\cos\psi_2} + \frac{N_3 m_n}{2\cos\psi_3} = \frac{m_n}{2}\left[\frac{N_2}{\cos\psi_2} + \frac{N_3}{\cos\psi_3}\right] \qquad \textbf{(9.24)}$$

This equation applies regardless of the hand of the gears.

Crossed helical gears are not used to transmit large amounts of power because the gears theoretically have only point contact where the teeth mesh. This is in contrast to parallel-shaft helical gears, which have line contact. For large power transfers between nonparallel

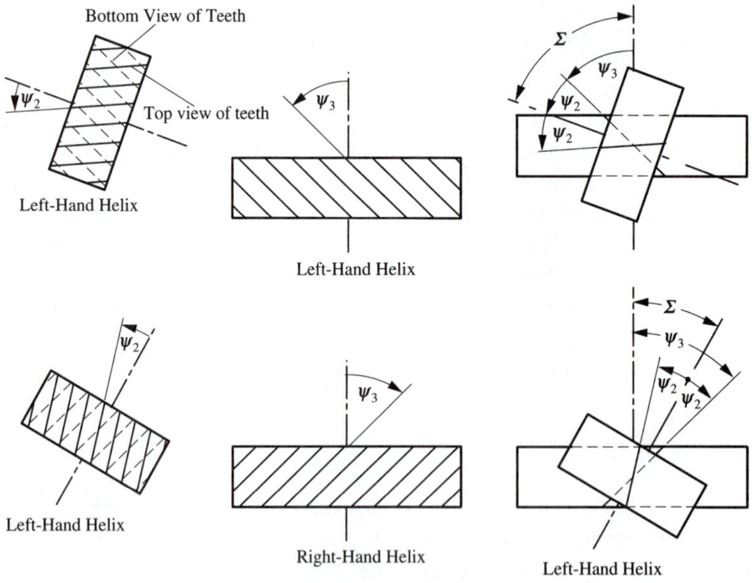

Figure 9.12 Relationship among shaft angle and helix angles for crossed helical gears.

shafts, bevel or hypoid gear sets are preferred. Worm gear sets can also be used when high velocity ratios are required.

EXAMPLE 9.3 **(*Crossed Helical Gear Geometry*)**

PROBLEM

Assume that two crossed helical gears have a shaft angle of 65 degrees and a velocity ratio of 2:1. The pinion (gear 2) has a normal diametral pitch of 10, a helix angle of 30°, and 70 teeth. Determine the helix angle, pitch diameter, and number of teeth on the meshing gears. Also find the center distance. Both gears are right handed, and both are cut with the same hob.

SOLUTION

First find the pitch diameter for gear 2 using

$$d_{p_2} = \frac{N_2}{P_{t_2}} = \frac{N_2}{P_n \cos\psi_2} = \frac{70}{10\cos 30°} = 8.083 \text{ in}$$

From the expression for the velocity ratio, the number of teeth on gear 3 is

$$N_3 = N_2\frac{\omega_2}{\omega_3} = 70\frac{2}{1} = 140$$

Next find the helix angle of the second gear, using Eq. (9.22) with a plus sign. Then,

$$\Sigma = \psi_2 + \psi_3$$

where $\Sigma = 65°$ and $\psi_2 = 30°$. Solving gives $\psi_3 = 35°$.

The pitch diameter for gear 3 is then given by

$$d_{t_3} = \frac{N_3}{P_n \cos\psi_3} = \frac{140}{10\cos 35°} = 17.091 \text{ in}$$

The center distance is given by

$$C = \frac{d_{t_2} + d_{t_3}}{2} = \frac{8.083 + 17.091}{2} = 12.587 \text{ in}$$

■

A summary of the equations specific to crossed helical gears is given in Table 9.3.

9.2 WORM GEARS

If the tooth of a helical gear makes a complete revolution on the pitch cylinder, the resulting gear is called a worm. This occurs when the helix angle and the gear face width are relatively large. In worms, the helix angle can exceed 80°. The mating gear is called a worm gear or worm wheel. Worm gear sets are used to connect nonintersecting shafts (usually at 90°) when the desired velocity reduction is large. A worm gear set is shown in Fig. 9.13.

Involute worms can be meshed with either spur or helical gears; however, point contact occurs and relatively small forces can be transmitted. Usually, the worm gear set is cut so that the gear partially envelops the worm. The design is either nonenveloping, single enveloping, or double enveloping. The three cases are illustrated schematically in Fig. 9.14. The worm and gear are nonenveloping (Fig. 9.14a) when the worm is meshed with a simple helical gear and point contact occurs. The worm and gear are single enveloping when the gear is cut to envelop the worm partially (Fig. 9.14b), and they are double enveloping when the pitch surface of the worm is cut in the shape of an hourglass and meshed with

Table 9.2 Summary of Helical Gear Formulas (Parallel Shafts)

Quantity	Formula
Transverse diametral pitch	$P_t = \frac{N}{d_t} = \frac{N}{2r_t} = \frac{\pi}{p_t} = P_n \cos\psi = \frac{1}{m_t} = \frac{\cos\psi}{m_n}$
Transverse circular pitch	$p_t = \frac{\pi}{P_t} = \frac{\pi d_t}{N} = \pi m_t = \frac{p_n}{\cos\psi} = p_a \tan\psi$
Normal diametral pitch	$P_n = \frac{P_t}{\cos\psi} = \frac{N}{2r_p \cos\psi} = \frac{\pi}{p_n} = \frac{1}{m_n} = \frac{1}{m_t \cos\psi}$
Normal circular pitch	$p_n = \frac{\pi}{P_n} = \pi m_n = p_t \cos\psi = p_a \sin\psi$
Axial pitch	$p_a = \frac{p_t}{\tan\psi} = \frac{p_n}{\sin\psi}$
Pitch cylinder diameter	$d_t = \frac{N}{P_t} = \frac{N}{P_n \cos\psi} = \frac{Np_t}{\pi} = \frac{Np_n}{\pi \cos\psi} = Nm_t$
Center distance	$C = r_{p_2} + r_{p_3} = \frac{d_{t_2} + d_{t_3}}{2} = \frac{N_2 + N_3}{2P_t} = \frac{p_t(N_2 + N_3)}{2\pi} = \frac{N_2 + N_3}{2P_n \cos\psi}$
Velocity ratio	$R = \frac{\omega_2}{\omega_3} = \frac{r_{p_3}}{r_{p_2}} = \frac{d_{t_3}}{d_{t_2}} = \frac{N_3}{N_2}$
Minimum face width	$F \geq \frac{1.15p_t}{\tan\psi}$ or $F \geq \frac{1.15\pi}{P_t \tan\psi}$
No. of teeth at undercutting	$N = \frac{2aP_t}{\sin^2\phi_t} = \frac{2kP_t}{P_n \sin^2\phi_t} = \frac{2k\cos\psi}{\sin^2\phi_t}$ ($k = 1$ for FD or 0.8 for ST)
Length of line of contact	$\lambda = -r_{p_2}\sin\phi_t + \sqrt{a_2^2 + 2a_2 r_{p_2} + r_{p_2}^2 \sin^2\phi_t}$ $-r_{p_3}\sin\phi_t + \sqrt{a_3^2 + 2a_3 p_{t_3} + r_{p_3}^2 \sin^2\phi_t}$
Transverse contact ratio	$m_{c_t} = \frac{\lambda}{p_t \cos\phi_t} = \frac{\lambda P_t}{\pi \cos\phi_t}$
Axial contact ratio	$m_{c_a} = \frac{F}{p_a} = \frac{F\tan\psi}{p_t}$
Total contact ratio	$m_c = m_{c_t} + m_{c_a}$
Transverse pressure angle	$\tan\phi_t = \frac{\tan\phi_n}{\cos\psi}$

Table 9.3 Summary of Crossed Helical Gear Formulas

Quantity	Formula
Shaft angle	$\Sigma = \psi_2 \pm \psi_3$
Pitch cylinder diameter	$d_t = \frac{N}{P_t} = \frac{N}{P_n \cos\psi} = \frac{Np_t}{\pi} = \frac{Np_n}{\pi\cos\psi} = Nm_t$
Center distance	$C = \frac{d_{t_2} + d_{t_3}}{2} = \frac{N_2}{2P_{t_2}} + \frac{N_3}{2P_{t_3}} = \frac{1}{2P_n}\left[\frac{N_2}{\cos\psi_2} + \frac{N_3}{\cos\psi_3}\right]$ $= \frac{m_n}{2}\left[\frac{N_2}{\cos\psi_2} + \frac{N_3}{\cos\psi_3}\right]$
Velocity ratio	$R = \frac{\omega_2}{\omega_3} = \frac{N_3}{N_2} = \frac{d_{t_3}P_{t_3}}{d_{t_2}P_{t_2}} = \frac{d_{t_3}P_n\cos\psi_3}{d_{t_2}P_n\cos\psi_2} = \frac{d_{t_3}\cos\psi_3}{d_{t_2}\cos\psi_2}$

Figure 9.13 Example of worm gear set.

a gear that also partially envelops the worm (Fig. 9.14c). The single and double enveloping worm gear sets will have line contact and can transmit considerably more power than can nonenveloping worm gear sets.

Note that in the cases of nonenveloping and single enveloping worm gear sets, the worm can drive the gear either by a rotation of the worm or by a translation of the worm along the axis tangent to the worm axis. Therefore, the alignment of the worm in the tangential direction is not critical. However, in double enveloping worm gear sets, the tangential alignment is critical. Both single and double enveloping worm gear sets must be accurately aligned in the axial and radial directions. Nonenveloping worm gears need to be accurately aligned in the radial direction. The shaft angles of all worm gear sets must be accurately aligned.

9.2.1 Worm Gear Nomenclature

A schematic drawing of a single enveloping worm gear set is shown in Fig. 9.15. The two gears have the same hand. The helix angle is defined in the same manner as was done for helical gears; however, as indicated in Fig. 9.15, the helix angle is usually quite large. The lead angle, λ, is the complement of the helix angle, ψ, that is,

$$\lambda + \psi = \frac{\pi}{2} \tag{9.25}$$

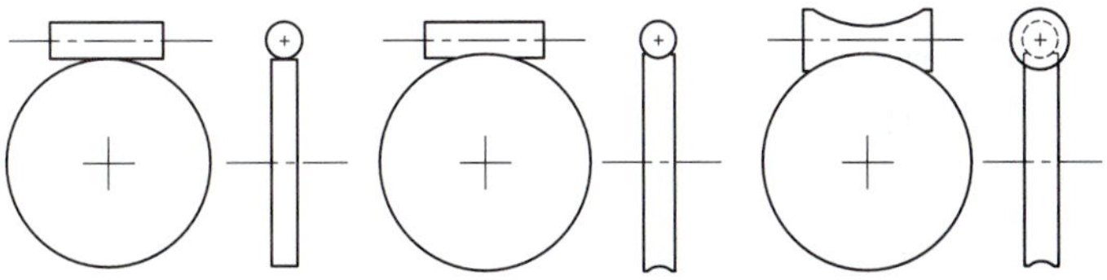

Figure 9.14 Types of worm gear sets.

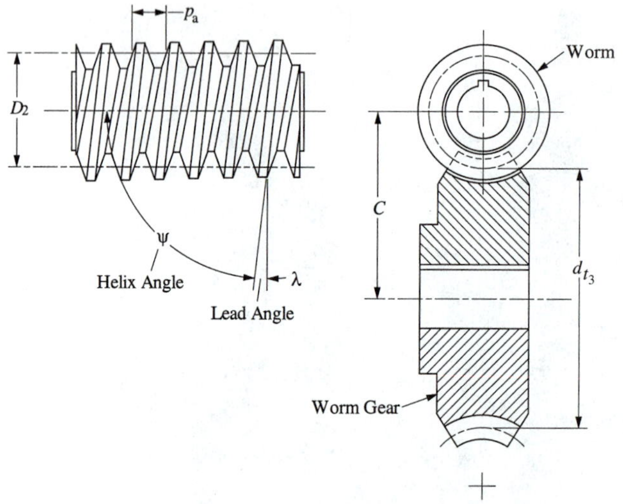

Figure 9.15 Worm gear nomenclature.

The lead, L, is the amount that the worm will advance with one revolution of the worm gear. If there is only one tooth, the lead is equal to the axial pitch, p_a. In general, if N_2 is the number of teeth on the worm,

$$L = N_2 p_a \tag{9.26}$$

If one tooth of the worm is unwrapped from the pitch cylinder, the lead of the worm is related to the lead angle, λ, as shown in Fig. 9.16. The relationship is

$$\tan \lambda = \frac{L}{\pi d_{t_2}} \tag{9.27}$$

The pitch diameter of the worm gear is the same as that for helical gears, that is,

$$d_{t_3} = \frac{p_{t_3} N_3}{\pi}$$

where N_3 is the number of teeth on the gear. The velocity ratio is given by Eq. (9.21) as

$$\frac{\omega_2}{\omega_3} = \frac{N_3}{N_2} = \frac{d_{t_3} \cos \psi_3}{d_{t_2} \cos \psi_2} \tag{9.28}$$

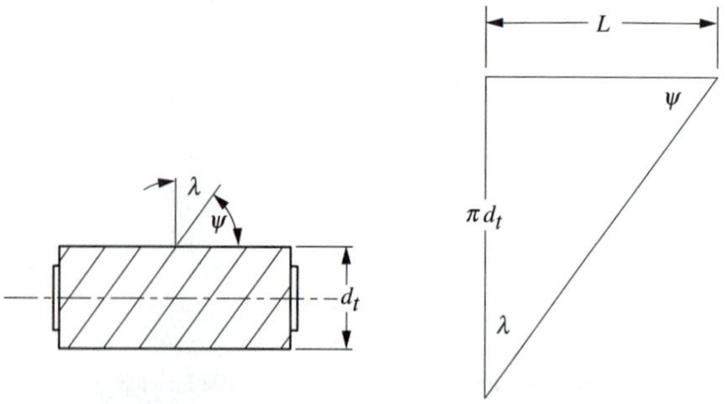

Figure 9.16 Representation of worm gear lead.

For a worm and worm gear to mesh properly with shafts at right angles, the following must be satisfied:

1. The helix angle of the gear must equal the lead angle of the worm.
2. The transverse circular pitch of the gear must equal the axial pitch of the worm.

If the shafts of the worm and gear are at 90° to each other, then

$$\psi_2 + \psi_3 = \frac{\pi}{2} \tag{9.29}$$

and $\psi_3 = \lambda$. Equation (9.28) can then be written as

$$\frac{\omega_2}{\omega_3} = \frac{d_{t_3} \cos \psi_3}{d_{t_2} \cos \psi_2} = \frac{d_{t_3} \cos \psi_3}{d_{t_2} \cos(\pi/2 - \psi_3)} = \frac{d_{t_3} \cos \psi_3}{d_{t_2} \sin \psi_3} = \frac{d_{t_3}}{d_{t_2} \tan \psi_3} = \frac{d_{t_3}}{d_{t_2} \tan \lambda}$$

Therefore, from Eq. (9.27),

$$\frac{\omega_2}{\omega_3} = \frac{d_{t_3}}{d_{t_2}} \frac{\pi d_{t_2}}{L} = \frac{\pi d_{t_3}}{L} \tag{9.30}$$

Worm gear sets may or may not be driven from both the worm and gear. In some applications, it is necessary for the drive to be self-locking. In such cases, only the worm can be the driver. In other cases, it must be possible to drive the worm gear set from either the worm or gear. Worms usually have relatively few teeth (one to eight), and when the number of teeth is small, the worm must be the driver. If the lead angle of the worm is greater than the friction angle of the surfaces in contact, the drive will be reversible. The coefficient of friction, μ, is related to the friction angle, β, by $\mu = \tan \beta$. Therefore,

Table 9.4 Summary of Worm Gear Formulas (Gear 2 is the Worm and 3 is the Gear)

Quantity	Formula
Lead	$L = N_2 p_a = \pi d_{t_2} \tan \lambda$
Lead angle	$\tan \lambda = \dfrac{L}{\pi d_{t_2}} = \dfrac{N_2 p_a}{\pi d_{t_2}}$
Minimum length of worm[6]	$f = 2\sqrt{\left(\dfrac{d_{t_3} + 2a}{2}\right)^2 - \left(\dfrac{d_{t_3} - 2a}{2}\right)^2}$
Normal circular pitch	$p_n = \dfrac{\pi}{P_n} = \pi m_n = p_t \cos \psi = p_a \sin \psi = p_a \cos \lambda$
Shaft angle	$\Sigma = \psi_2 \pm \psi_3$
Pitch cylinder diameter	$d_t = \dfrac{N}{P_t} = \dfrac{N}{P_n \cos \psi} = \dfrac{N p_t}{\pi} = \dfrac{N p_n}{\pi \cos \psi} = N m_t$
Effective face width of gear[6]	$F_e = \sqrt{(d_{t_2} + a + b)^2 - (d_{t_2})^2}$
Center distance	$C = \dfrac{d_{t_2} + d_{t_3}}{2} = \dfrac{N_2}{2P_{t_2}} + \dfrac{N_3}{2P_{t_3}} = \dfrac{1}{2P_n}\left[\dfrac{N_2}{\cos \psi_2} + \dfrac{N_3}{\cos \psi_3}\right]$ $= \dfrac{m_n}{2}\left[\dfrac{N_2}{\cos \psi_2} + \dfrac{N_3}{\cos \psi_3}\right]$
Velocity ratio	$R = \dfrac{\omega_2}{\omega_3} = \dfrac{N_3}{N_2} = \dfrac{d_{t_3} P_{t_3}}{d_{t_2} P_{t_2}} = \dfrac{d_{t_3} P_n \cos \psi_3}{d_{t_2} P_n \cos \psi_2} = \dfrac{d_{t_3} \cos \psi_3}{d_{t_2} \cos \psi_2}$

the worm gear set will be self-locking if the lead angle satisfies

$$\lambda < \tan^{-1} \mu \tag{9.31}$$

In general, a worm gear set cannot be back driven if the lead angle is less than 5° (helix angle > 85°).

A summary of the equations for worm gears is given in Table 9.4. A more complete list of equations is given by Townsend.[6]

EXAMPLE 9.4 ***(Worm and Gear Geometry)***

PROBLEM

Assume that a worm has three teeth and is driving a gear with 60 teeth. The shaft angle is 90°, the gear transverse circular pitch is 1.25 in, and the pitch diameter of the worm is 3.8 in. Find the helix angle of the gear, the center distance, and the lead angle of the worm. If the coefficient of friction between the worm and gear is 0.1, estimate whether or not the worm gear set can be back driven.

SOLUTION

The worm lead is given by Eq. (9.26) as

$$L = N_2 p_a = 3(1.25) = 3.75 \text{ in}$$

And from Eq. (9.30) the pitch diameter of the gear is

$$d_{t_3} = \frac{\omega_2}{\omega_3}\frac{L}{\pi} = \frac{60}{3}\frac{3.75}{\pi} = 23.87 \text{ in}$$

The lead angle, λ, of the worm can be computed from Eq. (9.27) as

$$\lambda = \tan^{-1}\left(\frac{L}{\pi d_{t_2}}\right) = \tan^{-1}\left(\frac{3.75}{\pi 3.8}\right) = 17.44°$$

Therefore, the helix angle of the gear is

$$\psi_3 = \lambda = 17.44°$$

and the helix angle of the worm is given by Eq. (9.29) as

$$\psi_2 = 90° - \psi_3 = 90° - 17.44° = 72.56°$$

The center distance is given by

$$C = r_{p_2} + r_{p_3} = \frac{d_{t_2} + d_{t_3}}{2} = \frac{3.8 + 23.87}{2} = 13.84 \text{ in}$$

If the worm gear set is to be reversible, $\lambda > \tan^{-1} \mu$. For the conditions given,

$$\tan^{-1} \mu = \tan^{-1}(0.1) = 5.71°$$

Therefore, the worm gear set is reversible. ■

9.3 INVOLUTE BEVEL GEARS

When power must be transferred between nonparallel intersecting shafts, bevel gears are usually used. In bevel gears, the pitch surfaces are cones. The shafts must have intersecting centerlines, but the intersection can be at any angle, although 90° (Fig. 9.17) is the most common angle.

The shafts have to be mounted so that the apexes of the pitch cones of the mating gears

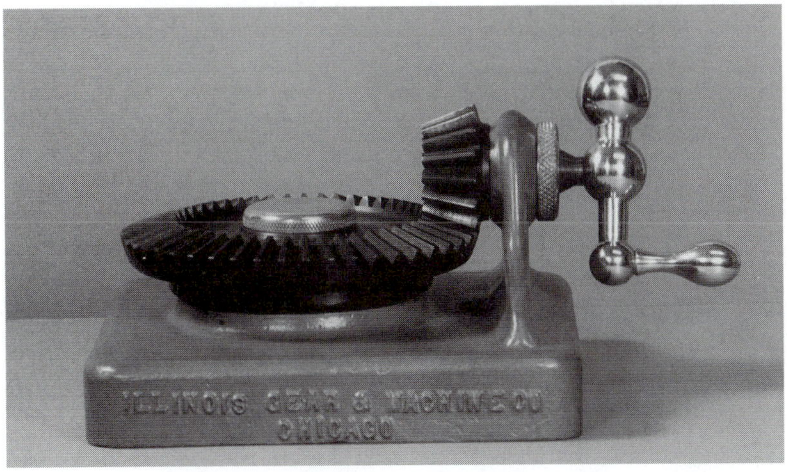

Figure 9.17 Bevel gears with 90° shaft angle.

are coincident (Fig. 9.18). The cones roll on each other without slipping and have spherical motion. Each point on each gear remains at a constant distance from the common apex.

The pitch diameter for bevel gears is the pitch cone diameter at the larger end (Fig. 9.19). The meshing bevel gears are contained within a sphere of radius r_o, as shown in Fig. 9.19, where the bases of the cones are contained on the surface of the sphere. The pitch cone angles, γ_i, determine the shaft angle, Σ, as shown in Fig. 9.19. These angles are related by

$$\Sigma = \gamma_2 + \gamma_3 \tag{9.32}$$

The velocity ratio for bevel gears is similar to that for spur and helical gears, that is,

$$\frac{\omega_2}{\omega_3} = \frac{N_3}{N_2} = \frac{d_3}{d_2} = \frac{r_3}{r_2} \tag{9.33}$$

where N_2 and N_3 are the numbers of teeth on gears 2 and 3, respectively, and d_2 and d_3 are the diameters at the large end of the pitch cones of gears 2 and 3, respectively (see Fig. 9.19). r_2 and r_3 are the corresponding pitch radii.

The pitch diameters can be related to the cone angles by considering the geometry represented in Fig. 9.19. In particular,

$$\sin \gamma_2 = \frac{d_2}{2r_o} = \sin(\Sigma - \gamma_3) = \sin \Sigma \cos \gamma_3 - \cos \Sigma \sin \gamma_3$$

Dividing by $\sin \Sigma \sin \gamma_3$ gives

$$\frac{\sin \gamma_2}{\sin \Sigma \sin \gamma_3} = \frac{\cos \gamma_3}{\sin \gamma_3} - \frac{\cos \Sigma}{\sin \Sigma}$$

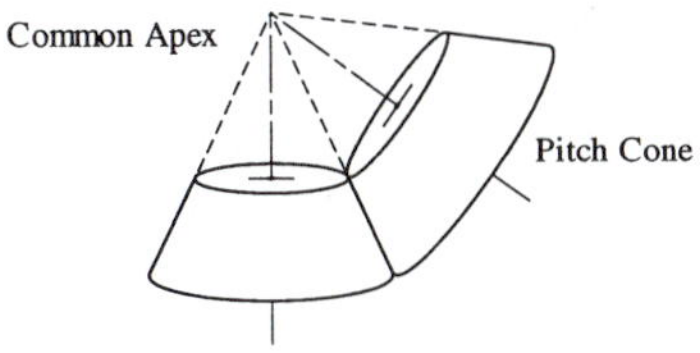

Figure 9.18 Pitch cones for bevel gears.

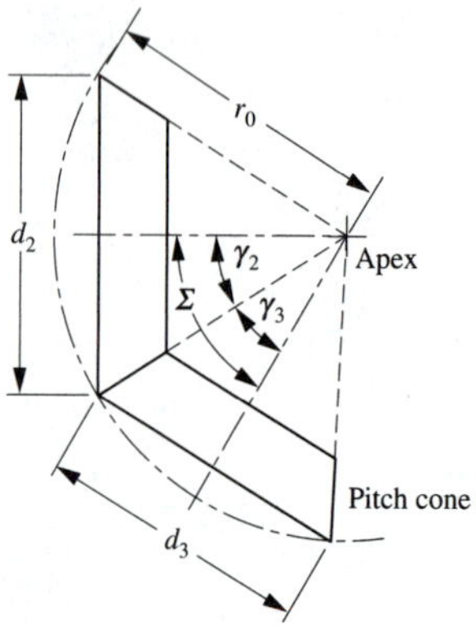

Figure 9.19 Relationship among pitch-cone angles.

or

$$\frac{1}{\sin\Sigma}\left[\frac{\sin\gamma_2}{\sin\gamma_3}+\cos\Sigma\right]=\frac{1}{\tan\gamma_3} \tag{9.34}$$

From the geometry,

$$\frac{\sin\gamma_2}{\sin\gamma_3}=\frac{d_2}{d_3} \tag{9.35}$$

From Eqs. (9.34) and (9.35),

$$\tan\gamma_3=\frac{\sin\Sigma}{[(d_2/d_3)+\cos\Sigma]}=\frac{\sin\Sigma}{[(N_2/N_3)+\cos\Sigma]} \tag{9.36}$$

Similarly,

$$\tan\gamma_2=\frac{\sin\Sigma}{[(d_3/d_2)+\cos\Sigma]}=\frac{\sin\Sigma}{[(N_3/N_2)+\cos\Sigma]} \tag{9.37}$$

When the shaft angle is 90°, which is the most common case, Eqs. (9.36) and (9.37) reduce to

$$\tan\gamma_3=\frac{d_3}{d_2}=\frac{N_3}{N_2} \tag{9.38}$$

and

$$\tan\gamma_2=\frac{d_2}{d_3}=\frac{N_2}{N_3} \tag{9.39}$$

9.3.1 Tredgold's Approximation for Bevel Gears

Because of the spherical geometry of bevel gears, it is difficult to draw and evaluate bevel gear properties such as contact ratio. Tredgold's approximation lets us approximate the bevel gears as equivalent spur gears. This approximation is used extensively, and the terminology of bevel-gear teeth has evolved around it.[2] The approximation is accurate enough for most practical purposes as long as the gear has eight or more teeth.

Tredgold's approximation involves the concept of a back cone for both meshing gears as shown in Fig. 9.20. The approximation recognizes that the action of the gear teeth in the vicinity of the contact location at the large end of the gear teeth is very similar to the action between two spur gears that have pitch radii r_{p_2} and r_{p_3} equal to the back cone radii, r_{e_2} and r_{e_3}.

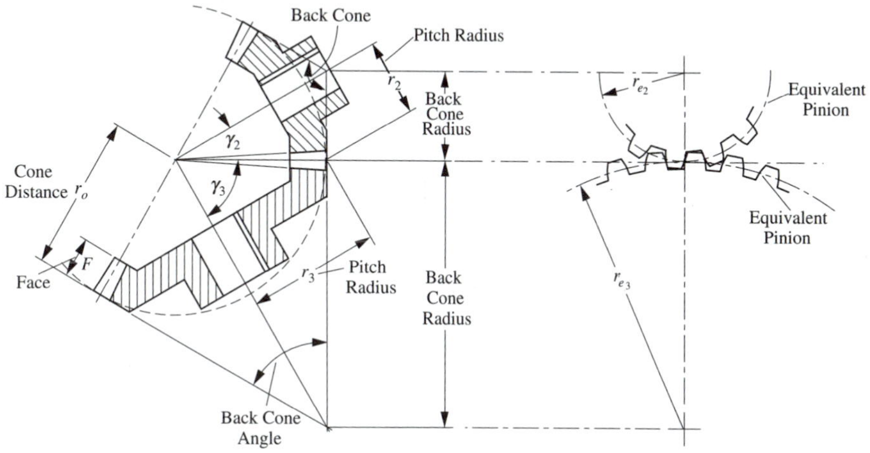

Figure 9.20 Approximation of bevel gears as equivalent spur gears (Tredgold's approximation).

Based on the geometry in Fig. 9.20, the pitch radii for the equivalent spur gears are given by

$$r_{e_2} = \frac{r_2}{\cos \gamma_2} \tag{9.40}$$

and

$$r_{e_3} = \frac{r_3}{\cos \gamma_3} \tag{9.41}$$

If p_c is the circular pitch at the large end of the bevel gears, the numbers of teeth on the equivalent spur gears are given by

$$N_{e_2} = \frac{2\pi r_{e_2}}{p_c} \tag{9.42}$$

and

$$N_{e_3} = \frac{2\pi r_{e_3}}{p_c} \tag{9.43}$$

Note that N_{e_2} and N_{e_3} need not be integers in Eqs. (9.42) and (9.43).

9.3.2 Additional Nomenclature for Bevel Gears

Figure 9.21 shows some of the additional terms used with bevel gears. Notice that most of the information characteristic of teeth sizes is defined for the large end of the gear teeth. Also notice that the apex of the pitch cone is not coincident with the apex of the face cone. This is to ensure that there is a constant clearance between the addendum of the given gear and the dedendum of the meshing gear. This also allows a larger fillet at the small end of the tooth than would otherwise be possible.

The AGMA[4] recommends a 20° pressure angle for bevel gears with 14 or more teeth and 25° for gears with 13 or fewer teeth. The minimum pressure angle is determined by undercutting on the pinion.

Bevel gears are usually mounted in a cantilever fashion. Because of this, the meshing gears will deflect away from each other, and the smaller end of the gears will carry even

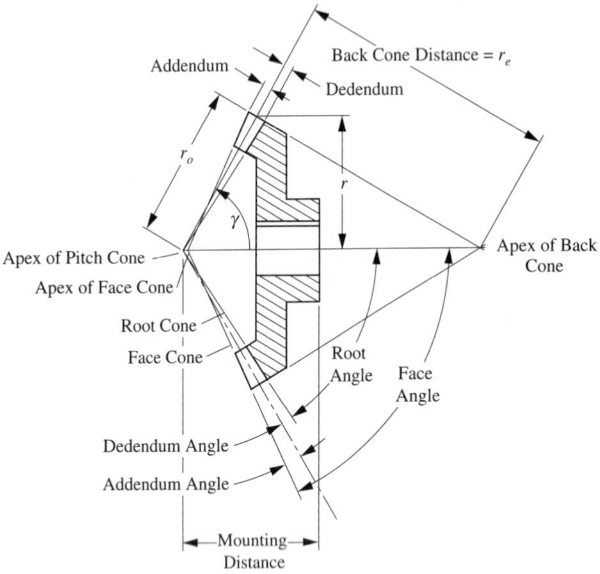

Figure 9.21 Nomenclature for bevel gears.

less load than they would without deflections. Because the large end of the gear tooth will carry most of the load, the face width of the teeth will be relatively small. Typical design practices are to limit the face width to

$$F < 0.3r_o \quad \text{or} \quad \frac{10.0}{P_d}\,(\text{or } 10.0\,m) \tag{9.44}$$

whichever is smaller.[3] This guideline is beneficial from a manufacturing standpoint because simpler tooling can be used with a narrower face width.

In addition to the general type shown in Fig. 9.18, there are three special types of bevel gears:

1. Crown bevel gears
2. Miter gears
3. Angular bevel gears

Each of these is discussed briefly in the following.

9.3.3 Crown Bevel Gears and Face Gears

Crown bevel gears are the bevel gear equivalent of a rack. A crown gear has a 90° pitch angle, and its teeth profiles are theoretically parts of great circles (the spherical equivalent of a straight line in the plane). The pitch cone of a crown gear is a cylinder of infinite radius, and the resulting involute teeth have straight sides. The pitch surface is a plane. Crown gears are shown in Figs. 9.22 and 9.23.

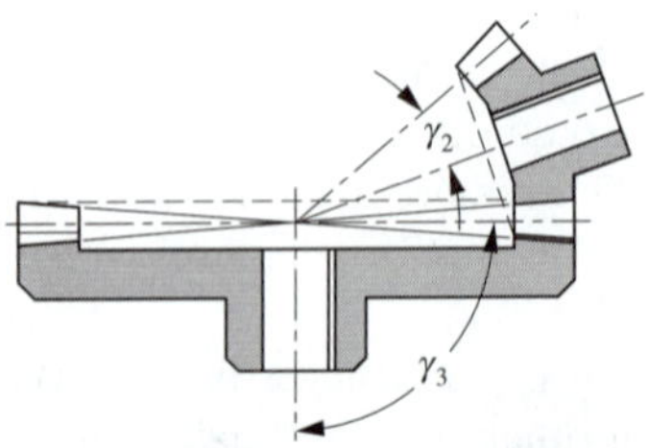

Figure 9.22 Crown gear geometry.

Figure 9.23 Crown gear model.

Face gears consist of a spur or helical pinion in combination with a conjugate gear of disk form. The face gear has the appearance of a crown gear except that the addendum and dedendum angles are zero. The gear is generated with a reciprocating pinion-shaped cutter that has the same diametral pitch and pressure angle as the mating pinion and is substantially the same size.[5] Face gears are shown in Figs. 9.24 and 9.25.

9.3.4 Miter Gears

Miter gears are mating bevel gears with equal numbers of teeth and with axes at right angles. These gears are used to transmit power around a 90° corner. Miter gears are shown in Figs. 9.26 and 9.27.

9.3.5 Angular Bevel Gears

Angular bevel gears have a shaft angle different from 90°. This is the most general type of bevel gear. Examples of angular bevel gears without a 90° shaft angle are shown in Figs. 9.28 and 9.29.

The AGMA[4] gives a standardized approach to determining bevel gear tooth proportions. Bevel gears in this system have unequal addendums for mating teeth. Tables are provided for 20° pressure angle gears with a 45° pitch angle. For angular bevel gears with a shaft angle different from 90°, a calculation procedure is given.

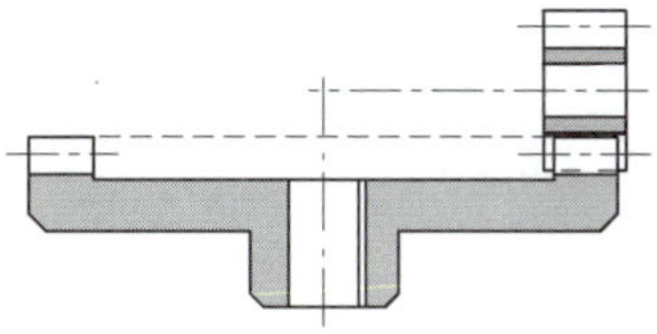

Figure 9.24 Face gear geometry.

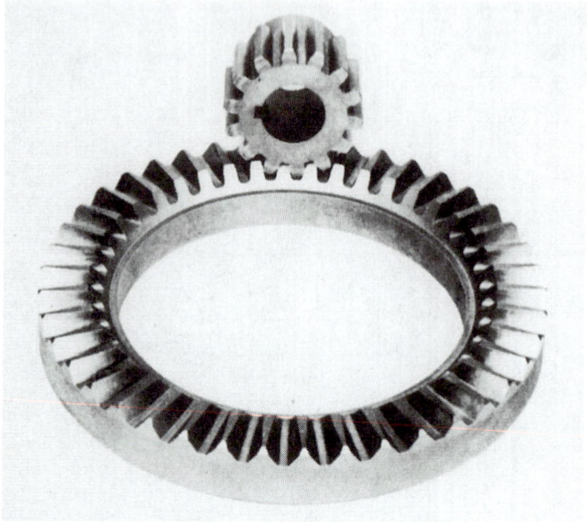

Figure 9.25 Face gear and pinion.[5]

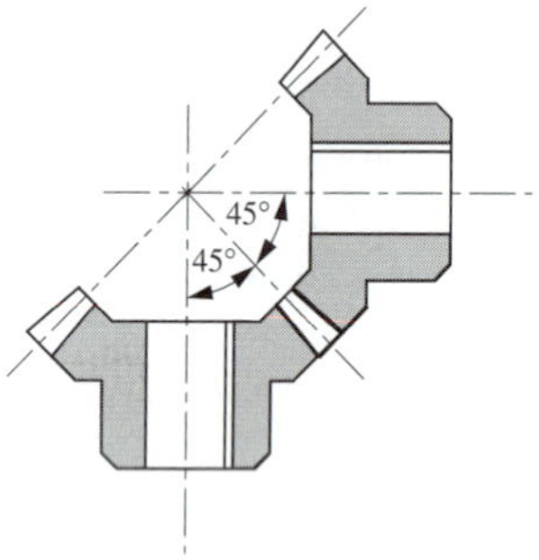

Figure 9.26 Miter gear geometry.

Figure 9.27 Miter gear model.

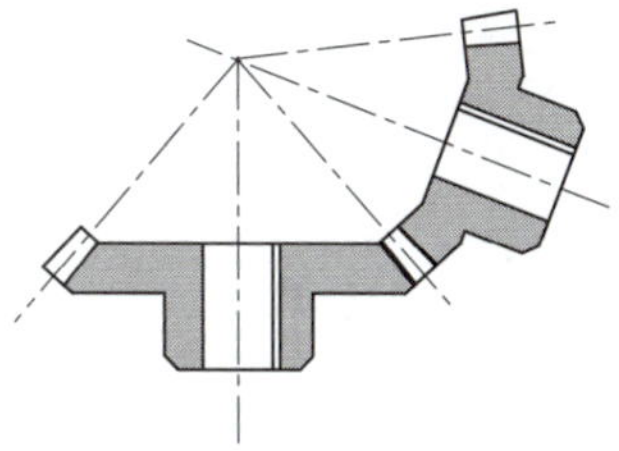

Figure 9.28 Angular bevel gear geometry.

A summary of the equations for bevel gears is given in Table 9.5. A more complete list of equations is given by Townsend.[6]

9.3.6 Zerol Bevel Gears

The previous examples of bevel gears have straight teeth. In addition to these gears, there are two types of bevel gears that have curved teeth. Zerol gears are patented bevel gears that have curved teeth but zero spiral angle at the middle of the teeth. Zerol gears have the same general tooth actions as straight bevel gears, and they can be used to replace straight-toothed bevel gears. The primary advantage over straight-toothed bevel gears is that they can be cut in the same machines as are used for spiral bevel gears, and the teeth can be easily ground.

Most modern straight-toothed bevel gears are also slightly convex, so the contact occurs in the central section of the teeth. This avoids contact on the narrower (and weaker) part of the teeth and allows a slight amount of adjustment when the gears are installed.

Table 9.5 Summary of Straight Bevel Gear Formulas (Pitch Information is for the Large End of Gear)

Quantity	Formula
Pitch diameter	$d = \dfrac{N}{P} = \dfrac{Np}{\pi} = Nm$
Pitch angle of gear	$\tan\gamma_2 = \dfrac{\sin\Sigma}{[d_3/d_2 + \cos\Sigma]} = \dfrac{\sin\Sigma}{[N_3/N_2 + \cos\Sigma]}$
Pitch angle of pinion	$\tan\gamma_3 = \dfrac{\sin\Sigma}{[d_2/d_3 + \cos\Sigma]} = \dfrac{\sin\Sigma}{[N_2/N_3 + \cos\Sigma]}$
Circular pitch	$p = \dfrac{\pi}{P} = \pi m$
Shaft angle	$\Sigma = \gamma_2 + \gamma_3$
Equivalent spur gear radius	$r_{e_i} = \dfrac{r_i}{\cos\gamma_i}, i = 2, 3$
No. of teeth on equivalent spur gear	$N_{e_i} = \dfrac{2\pi r_{e_i}}{p_c}, i = 2, 3$
Outer cone distance	$r_o = \dfrac{d_2}{2\sin\gamma_2} = \dfrac{d_3}{2\sin\gamma_3}$
Minimum face width	$F < 0.3r_o$ or $\dfrac{10.0}{P_d}$ (or $10.0\,m$)
Velocity ratio	$\dfrac{\omega_2}{\omega_3} = \dfrac{N_3}{N_2} = \dfrac{d_3}{d_2} = \dfrac{r_3}{r_2}$

Figure 9.29 Angular bevel gear model.

9.3.7 Spiral Bevel Gears

An example of spiral bevel miter gears is shown in Fig. 9.30. Spiral bevel gears have obliquely curved teeth. As in the case of straight-toothed bevel gears, the pitch cones of spiral bevel gears intersect and have a common apex. Spiral bevel gears are analogous to helical gears and are used for the same purpose. Spiral bevel gears have a gradual load transfer as the teeth engage, and they are much quieter and stronger than straight-toothed bevel gears. Therefore, spiral bevel gears are used for essentially all high-speed applications.

Figure 9.30 Spiral bevel miter gears.

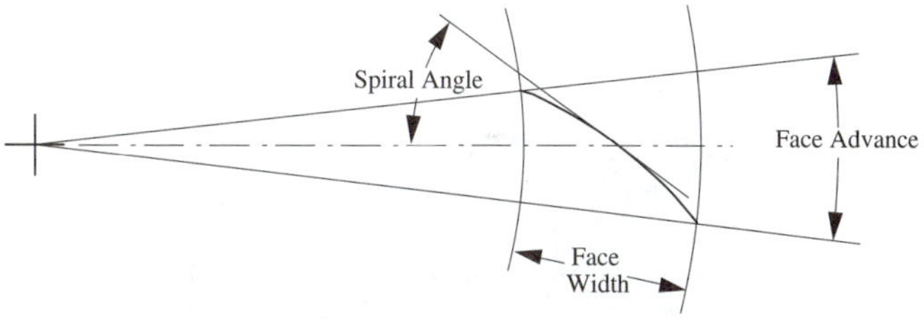

Figure 9.31 Geometry of one spiral gear tooth.

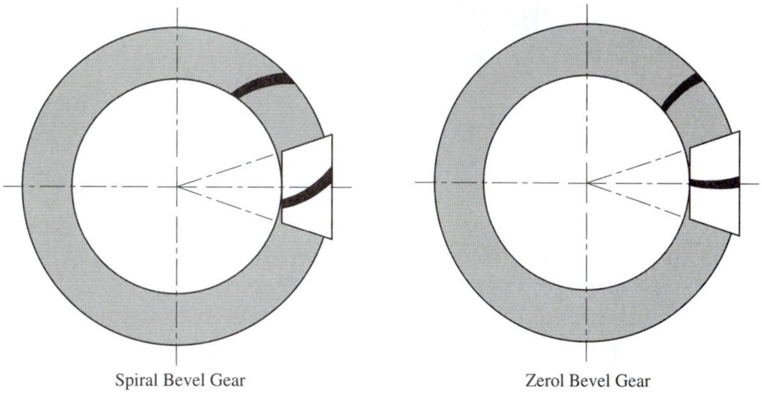

Figure 9.32 Comparison of spiral bevel gear and Zerol bevel gear.

Figure 9.33 Hypoid gear set.

Figure 9.34 The pitch surfaces of hypoid gears as hyperboloids of revolution.

The spiral angle (Fig. 9.31) is such that the tooth face advances due to the spiral exceeding the circular pitch. This provides a smooth transition of load from tooth to tooth and minimizes much of the tooth vibration that creates gear whine. This is in contrast to straight and Zerol bevel gears, with which the load is transferred from tooth to tooth immediately across the entire contact area. The geometries of Zerol and spiral bevel gears are compared to Fig. 9.32. The AGMA[5] gives standardized equations and graphs for determining spiral bevel tooth profiles.

9.3.8 Hypoid Gears

Hypoid gears are similar in appearance to spiral bevel gears; however, the shaft axes do not intersect (see Fig. 9.33). The pitch surfaces of hypoid gears are not cones as in the case of spiral bevel gears but are hyperboloids of revolution (see Fig. 9.34). This permits the teeth to maintain line contact.

Hypoid gears are used in the differentials of rear wheel–driven automobiles because the geometry lowers the level of the drive shaft. This permits a lower body. Another advantage of hypoid gears over spiral bevel gears is that the hypoid pinion teeth are stronger. This is because the hypoid pinion can be designed so that the spiral angle of the pinion is larger than that of the gear. This results in a larger and stronger pinion tooth than that of a spiral bevel pinion. Because of the length of contact on each gear tooth, hypoid gears are quieter than spiral bevel gears. They can also be used with higher reduction ratios.

9.4 REFERENCES

1. American Gear Manufacturers' Association (AGMA), "Tooth Proportions for Coarse Pitch Involute Spur Gears," AGMA Publication 201.02 and 201.02A 1968.
2. Shigley, J. E., *Kinematic Analysis of Mechanisms,* McGraw-Hill Book Co., New York (1975).
3. Mabie, H. H., and C. F. Reinholtz, *Mechanisms and Dynamics of Machinery,* 4th ed. John Wiley & Sons, New York (1987).
4. American Gear Manufacturers' Association (AGMA), "Design Manual for Bevel Gears," AGMA Publication 2005-B88, May, 1988.
5. American Gear Manufacturers' Association (AGMA), "Fine-Pitch On-Cutter Face Gears for 20-Degree Involute Spur Pinions," AGMA Publication 203.03, May, 1973.
6. Townsend, D. P., *Dudley's Gear Handbook,* McGraw-Hill Book Co., New York (1992).

9.5 CHAPTER 9 *Exercise Problems*

PROBLEM 9.1

Two helical gears are cut with a spur gear hob that has a diametral pitch of 4 and a pressure angle of 20°. The pinion has 15 teeth, the gear has 35 teeth, and the helix angle is 30°. Determine the minimum recommended face width. Using the minimum face width, find the transverse diametral pitch, the pitch cylinder radii, and the axial, transverse, and total contact ratios.

PROBLEM 9.2

Two helical gears are cut with the same tooth numbers and with the same cutter as given in Problem 9.1. The helix angle is 30°. Find the transverse pressure angle, the transverse diametral pitch, and the axial pitch.

PROBLEM 9.3

Two parallel helical gears are cut with a 20° normal pressure angle and a 45° helix angle. They have a diametral pitch of 12 in the normal plane and have 10 and 41 teeth, respectively. Find the transverse pressure angle, transverse circular pitch, and transverse diametral pitch. Also determine the minimum face width, and using that face width, determine the total contact ratio.

PROBLEM 9.4

A helical gear has 18 teeth and a transverse diametral pitch of 6. The face width is 1.5, and the helix angle is 25°. Determine the axial pitch, normal pitch, lead, transverse pitch diameter, and minimum face width.

PROBLEM 9.5

Two helical gears have 20 and 34 teeth and a normal diametral pitch of 8. The left-handed pinion has a helix angle of 40° and a rotational speed of 1000 rpm. The gear is also left-handed and has a helix angle of 40°. Determine the angular velocity of the gear, transverse diametral pitch of each gear, and pitch diameters.

PROBLEM 9.6

Two standard spur gears have a diametral pitch of 10, a pressure angle of 20°, and a velocity ratio equal to 3.5 : 1. The center distance is 8.55 in. Two helical gears are to be used to replace the two spur gears such that the center distance and angular velocity ratio remain unchanged. The helical gears are also to be cut with the same hob as that used to cut the spur gears. Determine the helix angle, tooth numbers, and minimum face width for the new gears if the helix angle is kept to a minimum.

PROBLEM 9.7

Two standard spur gears have a diametral pitch of 16 and a pressure angle of 20°. The tooth numbers are 36 and 100, and the gears were meshed at a standard center distance. After the gear reducer was designed and tested, the noise of the drive was found to be excessive. Therefore, the decision was made to replace the spur gears with helical gears. The helix angle chosen was 22°, and the tooth numbers were to remain unchanged. Determine the change in center distance required.

PROBLEM 9.8

A spur gear transmission consists of a pinion that drives two gears. The pinion has 24 teeth and a diametral pitch of 12. The velocity ratio for the pinion and one gear is 3 : 2 and for the pinion and the other gear is 5 : 2. To reduce the noise level, all three gears are to be replaced by helical gears such that the center distances and velocity ratios remain the same. The helical gears will be cut with a 16 pitch, 20° hob. If the helix angle is kept as low as possible, determine the number of teeth, face width, hand, helix angle, and outside diameter for each of the gears.

PROBLEM 9.9

A pair of helical gears have a module in the normal plane of 3 mm and a normal pressure angle of 20°. The gears mesh with parallel shafts and have 30 and 48 teeth. Determine the transverse module, the pitch diameters, the center distance, and the minimum face width.

PROBLEM 9.10

Two 20° spur gears have 36 and 90 teeth and a module of 1.5. The spur gears are to be replaced by helical gears such that the center distance and velocity ratio are not changed. The maximum allowed face width is 12.7 mm, and the hob module is 1.5 mm. Design the helical gear pair that has the smallest helix angle possible. Determine the numbers of teeth, the face width, the helix angle, and the outside diameters of the gears.

PROBLEM 9.11

Two helical gears are cut with a 20° hob with a module of 2. One gear is right handed, has a 30° helix angle, and has 36 teeth. The second gear is left handed, has a 40° helix angle, and has 72 teeth. Determine the shaft angle, the angular velocity ratio, and the center distance.

PROBLEM 9.12

Two crossed shafts are connected by helical gears such that the velocity ratio is 3:1, and the shaft angle is 60°. The center distance is 10 in, and the normal diametral pitch is 8. The pinion has 35 teeth. Assume that the gears are the same hand and determine the helix angles, pitch diameters, and recommended face widths.

PROBLEM 9.13

Two crossed shafts are connected by helical gears such that the velocity ratio is 3:2, and the shaft angle is 90°. The center distance is 5 in. Select a pair of gears that will satisfy the design constraints. What other information might be considered to reduce the number of arbitrary choices for the design?

PROBLEM 9.14

A helical gear with a normal diametral pitch of 8 is to be used to drive a spur gear at a shaft angle of 45°. The helical gear has 21 teeth, and the velocity ratio is 2:1. Determine the helix angle for the helical gear and the pitch diameter of both gears.

PROBLEM 9.15

Two crossed helical gears connect shafts making an angle of 45°. The pinion is right handed, has a helix angle of 20°, and contains 30 teeth. The gear is also right handed and contains 45 teeth. The transverse diametral pitch of the gear is 5. Determine the pitch diameter, the normal pitch, and the lead for each gear.

PROBLEM 9.16

The worm of a worm gear set has 2 teeth, and the gear has 58 teeth. The worm axial pitch is 1.25 in, and the pitch diameter is 3 in. The shaft angle is 90°. Determine the center distance for the two gears, the helix angle, and the lead for the worm.

PROBLEM 9.17

The shaft angle between two shafts is 90°, and the shafts are to be connected through a worm gear set. The center distance is 3 in, and the velocity ratio is 30:1. Determine a worm and gear that will satisfy the design requirements. Specify the number of teeth, lead angle, and pitch diameter for each gear. Also, determine the face width for the gear.

PROBLEM 9.18

A worm with two teeth drives a gear with 50 teeth. The gear has a pitch diameter of 8 in and a helix angle of 20°. The shaft angle between the two shafts is 80°. Determine the lead and pitch diameter of the worm.

PROBLEM 9.19

Two straight-toothed bevel gears mesh with a shaft angle of 90° and a diametral pitch of 5. The pinion has 20 teeth, and the gear ratio is 2:1. The addendum and dedendum are the same as for 20° stub teeth. For the gear, determine the pitch radius, cone angle, outside diameter, cone distance, and face width.

PROBLEM 9.20

A pair of straight-toothed bevel gears mesh with a shaft angle of 90° and a diametral pitch of 6. The pinion has 18 teeth, and the gear ratio is 2:1. The addendum and dedendum are the same as for 20° full-depth spur gear teeth. Determine the number of teeth on the gear and the pitch diameters of both the pinion and gear. Also, for the gear, determine the pitch angle, cone angle, outside diameter, cone distance, and face width.

PROBLEM 9.21

A pair of straight-toothed bevel gears mesh with a shaft angle of 80° and a diametral pitch of 7. The pinion has 20 teeth and a pitch cone angle of 40°. The gear ratio is 3:2. Determine the number of teeth on the gear and the pitch diameters of both the pinion and gear. Also determine the equivalent spur gear radii for both the pinion and the gear.

PROBLEM 9.22

A pair of straight-toothed bevel gears mesh with a shaft angle of 45° and a module of 5.08. The pinion has 16 teeth and a pitch cone angle of 20°. The gear ratio is 3:2. Determine the number of teeth on the gear and the pitch diameters of both the pinion and gear. Also determine the back-cone distance and the back-cone angle for the gear.

Chapter 10

Gear Trains

10.1 GEAR TRAINS

In Chapters 8 and 9, the characteristics of individual gears were discussed. However, in general, gears are of interest to designers only when they are used in pairs as motion and/or force transducers. These gear pairs can be combined in many ways to achieve desired input/output relationships. A combination of one or more gear pairs that are interrelated is called a gear train. All complex gear trains are combinations of the simple, compound, and planetary gear trains, discussed in this chapter.

10.2 DIRECTION OF ROTATION

As discussed in Chapters 8 and 9, the velocity ratio for two meshing gears (2 and 3) is

$$R = \pm\frac{\omega_2}{\omega_3} = \pm\frac{r_3}{r_2} = \pm\frac{N_3}{N_2} \tag{10.1}$$

where r_i and N_i are the pitch radii and number of teeth on gear i ($i = 2, 3$), the plus sign goes with an external gear meshing with an internal gear, and the minus sign goes with two external gears meshing.

When planar gears are involved, Eq. (10.1) can be used directly because all of the vectors are parallel. However, when bevel and crossed helical gears are involved, the angular velocities must be treated as vectors. For bevel gears, a relatively simple way to do this is to recognize that at the pitch point (the end of the tangent line to the pitch surfaces), rolling occurs, and the velocity at the pitch point on both gears is the same. From this,

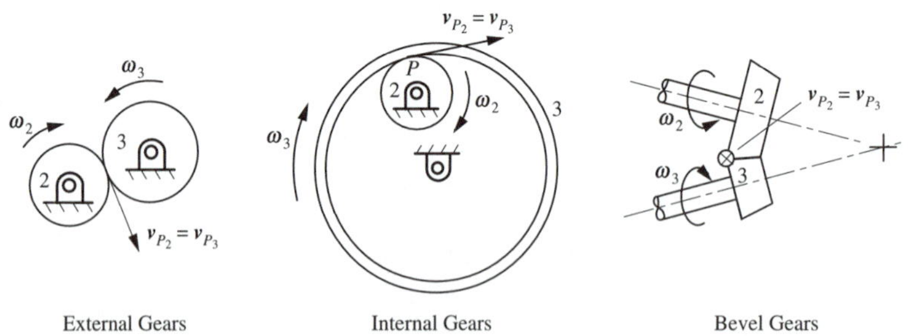

Figure 10.1 The direction of the angular velocities of two meshing gears can be determined from the direction of the velocity of the pitch point.

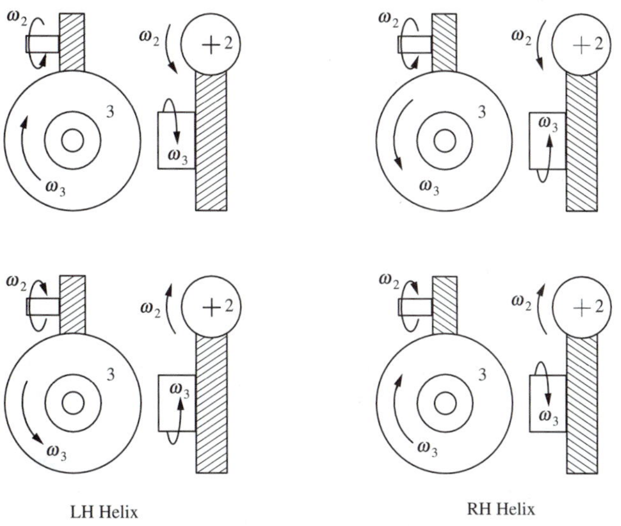

Figure 10.2 The direction of the angular velocities for crossed helical gears.

the direction of rotation can be inferred from the simple velocity relationship

$$\boldsymbol{v}_{P_2} = \boldsymbol{v}_{P_3} = \boldsymbol{\omega}_2 \times \boldsymbol{r}_2 = \boldsymbol{\omega}_3 \times \boldsymbol{r}_3 \tag{10.2}$$

where the pitch radius vectors ($\boldsymbol{r}_2$ and $\boldsymbol{r}_3$) for gears 2 and 3, respectively, are directed from the rotation axis to the pitch point. For bevel gears, the large end of the gear is used for the measurement of the pitch radii. This is shown in Fig. 10.1.

When crossed helical gears are involved, the process becomes a little more complicated because the hand of the gears, that is, the direction of twist of the helical teeth, affects the direction of rotation of the driven gear as shown in Fig. 10.2. To determine the direction of rotation of the pinion relative to the gear, treat the pinion as a screw and the gear as fixed. Observe the motion of the pinion relative to the gear as the pinion is rotated and the gear is viewed along the gear axis. If the pinion appears to advance toward the gear when the pinion is rotated, in reality the gear would rotate counterclockwise. If the pinion appears to withdraw from the gear when the pinion is rotated, the gear would rotate clockwise.

10.3 SIMPLE GEAR TRAINS

Simple gear trains can be divided into two types depending on whether idler gears are involved or not. Simple gear trains have only one gear on each shaft. These shafts rotate on bearings that are attached to the same frame. The gears may be of any type, for example, spur, bevel, hypoid, and worm. Figures 10.3–10.5 show various simple gear trains.

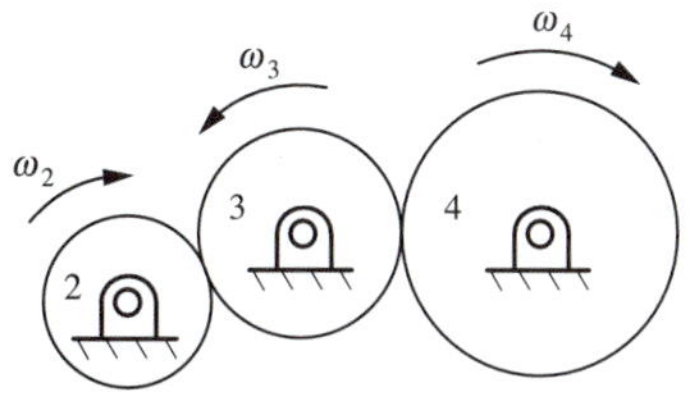

Figure 10.3 A simple gear train with all external gears and one idler (gear 3).

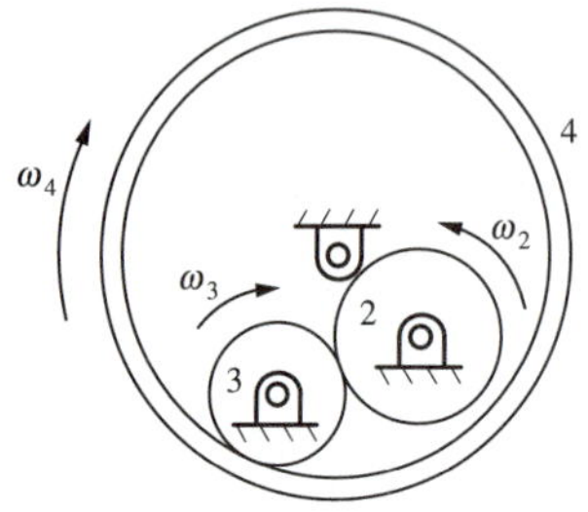

Figure 10.4 A simple gear train with external gear and one idler (gear 3).

The idler gears in simple gear trains can serve two purposes in design. One is to change the direction of motion of the output gear, and the second is to provide a spacer when two gears cannot be directly meshed because of the shaft locations. This occurs when there is a limit to the sizes that two gears can be but the shaft location is specified for reasons other than kinematics.

In gear trains, the overall gear reduction ratio for the gear box is usually of interest. This can be determined by writing the velocity ratio in terms of the tooth numbers at each mesh. For example, in Fig. 10.3,

$$\frac{\omega_2}{\omega_3} = -\frac{N_3}{N_2} \tag{10.3}$$

and

$$\frac{\omega_3}{\omega_4} = -\frac{N_4}{N_3} \tag{10.4}$$

Multiplying Eq. (10.4) by Eq. (10.3) gives

$$\frac{\omega_2}{\omega_3}\frac{\omega_3}{\omega_4} = \left(-\frac{N_3}{N_2}\right)\left(-\frac{N_4}{N_3}\right)$$

or

$$\frac{\omega_2}{\omega_4} = (-1)^2\frac{N_4}{N_2} \tag{10.5}$$

If we analyze the gear train in Fig. 10.4, we will get

$$\frac{\omega_2}{\omega_4} = (-1)^1\frac{N_4}{N_2} \tag{10.6}$$

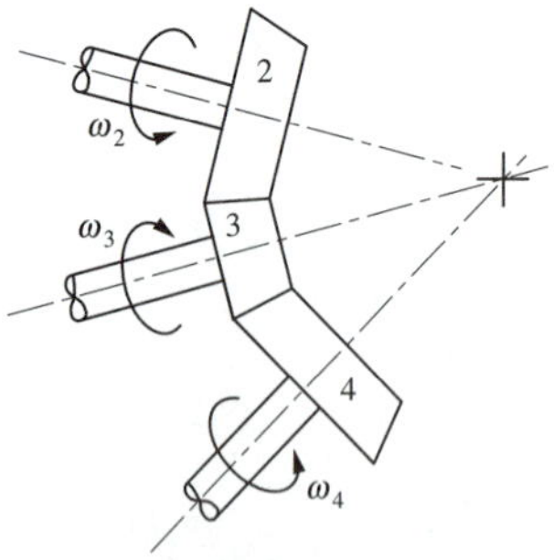

Figure 10.5 A simple gear train with external bevel gears and one idler (gear 3).

The analysis of the gear train in Fig. 10.5 is more difficult because we must treat the angular velocities as vectors to determine the directions mathematically. However, if we trace the angular-velocity directions using the procedure indicated in Fig. 10.1, we can determine the direction and compute the magnitude of the angular velocities separately. If we do this, we will find that the magnitude of the overall velocity ratio is given by

$$\frac{\omega_2}{\omega_4} = \frac{N_4}{N_2}$$

If we add more idler gears, the effect will be the same. Namely, the magnitude of the velocity ratio between the input and output shafts is a function of the numbers of teeth on the input and output gears only. The magnitude of the velocity ratio is independent of the size and number of idler gears. The sign of the train ratio for parallel-shaft gears, however, does depend on the number of idler gears. In particular, at each mesh between external gears, the velocity ratio changes sign. For internal gears, the velocity ratio remains the same sign. Therefore, if n is the number of meshes between *external* gears, the sign of the velocity ratio is given by $(-1)^n$. Note that each idler gear will have at least two mesh points.

A simple gear train can involve any number and types of gears. However, each gear in the gear train must be able to mesh with any other gear. Therefore, each gear must have the same normal pitch if the gears are to mesh properly.

10.3.1 Simple Reversing Mechanism

An idler gear can be used in a simple reversing mechanism shown in Fig. 10.6. This is a procedure commonly used to reverse the direction of rotation of the lead screw on small metal lathes. The procedure adds an extra idler to the simple gear train when the direction of rotation is to be reversed. The mechanism works well only when the gears are slowly moving or at rest, since there is no provision for ensuring that the gears will mesh easily when the direction change is made.

10.4 COMPOUND GEAR TRAINS

For all types of gears, the velocity ratio is limited for each mesh by practical considerations. For example, in spur gears, the velocity ratio at any mesh should not exceed 1:5. For larger reductions, compound gear trains should be used. Compound gear trains are characterized by the presence of two or more gears attached to the same shaft. The shafts,

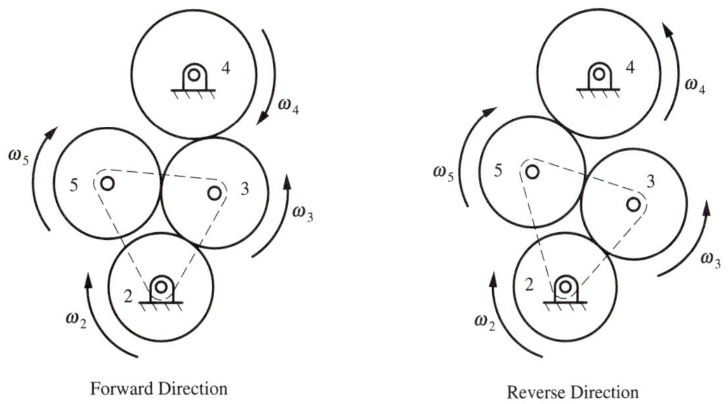

Figure 10.6 A simple reversing mechanism using an extra idler gear.

Figure 10.7 A compound gear train.

however, still rotate on bearings that are fixed to the frame. Unlike simple gear trains, the gears in a compound gear train need not and generally will not be of the same type. This is evident in Fig. 10.7, which shows an example of a commercial gear reducer.

The velocity ratios attainable in a compound gear train can be any size, with ratios in the thousands being possible. There is no theoretical limit to the number of passes (gear meshes) that can be made; however, practical issues such as friction and the functional need restrict the number in most applications.

A compound gear train is shown in Fig. 10.8. The symbolism often used for gear trains

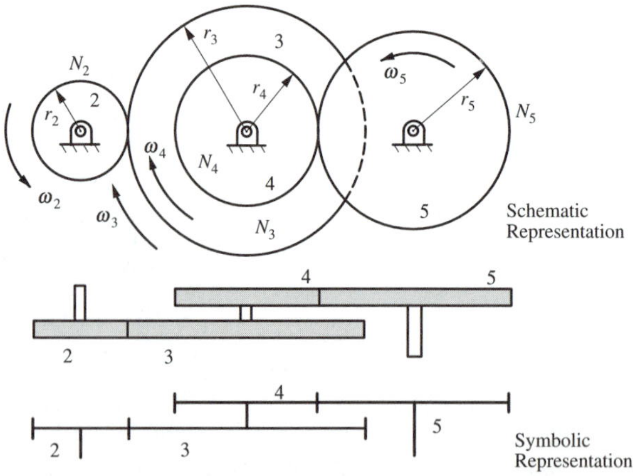

Figure 10.8. A compound gear train.

is also illustrated in the figure. The velocity ratio for the gear mesh can be written as

$$\frac{\omega_2}{\omega_3} = -\frac{N_3}{N_2} \tag{10.7}$$

and

$$\frac{\omega_4}{\omega_5} = -\frac{N_5}{N_4} \tag{10.8}$$

Because gears 3 and 4 are rigidly attached to the same shaft, we know that $\omega_3 = \omega_4$. The overall gear train velocity ratio is given by ω_2/ω_5.

If we solve Eq. (10.7) for ω_2 and Eq. (10.8) for ω_5 and use $\omega_3 = \omega_4$, the overall velocity ratio can be written as

$$\frac{\omega_2}{\omega_5} = \frac{-\omega_3(N_3/N_2)}{-\omega_4(N_4/N_5)} = \frac{(N_3/N_2)}{(N_4/N_5)} = \frac{N_3N_5}{N_2N_4} \tag{10.9}$$

or

$$\frac{\omega_2}{\omega_5} = \frac{\omega_2}{\omega_3}\frac{\omega_4}{\omega_5} = \frac{N_3N_5}{N_2N_4} \tag{10.10}$$

From Eq. (10.10), it is clear that we can compute the overall velocity ratio from the product of the velocity ratios at each mesh. We can do this either in terms of the velocities directly or, more beneficially, in terms of the tooth numbers. Notice that when parallel shaft gearing is involved, we can also use the ratios of the pitch circle radii, since these will be directly proportional to the tooth numbers. Equation (10.10) can then be extended to

$$\frac{\omega_2}{\omega_5} = \frac{\omega_2}{\omega_3}\frac{\omega_4}{\omega_5} = \frac{N_3N_5}{N_2N_4} = \frac{r_3r_5}{r_2r_4} \tag{10.11}$$

However, the most convenient parameter to use when computing velocity ratios is the tooth number on each gear. This is because the velocity ratios can be directly equated to the tooth ratios for all types of gearing, whereas the ratios of the pitch cylinder radii alone are not valid for gears with nonparallel shafts.

If we start with gear 2 as the input gear, we can treat each mesh of the gear train as having an input side and an output side. For example, in Fig. 10.8, at the mesh between gears 2 and 3, gear 2 would be the driver and gear 3 would be the driven gear. At the mesh between gears 4 and 5, gear 4 would be the driver and gear 5 would be the driven gear. Therefore, N_3 and N_5 would be associated with driven gears, and N_2 and N_4 would be associated with driver gears. In Eq. (10.11), the velocity ratio can be represented as the product of the driven gear numbers divided by the product of the driver gear numbers. This situation holds in general for compound gear trains. Mathematically, if n is the number of gear meshes (including idlers that each have two meshes), a general expression for the magnitude of the velocity ratio can be written as

$$\frac{\omega_{\text{input}}}{\omega_{\text{output}}} = \frac{\prod_{i=3}^{n} N_i}{\prod_{j=2}^{n} N_j} = \frac{\text{product of drive}n\text{ tooth numbers}}{\text{product of drive}r\text{ tooth numbers}} \tag{10.12}$$

Assuming that the gears are numbered sequentially, in Eq. (10.12), i includes only the tooth numbers for the odd gear numbers, and j includes only the tooth numbers for the even gear numbers. The sign of the gear ratio depends on the type of gears. If all parallel shaft gears are involved, we can use an extension of Eq. (10.4). Then,

$$\frac{\omega_{\text{input}}}{\omega_{\text{output}}} = (-1)^m \frac{\prod_{i=3}^{n} N_i}{\prod_{j=2}^{n} N_j} = (-1)^m \frac{\text{product of drive}n\text{ tooth numbers}}{\text{product of drive}r\text{ tooth numbers}}$$

where m is the number of meshes involving external gears.

EXAMPLE 10.1 ***(Analysis of Compound Gear Train)***

PROBLEM

Assume that the compound gear train in Fig. 10.9 has the tooth numbers given in parentheses. The angular velocity of gear 2 is 200 rpm in the direction shown. Find the magnitude and direction of the angular velocity of gear 10 and the velocity (magnitude and direction) of the rack that is gear 11.

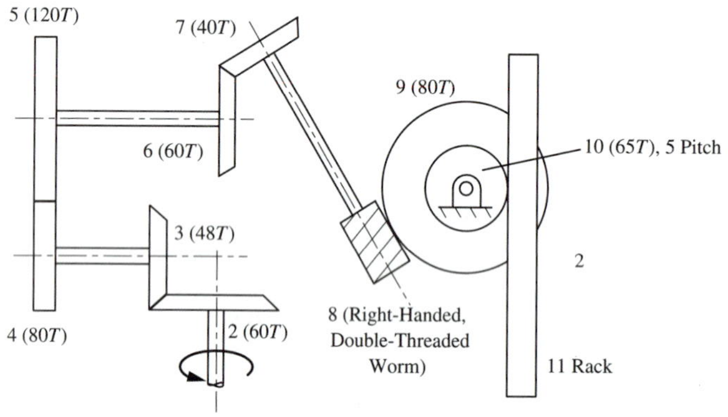

Figure 10.9 The gear train for Example 10.1.

SOLUTION

The velocity ratio for the gear drive between gears 2 and 9 is given by

$$\frac{\omega_9}{\omega_2} = \frac{N_2}{N_3}\frac{N_4}{N_5}\frac{N_6}{N_7}\frac{N_8}{N_9}$$

Therefore,

$$\omega_9 = \omega_2 \frac{N_2}{N_3}\frac{N_4}{N_5}\frac{N_6}{N_7}\frac{N_8}{N_9}$$

and

$$\omega_9 = 200\frac{60}{48}\frac{80}{120}\frac{60}{40}\frac{2}{80} = 6.25 \text{ rpm}$$

The velocity of the rack will be equal to the linear velocity of the pitch point on gear 10. The angular velocity of gear 10 is equal to the angular velocity of gear 9. The pitch diameter of gear 10 is given by

$$d_{10} = \frac{N_{10}}{P_{10}} = \frac{65}{5} = 13 \text{ in}$$

Therefore, the pitch velocity is given by

$$v = (\omega_{10})(d_{10}/2) = (6.25)(13/2)(2\pi/60) = 4.255 \text{ in/s}$$

However, we must now determine in which direction the rack moves (up or down). To do this, trace the pitch point velocities at each mesh. This is shown in Fig. 10.10. Gear 8 is a right-handed worm gear. Therefore, it will advance relative to gear 9 for a clockwise rotation. Consequently, gear 9 will rotate clockwise relative to the frame (link 1). If gear 9 and 10 rotate clockwise, then the rack will move down as shown. ■

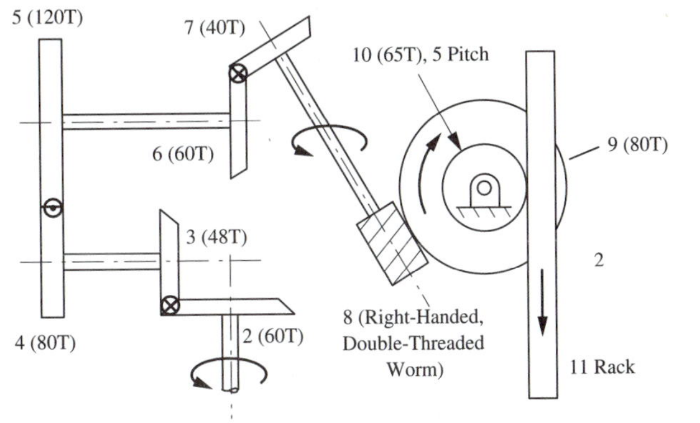

Figure 10.10 The directions of gear motion for Example 10.1.

10.4.1 Concentric Gear Trains

In a concentric gear train, the input and output shafts are collinear. An example of a concentric gear train is shown in Fig. 10.11. These gear trains are analyzed in much the same way as any compound gear train; however, the design is somewhat more complex. In the following, we will restrict the discussion to parallel-shaft gearing with a double reduction to illustrate a possible design procedure.

A concentric gear train with a two-stage reduction is shown in Fig. 10.12. A principal requirement for a concentric gear reducer is that

$$r_2 + r_3 = r_4 + r_5 \tag{10.13}$$

Figure 10.11 A concentric gear reducer.

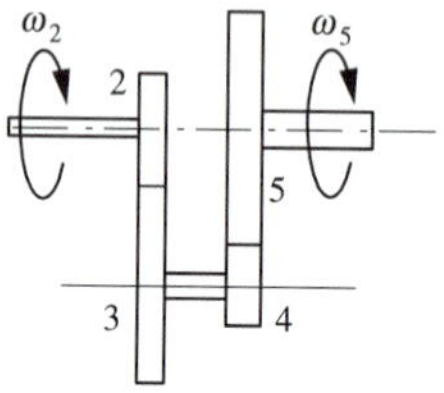

Figure 10.12 A concentric gear reducer with a two-stage reduction.

In addition, there may be a precise requirement for the overall reduction ratio. From before, the reduction ratio is given by

$$R = \frac{\omega_2}{\omega_5} = \frac{r_3}{r_2}\frac{r_5}{r_4} = \frac{N_3}{N_2}\frac{N_5}{N_4} \tag{10.14}$$

To mesh properly, gears 2 and 3 must have the same normal pitch, and gears 4 and 5 must have the same normal pitch. If helical gears are involved, we must select the helix angles, and this will give us some latitude in the design. Then,

$$\begin{aligned} \frac{N_2}{2r_2} &= P_{n_2} \cos \psi_2 \\ \frac{N_3}{2r_3} &= P_{n_2} \cos \psi_2 \end{aligned} \tag{10.15}$$

and

$$\begin{aligned} \frac{N_4}{2r_4} &= P_{n_4} \cos \psi_4 \\ \frac{N_5}{2r_5} &= P_{n_4} \cos \psi_4 \end{aligned} \tag{10.16}$$

In Eqs. (10.13)–(10.16) there are 12 unknowns and 6 equations. In addition, there is the constraint that the tooth numbers must be integers. Therefore, we can select six of the variables to solve the equations subject to the constraint that the tooth numbers are integers.

One design approach is to select first the tooth numbers to satisfy Eq. (10.14), which is typically the most difficult equation to satisfy. This is equivalent to selecting three of the variables. It may not always be easy or even possible to select tooth numbers in a practical range to solve Eq. (10.14) exactly. If the values of R are formed by ratios of small whole numbers, for example 1/2, 7/4, 4/9, many choices of whole numbers will satisfy the problem. In such cases, the best choice can be selected on the basis of criteria other than kinematics. However, other values of R are impossible to generate with simple gears. Examples are the square root of 2 and the ratio of two prime numbers (e.g., 503/2003). In such cases, it may be possible only to approximate the value for R.

When the machine function does not require an exact ratio, it is usual to select tooth numbers for a meshing gear pair that do not have common factors. This improves wear performance, because a defect on a gear tooth of one gear will make contact with all of the teeth on the mating gear equally rather than selectively making contact with a small number of teeth.

A number of elegant procedures are available for selecting the tooth numbers given R,[1] especially when R is given as a fraction where the numerator and denominator are whole numbers. Such procedures may be aided by tables of factors. For example, if $R = p/q$, we would look for values of N_3 and N_5 such that $N_3N_5 = p$ and values of N_2 and N_4 such that $N_2N_4 = q$. Alternatively, if the ranges for the tooth numbers are limited, we might simply conduct an exhaustive search for all possible combinations of tooth numbers that satisfy the condition for R. On modern computers, such a search is easy to program and takes very little time to conduct. A simple MATLAB program, *factor.m*, for finding factors of any integer is included on the disk with this book.

After the tooth numbers are established, we can select one of the normal diametral pitches, for example, P_{n_2}, and the corresponding helix angle (ψ_2). Then solve for r_2 and r_3. Given r_2 and r_3, Eqs. (10.13) and (10.16) can be solved for r_4, r_5, and $P_{n_4} \cos \psi_4$. Pick a standard value for P_{n_4}, and solve for the helix angle, ψ_4.

This discussion deals with kinematics alone. Obviously, other very important aspects of gear design are stress and wear considerations. The topic is properly treated in almost any book on machine design. Therefore, we will limit our discussion to kinematics with the assumption that sizing the teeth to carry the loading will be addressed elsewhere.

EXAMPLE 10.2 ***(Concentric Gear Box Design)***

PROBLEM

Assume that a concentric gear box is to be designed for a velocity ratio of $R = 20:1$. The first-stage reduction is to have a helix angle of 30° and normal diametral pitch of 8. Both sets of gears will have a normal pressure angle of 20°. Find values for the tooth numbers, pitch cylinder radii for all of the gears, and the diametral pitch and helix angle for gears 4 and 5.

SOLUTION

To avoid undercutting, we will limit the tooth numbers for N_2 and N_3 to 12 teeth (see Table 9.1). For the smallest possible gear box, assume that N_2 is 12. Initially, select N_4 to be 12 also. If the minimum helix angle for gears 4 and 5 is found to be less than 30°, we can specify it to be 30°, which will make 12 an acceptable tooth number. If the helix angle must be larger than 30°, the minimum value for N_4 could be smaller than 12. The velocity ratio can be written as

$$R = \frac{20n}{n}$$

where n is any integer. From Eq. (10.14), $n = N_2N_4$. Therefore, $n = 144$, and $20n$ is 2880.

To determine the factors for 2880, the MATLAB factor program was used. There are 42 factors for 2880. These are 1, 2, 3, 4, 5, 6, 8, 9, 10, 12, 15, 16, 18, 20, 24, 30, 32, 36, 40, 45, 48, 60, 64, 72, 80, 90, 96, 120, 144, 160, 180, 192, 240, 288, 320, 360, 480, 576, 720, 960, 1440, 2880. When designing the two-stage gear reducer, it is generally desirable to make the two gear reductions about the same. This avoids making one gear significantly larger than the others. Of the factors, 48 and 60 will give gear reductions of 4 and 5 for the two stages. Let us select 60 for the first stage (N_3) and 48 for the second stage (N_5). This will permit larger teeth (lower P_n) on the low-speed end of the gear reducer without making the gear diameters significantly larger than those for the high-end gears.

From Eqs. (10.15),

$$r_2 = \frac{N_2}{2P_{n_2} \cos \psi_2} = \frac{12}{2(8) \cos 30°} = 0.866 \text{ in}$$

and

$$r_3 = \frac{N_3}{2P_{n_3} \cos \psi_3} = \frac{N_3}{2P_{n_2} \cos \psi_2} = \frac{60}{2(8) \cos 30°} = 4.330 \text{ in}$$

From Eqs. (10.16),

$$\frac{N_4}{N_5} = \frac{r_4}{r_5} = \frac{12}{48} = \frac{1}{4}$$

Therefore,

$$r_5 = 4r_4$$

Substituting this expression into Eq. (10.13),

$$5r_4 = r_2 + r_3 = 0.866 + 4.330 = 5.196 \text{ in}$$

Therefore,

$$r_4 = \frac{5.196}{5} = 1.039 \text{ in}$$

and

$$r_5 = 4r_4 = 4(1.039) = 4.156 \text{ in}$$

The only unknowns are P_{n_4} and ψ_4. From Eqs. (10.16).

$$P_{n_4} \cos \psi_4 = \frac{N_4}{2r_4} = \frac{12}{2(1.039)} = 5.775$$

To illustrate the procedure, select a normal diametral pitch of 7. Then

$$\psi_4 = \cos^{-1}\left(\frac{5.775}{7}\right) = \cos^{-1}(0.825) = 34.4°$$

Note that in this example we have arbitrarily selected the diametral pitches. In an actual problem, these, along with the face widths, would be selected in part to accommodate the torque and speed requirements. ■

10.5 PLANETARY GEAR TRAINS

Both simple and compound gear trains have the restriction that their gear shafts must rotate in bearings fixed to the frame. However, this is a requirement that limits the versatility of the gear train. If one or more shafts rotate around another shaft as well as spinning about their own axes, the gear train is called a planetary or epicyclic gear train. Planetary gear trains are used extensively for compact gear reducers (Fig. 10.13). Also, because they are basically devices with multiple degrees of freedom, they are used in automatic transmissions for automobiles and trucks (Fig. 10.14).

Determining the velocity ratio for planetary gear trains is more difficult than for plain simple and compound gear trains. The motion of the gears involves the motion of the moving shaft or carrier along with the motion of the gears with respect to the carrier.

10.5.1 Planetary Gear Nomenclature

A simple planetary drive is shown schematically in Fig. 10.15. Planetary gear trains are typically made up of the following:

1. Sun gear (may or may not be fixed)
2. Planet gears (one or more)
3. Planet carrier
4. Internal ring gear (not used in all planetary gear trains)

Figure 10.13 A planetary gear reducer.

Figure 10.14 Planetary gears in automatic transmission.

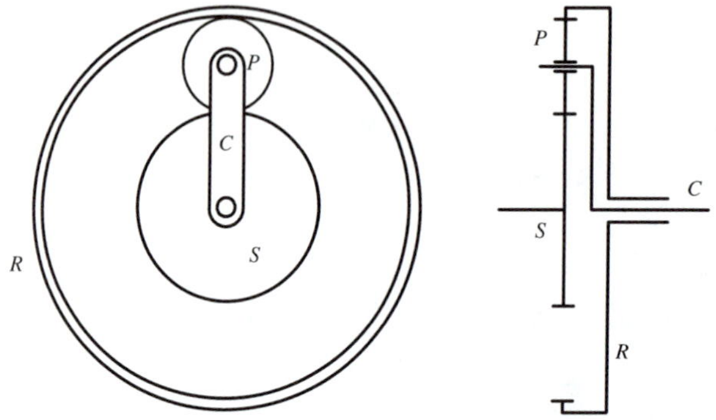

Figure 10.15 A simple planetary gear train.

The symbolism used to represent planetary drives is also shown in Fig. 10.15. This symbolism allows the designer to represent the structure of the planetary drive simply.

In Fig. 10.15, note that the carrier, ring gear, and sun gear all rotate about concentric axes. Also, three axes are evident from the simple figure. As shown below, the planetary gear train has two degrees of freedom, and the angular motion of two of the axes must be specified before the angular motion of the third can be determined.

Very high velocity reductions can be achieved with compound planetary gear trains. These systems involve compound planetary gears as shown in Figs. 10.16a. This also permits the ring gear in Fig. 10.15 to be replaced by another sun gear. The carrier can involve several shafts containing four or more planetary gears as shown in Fig. 10.16b for still greater reductions. And finally, it is possible to connect planetary gears in series as shown in Fig. 10.16c. In Fig. 10.16, the bearings associated with the frame link are not shown. It is understood that frame bearings will be required for all of the shafts rotating with fixed axes.

When planetary gear trains are connected in series, very high gear reductions are possible. Such gear reducers are common in small power tools. Figure 10.17 shows a small planetary gear reducer made up of two planetary systems in series used in a power screwdriver.

In the planetary gear trains indicated in Fig. 10.16, two of the shafts are inputs and one is the output. Typically, the angular velocity of one of the bodies is zero, but this is not required.

In Figs. 10.13–10.17, all of the gears in the planetary drives are parallel-shaft gears. However, this is not necessary. Perhaps the most common planetary drive is the differential (Fig. 10.18) in rear wheel–driven automobiles. This is a right-angle drive that involves a hypoid ring gear and pinion and bevel gear planets.

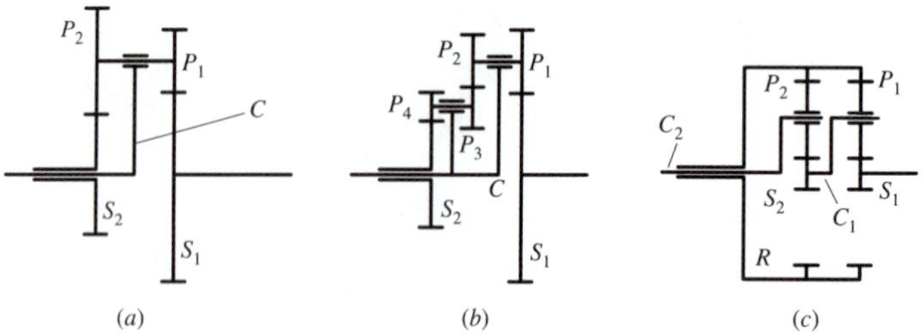

Figure 10.16 Complex planetary gear trains.

Figure 10.17 Planetary gear trains used in a power screwdriver.

Planetary gear trains are commonly analyzed using either the equation method or the tabular method. We will look at each procedure separately by analyzing example gear trains.

10.5.2 Analysis of Planetary Gear Trains Using Equations

In the equation method, the procedure is to write relative angular velocity equations (relative to the frame) for each of the gears with fixed rotation axes. Also, write relative velocity equations for the same gears relative to the carrier. If the angular velocities of two of the shafts are given, this procedure will always yield enough equations to solve for the angular velocities of all of the members in the system.

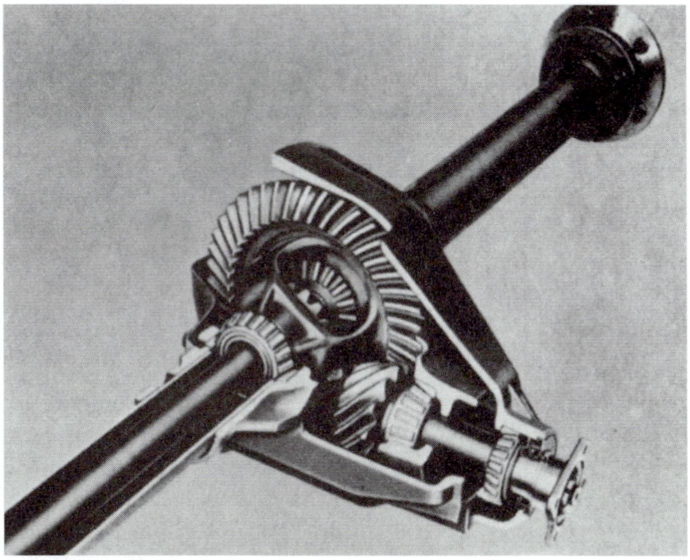

Figure 10.18 A planetary gear train used in the differential of a rear wheel–driven vehicle.

EXAMPLE 10.3 ***(Basic Procedure Using Equation Method)***

PROBLEM

Assume that the gear train in Fig. 10.15 has the frame as member 1, the sun gear as member 2, the planet as member 3, and the ring gear as member 4. Gear 2 is the input and rotates clockwise with an angular velocity of ${}^1\boldsymbol{\omega}_2$. Find an equation that involves the angular velocity of the carrier, ${}^1\boldsymbol{\omega}_2$, ${}^1\boldsymbol{\omega}_4$, and the tooth numbers for the individual gears.

SOLUTION

There are two gears, (2 and 4) that rotate about fixed axes in the system, and we can write the following angular velocity relationships for these gears using the chain rule for angular velocities:

$$ {}^1\omega_2 = {}^C\omega_2 + {}^1\omega_C \tag{10.17} $$

and

$$ {}^1\omega_4 = {}^C\omega_4 + {}^1\omega_C \tag{10.18} $$

If we make the carrier the reference link, the gears will move as an ordinary gear train in which the planet gear acts as an idler. Therefore, we can compute the velocity ratio relative to the carrier as

$$ \frac{{}^C\omega_2}{{}^C\omega_4} = -\frac{N_4}{N_2} \tag{10.19} $$

Next solve Eqs. (10.17) and (10.18) for ${}^C\omega_2$ and ${}^C\omega_4$, respectively. Then

$$ {}^C\omega_2 = {}^1\omega_2 - {}^1\omega_C \tag{10.20} $$

and

$$ {}^C\omega_4 = {}^1\omega_4 - {}^1\omega_C \tag{10.21} $$

Now divide Eq. (10.20) by (10.21) and equate the result to Eq. (10.19). The result is

$$ \frac{{}^1\omega_2 - {}^1\omega_C}{{}^1\omega_4 - {}^1\omega_C} = -\frac{N_4}{N_2} \tag{10.22} $$

Equation (10.22) gives the relationship for the velocities of the shafts coming from the gear train. Given any two of the angular velocities, the third can be determined. Note that it is important to identify the direction of the angular velocities with a plus or minus sign. Typically, we could select counterclockwise (CCW) as plus and clockwise (CW) as minus. ■

EXAMPLE 10.4 ***(Analysis of Planetary Gear Train Using Equation Method)***

PROBLEM

Assume that the carrier in Fig. 10.19 is member 6 and that it and gear 5 are driven clockwise at 150 and 30 rpm, respectively, when viewed from the right end. Find the magnitude and direction of the angular velocity of gear 2.

SOLUTION

There are two gears (2 and 5) that rotate about fixed axes in the system. As in the case of the previous example, we can write the angular velocity relationships for these gears using the chain rule for angular velocities. The equations are

$$ {}^1\omega_2 = {}^C\omega_2 + {}^1\omega_C \tag{10.23} $$

and

$$ {}^1\omega_5 = {}^C\omega_5 + {}^1\omega_C \tag{10.24} $$

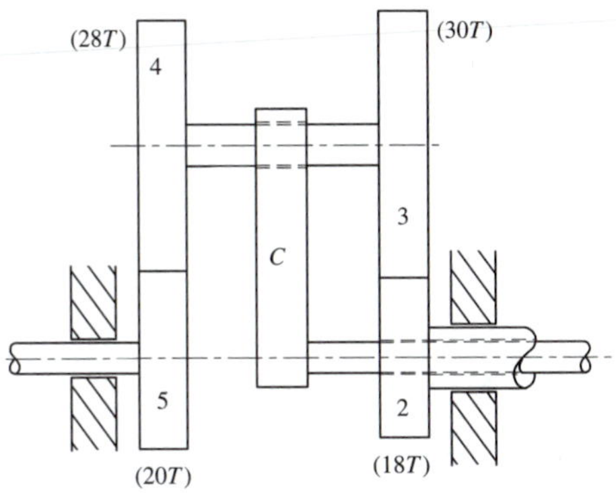

Figure 10.19 The planetary gear train for Example 10.4.

The angular velocity ratio of gears 2 and 5 relative to the carrier is

$$\frac{{}^C\omega_2}{{}^C\omega_5} = \frac{N_5 N_3}{N_4 N_2} \tag{10.25}$$

Notice that the velocity ratio is positive because both gears rotate in the same direction relative to the arm. Next solve Eqs. (10.23) and (10.24) for ${}^C\omega_2$ and ${}^C\omega_5$, respectively. Then,

$${}^C\omega_2 = {}^1\omega_2 - {}^1\omega_C \tag{10.26}$$

and

$${}^C\omega_5 = {}^1\omega_5 - {}^1\omega_C \tag{10.27}$$

Now divide Eq. (10.26) by (10.27) and equate the result to Eq. (10.25). This gives

$$\frac{{}^1\omega_2 - {}^1\omega_C}{{}^1\omega_5 - {}^1\omega_C} = \frac{N_5 N_3}{N_4 N_2} \tag{10.28}$$

Assuming CCW as positive, from the problem statement, ${}^1\omega_C = -150$ rpm and ${}^1\omega_4 = -50$ rpm. The tooth numbers are given in Fig. 10.19. Substituting the known values into Eq. (10.28) gives

$$\frac{{}^1\omega_2 + 150}{-50 + 150} = \frac{20}{28}\frac{30}{18}$$

or

$${}^1\omega_2 + 150 = 119.04$$

or

$${}^1\omega_2 = -30.95 \text{ rpm}$$

Therefore, the velocity of gear 2 is 30.95 rpm in the clockwise direction. ■

EXAMPLE 10.5 (*Analysis of Planetary Gear Train Using Equation Method*)

PROBLEM

Assume that gear 2 in Fig. 10.20 is driven at a speed of 60 rpm in the counterclockwise direction viewed from the right end. Gear 4 meshes with a fixed ring gear and with gear 5 as shown. Find the magnitude and direction of the angular velocity of gear 5.

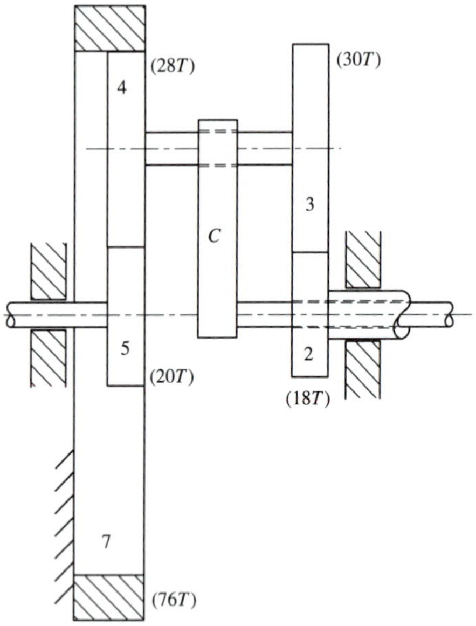

Figure 10.20 The planetary gear train for Example 10.5.

SOLUTION

There are three gears (2, 5, and 7) that can rotate about fixed axes in the system. We will include gear 7 in this list but ultimately will use the fact that its velocity is zero. As in Examples 10.3 and 10.4, we will solve the problem by writing relative velocity equations for all of the gears that have shafts that can rotate in fixed bearings. After rearranging, the resulting equations are

$$^{C}\omega_2 = {}^{1}\omega_2 - {}^{1}\omega_C \tag{10.29}$$

$$^{C}\omega_5 = {}^{1}\omega_5 - {}^{1}\omega_C \tag{10.30}$$

and

$$^{C}\omega_7 = {}^{1}\omega_7 - {}^{1}\omega_C \tag{10.31}$$

The angular velocity ratio of gears 2 and 5 relative to the carrier is

$$\frac{^{C}\omega_2}{^{C}\omega_5} = \frac{N_5 N_3}{N_4 N_2} \tag{10.32}$$

and that of gears 2 and 7 relative to the carrier is

$$\frac{^{C}\omega_2}{^{C}\omega_7} = -\frac{N_7 N_3}{N_4 N_2} \tag{10.33}$$

Now, divide Eq. (10.29) by Eq. (10.30) and equate the result with Eq. (10.32). This gives

$$\frac{^{1}\omega_2 - {}^{1}\omega_C}{^{1}\omega_5 - {}^{1}\omega_C} = \frac{N_5 N_3}{N_4 N_2} \tag{10.34}$$

Similarly, divide Eq. (10.29) by (10.31) and equate the result with Eq. (10.33) to get

$$\frac{^{1}\omega_2 - {}^{1}\omega_C}{^{1}\omega_7 - {}^{1}\omega_C} = -\frac{N_7 N_3}{N_4 N_2} \tag{10.35}$$

Equations (10.34) and (10.35) are the equations necessary for analyzing the planetary gear train. From the problem statement, we know that ${}^1\omega_2 = 60$ rpm and ${}^1\omega_7 = 0$. With these known values, only ${}^1\omega_C$ is unknown in Eq. (10.35). Substituting the known values into Eq. (10.35) gives

$$\frac{60 - {}^1\omega_C}{0 - {}^1\omega_C} = -\frac{76}{28}\frac{30}{18} = -4.5238$$

Solving gives

$${}^1\omega_C(1 + 4.5238) = 60 \Rightarrow {}^1\omega_C = 10.862 \text{ rpm}$$

Given ${}^1\omega_C$ and ${}^1\omega_2$, we can solve Eq. (10.34) for ${}^1\omega_5$. Substituting the known values into Eq. (10.34) gives

$$\frac{60 - 10.862}{{}^1\omega_5 - 10.862} = \frac{20}{28}\frac{30}{18} = 1.1905$$

Solving for ${}^1\omega_5$,

$${}^1\omega_5 - 10.862 = \frac{60 - 10.862}{1.1905} = 41.275 \Rightarrow {}^1\omega_5 = 52.137 \text{ rpm}$$

The value is positive, so ${}^1\omega_5$ is rotating counterclockwise when viewed from the right. ■

EXAMPLE 10.6 *(Analysis of Planetary Gear Trains in Series)*

PROBLEM

A two-stage planetary gear drive is represented in Fig. 10.21. Gear 2 is the input member, and carrier 7 is the output member. Gear 4 is a ring gear and is fixed. The carrier of the first stage is member 6, and it is rigidly connected to the gear that drives the second stage. Determine the velocity ratio of the gear drive.

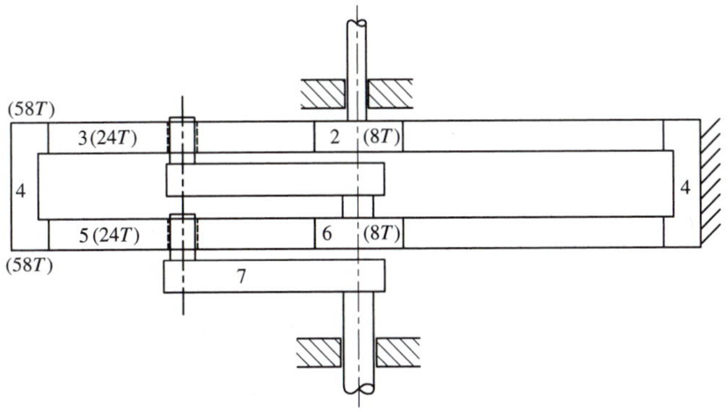

Figure 10.21 The planetary gear train for Example 10.6.

SOLUTION

There are three gears (2, 4, and 6) that can rotate about fixed axes in the system. Again, we will include the fixed ring gear in the equations and will set the velocity to zero once the equations are developed. As in the previous examples, we will solve the problem by writing relative velocity equations for all of the gears that have shafts that can rotate in fixed bearings. However, we must separate the two stages of the planetary drives when we write the equations. The first stage includes gears 2, 3, and 4 and the carrier is member 6. The second stage includes gears 6, 5, and 4 and carrier 7. The first stage can be analyzed independently of the second stage to determine the velocity of gear 6 in terms of the velocity of gear 2. The second stage can then be analyzed to determine the velocity of the second carrier (7) in terms of the velocity of gear 6. By combining the results of both stages, the velocity of link 7 can be determined as a function of gear 2 to determine the overall velocity ratio of the gear train.

After rearranging, the first-stage relative velocity equations are

$$^{6}\omega_2 = {}^{1}\omega_2 - {}^{1}\omega_6 \tag{10.36}$$

$$^{6}\omega_4 = {}^{1}\omega_4 - {}^{1}\omega_6 \tag{10.37}$$

The angular velocity ratio of gears 2 and 4 relative to the carrier (member 6) is

$$\frac{^{6}\omega_2}{^{6}\omega_4} = -\frac{N_4}{N_2} \tag{10.38}$$

Now, divide Eq. (10.36) by Eq. (10.37) and equate the result with Eq. (10.38). This gives

$$\frac{^{1}\omega_2 - {}^{1}\omega_6}{^{1}\omega_4 - {}^{1}\omega_6} = -\frac{N_4}{N_2} \tag{10.39}$$

Because $^{1}\omega_4 = 0$, this equation can be rewritten for $^{1}\omega_6$ as a function of $^{1}\omega_2$. The result is

$$^{1}\omega_6 = \frac{^{1}\omega_2}{1 + (N_4/N_2)} \tag{10.40}$$

We can now analyze the second stage in exactly the same manner as the first stage except that now the gears are 6, 5, and 4, and the carrier is 7. After rearranging, the second-stage relative velocity equations are

$$^{7}\omega_6 = {}^{1}\omega_6 - {}^{1}\omega_7 \tag{10.41}$$

$$^{7}\omega_4 = {}^{1}\omega_4 - {}^{1}\omega_7 \tag{10.42}$$

The angular velocity ratio of gears 4 and 6 relative to the carrier (member 7) is

$$\frac{^{7}\omega_6}{^{7}\omega_4} = -\frac{N_4}{N_6} \tag{10.43}$$

Now, divide Eq. (10.41) by Eq. (10.42) and equate the result with Eq. (10.43). This gives

$$\frac{^{1}\omega_6 - {}^{1}\omega_7}{^{1}\omega_4 - {}^{1}\omega_7} = -\frac{N_4}{N_6} \tag{10.44}$$

Because $^{1}\omega_4 = 0$, this equation can be rewritten for $^{1}\omega_7$ as a function of $^{1}\omega_6$. The result is

$$^{1}\omega_7 = \frac{^{1}\omega_6}{1 + (N_4/N_6)} \tag{10.45}$$

Combining Eqs. (10.45) and (10.40) and substituting the known tooth numbers gives

$$\frac{^{1}\omega_2}{^{1}\omega_7} = \left(1 + \frac{N_4}{N_2}\right)\left(1 + \frac{N_4}{N_6}\right) = \left(1 + \frac{58}{8}\right)\left(1 + \frac{58}{8}\right) = 68.06$$

Notice that in this example, the size of the planet gears does not affect the velocity ratio. They will have an impact on the size of the gear box, however. Also notice that members 2 and 7 both rotate in the same direction. ■

EXAMPLE 10.7 (*Analysis of Planetary Gear Train with Bevel Gears*)

PROBLEM

All of the planetary gear trains considered in the previous examples involved only parallel-shaft gears. Here we will analyze an example with bevel gears. The gear train is shown in

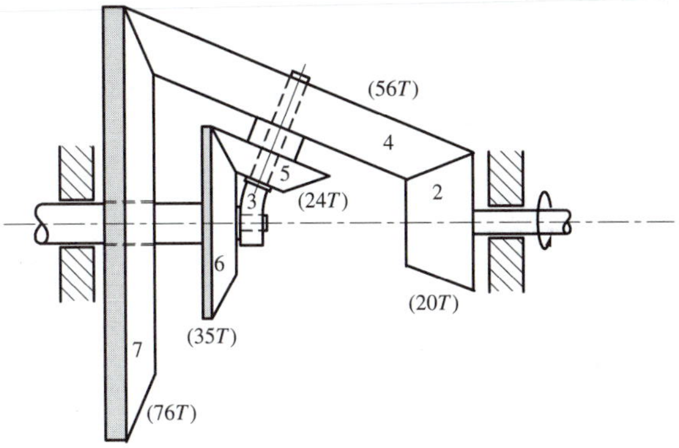

Figure 10.22 The planetary gear train for Example 10.7.

Fig. 10.22. The input to the gear train is gear 2 and the output is gear 6. The carrier is member 3, that rotates freely about the shaft on gear 6. Also, the compound planet gears (4 and 5) rotate about the axis of the carrier. Gear 7 is fixed to the frame. Assume that gear 2 rotates with an angular velocity of 100 rpm in the counterclockwise direction viewed from the right. Find the angular velocity of gear 7.

SOLUTION

Even though bevel gears are involved, we can approach the analysis in exactly the same way that was used for the parallel-shaft gears. There are three gears (2, 6, and 7) that can rotate about fixed axes in the system. Again, we will include gear 7 in this list but ultimately will use the fact that its velocity is zero. As in the previous examples, we will solve the problem by writing relative velocity equations for all of the gears that have shafts that can rotate in fixed bearings. After rearranging, the resulting equations are

$$^3\omega_2 = {}^1\omega_2 - {}^1\omega_3 \tag{10.46}$$

$$^3\omega_6 = {}^1\omega_6 - {}^1\omega_3 \tag{10.47}$$

and

$$^3\omega_7 = {}^1\omega_7 - {}^1\omega_3 \tag{10.48}$$

The angular velocity ratio of gears 2 and 6 relative to the carrier (3) is

$$\frac{^3\omega_2}{^3\omega_6} = -\frac{N_4 N_6}{N_2 N_5} \tag{10.49}$$

In Eq. (10.49) we must determine the direction by inspection. This will show that if the carrier is fixed and the motion is inverted so that all of the other links and gears (including 4) can move relative to the carrier, gears 2 and 6 will move in opposite directions. Similarly, the motion of gears 2 and 7 relative to the carrier is

$$\frac{^3\omega_2}{^3\omega_7} = -\frac{N_7}{N_2} \tag{10.50}$$

Again, relative to the carrier, gears 2 and 7 are seen to move in opposite directions. Now, divide Eq. (10.46) by Eq. (10.47) and equate the result with Eq. (10.49). This gives

$$\frac{^1\omega_2 - {}^1\omega_3}{^1\omega_6 - {}^1\omega_3} = -\frac{N_4 N_6}{N_2 N_5} \tag{10.51}$$

Also divide Eq. (10.46) by Eq. (10.48) and equate the result to Eq. (10.50). This gives

$$\frac{^1\omega_2 - {}^1\omega_3}{^1\omega_7 - {}^1\omega_3} = -\frac{N_7}{N_2} \tag{10.52}$$

Equations (10.51) and (10.52) are the equations necessary for analyzing the planetary gear train. From the problem statement, we know that ${}^1\omega_2 = 100$ and ${}^1\omega_7 = 0$. With these known values, only ${}^1\omega_3$ is unknown in Eq. (10.52). Substituting the known values into Eq. (10.52) gives

$$\frac{100 - {}^1\omega_3}{0 - {}^1\omega_3} = -\frac{76}{20} = -3.8$$

Solving gives

$${}^1\omega_3(1 + 3.8) = 100 \Rightarrow {}^1\omega_3 = 20.833 \text{ rpm}$$

Given ${}^1\omega_3$ and ${}^1\omega_2$, we can solve Eq. (10.51) for ${}^1\omega_6$. Substituting the known values into Eq. (10.51) gives

$$\frac{100 - 20.833}{{}^1\omega_6 - 20.833} = -\frac{56\ 35}{20\ 24} = -4.0833$$

Solving for ${}^1\omega_6$,

$${}^1\omega_6 = \frac{100 - 20.833}{-4.0833} + 20.833 = 1.444 \text{ rpm}$$

The value is positive, so ${}^1\omega_6$ is rotating in the same direction as ${}^1\omega_2$. Therefore, ${}^1\omega_6$ is rotating counterclockwise when viewed from the right. The overall velocity ratio for the gear box is

$$\frac{{}^1\omega_2}{{}^1\omega_6} = \frac{100}{1.444} = 69.2$$

■

10.5.3 Analysis of Planetary Gear Trains Using Tabular Method

10.5.3.1 Overview

The tabulation method is based on the knowledge that a planetary gear train is a linear system. The absolute angular velocity of any gear x that rotates about an axis fixed to the frame can be written as

$${}^1\omega_x = {}^C\omega_x + {}^1\omega_C \qquad \textbf{(10.53)}$$

where ${}^1\omega_C$ is the absolute angular velocity of the carrier, and ${}^C\omega_x$ is the angular velocity of the gear relative to the carrier. Also, because Eq. (10.53) is linear, we can multiply the input values by a constant, and the output value will be multiplied by the same constant. The tabular method is based on the idea of the linear relationship shown in Eq. (10.53) and superposition.

A simple planetary gear train is fundamentally a two-degree-of-freedom device. Therefore, we must specify two input velocities or displacements to compute the unknown displacement. For discussion, consider the planetary gear train shown in Fig. 10.23. There

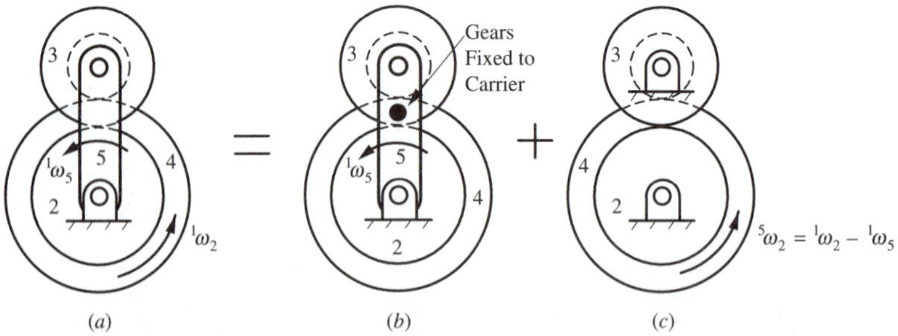

Figure 10.23 Analysis of a planetary drive using superposition.

is a compound gear for the planets and two sun gears, and the members are each assigned a number. Assume that the carrier is one of the input members, gear 2 is the other input member, and the known velocities are ${}^1\omega_2$ and ${}^1\omega_5$, respectively.

Now let us fix all of the gears to the carrier and rotate all of the gears by the velocity ${}^1\omega_5$. Then all of the gears, including gear 2, will have an initial angular velocity of ${}^1\omega_5$. To correct the angular velocity of gear 2 without changing the angular velocity of the carrier, let us fix the angular velocity of the carrier and move all of the gears relative to the carrier such that angular velocity of gear 2 ends up with the correct value when added to the velocity (${}^1\omega_5$) from step 1. This will also change all of the angular velocities in the gear train. The vector sum of the angular velocities from steps 1 and 2 for each gear will give the correct values. The vector sum is simply an algebraic sum because all of the gears have parallel shafts. The procedure, shown schematically in Fig. 10.23, is formalized in the tabular method of analysis as follows.

10.5.3.2 Procedure

The tabulation method begins with a table in which there is one column for each member in the gear train and a row for each of the following three steps.

1. Assume that all of the gears are locked to the carrier, and rotate the assembly with an angular velocity equal to the angular velocity of the carrier member. Tabulate this velocity under each member in the train.
2. Fix the arm, and rotate the second input member such that it ends up with the proper input velocity when steps 1 and 2 are added together. Tabulate the resulting velocity for each member in the train.
3. Add the results from steps 1 and 2 for each member in the train.

We will illustrate the procedure on three examples.

EXAMPLE 10.8 (*Tabulation Method for Simple Planetary Gear Train*)

PROBLEM

Assume that the planetary gear train in Fig. 10.24 has the frame as member 1, the sun gear as member 2, the planet gear as member 3, the ring gear as member 4, and the carrier as member 5. Gear 2 is the input and rotates counterclockwise with an angular velocity of 100 rpm, and the carrier rotates counterclockwise with an angular velocity of 200 rpm. Find the angular velocities of gears 3 and 4.

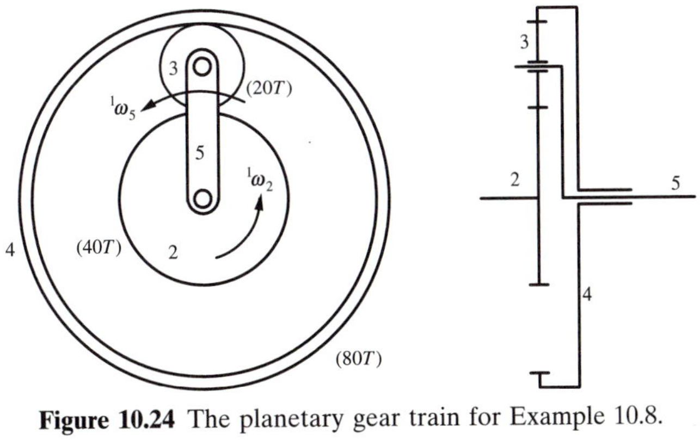

Figure 10.24 The planetary gear train for Example 10.8.

Table 10.1 Results for Example 10.8, rpm

Step	Carrier 5	Gear 2	Gear 3	Gear 4
1. Gears locked	200	200	200	200
2. Carrier fixed	0	−100	200	50
3. Total	200	100	400	250

SOLUTION

Following the procedure given above, the solution table is shown in Table 10.1. The four members of the system are associated with the columns, and the steps are associated with the rows. Step 1 of the procedure is to rotate the entire assembly by +200 rpm, which is the velocity of the carrier. Next, we will fix the carrier and rotate gear 2 by the velocity required to make its total angular velocity 100 rpm. To do this, we need to rotate gear 2 by −100 rpm relative to the arm.

When gear 2 rotates −100 rpm relative to the carrier, gears 3 and 4 will also rotate. The angular velocity of gear 3 will be

$$^1\omega_3 = -\frac{N_2}{N_3}\,^1\omega_2 = -\frac{40}{20}(-100) = +200$$

and

$$^1\omega_4 = -\frac{N_2}{N_4}\,^1\omega_2 = -\frac{40}{80}(-100) = +50$$

These values are entered into Table 10.1. The results are obtained by adding the results from the first two steps. From this, it is apparent that gear 3 rotates 400 rpm and gear 4 rotates 250 rpm, both in the counterclockwise (+) direction. ■

EXAMPLE 10.9 ***(Solution to Example 10.7 Using Tabulation Method)***

PROBLEM

Analyze the planetary gear train in Example 10.7 using the tabulation method.

SOLUTION

The problem is to determine the velocity ratio for the gear train, where the velocity ratio is defined by $^1\omega_2/^1\omega_6$. If we set $^1\omega_6 = 1$ and solve for the velocity of $^1\omega_2$, the velocity ratio will be given directly by the value for $^1\omega_2$. From the problem statement, we also know that $^1\omega_7 = 0$.

In the table, we will include only gears that rotate about axes that have bearings fixed to the frame. The planet gears rotate about the arm axis, which is skewed relative to the axis of the other gears. Therefore, the angular velocity of the planets is not obtained by a simple algebraic addition of the values from steps 1 and 2.

In the tabulation procedure, we assume that the velocity of the carrier is known. However, in this problem, the velocities of gears 6 and 7 are known. Therefore, we must treat the velocity of the carrier as unknown and solve for it. For step 1, assume that the gears are locked to the carrier and the assembly is turned by $+x$ rpm. This is shown in Table 10.2. In step 2, we must rotate gear 7 such that when the results of the first two steps are added, the final velocity for gear 7 will be zero. Clearly, then, gear 7 must be rotated by $-x$ rpm relative to the carrier. The remaining values in the second row of the table are determined by analyzing the gear train relative to the carrier.

Next sum the results from steps 1 and 2. From the problem statement, we know that $^1\omega_6 = 1$, and in the table, we can see that the velocity of gear 6 is also given by

$$^1\omega_6 = x(0.0694) = 1$$

Table 10.2 Results for Example 10.9, rpm

Step	Carrier 3	Gear 2	Gear 6	Gear 7
1. Gears locked	x	x	x	x
2. Carrier fixed	0	$x\frac{N_7}{N_2} = x\frac{76}{20}$	$-x\frac{N_7 N_5}{N_4 N_6} = -x\frac{76}{56}\frac{24}{35}$	$-x$
3. Total	x	$x\left[1 + \frac{76}{20}\right] = x(4.8)$	$x\left[1 - \frac{76}{56}\frac{24}{35}\right] = x(0.0694)$	0

Therefore,

$$x = 14.412$$

and

$$^1\omega_2 = x(4.8) = 14.412(4.8) = 69.2$$

Therefore the velocity ratio for the planetary gear train is

$$^1\omega_2/^1\omega_6 = 69.2$$

which is the same as that computed in Example 10.7. ■

EXAMPLE 10.10 (*Solution to Ferguson's Paradox Using Tabulation Method*)

PROBLEM

This is an interesting application of planetary gearing, and the planetary gear system employed is called Ferguson's paradox.[3] The mechanism is shown in Fig. 10.25. Gear 2 is fixed to the frame, and gear 5 is a planet that rotates relative to the carrier, which is member 6. Gears 2, 3, and 4 have tooth numbers 99, 100, and 101, respectively. All of the gears are cut from the same blank so that they will all mesh with gear 5, which has 20 teeth. If the arm makes 100 revolutions CCW, determine the number of revolutions made by gears 3, 4, and 5.

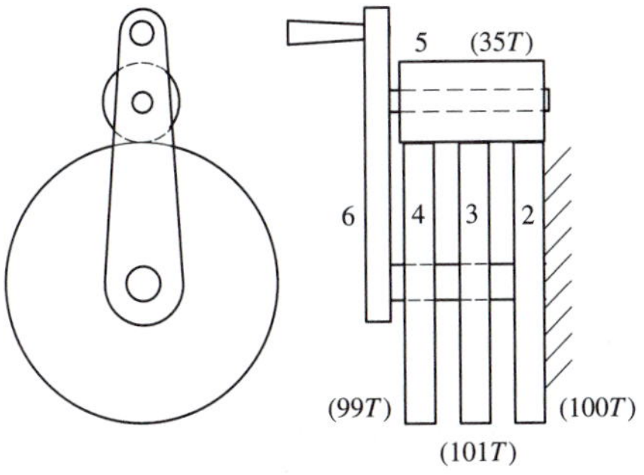

Figure 10.25 The planetary gear train for Example 10.10.

SOLUTION

Based on the problem statement, we know that the carrier moves 100 revolutions and gear 2 is fixed. As in the previous examples, the first step is to fix all of the gears to the arm and rotate the assembly by 100 turns. This is shown in Table 10.3. In step 2, we must rotate gear 2 such that when the results of the first two steps are added, the final velocity for gear 2 is zero. Therefore, gear 2 must be rotated by −100 rpm relative to the carrier. The remaining values in the second row of the table are determined by analyzing the gear train relative to the carrier.

Next sum the results from steps 1 and 2. The number of turns made by each gear in the mechanism is shown in the third row of Table 10.3. Notice that gear 3 makes 1 revolution in the direction of the motion of the carrier while gear 4 makes 1 revolution in the opposite

Table 10.3 Results for Example 10.10, revolutions

Step	Carrier 6	Gear 2	Gear 3	Gear 4	Gear 5
1. Gears locked	100	100	100	100	100
2. Carrier fixed	0	−100	−100/1.01	−100/0.99	−100/35
3. Total	100	0	1	−1	97.14

direction. Gear 2 is fixed by design. Therefore, as the carrier is turned, gear 3 will rotate very slowly in the direction of the carrier and gear 4 will rotate very slowly in the opposite direction. ■

10.6 REFERENCES

1. Merritt, H. E., *Gear Engineering,* John Wiley & Sons, New York (1971).
2. Mabie, H. H., and C. F. Reinholtz, *Mechanisms and Dynamics of Machinery,* 4th ed. John Wiley & Sons, New York (1987).
3. Shigley, J. E., *Kinematic Analysis of Mechanisms,* McGraw-Hill Book Co., New York (1975).

10.7 CHAPTER 10 Exercise Problems

PROBLEM 10.1

Find the angular velocity of gear 8 if the angular velocity of gear 2 is 800 rpm in the direction shown.

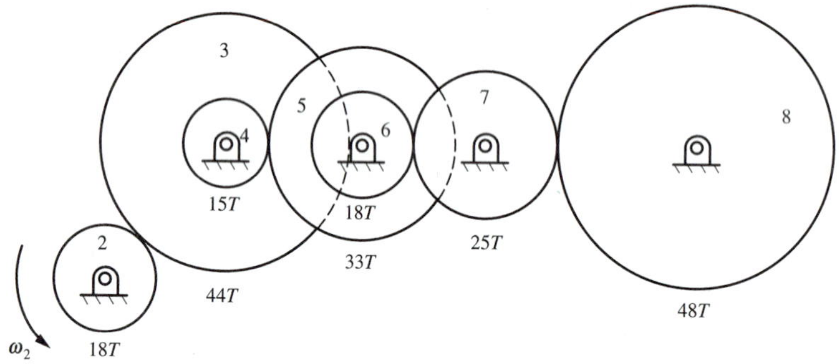

PROBLEM 10.2

Find the velocity of gear 8 in Problem 10.1 if the angular velocity of the driver (gear 2) is 300 rpm in the clockwise direction.

PROBLEM 10.3

The gear train given is for a machine tool. Power is input to the gear box through the pulley indicated, and the output power to the machine table is through gear 13. Gears 2 and 3, 4 and 5, and 11 and 12 are compound gears that can move axially on splined shafts to mesh with various different gears so that various combinations of overall gear ratios (ω_{13}/ω_2) can be produced. Determine the number of ratios possible and the overall gear ratio for each possibility.

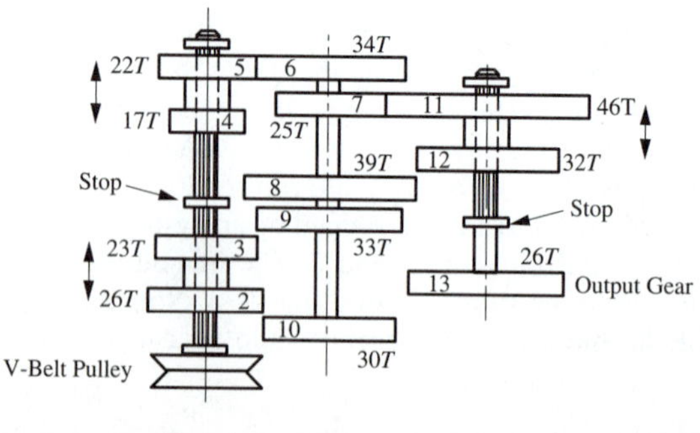

PROBLEM 10.4

A simple three-speed transmission is shown. The power flow is as follows: (a) First gear: gear 4 is shifted to mesh with gear 7; power flows through gears 2, 5, 7, 4. (b) Second gear: gear 3 is shifted to mesh with gear 6; power flows through gears 2, 5, 6, 3. (c) Third gear: gear 3 is shifted so that the clutch teeth on gear 3 mesh with those on gear 2; a direct drive results. (d) Reverse gear: gear 4 is shifted to mesh with gear 9; power flows through gears 2, 5, 8, 9, 4. An automobile with this transmission has a differential ratio of 3:1 and a tire outside diameter of 24 in. Determine the engine speed for the car under the following conditions: (i) first gear and the automobile is traveling at 15 mph; (ii) third gear and the automobile is traveling at 55 mph; (iii) reverse gear and the automobile is traveling at 3.5 mph.

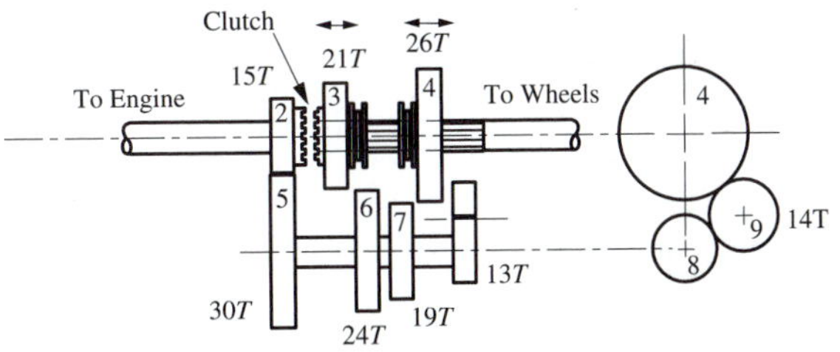

PROBLEM 10.5

Part of the gear train for a machine tool is shown. Compound gears 2 and 3 slide on a splined shaft so that gear 3 can mesh with gear 4 or gear 2 can mesh with gear 6. Also, compound gears 14 and 15 slide on a splined shaft so that gear 14 can mesh with gear 16 or gear 15 can mesh with gear 17. (a) If gear 3 meshes with gear 4, what are the two possible spindle speeds for a motor speed of 1800 rpm? (b) Now assume that gear 14 meshes with gear 16, and gear 2 meshes with gear 6. Gears 2, 3, 4, and 6 are standard and have the same diametrical pitch. What are the tooth numbers on gears 2 and 6 if the spindle speed is 130 ± 3 rpm?

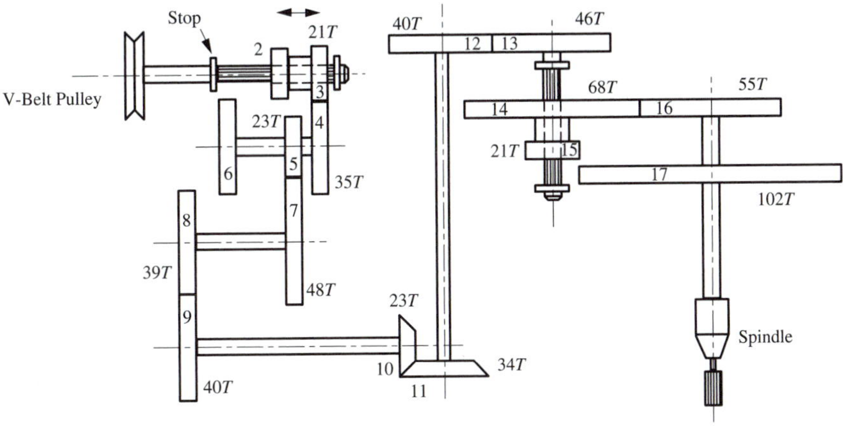

PROBLEM 10.6

An internal gear having 160 teeth and rotating counterclockwise at 30 rpm is connected through a gear train to an external gear, which rotates at 120 rpm in the counterclockwise direction. Using the minimum number of gears, select gears from the following list that will satisfy the design requirements. Tooth numbers for the available gears are 20, 22, 25, 30, 32, 34, 35, 40, 50, 55, 60, and 64. There is only one gear with each tooth number, and each gear has the same diametral pitch.

PROBLEM 10.7

Resolve Problem 10.6 if the external gear is concentric with the internal gear (the rotation axis is the same for both gears) and the external gear rotates clockwise.

PROBLEM 10.8

Resolve Problem 10.6 if the external gear is concentric with the internal gear and the external gear rotates counterclockwise.

PROBLEM 10.9 Resolve Problem 10.6 if the external gear rotates at 50 rpm.

PROBLEM 10.10 A gear reducer is to be designed as shown in the figure. Determine the diametral pitch and number of teeth on gears 4 and 5 if the speed of gear 2 (ω_2) is to be 10 times the speed of gear 5 (ω_5). The pitches of the two gears should be as nearly equal as possible, and no gear should have fewer than 15 teeth.

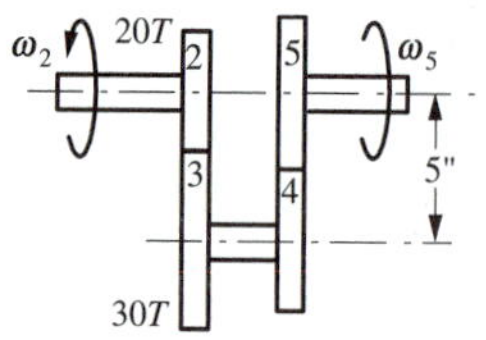

PROBLEM 10.11 Resolve Problem 10.10 if ω_2 is to be 8 times the speed of gear 5 (ω_5).

PROBLEM 10.12 Resolve Problem 10.10 if ω_2 is to be 6.5 times the speed of gears (ω_5).

PROBLEM 10.13 The gear train shown is a candidate for the spindle drive of a gear hobbing machine. The gear blank and the worm gear (gear 10) are mounted on the same shaft and rotate together. If the gear blank is to be driven clockwise, determine the hand of the hob. Also determine the velocity ratio (ω_8/ω_6) to cut 72 teeth on the gear blank.

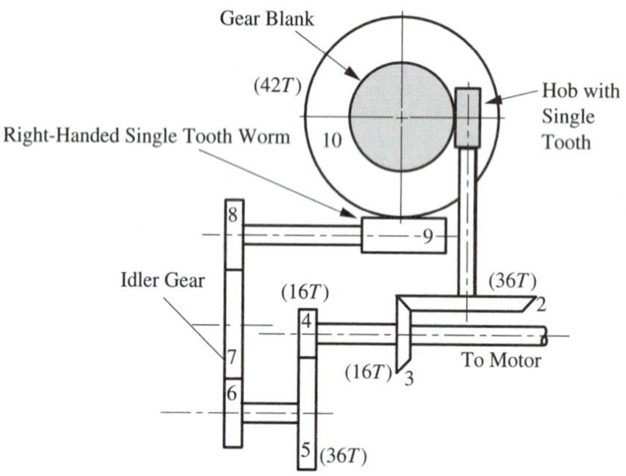

PROBLEM 10.14 Assume that the input shaft of a transmission rotates clockwise at 1800 rpm. The output shaft is driven at 160 rpm in the counterclockwise direction. None of the gears in the transmission is to be an idler, and the gear ratio at any given mesh is not to exceed 3 : 1. Gears are available that have all tooth numbers between 13 and 85; however, only one gear is available with each tooth number. Select the appropriate gears for the transmission, and sketch the configuration designed. Label the gears and tooth numbers.

PROBLEM 10.15 Resolve Problem 10.14 if the output shaft rotates at 210 rpm in the counterclockwise direction.

PROBLEM 10.16 Resolve Problem 10.14 if the output shaft rotates at 200 rpm in the *clockwise* direction.

PROBLEM 10.17 A simple gear reduction is to be used to generate a gear ratio equal to π. Make up a table of possible gear ratios where the maximum number of teeth on either gear is 100. This can be conveniently done using a simple computer program. Identify the gear set that most closely approximates the desired ratio. What is the error?

PROBLEM 10.18 A simple gear reduction is to be used to generate the gear ratio 0.467927. Make up a table of possible gear ratios where the maximum number of teeth on either gear is 100. Identify the gear set that most closely approximates the desired ratio. What is the error?

PROBLEM 10.19 A simple gear reduction is to be used to generate a gear ratio equal to $\sqrt{2}$. Make up a table of possible gear ratios where the maximum number of teeth on either gear is 100. Identify the gear set that most closely approximates the desired ratio. What is the error?

PROBLEM 10.20 In the gear train shown, gears 3 and 4 are integral. Gear 3 meshes with gear 2, and gear 4 meshes with gear 5. If gear 2 is fixed and $\omega_2 = 100$ rpm counterclockwise, determine the angular velocity of the carrier.

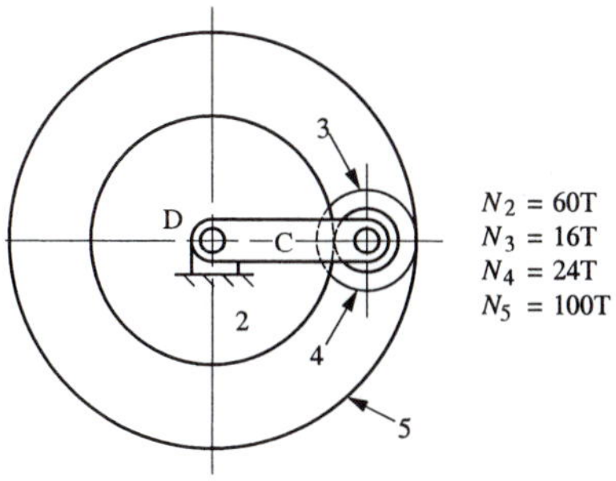

PROBLEM 10.21 Resolve Problem 10.20 if gear 5 is fixed and $\omega_5 = 100$ rpm counterclockwise.

PROBLEM 10.22 Resolve Problem 10.20 when $N_2 = 70T$, $N_3 = 35T$, $N_4 = 15T$, and $N_5 = 120$.

PROBLEM 10.23 In the figure given, axis *y-y* is fixed while axes *x-x* and z-z move with the arm. Gear 7 is fixed to the carrier. Gears 3 and 4, 5 and 6, and 8 and 9 are fixed together, respectively. Gears 3 and 4 move with planetary motion. If the tooth numbers are $N_2 = 16T$, $N_3 = 20T$, $N_4 = 22T$, $N_5 = 14T$, $N_6 = 15T$, $N_7 = 36T$, $N_8 = 20T$, $N_9 = 41T$, and $N_{10} = 97T$, determine the speed and direction of the output shaft.

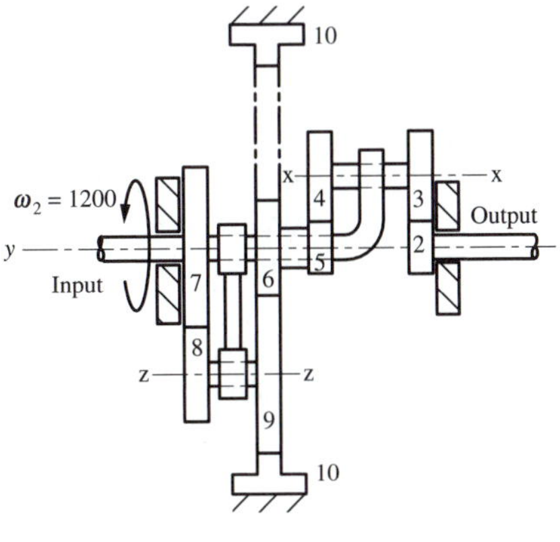

PROBLEM 10.24 Resolve Problem 10.23 when $N_2 = 16T$, $N_3 = 20T$, $N_4 = 16T$, $N_5 = 20T$, $N_6 = 15T$, $N_7 = 40T$, $N_8 = 18T$, $N_9 = 60T$, and $N_{10} = 135T$.

PROBLEM 10.25 Resolve Problem 10.23 when $N_2 = 14T$, $N_3 = 30T$, $N_4 = 14T$, $N_5 = 30T$, $N_6 = 15T$, $N_7 = 60T$, $N_8 = 15T$, $N_9 = 60T$, and $N_{10} = 135T$.

PROBLEM 10.26 In the gear train shown, gears 2 and 4, 6 and 7, and 3 and 9 are fixed together. If the angular velocity of the carrier is given, determine the angular velocity of gear 9.

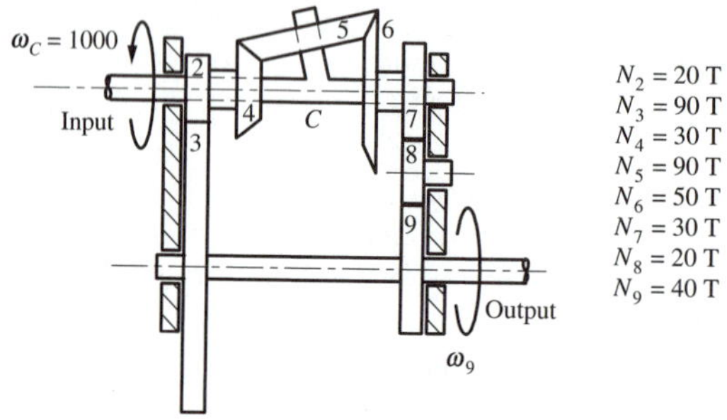

PROBLEM 10.27

Resolve Problem 10.26 if $N_2 = 10$T, $N_3 = 100$T, $N_7 = 20$T, $N_8 = 10$T, and $N_9 = 50$T.

PROBLEM 10.28

Resolve Problem 10.26 but assume that the shaft connecting gears 3 and 9 is the input shaft and the shaft of the carrier is the output shaft. Assume $\omega_9 = 500$ rpm counterclockwise and compute ω_C.

PROBLEM 10.29

The differential for a rear wheel–driven vehicle is shown schematically. If the drive shaft turns at 900 rpm, what is the speed of the vehicle if neither wheel slips and the outside diameter of the wheels is 24 in?

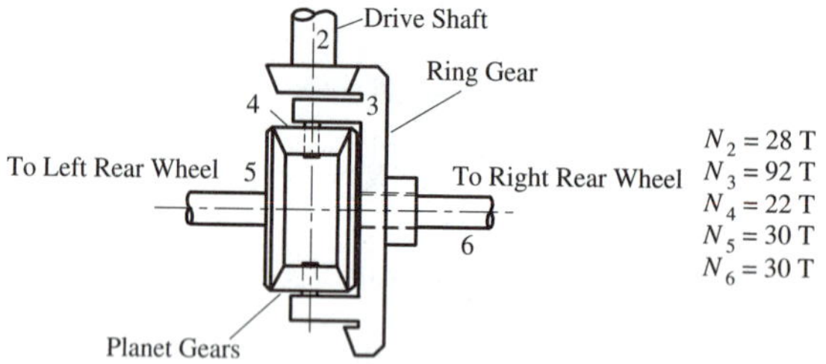

PROBLEM 10.30

Assume that the vehicle in Problem 10.29 is stopped so that the right wheel sits on a small icy patch and can spin freely while the left wheel does not spin. Determine the angular velocity of the right wheel if the angular speed of the drive shaft is 500 rpm.

PROBLEM 10.31

Assume that the vehicle in Problem 10.29 is traveling at 35 mph and turns around a curve with a radius of 50 ft from the centerline of the vehicle. The center-to-center distance between the treads of the right and left wheels is 60 in. Compute the rotational speed of each rear wheel, the rotational speed of the ring gear, and the rotational speed of the drive shaft.

PROBLEM 10.32

In the mechanism shown, derive an expression for the angular velocity of gear 7 (ω_7) in terms of ω_2 and ω_5 and the tooth numbers N_2, N_3, N_4, N_5, N_6, and N_8. Take counterclockwise viewing from the left as positive for the rotation of gears 2, 3, 4, 5, and 7. Viewed from the front of the page, take counterclockwise as the positive direction for gear 7.

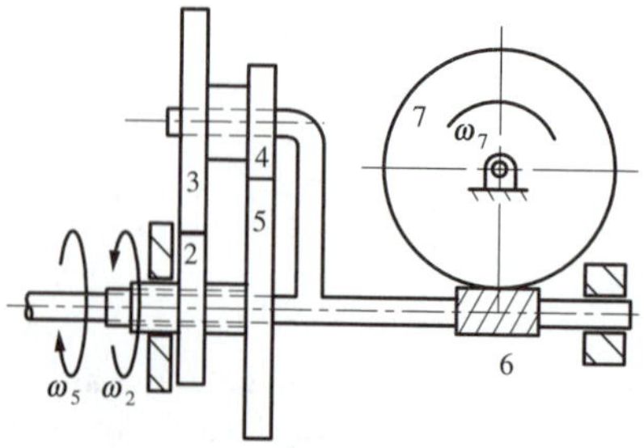

PROBLEM 10.33

In Problem 10.32, assume that $\omega_2 = 100$ rpm, $\omega_5 = -60$ rpm, $N_2 = 40T$, $N_3 = 60T$, $N_4 = 30T$, $N_5 = 70T$, $N_6 = 8T$, and $N_7 = 50T$. Determine the angular velocity of both gears 6 and 7.

PROBLEM 10.34[1]

The figure shows a schematic diagram of a semiautomatic transmission from the Model-T automobile. This was the forerunner of today's automatic transmission. A plate clutch, two banded clutches, and a system of pedals and levers (used to engage and disengage these plate and band clutches) operated in the proper sequence is shown in the table below. Determine the output/input speed ratio for each condition.

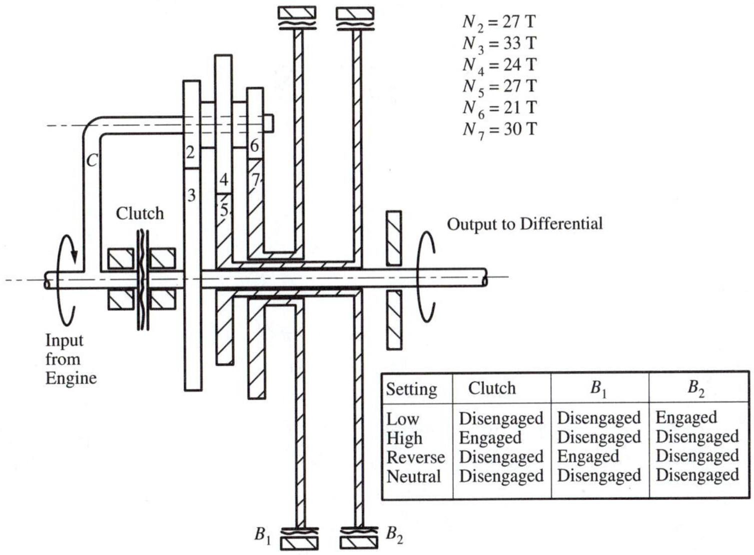

Setting	Clutch	B_1	B_2
Low	Disengaged	Disengaged	Engaged
High	Engaged	Disengaged	Disengaged
Reverse	Disengaged	Engaged	Disengaged
Neutral	Disengaged	Disengaged	Disengaged

PROBLEM 10.35

In Problem 10.34, if the engine rotates at 400 rpm, determine the angular velocity of gear 5 when the transmission is in low gear.

[1] Problem courtesy of Dr. Michael Stanisic, Notre Dame University.

Chapter 11

Static Force Analysis of Mechanisms

11.1 INTRODUCTION

Machines are used to apply mechanical force, energy, or power for useful purposes. So far, in this book, we have been concerned only with motion. However, mechanisms are used to transmit and apply force, as well as to generate desired motions. In many applications, the motion is per se unimportant. What is important is the application of force. The simplest and original "machines" transmitted force from an input location to an output location while magnifying or diminishing it. Examples are the first- and second-order levers, wedges, and pulley mechanisms. These are usually regarded as static machines. Many other forms of machine can be analyzed by the methods of statics because they function without motion or because the velocities of their motions are small enough that dynamic effects can be neglected. Examples are many types of clamps, pliers, and cutters; jacks, winches, and other heavy lifting devices; and many kinds of latches and toggles. Further, many machines require structural supports such as brackets, beams, frames, and trusses.

The design analysis of these mechanisms and their associated structures depends on static force analysis. In many other cases, mechanisms must deliver forces of controlled magnitude while generating a specified motion. If the motion is sufficiently slow for inertial forces to be neglected in comparison with the applied loads, the techniques of static force analysis can be applied. Further, even when inertial forces must be included, in the common case in which the motion can be considered to be known, D'Alembert's principle can be used to convert dynamic force analysis problems into the forms of static force analysis problems. This will be discussed in detail in Chapter 12.

The techniques used are fundamentally similar to those taught in introductory engineering mechanics courses. However, mechanisms and machines commonly have a relatively large number of members, some of which may have relatively complex geometries. Also, there are some techniques that lead to especially efficient solutions for some kinds of static machine problems. The basis of all static analysis is Newton's third law, as embodied in the concept of static equilibrium. The material in this chapter is composed of systematic ways to apply static equilibrium so that correct solutions to relatively complex problems can be reliably generated.

A further powerful motivation for pursuing static or dynamic force analysis is that conversion of a kinematic design into a real, physical mechanism design requires consideration of the loads on components and the stresses and deflections of those components. The loads on each member of a machine are usually of great interest to the machine designer because it is the engineer's responsibility to select the material to be used and to size the component so that it can safely resist those loads. That is, computation of the complete set of loads acting on a member in any critical situation provides essential initial data for a stress or deflection analysis of that member. Up to now in this book, we have usually indicated members by lines or simple polygons. It must always be remembered that this is a geometric convenience and that the lines and polygons represent physical objects with additional dimensions made of real, engineering materials. Analysis of the worst-case stresses or deflections and application of a relevant failure theory allow selection of materials and dimensions with the assurance that the part will withstand that loading condition without failing.

11.2 FORCES, MOMENTS, AND COUPLES

A force is a vector that has a definite line of action on a given link of the mechanism but not necessarily a definite point of application. If $\boldsymbol{i}, \boldsymbol{j}, \boldsymbol{k}$ are unit vectors respectively parallel to the x, y, and z axes of a Cartesian reference frame, the components of $\boldsymbol{F}$ in the directions of those axes are given by the dot products:

$$F_x = \boldsymbol{F}\cdot\boldsymbol{i}, \qquad F_y = \boldsymbol{F}\cdot\boldsymbol{j}, \qquad F_z = \boldsymbol{F}\cdot\boldsymbol{k} \tag{11.1}$$

and

$$\boldsymbol{F} = F_x\boldsymbol{i} + F_y\boldsymbol{j} + F_z\boldsymbol{k}$$

Forces may be internally applied forces such as a force due to a gas acting on a piston or body forces such as a weight or a magnetic force.

Two forces that have intersecting lines of action can be summed into a single, equivalent force as shown in Fig. 11.1. The resultant force will act along a line that passes through the point of intersection and lies in the plane defined by the two force vectors.

If two forces are equal and opposite but not collinear, the two forces cannot be resolved into a single force. However, the vector sum of the two equal and opposite forces will be zero. The forces constitute a couple as shown in Fig. 11.2. The moment of the couple is defined by

$$\boldsymbol{M} = \boldsymbol{r}\times\boldsymbol{F} = \begin{vmatrix} \boldsymbol{i} & \boldsymbol{j} & \boldsymbol{k} \\ r_x & r_y & r_z \\ F_x & F_y & F_z \end{vmatrix} = (r_yF_z - r_zF_y)\boldsymbol{i} + (r_zF_x - r_xF_z)\boldsymbol{j} + (r_xF_y - r_yF_x)\boldsymbol{k} \tag{11.2}$$

where

$$\boldsymbol{r} = r_x\boldsymbol{i} + r_y\boldsymbol{j} + r_z\boldsymbol{k}$$

The moment of a couple is independent of the point of application. It is a free vector that can be assumed to be applied anywhere on the body of interest. Also, the magnitude and direction are independent of how $\boldsymbol{r}$ is chosen. Two couples are equal if their moments [$\boldsymbol{M}$ vector in Eq. (11.2)] are equal.

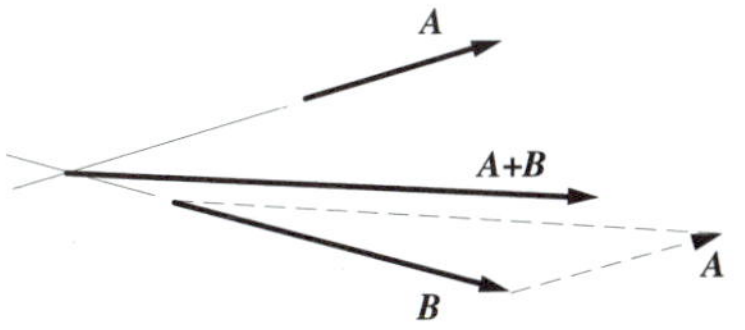

Figure 11.1 The resultant ($\boldsymbol{A} + \boldsymbol{B}$) is equivalent to the forces $\boldsymbol{A}$ and $\boldsymbol{B}$ acting simultaneously.

Figure 11.2 Two equal and opposite but noncollinear forces form a couple.

The moment generated by a single force about a point as shown in Fig. 11.3 is also given by Eq. (11.2). It is sometimes convenient to represent the moment in terms of the normal vector, $\boldsymbol{r}_n$, from the point to the force. Then,

$$\boldsymbol{M} = \boldsymbol{r}_n \times \boldsymbol{F} \tag{11.3}$$

When this is done, we can compute the moment magnitude directly from the magnitudes of $\boldsymbol{r}_n$ and $\boldsymbol{F}$. That is, $M = r_n(F)$. When this procedure is done, we must determine the direction of the moment by inspection.

The moment will be perpendicular to both $\boldsymbol{F}$ and $\boldsymbol{r}$ or $\boldsymbol{r}_n$. In planar problems, the moment will be normal to the plane containing $\boldsymbol{F}$ and $\boldsymbol{r}$. In this chapter, we will emphasize planar problems. Notice that the moment of a force about a point is very much dependent upon the location of the point.

If a force and a couple ($\boldsymbol{F}$ and $\boldsymbol{M}$) are applied to a rigid body, the system can be replaced by a single force such that the force will have the same effect on the system as the original force and couple. The new force vector will be equal to the original force vector but offset relative to the line of action of the original force vector by the normal distance $\boldsymbol{h}$. Here $\boldsymbol{h}$ is computed in such a way that $\boldsymbol{M} = \boldsymbol{h} \times \boldsymbol{F}$. Therefore,

$$|\boldsymbol{h}| = \frac{|\boldsymbol{M}|}{|\boldsymbol{F}|} \tag{11.4}$$

The vector $\boldsymbol{h}$ can be drawn from either side of the line of action of $\boldsymbol{F}$. However, the proper side of the line of action is the one that will make the sign of $\boldsymbol{h} \times \boldsymbol{F}$ the same as the sign of $\boldsymbol{M}$. An example is shown in Fig. 11.4. This procedure is used extensively when graphical analyses are conducted.

11.3 STATIC EQUILIBRIUM

A direct consequence of Newton's first two laws of motion is that, if a body is at rest, the sum of all forces acting on the body must be zero. Further, the sum of the moments of those forces about any point must also be zero. Thus, for any member of a structure:

$$\sum \boldsymbol{F} = 0 \tag{11.5}$$

$$\sum \boldsymbol{M}_0 = 0 \tag{11.6}$$

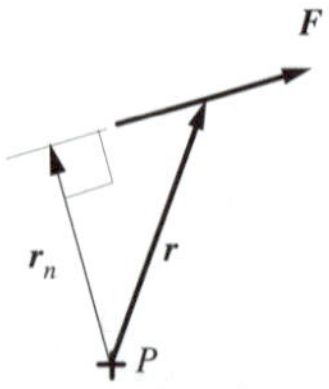

Figure 11.3 The moment of a force about a point.

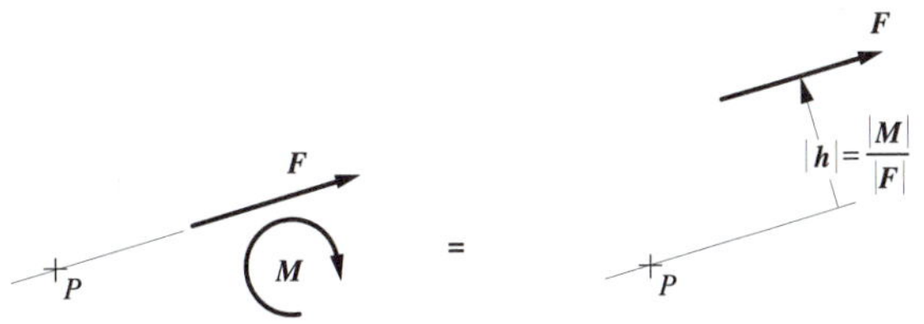

Figure 11.4 Replacing a force-moment system by a single force that is offset by the distance h.

where $\sum \boldsymbol{F}$ is the vector sum of all forces acting on the body, and $\sum \boldsymbol{M}_0$ is the vector sum of the moments of those forces about any chosen point, O. Equations (11.5) and (11.6) are called the equations of *static equilibrium.*

The basis of the static force analysis of any structure is the algebraic solution of the static equilibrium equations written for every member in the system. Equations (11.5) and (11.6) are typically associated with free-body diagrams. When the equations are written, they include all of the forces and moments associated with a given free-body diagram.

11.4 FREE-BODY DIAGRAMS

The first step in any static analysis of a mechanism is the construction or, more usually, sketching of the free-body diagram of each member of the mechanism. All forces acting on each member, including the forces of action and reaction between members, as well as externally applied loads, must be indicated on the free-body diagram.

A free-body diagram is a bookkeeping device to ensure that all relevant force components are included in the static equilibrium equations. The omission of *any* force component will lead to totally incorrect results. Experience has shown that disregard of the step of drawing free-body diagrams almost always results in incorrect equations and results. As a consequence, experienced machine designers never skip this step, although students often do!

The forces that can be transmitted across an ideal (frictionless) kinematic joint are related to the motions permitted by that joint. Basically, the work done by the transmitted forces in the directions of permitted motion must be zero. For example, a revolute joint permits rotation about its axis. Any force that is normal to that axis and whose line of action intersects it does no rotational work. Therefore, any force component in the plane of motion passing through the joint axis is transmitted. It is usually convenient to represent this set of possible forces by two components parallel to the x and y axis directions of the fixed reference frame, as shown in Figure 11.5a. In accordance with Newton's third law, the force components transmitted to one of the bodies joined from the other are equal and opposite to the force components received by the second body from the first. Therefore, when the free-bodies are drawn, these force components appear in equal and opposite pairs.

Similarly, a prismatic joint permits linear motion in one direction. A force normal to that direction does no work in the direction of motion. Also, a torque in the direction normal to the plane of motion will do no work in the direction of permitted motion. Therefore, these components will be transmitted by the prismatic joint, as indicated in Fig. 11.5b. Once again, the transmitted force and torque components will appear as equal and opposite pairs when the free bodies are drawn.

A rolling contact joint is similar to a revolute joint in permitting only pure rotation about the point of contact. The difference is that the axis of a revolute joint is fixed relative to each of the bodies that it joins, whereas the point of rolling contact migrates along the profiles of both bodies. The transmitted force system is also similar to that of a revolute joint, as indicated in Fig. 11.5c. A rolling and sliding contact permits both rotation in the

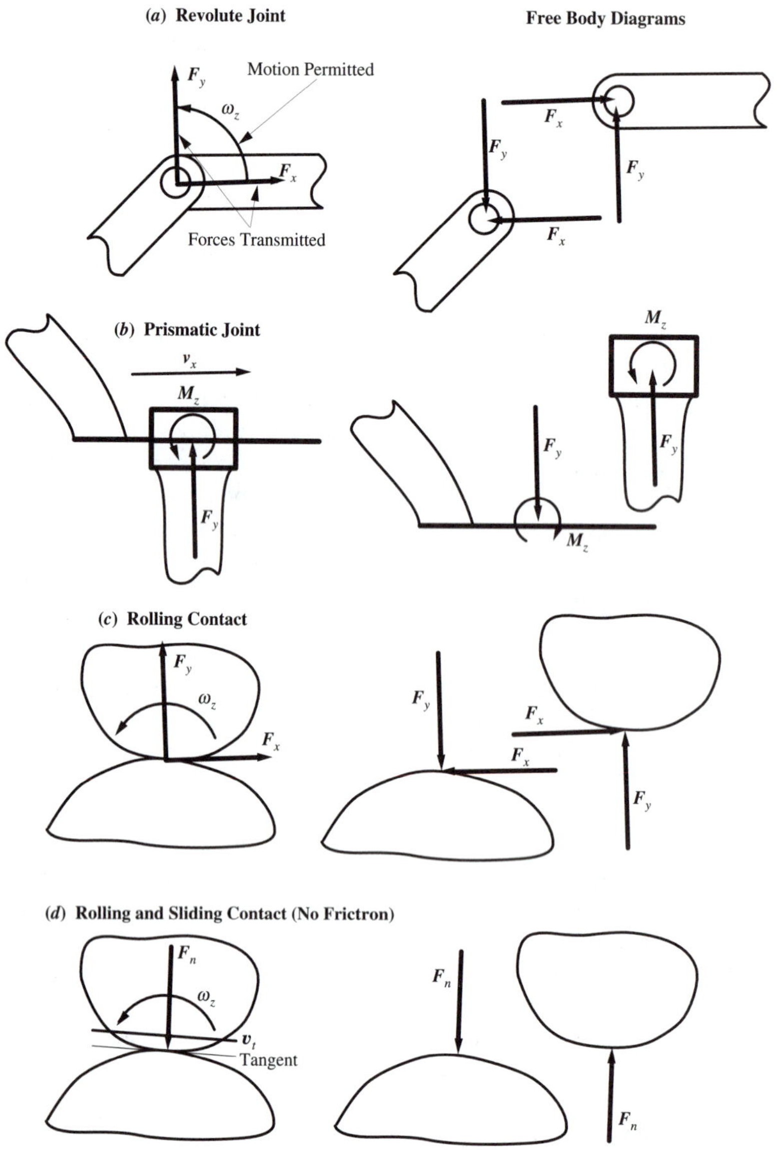

Figure 11.5 Force components transmitted by different types of kinematic joints in planar linkages.

direction normal to the plane of motion and sliding in the direction of the tangent at the point of contact. When friction is neglected, the only force component at the joint is a force along the normal at the point of contact. This is the only force component that can be transmitted by such a joint, as indicated in Fig. 11.5d.

As has been indicated, the forces transmitted by kinematic joints appear in equal and opposite pairs acting on the two members joined in the free-body diagrams. These are seldom the only forces acting in the system. Usually there will also be loads imposed on the system by external agencies. These external loads do not usually appear as equal and opposite pairs.

In addition to indicating all active force components on a set of free-body diagrams, it is good practice to include the dimensional information needed to locate the lines of action of the forces. This information will be needed when writing the static equilibrium equations.

EXAMPLE 11.1 ***(Drawing Free-Body Diagrams)***

PROBLEM

Draw free-body diagrams of all members of the vice-grip pliers in the position shown in Fig. 11.6. The objective of analysis will be to relate the forces, $\boldsymbol{F}_H$, exerted by the user's hand to the gripping forces, $\boldsymbol{F}_G$, exerted on the workpiece. The pliers mechanism can be regarded as being parallel to the horizontal plane, so there are no gravity loads in the plane of action.

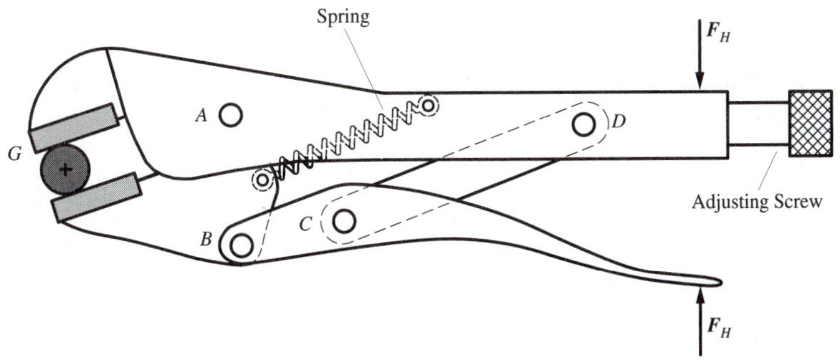

Figure 11.6 The vice-grip pliers of Example 11.1.

SOLUTION

Free-body diagrams of the individual components in the pliers are shown in Fig. 11.7. The only external loads on the system are the equal and opposite forces $\boldsymbol{F}_H$ exerted on the handles by the user's hand. The system as a whole can be regarded as a single free body (Fig. 11.6) expressing the requirement that the system as a whole be in static equilibrium. That, in turn, implies that if there are only two external forces acting on the system, those forces must be equal and opposite and must act along the same line of action.

In Fig. 11.7, the force between links is represented by the letter $\boldsymbol{F}$ with a subscript including the numbers for the two links. The force $\boldsymbol{F}_{23}$ is interpreted as the force that link 2 exerts on link 3. Similarly, $\boldsymbol{F}_{32}$ is the force that link 3 exerts on link 2. Clearly, $\boldsymbol{F}_{23} = -\boldsymbol{F}_{32}$. If a force is not the internal force between two rigid bodies, the subscript will correspond to the location where the force is applied or to the type of force (e.g., $\boldsymbol{F}_S$ or $\boldsymbol{F}_H$).

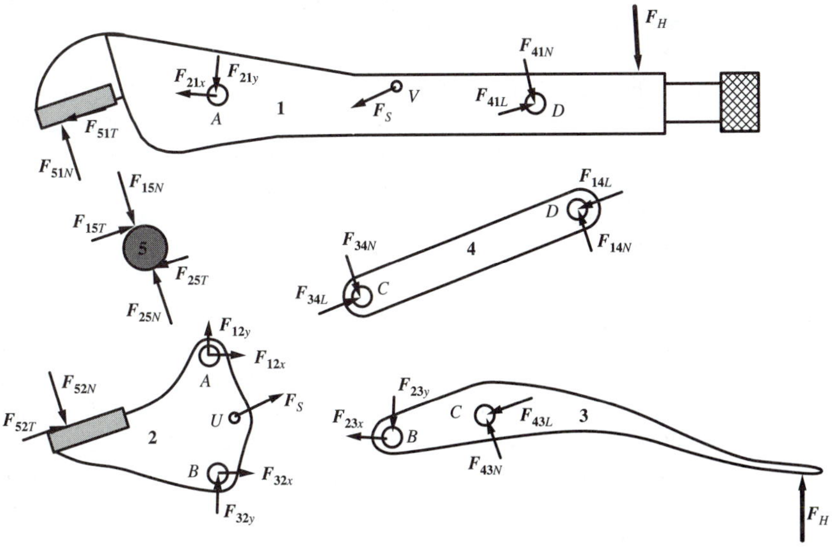

Figure 11.7 The free-body diagrams of Example 11.1.

The forces, $\boldsymbol{F}_S$, exerted on the system by the spring can also be treated as external forces for the purposes of the present analysis. Once again, these two forces are equal and opposite and act along the spring axis. The magnitude of $\boldsymbol{F}_S$ would be computed based on the current length of the spring, the free length, and the spring constant.

The two contacts between the workpiece and the jaws of the vice grip are modeled as rolling contacts. That is, they transmit only forces whose lines of action pass through the contact points. These contact forces are, in principle, unknown in both magnitude and direction. Therefore, they are represented by two orthogonal force components $\boldsymbol{F}_{25N}$ and $\boldsymbol{F}_{25T}$ (or by $\boldsymbol{F}_{15N}$ and $\boldsymbol{F}_{15T}$). In fact, since there are only two forces acting on the workpiece, those two forces must be equal, opposite, and collinear. That is, $\boldsymbol{F}_{25T} = \boldsymbol{F}_{15T} = 0$.

The joints connecting the members of the mechanism are all revolutes. That is, each of them transmits a force whose line of action passes through the revolute axis. These forces are initially unknown in both magnitude and direction. It is convenient to represent them by means of two components with specified directions but unknown magnitudes. In many cases those directions are arbitrarily assigned to be parallel to the x and y axes of the fixed reference frame, but it is not necessary that they have those directions. In the case of joints C and D, the force components are taken to be along the line CD and normal to that line. This is helpful in this case because the link CD actually has only two forces acting on it, one at C and one at D, so equilibrium of that link requires that those forces be equal, opposite, and collinear. Therefore in link 4, $\boldsymbol{F}_{14N} = \boldsymbol{F}_{34N} = 0$, and $\boldsymbol{F}_{14L} = -\boldsymbol{F}_{34L}$. Note, however, that $\boldsymbol{F}_{14L}$ and $\boldsymbol{F}_{34L}$ may be either compressive or tensile.

Notice that every internal force component appears in two of the free-body diagrams and that the senses of the two equal components are opposed. It is convenient to label the two components of each pair identically rather than putting a negative sign on one of them. This is consistent if we regard the labels as representing their magnitudes and the vectors drawn as representing their directions, but it can be confusing because it seems different from the conventions used earlier in this book. It is, of course, necessary to use positive and negative signs on the components when writing equilibrium equations. ■

11.5 GRAPHICAL FORCE ANALYSIS

The force analysis of mechanisms can be conducted either graphically or analytically. The graphical approach is the easiest to perform if only one position is of interest. Also, in the design stage, the graphical procedure allows us to visualize the effect of the location of joints and force application points on the forces that are transmitted in the mechanism. However, for several positions, the graphical procedure is usually inefficient and analytical methods that can be computerized are preferred. Also, if extreme accuracy is required, the analytical procedure is necessary.

Static equilibrium of any free body requires that all forces acting on it sum to zero and that the moments of all those forces about any chosen point should also sum to zero. The force balance can be accomplished graphically by closing a force polygon. However, this does not ensure that the moment equation is satisfied. There are graphical methods for summing moments, but with an electronic calculator it is simpler to measure the normal distances to the lines of action of the forces from the point about which moments are being taken and to compute the moments.

Two special cases are very useful, regardless of whether graphical or analytical solution methods are used. They are shown in Fig. 11.8. The first of these is the case of a member with only two forces (and no couples or moments) acting on it. This case has already been referred to in Example 11.1. To satisfy the force equilibrium conditions, the two forces must be equal and opposite. However, a pair of equal and opposite forces constitutes a couple with moment about any point equal to the magnitude of one of the forces times the normal distance between them. This moment can be zero only if the two forces have the same line of action. Thus, if only two forces act on a member, the forces must be equal, opposite, and collinear.

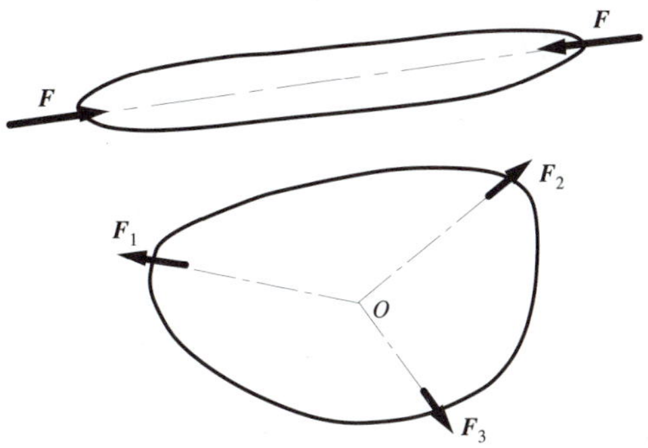

Figure 11.8 Two and three force members.

The second case occurs when only three, nonparallel forces act on a given member. In this case, their lines of action must intersect at a single point. This follows because the lines of action of any two nonparallel planar forces always intersect. If moments are taken about that point of intersection (O in Fig. 11.8), the moment of the system is the magnitude of the third force times the normal distance of its line of action from the point O. Therefore that moment can be zero only if the normal distance is zero; in other words, the system can be in equilibrium only if the lines of action of all three forces pass through a single point.

The result for a two-force member greatly simplifies the analysis because the line of action of the force pair of such a member is automatically the line joining the points of application of the forces. The result for a three-force member allows identification of the line of action of the third force if the lines of action of the other two are known.

The procedure used when graphically solving for the unknown forces in a mechanism is first to draw free-body diagrams of the members. The next step is to look for a solvable member. Since we have, effectively, three scalar equations for each member, a solvable member has three (or fewer) unknown force components. These might be a force that is unknown in both magnitude and direction, which is equivalent to two unknown components, plus a force with known direction but unknown magnitude or a force unknown in both magnitude and direction plus an unknown torque. There may also be other, known forces acting on the member.

It might seem that there are numerous possible combinations of three unknown force components. However, they all reduce to only two cases: (a) three forces on known lines of action with unknown magnitudes, or (b) two forces on known lines of action with unknown magnitudes plus an unknown torque. The first case cited in the previous paragraph—a force of unknown magnitude and direction applied at a known point plus a force with known line of action but unknown magnitude—is equivalent to case (a) stated here. If the force with unknown magnitude and direction is represented by two unknown components on lines of action passing through the point of application, the two cases become identical.

In both case (a) and case (b) the solution procedure can be started by taking moments about the point of intersection of two of the unknown forces. If two of those forces are components of a force with unknown magnitude and direction applied at a point, that point is the point of intersection and is a convenient point about which to take moments. Because the two forces whose lines of action intersect at the point about which moments are taken have zero moments about that point, the moment balance involves only one unknown force magnitude, or torque. This unknown can be solved for. A force polygon can then be drawn to solve for the remaining unknown force components. The procedure is illustrated in the following example.

EXAMPLE 11.2 ***(Graphical Force Analysis of Vice-Grip Pliers)***

PROBLEM

Find the grip force F_G exerted on the workpiece by the vice-grip pliers of Example 11.1, shown in Fig. 11.7, if the forces F_H exerted by the user are 25 lb in the location shown in Fig. 11.9. The spring force, F_S is 10 lb. Also find the forces transmitted by the revolute joints at A, B, and C.

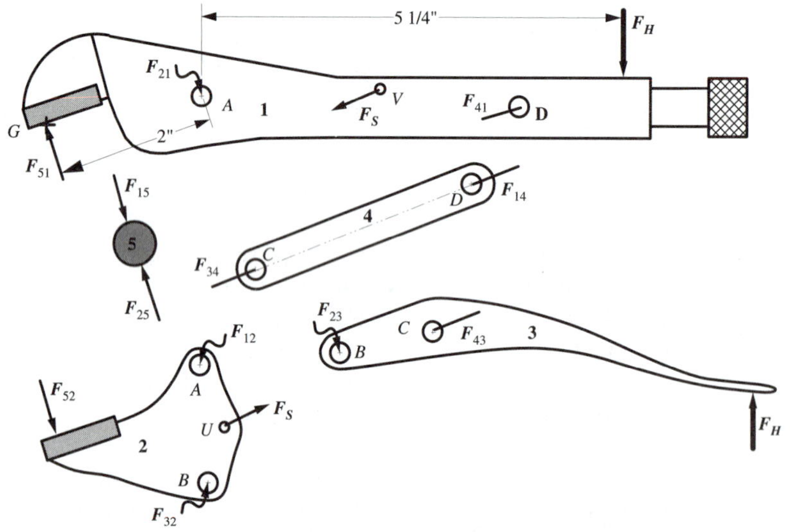

Figure 11.9 The free-body diagrams of Examples 11.1 and 11.2 simplified by taking advantage of the fact that members 4 and 5 are two-force members.

SOLUTION

This is an example of a mechanism that can be analyzed graphically without making any calculations beyond scaling and unscaling vectors drawn in polygons. The free-body diagrams of all members are shown in Fig. 11.7. However, those free-body diagrams can be simplified by taking advantage of the two-force members 4 and 5. The result is the set of free-body diagrams of Fig. 11.9. In Fig. 11.9 we have used the convention often used in graphical analyses of representing forces of unknown direction and magnitude by a wavy line. A force with a known direction but unknown magnitude is represented by a line without an arrowhead.

The first step in the analysis is to identify the free-body diagrams that have no more than three unknown quantities and that have at least one known vector (direction and magnitude). Member 3 is the only member satisfying this condition. Member 3 is a three-force member with the known force $\boldsymbol{F}_H$, the force $\boldsymbol{F}_{43}$ whose line of action is known but whose magnitude is unknown, and the force $\boldsymbol{F}_{23}$ applied at point B, whose magnitude and direction are both unknown. Thus there are three unknown force magnitudes, two components of $\boldsymbol{F}_{23}$ and the magnitude of $\boldsymbol{F}_{43}$.

Because member 3 is a three-force member with no applied couples, all of the forces must intersect at a point. We know the magnitude and direction of $\boldsymbol{F}_H$ and the direction of $\boldsymbol{F}_{43}$. The lines of action of these two forces intersect at point O in Fig. 11.10. If we sum moments about point O, the only force contributing to the moment summation is $\boldsymbol{F}_{23}$. Therefore, the line of action of $\boldsymbol{F}_{23}$ must pass through O. Now if we sum forces vectorially, we know that

$$\sum \boldsymbol{F} = \boldsymbol{0} = \boldsymbol{F}_H + \boldsymbol{F}_{23} + \boldsymbol{F}_{43} \tag{11.7}$$

The force polygon corresponding to Eq. (11.7) is shown in Fig. 11.10. Note that the polygon is drawn so that the vectors add together in a head-to-tail fashion because all of the vectors in Eq. (11.7) are positive, and they sum to zero. From the polygon,

$$\boldsymbol{F}_{23} = 709\,\text{lb} \quad \text{and} \quad \boldsymbol{F}_{43} = 716\,\text{lb}$$

Next consider link 2. Initially, the free-body diagram associated with link 2 has five unknowns. However, we know the magnitude and direction for $\boldsymbol{F}_{32}$. Also, there are no concentrated moments on link 2. Therefore, link 2 has a total of three unknowns, and these unknowns can be determined. When we begin, we have two known forces ($\boldsymbol{F}_{32}$ and $\boldsymbol{F}_S$) applied to link 2. The first operation is to replace these two forces by an equivalent force. The equivalent force will be the vector sum of $\boldsymbol{F}_{32}$ and $\boldsymbol{F}_S$ and will create the same moment about an arbitrary point. For this to be the case, the resultant of $\boldsymbol{F}_{32}$ and $\boldsymbol{F}_S$ must pass through point M, the intersection of $\boldsymbol{F}_{32}$ and $\boldsymbol{F}_S$. The resultant of $\boldsymbol{F}_{32}$ and $\boldsymbol{F}_S$ is approximately equal to $\boldsymbol{F}_{32}$ because $\boldsymbol{F}_{32}$ is so much larger than $\boldsymbol{F}_S$. The resultant ($\boldsymbol{F}_{32} + \boldsymbol{F}_S$) is known in both direction and magnitude. Link 2 then reduces to another three-force member with one unknown magnitude ($\boldsymbol{F}_{52}$) and one unknown direction and magnitude ($\boldsymbol{F}_{12}$). The force $\boldsymbol{F}_{12}$ must pass through the intersection of the force $\boldsymbol{F}_{32} + \boldsymbol{F}_S$ and $\boldsymbol{F}_{52}$. This occurs at N. Knowing the direction of $\boldsymbol{F}_{12}$, we can sum forces on the free-body diagram and get

$$\sum \boldsymbol{F} = \boldsymbol{0} = \boldsymbol{F}_{32} + \boldsymbol{F}_S + \boldsymbol{F}_{12} + \boldsymbol{F}_{52} \tag{11.8}$$

This equation can be solved using the force polygon in Fig. 11.10. Again, we add the vectors head to tail. This will give us both the magnitude and sense (i.e., the end that contains the arrowhead) of the unknown vectors. From the polygon,

$$F_{12} = 844 \text{ lb}$$

and

$$F_{52} = 489 \text{ lb}$$

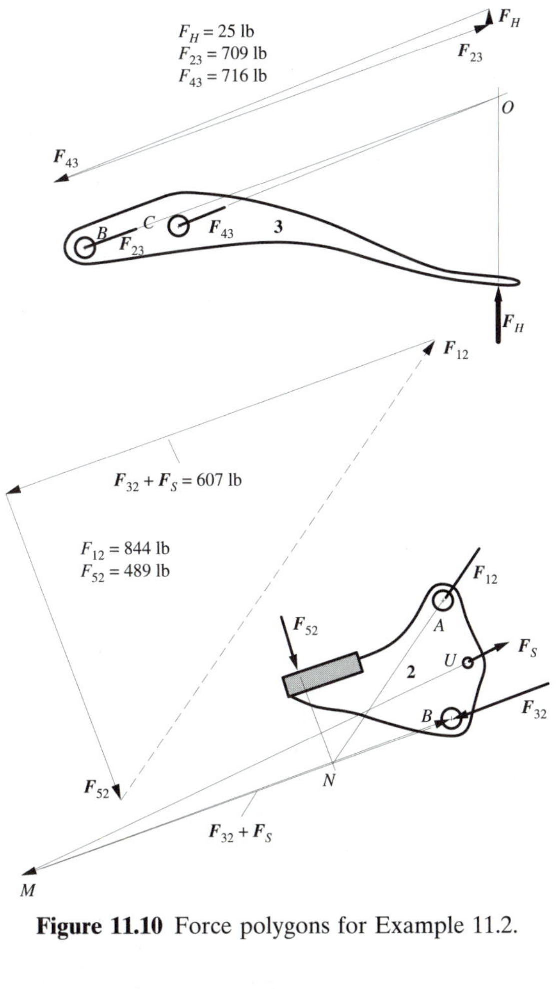

Figure 11.10 Force polygons for Example 11.2.

Notice that free body 1 was not used. It might appear that 4 and 5 were also not used. However, these are two-force members. By identifying the lines of action of the respective force pairs, we have ensured their static equilibrium, so they have, in fact, been used. We would have needed to use free body 1 if we had not noted that the whole system is subject to only two external forces and required those two forces to be equal, opposite, and collinear. Because we effectively used the whole system as one free body, we could solve for the forces without considering one member as a free body. In general, if the linkage obeys the planar constraint criterion, the number of free bodies needed for complete soution is equal to the number of members in the linkage.

Notice also that this linkage converts the modest force of 25 lb applied to the handles by the user to a gripping force of 489 lb. That is, it magnifies the force by a factor of about 19.6. You may notice that BC is nearly aligned with CD. This is an example of a toggle linkage. As the linkage moves into the toggle position (B, C, and D collinear), the mechanism can generate a very large force (theoretically infinite) at G. Toggle linkages like this can not only be used to generate large forces but also can be snapped over center into a statically stable position allowing the workpiece to be clamped indefinitely. This is, in fact, the way in which the vice-grip pliers are intended to be used.

This problem illustrates one of the reasons for conducting the force analysis graphically. By observing the force polygon associated with link 3, we can visualize how the forces increase as the linkage moves into the toggle position. Such visual feedback often provides valuable insight during the design stage of the product development into how a given linkage will behave. ■

EXAMPLE 11.3 ***(Graphical Force Analysis of Four-Bar Linkage)***

PROBLEM

For the linkage shown in Fig. 11.11, find the forque $\boldsymbol{T}_{12}$ required if $P = 120$ lb in the direction shown. The linkage dimensions are as follows:

$$AB = 6\text{ in} \quad EC = 12\text{ in} \quad AE = 8\text{ in}$$
$$BC = 18\text{ in} \quad ED = 5\text{ in}$$

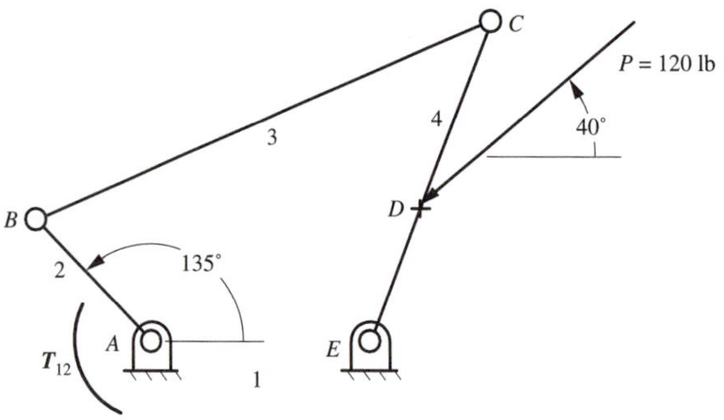

Figure 11.11 Mechanism for Example 11.3.

SOLUTION

This mechanism can be analyzed graphically for most of the procedure; however, ultimately the unknown torque, $\boldsymbol{T}_{12}$, must be computed. To begin the analysis, draw the general free-body diagrams for each link. We can simplify the free-body diagram for link 3 by noting that it is a two-force member. Therefore, $\boldsymbol{F}_{23}$ and $\boldsymbol{F}_{43}$ must be equal, opposite, and collinear as shown in Fig. 11.12. Knowing the direction of $\boldsymbol{F}_{23}$ (and therefore $\boldsymbol{F}_{32}$), it is apparent that link 2 has only two forces also. Therefore, because these two forces must sum to zero for static equilibrium, they are equal and opposite; however, they are *not* collinear because of the torque $\boldsymbol{T}_{12}$ applied to link 2. For the forces in a two-force member to be collinear, there must be no applied torques or moments.

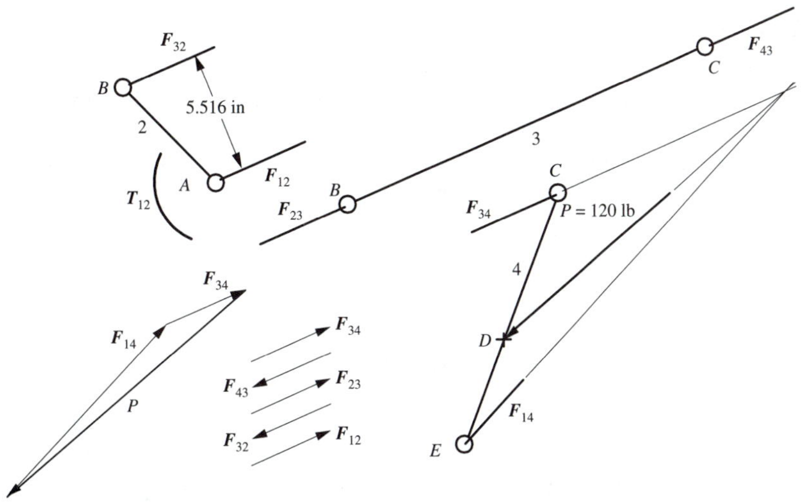

Figure 11.12 Free-body diagrams and force polygon for mechanism in Example 11.3.

Because the direction of $\boldsymbol{F}_{43}$ and therefore that of $\boldsymbol{F}_{34}$ is known, link 4 becomes a three-force member. The lines of action of the three forces must intersect as shown in Fig. 11.12, and this gives us the direction of $\boldsymbol{F}_{14}$. Knowing the direction of $\boldsymbol{F}_{14}$, we can sum forces on link 4 and solve for the unknowns. That is,

$$\sum \boldsymbol{F} = \boldsymbol{0} = \boldsymbol{P} + \boldsymbol{F}_{14} + \boldsymbol{F}_{34}$$

From the force polygon in Fig. 11.12, $\boldsymbol{F}_{34} = 33.6$ lb in the direction shown. From the free-body diagrams and force equilibrium considerations (see Fig. 11.12), we know that $\boldsymbol{F}_{34} = -\boldsymbol{F}_{43} = \boldsymbol{F}_{23} = -\boldsymbol{F}_{32} = \boldsymbol{F}_{12}$. Knowing $\boldsymbol{F}_{32}$ and $\boldsymbol{F}_{12}$, we can compute the torque $\boldsymbol{T}_{12}$. Summing moments about point A gives

$$\sum \boldsymbol{M}_A = \boldsymbol{T}_{12} + \boldsymbol{h} \times \boldsymbol{F}_{32} = 0$$

For equilibrium, the torque $\boldsymbol{T}_{12}$ must be equal and opposite to $\boldsymbol{h} \times \boldsymbol{F}_{32}$. This is shown in Fig. 11.13. Because the cross-product $\boldsymbol{h} \times \boldsymbol{F}_{32}$ is counterclockwise, the torque must be clockwise. Therefore, we need to compute only the magnitude of the torque. This is given by

$$T_{12} = h(F_{32}) = 5.516(33.6) = 185.3 \text{ in-lb}$$

and

$$\boldsymbol{T}_{12} = 185.3 \text{ in-lb CW}$$

■

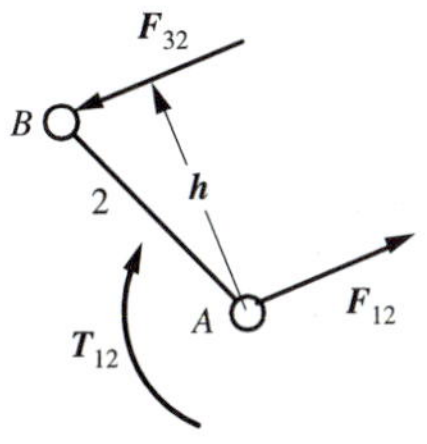

Figure 11.13 Free-body diagram for link 2.

EXAMPLE 11.4 (*Graphical Force Analysis of Cam Mechanism*)

PROBLEM

As an example of a mechanism with higher pairs, analyze the cam mechanism in Fig. 11.14 for the torque $\boldsymbol{T}_{12}$. Assume that the kinetic coefficient of friction is 0.15 between the follower and cam, and the torque $\boldsymbol{T}_{13}$ on the follower is 60 in-lb. The weight of the follower is 20 lb, and the weight of the cam is 30 lb.

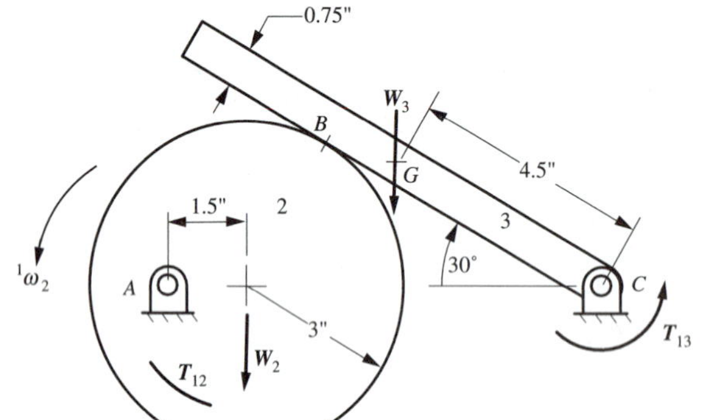

Figure 11.14 Cam mechanism for Example 11.5.

SOLUTION

The cam mechanism involves only two moving links, and therefore only two free-body diagrams are necessary. These are shown in Fig. 11.15. The force components will then be as shown in Fig. 11.15.

To begin the analysis, start with link 3, where the known forces are located. Initially, link 3 does not appear to be a three-force member that can be analyzed easily using graphical procedures because of the torque $\boldsymbol{T}_{13}$. However, as discussed in Section 11.2, we can replace the force $\boldsymbol{W}_3$ and torque $\boldsymbol{T}_{13}$ with an equivalent system if we translate the force $\boldsymbol{W}_3$ by an amount h, where h is given by

$$h = \frac{|\boldsymbol{T}_{13}|}{|\boldsymbol{W}_3|} = \frac{60}{20} = 3 \text{ in}$$

Note that the force is translated so as to produce the torque $\boldsymbol{T}_{13}$. Therefore, if we sum moments about any point, the force $\boldsymbol{W}_3$ alone will produce the same moment as the original set of $\boldsymbol{T}_{13}$ and $\boldsymbol{W}_3$.

The next operation is to replace the normal force ($\boldsymbol{F}_{23_n}$) and friction force ($0.15\boldsymbol{F}_{23_n}$) at B with the total resultant force $\boldsymbol{F}_{23}$. Before this is done, however, we must determine the direction for the friction force. Friction makes the mechanism analysis nonlinear, and we must know the correct sign for the friction force or we will not compute the correct answers. The friction force is in the direction of the relative motion or impending motion of the link contacting the body associated with the free-body diagram. In this problem, the cam rotates counterclockwise relative to the frame and the follower. Therefore, the motion of the cam relative to the follower at the point of contact is away from the follower pivot as shown in Fig. 11.14. After the direction of the friction force is known, we can determine the line of action of the resultant force $\boldsymbol{F}_{23}$ as shown in Fig. 11.15. The resultant force is applied at B.

We now have a three-force member that involves $\boldsymbol{F}_{23}$, $\boldsymbol{F}_{13}$, and $\boldsymbol{W}_3$ (translated by the amount h). The lines of action of these three forces must intersect at a point as shown in Fig. 11.15. This identifies the line of action of force $\boldsymbol{F}_{13}$, which was initially unknown. We can now sum forces on the free-body diagram using

$$\sum \boldsymbol{F} = \boldsymbol{0} = \boldsymbol{W}_3 + \boldsymbol{F}_{13} + \boldsymbol{F}_{23}$$

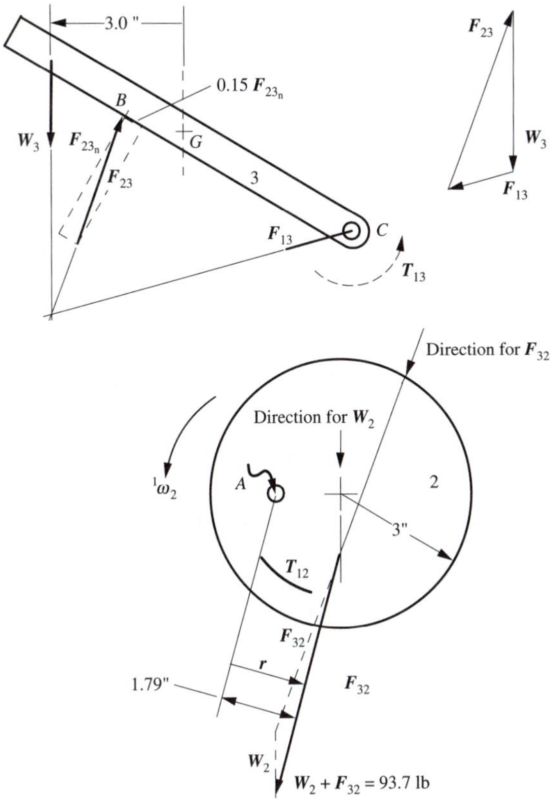

Figure 11.15 Free-body diagrams and force polygon for mechanism for Example 11.5.

The force polygon is shown in Fig. 11.15. The force $\boldsymbol{F}_{23}$ can be scaled from the force polygon to be 71.18 lb. We know that $\boldsymbol{F}_{23} = -\boldsymbol{F}_{32}$; therefore, we can apply the known force to the free-body diagram for link 2. On link 2, there are two known forces and an unknown torque. We can proceed in several ways; however, an especially simple way is to find the resultant of $\boldsymbol{W}_2 + \boldsymbol{F}_{32}$. We will then have only one force other than $\boldsymbol{F}_{12}$ and a torque applied to link 2. If we sum moments about point A, only the resultant $\boldsymbol{W}_2 + \boldsymbol{F}_{32}$ and $\boldsymbol{T}_{12}$ will enter into the calculations. The resulting equation is

$$\sum \boldsymbol{M}_A = \boldsymbol{0} = \boldsymbol{T}_{12} + \boldsymbol{r} \times (\boldsymbol{W}_2 + \boldsymbol{F}_{32})$$

This equation can be simplified because we can measure directly the perpendicular distance from point A to the line of action of the resultant $\boldsymbol{W}_2 + \boldsymbol{F}_{32}$. Then,

$$|\boldsymbol{T}_{12}| = |\boldsymbol{r}|\,|\boldsymbol{W}_2 + \boldsymbol{F}_{32}| = 1.79(93.7) = 168 \text{ in-lb}$$

We can identify the direction of $\boldsymbol{T}_{12}$ by inspection or by actually performing the cross-product indicated above. The result is

$$\boldsymbol{T}_{12} = 168 \text{ in-lb (CCW)}$$

■

11.6 ANALYTICAL APPROACH TO FORCE ANALYSIS

A body is in static equilibrium when the resultant (sum) of all the forces acting on it is zero and the resultant of all their moments about any point is zero.

If a system of rigid bodies, such as a mechanism, is in static equilibrium, then each individual body is in equilibrium under the action of all the forces acting on it, including those exerted on it by other members. A free-body diagram should be drawn for each member, and the appropriate force and moment equilibrium equations should then be written for each free body and solved for the unknown forces or moments.

For an analytical force analysis, we must know the coordinates of all points involved in the analysis. Therefore, before the force analysis is conducted, a position analysis must be conducted. As indicated in Chapters 2 and 3, the position equations will be nonlinear. After the position analysis is conducted, however, the force equations will be linear (unless friction is involved), and therefore they can be easily solved. We will illustrate the procedure in three examples.

EXAMPLE 11.5 (*Analytical Force Analysis of Four-Bar Linkage*)

PROBLEM

Analyze the linkage in Example 11.3 analytically to find the torque $\boldsymbol{T}_{12}$ required if $P = 120$ lb and the driver link 2 is at an angle of 135° with the horizontal axis.

SOLUTION

Before the force analysis can be conducted, we must determine the coordinates of the points relative to the coordinate system shown in Fig. 11.16. This can be done using the procedures discussed in Chapter 3, and all vectors will be defined using the sign convention and nomenclature given in Chapter 3 (see Fig. 3.3). This will let us determine the angular orientation of each link and each force. After the position analysis, we can summarize the known position and force information as follows:

$$\boldsymbol{P} = 120 \angle 220° = -91.9\boldsymbol{i} - 77.1\boldsymbol{j}$$
$$\boldsymbol{F}_{34} = F_{34} \angle 22.65° = F_{34}\,(0.9228)\boldsymbol{i} + F_{34}\,(0.3851)\boldsymbol{j}$$
$$\boldsymbol{r}_{B/A} = r_{B/A} \angle 135° = 6 \angle 135° = -4.243\boldsymbol{i} + 4.243\boldsymbol{j}$$
$$\boldsymbol{r}_{C/B} = r_{C/B} \angle 22.65° = 18 \angle 22.65° = 16.61\boldsymbol{i} + 6.93\boldsymbol{j}$$
$$\boldsymbol{r}_{C/E} = r_{C/E} \angle 68.65° = 12 \angle 68.65° = 4.368\boldsymbol{i} + 11.18\boldsymbol{j}$$

and

$$\boldsymbol{r}_{D/E} = r_{D/E} \angle 68.65° = 5 \angle 68.65° = 1.820\boldsymbol{i} + 4.657\boldsymbol{j}$$

To begin the analysis, draw the general free-body diagrams for each link as shown in Fig. 11.17. Again, we can simplify the free-body diagram for link 3 by noting that it is a two-force member with no applied moments. However, in an analytical force analysis, it is usually not efficient to simplify three-force members as was done in the graphical analysis. Therefore, we will represent $\boldsymbol{F}_{14}$ by its x and y components. For the analysis, begin with link 4, because this is the only location where a force magnitude is known. There are only three unknowns associated with this link. These are the magnitude of $\boldsymbol{F}_{34}$ and the magnitude of the two components of

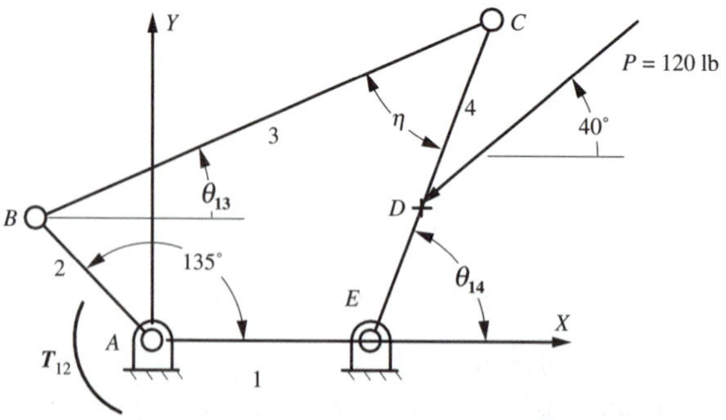

Figure 11.16 Coordinate axes for mechanism in Example 11.5.

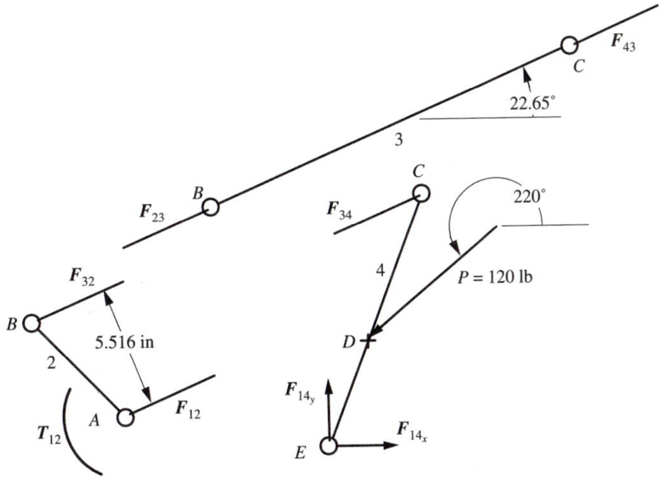

Figure 11.17 Free-body diagrams for analytical force analysis for Example 11.5.

$\boldsymbol{F}_{14}$. We do not need to determine $\boldsymbol{F}_{14}$ to solve the problem. To compute $\boldsymbol{F}_{34}$, sum moments about point E in the free-body diagram for link 4. Then,

$$\sum \boldsymbol{M}_E = \boldsymbol{r}_{C/E} \times \boldsymbol{F}_{34} + \boldsymbol{r}_{D/E} \times \boldsymbol{P} = \boldsymbol{0} \tag{11.9}$$

where

$$\boldsymbol{r}_{C/E} \times \boldsymbol{F}_{34} = \begin{vmatrix} \boldsymbol{i} & \boldsymbol{j} & \boldsymbol{k} \\ 4.368 & 11.18 & 0 \\ 0.9228F_{34} & 0.3851F_{34} & 0 \end{vmatrix} = -8.635F_{34}\boldsymbol{k}$$

and

$$\boldsymbol{r}_{D/E} \times \boldsymbol{P} = \begin{vmatrix} \boldsymbol{i} & \boldsymbol{j} & \boldsymbol{k} \\ 1.820 & 4.657 & 0 \\ -91.9 & -77.1 & 0 \end{vmatrix} = 287.6\boldsymbol{k}$$

Therefore,

$$F_{34} = 287.6/8.635 = 33.31 \text{ lb}$$

Because the force is positive,

$$\boldsymbol{F}_{34} = 33.31 \angle 22.65^\circ$$

From the free-body diagrams,

$$\boldsymbol{F}_{32} = -\boldsymbol{F}_{23} = \boldsymbol{F}_{43} = -\boldsymbol{F}_{34}$$

Therefore,

$$\boldsymbol{F}_{32} = -33.31 \angle 22.65^\circ = -30.74\boldsymbol{i} - 12.83\boldsymbol{j}$$

Now considering the free-body diagram for link 2, we can sum moments about point A. The result of this is

$$\sum \boldsymbol{M}_A = \boldsymbol{r}_{B/A} \times \boldsymbol{F}_{32} + \boldsymbol{T}_{12} = \boldsymbol{0} \tag{11.10}$$

Here we are using the convention that a positive value for T_{12} corresponds to a counterclockwise torque. From Eq. (11.9),

$$T_{12} = -r_{B/A} \times F_{32} = -\begin{vmatrix} i & j & k \\ -4.243 & 4.243 & 0 \\ -30.74 & -12.83 & 0 \end{vmatrix} = -184.8k \text{ in-lb}$$

or

$$T_{12} = 184.8 \text{ in-lb CW}$$

Note that this result is very close to that computed in the graphical analysis. In general, some error must be expected in the graphical analysis because of the graphical constructions. ■

11.6.1 Transmission Angle in a Four-Bar Linkage

In Chapter 4, the transmission angle in a four-bar linkage was mentioned in conjunction with the design of optimal linkages for function generation. The transmission angle is the angle η in Fig. 11.16. If we reexamine Eq. (11.10) we find that

$$|T_{12}| = |r_{B/A} \times F_{32}| \tag{11.11}$$

For a given position of the linkage, the magnitude and orientation of $r_{B/A}$ and F_{32} will be fixed, and the magnitude of the torque required for equilibrium will be directly proportional to the magnitude of F_{32}. Therefore, the torque required to maintain equilibrium will increase with the magnitude of F_{32}. Now examining Eq. (11.9) and using an alternative expression for the cross-product gives

$$|r_{C/E} \times F_{34}| = |r_{C/E}|\,|F_{34}| \sin\eta = |r_{D/E} \times P|$$

or

$$|F_{34}| = \frac{|r_{D/E} \times P|}{|r_{C/E}| \sin\eta} = |F_{32}| \tag{11.12}$$

From Eq. (11.12) it is clear that the magnitude of the force F_{32} increases rapidly in a nonlinear fashion as the transmission angle η approaches 0 or π, and F_{32} reaches a minimum value for $\pi/2$. To reduce the torque required for equilibrium, we need to reduce the magnitude of F_{32}, which means that we want to make the transmission angle as close to $\pi/2$ as possible. This is consistent with the optimization procedure discussed in Chapter 4, where the objective was to minimize $|\pi/2 - \eta|$.

In general, the transmission angle should be held between 30° and 150°. If the transmission angle approaches 0 or 180°, high bearing loads, excessive wear, and binding in the joints can be expected.

EXAMPLE 11.6 ***(Analytical Force Analysis of Drilling-Mud Pump)***

PROBLEM

A pump used for pumping drilling mud in oil well drilling has two double-acting cylinders. The linkage of the piston to the crankshaft for each cylinder is arranged as shown in Fig. 11.18.

On the upstroke, the gage pressure in the cylinder above the piston is 750 psi above atmospheric, and on the bottom side of the piston the pressure is 5 psi *below* atmospheric. The bore of the cylinder has a diameter of 6 in, and the piston rod is 1.5 in in diameter. The weight of the connecting rod, which can be regarded as a uniform rod, is 50 lb. The weight of the piston, piston rod, and crosshead assembly is 30 lb. The crank weighs 45 lb and has its center of mass 3 in from the crankshaft axis as shown. The frictional resistance from the piston and gland seals and the crosshead is estimated to total 12 lb. In the position shown, find the torque that must

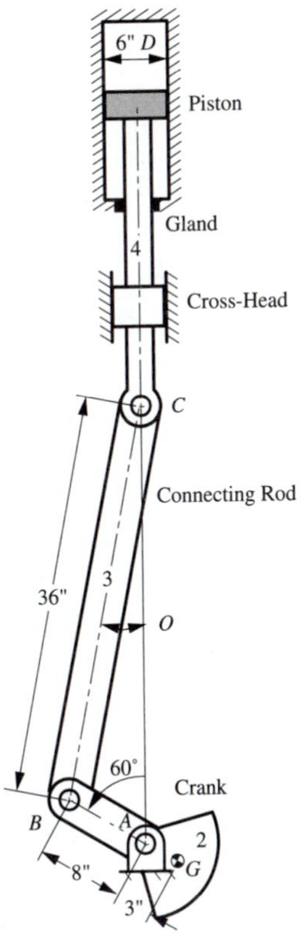

Figure 11.18 The mechanism of an oil drilling mud pump analyzed in Example 11.6.

be applied to the crank by the motor, the axial loads in the connecting rod and piston rod, the loads on all three bearings, and the side thrust resisted by the crosshead.

SOLUTION

Before the force analysis can be conducted, we need to determine the orientation of the connecting rod and the force on the piston. The angle ACB can be calculated using the sine rule:

$$\frac{\sin\theta}{8} = \frac{\sin 60^\circ}{36}$$

or

$$\theta = 11.096^\circ$$

The area of the top face of the piston is

$$A_T = \pi \times 3^2 = 28.27 \text{ in}^2$$

Therefore the downward force on the top of the piston is

$$750 \times A_T = 21{,}210 \text{ lb}$$

The net area of the bottom face of the piston is

$$A_B = A_T - \pi \times 0.75^2 = 26.51 \text{ in}^2$$

Therefore the force on the bottom of the piston is

$$5 \times A_B = 130 \text{ lb}$$

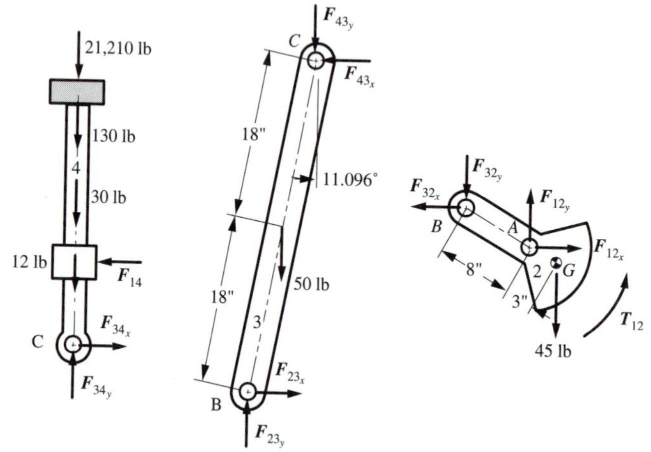

Figure 11.19 The free-body diagrams for the mechanism of Fig. 11.18 and Example 11.6.

This is also downward because the pressure is lower than atmospheric below the piston. The free-body diagrams for the pump members are shown in Fig. 11.19. Note that we have assumed the directions for the force components. If an assumed direction is wrong, the force component will be negative.

Applying static equilibrium to the piston (link 4):

$$\sum F_x = 0 = F_{34_x} - F_{14}: \qquad \Rightarrow F_{34_x} = F_{14}$$

$$\sum F_y = 0 = F_{34_y} - 21{,}210 - 30 - 12 - 130: \qquad \Rightarrow F_{34_y} = 21{,}380$$

Here we have rounded the result to four significant figures. Hence the piston rod axial load is 21,380 lb.

Applying static equilibrium to the connecting rod:

$$\sum F_x = 0 = -F_{43_x} + F_{23_x}: \qquad \Rightarrow F_{23_x} = F_{43_x} = F_{34_x} = F_{14}$$

Again note that we have assumed the direction for the force components and are dealing only with magnitudes.

$$\sum F_y = 0 = F_{23_y} - 21380 - 50: \qquad \Rightarrow F_{23_y} = 21{,}430$$

For the moment summation, we will use the forces and the normal distances from their lines of action to the point about which the summation of moments is made. For link 3, we will sum moments about point B.

$$\sum M_B = \sum R_n(F) = F_{43_x}(36)(\cos 11.096°) - 21380(36)(\sin 11.096°) - 50(18)(\sin 11.096°)$$

giving

$$F_{43_x} = 4{,}200 \text{ lb}$$

Hence the connecting rod load at C is $\sqrt{21430^2 + 4200^2} = 21{,}840$ lb. This is also the load on the bearing at C.

Applying static equilibrium to the crank and noting that $F_{23_x} = F_{32_x}$ and $F_{23_y} = F_{32_y}$,

$$\sum F_x = 0 = F_{12_x} - 4200 \Rightarrow F_{12_x} = 4200 \text{ lb}$$

$$\sum F_y = 0 = F_{12_y} - 21430 - 45 \Rightarrow F_{12_y} = 21{,}480 \text{ lb}$$

$$\sum M_A = T_{12} + 21430(8)(\sin 60°) + 4200(8)(\cos 60°) - 45(3)(\sin 60°)$$

So

$$T_{12} = -165{,}200 \text{ lb-in}$$

or

$$\boldsymbol{T}_{12} = 13{,}800 \text{ ft-lb CW}$$

The bearing load at A is

$$F_{12} = \sqrt{21480^2 + 4200^2} = 21{,}900 \text{ lb}$$

This completes the analysis. ■

11.7 FRICTION CONSIDERATIONS

In the previous examples, we have considered friction briefly. In this section, we will investigate the effect of friction more formally for various types of joints. Friction can be extremely important in the design of mechanisms. If not properly handled, friction can greatly reduce the efficiency of a mechanism and increase power requirements. Because of friction, mechanical work is converted to heat, and the resulting heat buildup can degrade the materials in the mechanism, especially when polymers are used. Friction also contributes to wear, and in extreme cases friction can cause a mechanism to seize up.

The friction force is perpendicular to the contact force at a joint, and assuming Coulomb friction, the friction force will be proportional to the normal contact force. The direction of the friction force is determined by the direction of the relative motion or impending relative motion. Therefore, before a friction force analysis is conducted, it is important to conduct a velocity analysis in enough detail to determine the direction of the relative motion between different links. The magnitudes of the relative velocities are not required if Coulomb friction is assumed. The relative velocity magnitudes will be needed, however, if viscous friction is assumed. In the discussions here, only Coulomb friction will be considered. In the following, we will consider friction in cam joints, sliding or prismatic joints, and revolute joints.

11.7.1 Friction in Cam Contact

Two links (2 and 3) in cam contact are shown in Fig. 11.20. If we treat the problem as planar, the two links will have point contact at the cam joint, and the contact force can be resolved into two components, one normal to the common tangent and one along the common tangent. The component in the direction of the common tangent will be the friction force, and it can be related to the normal force by

$$F_{32_t} = \mu F_{32_n}$$

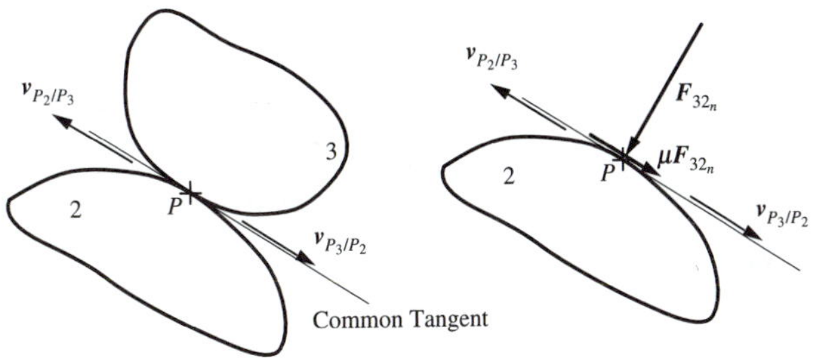

Figure 11.20 Friction forces in a cam joint.

where μ is the coefficient of friction. Note that when Coulomb friction is assumed, the friction force depends on μ and F_{32_n} only; that is, F_{32_t} is independent of the area at the contact location. If the bodies have relative motion, μ will be equal to the kinetic coefficient of friction. If the bodies are at rest but the motion is impending, then the value of μ can range from 0 to μ_s, the static coefficient of friction. The actual coefficient of friction is equal to μ_s only when motion is about to occur.

To determine the direction of the friction force on link 2, we must determine the direction for the velocity ${}^3\boldsymbol{v}_{P_2/P_3} = {}^3\boldsymbol{v}_{P_2}$. This is the relative velocity of the contact point on link 2 relative to link 3. The friction force will be opposite to this relative velocity vector. Alternatively, we can determine the velocity of P_3 relative to link 2 or ${}^2\boldsymbol{v}_{P_3/P_2} = {}^2\boldsymbol{v}_{P_3}$. This is the point on link 3 that is causing the force on link 2. The friction force on link 2 will be in the same direction as the velocity ${}^2\boldsymbol{v}_{P_3/P_2} = {}^2\boldsymbol{v}_{P_3}$. Simplistically, we can think of P_3 dragging against link 2 to cause the friction force. In this case, the friction force will be in the direction of the dragging motion, that is, in the direction of ${}^2\boldsymbol{v}_{P_3}$.

11.7.2 Friction in Slider Joints

Two links (3 and 5) in sliding contact are shown in Fig. 11.21a. The slider (link 3) is assumed to have two loads $\boldsymbol{P}$ and $\boldsymbol{W}$ applied to it from links other than link 5. If we resolve the force from link 5 on link 3 into two components, the normal force will be perpendicular to the contact surface and the friction force will be tangent to it as shown in Fig. 11.21b. The direction of the friction force will be opposite to the velocity of any point on the slider relative to link 5. In Fig. 11.21b, point A_3 is chosen as a typical point.

The two components of the force $\boldsymbol{F}_{53}$ can be resolved into a single force if the friction angle ϕ is known. This angle is given by

$$\phi = \tan^{-1}\left(\frac{\mu F_{53_n}}{F_{53_n}}\right) = \tan^{-1}(\mu) \tag{11.13}$$

Note that, from force equilibrium on link 3, $W = F_{53_n}$ and $P = \mu F_{53_n}$. If both P and W are given and the block is in static equilibrium, then μ cannot be specified. It will assume whatever value is required for equilibrium. However, μ cannot be larger than the static coefficient of friction if the mechanism is not moving or the kinetic coefficient of friction if the mechanism is moving. If μ is given, then P can be computed if F_{53_n} is known.

If the normal force and friction force are resolved into a single force oriented at an angle of ϕ relative to the normal direction, link 3 becomes a three-force member. Therefore, the forces must intersect at a point as shown in Fig. 11.21b. This will indicate the location where the total force $\boldsymbol{F}_{53}$ is applied to link 3. It is possible that the point of application of the force must be beyond the physical limits of the block. When this occurs, the block will

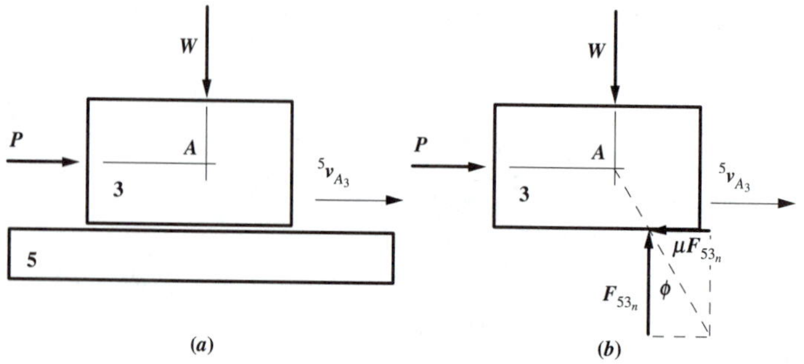

Figure 11.21 Friction forces in a slider joint.

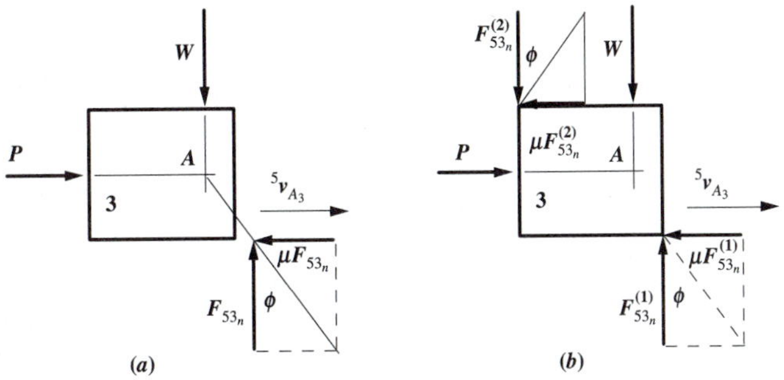

Figure 11.22 Condition indicating that the block must contact the slide at two locations.

tend to rotate, and if the block is not supported on both top and bottom, static equilibrium cannot be maintained. This condition is indicated in Fig. 11.22a. When designing a mechanism, we must provide for a support on both sides of the slider whenever there is a possibility that the friction force will become large enough to cause tipping.

If the location for contact from one side of the block is beyond the physical boundary of the block, the block will tip slightly and bring the opposite side into contact. There will then be a normal and friction force on both sides of the block. When this occurs, we must treat the force $\boldsymbol{F}_{53}$ as two components ($\boldsymbol{F}^{(1)}_{53}$ and $\boldsymbol{F}^{(2)}_{53}$) as shown in Fig. 11.22b. The two components will be oriented at an angle $\pm\phi$ with respect to the normal direction; however, one component will be oriented at an angle of $+\phi$, and the other component will be oriented at an angle of $-\phi$ as shown in Fig. 11.22b.

When the condition indicated in Fig. 11.22 is satisfied, we can analyze the problem directly using analytical techniques by summing forces and moments. For a graphical solution, the problem can be simplified by resolving $\boldsymbol{F}^{(1)}_{53}$ and $\boldsymbol{F}^{(2)}_{53}$ into a single force that acts through the intersection of the lines of action of the two components. This is shown in Fig. 11.23a. Initially, we will not know the direction for the resultant; however, we now have a three-force member, and the lines of action of all of the forces must intersect at a point. The directions of P and W are already known, and their lines of action intersect at A. Therefore, the line of action of the resultant $\boldsymbol{F}_{53}$ must be along the line AB as shown in Fig. 11.23b. The individual components can then be determined by graphically summing forces.

The procedure indicated here can be used regardless of how the block is actually constrained; for example, it can be mounted over a rod or captured within a cylinder.

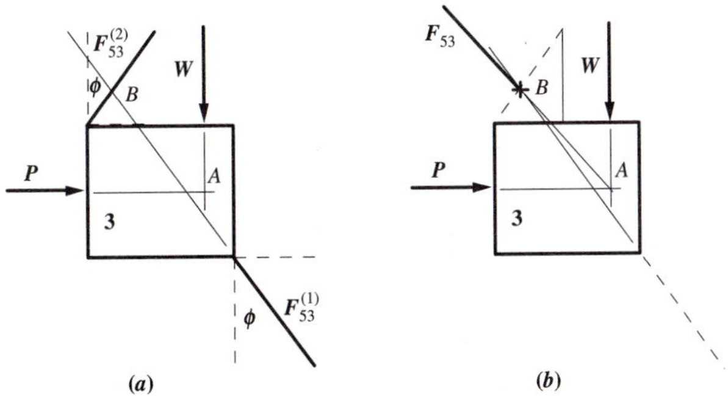

Figure 11.23 Resolving the two components of the slider force into a single component.

11.7.3 Friction in Revolute Joints

Although slider friction is probably the most significant problem area in mechanisms when friction is addressed, pin friction is also important, especially when transmission angles are poor.

Pin friction occurs in a revolute joint at the contact point between the pin and bearing. The effect is to create a torque called a friction torque in the joint. The magnitude of the torque can be determined by considering in detail how the forces are transmitted through the joint. A revolute joint between two arbitrary links (2 and 4) is represented in Fig. 11.24. In the figure, the clearance in the joint is greatly exaggerated. The nominal radius of the pin in the joint is R, and the friction coefficient is again taken as μ.

Figure 11.24b shows the forces that act on the pin part of the joint. The friction force is again equal to μ times the normal force, and the friction angle, ϕ, is computed as was done in Eq. (11.13). That is, $\phi = \tan^{-1}(\mu)$. Because of the friction force, a torque must be applied to link 2 for static equilibrium. This torque caused by the friction force is called the friction torque, and its magnitude is equal to

$$T_F = \mu F_{42_n} R \tag{11.14}$$

The friction force opposes the motion of link 2 relative to link 4, and therefore the torque resulting from the friction force also opposes the relative motion. The total contact force is the resultant of the friction force and the normal force. As shown in Fig. 11.24b, the line of action of the resultant force $\boldsymbol{F}_{42}$ will be tangent to a circle called the friction circle, and the radius of the friction circle is given by

$$R_F = R \sin \phi \tag{11.15}$$

There will be a friction circle associated with each revolute joint, and in each case, the joint force will be tangent to the friction circle. The joint force must be located on the proper side of the friction circle to oppose the relative motion. The effect of the friction circles is to alter slightly the line of action of the joint forces. A binary link with a revolute joint at both ends will still be a two-force member, but the line of action of the forces will not pass through the rotation axes. As shown in Fig. 11.25, there are four possible lines

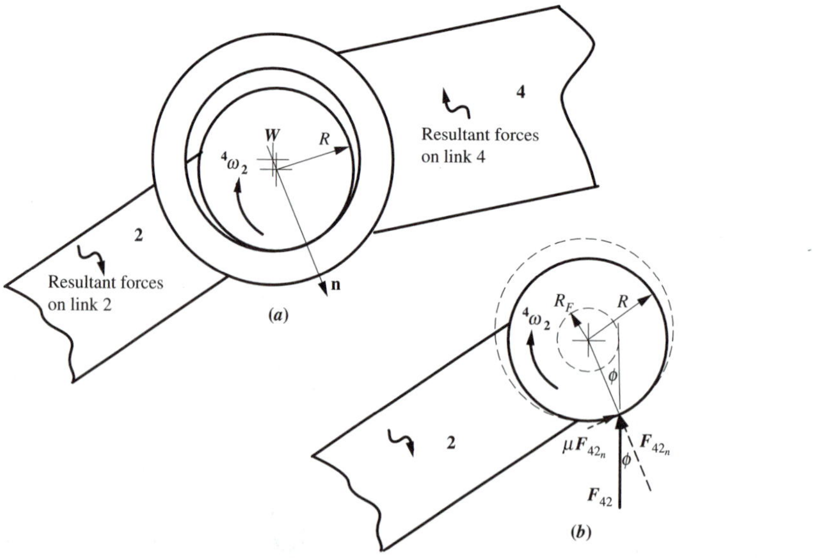

Figure 11.24 A revolute joint with friction.

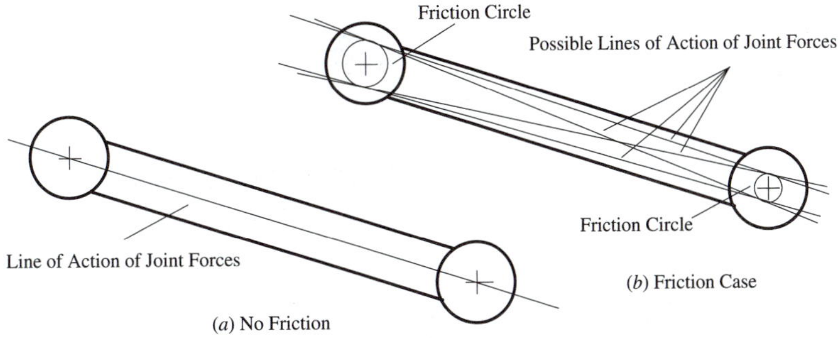

Figure 11.25 Possible lines of action of joint forces when pin friction is involved.

of action for the joint forces, depending on which side of the friction circles the joint forces are located.

To determine the proper side of the friction circles for the joint forces, it is necessary to know the general direction of the forces and the relative motion at the joints before the analysis is begun. This is usually done by conducting a frictionless analysis first and by conducting a velocity analysis. If the direction is chosen incorrectly, the wrong side of the friction circle will be indicated to oppose the relative motion in the joint. This in turn will slightly alter the exact direction of the joint force. Therefore, if any of the joint forces are negative (wrong direction assumed), the analysis must be redone using the proper directions.

EXAMPLE 11.7 *(Analysis of Slider-Crank Mechanism with Friction)*

PROBLEM

A slider-crank mechanism has a piston load of 200 lb in the direction shown in Fig. 11.26, and the crank angle (θ_2) is 120°. The coefficient of friction in each pin and between the slider and frame is 0.2, and the diameter of the pin at each revolute joint is 2 in. Find the torque $\boldsymbol{T}_{12}$ required for static equilibrium with and without friction.

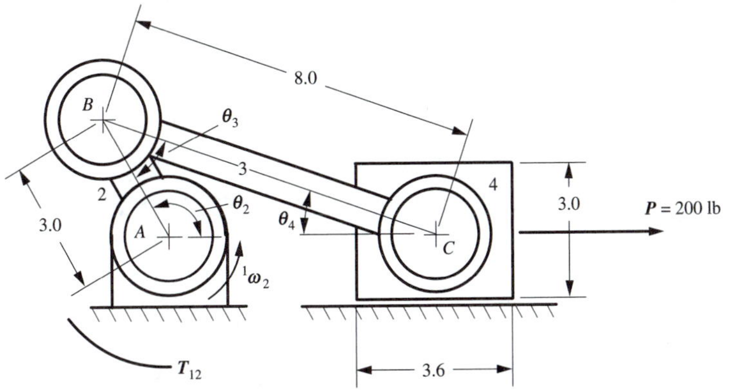

Figure 11.26 The mechanism for Example 11.7.

SOLUTION

We will perform the analysis graphically. Before the friction analysis is conducted, we need to determine the relative motion at each joint and the general direction of the forces at each joint. We can determine the relative motion by inspection because we need only to identify whether

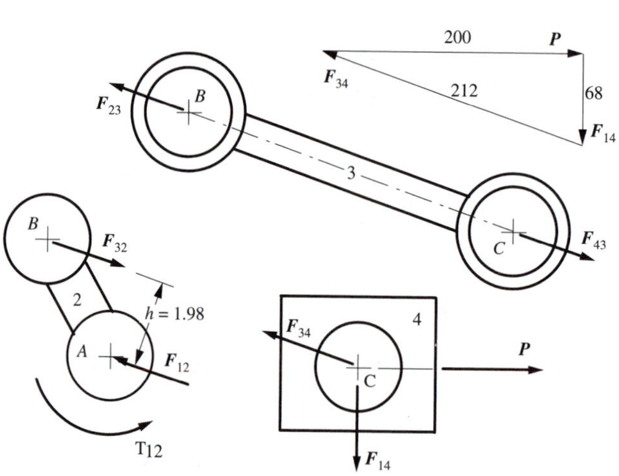

Figure 11.27 Free-body diagrams and force polygon for zero-friction case in Example 11.7.

individual angles are increasing or decreasing. For the direction of the crank rotation given:

θ_2 is increasing
θ_3 is decreasing
θ_4 is decreasing.

To determine the general directions of the joint forces, a zero-friction analysis can be conducted. For this analysis, the free-body diagrams and force polygons are shown in Fig. 11.27. To begin this analysis, we recognize that link 3 is a two-force member. Therefore, the line of action of the joint forces is along the line *BC*. Link 4 is then a three-force member, and the lines of action of the three forces intersect at point *C*. By summing forces vectorially on link 4, the magnitudes of all of the forces can be determined. The force summation equation is

$$\Sigma \boldsymbol{F} = \boldsymbol{0} = \boldsymbol{F}_{14} + \boldsymbol{P} + \boldsymbol{F}_{34} \tag{11.16}$$

The force polygon gives the magnitude and direction for each of the vectors. From equilibrium considerations at each joint, we know

$$\boldsymbol{F}_{32} = -\boldsymbol{F}_{23} = \boldsymbol{F}_{43} = -\boldsymbol{F}_{34} \tag{11.17}$$

and

$$\boldsymbol{F}_{12} = -\boldsymbol{F}_{32} \tag{11.18}$$

This gives us the general direction of all forces at the joints. To determine the torque $\boldsymbol{T}_{12}$ for equilibrium in the nonfriction case, sum moments about point *A* of the free-body diagram for link 2. From this we get

$$T_{12} = hF_{32} = 212(1.98) = 420 \text{ in-lb}$$

By inspection, the torque must be counterclockwise. Therefore,

$$\boldsymbol{T}_{12} = 420 \text{ in-lb CCW (no-friction case)}$$

To analyze the system with friction, we need to compute the friction angle and friction circle radius for each joint. The friction angle is

$$\phi = \tan^{-1}(\mu) = \tan^{-1}(0.2) = 11.31°$$

From Eq. (11.15), the friction circle radius at each joint is

$$R_f = R \sin \phi = 1 \sin 11.31° = 0.20 \text{ in}$$

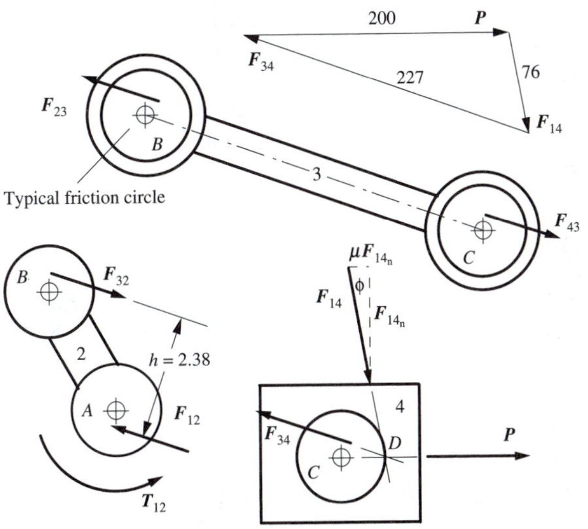

Figure 11.28 Free-body diagrams and force polygon for friction case in Example 11.7.

The friction circles are shown in Fig. 11.28. We must now determine at which side of the friction circles the forces are located. Considering joint C on link 3 first, the force F_{43} is tensile and link 3 is rotating counterclockwise relative to link 4 (θ_4 is getting smaller). We must locate $\boldsymbol{F}_{43}$ tangent to the friction circle such that it opposes this counterclockwise rotation. This means that $\boldsymbol{F}_{43}$ must be located on the upper side of the friction circle. Then $\boldsymbol{F}_{34}$ will be located on the upper side of the corresponding friction circle on link 4.

Next consider joint B on link 3. Force $\boldsymbol{F}_{43}$ is tensile, and link 3 rotates clockwise relative to link 2 (θ_3 is decreasing). We must locate $\boldsymbol{F}_{23}$ tangent to the friction circle such that it opposes this clockwise rotation. This means that $\boldsymbol{F}_{23}$ must be located on the upper side of the friction circle. Similarly, $\boldsymbol{F}_{32}$ will be located on the upper side of the corresponding friction circle on link 2.

Moving to joint A on link 2, the force $\boldsymbol{F}_{12}$ is oriented approximately as shown in Fig. 11.27. Also, link 2 is rotating counterclockwise relative to link 1. If $\boldsymbol{F}_{12}$ is to provide a torque that opposes this motion, it must be located on the lower part of the friction circle as shown in Fig. 11.28.

On link 4, we must orient the force $\boldsymbol{F}_{14}$ relative to the block. Because link 4 is a three-force member, all of the forces must intersect at a point. Because we know the lines of action of $\boldsymbol{P}$ and $\boldsymbol{F}_{34}$, we know the location of the point of intersection (D in Fig. 11.28). We also know that the block is moving to the left relative to the frame so that the friction force must be directed toward the right. Therefore, $\boldsymbol{F}_{14}$ must be inclined as shown in Fig. 11.28, and its line of action must intersect the other two forces on link 4 at D. Note that $\boldsymbol{F}_{14}$ makes contact within the physical limits of the block, and therefore the block does not tip. If contact were indicated beyond the physical limits of the block, we would have to use the procedure indicated in Fig. 11.22.

Now the directions of the forces on link 4 are known. We can sum forces (Eq. 11.16) and solve for the unknowns. The force polygon is shown in Fig. 11.28. From the force polygon and Eqs. (11.17) and (11.18), we can determine the forces on link 2. The torque $\boldsymbol{T}_{12}$ can then be computed. The magnitude of $\boldsymbol{T}_{12}$ is give by

$$T_{12} = hF_{32} = 227(2.38) = 540 \text{ in-lb}$$

By inspection, the torque must be counterclockwise. Therefore,

$$\boldsymbol{T}_{12} = 540 \text{ in-lb CCW}$$

Because of friction, the torque needed for equilibrium has increased by almost 28 percent. The friction coefficients used in this example are high in order to illustrate the procedure. However, the torque requirements to drive a mechanism can increase significantly when friction is involved. Therefore, the designer should reduce the effects of friction either by careful bearing design and lubrication or by altering the linkage geometry (when possible) to reduce the sensitivity of the mechanism to friction. ■

11.8 IN-PLANE AND OUT-OF-PLANE FORCE SYSTEMS

It is important to remember, when analyzing planar mechanisms, that although motion occurs only parallel to a single plane, it is possible to apply fully three-dimensional force systems to the mechanism. Only the components of force and moment that act in the plane of motion affect the action of the mechanism. These *in-plane* force components are the two components of force parallel to the plane of motion and the moment vector normal to that plane. If the force system is resolved into Cartesian components, the remaining three components—the force normal to the plane of motion and the two components of moment parallel to the plane of motion—do not affect the action of the mechanism. Nevertheless, they apply loads to the components and contribute to the stresses in those components. These *out-of-plane* force components act on the mechanism as a structure.

Deflections due to out-of-plane forces are particularly important in planar mechanisms since an excessive deflection may result in significant deviations from parallelism of the joint axes, which can result in binding of the mechanism or generation of excessive internal stresses. For this reason, the deflection of members is often a concern even when strength is more than adequate.

Since in-plane and out-of-plane force systems affect a mechanism in quite different ways, it is convenient to decompose the force analysis of a planar mechanism into separate analyses of the in-plane and out-of-plane force systems. However, this must be done with some care, since an in-plane force may contribute to an out-of-plane moment and vice versa.

EXAMPLE 11.8 ***(Analysis of Out-of-Plane Forces)***

PROBLEM

The SCARA robot shown in Figure 11.29 has vertical revolute axes 1, 2, and 3. It carries a payload of 50 lb with center of mass on the axis of joint 3. In addition, the members A and B

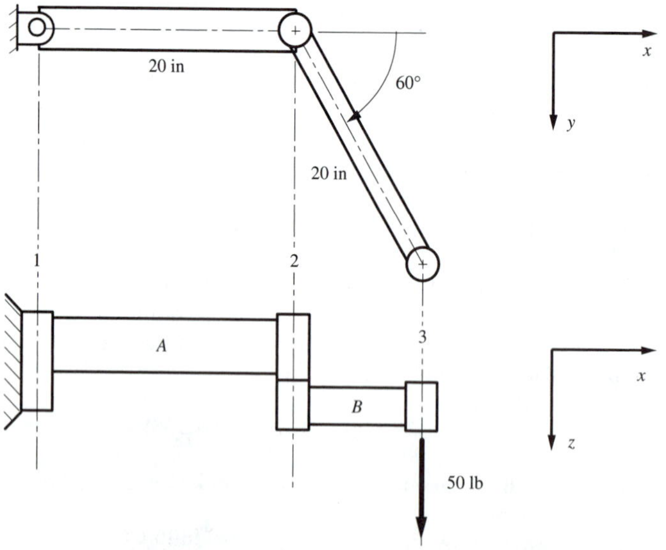

Figure 11.29 The SCARA robot considered in Example 11.8.

can be regarded as uniform beams with respective weights 60 lb and 40 lb. Characterize all the loads on members A and B in the position shown.

SOLUTION

Here all the forces are out-of-plane forces. As far as a force analysis is concerned, the manipulator behaves as a cranked cantilever beam. To find the loads on the members A and B, it is first necessary to draw free-body diagrams of each of them. A free-body diagram of member B in the x-z plane is shown in Fig. 11.30.

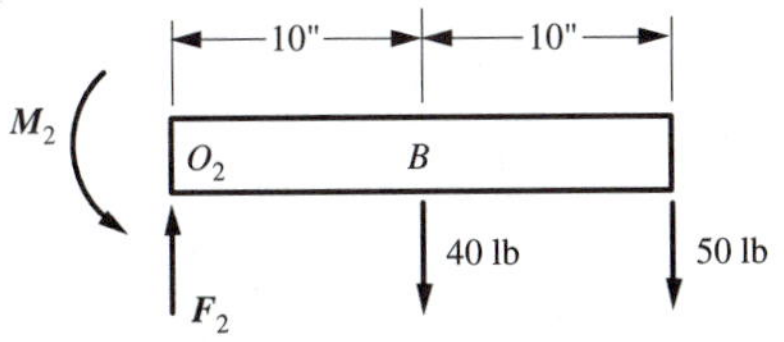

Figure 11.30 The free-body diagram of member B.

As may be seen, member B is a simple cantilever beam considered to be embedded at joint 2. The shear force F_2 and bending moment M_2 at joint 2 can be found by writing the vertical force equilibrium equation and the moment equation about O_2 .

$$\sum F_y = 0: \quad F_2 = 40 + 50 = 90 \text{ lb}$$

$$\sum M_2 = 0: \quad M_2 = 40 \times 10 + 50 \times 20 = 1400 \text{ in-lb}$$

The free-body diagram for member A is shown in Fig. 11.31.

The reaction force F_1 and moments M_{1y} and M_{1x} at joint axis 1 can now be computed using vertical force equilibrium together with moment balances about the y and x directions.

In the plane of the beam:

$$\sum F_y = 0: \quad F_1 = 60 + 90 = 150 \text{ lb}$$

$$\sum M_y = 0: \quad M_{1y} = 1400 \sin 30° + 90 \times 20 + 60 \times 10 = 3100 \text{ in-lb}$$

About the longitudinal axis of the beam:

$$\sum M_x = 0: \quad M_{1x} = -1400 \cos 30° = -1212 \text{ in-lb}$$

Notice that, although there are no forces in the plane of motion and no motion of the system, there are significant forces acting on the members. It is always necessary to consider out-of-plane forces, as well as in-plane forces. ■

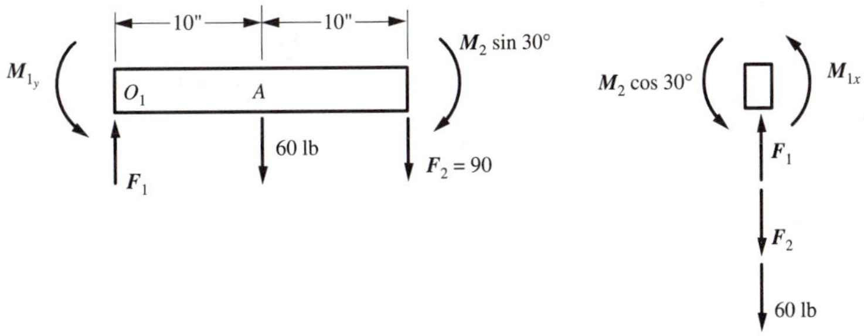

Figure 11.31 The free-body diagram of member A. This figure shows the view of the member in its x-y plane together with an end view showing the torsional components. The bending moment M_2 at joint 2 is applied, along with the shear force to member A. However, because of the 60° angle between the longitudinal axes of members A and B, it is necessary to resolve M_2 into a component $M_2 \sin 30°$ in the plane of member A, along with a component $M_2 \cos 30°$, which is a torsional moment about its longitudinal axis.

EXAMPLE 11.9 *(Three-Dimensional Force Analysis of Robot)*

PROBLEM

The SCARA robot of Fig. 11.29 and Example 11.8 is required to exert a 20-lb horizontal tool force on the line of action shown in Fig. 11.32. Find the torques that must be exerted by the actuators at joints 1 and 2 and the reaction forces and moments at the joints. The weights of the members and end effector remain the same as in Example 11.8.

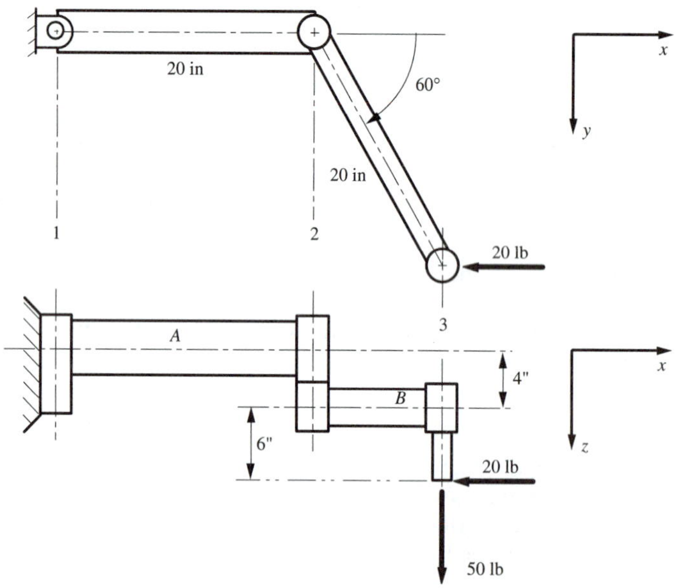

Figure 11.32 A 20-lb horizontal tool load has been added to the weight load in Example 11.8. In this example we have both in-plane and out-of-plane force systems.

SOLUTION

There are many different possible approaches to this problem. However, when dealing with fully three-dimensional systems like this, it is usually best to decide on three orthogonal coordinate axis directions and resolve all force and moment components into those directions. This may yield more equations than other choices of directions in which to resolve, but it has the advantage of being a very systematic procedure that reduces the chance of error.

It is first necessary to draw free-body diagrams of each of the two bodies in the system. Since the force system is three-dimensional, in Fig. 11.33 we have resorted to three orthogonal views of each member to ensure that every force component is visible. In fact, each component should appear on at least two of these views.

We can now proceed to write the six static equilibrium equations for each member by referring to the free-body diagrams in Fig. 11.33:
Member B:

$$\sum F_x = 0: \quad F_{2x} = 20$$

$$\sum F_y = 0: \quad 0 = 0$$

$$\sum F_z = 0: \quad F_{2z} + 40 + 50 = 0$$

Taking moments about point O_2:

$$\sum M_x = 0: \quad M_{2x} + 50 \times 20 \sin 60^\circ + 40 \times 10 \sin 60^\circ = 0$$

$$\sum M_y = 0: \quad M_{2y} = 10 \times 20 + 50 \times 20 \cos 60^\circ + 40 \times 10 \cos 60^\circ$$

$$\sum M_z = 0: \quad M_{2z} + 20 \times 20 \sin 60^\circ = 0$$

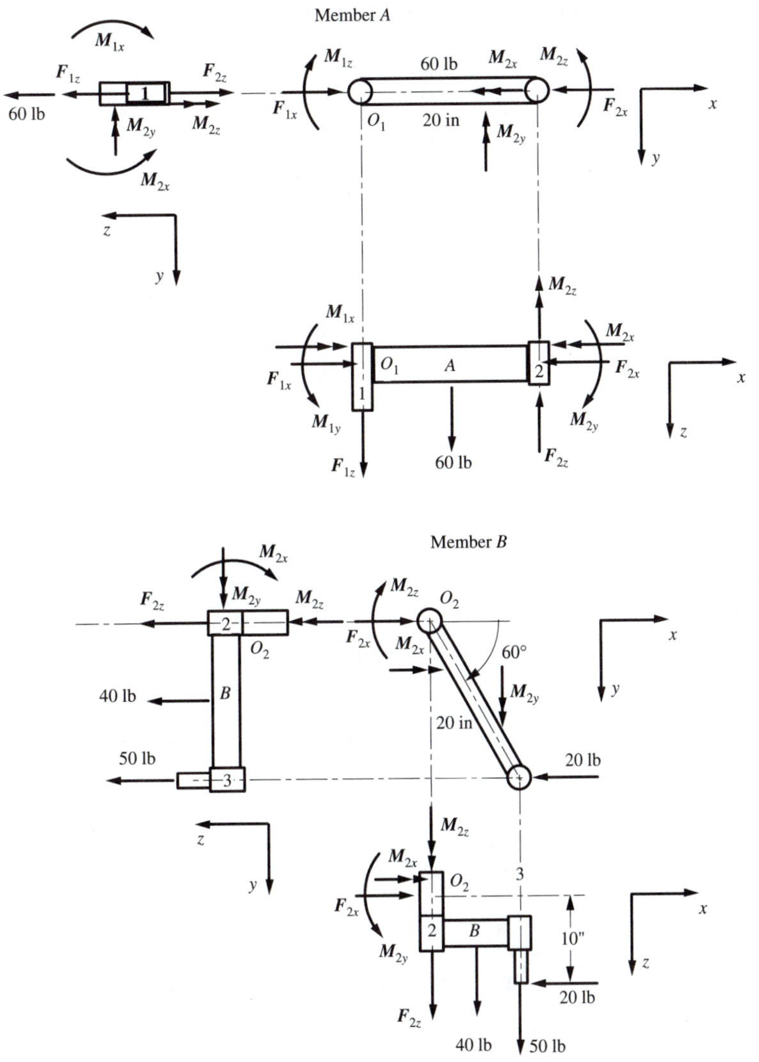

Figure 11.33 Free-body diagrams for Example 11.9. Each free body is shown in three orthogonal views: projected on the *x-y* plane, the *x-z* plane, and the *y-z* plane, respectively.

Solution gives:

$$F_{2x} = 20\text{ lb}$$
$$F_{2z} = -90\text{ lb}$$
$$M_{2x} = -1212\text{ lb-in}$$
$$M_{2y} = 900\text{ lb-in}$$
$$M_{2z} = -346\text{ lb-in}$$

Member *A*:

$$\sum F_x = 0: \quad F_{1x} = F_{2x}$$
$$\sum F_y = 0: \quad 0 = 0$$
$$\sum F_z = 0: \quad F_{1z} + 60 = F_{2z}$$

Taking moments about point O_1:

$$\sum M_x = 0: \qquad M_{1x} = M_{2x}$$
$$\sum M_y = 0: \qquad M_{1y} = M_{2y} + 60 \times 10 - F_{2z} \times 20$$
$$\sum M_z = 0: \qquad M_{1z} = M_{2z}$$

Substitution of the previously calculated values for F_{2x}, F_{2z}, M_{2x}, M_{2y}, and M_{2x} gives

$$F_{1x} = 20 \text{ lb}$$
$$F_{1z} = -150 \text{ lb}$$
$$M_{1x} = -1212 \text{ in-lb}$$
$$M_{1y} = 900 + 600 - (-90) \times 20 = 3300 \text{ in-lb}$$
$$M_{1z} = -346 \text{ in-lb}$$

This completes the solution. The torques M_{1z} and M_{2z} that must be produced by the actuators at joints 1 and 2, respectively, are both 346 in-lb in magnitude. The reaction forces and moments F_{1x}, F_{1z}, M_{1x}, M_{1y}, F_{2x}, F_{2z}, M_{2x}, M_{2y} may be compared with the results of Example 11.8. As may be seen, the vertical forces F_{1z} and F_{2z} remain the same, as does M_{1x}. The other forces and moments are changed by the presence of the horizontal tool load. ■

11.9 CONSERVATION OF ENERGY AND POWER

When a linkage is subjected to significant loads only on the input and output links and the system can be regarded as static, a quick and easy method of computing the ratio of input force/torque to output force/torque is available. This is particularly useful if a velocity analysis has already been performed for other reasons, because a velocity analysis is needed as part of the solution.

For a given mechanism, the input power equals the output power if the following are satisfied:

1. There are no energy losses in the mechanism. This means that there is no friction in the joints or no other form of energy dissipation.
2. There is no energy stored in the mechanism as it moves. This means that there are no springs in the system and that the energy associated with potential energy because of the weight of the links and the change in the centers of gravity of the links is negligible.
3. The velocities are small enough or the masses are small enough that changes in kinetic energy are negligible.

Thus, if the input and output loads are torques $\boldsymbol{T}_i$ and $\boldsymbol{T}_o$, respectively, and the angular velocities of the input and output links are $\boldsymbol{\omega}_i$ and $\boldsymbol{\omega}_o$, respectively:

$$\boldsymbol{T}_i \cdot \boldsymbol{\omega}_i + \boldsymbol{T}_o \cdot \boldsymbol{\omega}_o = 0$$

The + sign in the equation implies that if $\boldsymbol{T}_i$ is in the same direction as $\boldsymbol{\omega}_i$, as is often, but not necessarily, the case, then $\boldsymbol{T}_o$ will be in the opposite direction to $\boldsymbol{\omega}_o$.

Similarly, if the input and output are $\boldsymbol{F}_i$ and $\boldsymbol{F}_o$, and $\boldsymbol{v}_{pi}$ and $\boldsymbol{v}_{po}$ are the respective velocities of their points of application, then

$$\boldsymbol{F}_i \cdot \boldsymbol{v}_{pi} + \boldsymbol{F}_o \cdot \boldsymbol{v}_{po} = 0$$

Similarly, for a force input and torque output

$$\boldsymbol{F}_i \cdot \boldsymbol{v}_{pi} + \boldsymbol{T}_o \cdot \boldsymbol{\omega}_o = 0$$

and for a torque input and force output

$$\boldsymbol{T}_i \cdot \boldsymbol{\omega}_i + \boldsymbol{F}_o \cdot \boldsymbol{v}_{po} = 0$$

or, in general,

$$\sum_{k=1}^{m} \boldsymbol{T}_k \cdot \boldsymbol{\omega}_k + \sum_{j=1}^{n} \boldsymbol{F}_j \cdot \boldsymbol{v}_{pj} = 0 \qquad \textbf{(11.19)}$$

where $\boldsymbol{T}_k$, $k = 1, 2, \ldots, m$, are all torques acting on members of the system and $\boldsymbol{\omega}_k$, $k = 1, 2, \ldots, m$, are the angular velocities of the links to which they are respectively applied. $\boldsymbol{F}_j$, $j = 1, 2, \ldots, n$, are all of the external forces acting on the system, and $\boldsymbol{v}_{pj}$, $j = 1, 2, \ldots, n$, are the respective velocities of their points of applications. Thus, more generally, if the necessary velocity information for the system is available, a static force analysis can be performed using energy conservation as an alternative to static equilibrium. Depending on the system characteristics, this may or may not be more efficient than performing a force analysis using free-body diagrams and the equations for static equilibrium.

When computing the power, the dot product for the force-velocity expression is given by

$$\boldsymbol{F}_i \cdot \boldsymbol{v}_{pi} = F_i v_{pi} \cos \theta_i = F_x v_x + F_y v_y + F_z v_z$$

where the magnitudes of the force and torque can be both positive and negative and θ_i is the angle between $\boldsymbol{F}_i$ and v_{pi}. The dot product of the torque and angular velocity can be computed by multiplying the magnitudes of the two vectors, again realizing that the signs can be either positive or negative depending on whether the torque and angular velocity are in the same direction or opposite directions. Situations of positive and negative power are illustrated schematically in Fig. 11.34.

The velocity analysis can be conducted using any procedure that is convenient. Note, however, that only the velocities associated with the points and links associated with the application of external forces and torques are needed. Because only selected velocities are usually required in Eq. (11.19), force analyses using the conservation of energy are often done in conjunction with the use of instant centers for the velocity analysis. Of course, if the velocities are already known from some other type of analysis, these can be used directly.

Note that even though Eq. (11.19) involves the velocities, the results will not depend on the value chosen for the input velocity. This must be the case because we could also conduct the force analysis using free-body diagrams and the equations for force equilibrium, and these equations are a function of position and the applied forces only. Recall that the velocity problem is a linear problem. Therefore, if the input velocity is multiplied by a constant, then all of the velocities including the angular velocities will be multiplied by the same constant. Therefore, if we divide Eq. (11.19) by the magnitude of the input velocity, the force/torque results will be unchanged. Because of this, we need not know the actual value for the input velocity before conducting the force analysis. In fact, the mechanism need not be moving at all. Because of this, a common value to choose for the input velocity is 1.

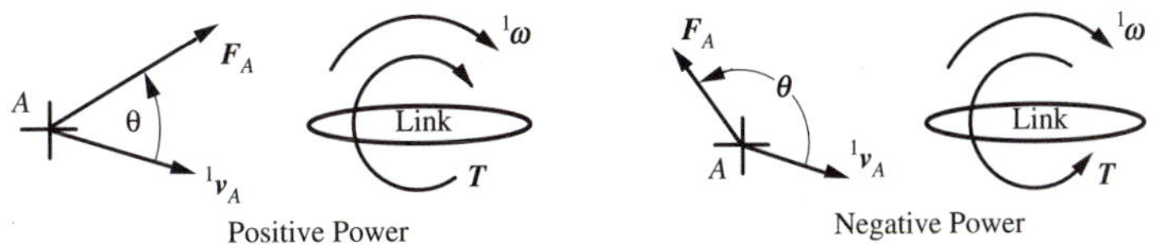

Figure 11.34 Situations for positive and negative power.

EXAMPLE 11.10 ***(Analysis of Slider-Crank Mechanism Using Conservation of Power)***

PROBLEM

Estimate the necessary driving torque for the mud pump of Example 11.6 using conservation of power.

SOLUTION

To obtain a quick estimate of the forces, we will neglect the effect of the weight of the components. This is justifiable because those weights are much smaller than the piston load of $P = 21{,}340$ lb (Example 11.6).

We will conduct the velocity analysis using instant centers because only two velocities are required. Since the piston load can be considered to act at point C on member 4 and the driving torque acts on member 2, we need I_{12}, I_{14}, and I_{24} to develop the necessary relationship between the input and output velocity using the instantaneous center method. I_{24} is located as shown in Fig. 11.35. The distance $I_{24}I_{12}$ can be scaled from the drawing or calculated. To calculate it, use angle ACB (11.096° from Example 11.6) and first find

$$\phi = \angle BI_{24}A = 90° - 11.096° = 78.904°$$

$$\beta = \angle BAI_{24} = 90° - 60° = 30°$$

and

$$\gamma = \angle ABI_{24} = 180° - \phi - \beta = 180° - 78.904° - 30° = 71.096°$$

Now use the sine rule to find the length $r = AI_{24}$

$$r = \sin\gamma \frac{AB}{\sin\phi} = \frac{8 \sin 71.096°}{\sin 78.904°} = 7.713 \text{ in.}$$

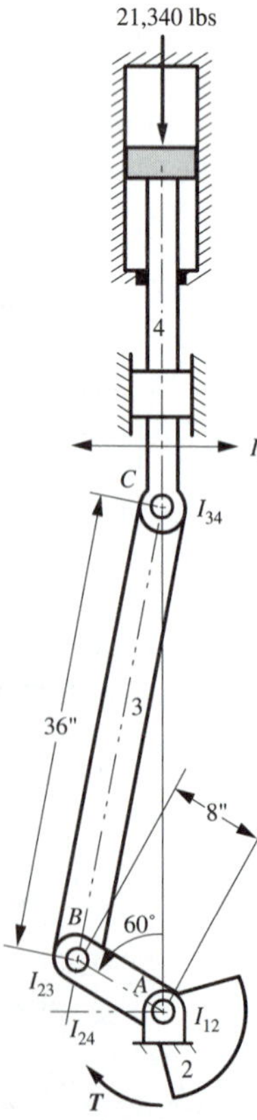

Figure 11.35 Location of instant centers for mud pump.

The velocity of I_{24} is equal to the velocity of C_4 where the force of 21,340 lb is applied. Also, if we consider I_{24} to be a point on link 2,

$$\boldsymbol{v}_{I_{24}} = {}^1\boldsymbol{\omega}_2 \times \boldsymbol{r}_{I_{24}/A}$$

and the magnitude is the velocity of I_{24},

$$v_{I_{24}} = {}^1\omega_2(\boldsymbol{r}_{I_{24}/A})$$

From conservation of power,

$$\boldsymbol{T} \cdot {}^1\boldsymbol{\omega}_2 + \boldsymbol{P} \cdot \boldsymbol{v}_{C_4} = 0 \qquad \textbf{(11.20)}$$

Let us assume that link 2 is rotating clockwise. Then $\boldsymbol{P}$ and $\boldsymbol{v}_{C_4}$ are in opposite directions. Therefore, power is being taken out at the piston and put into the system at the crank. Consequently, the torque, $\boldsymbol{T}$, must be in the same direction as the angular velocity of link 2 or clockwise. Therefore, we need only to use Eq. (11.20) to find the magnitude of $\boldsymbol{T}$. This gives

$$T = P\frac{v_{C_4}}{{}^1\omega_2} = P\frac{v_{I_{24}}}{{}^1\omega_2} = P\frac{{}^1\omega_2(r_{I_{24}/A})}{{}^1\omega_2} = P(r_{I_{24}/A})$$

Therefore,

$$T = 21340(7.713) = 164{,}600 \text{ in-lb or } 13{,}700 \text{ ft-lb clockwise}$$

This may be compared with the value of $T = 13{,}800$ ft-lb obtained from the full static equilibrium analysis. The difference between the values is accounted for by the fact that the weights of the linkage members were included in the former analysis. ■

EXAMPLE 11.11 *(Analysis of Four-Bar Linkage Using Conservation of Power)*

PROBLEM

In the four-bar linkage represented in Fig. 11.36, a torque $\boldsymbol{T}_{12}$ is applied to link 2, and it is resisted by a torque $\boldsymbol{T}_{14}$ on link 4. Find a relationship for the mechanical advantage of the mechanism in terms of the mechanism geometry.

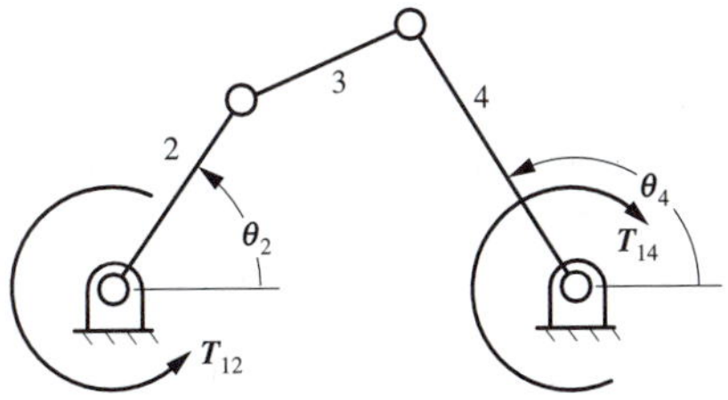

Figure 11.36 The linkage for Example 11.11.

SOLUTION

The mechanical advantage (MA) for the linkage is a function of position and is given by

$$MA = \frac{\text{output torque}}{\text{input torque}} = \frac{T_{14}}{T_{12}} \qquad \textbf{(11.21)}$$

Therefore, we must find a geometric relationship equal to the ratio of the output torque divided by the input torque. We will use conservation of power for the analysis and conduct the velocity analysis using instant centers of velocity. To begin, the power expression is

$$\boldsymbol{T}_{12} \cdot {}^1\boldsymbol{\omega}_2 + \boldsymbol{T}_{14} \cdot {}^1\boldsymbol{\omega}_4 = 0 \qquad \textbf{(11.22)}$$

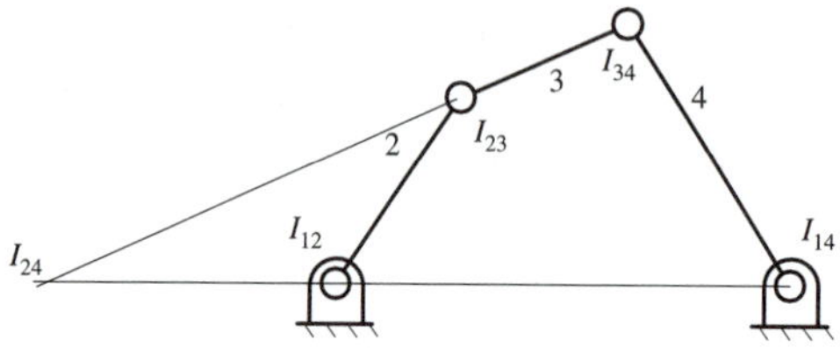

Figure 11.37 Instant center locations for the linkage in Fig. 11.36.

From this expression, if $\boldsymbol{T}_{12}$ and ${}^1\boldsymbol{\omega}_2$ are in the same direction, then $\boldsymbol{T}_{14}$ and ${}^1\boldsymbol{\omega}_4$ must be in opposite directions. Therefore, if we know the signs of $\boldsymbol{T}_{12}$, ${}^1\boldsymbol{\omega}_2$, and ${}^1\boldsymbol{\omega}_4$, we can determine the sign of $\boldsymbol{T}_{14}$ by inspection after a kinematic analysis is conducted. Consequently, we can concentrate on the magnitude of $\boldsymbol{T}_{14}$. We can compute the magnitude of $\boldsymbol{T}_{14}$ directly from Eq. (11.22). That is,

$$T_{14} = \frac{T_{12}{}^1\omega_2}{{}^1\omega_4} = T_{12}\frac{{}^1\omega_2}{{}^1\omega_4}$$

Therefore,

$$\frac{T_{14}}{T_{12}} = \frac{{}^1\omega_2}{{}^1\omega_4} \tag{11.23}$$

Based on our knowledge of instant centers, the ratio ${}^1\omega_2/{}^1\omega_4$ will be a function of geometry only. To find the ratio of angular velocities, we need first to find the instant centers I_{12}, I_{14}, and I_{24}. These are shown in Fig. 11.37. The velocity of the two points located at I_{24} is given by

$$\boldsymbol{v}_{I_{24}} = {}^1\boldsymbol{\omega}_2 \times \boldsymbol{r}_{I_{24}/I_{12}} = {}^1\boldsymbol{\omega}_4 \times \boldsymbol{r}_{I_{24}/I_{14}}$$

and

$${}^1\omega_2 r_{I_{24}/I_{12}} = {}^1\omega_4 r_{I_{24}/I_{14}}$$

Therefore,

$$\frac{{}^1\omega_2}{{}^1\omega_4} = \frac{r_{I_{24}/I_{14}}}{r_{I_{24}/I_{12}}} \tag{11.24}$$

Equation (11.24) shows that the ratio of the angular velocities is clearly a function of position, that is, of the location of the instant centers.

Now combining Eqs. (11.21), (11.23), and (11.24) gives

$$MA = \frac{T_{14}}{T_{12}} = \frac{{}^1\omega_2}{{}^1\omega_4} = \frac{r_{I_{24}/I_{14}}}{r_{I_{24}/I_{12}}}$$

Note that the sign of either ${}^1\omega_2$ or ${}^1\omega_4$ can be negative. Therefore, mechanical advantage may be positive or negative, although in many instances only the magnitude is of interest. When MA is positive, the torques are in the same direction, and when it is negative they are in opposite directions. In this particular example, $r_{I_{24}/I_{14}}$ and $r_{I_{24}/I_{12}}$ (and therefore ${}^1\omega_2$ and ${}^1\omega_4$) will be in the same direction when I_{24} is outside of I_{12} and I_{14}. They will be in opposite directions when I_{24} is between I_{12} and I_{14}.

■

11.10 VIRTUAL WORK

It is possible to apply the conservation of enegy method to obtain force relationships in truly static systems in which no real movement occurs. The technique is to imagine a small displacement of the point of application of the input force and compute the corresponding displacement of the point or points of application of the output force or forces. The sum of the work done by all external forces must be zero. This is the principle of *virtual work*. It is identical to conservation of energy except that the displacements are virtual or imaginary.

The problem of deriving a consistent set of displacements of the points of application of the external forces, which is the key to this technique, can be converted to a velocity analysis. If the velocity of each application point is multiplied by the same small time interval, the result is a set of consistent, small displacements. The time interval, being the same in all cases, will cancel from the energy equation so that it becomes, in fact, identical to Eq. (11.19). The way in which the method is used will be illustrated by the following example.

EXAMPLE 11.11 *(Analysis of Vice Grip Using Virtual Work)*

PROBLEM

Find the grip force, $\boldsymbol{F}_G$, when the force applied to the handles of the vice-grip pliers of Fig. 11.6 is 25 lb. The lines of action of the grip forces and hand forces are located as indicated in Fig. 11.7.

SOLUTION

Although the system depicted in Fig. 11.6 is truly static, we can still imagine a small displacement such as might occur if the workpiece were compliant and deflected a little. When using virtual work, we must account for all of the work done during the virtual displacement. In Fig. 11.38, work will be done by the forces at G and H and by the spring force. However, the spring force is so much less than the other forces that the energy stored in the spring can be ignored.

To obtain consistent displacements efficiently, we can again use the instantaneous center method. The six instantaneous centers of the mechanism are located as shown in Fig. 11.38. If member 1 is treated as the base and the point of application of $\boldsymbol{F}_H$ on the lower handle is displaced, there will also be a displacement of the point of application of $\boldsymbol{F}_G$ on the lower jaw. Since the former is on member 3 and the latter is on member 2, we need I_{13}, I_{12}, and I_{23} to obtain the necessary consistent pair of velocities (or displacements).

Scaling from the drawing shown in Fig. 11.38:

$$\overrightarrow{I_{12}I_{23}} = 1.469 \text{ in}; \qquad \overrightarrow{I_{13}I_{23}} = 0.186 \text{ in}$$

Hence

$$\omega_2/\omega_3 = 0.186/1.469$$

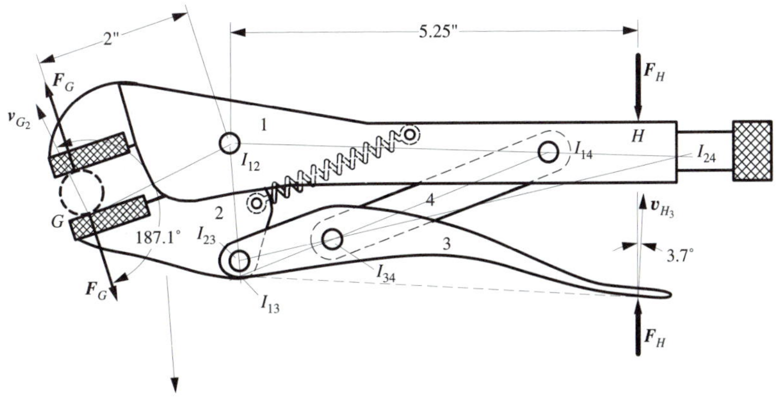

Figure 11.38 Locations of the instantaneous centers for the vice-grip pliers of Example 11.11.

Also

$$\overrightarrow{HI_{13}} = 5.129 \text{ in} \quad \text{and} \quad \overrightarrow{GI_{12}} = 2.019 \text{ in}$$

Now

$$v_G/v_H = (\overrightarrow{GI_{12}} \times \omega_2)/\overrightarrow{HI_{13}} \times \omega_3) = (2.019 \times 0.186)/(5.129 \times 1.469)$$

or

$$v_G = 0.0498 v_H$$

Also, the angle between $\boldsymbol{v}_H$ and $\boldsymbol{F}_H$ is 3.7°, while that between $\boldsymbol{v}_G$ and $\boldsymbol{F}_G$ is 187.1°. Hence:

$$F_H v_H \cos 3.7° + F_G v_G \cos 187.1° = 0$$

or

$$v_H \times 25 \cos 3.7° + F_G \times v_H \times 0.0498 \times \cos 187.1° = 0$$

Notice that the equation may be divided by v_H, indicating that the results are independent of the value chosen for v_H. The final result for F_G is

$$F_G = 505 \text{ lb}$$

This value may be compared with that obtained by a completely different method in Example 11.2. The small difference is due to graphical inaccuracies.

In the present instance we chose not to include the spring force because it proved to be negligible in Example 11.2. It could have been included, however, by finding the virtual velocity of its point of application on link 2 and including the scalar product of that velocity with the spring force in the conservation of energy equation (Eq. 11.19). The change in F_G would be insignificant, however. ■

11.11 GEAR LOADS

11.11.1 Spur Gears

In the analysis, we will assume that the spur gears have an involute profile. If friction is neglected, the force exerted by one of a pair of meshing gears on the other acts along the line normal to the teeth at the point of contact. That is, it acts along the line of action or pressure line shown in Fig. 11.39.

Free-body diagrams of the gear and pinion are as shown in Fig. 11.40. The normal force W is tangent to the base circles of both the pinion and gear, and the torque on the pinion is

$$T_1 = W r_{b_1} = W r_{p_1} \cos \psi \tag{11.25}$$

The torque on the gear is

$$T_2 = W r_{b_2} = W r_{p_2} \cos \psi \tag{11.26}$$

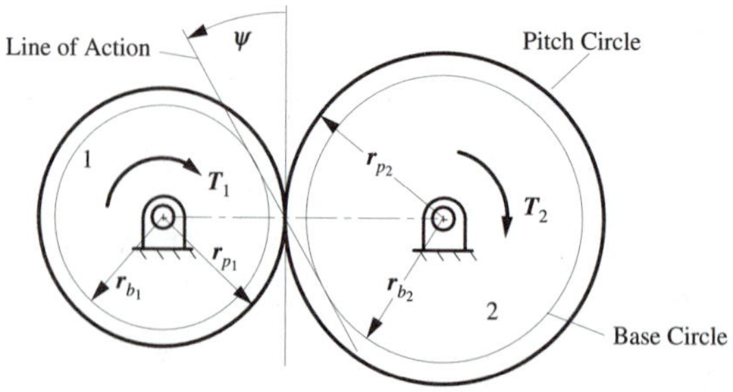

Figure 11.39 Torque transmitted through spur gears.

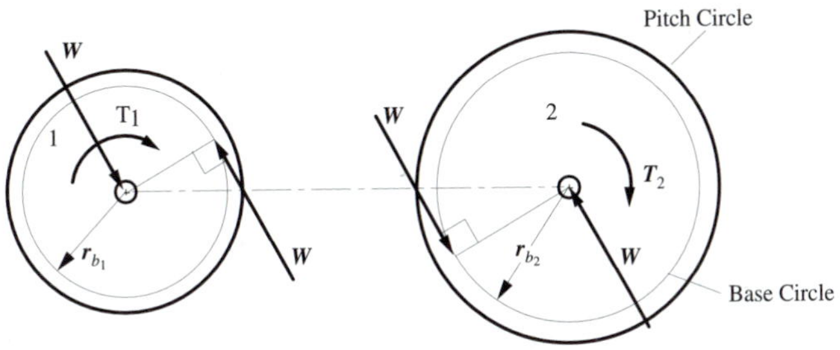

Figure 11.40 Free-body diagrams of spur gears.

Therefore

$$\frac{T_2}{T_1} = \frac{r_{p_2}}{r_{p_1}} = \frac{N_2}{N_1}$$

where N_1 and N_2 are the numbers of teeth on the pinion and gear, respectively.

Also, from Eqs. (11.25) and (11.26)

$$W = \frac{T_1}{r_{p_1} \cos \psi} = \frac{T_2}{r_{p_2} \cos \psi} \tag{11.27}$$

It may be seen that W is also the load on the bearings supporting the shafts of both the gear and the pinion.

If friction is considered, an additional force component of μW, where μ is the coefficient of friction, appears tangent to the teeth at the point of contact. This is shown in Fig. 11.41. Because W is constant, the friction force is constant in magnitude, and the bearing reactions (F_B) are also constant in magnitude and given by

$$F_B = \sqrt{W^2 + (\mu W)^2} = W\sqrt{1 + \mu^2}$$

However, the torques are

$$T_1 = Wr_{b_1} - \mu Wx \tag{11.28}$$

$$T_2 = Wr_{b_2} - \mu Wy \tag{11.29}$$

The distances x and y vary as the point of contact moves along the path of contact. However,

$$x + y = r_{b_1} \tan \psi + r_{b_2} \tan \psi = (r_{p_1} + r_{p_2}) \sin \psi = \frac{N_1 + N_2}{2P} \sin \psi \tag{11.30}$$

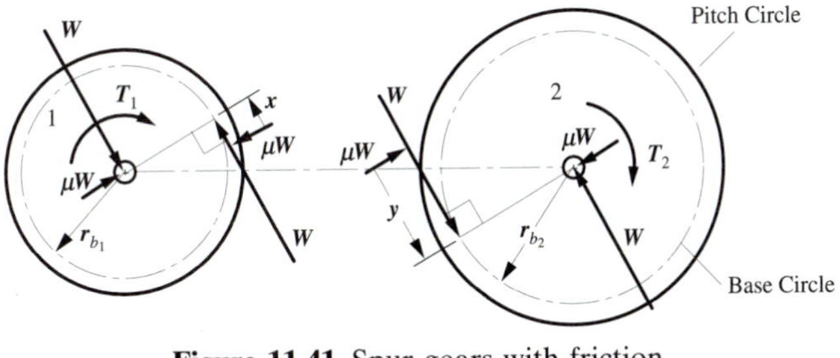

Figure 11.41 Spur gears with friction.

where P is the diametrical pitch. From Eqs. (11.28)–(11.30),

$$\frac{T_2}{T_1} = \frac{r_{b_2} - \mu y}{r_{b_1} - \mu x} = \frac{\dfrac{N_2}{2P}\cos\psi - \mu\left(\dfrac{N_1 + N_2}{2P}\sin\psi - x\right)}{\dfrac{N_1}{2P}\cos\psi - \mu x}$$

or

$$\frac{T_2}{T_1} = \frac{N_2\cos\psi - \mu(N_1 + N_2)\sin\psi + 2\mu P x}{N_1\cos\psi - 2\mu P x}$$

Thus, T_2/T_1 is no longer constant but is a function of the position x of the point of contact. This is a source of torsional vibration. The amount of variation is a function of the length of the path of contact. Note that T_2/T_1 is always less than N_2/N_1 when averaged over the contact cycle. The sharing of loads among two or more pairs of teeth in contact does not change the bearing loads, but it does affect the torque fluctuations and tends to diminish them. Since the load sharing depends on the tooth stiffnesses, modeling of this effect becomes complicated and is beyond the scope of this text.

11.11.2 Helical Gears

The teeth of an involute helical gear are similar in cross section to those of a spur gear. However, instead of being cut with their faces parallel to the gear axis, they are cut at an angle α to that axis where α is the helix angle of the gear. It is similar to the lead angle of a screw, except that the teeth of the helical gear are straight. That is, a straight line tangent to the pitch cylinder of the gear at its midplane can be drawn on the face of each tooth. The set of such lines from all teeth form the generators of a hyperboloid of one sheet. If this line is moved parallel to itself along an involute curve with the pitch circle radius equal to that of the pitch cylinder, it will generate the tooth face.

If friction is neglected, the force transmitted between contacting teeth is normal to the gear teeth at the point of contact. That is, it lies in the normal plane. It is inclined at angle ψ_n to the cylinder tangent at the point of contact as shown in Fig. 11.42. Here, ψ_n is the

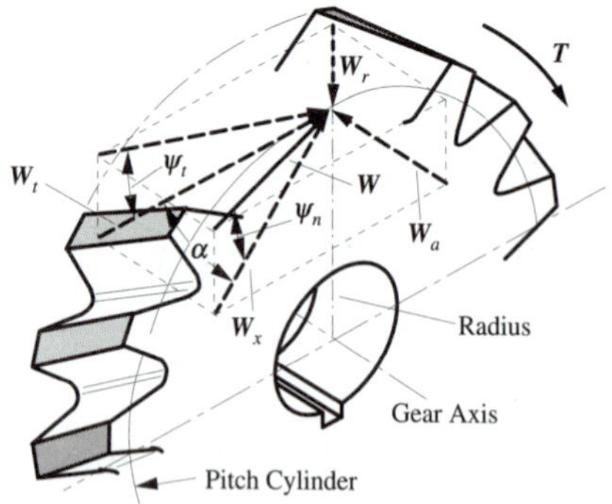

Figure 11.42 Resolution of contact force on tooth of helical gear. W is the contact force acting along the pitch line. W_r is the radial component of W. W_x is the component of W that is orthogonal to W_r. W_x is tangent to the pitch cylinder of the gear. W, W_r, and W_x lie in a plane normal to the gear tooth face at the point of contact. α is the helix angle of the teeth. W_a is parallel to the gear axis of rotation, and W_t is the tangent to the pitch cylinder normal to W_a and passing through the pitch point.

pressure angle for the hob used to cut the gear. Resolving the contact force W into the plane normal to the gear tooth gives

$$W_r = W \sin \psi_n$$

and

$$W_x = W \cos \psi_n$$

Resolving W_x in the plane tangent to the pitch cylinder

$$W_t = W_x \cos \alpha = W \cos \psi_n \cos \alpha$$

and

$$W_a = W_x \sin \alpha = W \cos \psi_n \sin \alpha$$

Also, from Fig. 11.42,

$$W_r \tan \psi_n = W_r \tan \psi_t \cos \alpha$$

and

$$\tan \psi_t = \frac{\tan \psi_n}{\cos \alpha} \tag{11.31}$$

Hence

$$\frac{W_r}{W_t} = \frac{\sin \psi_n}{\cos \psi_n \cos \alpha} = \frac{\tan \psi_n}{\cos \alpha} = \tan \psi_t$$

or

$$W_r = W_t \tan \psi_t$$

$$\frac{W_a}{W_t} = \tan \alpha \tag{11.32}$$

or

$$W_a = W_t \tan \alpha \tag{11.33}$$

The torque acting on the gear is represented in the free-body diagram in Fig. 11.43. From the figure,

$$T = W_t r_p \tag{11.34}$$

where r_p is the radius of the pitch circle. Using these relationships, the components W_t, W_a, and W_r can be computed in terms of the torque T.

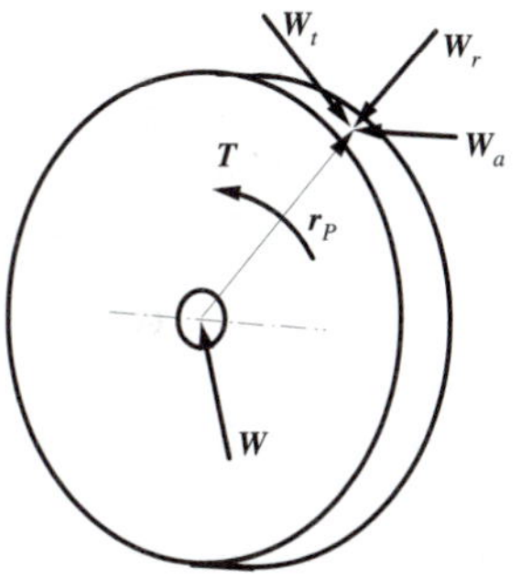

Figure 11.43 The free-body diagram of the gear. The only component of the contact force W that exerts a moment about the gear axis is W_t.

Note that the axial thrust W_a also produces a couple tending to rotate the gear in the plane of the shaft axis. This produces an asymmetry in the bearing loads. The couple has magnitude

$$T_a = W_a r_p = W_t r_p \tan\alpha \tag{11.35}$$

EXAMPLE 11.12 ***(Analysis of Helical Gear)***

PROBLEM

A helical gear is cut with a standard 25° pressure angle, diametrical pitch 4 spur gear cutter at a helix angle of 30°. If the gear has 40 teeth and transmits 100 hp at a speed of 500 rpm, calculate the radial and thrust loads to be carried by the mounting bearings. Also compute the couple due to the axial component. If the shaft bearings are to be 6 in apart, compute the additional radial bearing load from this source.

SOLUTION

To solve the problem, we must determine the tangential component, W_t, of the normal force, and from this we can determine the other components. Using Eq. (11.34), the relationship between W_t, the power P, and the angular velocity, ω, of the gear is

$$W_t = \frac{T}{r_p} = \frac{P}{\omega r_p} \tag{11.36}$$

We can compute the pitch radius from the number of teeth and the diametrical pitch. That is

$$r_p = \frac{N}{2P} = \frac{40}{2(4)} = 5 \text{ in}$$

Because the power is given in hp, r_p in inches, and ω in rpm, Eq. (11.36) becomes

$$W_t = \frac{(33{,}000)(12)(\text{hp})}{2\pi\omega r_p} = \frac{63{,}025(\text{hp})}{\omega r_p} = \frac{63{,}025(100)}{(500)(5)} = 2520 \text{ lb}$$

From Eqs. (11.31) and (11.32),

$$W_r = W_t \tan\psi_t = W_t \frac{\tan\psi_n}{\cos\alpha} = 2520\frac{\tan 25^\circ}{\cos 30^\circ} = 1357 \text{ lb}$$

And from Eq. (11.33),

$$W_a = W_t \tan\alpha = 2520 \tan 30^\circ = 1455$$

The couple due to the axial load is given by Eq. (11.35) as

$$T_a = W_a r_p = W_t r_p \tan\alpha = 2520(5)\tan 30^\circ = 7275 \text{ in-lb}$$

The shaft bearings must produce a couple that will react with this couple to place the shaft in static equilibrium. The couple produced by the bearings is $6F$, so the *increase* in the reaction forces to support the couple produced by the axial load is

$$F = \frac{7275}{6} = 1212 \text{ lb}$$

■

11.11.3 Worm Gears

Worm gears are a special case of crossed-helical gears, and the forces on the worm from the gear are usually resolved as shown in Fig. 11.44. The force equations are similar to those for helical gears except that in the case of worm gears, the lead angle, λ, is usually used instead of the helix angle. With worm gears, the worm is almost always the driver because in most cases friction makes it impossible to drive the worm from the gear.

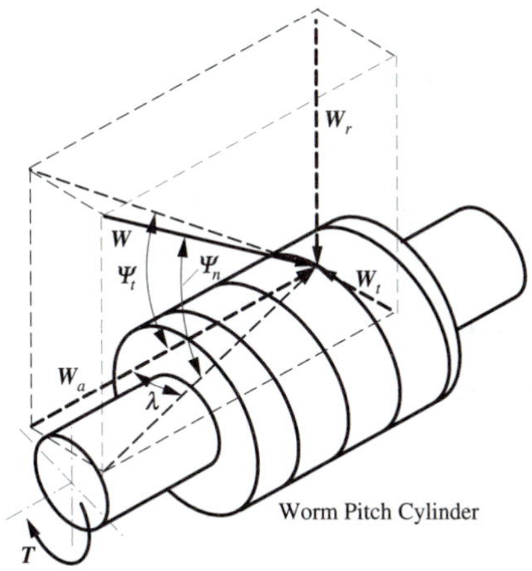

Figure 11.44 The forces acting on a worm gear.

Referring to Fig. 11.44, the force component equations for the worm can be written as

$$W_t = \frac{T}{r_p}$$

where r_p is the pitch radius and T is the input torque on the worm.

$$W_a = \frac{W_t}{\tan\lambda}$$

$$W_r = W_t \frac{\tan\psi_t}{\tan\lambda}$$

and

$$W = \sqrt{W_a^2 + W_r^2 + W_t^2}$$

In most worm gear drives, the gear axis is at right angles to that of the worm. Because of this, W_a for the worm becomes W_t for the gear, and W_t for the worm becomes W_a for the gear. The radial component is the same for both the worm and the gear.

For typical values for λ and ψ_t, the magnitude of W is much larger than the magnitude of W_t. This makes worm gears somewhat inefficient, because even small values of the coefficient of friction will produce large friction forces compared with W_t. This makes the equations in this section somewhat approximate for design purposes. For accurate designs, the contribution of friction must be included in the force calculations. If the friction is not included in the analysis, the effect of friction must be accounted for through a safety factor for the gear box.

It is also necessary to account for the frictional losses by providing a means for cooling the gear box. This is commonly done by providing fins on the gear box to improve convective heat transfer. In extreme cases, the oil is cooled separately. In general, most of the frictional losses will be converted to heat, and if the design does not properly account for this heat load, the oil temperature will rise and the gear box will fail prematurely.

11.11.4 Straight Bevel Gears

In computing the tooth forces on bevel gears, it is necessary to estimate the point of application of the force. It is typical to assume that the contact force acts at the middle of the tooth face. In practice, it is probably somewhat farther toward the thick end of the

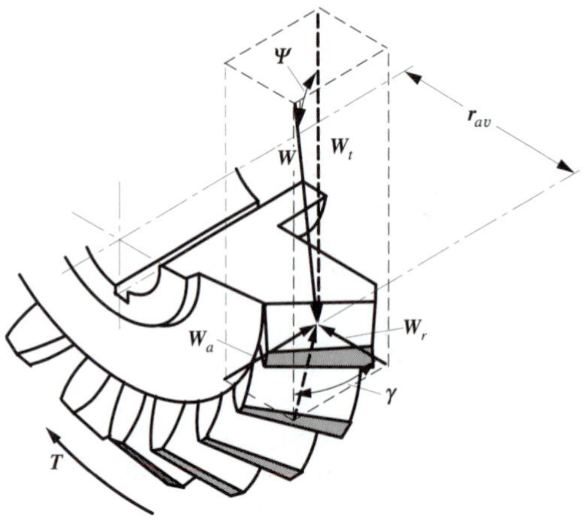

Figure 11.45 The force components for a straight-toothed bevel gear.

tooth because the tooth is stiffer in that direction. However, the error in assuming that the force contacts the middle of the tooth is small, and this will generally give slightly conservative results. In the absence of friction, the contact force, W, will be normal to the tooth, and as indicated in Fig. 11.45, the force can be resolved into three components. The tangential component, W_t, is the component that generates the shaft torque, and it is given by

$$W_t = W \cos \psi$$

where ψ is the pressure angle. The force component in the plane of the gear shaft is

$$W_y = W \sin \psi$$

and this component can be used to compute the radial and axial components. These are

$$W_r = W_y \cos \gamma = W \sin \psi \cos \gamma$$
$$W_a = W_y \sin \gamma = W \sin \psi \sin \gamma$$

where γ is the cone angle for the gear. The torque, T, transmitted by the gear is given by

$$T = W_t r_{av}$$

where r_{av} is the average radius of the gear pitch cone. Note that as in the case of a helical gear, the force component W_a produces a time-varying couple on the shaft, and the couple must be resisted by the bearings.

11.12 CHAPTER 11 *Exercise Problems*

PROBLEM 11.1

In the mechanism shown, sketch a free-body diagram of each link, and determine the force P that is necessary for equilibrium if $T_{12} = 90\ N - m$ and $\theta = 90°$.

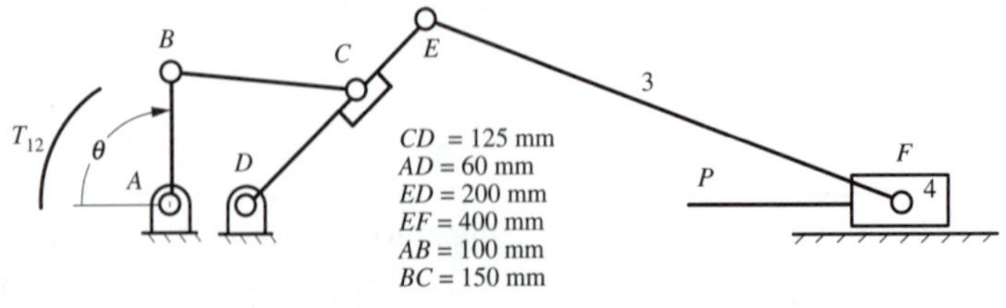

PROBLEM 11.2

Draw a free-body diagram for each of the members of the mechanism shown, and find the magnitude and direction of all the forces and moments. Compute the torque applied to link 2 to maintain static equilibrium. Link 2 is horizontal.

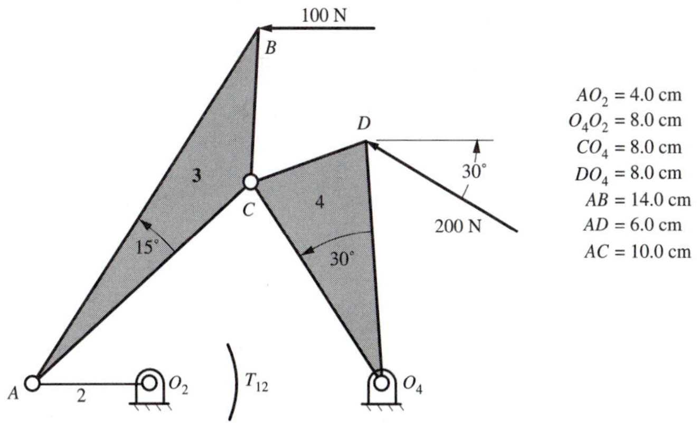

PROBLEM 11.3

If a force of 1000 lb is applied to the slider as shown, determine the force ***P*** required for static equilibrium.

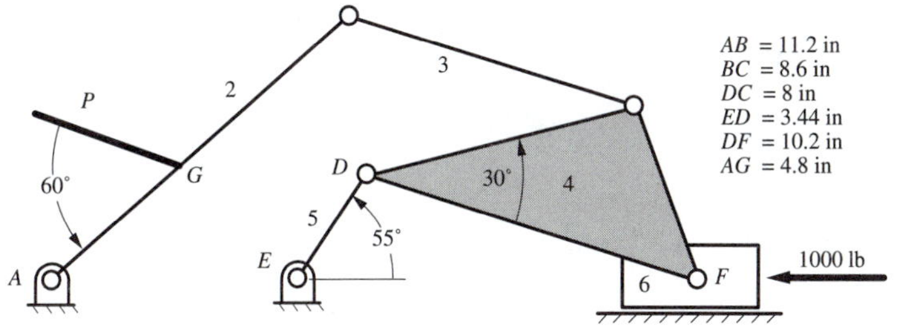

PROBLEM 11.4

For the mechanism and data given, determine the cam torque, $\boldsymbol{T}_{12}$, and the forces on the frame at points A and C ($\boldsymbol{F}_{21}$ and $\boldsymbol{F}_{31}$). Assume that there is friction between the cam and follower only.

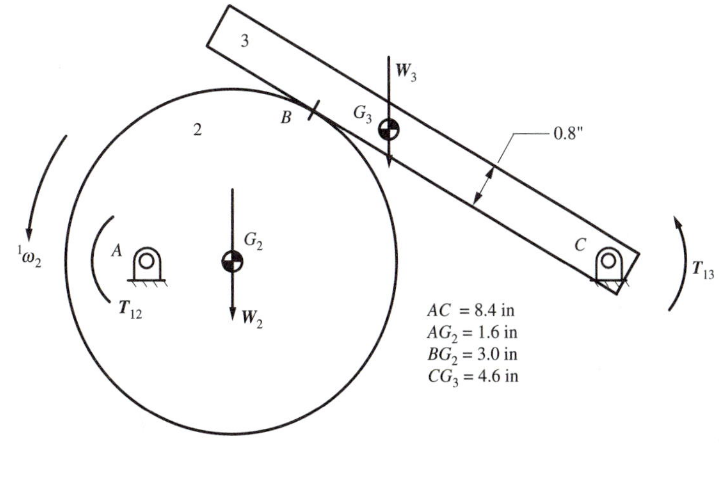

PROBLEM 11.5

In the mechanism shown, $\boldsymbol{P}$ = 100 lb. Find the value of the force $\boldsymbol{Q}$ on the block for equilibrium. Use energy methods.

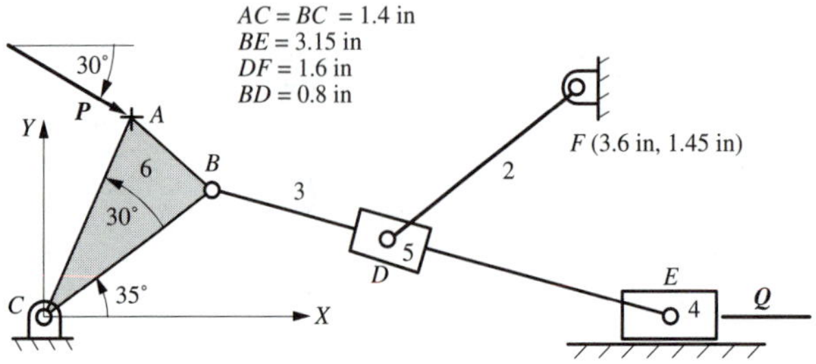

PROBLEM 11.6

If T_{12} is 1 in-lb, find $\boldsymbol{P}_{16}$ using energy methods.

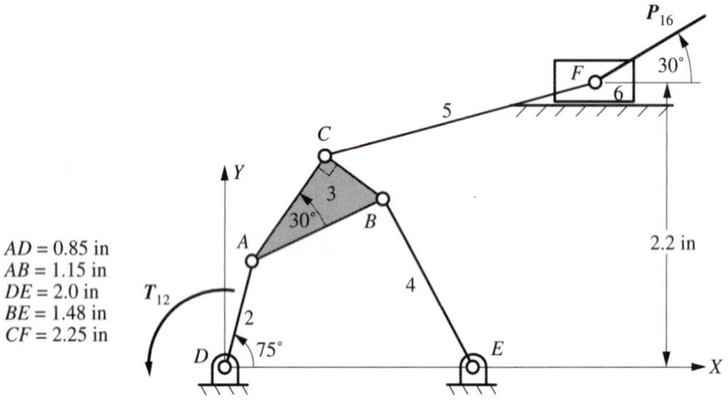

PROBLEM 11.7

Assume that the force $\boldsymbol{P}$ is 10 lb. Use energy methods to find the torque $\boldsymbol{T}_{12}$ required for equilibrium.

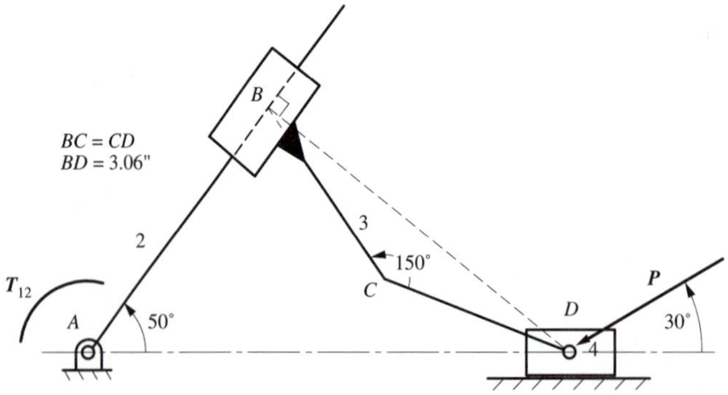

PROBLEM 11.8

In the four-bar linkage shown, the force $\boldsymbol{P}$ is 100 lb. Use energy methods to find the torque $\boldsymbol{T}_{12}$ required for equilibrium.

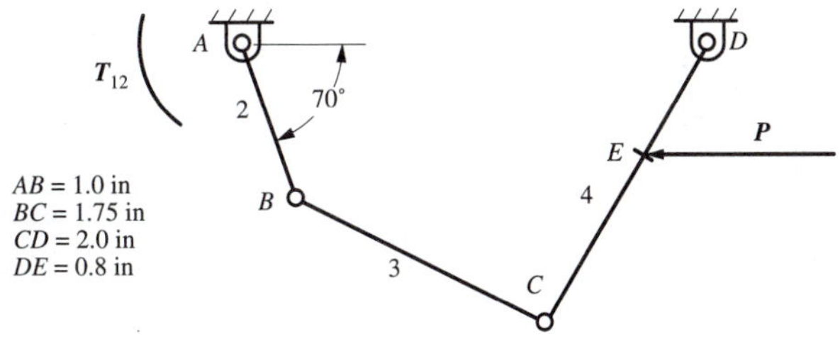

PROBLEM 11.9

If $\boldsymbol{F}_{14}$ is 100 lb, find the force $\boldsymbol{F}_{12}$ required for static equilibrium.

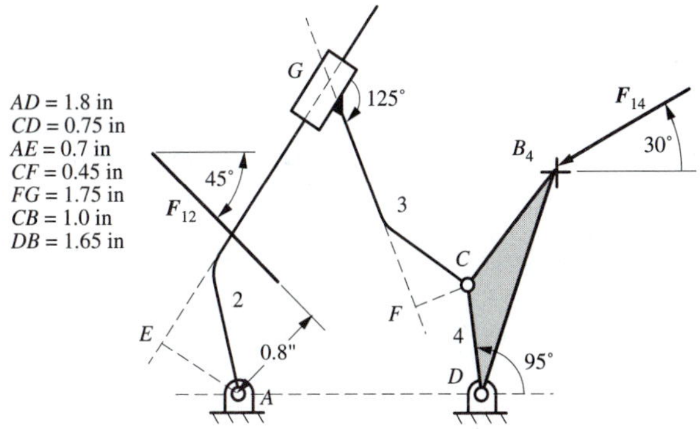

PROBLEM 11.10

In the eight-link mechanism, most of the linkage is contained in the black box and some of the instant centers are located as shown. The force $\boldsymbol{P}$ is 100 lb and is applied to point D on link 8. Compute the velocity of point C_8 and determine the torque $\boldsymbol{T}_{12}$ necessary for equilibrium.

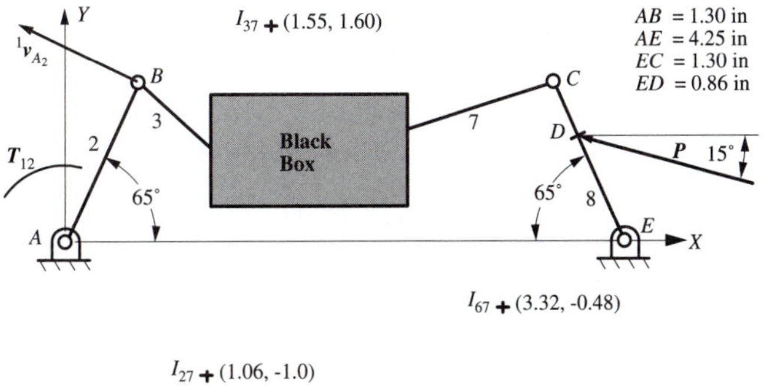

PROBLEM 11.11

The mechanism shown is called a vice grip because a very high force at E can be generated with a relatively small force at G when points A, B and C are collinear (toggle position). In the position shown, determine the ratio F_{14}/F_{13}.

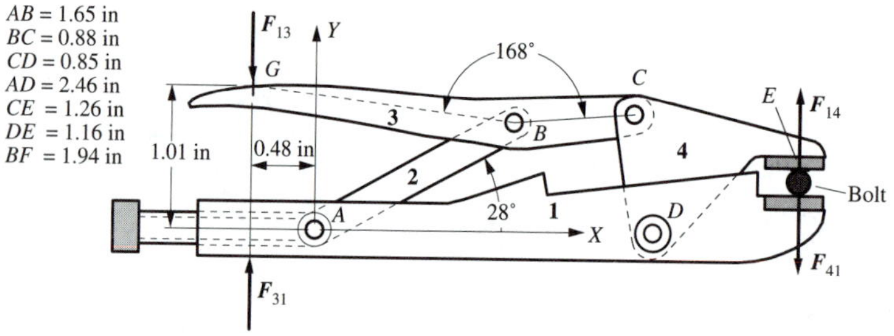

PROBLEM 11.12

If Q is 100 lb in the direction shown, use energy methods to find the torque T_{12} required for equilibrium.

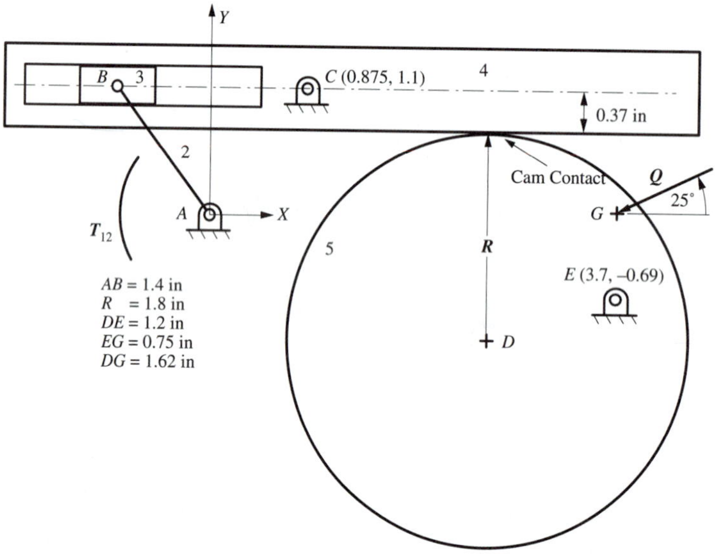

PROBLEM 11.13

If Q is 100 lb in the direction shown, use energy methods to find T_{12}.

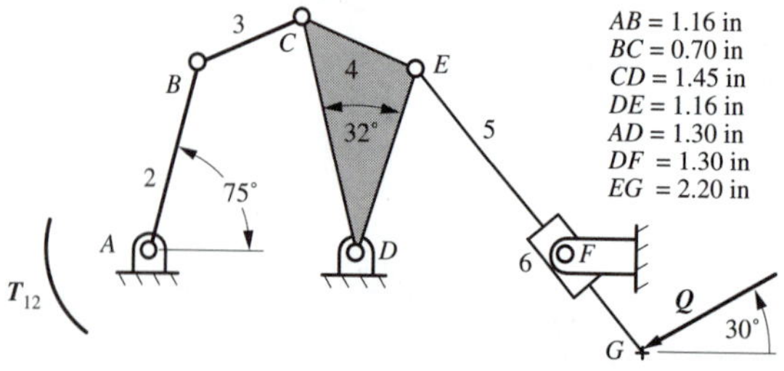

PROBLEM 11.14 If the velocity of link 2 is 10 in/s and the force on link 2 is 100 lb in the direction shown, find the torque on link 6 required to maintain equilibrium in the mechanism.

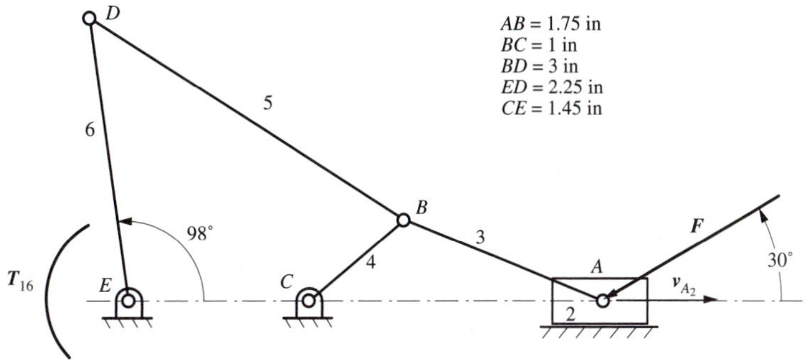

PROBLEM 11.15 In the mechanism shown, point F is a swivel at the midpoint of link 3 that carries link 5. The motion of the four-bar linkage causes arm 6 to oscillate. If link 2 rotates counterclockwise at 12 rad/s and is driven by a torque of 20 ft-lb, determine the resisting torque on link 6 required for equilibrium.

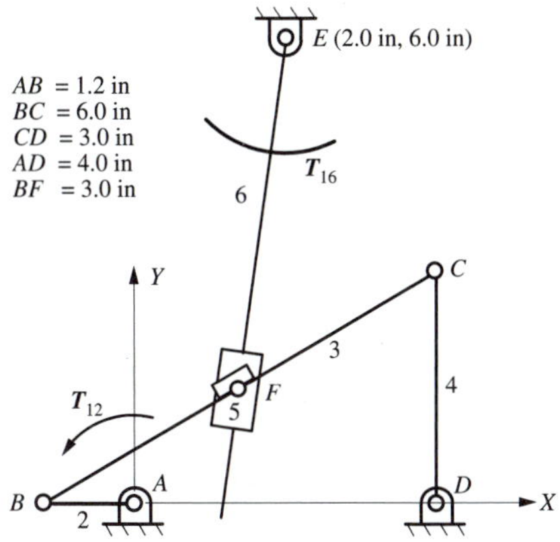

PROBLEM 11.16 Find the torque $\boldsymbol{T}_{12}$ for a coefficient of friction μ of 0.0 and 0.2. Assume that the radius of each pin is 1 in, and consider both pin and slider friction. Link 2 rotates CW.

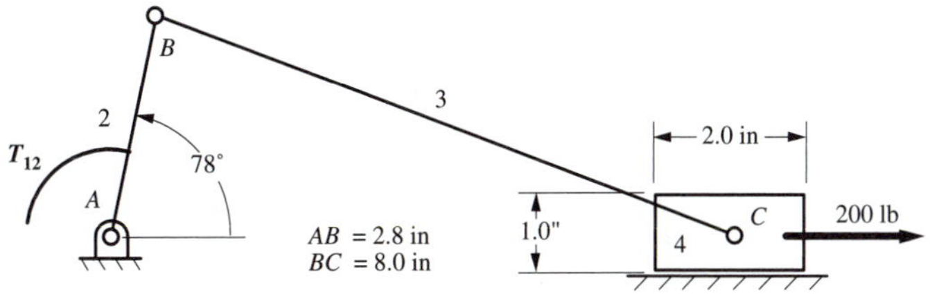

PROBLEM 11.17

For the position given for the slider-crank mechanism, find the torque T_{12} required for equilibrium. The radius of each pin is 1 in, and the friction coefficient at the pin between links 3 and 4 and between the block and the frame is 0.3. Elsewhere, the coefficient of friction is 0. Link 2 rotates CCW.

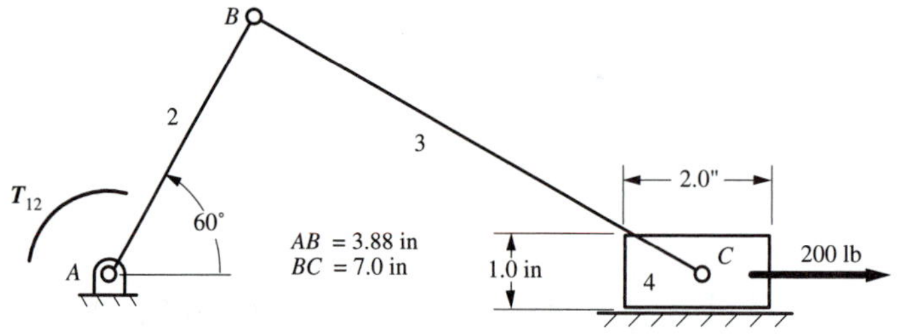

PROBLEM 11.18

If the radius of each pin is 0.9 in and the coefficient of friction at all joints is 0.15, find the torque T_{12} required for equilibrium in the position shown. Link 2 rotates CCW.

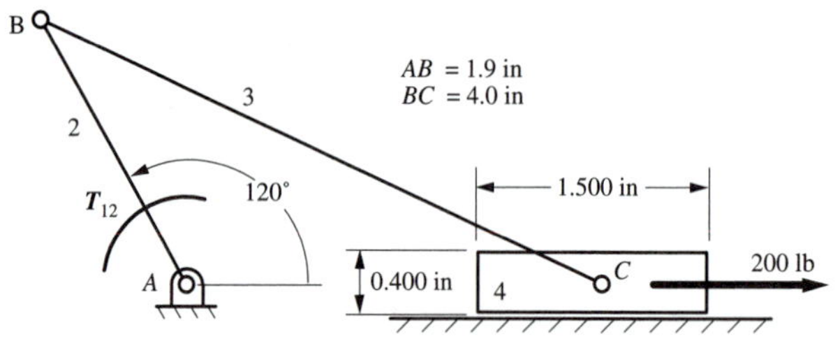

PROBLEM 11.19

If the coefficient of friction is 0.4 at each pin and each pin radius is 1 in, determine the force F required for equilibrium. Link 2 rotates CCW.

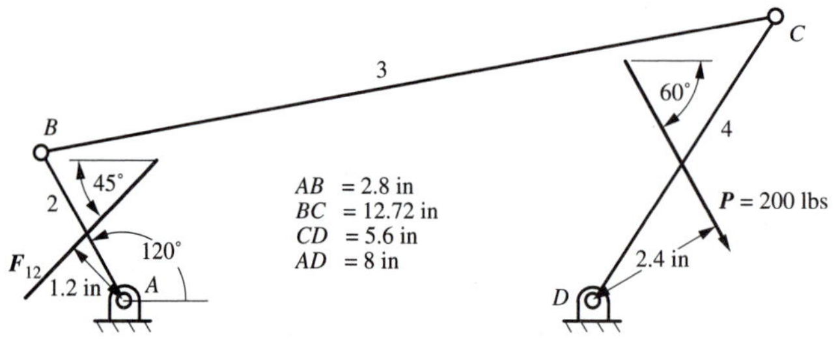

Chapter 12

Dynamic Force Analysis

12.1 INTRODUCTION

In the subject of machine dynamics we bring together *kinematics,* the study of motion and geometry, with *kinetics,* the study of the relationship between force and motion, to derive information on the forces and torques active in moving machinery. This information is, in turn, essential to the computation of the stresses internal to the members of a machine and the elastic deflections of the members. Because two major failure modes of machine members are deformation or fracture due to excessive stress and vibration and interference due to excessive elastic deflection, this information is obviously crucial for the purposes of machine design.

The computation of stresses and deflections in machine members is beyond the scope of this book. It is covered in texts on strength of materials, stress analysis, and design of machine elements. There are, of course, powerful computer tools such as finite-element analysis codes that can be used for this purpose. However, the input to such a program must include a complete description of the forces acting on the member. This includes both forces of reaction against the base or other members and active forces and torques produced by self-weight, by loads, including dynamic body forces such as centrifugal force, and by actuators or prime movers powering the machine.

The relationship between force and motion that is central to the study of kinetics is Newton's second law:

$$\boldsymbol{F} = m\boldsymbol{a} \tag{12.1}$$

where $\boldsymbol{F}$ is the force acting on a particle, m is its mass, and $\boldsymbol{a}$ is its acceleration.

Since this relationship refers to only a single particle and the machine members we want to deal with are bodies with distributed mass, it is necessary to derive appropriate equations relating force and torque for a rigid body with distributed mass from Newton's second law and the rigidity assumption. This results in the *Newton-Euler* equations describing the motion of a rigid body. The description of the motion of the *center of mass* of the body has exactly the same form as Eq. (12.1). That is, the center of mass moves as though the entire mass of the body were condensed into a particle at that point:

$$\sum \boldsymbol{F} = m\boldsymbol{a}_G \tag{12.2}$$

where $\boldsymbol{a}_G$ is the acceleration of the center of mass.

Rotary motion of the body is described by Euler's equation:

$$\sum \boldsymbol{M}_G = \boldsymbol{I}_G\boldsymbol{\alpha} \tag{12.3}$$

where $\sum \boldsymbol{M}_G$ is the resultant moment of the force system acting on the body *about the center of mass,* $\boldsymbol{I}_G$ is the inertia matrix based on fixed coordinate axes with origin at the center of mass, and $\boldsymbol{\alpha}$ is the angular acceleration of the body relative to that same fixed frame. Alternatively, if the body is rotating about a fixed point, P, moments may be taken about that point and the inertia matrix expressed in the same fixed coordinate frame with origin at P. Euler's equation then becomes

$$\sum \boldsymbol{M}_P = \boldsymbol{I}_P\boldsymbol{\alpha} \tag{12.4}$$

It is important to note that this special case, rotation about a fixed point, is the only case in which the inertia matrix may be based on a point other than the center of mass.

In practice, for any case except rotation of a symmetric body about a fixed axis of symmetry, it is much more convenient to express the inertia matrix about a coordinate frame *fixed in the moving body*. There is always a set of three orthogonal axes, fixed in the body, for which the inertia matrix becomes a diagonal matrix. These axes are called the *principal axes of inertia*. If the inertia matrix, $\boldsymbol{I}_G$, is expressed in the principal axes, and the angular velocity, $\boldsymbol{\omega}$, and angular acceleration, $\boldsymbol{\alpha}$, are expressed in the same frame, Eq. (12.3) becomes

$$\sum \boldsymbol{M}_G = \boldsymbol{I}_G\boldsymbol{\alpha} + \boldsymbol{\omega} \times \boldsymbol{I}_G\boldsymbol{\omega} \tag{12.5}$$

This is the form of Euler's equation most useful for spatial motion.

The methodology of dynamics problems is to use Eqs. (12.2)–(12.5) applied to each free body in the system in a manner exactly analogous to the use of static equilibrium equations in analyzing a structure. If the right-hand sides of Eqs. (12.2) and (12.3) are set to zero, they become identical to the static equilibrium equations. This is why Eqs. (12.2) and (12.3) are often called *dynamic equilibrium equations*.

Of course, the presence of the inertia terms on the right-hand sides of the dynamic equilibrium equations complicates their solution. Most dynamic problems in engineering fall into one of two classes. In the first of these, the motion of each body in the system is known. Therefore the acceleration of the center of mass and the angular acceleration of each member are known or can be computed by the use of kinematic techniques. Thus the right-hand sides of the dynamic equilibrium equations can be treated as known quantities, and the equations can be solved algebraically in a manner exactly analogous to the solution of the static equilibrium equations.

In the second class of problems, the motion of each body is not known a priori, and it is necessary to treat the accelerations on the right-hand sides of the equations as the second derivatives of position variables. The forces and moments acting on the member, which appear on the left-hand sides of the equations, must then also be related to the position variables to produce a set of differential equations in those position variables. These equations are called the *equations of motion* of the system. Description of the motion of the system then requires solution of the equations of motion.

In many machine design problems the motion of some input element is specified and dynamic analysis of these systems falls into the first of the two classes described above. These problems may be called *machine dynamics problems*. In this book we will confine ourselves to this class of dynamic problems.

Dynamic problems that must be solved by solution of the differential equations of motion are treated in courses and texts on vibrations or system dynamics. Another class of such problems is treated in the area of multibody system dynamics.

12.2 PROBLEMS SOLUBLE VIA PARTICLE KINETICS

Some machine design problems require only Eq. (12.1) or the energy and momentum relationships derived from it for solution. That is, it is not necessary to use rotary inertia. Effectively, the inertias can be modeled as particles with adequate accuracy.

12.2.1 Dynamic Equilibrium of Systems of Particles

First, it is necessary to relate the positions of the concentrated masses of the system kinematically in order to find their accelerations. In principle, this requires first relating their positions, then their velocities and accelerations. The way this works out is illustrated by the following examples.

EXAMPLE 12.1 ***(Vehicle Acceleration and Braking)***

PROBLEM

An automobile has a total weight of $W = 4000$ lb. As shown in Fig. 12.1, its wheel base (distance between the front and rear axle centers) is $b + c = 90$, in and its center of mass is $b = 40$ in behind the front axle center. When it is parked on a level surface, the center of mass is $h = 25$ in above the ground. The wheel radius is $r = 12.5$ in, and on a good surface, the coefficient of friction between the tires and the road is $\mu = 0.8$. If the weights and moments of inertia of the wheels are neglected, estimate the ratio of the maximum acceleration that the vehicle can achieve to the gravitational acceleration, g, if it is driven by

1. The rear wheels
2. The front wheels
3. Also find the maximum deceleration that can be achieved if the braking effort is optimally apportioned between front and rear wheels. What is the percentage of the braking force at the front wheels?

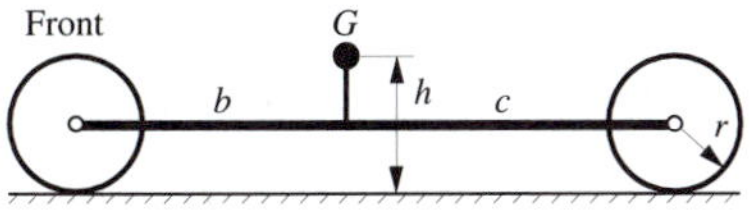

Figure 12.1 The vehicle model for Example 12.1.

SOLUTION

In this instance, the kinematic part of the solution is trivial. Since the wheel inertia is neglected, it is not even necessary to develop expressions for the angular velocities and angular accelerations of the wheels. We therefore start with step (3) in the procedure above.

1. Since the wheel weight and inertia will not be considered separately, we start with a free-body diagram of the whole vehicle shown in Fig. 12.2.

   There is no rotation of the vehicle mass, so the angular acceleration is zero. Therefore, the inertia torque term ($I_G\alpha$) in Eq. (12.3) is zero.

   The term $m\boldsymbol{a}_G$ in Eq. (12.2) can be computed and treated as a known force. If we move the term to the left-hand side of the equation, we can treat it as an applied force and solve the problem as a statics problem. As we will discuss in more detail later, when we treat the term, $m\boldsymbol{a}_G$, in this way, it is called an *inertia force,* $\boldsymbol{F}_I$, and it is in the direction opposite to that of the acceleration. The magnitude of F_I is given by

$$\boldsymbol{F}_I = -\frac{W\boldsymbol{a}}{g}$$

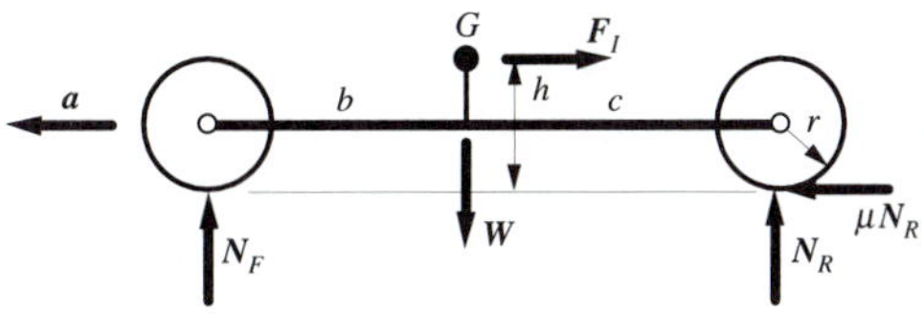

Figure 12.2 Free-body diagram when vehicle is driven via the rear wheels.

where a is the magnitude of the acceleration $\boldsymbol{a}$, which is positive in the forward direction, and W is the magnitude of the weight, $\boldsymbol{W}$. That is, the mass of the vehicle is W/g.

The dynamic equilibrium equations are

$$\sum \boldsymbol{F}_x = 0: \qquad -\boldsymbol{F}_I + \mu \boldsymbol{N}_R = 0 \qquad (a)$$

$$\sum \boldsymbol{F}_y = 0: \qquad \boldsymbol{W} = \boldsymbol{N}_R + \boldsymbol{N}_F \qquad (b)$$

$$\sum \boldsymbol{M}_G = 0: \qquad c\boldsymbol{N}_R = b\boldsymbol{N}_F + h\mu \boldsymbol{N}_R \qquad (c)$$

Moments are conveniently taken about G in order to eliminate W and a from the moment equation.

Using Eq. (a) together with the expression for F_I gives

$$N_R = \frac{Wa}{\mu g}$$

Applying this to Eq. (b),

$$N_F = W\left(1 - \frac{a}{\mu g}\right)$$

Substitution of these expressions into Eq. (c) gives, after some rearrangement,

$$\frac{a}{g}\left(\frac{b+c}{\mu} - h\right) = b$$

or

$$A = \frac{a}{g} = \frac{\mu b}{b + c - \mu h}$$

where A is the required ratio of the acceleration to the gravitational acceleration. Substitution of the values of μ, b, c, and h given in the problem statement results in

$$A = 0.457$$

That is, the maximum acceleration that can be achieved by the vehicle is 45.7% of the gravitational acceleration.

2. The free-body diagram for front wheel drive is as shown in Fig. 12.3.

The dynamic equilibrium equations are

$$\sum \boldsymbol{F}_x = 0: \qquad -\boldsymbol{F}_I + \mu \boldsymbol{N}_F = 0 \qquad (d)$$

$$\sum \boldsymbol{F}_y = 0: \qquad \boldsymbol{W} = \boldsymbol{N}_R + \boldsymbol{N}_F \qquad (e)$$

$$\sum \boldsymbol{M}_G = 0: \qquad c\boldsymbol{N}_R = b\boldsymbol{N}_F + h\mu \boldsymbol{N}_F \qquad (f)$$

Substitution for F_I in Eq. (d) gives

$$N_F = \frac{Wa}{\mu g}$$

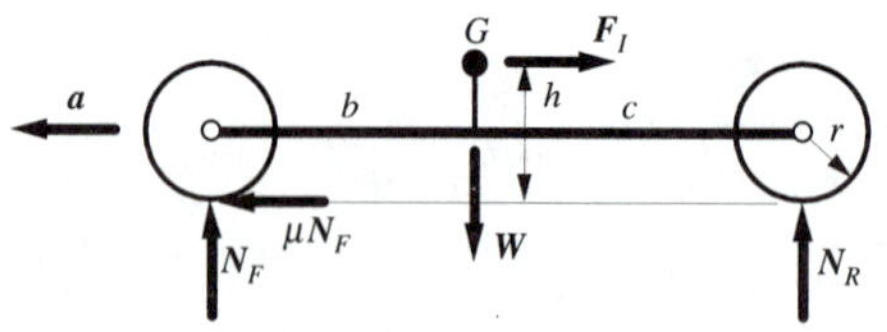

Figure 12.3 The free-body diagram when the vehicle is driven via the front wheels.

Substitution in Eq. (*e*) gives

$$N_R = W\left(1 - \frac{a}{\mu g}\right)$$

Substitution of these values into Eq. (*f*) gives, after rearrangement

$$\frac{a}{g}\left(\frac{b+c}{\mu} + h\right) = c$$

Hence

$$A = \frac{a}{g} = \frac{\mu c}{b + c + \mu h}$$

Substitution of the given values yields

$$A = 0.364$$

As can be seen, for the values given, the acceleration that can be achieved with front wheel drive is significantly lower than with rear wheel drive.

3. The free-body diagram for braking is as shown in Fig. 12.4.
In this case the inertia force is directed forward. The dynamic equilibrium equations are, in this instance,

$$\sum \boldsymbol{F}_x = 0: \qquad \boldsymbol{F}_I - \mu(\boldsymbol{N}_R + \boldsymbol{N}_F) = 0 \qquad (g)$$

$$\sum \boldsymbol{F}_y = 0: \qquad \boldsymbol{W} = \boldsymbol{N}_R + \boldsymbol{N}_F \qquad (h)$$

$$\sum \boldsymbol{M}_G = 0: \qquad c\boldsymbol{N}_R + h\mu(\boldsymbol{N}_R + \boldsymbol{N}_F) = b\boldsymbol{N}_F \qquad (i)$$

Substitution from Eq. (*h*) into Eq. (*g*), together with substitution for F_I, gives

$$\frac{Wa}{g} = \mu W$$

or $A = \mu$. For the values given, $A = 0.8$.
Substitution from Eq. (*h*) into Eq. (*i*) gives

$$N_F(b + c) = (c + h\mu)W$$

or

$$N_F = W\frac{c + h\mu}{b + c}$$

For the values given

$$N_F = 0.778W \quad \text{and} \quad N_R = 0.222W$$

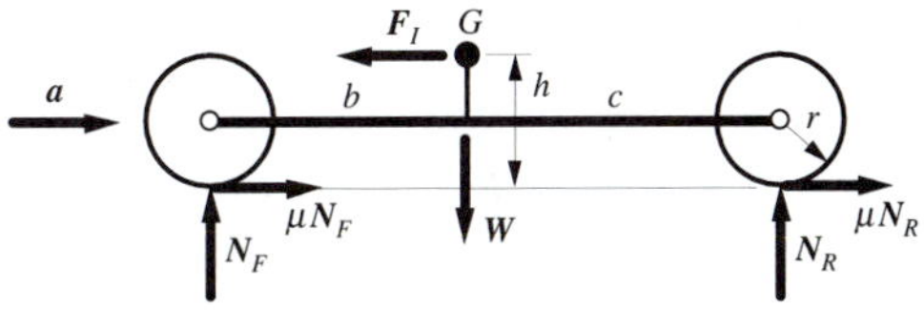

Figure 12.4 The free-body diagram during braking.

Therefore, 77.8% of the braking effort is at the front wheels, which is why the front brakes of automotive vehicles are usually much bigger than the rear brakes.

Notice also that, for this configuration, the deceleration available through optimal braking is considerably greater than the maximum possible acceleration. ■

EXAMPLE 12.2 (*Flyball Governor*)

PROBLEM

A flyball governor is arranged as shown in Fig. 12.5 to allow adjustment of the governed speed by adjusting the preload in the spring. As the weights swing outward under the influence of centrifugal force, the spring is compressed.

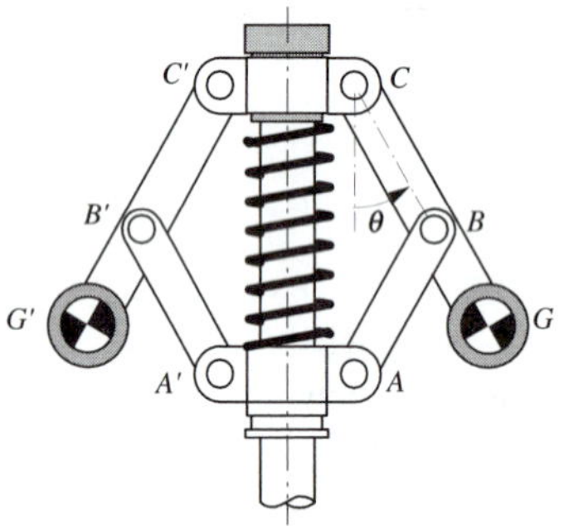

Figure 12.5 The flywheel governor discussed in Example 12.2 The dimensions are $AB = BC = A'B' = B'C' = 1.5$ in., $CG = C'G' = 2.438$ in., $CC' = 1.25$ in. G and G' are the effective centers of mass of the arms.

The governor operates a throttle valve to control the speed of an engine. The valve is fully closed when the angle θ is 75°. Compute the speed at which the valve is closed if the effective centers of mass of the arms are in the locations G and G' shown and other inertias can be neglected. The weight of each arm is 0.25 lb. The adjusting nut is set so that the spring is at its natural or free length when $\theta = 5°$. The stiffness of the spring is 20 lb/in. The length of the links are as indicated in the figure caption.

SOLUTION

The radius of rotation of the center of mass of each arm is

$$r = 2.438 \sin 75° + 0.625 = 2.979 \text{ in}$$

Hence the centrifugal acceleration of the point G is

$$a_G = r\omega^2 = (2.979/12)\omega^2 = 0.248\omega^2$$

The magnitude of the inertia force is

$$F_I = ma_G = (0.25/32.2) \times 0.248\omega^2 = 0.00193\omega^2$$

Free-body diagrams of links AB and BC are shown in Fig. 12.6. Since AB is a two-force member, the forces A and B are equal, opposite, and collinear. For link BC,

$$\sum \boldsymbol{F}_x = 0: \quad F_{CN} + F_I - F_{AB} \sin 75° = 0 \qquad (a)$$

$$\sum \boldsymbol{F}_y = 0: \quad F_S - F_{AB} \cos 75° = 0 \qquad (b)$$

where F_S is half the force from the spring.

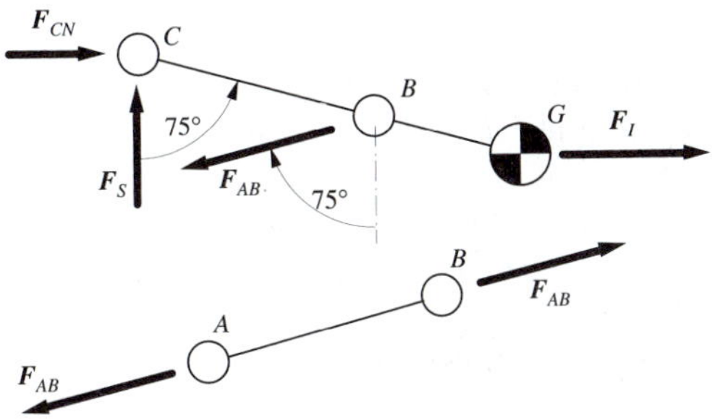

Figure 12.6 Free-body diagrams of links AB and BC used in the solution of Example 12.2.

Taking moments about point C in the free-body diagram for link BC,

$$\sum \boldsymbol{M}_C = 0: \qquad F_{AB}\overrightarrow{BC}\sin 30° - F_I\overrightarrow{CG}\sin 15° = 0 = F_{AB}1.5\sin 30° - F_I 2.438\sin 15° \qquad (c)$$

The distance AC in this position is

$$AC = 2 \times 1.5\cos 75° = 0.776 \text{ in}$$

When the spring is at its natural length

$$AC = 2 \times 1.5\cos 5° = 2.989 \text{ in}$$

Hence the spring is compressed the distance

$$x = 2.989 - 0.776 = 2.212 \text{ in}$$

and the spring force is

$$2F_S = 20 \times 2.212 = 44.24 \text{ lb}$$

or

$$F_S = 22.12 \text{ lb}$$

Hence, from Eq. (b)

$$F_{AB} = F_S/\cos 75° = 22.12/\cos 75° = 85.47 \text{ lb}$$

Using Eq. (c)

$$F_I = \frac{1.5\sin 30°}{2.438\sin 15°}F_{AB} = \frac{1.5\sin 30°}{2.438\sin 15°}85.47 = 101.58 \text{ lb}$$

Therefore

$$\omega^2 = F_I/0.00193 = 101.58/0.00193$$

or

$$\omega = 229.4 \text{ rad/s}$$

or

$$\omega = 2{,}191 \text{ rpm}$$

This is the speed at which the governor will cause the engine to operate. ■

12.2.2 Conservation of Energy

Conservation of energy is useful in mechanism problems in which there is interchange between kinetic and potential energy. Most often the potential energy is either gravitational potential energy or strain energy. If the total energy in the system can be calculated in some position of the system and the system has mobility one, then the kinetic energy and hence the velocity can be calculated in any other position of the system, provided joint friction can be neglected. This type of calculation is usually much more efficient than pursuing the same result via dynamic equilibrium. Conservation of energy provides information on only the variables that determine energy components. Thus, although positions and velocities can be related in this way, the method will yield no information on accelerations.

Conservation of energy can often be used in association with conservation of momentum to solve problems in which energy is stored in a spring, a raised weight, or a flywheel and then rapidly released, as in punching and stamping machines, jackhammers, and similar types of mechanisms.

12.2.3 Conservation of Momentum

Conservation of momentum is particularly useful in mechanism problems that involve impacts or other short-period, impulsive events. The methodology is to compare the momentum of the system immediately before and immediately after the impulsive event. In problems involving one-dimensional motion, only conservation of linear momentum is required. In problems involving two- or three-dimensional, rigid-body motion, conservation of angular momentum is also required.

The general methodology of problems involving conservation of energy and conservation of momentum can be studied in the following example.

EXAMPLE 12.3 **(*Hydraulic Impactor*)**

PROBLEM

A hydraulic impactor (jackhammer) uses a rotary hydraulic motor to turn a cylindrical cam, as shown in Fig. 12.7. The cam has a ramp, which causes the follower to extend a spring, and a step, which releases the follower and allows the spring to contract, accelerating a hammer. At the end of the stroke, the hammer strikes the impactor bit, driving it into the ground. If it

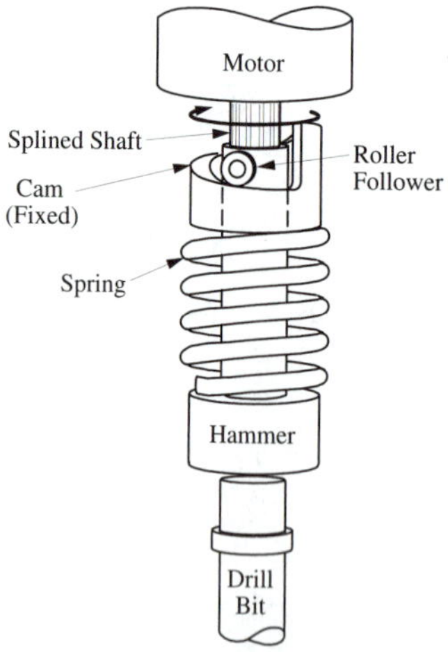

Figure 12.7 The drive mechanism of the hydraulic impactor discussed in Example 12.3.

is assumed that the impact between the hammer and the bit is perfectly elastic (no loss of energy), find the velocity and kinetic energy of the bit immediately after it is struck by the hammer. The bit weighs 20 lb, and the hammer weighs 17.5 lb. The spring has a rate of 2000 lb/in, and the cam lift is 2 in. At the end of the stroke, when the hammer strikes the bit, the spring is compressed 0.5 in. Also estimate the hydraulic power needed to drive the device if it delivers 5 strokes per second.

SOLUTION

The critical position and velocity parameters for this problem are shown in Fig. 12.8. We first use conservation of energy to find the velocity of the hammer at the instant of impact.

When the hammer is in the fully raised position (Fig. 12.8a), the hammer and bit are at rest, so kinetic energy is

$$K_0 = 0$$

The strain energy is

$$P_{0S} = \tfrac{1}{2} \times 2000 \times 2.5^2 = 6250 \text{ lb-in} = 520.8 \text{ ft-lb}$$

Here 2.5 in is used because that is the total extension of the spring in this position. Note that the spring is extended 0.5 in even when the cam follower is in its lowest position.

If the position at which the hammer strikes the bit is taken as the zero reference, the gravitational potential energy of the hammer is

$$P_{0G} = 17.5 \times 2/12 = 2.9 \text{ ft-lb}$$

Hence the total mechanical energy of the system when the hammer is fully raised is

$$U_0 = K_0 + P_{0s} + P_{0G} = 523.7 \text{ ft-lb}$$

Now consider the instant immediately before the impact of the hammer on the bit, which is illustrated by Fig. 12.8b. If the velocity of the hammer is $\boldsymbol{v}_1$ the kinetic energy of the system is

$$K_1 = \frac{1}{2} \times \frac{17.5}{32.2} \times v_1^2 = 0.272 v_1^2 \text{ ft-lb}$$

The strain energy is now

$$P_{1S} = \tfrac{1}{2} \times 2000 \times 0.5^2 = 250 \text{ lb-in} = 20.8 \text{ ft-lb}$$

and the gravitational potential energy is zero.

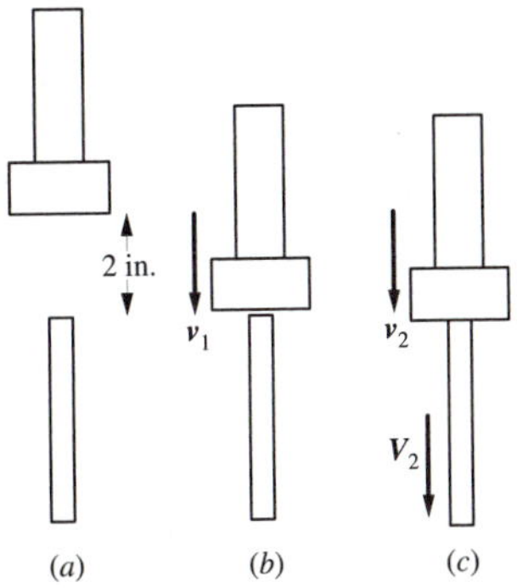

Figure 12.8 Three critical configurations of the system: (a) with the hammer fully raised and both hammer and bit at rest; (b) at the instant before impact with the hammer at velocity $\boldsymbol{v}_1$ and the bit at rest; and (c) at the instant after impact with both hammer and bit moving, the hammer at velocity $\boldsymbol{v}_2$ and the bit at velocity $\boldsymbol{V}_2$.

Therefore, the total mechanical energy is

$$U_1 = K_1 + P_{1S} = 0.272\, v_1^2 + 20.8 \text{ ft-lb}$$

Equating this expression to the system energy when the hammer is fully raised gives

$$U_1 = U_0 = 0.272 v_1^2 + 20.8 = 523.7 \text{ ft-lb}$$

Hence, solving for v_1,

$$v_1 = 43.0 \text{ ft/s}$$

and

$$K_1 = 523.7 - 20.8 = 502.9 \text{ ft-lb}$$

We now use both conservation of momentum and conservation of energy to determine the velocities of hammer and bit immediately after the impact, which is the condition shown in Fig. 12.8c.

Let v_2 be the velocity of the hammer immediately after impact, and let V_2 be the velocity of the bit at that instant. Both are taken to be positive in the downward direction. The momentum of the system before impact is

$$g_1 = \frac{17.5}{32.2} \times 43.0 = 23.4 \text{ lb-s}$$

Immediately after impact, the momentum is

$$g_2 = \frac{20}{32.2} \times V_2 + \frac{17.5}{32.2} \times v_2$$

Equating g_1 and g_2 gives

$$23.4 = \frac{20}{32.2} \times V_2 + \frac{17.5}{32.2} \times v_2$$

or

$$V_2 + 0.875 v_2 = 37.63 \qquad (a)$$

The kinetic energy of the system immediately after the impact is

$$K_2 = \frac{1}{2} \times \frac{20}{32.2} V_2^2 + \frac{1}{2} \times \frac{17.5}{32.2} v_2^2$$

Since the impact is elastic, K_1 and K_2 can be equated

$$502.9 = \frac{1}{2} \times \frac{20}{32.2} V_2^2 + \frac{1}{2} \times \frac{17.5}{32.2} v_2^2$$

or

$$1619 = V_2^2 + 0.875 v_2^2 \qquad (b)$$

Substitution for V_2 from Eq. (a) into Eq. (b) gives

$$1619 = (37.63 - 0.875 v_2)^2 + 0.875 v_2^2$$

Expansion and simplification of this equation give

$$1.641 v_2^2 - 65.85 v_2 - 203.0 = 0$$

Solution of this quadratic equation in v_2 gives the following values:

$$v_2 = 43.0 \quad \text{and } v_2 = -2.875 \text{ ft/s}$$

The first of these values is the velocity of the hammer before impact. That must be a solution because the values before impact give the same energy and momentum, but the second value is then the correct solution for the velocity of the hammer after impact. The negative value indicates that the hammer rebounds slightly in the upward direction.

Substitution back into Eq. (*a*) gives

$$V_2 = 37.63 - 0.875 \times (-2.875) = 40.1 \text{ ft/s}$$

The kinetic energy of the bit at this instant is

$$K_B = \frac{1}{2} \times \frac{20}{32.2} \times 40.1^2 = 499 \text{ ft-lb}$$

The strain energy put into the spring by each rotation of the cam is

$$P_{0S} - P_{1S} = 520.8 - 20.8 = 500 \text{ ft-lb}$$

The gravitational potential energy put in by raising the hammer is

$$P_{0G} = 2.9 \text{ ft-lb}$$

Hence the energy put in on each rotation of the cam is

$$u = 502.9 \text{ ft-lb}$$

If the cam rotates 5 times per second, the power put in is

$$P = 5 \times 502.9 = 2510 \text{ ft-lb/s} = 2510/550 = 4.57 \text{ hp}$$

This completes solution of the problem. ■

12.3 DYNAMIC EQUILIBRIUM OF SYSTEMS OF RIGID BODIES

Equation (12.3) is applicable for general spatial motion. The matrix $\boldsymbol{I}_G$ is a symmetric, 3×3 matrix.

In the present work, we will restrict consideration to planar motion with one of the principal axes normal to the plane of motion. In this case Eq. (12.3) becomes

$$\sum \boldsymbol{M}_G \cdot \boldsymbol{w} = I_G \alpha \qquad \textbf{(12.6)}$$

where I_G is a scalar quantity, being the moment of inertia of the body about an axis through the center of mass normal to the plane of motion, $\boldsymbol{w}$ is a unit vector normal to the plane of motion, and $\boldsymbol{\alpha}$ is the angular acceleration, taken to be positive in the $\boldsymbol{w}$ direction. That is,

$$\boldsymbol{\alpha} = \alpha \boldsymbol{w} \qquad \textbf{(12.7)}$$

We make this restriction because the vast majority of machine dynamics problems are planar motion problems. Solutions of problems involving general spatial motion follow the same lines as the examples studied below, but require use of the general forms of Euler's equation: Eq. (12.3) or Eq. (12.5).

If a body is in motion, the sum of the forces acting on it is equal to its mass multiplied by the acceleration of its center of mass, that is,

$$\sum \boldsymbol{F} = m\boldsymbol{a}_G \qquad \textbf{(12.8)}$$

If we introduce a force, $\boldsymbol{F}_I$, such that

$$\boldsymbol{F}_I = -m\boldsymbol{a}_G \qquad \textbf{(12.9)}$$

called the inertia force, which is applied on a line passing through the center of mass, and a couple, M_I,

$$M_I = -I_G \boldsymbol{\alpha} \qquad \textbf{(12.10)}$$

called the inertia torque, and treat these in the same way as any other external force and torque, then the dynamic equilibrium equations assume the same form as the static equilibrium equations:

$$\sum \boldsymbol{F} = 0 \qquad \textbf{(12.11)}$$

$$\sum \boldsymbol{M}_O = 0 \qquad \textbf{(12.12)}$$

where O is any point in the plane about which moments are taken. Notice that, since all moment vectors are normal to the plane of motion, the moment equation may be treated as a scalar equation. The summation of the forces $\sum \boldsymbol{F}$ now includes $\boldsymbol{F}_I$, and $\sum \boldsymbol{M}_O$ includes $\boldsymbol{M}_I$. Note that the point of application of the inertia force is defined to be the center of mass and that the moment of inertia used in computing the inertia couple is that about the center of mass, whether or not the moments are taken about the center of mass.

The replacement of the $-m\boldsymbol{a}_G$ term with the inertia force and the replacement of $-I_G\alpha$ with the inertia torque to convert the dynamic equilibrium equations into the same form as the static equilibrium equations is known as D'Alembert's principle. The general procedure for solving a dynamic equilibrium problem in machine dynamics is the following:

1. Solve the position, velocity, and acceleration kinematics of the system using the procedures of Chapters 2 and 3 to obtain the accelerations of all bodies with significant mass and the angular accelerations of all bodies with significant inertia.
2. Compute the inertia force and couple acting on each body according to D'Alembert's principle.
3. Apply the inertia force and couple as an external force and torque to each member. The line of action of the inertia force passes through the center of mass of each member.
4. Draw a free-body diagram of each member including all external forces acting on that member and all reaction forces from other members to which it is connected.
5. Write three force and moment equlibrium equations for each member.
6. Solve the equations for the unknown forces.

It may be seen that the last three steps in the process are identical to the solution of a static equilibrium problem.

EXAMPLE 12.4 (*Dynamic Force Analysis*)

PROBLEM

The members of the mechanism of Example 2.4 and Fig. 2.17 have the inertial properties tabulated and indicated in Fig. 12.9. Find the driving torque that must be applied to the crank, member 2, in order to maintain the constant angular velocity of 60 rpm in the clockwise direction.

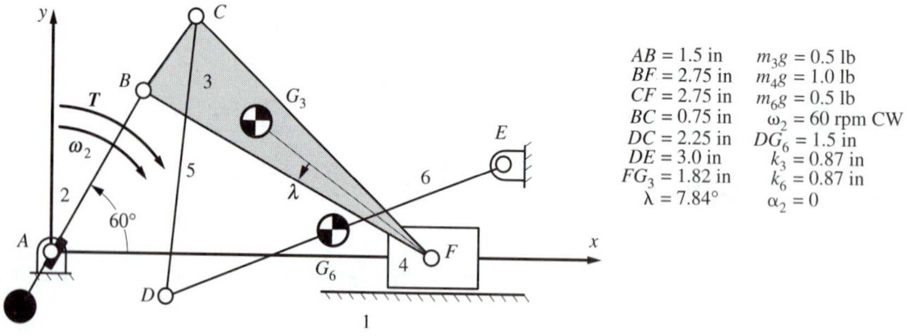

Figure 12.9 The mechanism of Example 12.4.

Friction in all joints (including the prismatic joint) and the mass and moment of inertia of link 5 may be neglected. The mechanism moves in the horizontal plane.

SOLUTION

Since the geometry and velocity and acceleration are identical to those of Example 2.4, that solution can be used for parts 1 and 2 of the solution procedure, with one modification. It is necessary to find the accelerations of points G_3 and G_6.

In both cases the acceleration image is used. G_6 is at the midpoint of DE. Since point E is a fixed point, it maps into point o' on the acceleration polygon, and hence the image of $\overrightarrow{DE}$ is $\overrightarrow{d'o'}$. Next g_6' is located at the midpoint of $\overrightarrow{d'o'}$, and the acceleration $\boldsymbol{a}_{G_6}$ is the vector $\overrightarrow{o'g_6'}$.

Similarly, point g_3' can be located on the image $b'c'f'$ of BCF by constructing $\angle b'f'g_3'$ equal to $\angle BFG_3 = \lambda$, and making

$$\frac{f'g_3'}{b'f'} = \frac{FG_3}{BF}$$

The accelerations of these two points can now be scaled from the acceleration diagram.

$$\boldsymbol{a}_{G_6} = 1.00 \times 20 = 20.0 \text{ in/s}^2 \text{ at } -63° \text{ to the } x \text{ axis.}$$

$$\boldsymbol{a}_{G_3} = 2.16 \times 20 = 43.2 \text{ in/s}^2 \text{ at } -132° \text{ to the } x \text{ axis.}$$

Scaling $\boldsymbol{a}^t_{F/B}$ from Fig. 12.10 gives

$$\boldsymbol{\alpha}_3 = \frac{\boldsymbol{a}^t_{F/B}}{BF} = \frac{2.63 \times 20}{2.75} = 19.1 \text{ rad/s}^2 \text{ CCW}$$

In addition, we need the angular acceleration information that was derived in Example 2.4:

$$\boldsymbol{\alpha}_6 = 40/3.0 = 13.3 \text{ rad/s}^2 \text{ CCW}$$

$$\boldsymbol{a}_F = 12.9 \text{ in/s}^2 \text{ (to the left)}$$

It may be noted that we have not attempted to find the acceleration of the center of mass of link 2, and in fact there is not sufficient information to find it. This is because it is not needed as long as we are seeking only the torque, T, and not the reaction force at point A, and as long as the angular velocity of member 2 is constant.

Step 3 in the procedure given above can now be addressed. The free-body diagrams for the mechanism are shown in Fig. 12.11.

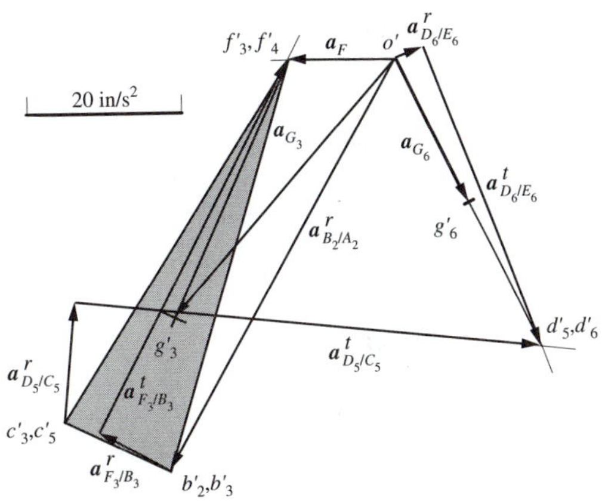

Figure 12.10 The acceleration diagram of Fig. 2.18 modified to allow computation of the acceleration of points G_3 and G_6. The acceleration image is used. It is necessary only to construct the position of point g_3' such that $\angle c'f'g_3' = \lambda$ and $f'g_3'/f'c' = FG_3/FC$, and of point g_6' at the midpoint of $o'd'$.

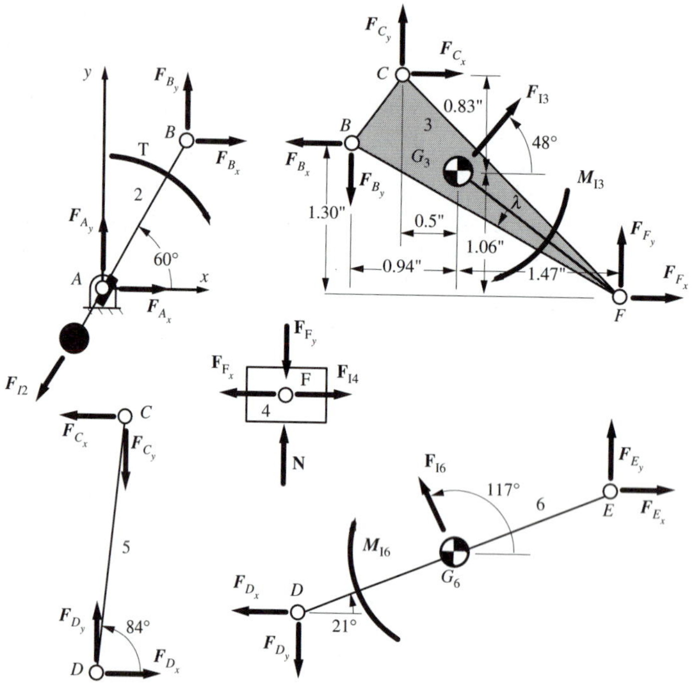

Figure 12.11 Free-body diagrams for Example 12.4.

The inertia force acting at G_3 is calculated as follows:

$$\boldsymbol{F}_{I3} = -m_3\boldsymbol{a}_{G_3} = \frac{0.5}{32.2} \times \frac{43.2}{12} = 0.056 \text{ lb at } 48° \text{ to the } x \text{ axis.}$$

Notice that the direction of $\boldsymbol{F}_{I3}$ is determined by simply adding 180° to the direction of $\boldsymbol{a}_{G_3}$. Similarly, the inertia force acting at G_6 is calculated:

$$\boldsymbol{F}_{I6} = -m_3\boldsymbol{a}_{G_6} = \frac{0.5}{32.2} \times \frac{20.0}{12} = 0.026 \text{ lb at } 117° \text{ to the } x \text{ axis}$$

The inertial force acting on the translating mass 4 is

$$\boldsymbol{F}_{I4} = -m_4\boldsymbol{a}_{G_4} = \frac{1.0}{32.2} \times \frac{13.9}{12} = 0.036 \text{ lb} \rightarrow$$

The inertial couple acting on member 3 is calculated as follows:

$$\boldsymbol{M}_{I3} = -I_3\boldsymbol{\alpha}_3 = m_3k_3^2\boldsymbol{\alpha}_3 = \frac{0.5}{32.2} \times \left(\frac{0.87}{12}\right)^2 \times 19.1 = 0.00156 \text{ lb-ft} = 0.0187 \text{ lb-in CW}$$

The inertia couple acting on member 6 is

$$\boldsymbol{M}_{I6} = I_6\boldsymbol{\alpha}_6 = -m_6\,k_6^2\boldsymbol{\alpha}_6 = \frac{0.5}{32.2} \times \left(\frac{0.87}{12}\right)^2 \times 13.3 = 0.00109 \text{ lb-ft} = 0.0130 \text{ lb-in CW.}$$

Step 4 in the procedure is the drawing of free-body diagrams of all members:

Step 5 of the solution procedure is writing dynamic equilibrium equations for each member. Starting with member 2:

$$\sum M_A = 0: \qquad T + F_{Bx} \times 1.5 \sin 60° = F_{By} \times 1.5 \cos 60°$$

Here we have chosen to take moments about point A because doing so eliminates the components of F_A and also F_{I2}. Force equilibrium would, in this case, give two equations that could ultimately be solved for the components of F_A. Since we are not interested in F_A, they are not written out.

Moving to member 3,

$$\sum F_x = 0: \qquad F_{C_x} + F_{F_x} + 0.056 \cos 48° = F_{B_x}$$

$$\sum F_y = 0: \qquad F_{C_y} + F_{F_y} + 0.056 \sin 48° = F_{B_y}$$

$$\sum M_{G_3} = 0: \qquad F_{C_x} \times 0.83 + F_{C_y} \times 0.5 + 0.0187 = F_{B_x} \times 0.24 + F_{B_y} \times 0.94 + F_{F_x} \times 1.06 + F_{F_y} \times 1.47$$

Here the choice of the point about which to take moments makes little difference. Use of G_3 slightly simplifies the equation since it eliminates $\boldsymbol{F}_{I3}$.

Continuing with member 4:

$$\sum F_x = 0: \qquad 0.036 = F_{F_x}$$

$$\sum F_y = 0: \qquad N = F_{F_y}$$

No moment equation is written because this member is restrained from rotation.

Member 5 is a two-force member because its mass and moment of inertia are neglected. This implies both that $F_C = F_D$ and that both forces are aligned and opposed along the axis of the member. The alignment along the axis is required by the moment equation. Consequently,

$$\sum F_x = 0: \qquad F_{C_x} = F_{D_x}$$

$$\sum F_y = 0: \qquad F_{C_y} = F_{D_y}$$

$$\sum M_D = 0: \qquad F_{C_y} \times 2.25 \cos 84° = F_{C_x} \times 2.25 \sin 84°$$

or

$$F_{C_y} = F_{C_x} \tan 84° \quad \text{and} \quad F_{D_y} = F_{D_x} \tan 84°$$

Finally, for member 6,

$$\sum M_E = 0: \qquad 0.0130 + F_{D_x} \times 3 \sin 21° + 0.026 \times 1.5 \sin 96° = F_{D_y} \times 3 \cos 21°$$

The force equilibrium equations would yield only expressions for the components of F_E. Since we are not interested in F_E and can eliminate it by taking moments about point E, the force equilibrium equations are not needed. Substitution of the previous relationship between F_{D_x} and F_{D_y} into this equation gives

$$0.0130 + F_{D_x} \times 3 \sin 21° + 0.026 \times 1.5 \sin 96° = F_{D_x} \tan 84° \times 3 \cos 21°$$

or

$$F_{C_x} = F_{D_x} = 0.0020 \text{ lb}$$

Then

$$F_{C_y} = F_{D_y} = 0.0020 \times \tan 84° = 0.0193 \text{ lb}$$

Substitution of these values into the force equilibrium equations of member 3 gives

$$F_{B_x} = 0.0395 + F_{F_x} = 0.0395 + 0.036 = 0.0755$$

$$F_{B_y} = 0.0609 + F_{F_y}$$

Substitution into the rotation equations gives

$$0.0300 = 0.24F_{B_x} + 0.94F_{B_y} + 1.06F_{F_x} + 1.47F_{F_y}$$

$$= 0.24 \times 0.0755 + 0.94F_{B_y} + 1.06 \times 0.036 + 1.47F_{F_y}$$

or

$$0 = 0.0263 + 0.94F_{B_y} + 1.47F_{F_y}$$

Elimination of F_{F_y} gives

$$0 = 0.0263 + 0.94F_{B_y} + 1.47(F_{B_y} - 0.0609)$$

or

$$F_{B_y} = 0.0262 \text{ lb}$$

The values obtained for F_{B_x} and F_{B_y} can be substituted into the moment equation of member 2 to give

$$T + 0.0755 \times 1.5 \sin 60° = 0.0262 \times 1.5 \cos 60°$$

or

$$T = -0.078 \text{ lb-in}$$

Therefore, when passing through this position a torque of 0.078 lb-in in the counterclockwise direction is needed to prevent member 2 from accelerating and to maintain its constant speed of rotation. ■

12.4 FLYWHEELS

Flywheels are used to store energy and to smooth speed fluctuations during a machine cycle. They serve a function similar to that of a capacitor in an electric circuit or an accumulator in a hydraulic circuit. In a typical situation, a machine that needs to operate at near-constant speed must be matched to one that produces significant torque fluctuations. For example, an electric motor that works best at near-constant speed may need to drive a punch press, or reciprocating compressor, that produces a strongly fluctuating load torque. Conversely, an internal combustion engine that produces strong torque fluctuations may be required to drive a load shaft at close to constant velocity.

Actually, a flywheel is usually an essential component of an internal combustion engine because the cycle includes strokes in which air is being compressed in the cylinder, so the engine is absorbing energy rather than producing it. The flywheel allows energy to be stored during the power stroke when the charge is burning and expanding and returned to the piston when it is compressing the charge. This is also a reason for the use of engines with multiple cylinders, since the torque fluctuations for different cylinders can be evenly distributed over the engine cycle, reducing the output torque fluctuations and reducing the size of the requisite flywheel.

If the machine-induced energy fluctuations are known and the allowable speed fluctuation is specified, the requisite flywheel inertia is readily calculated. The coefficient of speed fluctuation, c_δ, is defined as follows:

$$c_\delta = \frac{\omega_2 - \omega_1}{\omega} \tag{12.13}$$

where ω_2 is the maximum flywheel angular velocity, ω_1 is the minimum flywheel angular velocity, and ω is the average flywheel velocity. The allowable fluctuation coefficient varies by application, from 0.05 or more for agricultural and mining machinery down to about 0.003 for alternating-current generators. Values appropriate to given applications may be found in machine design handbooks.

If the moment of inertia of the flywheel is I_W, the change in flywheel energy is

$$\Delta E = \frac{I_W}{2}(\omega_2^2 - \omega_1^2) \tag{12.14}$$

Although speed may fluctuate asymmetrically over the cycle, it is often adequate to approximate the average angular velocity as the mean of ω_2 and ω_1:

$$\omega = \frac{\omega_2 + \omega_1}{2}$$

so

$$\omega_2^2 - \omega_1^2 = 2c_\delta \omega^2$$

and

$$I_W = \frac{\Delta E}{c_\delta \omega^2} \tag{12.15}$$

can be used to estimate the required flywheel inertia.

The following example illustrates the way in which a flywheel can be sized for a particular application

EXAMPLE 12.5 (*Punch Press*)

PROBLEM

A machine used to punch holes in metal plate is to be driven by an induction motor with rated speed 1700 rpm. The allowable drop in motor speed is 15%. The machine will be required to punch holes of diameter up to 1 in in steel plate up to 0.5 in thick with shear strength of 50,000 psi. The holes are to be punched at a rate of one every 2.5 s. Find the requisite motor power and flywheel inertia.

SOLUTION

The maximum punch force is

$$F = \pi d t \tau$$

where $\tau = 50{,}000$ psi is the shear strength of the material, d is the diameter of the hole, and t is the plate thickness. That is, the maximum punch force is simply the shear area multiplied by the shear strength. The force profile as the punch penetrates is irregular with the peak force occurring at a penetration distance of about 3/8 of plate thickness. A typical profile is sketched in Fig. 12.12.

The area under the curve of punch force versus depth of penetration is the energy used in the punching operation. It can be measured experimentally, but

$$\Delta E = \frac{Ft}{2} \tag{12.16}$$

is a frequently used approximation. In the present instance

$$F = \pi \times 1.0 \times 0.5 \times 50{,}000 = 78{,}540 \text{ lb}$$

So

$$\Delta E = 78{,}540 \times 0.5/2 = 19{,}600 \text{ lb-in} = 1640 \text{ lb-ft}$$

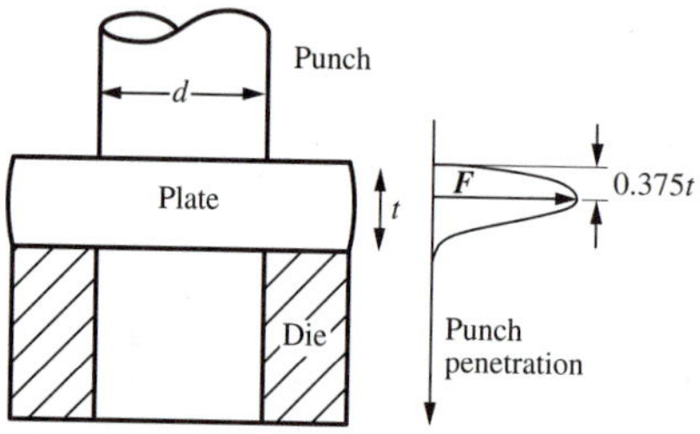

Figure 12.12 Punch force as a function of depth of penetration. The peak force, F, occurs at a penetration depth of about 0.375 t, where t is the plate thickness.

At a punching rate of once every 2.5 s the average power required is

$$P = 1640/2.5 = 654 \text{ ft-lb/s} = 1.19 \text{ hp}$$

This is the power for which the motor should be sized.

Now, if the rated motor speed is 1700 rpm we can assume that the maximum motor speed will be this value, so

$$\omega_2 = \frac{1700 \times 2\pi}{60} = 178 \text{ rad/s}$$

Also, we can assume that the motor speed quickly drops to its minimum value during the punch stroke and that it is then built back up approximately uniformly to the maximum value in the remainder of the cycle time. Therefore

$$\omega = \frac{\omega_1 + \omega_2}{2}$$

is an adequate approximation to the average motor speed, ω.

Since the allowable motor speed variation is 15%, $c_\delta = 0.15$ and so, applying Eq. (12.13),

$$0.15 = \frac{\omega_2 - \omega_1}{\omega} = \frac{2(\omega_2 - \omega_1)}{\omega_2 + \omega_1}$$

Substitution of $\omega_2 = 178$ gives

$$0.075(178 + \omega_1) = 178 - \omega_1$$

or

$$\omega_1 = 153 \text{ rad/s}$$

Also, applying Eq. (12.14),

$$1640 = \frac{I_W}{2}(178^2 - 153^2)$$

or

$$I_W = 0.396 \text{ lb-ft s}^2$$

■

Sizing of a flywheel for an internal combustion engine starts with the pressure-volume chart of the engine's combustion cycle. Multiplication of the pressure by the piston area produces the piston force. Force analysis of the slider-crank mechanism of the engine with the piston force as input produces crankshaft torque as a function of crank angle. The flywheel is then sized to bring the speed fluctuations produced by the torque variation within acceptable limits. Although this process is beyond the scope of this book, it is fully explained in texts on the dynamics of reciprocating machinery.

12.5 CHAPTER 12 *Exercise Problems*

PROBLEM 12.1

The four-wheeled vehicle shown slides down a steep slope with its rear wheels locked (not moving relative to the body) and its front wheels rolling freely. If M is the mass of the vehicle, h the normal distance from its center of mass, G, to the ground, r the wheel radius, and $2c$ the distance between the axles, find the acceleration of the vehicle. The angle of the slope is θ, and the coefficient of friction between the wheels and the ground is μ. The mass and moment of inertia of each wheel about its axle may be neglected. What is the largest value for the angle θ at which the vehicle will not slide?

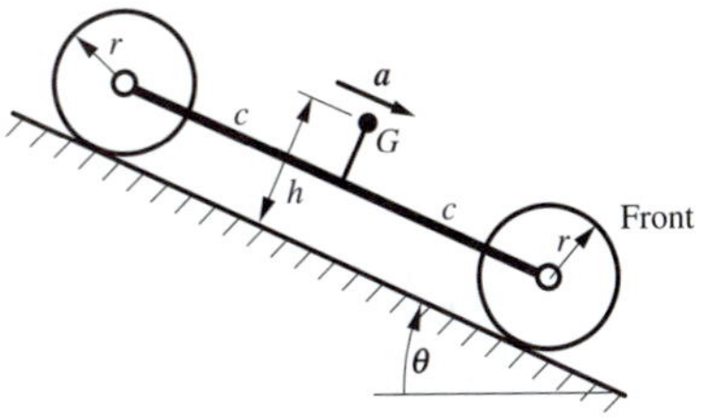

PROBLEM 12.2

The flyball governor shown is started from rest and accelerated slowly about the axis of rotation. At what speed of rotation will it be in the position shown? Friction may be neglected. Ignore the masses of the four links.

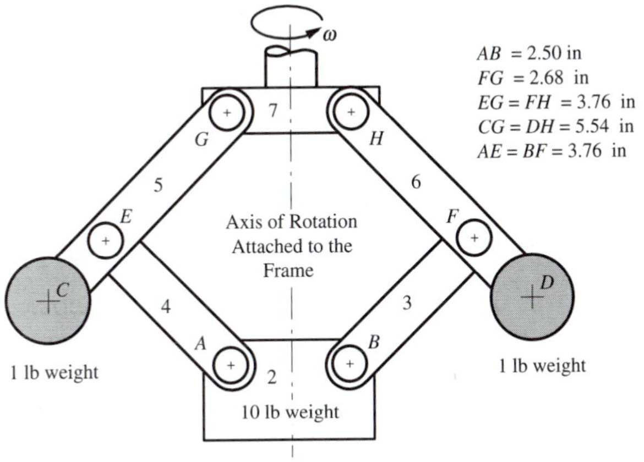

PROBLEM 12.3

Solve Problem 12.2 assuming a coefficient of friction of 0.3 at each of the six pin joints. The diameter of each joint is 0.8 in.

PROBLEM 12.4

A wheel, of mass m and radius r, rolls without slipping on a horizontal plane. It hits a step of height h. If the velocity of the center of the wheel before striking the step is V, directed as shown, find:

1. The magnitude and direction of the velocity of the center of the wheel immediately after the impact
2. The minimum value of V for which the wheel surmounts the step
3. The impulse exerted on the wheel by the edge of the step at impact

The impact may be considered to take place over a vanishingly small time interval. The wheel is assumed to remain in contact with the edge of the step after the impact. The wheel may be considered to have moment of inertia about its center in the direction of rotation $I = mr^2$.

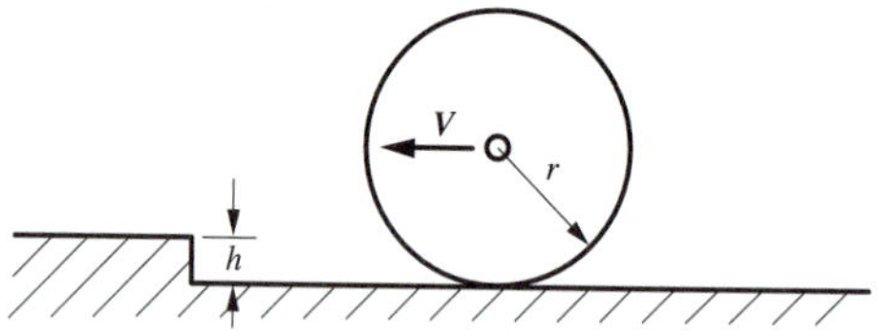

PROBLEM 12.5

In the mechanism shown below, link 2 rotates at an angular velocity of 20 rad/s (CW) and angular acceleration of 140 rad/s^2 (CW). Find the torque which must be applied to link 2 to maintain equlibrium. Link 2 is balanced so that its center of mass is at the pivot O_2. The center of mass of link 3 is at A, and the revolute joint axes are horizontal. Friction may be neglected.

$$O_2A = CA = 100 \text{ mm} \qquad m_3 = 0.74 \text{ kg} \qquad I_{G2} = .00205 \text{ kg-s}^2\text{-m}$$

$$m_4 = 0.32 \text{ kg} \qquad I_{G3} = .0062 \text{ kg-s}^2\text{-m}$$

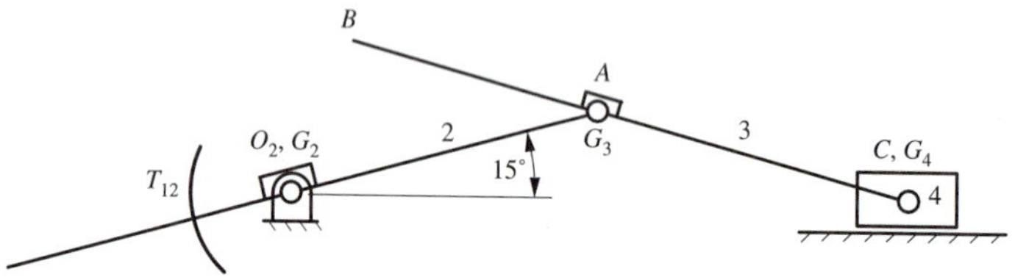

PROBLEM 12.6

Find the external torque (T_{12}) that must be applied to link 2 of the mechanism illustrated to drive it at $\omega_2 = 1800$ rad/s CCW and $\alpha_2 = 0$ rad/s^2. Link 2 is in a horizontal position, and it is balanced so that its center of mass is at the pivot O_2. The joint axes are horizontal and friction may be neglected.

$$W_3 = 0.708 \text{ lb} \qquad I_{G_3} = 0.0154 \text{ lb-s}^2\text{-in}$$

$$W_4 = 0.780 \text{ lb} \qquad I_{G_4} = 0.0112 \text{ lb-s}^2\text{-in}$$

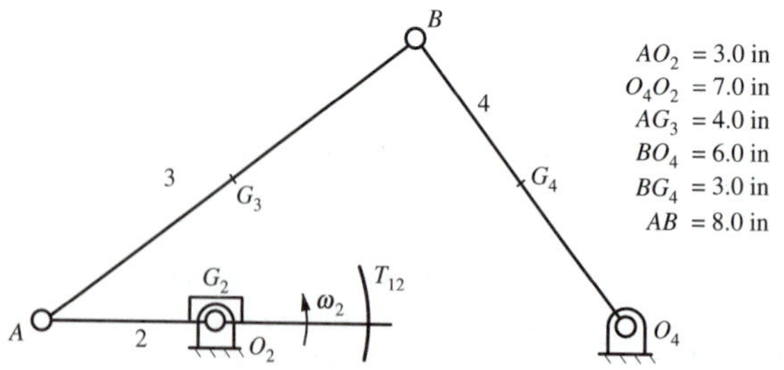

PROBLEM 12.7

Find the external torque (T_{12}) that must be applied to link 2 of the mechanism illustrated in order to drive it at $\omega_2 = 210$ rad/s CCW and $\alpha_2 = 0$ rad/s^2. Link 2 is balanced so that its center of mass is at the pivot O_2. The mechanism moves in the horizontal plane and friction may be neglected.

$$W_3 = 3.4 \text{ lb} \qquad I_{G_3} = 0.1085 \text{ lb-s}^2\text{-in}$$

$$W_4 = 2.86 \text{ lb}$$

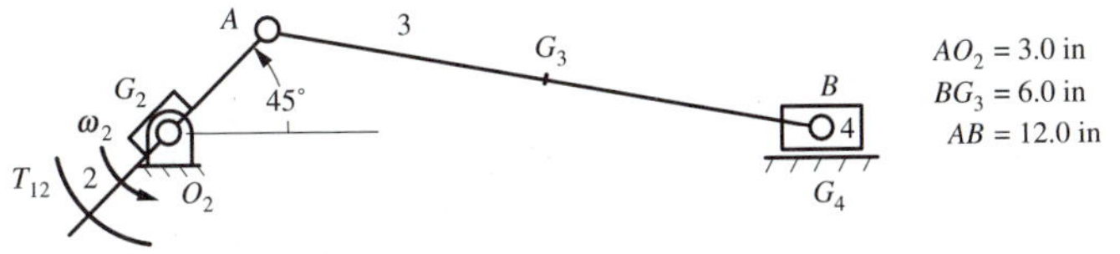

PROBLEM 12.8

In the mechanism shown, the center of mass of link 3 is at G_3, which is located at the center of link 3. The mass of link 3 is 0.5 kg. Its moment of inertia about G_3 is 0.0012 N − s^2 − m. The mechanism moves in the horizontal plane. The weights and moments of inertia of members 2 and 4 may be neglected. Link 2 is driven at a constant angular velocity of 50 rad/s CW by the torque applied to link 2. There are no other external forces or torques applied to the mechanism. Friction may be neglected.

1. Find the magnitudes and directions of the inertia force and inertia torque acting on link 3.
2. Hence find the magnitudes and directions of the forces exerted on link 3 by link 2, at A, and by link 4, at B. You may use either a graphical solution or numerical solution of the dynamic equilibrium equations.

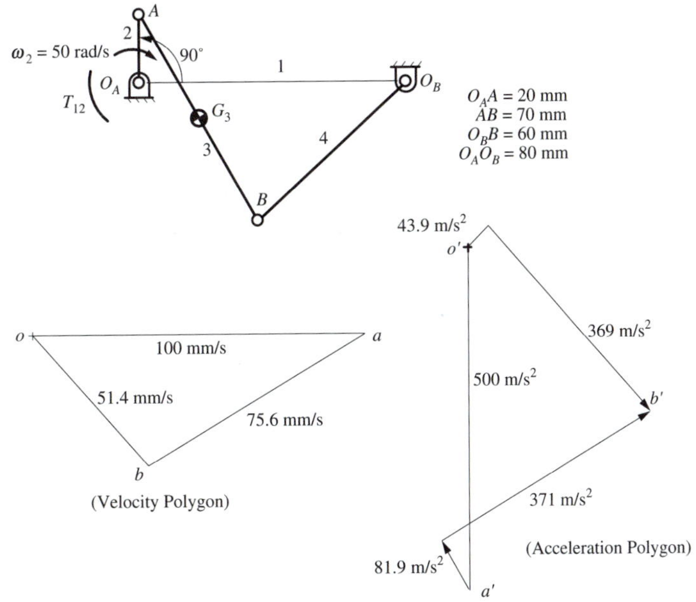

PROBLEM 12.9

Link AB of the geared five-bar linkage shown drives CCW against a load torque $T_{15} = 25$ in-lb. If ${}^1\boldsymbol{\omega}_2 = 0.001$ rpm CW, find the driving torque T_{12}. The mechanism moves in the horizontal plane, and friction may be neglected. The gears 2 and 3 are represented by their pitch circles. Both gears turn on bearings supported by the tie link, 4. The weight of link 2 is small and can be neglected. Gear 3 is 0.2 in thick and may be treated as a solid disk.

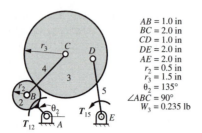

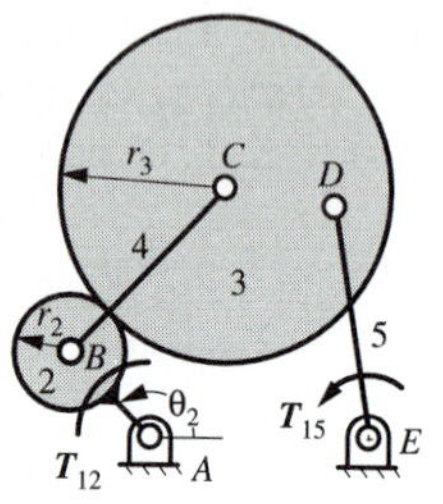

$AB = 1.0$ in
$BC = 2.0$ in
$CD = 1.0$ in
$DE = 2.0$ in
$AE = 2.0$ in
$r_2 = 0.5$ in
$r_3 = 1.5$ in
$\theta_2 = 135°$
$\angle ABC = 90°$
$W_3 = 0.235$ lb

PROBLEM 12.10

A punch press similar to that of Example 12.5 is to punch holes of diameter up to 0.75 in through steel plate up to 0.375 in thick. The shear strength of the steel will range up to 60,000 psi. The rated speed of the motor is 1500 rpm, and a 10% drop in motor speed is allowable. If holes are to be punched at a maximum rate of 1 per second, find the requisite motor power and flywheel inertia.

Chapter 13

Shaking Forces and Balancing

13.1 INTRODUCTION

Fast-moving machinery with rotating or reciprocating masses is a significant source of vibration excitation. A major theme in machine dynamics and machine design is seeking to minimize the fluctuating forces that such machinery applies to its environment via its mounts. Rapidly rotating masses such as those in electric motors and generators, steam and gas turbines, vehicle wheels, and many other situations can generate significant fluctuating forces with even tiny amounts of unbalance. Combinations of rotating and reciprocating masses are found in internal combustion engines, pumps, compressors, and many other types of machinery. They are strong generators of fluctuating forces, but those forces can be at least partially balanced by appropriately placed weights. It is necessary to discuss the procedures used for balancing in these somewhat diverse types of system.

When any mechanism is operated at high speeds, two types of forces must be considered. These are externally applied forces and inertial forces. Inertial forces arise when the individual members are subjected to large accelerations. In general, the inertial force system acting on a given member can be represented as an inertia force acting on a line through the center of mass, together with an inertia torque, as was described in Section 12.2. The force is given by $-m\boldsymbol{a}$ and the couple by $-I\boldsymbol{\alpha}$, where m is the mass of the member, I is the mass moment of inertia about the center of mass, $\boldsymbol{a}$ is the linear acceleration of the center of mass, and $\boldsymbol{\alpha}$ is the angular acceleration of the member. In high-speed machinery, the inertial forces may be larger in magnitude than the external forces. Consequently, when the mechanism is designed, both types of forces must be taken into account.

In general, the external forces will be associated with the useful function that the mechanism is to perform and with driving the mechanism. There is often little that can be done to alter their magnitudes. On the other hand, the inertial forces are due entirely to the mass and motion characteristics of the machine members. Therefore, prudent design practice dictates that the inertial forces be minimized. This can be done either by reducing the masses and moments of inertia of the moving members or by reducing the linear and angular accelerations. The masses may be reduced by using lighter materials and optimal geometries. For a given kinematic geometry, the angular accelerations cannot be reduced. However, the linear accelerations of the centers of mass can be reduced by moving the centers of mass toward points of zero acceleration. The way this is often done is to add mass in the form of balance weights to move the overall center of mass of a given member to a location of reduced acceleration.

13.2 SINGLE-PLANE (STATIC) BALANCING

In spite of all the care that may be taken in the design and manufacture of a rotating part, whether the part is completely machined, cast, or forged or is assembled from various parts as in the case of the armature of an electric motor, it is uncommon for it to run smoothly, particularly if the operating speed is high. Variations in dimensions due to machining, variations in homogeneity of the material, variations in the methods of assembly, and eccentricity of bearing surfaces all contribute to offsetting the center of mass from the axis of rotation.

The curve in Fig. 13.1 emphasizes the effect of a small amount of unbalance at high speeds. The curve shows the centrifugal force produced by an inch-ounce of unbalance at various angular speeds. (An inch-ounce is defined as 1 ounce of weight at 1 inch from the axis of rotation.) The centrifugal force due to 1 in-oz at 1000 rpm is 1.76 lb. At 10,000 rpm it is 176 lb. That is, it increases as the square of the speed. It is evident that the centrifugal force produced on a large rotor can be very large, even if the center of gravity is displaced only a small amount from the axis of rotation, and consequently large shaking forces will be produced on the structure. For example, consider the rotor of an aircraft gas turbine weighing 400 lb that operates at 16,000 rpm and suppose the center of mass is 0.001 in from the axis of rotation. The 6.4 in-oz of unbalance would cause a centrifugal force of

$$F = mR\omega^2 = \frac{400}{32.2} \cdot \frac{0.001}{12} \cdot \left(\frac{2\pi \cdot 16{,}000}{60}\right)^2 = 2924 \text{ lb}$$

Such a force could cause considerable damage to the machine. Since it is usually impossible to manufacture the rotor of a machine so that the center of gravity will lie within 0.001 in of the axis of rotation, the part must be balanced after manufacture, and the balancing is done experimentally.

To illustrate the principles that are involved in balancing a rotating mass, we will first consider a rod rotating at a constant angular velocity ω to which is attached a single

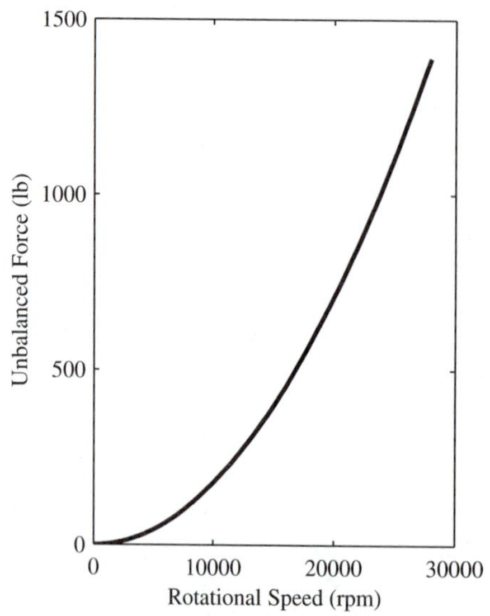

Figure 13.1 Effect of 1 oz-in of unbalance.

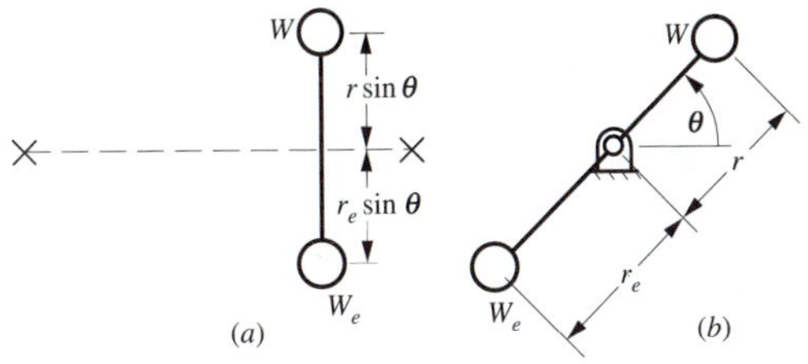

Figure 13.2 Single rotating mass.

concentrated mass of weight W at radius r. Let W_e be the weight (called the counterbalance weight) that must be added at some radius r_e in order to produce equilibrium as shown in Fig. 13.2. Static balance will be produced if the sum of the moments of the weights about the axis of rotation is zero. That is,

$$-Wr\cos\theta + W_e r_e \cos\theta = 0$$

or

$$W_e r_e = Wr \tag{13.1}$$

If the value of r_e is arbitrarily chosen, then the value of W_e can be found from Eq. (13.1). When the system is statically balanced, the shaft will not have any tendency to rotate in its bearings under the influence of gravity regardless of the position to which it is rotated. If the system is rotated with angular velocity ω, the static balance condition also ensures that the sum of the inertia forces is zero, as illustrated in Fig. 13.2. That is

$$\frac{W}{g} r\omega^2 - \frac{W_e}{g} r_e\omega^2 = 0$$

or

$$W_e r_e = Wr$$

which is identical to Eq. (13.1).

The simplest situation is a single rotor that can be regarded as rotating in a single plane, such as that just considered. This type of balancing is called static balancing because a static-type balance procedure in which the measurements are made on the rotor while it is stationary can be used. If the rotor cannot be regarded as spinning in a single plane the balance procedure involves measurements while the rotor is spinning. This procedure is called dynamic balancing.

Of course, there are situations in which an eccentric mass is deliberately spun to produce a rotating force vector. Example 13.1 concerns an eccentric rotor used to excite vibratory motion of a screen used to sort material particles by size.

EXAMPLE 13.1 ***(Eccentric Mass Out of Balance)***

PROBLEM

An eccentric mass is rotated on a shaft mounted on the frame of a vibrating screen used to sort iron ore into different sizes. The mass is shaped as a circular sector as shown in Fig. 13.3. The mass weighs 20 lb, and its mass center is 4 in from the shaft axis. If it is rotated at 600 rpm, what is the force that it exerts on the shaft and hence on the frame of the screen?

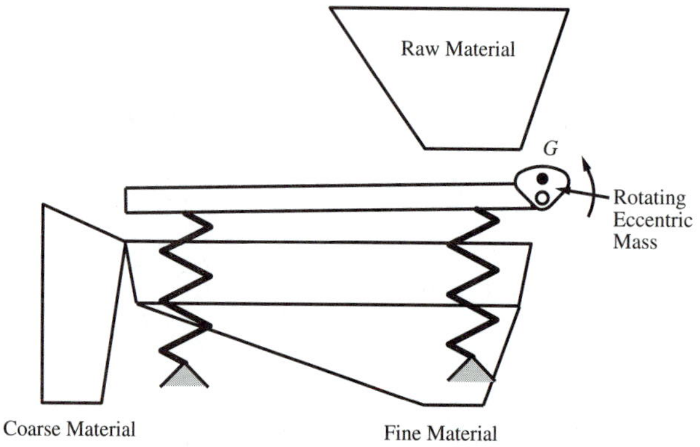

Figure 13.3 The vibrating screen arrangement discussed in Example 13.1. G is the center of mass of the rotor.

SOLUTION

The exciting force here is simply the centrifugal force of the rotating mass. The angular velocity is

$$\omega = 600 \times 2\pi/60 = 62.83 \text{ rad/s}$$

so the acceleration of the center of mass is

$$a = r\omega^2 = (4/12) \times 62.83^2 = 1316 \text{ ft/s}^2$$

Therefore, the magnitude of the rotating force is

$$F = ma = (20/32.2) \times 1316 = 817 \text{ lb}$$ ■

The product of the mass of a rotor and the distance from its center of mass to the shaft axis is called the unbalance. An unbalance of a given magnitude may be removed by adding a mass 180° out of phase with it or by subtracting mass with the same unbalance in phase with it. Referring to Fig. 13.4, if the mass of the rotor is m_A and located at point A, then its unbalance is $m_A r_A$. If a balancing mass is to be replaced at point B directly opposite point A, then its mass must be m_B such that $m_B r_B = m_A r_A$. Similarly, balance can also be achieved by *removing* mass m_C at point C on the same line with A and the shaft axis provided $m_C r_C = m_A r_A$.

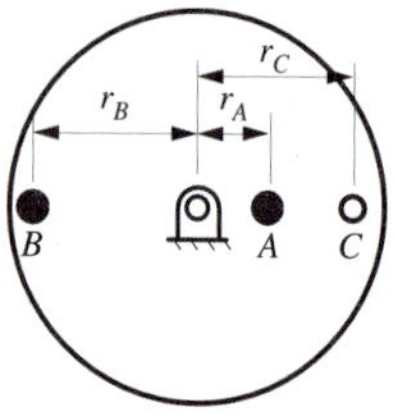

Figure 13.4 A rotor with an eccentric mass with mass center at point A. The unbalance can be removed by placing a mass at point B or removing a mass at point c.

EXAMPLE 13.2 (*Static Balancing*)

PROBLEM

A factory producing gear sets statically balances spur gear wheels using the following procedure (Fig. 13.5).

The wheel is mounted on a spindle that turns in very low friction bearings. The wheel is allowed to turn freely under gravity until it comes to rest. The low point, X, is marked. A small, known mass, m_y, is next removed from the wheel at point Y, which is 90° clockwise from X

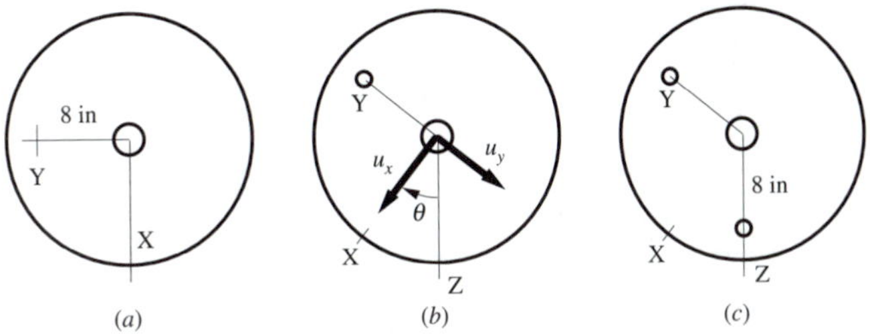

Figure 13.5 The static balance procedure used in Example 13.2. Three stages in the process are shown from left to right.

and at a set radius, r_y. The wheel is then replaced on the spindle and comes to rest with point Z at its lowest point. A calculated mass, m_z, is then drilled out at a specified radius, r_z, on the radial line through Z to balance the wheel.

If the mass drilled out of a given wheel at point Y weighs 0.1 lb at radius 8 in, and the angle θ, between the radial lines through X and Z is 35°, find the weight of the mass to be removed at Z if it is also drilled out at a radius of 8 in.

SOLUTION

Let the initial unbalance be u_x. The unbalance of the trial mass at point Y is

$$u_y = 0.1 \times 8 = 0.8 \text{ in-lb}$$

Now, when the wheel is in equilibrium after the trial mass is removed,

$$u_x \sin\theta = u_y \cos\theta$$

Note that the unbalance is in the *opposite* direction to Y since mass was removed at Y, not added.

$$u_x = 0.8 \cot 35° = 1.143 \text{ in-lb}$$

The total unbalance is then

$$u_z = u_x \cos\theta + u_y \sin\theta = 1.143 \cos 35° + 0.8 \sin 35° = 1.394 \text{ in-lb}$$

Hence the weight of the mass to be removed at Z is

$$m_z = 1.394/8 = 0.174 \text{ lb}$$

■

13.3 MULTIPLANE (DYNAMIC) BALANCING

If the inertial mass distribution of the rotor cannot be regarded as planar, fluctuating moments will be generated about axes normal to the axis of rotation in addition to the radial forces. Although the system may be statically balanced by eliminating the resultant radial forces, the shaking moments produced can still be potent sources of vibration excitation. Multiplane or dynamic balancing techniques allow elimination of both radial unbalance forces and moments.

To see how this works, consider the system shown in Fig. 13.6. The shaft shown has two rotors mounted on it, each with an unbalance. The shaft turns in two bearings, as shown. Let us first calculate the unbalance forces, F_A and F_C, exerted on the shaft by the bearings.

The unbalance of the left-hand rotor is

$$u_P = m_P r_P$$

where m_P is the mass of the left-hand rotor and r_P is the eccentricity of its center of mass. Hence, at a constant rotation speed ω, the unbalanced force magnitude is

$$F_P = u_P \omega^2$$

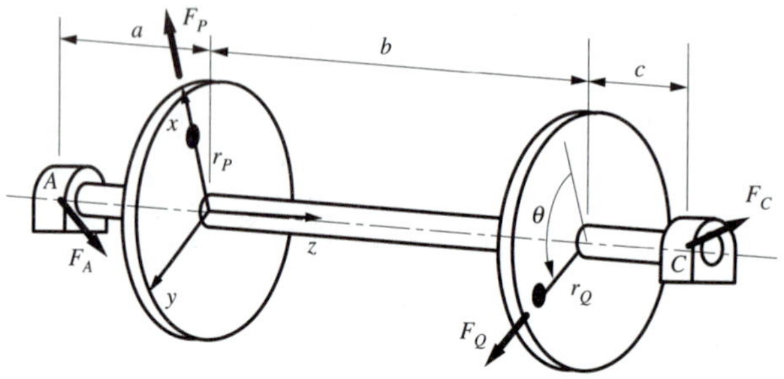

Figure 13.6 A shaft with two rotors, both of which have unbalances. The left rotor plane is designated by P and the right by Q.

Similarly, at the right-hand rotor, the unbalance is

$$u_Q = m_Q r_Q$$

where m_Q is the mass of the right-hand rotor and r_Q is the eccentricity of its center of mass. Hence, the unbalanced force magnitude is

$$F_Q = u_Q \omega^2$$

Using a reference frame fixed to the left-hand rotor as shown in Fig. 13.6 with the z axis along the shaft axis and the x axis aligned with the direction of the unbalance u_P, we can write the following force equilibrium equations:

$$\sum F_x = 0: \qquad F_P + F_Q \cos\theta + F_{A_x} + F_{C_x} = 0 \tag{13.2}$$

$$\sum F_y = 0: \qquad F_Q \sin\theta + F_{A_y} + F_{C_y} = 0 \tag{13.3}$$

It is convenient to take moments about A because then F_C is the only unknown force appearing in the equations.

$$\sum M_A = 0: \qquad a\boldsymbol{k} \times F_P \boldsymbol{i} + (a+b)\boldsymbol{k} \times F_Q(\cos\theta\, \boldsymbol{i} + \sin\theta\, \boldsymbol{j}) + (a+b+c)\boldsymbol{k} \times (F_{C_x}\boldsymbol{i} + F_{C_y}\boldsymbol{j}) = 0$$

or

$$aF_P\boldsymbol{j} + (a+b)F_Q(\cos\theta\,\boldsymbol{j} - \sin\theta\,\boldsymbol{i}) + (a+b+c)(F_{C_x}\boldsymbol{j} - F_{C_y}\boldsymbol{i}) = 0$$

Separating the component equations:

$$-(a+b)F_Q \sin\theta - (a+b+c)F_{C_y} = 0 \tag{13.4}$$

$$aF_P + (a+b)F_Q\cos\theta + (a+b+c)F_{C_x} = 0 \tag{13.5}$$

We can solve for the components of F_C from Eqs. (13.4) and (13.5) and then solve for the components of F_A from Eqs. (13.2) and (13.3). From Eq. (13.4)

$$F_{C_y} = -\frac{(a+b)F_Q \sin\theta}{a+b+c}$$

and from Eq. (13.5)

$$F_{C_x} = -\frac{aF_P + (a+b)F_Q\cos\theta}{a+b+c}$$

From Eq. (13.2)

$$F_{A_x} = -F_P - F_Q \cos\theta - F_{C_x} = -F_P - F_Q \cos\theta + \frac{aF_P + (a+b)F_Q \cos\theta}{a+b+c}$$

or

$$F_{A_x} = -\frac{(b+c)F_P + cF_Q \cos\theta}{a+b+c}$$

From Eq. (13.3)

$$F_{A_y} = -F_Q \sin\theta - F_{C_y} = -F_Q \sin\theta + \frac{(a+b)F_Q \sin\theta}{a+b+c}$$

or

$$F_{A_y} = -\frac{cF_Q \sin\theta}{a+b+c}$$

Thus the forces applied to the bearings by the shaft are

$$\boldsymbol{F}_A^* = -\boldsymbol{F}_A = \frac{\{(b+c)F_P + cF_Q \cos\theta\}\boldsymbol{i} + cF_Q \sin\theta\boldsymbol{j}}{a+b+c}$$

and

$$\boldsymbol{F}_C^* = -\boldsymbol{F}_C = \frac{\{aF_P + (a+b)F_Q \cos\theta\}\boldsymbol{i} + (a+b)F_Q \sin\theta\boldsymbol{j}}{a+b+c}$$

These forces rotate with the shaft, so they fluctuate sinusoidally in any given direction.

If the system is statically balanced by adding a balance weight in the plane P, the required unbalance is

$$\boldsymbol{u} = -\boldsymbol{u}_P - \boldsymbol{u}_Q = -u_P\boldsymbol{i} - u_Q(\cos\theta\,\boldsymbol{i} + \sin\theta\,\boldsymbol{j})$$

Addition of an unbalance of this magnitude and direction in plane P adds a force

$$\boldsymbol{F} = \omega^2\boldsymbol{u} = -\boldsymbol{F}_P - \boldsymbol{F}_Q$$

to the system. The dynamic equilibrium equations become

$$\Sigma F_x = 0: \quad -F_P - F_Q \cos\theta + F_P + F_Q \cos\theta + F_{A_x} + F_{C_x} = F_{A_x} + F_{C_x} = 0 \quad \textbf{(13.6)}$$

$$\Sigma F_y = 0: \quad -F_Q \sin\theta + F_Q \sin\theta + F_{A_y} + F_{C_y} = F_{A_y} + F_{C_y} = 0 \quad \textbf{(13.7)}$$

Once again, taking moments about A:

$$\Sigma M_A = 0: \quad -a\boldsymbol{k} \times \{F_P\boldsymbol{i} + F_Q(\cos\theta\,\boldsymbol{i} + \sin\theta\,\boldsymbol{j})\} + a\boldsymbol{k} \times F_P\boldsymbol{i} + (a+b)\boldsymbol{k} \times F_Q(\cos\theta\,\boldsymbol{i} + \sin\theta\,\boldsymbol{j}) + (a+b+c)\boldsymbol{k} \times (F_{C_x}\boldsymbol{i} + F_{C_y}\boldsymbol{j}) = 0$$

or

$$bF_Q(\cos\theta\,\boldsymbol{j} - \sin\theta\,\boldsymbol{i}) + (a+b+c)(F_{C_x}\boldsymbol{j} - F_{C_y}\boldsymbol{i}) = 0$$

Therefore

$$F_{C_x} = -\frac{bF_Q \cos\theta}{a+b+c}$$

$$F_{C_y} = -\frac{bF_Q \sin\theta}{a+b+c}$$

and, from Eqs. (13.6) and (13.7),

$$F_{A_x} = -F_{C_x} = \frac{bF_Q \cos\theta}{a+b+c}$$

$$F_{A_y} = -F_{C_y} = \frac{bF_Q \sin\theta}{a+b+c}$$

As may be seen, although the system is now statically balanced in that the resultant radial force is zero, the forces exerted on the bearings are not zero, and so a vibration excitation effect is still present. Since $F_A = -F_C$, the bearing forces form a couple with moment

$$\boldsymbol{M} = (a+b+c)\boldsymbol{k} \times (F_{A_x}\boldsymbol{i} + F_{A_y}\boldsymbol{j}) = (a+b+c)(F_{A_x}\boldsymbol{j} - F_{A_y}\boldsymbol{i})$$
$$= (a+b+c)F_Q(\cos\theta\,\boldsymbol{j} - \sin\theta\,\boldsymbol{i})$$

This couple rotates with the shaft.

EXAMPLE 13.3 (*Two-Plane Balance*)

PROBLEM

A small gas turbine rotor carries four blade disks positioned in planes *A, B, C, D* as shown in Fig. 13.7. The shaft turns in bearings in planes *P* and *Q*. Relative to the reference frame shown, the unbalances of the blade disks are, respectively,

$$u_A = 0.08 \text{ lb-in at } 0°; \qquad u_B = 0.03 \text{ lb-in at } 70°;$$
$$u_C = 0.04 \text{ lb-in at } -115°; \qquad u_D = 0.03 \text{ lb-in at } 160°$$

1. Compute the rotating radial forces that the rotor would apply to the bearings when spinning at the engine's normal operating speed of 15,000 rpm.
2. The rotor is to be balanced by removing material from disk *A* at radius 3.5 in and from disk *D* at radius 2.0 in. Calculate the mass that must be removed at each location and the angular position at which it must be removed.

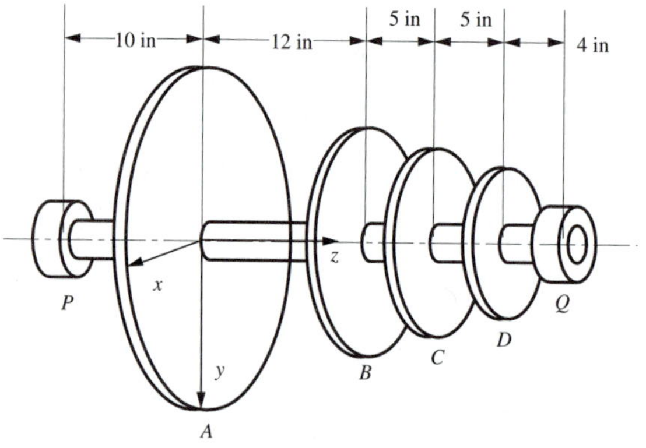

Figure 13.7 The turbine rotor whose balance is analyzed in Example 13.3.

SOLUTION

1. $\omega = 15{,}000 \times 2\pi/60 = 1571$ rad/s. Therefore, the unbalance forces at the respective blade disks are

$$\boldsymbol{F}_A = \frac{0.08 \times 1571^2}{32.2 \times 12} = 511 \text{ lb } \angle 0°$$

$$\boldsymbol{F}_B = \frac{0.03 \times 1571^2}{32.2 \times 12} = 192 \text{ lb } \angle 70°$$

$$\boldsymbol{F}_C = \frac{0.04 \times 1571^2}{32.2 \times 12} = 255 \text{ lb} \angle -115°$$

$$\boldsymbol{F}_D = \frac{0.03 \times 1571^2}{32.2 \times 12} = 192 \text{ lb} \angle 160°$$

Let $\boldsymbol{F}_P = F_{P_x}\boldsymbol{i} + F_{P_y}\boldsymbol{j}$ be the force exerted on the shaft by the bearing at P, and let $\boldsymbol{F}_Q = F_{Q_x}\boldsymbol{i} + F_{Q_y}\boldsymbol{j}$ be the force exerted on the shaft by the bearing at Q. The force equilibrium equations are:

$$\Sigma F_x = 0: \qquad F_{P_x} + F_{Q_x} + F_A \cos 0° + F_B \cos 70° + F_C \cos(-115°) + F_D \cos 160° = 0$$

or

$$F_{P_x} + F_{Q_x} + 511 + 192 \cos 70° + 255 \cos(-115°) + 192 \cos 160° = 0$$

giving

$$F_{P_x} + F_{Q_x} = -288 \text{ lb} \tag{13.8}$$

$$\Sigma F_y = 0: \qquad F_{P_y} + F_{Q_y} + F_A \sin 0° + F_B \sin 70° + F_C \sin(-115°) + F_D \sin 160° = 0$$

or

$$F_{P_y} + F_{Q_y} + 192 \sin 70° + 255 \sin(-115°) + 192 \sin 160° = 0$$

giving

$$F_{P_y} + F_{Q_y} = -15 \text{ lb} \tag{13.9}$$

Taking moments about P:

$$\Sigma M_P = 0: \qquad 10\boldsymbol{k} \times 511\boldsymbol{i} + 22\boldsymbol{k} \times (65.7\boldsymbol{i} + 180\boldsymbol{j}) + 27\boldsymbol{k} \times (-107.8\boldsymbol{i} - 231\boldsymbol{j}) + 32\boldsymbol{k} \times (-180\boldsymbol{i} + 65.7\boldsymbol{j}) + 36\boldsymbol{k} \times (F_{Q_x}\boldsymbol{i} + F_{Q_y}\boldsymbol{j}) = 0$$

Therefore

$$F_{Q_y} = \frac{1}{36}(-3960 + 6237 - 2102) = 4.9 \text{ lb}$$

$$F_{Q_x} = -\frac{1}{36}(5110 + 1445 - 2911 - 5760) = 58.8 \text{ lb}$$

and substitution back into Eqs. (13.8) and (13.9) gives

$$F_{P_x} = -288 - 59 = -347 \text{ lb}$$

$$F_{P_y} = -15 - 5 = -20 \text{ lb}$$

Consequently, the magnitude and direction of the bearing force at P is

$$F_P = \sqrt{347^2 + 20^2} = 348 \text{ lb} \angle 183.3°$$

Similarly, the bearing force at Q is

$$F_Q = \sqrt{4.9^2 + 58.8^2} = 59 \text{ lb} \angle 4.8°$$

2. Since the unbalance force is simply the unbalance multiplied by ω^2 and has the same direction as the unbalance, it is convenient simply to work with the unbalances. The u_A is the required unbalance to be added to disk A, and u_D is the unbalance to be added to disk D. If the rotor is fully dynamically balanced, the forces at the bearings P and Q are both zero. The x direction force equilibrium equation gives

$$u_{A_x} + 0.08 + 0.03 \cos 70° + 0.04 \cos(-115°) + 0.03 \cos 160° + u_{D_x} = 0$$

The y direction equation is

$$u_{A_y} + 0 + 0.03 \sin 70° + 0.04 \sin(-115°) + 0.03 \sin 160° + u_{D_y} = 0$$

These equations give

$$u_{A_x} + u_{D_x} = -0.0452 \text{ lb-in} \quad \textbf{(13.10)}$$

$$u_{A_y} + u_{D_y} = -0.0022 \text{ lb-in} \quad \textbf{(13.11)}$$

Since the bearing forces are zero, it is better to take moments about A, so only the components of u_D appear as unknowns in the moment equations. Hence

$$12\boldsymbol{k} \times (0.03 \cos 70°\boldsymbol{i} + 0.03 \sin 70°\boldsymbol{j}) + 17\boldsymbol{k} \times (0.04 \cos(-115°)\boldsymbol{i} + 0.04 \sin(-115°)\boldsymbol{j}) + 22\boldsymbol{k} \times ((0.03 \cos 160° + u_{D_x})\boldsymbol{i} + (0.03 \sin 160° + u_{D_y})\boldsymbol{j}) = 0$$

This reduces to

$$0.0523 - 22u_{D_y} = 0$$

$$-0.7845 + 22u_{D_x} = 0$$

or

$$u_{D_x} = 0.0357 \text{ lb-in}$$

$$u_{D_y} = 0.0024 \text{ lb-in}$$

so the magnitude of the required additional unbalance at disk D is

$$u_D = \sqrt{0.0357^2 + 0.0024^2} = 0.0358 \text{ lb-in}$$

and its direction is

$$\angle u_D = \tan^{-1}(0.0024/0.0357) = 3.8° \text{ relative to the } x \text{ axis}$$

Substitution back into Eqs. (13.10) and (13.11) gives

$$u_{A_x} = -0.0452 - 0.0357 = -0.0809 \text{ lb-in}$$

$$u_{A_y} = -0.0022 - 0.0024 = -0.0046 \text{ lb-in}$$

so the magnitude of the required additional unbalance at disk A is

$$u_A = \sqrt{0.0809^2 + 0.0046^2} = 0.0810 \text{ lb-in}$$

and its direction is

$$\angle u_A = 180° + \tan^{-1}(0.0046/0.0809) = 183.3° \text{ relative to the } x \text{ axis}$$

Now the unbalance at A is to be created by *removing* mass at the radius 3.5 in. The weight of the mass to be removed is

$$m_A = 0.0810/3.5 = 0.023 \text{ lb}$$

Since the material will be removed, it should be removed at the angle 180° + 183.3° = 363.3°, that is, at angle 3.3° to the x axis in the positive rotation direction about the z axis.

Similarly, at disk D the material is to be removed at the radius 2.0 in. The weight to be removed is

$$m_D = 0.0358/2.0 = 0.0179 \text{ lb}$$

and it must be removed at the angle 180° + 3.8° = 183.8° measured from the positive x axis direction. ■

13.4 BALANCING RECIPROCATING MASSES

A second major source of vibration excitation in machinery is the presence of masses that perform oscillatory motions. Of course, the classic case is the reciprocating piston mass in an internal combustion engine, or a reciprocating pump or compressor.

Although the motion of the reciprocating mass in one of these machines is not strictly harmonic, it is customary to treat it as being so for the purposes of trying to reduce the tendency of the machine to excite vibration. There are several reasons for making this approximation. One is that a simple harmonic oscillation is easy to model and conceptualize. Another is that the fundamental frequency of oscillation is really most important. If we imagine the oscillatory acceleration decomposed into Fourier components, there will be a large-amplitude fundamental with the period of the overall oscillation and a train of higher harmonics with much smaller amplitudes. Not only are these harmonics less effective in exciting vibration because of their smaller amplitudes, but their frequencies are much higher than the fundamental. Since higher frequency vibrations are much more effectively damped out in most structures, the fundamental frequency tends to dominate the transmitted vibration.

Another approximation that is commonly used when dynamically modeling reciprocating machines is to model the mass of the connecting rod as two equivalent masses, one located on the crank pin axis and the other at the wrist pin axis. That is, the distributed mass of the connecting rod is split into a reciprocating point mass, at the wrist pin, and a pure rotating point mass, at the crank pin. The reciprocating component then simply becomes a part of the piston mass, and the rotating component becomes a part of the mass of the crank pin.

This decomposition can be done to preserve the correct location of the center of mass of the connecting rod. However, the moment of inertia of the two mass elements about the center of mass will not be exactly the same as that of the connecting rod. The error introduced in this manner is minimal in most analyses.

13.4.1 Expression for Lumped Mass Distribution

To develop an expression for discretizing the distribution of the masses of the crank and connecting rod consider the slider-crank mechanism shown in Fig. 13.8. The crank is

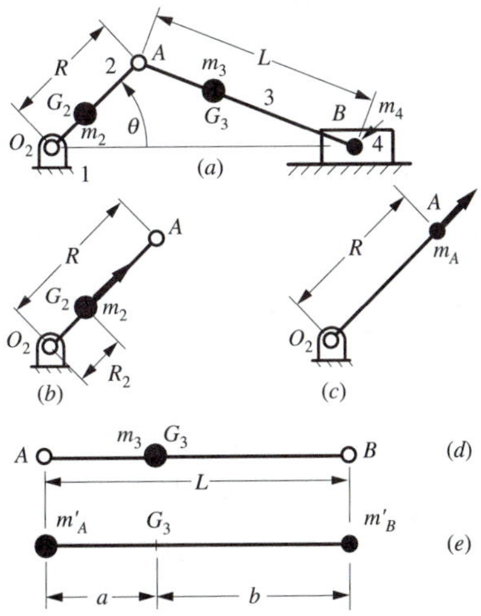

Figure 13.8 A slider-crank mechanism: replacement of connecting rod and crank pin inertias by approximately equivalent lumped masses located at points A and B.

assumed to rotate with a constant angular velocity ω. Points A and B are the crank pin and wrist pin, respectively, and G_2 and G_3 are the centers of mass of links 2 and 3 respectively. The sum of the weights of the crank and crank pin is m_2g, and the centrifugal force acts outward along line G_2A as shown in Fig. 13.8b. In Fig. 13.8c, the masses of the crank and crank pin have been replaced by a concentrated mass m_A located at A such that the centrifugal forces shown in Figs. 13.8b and c will be equal. Thus,

$$m_A R\omega^2 = m_2 R_2 \omega^2$$

or

$$m_A = \frac{R_2}{R} m_2 \tag{13.12}$$

Note that in Eq. (13.12) it is assumed that the center of mass of the crank and point A are on the same side of the rotation axis. If G_2 is on the opposite side, the effective mass of the crank will be *negative.* This is because the inertial force due to G_2 will be in the opposite direction from that due to a mass at A.

The connecting rod of mass m_3 in Fig. 13.8d can be approximated by the system shown in Fig. 13.8e, which consists of the two concentrated weights m'_A and m'_B connected by a weightless rod. For the center of mass of the substitute masses m'_A and m'_B to remain at G_3

$$m'_B(a + b) = (m'_A + m'_B)a$$

or

$$m'_B = \frac{m_3 a}{a + b} = \frac{m_3 a}{L} \tag{13.13}$$

where $m_3 = (m'_A + m'_B)$ is the mass of the connecting rod, b is the distance from the center of mass of the connecting rod to the center of the piston pin, and a is the distance from the center of mass of the connecting rod to the center of the crank pin, Similarly,

$$m'_A = \frac{m_3 b}{a + b} = \frac{m_3 b}{L} \tag{13.14}$$

The replacement system in Fig. 13.8e will have the same inertia forces as the actual connecting rod but its inertia torque will be somewhat different. The total equivalent masses at A and B can be represented by

$$\overline{m}_A = m'_A + m_A \tag{13.15}$$

and

$$\overline{m}_B = m'_B + m_4 \tag{13.16}$$

where m_4 is the weight of the piston.

It should be noted that this replacement is an approximation. It almost always increases the effective moment of inertia of the connecting rod about its mass center. This is because the radius of gyration of the central bar portion of the member is actually relatively small: $L/\sqrt{12}$ if the bar is uniform. Modeling it as lumped masses at its ends always increases the effective moment of inertia.

The shaking forces generated by the reciprocating and rotating masses of a reciprocating machine can now be estimated as in the following example.

EXAMPLE 13.4 (*Shaking Force Calculation*)

PROBLEM

A single-cylinder reciprocating air compressor has a piston that weighs 2.5 lb and a connecting rod of length 12 in weighing 2.0 lb, with its center of mass 4 in from the center of the crank pin. The crank radius is 3 in with a counterbalance weight 4.0 lb and center of mass 2 in on the reverse side of the crankshaft axis. Estimate the shaking force for a constant crank angular velocity of 300 rpm (cw) at 30° after top dead center (T.D.C.).

SOLUTION

Figure 13.9 shows the linkage geometry in the specified position, together with the corresponding velocity and acceleration diagrams.

The angular velocity of the crank (member 2) is

$$\omega_2 = 300 \times 2\pi/60 = 31.42 \text{ rad/s CW}$$

Hence the velocity of point A is

$$\mathbf{v}_A = 0.25 \times 31.42 = 7.85 \text{ ft/s} \angle -30°$$

which is plotted as $\overrightarrow{oa}$ on the velocity diagram. The velocity of point B is given by $\mathbf{v}_B = \mathbf{v}_A + \mathbf{v}_{B/A}$. The direction of both $\mathbf{v}_B$ and $\mathbf{v}_{B/A}$ are known.

Completion of the triangle by drawing $\overrightarrow{ob}$ parallel to the slide and $\overrightarrow{ab}$ normal to $\overrightarrow{AB}$ gives

$$\mathbf{v}_{B/A} = \overrightarrow{ab} = 6.87 \text{ ft/s in the direction shown}$$

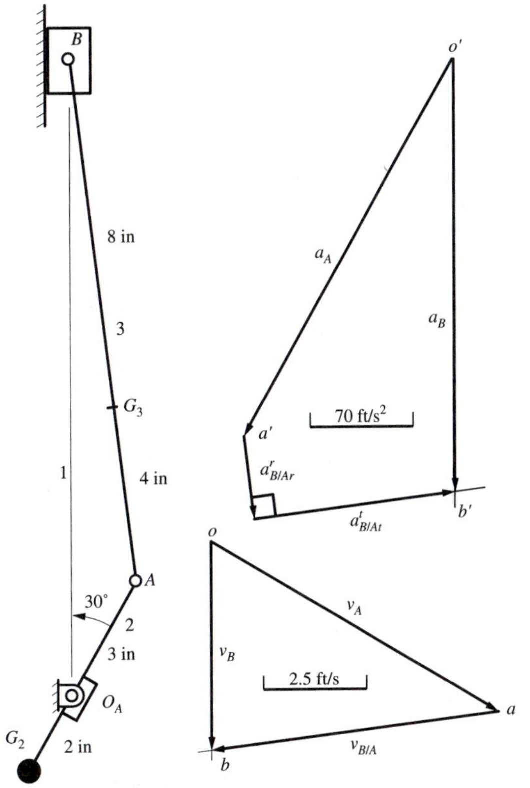

Figure 13.9 Position, velocity, and acceleration diagrams for the air compressor mechanism of Example 13.4.

Hence, the angular velocity of the connecting rod, member 3, is

$$\omega_3 = \frac{ab}{AB} = \frac{6.87}{1.0} = 6.87 \text{ rad/s CCW}$$

Since the angular velocity of member 2 is constant

$$\boldsymbol{a}_A = \boldsymbol{a}_A^r = 0.25 \times 31.42^2 = 246.8 \text{ ft/s}^2 \angle -120°$$

This is plotted on the acceleration diagram as $\overrightarrow{o'a'}$. The acceleration of B is given by $\boldsymbol{a}_B = \boldsymbol{a}_A + \boldsymbol{a}_{B/A} = \boldsymbol{a}_A + \boldsymbol{a}_{B/A}^r + \boldsymbol{a}_{B/A}^t$. The radial component of $\boldsymbol{a}_{B/A}^r$ can be calculated and plotted

$$\boldsymbol{a}_{B/A}^r = 1.0 \times 6.87^2 = 47.20 \text{ ft/s}^2$$

parallel to $\overrightarrow{BA}$ as shown in the figure. The direction of $\boldsymbol{a}_{B/A}^t$ can now be plotted through the tip of this arrow. Its intersection with the direction of $\boldsymbol{a}_B$, parallel to the slide, gives the point b' and completes the polygon. Scaling from the diagram gives

$$\boldsymbol{a}_B = 248 \text{ ft/s}^2 \text{ (down)}$$

This is actually all the acceleration information needed to solve the problem because the distributed mass of the connecting rod will be approximated by lumped masses at points A and B.

Applying Eqs. (13.13) and (13.14), the equivalent weights for the connecting rod are

$$m'_A = (8/12)\,2.0 = 1.333 \text{ lb}$$

$$m'_B = (4/12)\,2.0 = 0.667 \text{ lb}$$

Therefore, the total effective weight at point B is

$$\overline{m}_B = 0.667 + 2.5 = 3.167 \text{ lb}$$

The reciprocating shaking force component is, therefore,

$$\boldsymbol{F}_B = -\overline{m}_B \boldsymbol{a}_B = 248 \times 3.167/32.2 = 24.4 \text{ lb (up)}$$

The rotating mass at A in Fig. 13.10 will be the sum of m'_A and the effective mass m_A of the crank referred to point A. Note that points A and G_2 are on opposite sides of the rotation axis. Therefore, the effective mass of the crank alone will be negative. From Eq. (13.12) the effective mass m_A of the crank is given by

$$m_A = \frac{R_2}{R} m_2 = -\frac{2}{3} 4.0 = -2.666 \text{ lb}$$

The total equivalent mass at A is given by Eq. (13.15) as

$$\overline{m}_A = m'_A + m_A = 1.333 - 2.666 = -1.333 \text{ lb}$$

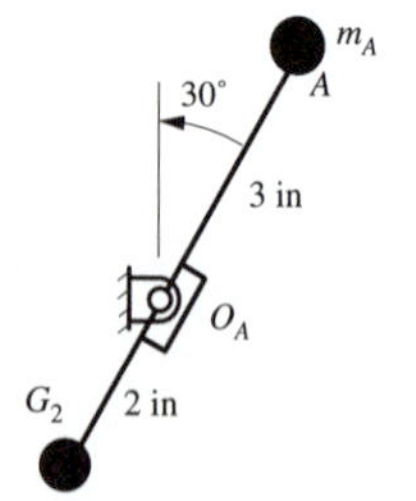

Figure 13.10 Effective mass distribution of the continuously rotating crank of Example 13.4. The mass m_A is the equivalent rotating mass of the connecting rod.

Therefore, the center of gravity for M_A is on the same side of the rotation axis as G_2. Consequently, the inertia force due to the effective rotating mass is

$$\boldsymbol{F}_A = \overline{m}_A R\omega^2 = \frac{-1.333(3)(31.42)^2}{12 \times 32.2} = -10.220 \angle 60° = 10.220 \angle -120°$$

If the effective mass at A were positive, the direction of the inertial force would be opposite to the acceleration of point A or at an angle of 60° with the horizontal axis. In this system, the rotating masses contribute very little to the shaking force. The reciprocating masses are dominant in the generation of the shaking force. ■

13.4.2 Analytical Approach to Balancing a Slider-Crank Mechanism

As already discussed, the inertia of the connecting rod can be approximated by two concentrated masses, one at the crank pin and the other at the wrist pin. The slider-crank system them becomes a combination of a mass assumed to be rotating at constant angular velocity (the crank mass together with the equivalent mass at the crank pin from the connecting rod) and a reciprocating mass (the piston and wrist pin masses, together with the equivalent mass from the connecting rod concentrated at the wrist pin). To a first approximation, the piston motion can be regarded as a simple harmonic motion. This is equivalent to replacing a Fourier series by its fundamental term; it is an approximation, but a useful one. The inertia force from the reciprocating mass is then a simple harmonic force acting along the cylinder axis. The inertia force from the rotating mass rotates with the crank. Its component along the cylinder axis is also harmonic, with the same period as the inertia force from the reciprocating mass. The rotating inertia force can be completely removed by counterweighting the crank so that the center of mass of the equivalent mass system is at the crankshaft axis. The inertia force of the reciprocating mass cannot be completely removed. However, it can be offset by increasing the mass of the counterweight on the crank so that the axial component of the crank inertia force opposes the inertia force from the reciprocating mass. This reduces the magnitude of the axial inertia force from the reciprocating mass, but at the cost of reintroducing a lateral shaking force from the counterweighted crank. A popular arrangement for single-cylinder engines is to use a crank counterweight that is large enough to halve the amplitude of the reciprocating inertia force. This results in a rotating inertia force of half the amplitude of the original reciprocating unbalance. The resultant inertia force rotates in the *opposite* direction to the crank. If one is willing to tolerate the extra complexity of arranging a counterrotating counterweight on the crankshaft axis, this remaining unbalance force can also be removed.

The shaking force is the *resultant of all the forces acting on the frame of a mechanism due to inertia forces only.* Thus, if the resultant of all the forces due to inertia effects acting on the frame is zero, there is no *shaking force*. There may, nevertheless, be a *shaking couple* present. Balancing a mechanism consists of eliminating the shaking force and shaking couple. In some instances we can accomplish both. We shall discover that in most mechanisms, by adding appropriate balancing weights, we can reduce the shaking force and shaking couple, but it is usually not practical to provide a means of completely eliminating them.

Since the slider-crank mechanism is so widely used in such machines as internal-combustion engines and compressors, considerable work has been done on the development of techniques for balancing these mechanisms. The approach to balance the slider-crank mechanism will use the approximate mass distribution developed in Section 13.4.1 along with an approximation for the acceleration of point B at the piston. An expression for the acceleration of the piston is developed in the following section.

13.4.2.1 Approximate Expression for Piston Acceleration

If the crank moves with constant angular velocity ω, the acceleration of any point on the crank is given by

$$a_c = r\omega^2 \tag{13.17}$$

where r is the distance from O_2 to the point of interest. In Fig. 13.11, point A is the primary point of interest. If the lumped-mass model is used for the connecting rod, the only other point of interest is point B at the wrist pin. The acceleration of this point can be determined using Fig. 13.11. A local x, y coordinate system is located at point O_2 as shown such that the x axis is along the line O_2B.

The position of the point B relative to O_2, the origin of the x and y coordinate axes, is represented by x. Link 2 is at an angle θ from the x axis for the position shown. The connecting rod makes an angle of ϕ with the x axis. The distance x is given by the sum of the projection of the crank and connecting rod on the x axis:

$$x = R\cos\theta + L\cos\phi \tag{13.18}$$

It is desirable to represent x as a function of R, L, and θ. To do this ϕ may be eliminated by noting that the vertical projections of R and L are equal, or

$$R\sin\theta = L\sin\phi$$

or

$$\sin\phi = (R/L)\sin\theta \tag{13.19}$$

but

$$\cos\phi = \sqrt{1 - (\sin\phi)^2} \tag{13.20}$$

Therefore, substitution of Eq. (13.19) into Eq. (13.20) gives

$$\cos\phi = \sqrt{1 - [(R/L)\sin\theta]^2} \tag{13.21}$$

Substitute Eq. (13.21) into Eq. (13.18) to get

$$x = R\cos\theta + L\sqrt{1 - [(R/L)\sin\theta]^2} \tag{13.22}$$

Equation (13.22) is an exact expression for the location of the piston relative to the center of the crank bearing.

The expression for the velocity may be obtained by differentiating x with respect to t. Noting that $d\theta/dt = \omega$, the angular velocity of the piston is

$$v_B = \frac{dx}{dt} = -R\omega\left[\sin\theta + \frac{R}{2L}\sin 2\theta \Big/ \sqrt{1 - \left(\frac{R}{L}\sin\theta\right)^2}\right] \tag{13.23}$$

The expression for the acceleration of the piston may be obtained by differentiating v with respect to t and remembering that ω, the angular speed of the crank, is constant. Then,

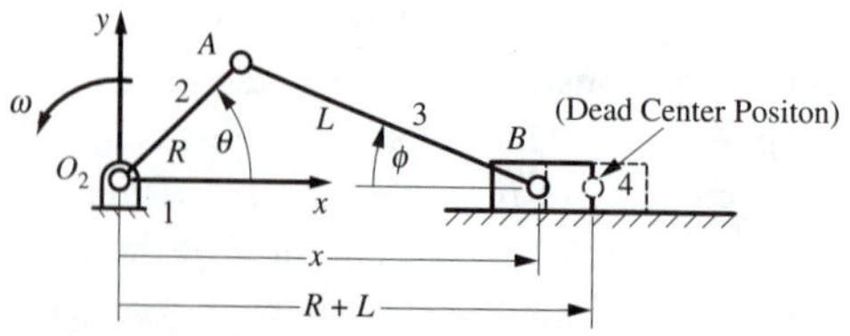

Figure 13.11 Slider crank mechanism coordinate system.

$$a_B = \frac{dv_B}{dt} = -R\omega^2 \left[\cos\theta + \frac{\frac{R}{L}\cos 2\theta \left\{1 - \left(\frac{R}{L}\right)^2\right\} + \left(\frac{R}{L}\right)^3 \cos^4\theta}{\left\{1 - \left(\frac{R}{L}\sin\theta\right)^2\right\}^{3/2}} \right] \tag{13.24}$$

If the velocity or acceleration is negative, it is directed from the piston to the crank bearing. The angular velocity of the crank is positive if it is counterclockwise.

For most practical engines, the value of R/L will be less than 1/4 so that $(R/L)^2$ will be small compared with 1. With this approximation, the expression for the acceleration can be simplified with little loss of accuracy to

$$a_B = -R\omega^2 \left[\cos\theta + \frac{R}{L}\cos 2\theta\right] \tag{13.25}$$

The maximum error in the approximate expressions above is around 0.5% of the maximum values for $(R/L = 1/4)$. Therefore, because of the form for the equations, the approximate expressions are sometimes used in analytical work. However, when a computer is used, the simplification is not necessary because it is only slightly more difficult to compute the exact value for the acceleration of the piston.

13.5 EXPRESSIONS FOR INERTIAL FORCES

We can represent the inertia forces as the vectors

$$\boldsymbol{f}_A = -\overline{m}_A \boldsymbol{a}_A = \overline{m}_A R\omega^2 \angle\theta \tag{13.26}$$

and

$$\boldsymbol{f}_B = -\overline{m}_B \boldsymbol{a}_B \cong \overline{m}_B R\omega^2 \left(\cos\theta + \frac{R}{L}\cos 2\theta\right) \angle 0^\circ \tag{13.27}$$

Therefore, in terms of x and y components, the total shaking force would be

$$\boldsymbol{f}_S = \boldsymbol{f}_A + \boldsymbol{f}_B = R\omega^2 \left[(\overline{m}_A + \overline{m}_B)\cos\theta + \overline{m}_B \left(\frac{R}{L}\right)\cos 2\theta\right] \boldsymbol{i} + R\omega^2(\overline{m}_A \sin\theta)\boldsymbol{j} \tag{13.28}$$

If a counterbalance mass is added to the crank, the mass would be added to the side of the crank that is opposite A. As discussed in Section 13.2, this is equivalent to subtracting mass from the side of the crank where A is located. Regardless of whether we add or subtract mass, it is convenient to represent the "added" mass as an equivalent mass at A by using Eq. (13.12). If the counterbalance mass is m_{cb}, at a distance R_c from the crank

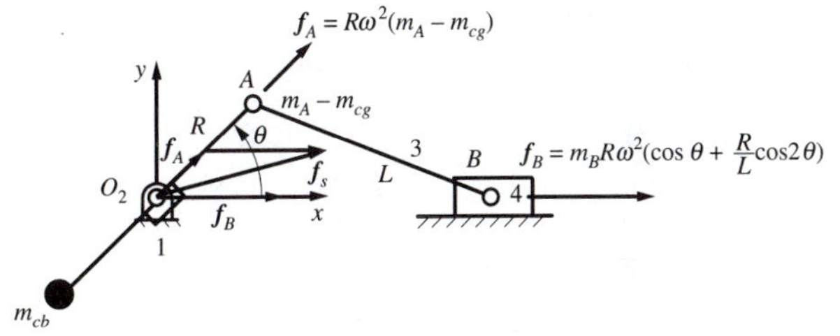

Figure 13.12 Equivalent slider-crank mechanism with inertial forces (ω = constant).

axis, the equivalent mass $\overline{m}_{cb}$ at point A would be

$$\overline{m}_{cb} = \frac{R_c}{R} m_{cb}$$

The total inertial force at point A would then be

$$f_A = (\overline{m}_A - \overline{m}_{cb})R\omega^2 \angle\theta$$

and Eq. (13.28) can be rewritten as

$$\begin{aligned} f_S &= f_A + f_B \\ &= R\omega^2 \left[(\overline{m}_A + \overline{m}_B - \overline{m}_{cb}) \cos\theta + \overline{m}_B \left(\frac{R}{L}\right) \cos 2\theta \right] \boldsymbol{i} + R\omega^2 (\overline{m}_A - \overline{m}_{cb}) \sin\theta \boldsymbol{j} \end{aligned} \tag{13.29}$$

The results are shown schematically in Fig. 13.12. Equation (13.29) can be easily programmed using MATLAB. This permits the user to compute the shaking force as a function of θ and to determine the best choice of counterbalance weight to minimize the maximum value for the shaking force. The program *shake.m,* which is included on the disk with this book, performs the necessary calculations and plots the results.

Notice that $\overline{m}_{cb}$ appears in both components of f_S. The objective of the balancing procedure is to adjust $\overline{m}_{cb}$ to reduce the magnitude of f_S that varies directly with θ and in part with 2θ. It is customary to refer to the portion of the force occurring at the circular frequency ω rad/s as the primary inertia force and the portion occurring at 2ω rad/s as the secondary inertia force. We note that the vertical component has only a primary part and that it therefore varies directly with the crankshaft speed. On the other hand, the horizontal component, which is in the direction of the cylinder axis, has a primary part varying directly with the crankshaft speed and a secondary part that varies at twice the crankshaft speed.

Note also that the equations derived are in terms of mass. In general, we will measure directly the weights for the crank and connecting rod and not the mass. However, as used in the earlier examples, the mass can be conveniently determined from the weight by dividing the weight by the acceleration of gravity.

EXAMPLE 13.5 (*Balancing a Slider-Crank Mechanism*)

The purpose of this example is to introduce the student to a numerical technique for balancing a model of a slider-crank mechanism and to give an example of the use of a computer program for balancing a slider-crank mechanism. The computer program *shake.m* will be used to determine the "optimum" counterbalance value for a slider-crank mechanism that might be used in a small engine.

The program computes the maximum value of the shaking force for a specified value of the counterbalance weight (assumed to be located at a distance equal to the crank radius). The program also computes the maximum shaking force value for no counterbalance weight and for the optimum counterbalance weight. The maximum value is found through a simple search over the crank angle range of 0 to 360°. Given the counterbalance weight W_{cb}, the program increments the crank angle θ in 1-deg increments. The shaking force is computed at each of these θ values, and the maximum value is selected.

PROBLEM

For the analysis, assume that the following data have been recorded:

Piston	$W_4 = 20$ lb
Rod	$L = 14$ in
	$W'_A = 24.29$ lb

	$W'_B = 9.71$ lb
Crank	$R = 4.0$ in
	$W_A = 3.75$ lb
Rotation speed	1000 rpm
Gravity	$g = 386$ in/s^2

Use the program *shake.m* given on the disk with this book to find the optimum counterbalance weight for the mechanism. Note that the weights of the crank and the connecting rod mentioned are already referred to the crank pin and the piston pin.

SOLUTION Begin the optimization with an initial value for the counterbalance mass as $\overline{m}_{cb} = \overline{m}_A + 2\overline{m}_B/3$. This value is often suggested as the optimum counterbalance mass. The counterbalance weight corresponding to this mass is

$$W_{cb} = (W_A + W'_A) + 2(W'_B + W_4)/3 = (3.75 + 24.29) + 2(9.71 + 20)/3 = 47.85 \text{ lb}.$$

The program *shake.m* was run with three values for the counterbalance force. The first value was the given value (47.85 lb), the second value was no counterbalance force, and the third value was the optimum value (45.87 lb) determined by the program. The corresponding maximum shaking force values are given in Table 13.1. The program also plots the corresponding shaking force for each of the values of the counterbalance force. These are shown in Fig. 13.13. Notice that the shaking force can be reduced by almost 70 percent by the addition of the simple counterbalance weights. ■

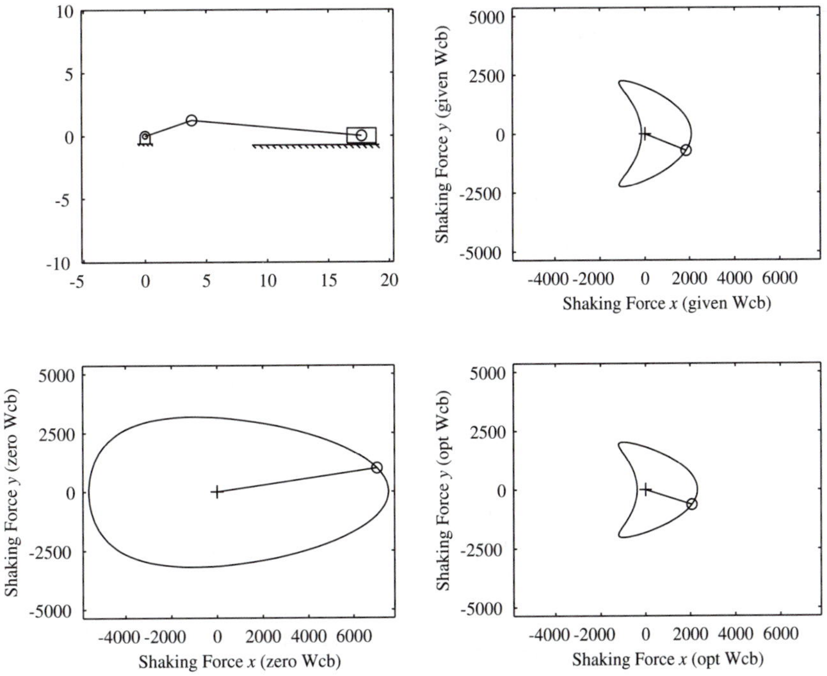

Figure 13.13 Shaking force diagrams for Example 13.5. The top right diagram is for the given counterbalance weight. The bottom left diagram is for no counterbalance weight, and the bottom right diagram is for the optimum counterbalance weight.

Table 13.1 Summary of Shaking Force Values for Example 13.5

Counterbalance Weight (lb)	Maximum Shaking Force (lb)	Crank Angle
0	7527.4	0°
47.85	2495.4	98.18°
45.87	2313.7	258.18°

13.6 BALANCING MULTICYLINDER MACHINES

The approximation of the connecting rod and crank pin masses used above is also routinely used when discussing multicylinder engine balancing. Because of these approximations and manufacturing variations, even a perfectly balanced multicylinder engine, pump, or compressor will generate some residual vibration. For this reason, vibration isolating mounts are always necessary. In many situations it is necessary to trade off the complexity involved in achieving a higher level of balance against using a simpler scheme and relying on vibration mounts to further reduce the remaining vibration.

Balance is not the only constraint when selecting an automotive engine configuration. The firing sequence of the cylinders must be even, and engine designers try to optimize the firing sequence to minimize the elastic windup of the crankshaft. Two-stroke and diesel engines, which fire every piston stroke, require different crankshaft geometries and balance arrangements from four-stroke engines that fire every second piston stroke.

In multicylinder engines and compressors, the weight of the crankshaft and the part of the weight of the connecting rod that is concentrated at the crank pin can be counterbalanced exactly using the procedures discussed earlier. Therefore, only the balancing of the reciprocating weights will be considered here.

If we have n cylinders, the inertial force associated with each piston is in the direction of the piston travel and is given by

$$f_{B_i} = \overline{m}_{B_i} R_i \omega^2 \left[\cos(\theta + \phi_i - \psi_i) + \frac{R_i}{L_i} \cos 2(\theta + \phi_i - \psi_i) \right], \qquad i = 1, 2, \ldots, n \tag{13.30}$$

where $\overline{m}_{B_i}$, R_i, L_i, and ϕ_i are the reciprocating mass, the crank radius, the connecting rod length, and the phase angle, respectively, for the ith cylinder. The angle θ gives the position of the piston axis for cylinder 1 relative to the x axis, and each cylinder axis is oriented at an angle of ψ_i relative to the axis of cylinder 1. If the first crank starts in the direction of the x axis, the angle θ is equal to the crank angular velocity multiplied by the time. In most engines and compressors, $\overline{m}_{B_i}$, R_i, and L_i will be the same for all cylinders. Therefore, the subscript i must be maintained for the angles ϕ_i and ψ_i only. The phase angle ϕ_i gives the angle from crank 1 to crank i as shown in Fig. 13.14. Both $\phi_1 = 0$ and $\psi_1 = 0$, although both angles are usually included in the equations for symmetry.

The total shaking force will be the vector sum of the forces from all the cylinders. For the analysis, the coordinate axes will be oriented as shown in Fig. 13.14. The z axis is along the rotation axis of the crankshaft, the x axis is in the direction of the piston travel of the first cylinder, and the y axis is defined such that a right-handed coordinate system results. Then, the x component of the shaking force from the cylinder is

$$f_{B_{x_i}} = f_{B_i} \cos \psi_i = \overline{m}_B R \omega^2 \left[\cos(\theta + \phi_i - \psi_i) + \frac{R}{L} \cos 2(\theta + \phi_i - \psi_i) \right] \cos \psi_i \tag{13.31}$$

and the y component is

$$f_{B_{y_i}} = f_{B_i} \sin \psi_i = \overline{m}_B R \omega^2 \left[\cos(\theta + \phi_i - \psi_i) + \frac{R}{L} \cos 2(\theta + \phi_i - \psi_i) \right] \sin \psi_i \tag{13.32}$$

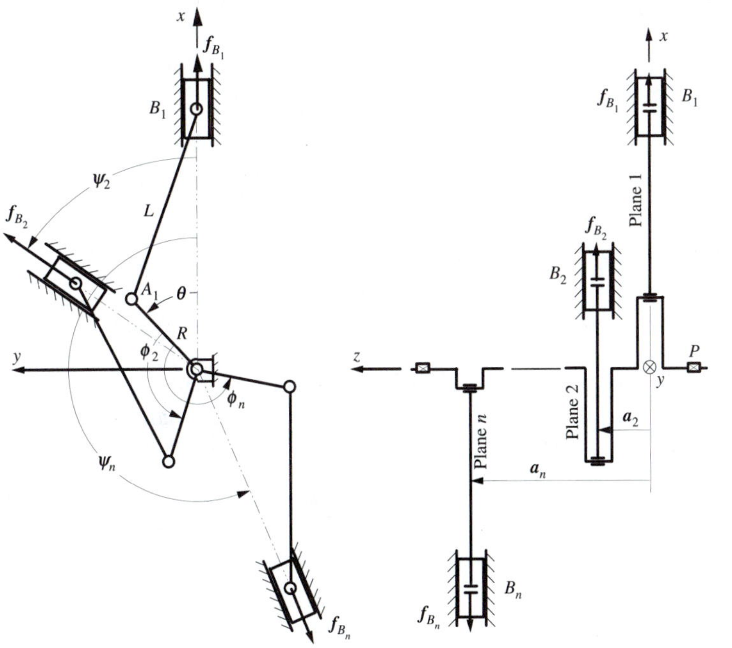

Figure 13.14 Schematic diagram of shaking forces in multicylinder engine.

The total inertial force vector due to the reciprocating masses is given by summing the vector contribution from each cylinder. It is convenient to sum the x and y components separately when doing this. The result is

$$f_x = \overline{m}_B R\omega^2 \left[\sum_{i=1}^{n} \cos(\theta + \phi_i - \psi_i) \cos \psi_i + \frac{R}{L} \sum_{i=1}^{n} \cos 2(\theta + \phi_i - \psi_i) \cos \psi_i \right] \tag{13.33}$$

and

$$f_y = \overline{m}_B R\omega^2 \left[\sum_{i=1}^{n} \cos(\theta + \phi_i - \psi_i) \sin \psi_i + \frac{R}{L} \sum_{i=1}^{n} \cos 2(\theta + \phi_i - \psi_i) \sin \psi_i \right] \tag{13.34}$$

Equations (13.33) and (13.34) can be simplified using the following trigonometric identities:

$$\cos(\theta + \phi_i - \psi_i) = \cos\theta \cos(\phi_i - \psi_i) - \sin\theta \sin(\phi_i - \psi_i) \tag{13.35}$$

and

$$\cos 2(\theta + \phi_i - \psi_i) = \cos 2\theta \cos 2(\phi_i - \psi_i) - \sin 2\theta \sin 2(\phi_i - \psi_i) \tag{13.36}$$

Then

$$\begin{aligned} f_x &= \overline{m}_B R\omega^2 \left[\sum_{i=1}^{n} \cos(\theta + \phi_i - \psi_i) \cos \psi_i + \frac{R}{L} \sum_{i=1}^{n} \cos 2(\theta + \phi_i - \psi_i) \cos \psi_i \right] \\ &= \overline{m}_B R\omega^2 \left[\cos\theta \sum_{i=1}^{n} \cos(\phi_i - \psi_i) \cos \psi_i - \sin\theta \sum_{i=1}^{n} \sin(\phi_i - \psi_i) \cos \psi_i \right. \\ &\qquad \left. + \frac{R}{L} \cos 2\theta \sum_{i=1}^{n} \cos 2(\phi_i - \psi_i) \cos \psi_i - \frac{R}{L} \sin 2\theta \sum_{i=1}^{n} \sin 2(\phi_i - \psi_i) \cos \psi_i \right] \end{aligned} \tag{13.37}$$

and

$$f_y = \overline{m}_B R\omega^2 \left[\sum_{i=1}^{n} \cos(\theta + \phi_i - \psi_i) \sin \psi_i + \frac{R}{L} \sum_{i=1}^{n} \cos 2(\theta + \phi_i - \psi_i) \sin \psi_i \right]$$

$$= \overline{m}_B R\omega^2 \left[\cos \theta \sum_{i=1}^{n} \cos(\phi_i - \psi_i) \sin \psi_i - \sin \theta \sum_{i=1}^{n} \sin(\phi_i - \psi_i) \sin \psi_i \right.$$

$$\left. + \frac{R}{L} \cos 2\theta \sum_{i=1}^{n} \cos 2(\phi_i - \psi_i) \sin \psi_i - \frac{R}{L} \sin 2\theta \sum_{i=1}^{n} \sin 2(\phi_i - \psi_i) \sin \psi_i \right] \tag{13.38}$$

To balance the forces, it is necessary for f_x and f_y to be zero. This can be achieved by making the resultant of the terms on the right-hand side of Eqs. (13.37) and (13.38) equal to zero. This occurs for all values of θ when

$$\sum_{i=1}^{n} \cos(\phi_i - \psi_i) \cos \psi_i = 0 \tag{13.39}$$

$$\sum_{i=1}^{n} \cos(\phi_i - \psi_i) \sin \psi_i = 0 \tag{13.40}$$

$$\sum_{i=1}^{n} \sin(\phi_i - \psi_i) \sin \psi_i = 0 \tag{13.41}$$

$$\sum_{i=1}^{n} \sin(\phi_i - \psi_i) \cos \psi_i = 0 \tag{13.42}$$

$$\sum_{i=1}^{n} \cos 2(\phi_i - \psi_i) \cos \psi_i = 0 \tag{13.43}$$

$$\sum_{i=1}^{n} \cos 2(\phi_i - \psi_i) \sin \psi_i = 0 \tag{13.44}$$

$$\sum_{i=1}^{n} \sin 2(\phi_i - \psi_i) \sin \psi_i = 0 \tag{13.45}$$

$$\sum_{i=1}^{n} \sin 2(\phi_i - \psi_i) \cos \psi_i = 0 \tag{13.46}$$

Equations (13.39)–(13.42) are necessary to balance the primary shaking forces, and Eqs. (13.43)–(13.46) are necessary to balance the secondary shaking forces. In some crank arrangements, it will be possible to satisfy all eight equations. In other cases, it may be possible to satisfy either Eqs. (13.39)–(13.42) or Eqs. (13.43)–(13.46). If there is a choice, it is generally more important to balance the primary shaking forces rather than the secondary shaking forces because the factor (R/L) will always be less than 1, making the primary shaking forces larger than the secondary shaking forces.

In most cases, the cylinders will be offset along the z axis. Therefore, the individual shaking forces will not be in a single plane, and there is a possibility of a shaking moment. The components of the shaking moment can be determined by summing moments about the x and y axes, which are in the plane of cylinder 1. This will give two components of the shaking moment. If a_i is the distance from the plane of cylinder 1 to the plane of cylinder i ($a_1 = 0$), the x component of the moment is given by

$$\begin{aligned} M_x &= -\sum_{i=1}^{n} f_{B_{y_i}} a_i \\ &= -\overline{m}_B R\omega^2 \Bigg[\cos\theta \sum_{i=1}^{n} a_i \cos(\phi_i - \psi_i) \sin\psi_i - \sin\theta \sum_{i=1}^{n} a_i \sin(\phi_i - \psi_i) \sin\psi_i \\ &\qquad + \frac{R}{L}\cos 2\theta \sum_{i=1}^{n} a_i \cos 2(\phi_i - \psi_i) \sin\psi_i - \frac{R}{L}\sin 2\theta \sum_{i=1}^{n} a_i \sin 2(\phi_i - \psi_i) \sin\psi_i \Bigg] \end{aligned} \tag{13.47}$$

and the y component is given by

$$\begin{aligned} M_y &= \sum_{i=1}^{n} f_{B_{x_i}} a_i \\ &= \overline{m}_B R\omega^2 \Bigg[\cos\theta \sum_{i=1}^{n} a_i \cos(\phi_i - \psi_i) \cos\psi_i - \sin\theta \sum_{i=1}^{n} a_i \sin(\phi_i - \psi_i) \cos\psi_i \\ &\qquad + \frac{R}{L}\cos 2\theta \sum_{i=1}^{n} a_i \cos 2(\phi_i - \psi_i) \cos\psi_i - \frac{R}{L}\sin 2\theta \sum_{i=1}^{n} a_i \sin 2(\phi_i - \psi_i) \cos\psi_i \Bigg] \end{aligned} \tag{13.48}$$

For dynamic balance, it is important to balance both the shaking force and the shaking moment. Both will cause undesirable vibrations at the engine base. To balance the shaking moment, we want the moment components in Eqs. (13.47) and (13.48) to be zero. This occurs when

$$\sum_{i=1}^{n} a_i \cos(\phi_i - \psi_i) \cos\psi_i = 0 \tag{13.49}$$

$$\sum_{i=1}^{n} a_i \cos(\phi_i - \psi_i) \sin\psi_i = 0 \tag{13.50}$$

$$\sum_{i=1}^{n} a_i \sin(\phi_i - \psi_i) \sin\psi_i = 0 \tag{13.51}$$

$$\sum_{i=1}^{n} a_i \sin(\phi_i - \psi_i) \cos\psi_i = 0 \tag{13.52}$$

$$\sum_{i=1}^{n} a_i \cos 2(\phi_i - \psi_i) \cos\psi_i = 0 \tag{13.53}$$

$$\sum_{i=1}^{n} a_i \cos 2(\phi_i - \psi_i) \sin\psi_i = 0 \tag{13.54}$$

$$\sum_{i=1}^{n} a_i \sin 2(\phi_i - \psi_i) \sin\psi_i = 0 \tag{13.55}$$

$$\sum_{i=1}^{n} a_i \sin 2(\phi_i - \psi_i) \cos\psi_i = 0 \tag{13.56}$$

As in the case of the shaking force, Eqs. (13.49)–(13.52) are necessary to balance the primary shaking moments, and Eqs. (13.53)–(13.56) are necessary to balance the secondary shaking moments.

Notice that it may be possible to reduce the shaking forces using counterweights even though Eqs. (13.39)–(13.46) and (13.49)–(13.56) are not satisfied. However, it is much

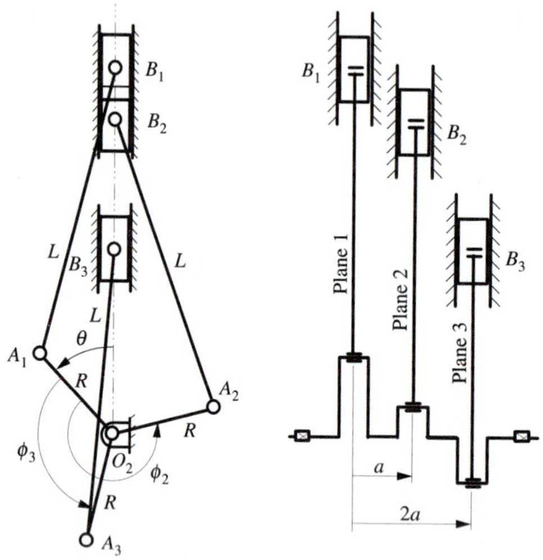

Figure 13.15 Schematic diagram of three-cylinder in-line engine.

more efficient if the shaking forces can be internally balanced by a careful design of the crank phase angles. Even in cases in which it is not possible to balance the shaking forces perfectly by satisfying all of the equations directly, if the equations are satisfied approximately, the shaking forces and moments will be reduced, and the sizes of any additional counterbalance weights will also be reduced.

To illustrate the use of the equations, we will investigate two examples. For a more extensive consideration of the topic, the reader is referred to the work by Holowenko.[1]

13.6.1 Balancing a Three-Cylinder In-Line Engine

Three-cylinder in-line engines are commonly used in small utility tractors. An example of this cylinder arrangement is shown in Fig. 13.15. Notice that ψ_i is zero for each cylinder, and the crank phase angles are distributed symmetrically about the crank shaft axis. Because $\psi_i = 0$, $\sin \psi_i = 0$ and $\cos \psi_i = 1$. Therefore, Eqs. (13.40), (13.41), (13.44), and (13.45) and Eqs. (13.50), (13.51), (13.54), and (13.55) are satisfied identically. The left-hand sides of the remaining equations are given in the following:

$$\sum_{i=1}^{3} \cos \phi_i = 1 + \cos 240° + \cos 120° = 1 - \frac{1}{2} - \frac{1}{2} = 0$$

$$\sum_{i=1}^{3} \sin \phi_i = 0 + \sin 240° + \sin 120° = 0 - \frac{\sqrt{3}}{2} + \frac{\sqrt{3}}{2} = 0$$

$$\sum_{i=1}^{3} \cos 2\phi_i = 1 + \cos 480° + \cos 240° = 1 - \frac{1}{2} - \frac{1}{2} = 0$$

$$\sum_{i=1}^{3} \sin 2\phi_i = 1 + \sin 480° + \sin 240° = 1 + \frac{\sqrt{3}}{2} - \frac{\sqrt{3}}{2} = 0$$

[1] Holowenko, A. R., *Dynamics of Machinery,* John Wiley & Sons, New York (1955).

$$\sum_{i=1}^{3} a_i \cos\phi_i = 0 + a\cos 240° + 2a\cos 120° = 0 - \frac{a}{2} - \frac{2a}{2} = -\frac{3a}{2}$$

$$\sum_{i=1}^{3} a_i \sin\phi_i = 0 + a\sin 240° + 2a\sin 120° = 0 - \frac{a\sqrt{3}}{2} + \frac{2a\sqrt{3}}{2} = \frac{a\sqrt{3}}{2}$$

$$\sum_{i=1}^{3} a_i \cos 2\phi_i = 0 + a\cos 480° + 2a\cos 240° = 0 - \frac{a}{2} - \frac{2a}{2} = -\frac{3a}{2}$$

$$\sum_{i=1}^{3} a_i \sin 2\phi_i = 0 + a\sin 480° + 2a\sin 240° = 0 + \frac{a\sqrt{3}}{2} - \frac{2a\sqrt{3}}{2} = -\frac{a\sqrt{3}}{2}$$

In this engine, the shaking forces are balanced exactly, but the primary and secondary moments are not balanced. The moments form a shaking couple on the engine. The magnitude of the couple is given by Eq. (13.48) as

$$\begin{aligned} M &= \overline{m}_B R\omega^2\left[-\frac{3a}{2}\cos\theta - \frac{a\sqrt{3}}{2}\sin\theta - \frac{R}{L}\frac{3a}{2}\cos 2\theta + \frac{R}{L}\frac{a\sqrt{3}}{2}\sin 2\theta\right] \\ &= \frac{\overline{m}_B aR\omega^2}{2}\left[-3\cos\theta - \sqrt{3}\sin\theta - 3\frac{R}{L}\cos 2\theta + \frac{R}{L}\sqrt{3}\sin 2\theta\right] \end{aligned} \quad \textbf{(13.57)}$$

This equation can be simplified by the trigonometric relationship

$$p\cos\beta + q\sin\beta = \sqrt{p^2+q^2}\sin(\beta+\gamma) \quad \textbf{(13.58)}$$

where $\tan\gamma = p/q$. Equation (13.57) can then be written as

$$M = \frac{\overline{m}_B aR\omega^2}{2}\left[-2\sqrt{3}\sin(\theta+60°) - 2\frac{R}{L}\sqrt{3}\sin(2\theta-60°)\right] \quad \textbf{(13.59)}$$

The first term is the primary shaking moment and the second term is the secondary shaking moment. The primary shaking moment can be balanced by counterweights that rotate at the engine speed but are out of phase with the first crank by positive 60°. The secondary shaking moment can be balanced by counterweights that rotate at two times the engine speed and are out of phase with the first crank by negative 60°. Figure 13.16 shows one

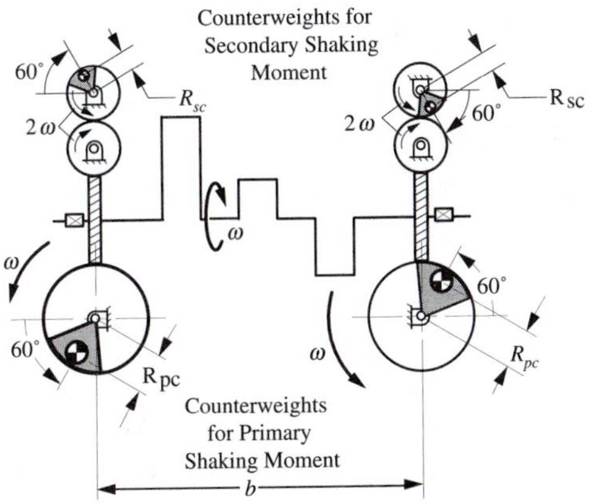

Figure 13.16 Arrangements to balance primary and secondary shaking moments. The counterbalance weights are shown for the position when $\theta = 0$.

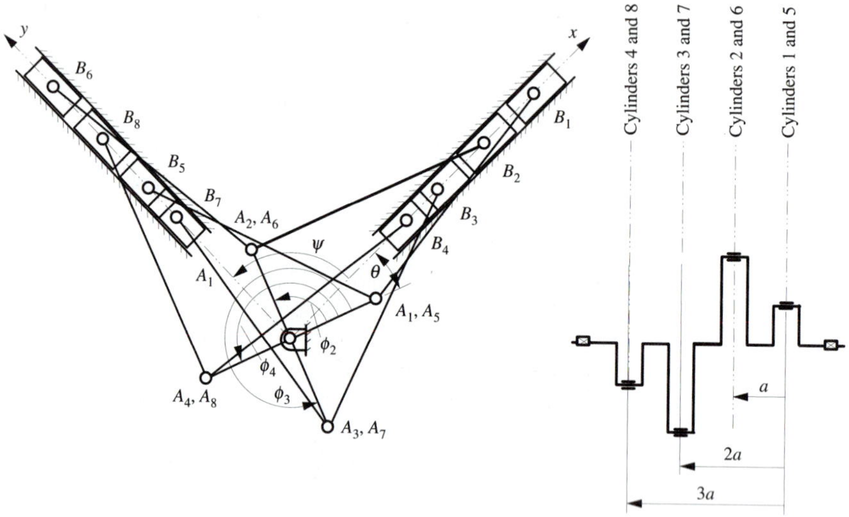

Figure 13.17 Eight-cylinder engine with 90° V.

balancing scheme for the shaking moments. The mass of the counterbalances for the primary moment are[1]

$$m_{cb_{\text{primary}}} = 2\sqrt{3}\overline{m}_B \frac{R}{R_{\text{pc}}} \frac{a}{b}$$

and for the secondary moment

$$m_{cb_{\text{secondary}}} = \frac{3}{4}\overline{m}_B \frac{R}{R_{\text{sc}}} \frac{a}{b}$$

where R_{pc} is the radial location of the center of mass of the counterweights for the primary shaking moment, R_{sc} is the radial location of the center of mass of the counterweights for the secondary shaking moment, and b is the distance between the rotation axes of the counterweights. The counterweights for the primary shaking moment rotate at the same velocity as the crankshaft, and the counterweights for the secondary shaking moment rotate at twice the velocity of the crankshaft.

13.6.2 Balancing an Eight-Cylinder V Engine

Eight-cylinder V engines (V-8) are commonly used in high-performance automobiles. An example of this cylinder arrangement is shown in Fig. 13.17. The engine shown is assumed to have the cylinder banks oriented at 90° to each other. Cylinders 1 and 5, 2 and 6, 3 and 7, and 4 and 8 are symmetrically located relative to the vertical axis. Then ψ_i is 90° for cylinders 5, 6, 7, and 8 and 0° for cylinders 1, 2, 3, and 4. The phase angles are $\phi_1 = \phi_5 = 0°$, $\phi_2 = \phi_6 = 90°$, $\phi_3 = \phi_7 = 270°$, and $\phi_4 = \phi_8 = 180°$. Substitution of the known angles into Eqs. (13.39)–(13.46), and (13.53)–(13.56) will show that the shaking force equations and the secondary shaking moment equations are satisfied. However, the primary

[1] Holowenko, A. R., *Dynamics of Machinery,* John Wiley & Sons, New York (1955).

shaking moment equations are not satisfied. The left-hand sides of Eqs. (13.49)–(13.52) give

$$\sum_{i=1}^{n} a_i \cos(\phi_i - \psi_i) \cos \psi_i = -3a \qquad \sum_{i=1}^{n} a_i \cos(\phi_i - \psi_i) \sin \psi_i = -a$$

$$\sum_{i=1}^{n} a_i \sin(\phi_i - \psi_i) \sin \psi_i = 3a \qquad \sum_{i=1}^{n} a_i \sin(\phi_i - \psi_i) \cos \psi_i = -a$$

The x component of the shaking moment is given by Eq. (13.47). Substitution of numbers into that equation gives

$$M_x = -\overline{m}_B R\omega^2 \left[\cos\theta \sum_{i=1}^{n} a_i \cos(\phi_i - \psi_i) \sin \psi_i - \sin\theta \sum_{i=1}^{n} a_i \sin(\phi_i - \psi_i) \sin \psi_i \right.$$
$$\left. + \frac{R}{L} \cos 2\theta \sum_{i=1}^{n} a_i \cos 2(\phi_i - \psi_i) \sin \psi_i - \frac{R}{L} \sin 2\theta \sum_{i=1}^{n} a_i \sin 2(\phi_i - \psi_i) \sin \psi_i \right]$$
$$= \overline{m}_B R\omega^2 a[\cos\theta + 3\sin\theta]$$

or using Eq. (13.58) with the trigonometric identity $\sin\theta = \cos(\theta - 90°)$,

$$M_x = \sqrt{10}\,\overline{m}_B R\omega^2 a \cos(\theta - 71.56°) \tag{13.60}$$

The y component of the shaking force is given by Eq. (13.48). Substitution of numbers into that equation gives

$$M_y = \overline{m}_B R\omega^2 \left[\cos\theta \sum_{i=1}^{n} a_i \cos(\phi_i - \psi_i) \cos \psi_i - \sin\theta \sum_{i=1}^{n} a_i \sin(\phi_i - \psi_i) \cos \psi_i \right.$$
$$\left. + \frac{R}{L} \cos 2\theta \sum_{i=1}^{n} a_i \cos 2(\phi_i - \psi_i) \cos \psi_i - \frac{R}{L} \sin 2\theta \sum_{i=1}^{n} a_i \sin 2(\phi_i - \psi_i) \cos \psi_i \right]$$
$$= \overline{m}_B R\omega^2 a[-3\cos\theta + \sin\theta]$$

or using Eq. (13.58),

$$M_y = \sqrt{10}\,\overline{m}_B R\omega^2 a \sin(\theta - 71.56°)$$

The x and y components of the shaking moment can be combined vectorially to form the vector

$$\boldsymbol{M} = \sqrt{10}\,\overline{m}_B R\omega^2 a[\cos(\theta - 71.56°)\boldsymbol{i} + \sin(\theta - 71.56°)\boldsymbol{j}]$$

Therefore, the shaking moment is of constant magnitude and can be balanced by the couple formed by rotating counterweights. The magnitude of the couple must be equal to $\sqrt{10}\,\overline{m}_B R\omega^2 a$. A possible counterbalance arrangement for the crankshaft position corresponding to $\theta = 0$ is given in Fig. 13.18.

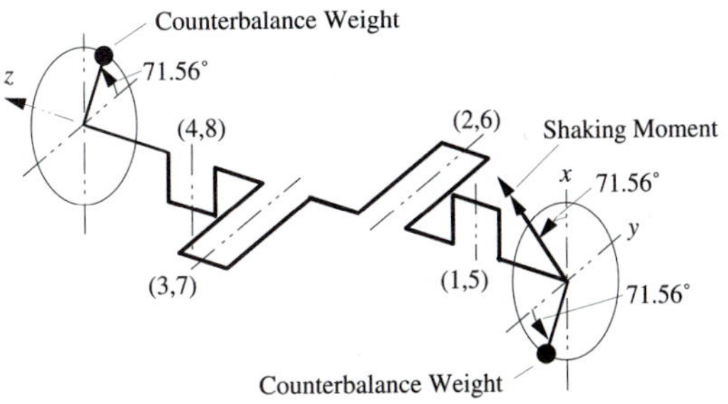

Figure 13.18 Rotating counterweights used to balance shaking forces in an eight-cylinder V engine.

13.7 CHAPTER 13 Exercise Problems

PROBLEM 13.1

The figure below shows a system with two weights, W_A and W_B, which have been found to balance a system of weights (not shown) on the shaft. The weights for W_A and W_B are 4 and 8 lb, respectively, and the radii, r_A and r_B, are both 6 inches. Later, it is decided to replace W_A and W_B by two weights, W_C and W_D, where the planes for the two weights are as shown. What are the magnitudes and angular locations of W_C and W_D if the radius of the center of gravity for both links is 5 in?

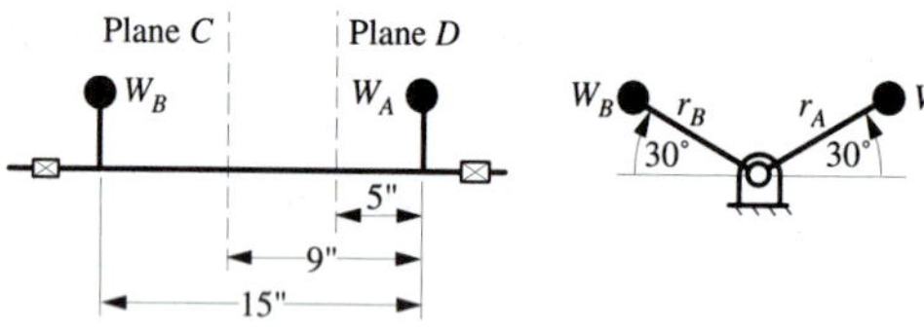

PROBLEM 13.2

The figure below shows a system with two weights, W_A and W_B, which have been found to balance a system of weights (not shown) on the shaft. The weights for W_A and W_B are 6 and 8 lb, respectively, and the radii, r_A and r_B, are both 5 inches. It is decided to replace W_A and W_B by two weights, W_C and W_D, where the planes for the two weights are as shown. What are the magnitudes and angular locations of W_C and W_D if the radius of the center of gravity for both links is 6 in?

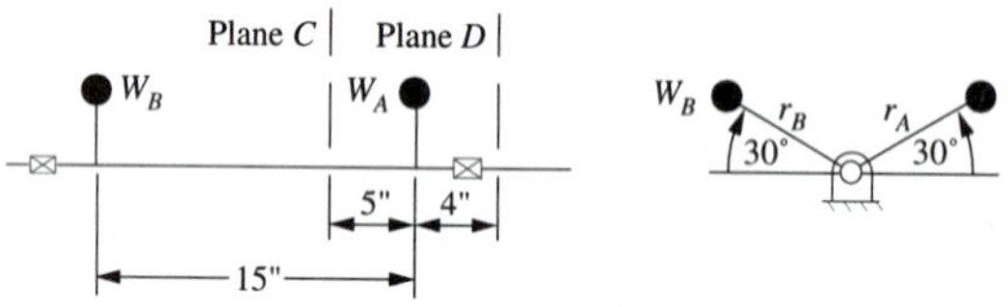

PROBLEM 13.3

Three rotating weights W_1, W_2, W_3 are to be balanced by two weights W_A and W_B in planes A and B. Determine the magnitudes and angular locations of the counterbalance weights necessary to balance the rotating weights.

$W_1 = 12$ lb	$r_1 = 2.5$ in	$\theta_1 = 30°$	$z_A = 4''$
$W_2 = 9$ lb	$r_2 = 2.5$ in	$\theta_2 = 150°$	$z_B = 14''$
$W_3 = 8$ lb	$r_3 = 2.5$ in	$\theta_3 = 270°$	
$W_A = ?$ lb	$r_A = 2.5$ in	$\theta_A = ?$	
$W_B = ?$ lb	$r_B = 2.5$ in	$\theta_B = ?$	

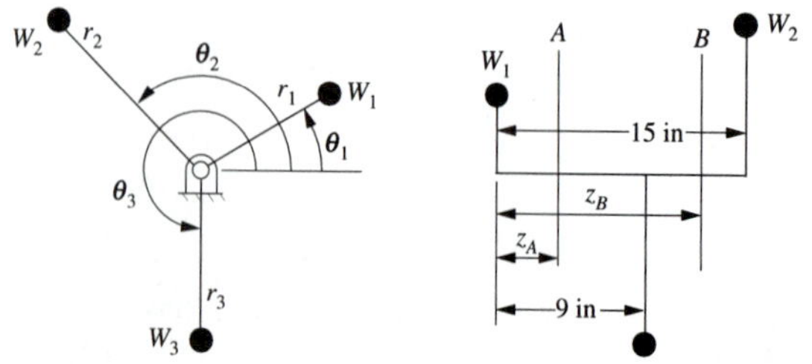

PROBLEM 13.4 Resolve Problem 13.3 if z_A and z_B are 4.5 in and 12 in, respectively.

PROBLEM 13.5 Four weights, W_1, W_2, W_3, and W_4, are all rotating in a single plane. Determine the magnitude and angular location of the single weight necessary to balance the four rotating weights. Assume that the radius to the center of gravity of the balancing weight is 9 in. The shaft is rotating at 1800 rpm.

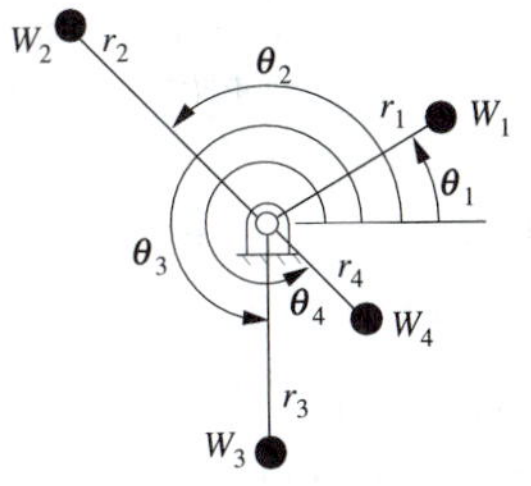

$W_1 = 12$ lb	$r_1 = 9$ in	$\theta_1 = 30°$
$W_2 = 9$ lb	$r_2 = 12$ in	$\theta_1 = 135°$
$W_3 = 8$ lb	$r_3 = 10$ in	$\theta_1 = 270°$
$W_4 = 5$ lb	$r_A = 8$ in	$\theta_1 = 315°$

PROBLEM 13.6 Resolve Problem 13.5 for the following set of data.

$W_1 = 20$ lb	$r_1 = 4$ in	$\theta_1 = 45°$
$W_2 = 10$ lb	$r_2 = 12$ in	$\theta_2 = 135°$
$W_3 = 8$ lb	$r_3 = 12$ in	$\theta_3 = 180°$
$W_4 = 6$ lb	$r_4 = 10$ in	$\theta_4 = 270°$

PROBLEM 13.7 For the mechanism shown, determine the magnitude and location of the shaking force acting on the frame. Determine the location with respect to point A. Also find the magnitude of the reaction force at point A and at point C. Assume that $W_3 \gg W_2$ and W_4.

$W_3 = 2.0$ lb $\quad I_3 = 0.1$ lb-s²-in $\quad {}^1\omega_2 = 6.28$ rad/s CCW (constant)

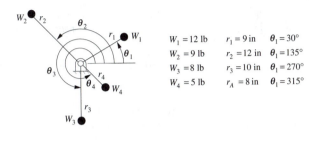

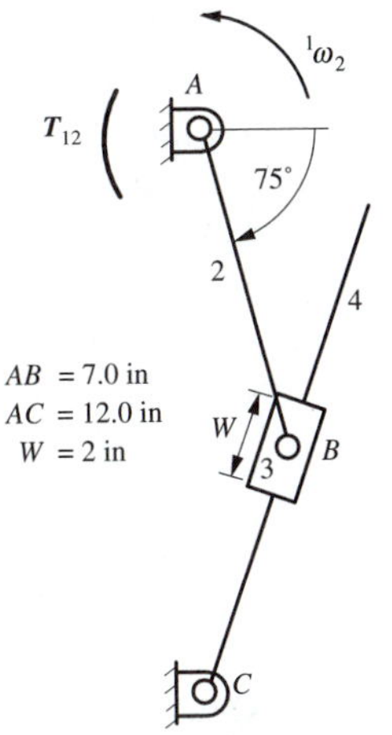

PROBLEM 13.8 For the mechanism and data given, determine the shaking force and its location relative to point A. Draw the shaking force vector on the figure. The force F_B is 10 lb in the direction shown. For the moments of inertia of link 3, use $g = 386$ in/s² and $I_G = ml^2/12$.

${}^1\omega_2 = 160$ rad/s CCW	${}^1\alpha_2 = 0$ rad/s²	$W_2 = 0.95$ lb
$I_{G_2} = 0.00369$ lb-s²-in	$W_3 = 3.5$ lb	$W_4 = 2.5$ lb

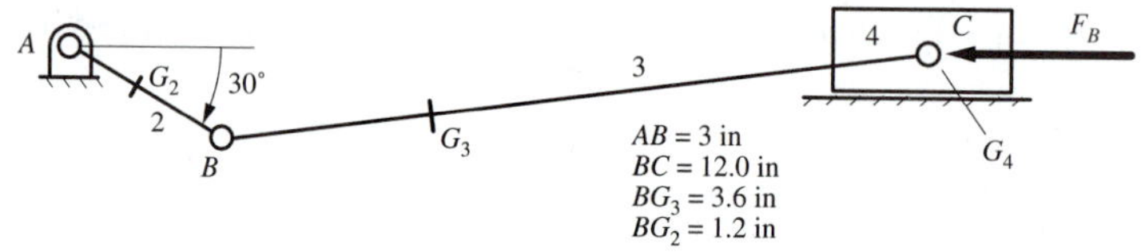

PROBLEM 13.9

For the mechanism shown, determine the magnitude and location of the shaking force acting on the frame. Determine the location with respect to point A. Draw the shaking force vector on the figure.

$$\omega_2 = 12 \text{ rad/s CCW} \qquad \alpha_2 = 0$$

$$W_2 = 0.5\,N \qquad W_3 = 2.5\,N \qquad W_4 = 1.5\,N$$

For the moments of inertia for the links, use $g = 9.81 \text{ m/s}^2$ and $I_G = ml^2/12$, that is, assume that each link is a bar with a uniform cross-section.

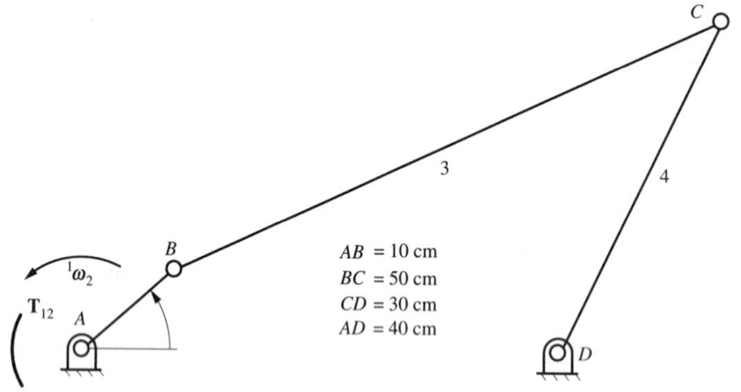

PROBLEM 13.10

For the mechanism given, assume that $^1\omega_2$ is 200 rad/s CCW (constant), and link 2 is balanced so that its center of mass is located at the pivot at point A. Also assume that I_{G_2} is small enough to be neglected. For the data given, determine the shaking force and its location relative to point A. Draw the shaking force vector on the figure.

$$I_{G_3} = 0.0106 \text{ lb-s}^2\text{-in} \qquad W_3 = 2.65 \text{ lb}$$

$$I_{G_4} = 0.0531 \text{ lb-s}^2\text{-in} \qquad W_4 = 6.72 \text{ lb}$$

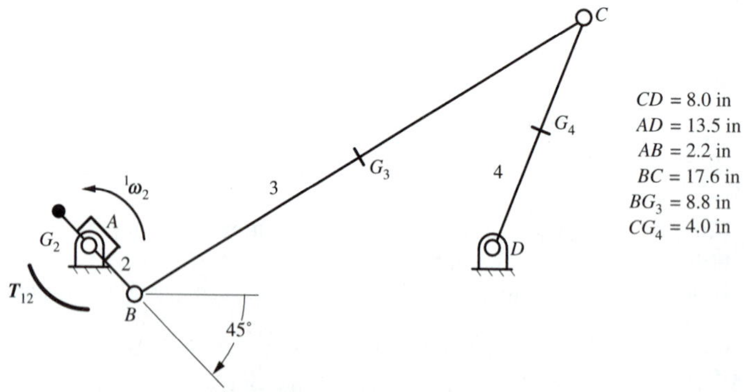

PROBLEM 13.11 A single cylinder engine is mounted so that the crankshaft is horizontal as shown in Fig. 13.12. The engine is characterized by the following data.

Rotational speed	1200 rpm
Stroke	6 in
Length of connecting rod	12 in
Distance from crank pin to *CG* of connecting rod	4 in
Equivalent unbalanced weight of crank at a 3 in radius	6 lb
Weight of piston	7 lb
Weight of connecting rod	15 lb

Determine the magnitude of the shaking force when the crank angle is 120° if there is no counterbalance weight. Then determine the shaking force at the crank location if a counterbalancing weight is added that is equal to $\overline{m}_{cb} = \overline{m}_A + 2\overline{m}_B/3$.

PROBLEM 13.12 Resolve Problem 13.11 if the stroke is 4 in, the engine speed is 1800 rpm, and the equivalent unbalanced weight of the crank at a 2 in radius is 5 lb.

PROBLEM 13.13 For the engine given in Problem 13.11, lump the weight of the connecting rod at the crank pin and piston pin and draw the polar shaking force diagram for the following three cases.

1. No counterbalancing weights
2. A counterbalancing weight equal to the sum of the crank weight at the crank radius, the part of the connecting rod weight assumed to be concentrated at the crank pin, the weight of the piston, and the part of the connecting rod weight concentrated at the piston pin.
3. A counterbalancing weight equal to the sum of the crank weight at the crank radius, the part of the connecting rod weight assumed to be concentrated at the crank pin, and half of the weight concentrated at the piston pin (weight of the piston and part of the connecting rod weight concentrated at the piston pin).

PROBLEM 13.14 The two-cylinder engine shown below has identical cranks, connecting rods, and pistons. The rotary masses are perfectly balanced. Derive an expression for the shaking forces and shaking moments using the symbols indicated if $\phi_2 = 90°$. Are the primary or secondary shaking forces balanced? What about the primary and secondary shaking moments?

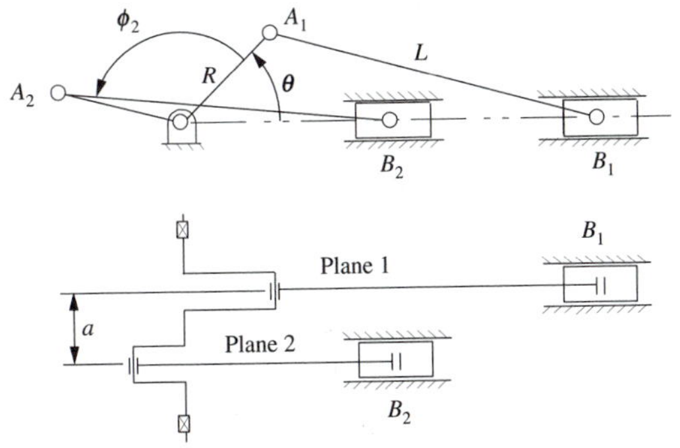

PROBLEM 13.15 Resolve Problem 13.14 when $\phi_2 = 180°$.

PROBLEM 13.16 The four-cylinder engine shown below has identical cranks, connecting rods, and pistons. The rotary masses are perfectly balanced. Derive an expression for the shaking forces and shaking

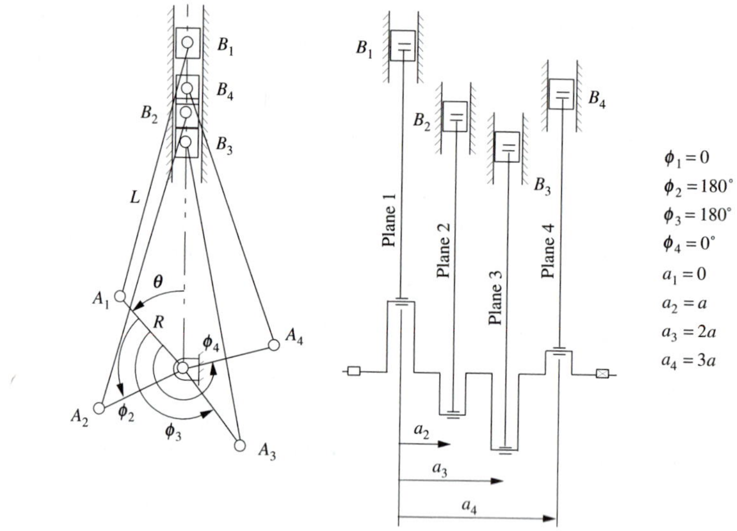

moments for the angles and offset values indicated. Are the primary or secondary shaking forces balanced? What about the primary and secondary shaking moments?

PROBLEM 13.17

Resolve Problem 13.16 for the following values for the phase angles and offset distances.

$$\phi_1 = 0° \qquad \phi_2 = 90° \qquad \phi_3 = 270° \qquad \phi_4 = 180°$$
$$a_1 = 0 \qquad a_2 = a \qquad a_3 = 2a \qquad a_4 = 3a$$

PROBLEM 13.18

The six-cylinder engine shown below has identical cranks, connecting rods, and pistons. The rotary masses are perfectly balanced. Derive an expression for the shaking forces and shaking moments for the angles and offset values indicated. Are the primary or secondary shaking forces balanced? What about the primary and secondary shaking moments?

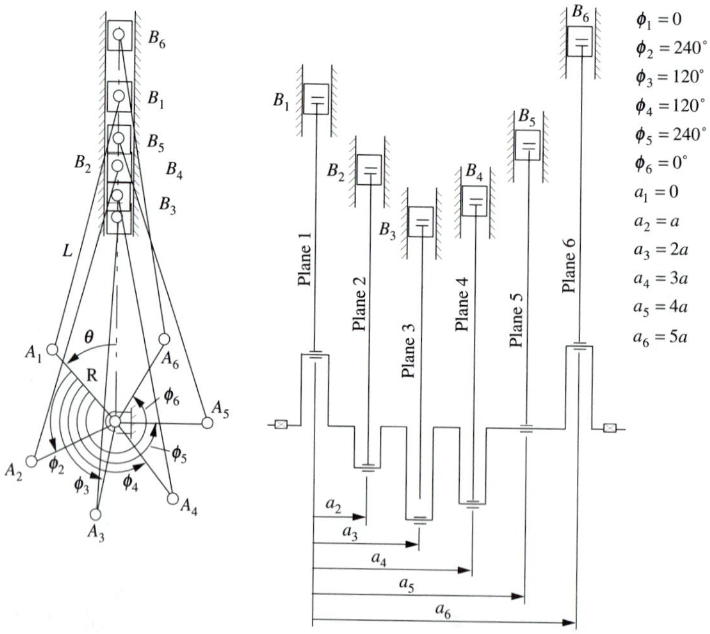

PROBLEM 13.19

Resolve Problem 13.18 for the following values for the phase angles and offset distances.

$\phi_1 = 0°$	$\phi_2 = 120°$	$\phi_3 = 240°$	$\phi_4 = 60°$	$\phi_5 = 300°$	$\phi_6 = 180°$
$a_1 = 0$	$a_2 = a$	$a_3 = 2a$	$a_4 = 3a$	$a_5 = 4a$	$a_6 = 5a$

PROBLEM 13.20

The two-cylinder V engine shown below has identical cranks, connecting rods, and pistons. The rotary masses are perfectly balanced. Derive an expression for the shaking forces and shaking moments for the angles and offset values indicated. The V angle is $\psi = 60°$, and the phase angle is $\phi_2 = 90°$. Are the primary or secondary shaking forces balanced? What about the primary and secondary shaking moments?

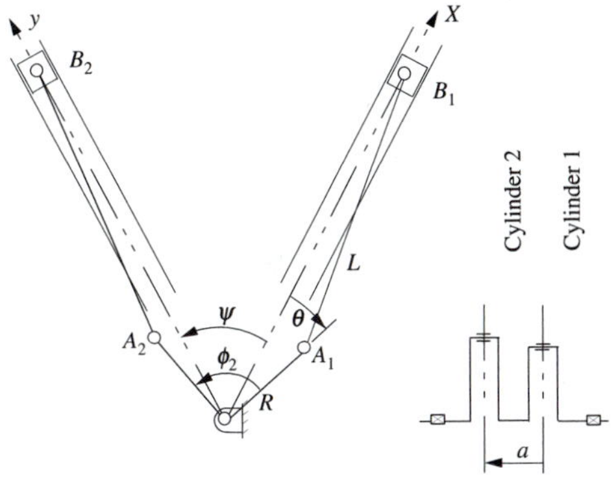

PROBLEM 13.21

Resolve Problem 13.20 if $\psi = 180°$ and $\phi_2 = 180°$.

PROBLEM 13.22

The six-cylinder V engine shown below has identical cranks, connecting rods, and pistons. The rotary masses are perfectly balanced. Derive an expression for the shaking forces and shaking moments for the angles and offset values indicated. The *V* angle is $\psi = 60°$, and the phase angles ϕ_1, ϕ_2, and ϕ_3 are 0°, 120°, and 240°, respectively. Are the primary or secondary shaking forces balanced? What about the primary and secondary shaking moments?

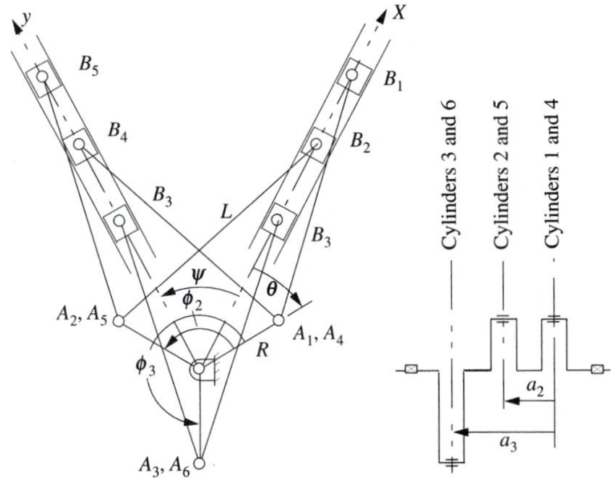

Index

D

E

Title page Courtesy Gary L. Kinzel.

Chapter 1 Page 1: Courtesy Ken Waldron. Page 7, 11, and 12: Courtesy Gary L. Kinzel. Page 27: Courtesy John Deere & Company. Page 31: Courtesy Gary L. Kinzel.

Chapter 2 Page 121: Courtesy Gary L. Kinzel.

Chapter 3 Page 170: Courtesy Caterpillar, Inc. Page 171: Courtesy Power Team, a Division of SPX Corporation, Owatonna, MN. Page 172: Courtesy Gary L. Kinzel. Page 182: Courtesy John Deere & Company. Pages 183, 189, 233 (top), 233 (bottom), 234 (top), 234 (bottom) and 235: Courtesy Gary L. Kinzel.

Chapter 5 Pages 300 and 318: Courtesy Gary L. Kinzel. Page 304: Courtesy Ken Waldron.

Chapter 8 Page 428 (left) and 428 (right): Courtesy Gary L. Kinzel. Page 444: Courtesy Horsburgh & Scott Company. Page 445: Courtesy Fellows Corporation. Page 446: Courtesy Bourn & Koch Machine Tool Company.

Chapter 9 Page 463, 464 (top), 464 (bottom), 465, 479, 483, 487, 488 (bottom), 490 (top), 490 (bottom), 491, and 492: Courtesy Gary L. Kinzel. Page 468 (top): Courtesy Bourn & Koch Machine Tool Company. Page 468 (bottom): Courtesy Fellows Corporation. Page 488 (top): Extracted from AGMA 203.03, Fine Pitch, On-Center Face Gears for 20-Degree Involute Spur Pinions, with the permission of the publisher, the American Gear Manufacturers Association, Alexandria, VA.

Chapter 10 Page 500: Courtesy Rexnord Corporation. Page 503: Courtesy Rockwell Automation Dodge. Page 507: Courtesy ANDANTEX USA, Inc. Page 508: Courtesy Zahnradfabrik Friedrichshafen AG. Page 509 (top): Courtesy Gary L. Kinzel. Page 509 (bottom): Courtesy The Gleason Works.